James Stormonth

A Manual of Scientific Terms

Pronouncing, Etymological and Explanatory

James Stormonth

A Manual of Scientific Terms
Pronouncing, Etymological and Explanatory

ISBN/EAN: 9783337337995

Printed in Europe, USA, Canada, Australia, Japan

Cover: Foto ©berggeist007 / pixelio.de

More available books at **www.hansebooks.com**

A

MANUAL OF SCIENTIFIC TERMS:

Pronouncing, Etymological, and Explanatory;

CHIEFLY COMPRISING

TERMS IN BOTANY, NATURAL HISTORY, ANATOMY,
MEDICINE, AND VETERINARY SCIENCE:

WITH AN APPENDIX OF SPECIFIC NAMES.

Designed for the use of Junior Medical Students, and others
studying one or other of these Sciences.

By REV. JAMES STORMONTH,

AUTHOR OF 'THE ETYMOLOGICAL AND PRONOUNCING DICTIONARY OF THE ENGLISH
LANGUAGE, 'THE HANDY ENGLISH WORD-BOOK, AND COMPLETE
ICTIONARY APPENDIX,' ETC.

Second Edition.

EDINBURGH:
MACLACHLAN AND STEWART.
LONDON: SIMPKIN, MARSHALL, & CO.
1885.

CONTENTS.

PREFACE.

THE increasingly numerous class of learners and junior students in our higher class schools and colleges, as well as of general readers, experience the want of a 'Manual of Scientific Terms' specially suited for the particular stage of their studies. The present work is an attempt to furnish such Lists of Terms in Botany, Natural History, Chemistry, Anatomy, Medicine, and Veterinary Science, as may be met with in the student's ordinary text-books, and in the current literature of the day. No attempt has been made to supply an exhaustive vocabulary. The *selected words* have been taken from the more popular text-books ; and the definitions are very generally given in the language employed in them, though often simplified. The Author feels assured that the special end in view, in the preparation of the present work, has been best served by making this legitimate use of such text-books ; for the language and phraseology employed in his text-books must become familiar to the student in the course of his studies.

The general terms in Botany will be found a pretty exhaustive list. All the names of Orders, and generally Sub-orders, as found in *Balfour's Manual of Botany*, have been inserted. Only, however, such names of genera and species are given as have some noteworthy particulars or qualities affirmed of them. The list of Latin Anatomical terms will be found a very full one ; but only such compound terms are given as are deemed most useful, inasmuch as the Latin compounds are now very often laid aside, and their English equivalents employed instead. The terms in Natural History, Chemistry, Medicine, and Veterinary Science, will be found in sufficient numbers to meet ordinary requirements.

Each term is followed by its re-spelling in simple phonetic characters, and accented, while the syllabication employed is that laid down in the Author's *Handy English Word-Book and Complete Dictionary Appendix.*[1] By using these methods, the pronunciation has been indicated with very great precision. A system of re-spellings with the view of indicating the correct pronunciation of scientific terms was much needed, for the pronunciation of many of them sadly puzzled most persons, and

[1] Edinburgh : W. P. Nimmo.

vii

even scientific men were frequently not agreed as to the correct pronunciation of many terms which they were constantly using ; indeed, they
were not unseldom inconsistent with themselves in the pronunciation of
many scientific terms. The endeavour is here made, and it is hoped
not unsuccessfully, to settle the pronunciation of terms in regard to
which there has hitherto been no certain agreement. Where the
pronunciation of a term has been fixed by usage, such of course has
been retained. In every case, the analogy of the English language,
and the etymology of the term, has been considered in fixing its
pronunciation. The student must particularly remember that the
symbols (– ◡) here employed to indicate English pronunciation
regard the *quality* of the vowel-sounds only, and not *quantity*, as in
Latin and Greek. Hence it often happens that the root-words, or
Latin words within the brackets, may have such a symbol as (◡),
while the corresponding syllable in the Latin word, re-spelt for pronunciation, may have the symbol (–), and *vice versâ*.

The student or learner should, in the case of a difficult or uncertain
word, enunciate aloud each syllable of the re-spelling, distinctly and
firmly, pronouncing each syllable and word repeatedly, always taking
care to place a halt or stress on the accents marked thus (ʹ or ʹ). By
such practice, the pronunciation of the terms will be well and correctly
done ; and repetition, first slowly, and then more rapidly, will tend to
fix their pronunciation in the memory. The re-spellings is an important
feature in this work, and its importance will readily be seen and
understood. The table of sound symbols on page x. should be carefully studied.

The root-words or etymologies, and Latin words with literal meanings
and their quantities, are placed within brackets. It is of great importance to exhibit the primary meanings of the terms by means of root-
words, and the simpler forms of Latin terms. A root-word not only
exhibits the origin of a particular term, but very frequently supplies
a key to the primary meanings of a large number of other related
terms. In the case of specific and anatomical terms, the meaning of
each Latin word is given clearly and literally.

In the study of Botany, Anatomy, and Geology, *specific names* are
apt to be misunderstood. At any rate, to the learner and young
student, they are often mysterious and puzzling. Accordingly, a considerable list of the second names of the binomials, used as specific
names, has been given by way of an Appendix, beginning on page 448.
In order to smooth the way to many not over familiar with Latin, or
entirely ignorant of that language, the list of specific names has been
prefaced by a short introduction ; and a considerable number of
examples of Latin nouns and adjectives have been declined, marked
for pronunciation, and defined. This will be found no less useful
to the student in the study of anatomical terms.

Following the specific names, there will be found a pretty full list of prefixes, including those used in connection with scientific terms. The list of postfixes contains only the common terminations of scientific terms. These are not only explained, but also illustrated by examples. In the examples the roots are printed in black letters, and the prefixes and postfixes in italics. A list of abbreviations in use by medical practitioners, with their unabbreviated forms and meanings, concludes the work.

Students or learners are recommended to use the present work before commencing their special studies, or, at least, in the earlier stages of them, as a daily lesson-book. Let a certain portion be accurately committed to memory daily, or frequently, and the very great advantage of such a course will be seen and felt on its accomplishment, for by so doing students will attain a competent knowledge of the spellings of the terms, their pronunciation, their root-words, and their definitions. Equipped with this knowledge, they will be able to follow the lectures and instructions of their teachers and professors both with freshness and intelligence. In short, by a little hard work to begin with, their after-studies will become very much more pleasant and profitable.

Though great care has been exercised in the preparation of the work, it is scarcely to be expected that it will be free from error. It is hoped, however, that errors will be unimportant, and few in number.

The Author has here to acknowledge the deep obligation under which he lies to G. W. Balfour, Esq., M.D., F.R.C.P., of Edinburgh, and to Alexander Morison, Esq., M.D., of Canonbury, London, for many excellent suggestions. The whole proof sheets were carefully read by Dr. Balfour, and a similar service was rendered by Dr. Morison on the MS. Of course, the suggestions and corrections made by these gentlemen chiefly regarded terms in those departments of science with which they were most conversant. The Author also gladly acknowledges similar favours from other friends. While thus gratefully acknowledging his obligation to these gentlemen, and other friends, it must be distinctly understood that the Author is alone responsible for any deficiencies which may be found in the work.

The Author of this compilation now submits his work to the judgment of professional men, and the general public, and he will be glad to learn that their judgment in regard to the objects of the work, as well as the manner of its execution, is a favourable one.

EDINBURGH, 15th March 1879.

THE FOLLOWING GENERAL RULES AS AFFECTING THE PRONUNCIATION OF A LARGE CLASS OF WORDS HAVING *c* OR *g* AS ONE OF THEIR ELEMENTS, CANNOT BUT PROVE USEFUL TO THE READER IN CONSULTING THE FOLLOWING PAGES. See list in *English Spellings and Spelling Rules*, p. 70.[1]

I. *c* is generally pronounced as *k* (1) when followed by one of the vowels *a, o, u*, as in 'cake,' 'becoming,' 'concuss'; (2) when followed by a consonant, except *h*, as in 'accord,' 'clime'; (3) when it terminates a word, as in 'physic,' 'music,' 'zinc.'

c is generally pronounced *s* when it comes before one of the vowels *e, i, y*, as in 'avarice,' 'cipher,' 'fancy.'

II. *g* is generally pronounced as *dj* before *e, i, y, œ*, as in 'page,' 'pageantry,' 'rage,' 'origin,' 'oxygen,' 'regent,' 'pugilism,' 'rugæ,' 'monogyn.'

g is generally hard (1) when it comes before the vowels *a, o, u*, as in 'prefigure,' 'regulate,' 'organ,' 'regard,' 'legume,' 'rigorous'; (2) when it comes before any consonant, except *h*, as in 'progress,' 'quagmire,' 'pugnacious'; (3) when it terminates a word, as in 'rag,' 'ring,' 'rung,' 'strong.'

Edinburgh : W. P. Nimmo.

THE SOUND SYMBOLS USED IN THE RE-SPELLINGS

FOR PRONUNCIATION IN

THE STUDENT'S MANUAL OF SCIENTIFIC TERMS.

a,	*ā* as in	mate, fate, fail, aye.	ou, ow, *ŏw*	as in	noun, bough, cow.
a,	*ă* ,,	mat, fat.	oi, oy, *ŏў*	,,	boy, soil.
a,	*â* ,,	far, calm, father.	u, ew, *ū*	,,	pure, due, few.
a, aw, *aw* ,,		awl, fall, law.	c,	*s* ,,	acid, cell, face.
e, ee, *ē* ,,		mete, meet, feet, free.	c,	*k* ,,	cone, colic, tract.
e,	*ĕ* ,,	met, bed.	ch,	*tsh* ,,	chair, larch, church.
e,	*ė* ,,	her, fern, heard.	ch,	*sh* ,,	chemise, drench, match.
i,	*ī* ,,	pine, height, sigh, tie.	ch,	*k* ,,	chaos, anchor, scholar.
i,	*ĭ* ,,	pin, tin, ability.	g,	*g* ,,	game, gone, gun.
o,	*ō* ,,	note, toll, soul.	g,	*j* ,,	George, gem, gin.
o,	*ŏ* ,,	not, plot.	g, dg, *dj* ,,		judge, ledge, rage.
o, oo, *ŏ* ,,		move, smooth.	th,	*th* ,,	thing, breath.
o, oo, *ŏŏ* ,,		woman, foot, soot.	th,	*tħ* ,,	there, breathe.

NOTE.—Among well-educated people, in Scotland at least, *ī* seems to have two sounds—

1. *ī* in its proper name-sound, as in sigh, try, high, my, tie, liar, bye, hire, sire.
2. *ĕĭ* forming a sound resulting from the combined sounds of *e* and *i*, as in height, pine, mine, sight, write, white, flight, fright, might, trite.

COMMON ABBREVIATIONS USED IN THE WORK.

a. or adj.,	.	adjective.
anat.,	. .	anatomy.
anc.,	. .	ancient.
bot.,	. .	botany.
chem.,	. .	chemistry.
dim.,	. .	diminutive.
E.,	. . .	East.
entom.,	. .	entomology.
far.,	. . .	farriery.
fem.,	. . .	feminine.
gen.,	. . .	gender, genitive.
geol.,	. .	geology.
hort.,	. .	horticulture.
instr.,	. .	instrument.
masc.,	. .	masculine.
med.,	. .	medicine.
min.,	. .	mineralogy.
N.,	. . .	North.
n.,	. . .	noun.

nat. hist.,	.	natural history.
neut.,	. .	neuter.
nom.,	. .	nominative.
obj.,	. . .	objective.
ornith.,	. .	ornithology.
palaeon.,	.	palaeontology.
path.,	. .	pathology.
pert.,	. .	pertaining.
phren.,	. .	phrenology.
phys.,	. .	physiology, physics.
plu.,	. . .	plural.
poss.,	. .	possessive.
pref.,	. .	prefix.
S.,	. . .	South.
sing.,	. .	singular.
superl.,	. .	superlative.
surg.,	. .	surgery.
W.,	. . .	West.
zool.,	. .	zoology.

ABBREVIATIONS OF PROPER NAMES, FORMING NAMES OF LANGUAGES, WHICH DESIGNATE ROOT-WORDS.

Ar.,	. . .	Arabic.
AS.,	. . .	Anglo-Saxon.
Celt.,	. .	Celtic.
Chin.,	. .	Chinese.
Dan.,	. .	Danish.
Dut.,	. .	Dutch.
Eng.,	. .	English.
F.,	. . .	French.
Gael.,	. .	Gaelic.
Ger.,	. .	German.
Gr.,	. . .	Greek.

Icel.,	. .	Icelandic.
It.,	. . .	Italian.
L.,	. . .	Latin.
mid. L.,	.	Latin of the Middle Ages, late Latin, or Latin not classical.
Norm. F.,		Norman French.
Old Eng.,	.	Old English.
Sans.,	. .	Sanscrit.
Scot.,	. .	Scotch.
Sp.,	. .	Spanish.

NOTE.—For Abbreviations of Medical Terms, see page 483.

THE STUDENT'S MANUAL

OF

SCIENTIFIC TERMS.

———◆———

abaxial, a., *ăb·ăks'·ĭ·ăl* (*ab*, from, and *axial*), not in the axis; in *bot.*, applied to the embryo which is out of the axis of the seed: **abaxile**, a., *ăb·ăks'·ĭl*, in same sense.

abbreviated, a., *ăb·brēv'·ĭ·āt·ĕd*, also **abbreviate**, a., *ăb·brēv'·ĭ·āt* (L. *ab*, from; *brevis*, short), in *bot.*, applied to one part when shorter than another.

abdomen, n., *ăb·dōm'·ĕn* (L. *abdōmen*, the belly—from *abdo*, I conceal), the lower belly : **abdominal**, a., *ăb·dŏm'·ĭn·ăl*, belonging to the lower belly.

abducent, a., *ăb·dūs'·ĕnt* (L. *abducens*, leading away or from), separating ; drawing back : **abducens**, n., *ăb·dūs'·ĕnz* (L.), applied to the sixth cranial nerve, which, distributed to the external rectus muscle, turns the eyeball outwards ; hence it is called the **abducens oculi**, *ŏk'·ūl·ī* (L. *oculi*, of the eye).

abduction, n., *ăb·dŭk'·shŭn* (L. *ab*, from ; *duco*, I lead, *ductus*, led), the act of drawing away from ; the movement of a limb from the median line—that is, the middle line of the body ; see ' median line.'

abductor, n., *ăb·dŭkt'·ŏr* (L. *abductor*, that which draws out-

A

wards), a muscle that draws a limb or part outwards : **abductor indicis manus**, *ĭn'·dĭs·ĭs măn'·ŭs* (L. *index*, an index, *indĭcis*, of the index ; and *manŭs*, the hand, *manŭs*, of the hand), the muscle that puts outwards the index finger of the hand : **abductor minimi digiti**, *mĭn'·ĭm·ī dĭdj'·ĭt·ī* (L. *minimus*, the least ; *digitus*, the finger), the abductor of the least finger ; the muscle that draws away the little finger from the ring finger : **abductor minimi digiti pedis**, *pĕd'·ĭs* (L. *pēs*, a foot, *pĕdis*, of a foot), the abductor of the least finger of the foot ; the muscle that draws the little toe away from its neighbour : **abductor oculi**, *ŏk'·ūl·ī* (L. *oculus*, the eye, *oculi*, of the eye), the muscle that extends or expands the eye : **abductor pollicis manus**, *pŏl'·lĭs·ĭs măn'·ūs* (L. *pollex*, the thumb, *pollicis*, of the thumb ; *manŭs*, the hand, *manŭs*, of the hand), the muscle that draws outwards the thumb of the hand : **abductor pollicis pedis**, *pĕd'·ĭs* (L. *pēs*, a foot, *pĕdis*, of a foot), the muscle that extends the thumb or great toe of the foot.

aberrant, u., *ăb·ĕr'·ănt* (L. *ab*, from ; *errans*, wandering, gen. *errantis*), departing from the regular type: **aberration, n.,** *ăb·ĕr·ā'shŭn,* a disordered state of the intellect; any deviation from the usual and natural appearance.

Abies, n., *ăb'ĭ·ēz* (L. *abiēs,* the silver fir, *abiĕtis,* of the silver fir), the fir tree; a genus of trees: **Abietineæ, n.** plu., *ăb'ĭ·ĕt·ĭn'ĕ·ē,* the sub-order of the Coniferæ or cone-bearing family, including the fir and spruce: **abietic, a.,** *ăb'ĭ·ĕt'ĭk,* belonging to the fir tree: **Abies picea,** *pĭs'ĕ·ă* (Gr. *peukē,* the pine or pitch tree), the silver fir which furnishes turpentine: **A. balsamea,** *băl·săm'ĕ·ă* (L. *balsameus,* balsamic—from *balsămum,* balsam), a species which produces Canada balsam; balm of Gilead fir: **A. Canadensis,** *kăn'ă·dĕns'ĭs* (from Canada), hemlock spruce, which furnishes a balsam: **A. excelsa,** *ĕk·sĕls'ă* (L. *excelsus,* lofty, high), the Norway spruce, producing Burgundy pitch: **A. nigra,** *nĭg'ră* (L. *nĭger* or *nĭgra,* black), the black spruce, from which, and other species, spruce beer is made.

abiogenesis, n., *ăb'ĭ·ō·jĕn'ĕs·ĭs* (Gr. *a,* without; *bios,* life; *genesis,* origin, source), the doctrine that living bodies may be evolved from inorganic matter; spontaneous generation.

ablactation, n., *ăb'lăkt·ā'shŭn* (L. *ab,* from; *lacto,* I suckle—from *lac,* milk), weaning of a child from the breast.

ablation, n., *ăb·lā'shŭn* (L. *ablātum,* to take or bear away—from *ab,* from or away; *latum,* to carry or bear), a taking away; the removal of a part from a body by excision, extirpation, or amputation.

ablepsia, n., *ă·blĕps'ĭ·ă,* also **ablepsy, n.,** *ă·blĕps'ĭ* (Gr. *a,* without, not; *blepo,* I see, *blepso,* I shall see), want of sight; blindness.

abluent, n., a., *ăb'lŏ·ĕnt* (L. *ablŭo,* I wash off or away—from *ab,* from;

lŭo, I wash, *luens,* washing), a substance which carries off impurities; same as 'abstergent'and 'detergent': **ablution, n.,** *ăb·lō'shŭn,* a cleansing or purification.

abnormal, a., *ăb·nŏrm'ăl* (L. *ab,* from; *norma,* a rule), anything out of the usual or natural course; irregular.

abomasum, n., *ăb'ŏm·āz'ŭm,* also **abomas'us,** *-āz'ŭs* (L. *ab,* from; *ŏmāsum,* tripe, the paunch), the fourth cavity of the stomach of ruminant animals, as the cow.

aborticide, n., *ăb·ŏrt'ĭ·sīd* (L. *abortus,* an untimely birth; *cædo,* I kill), the destroying of the fœtus in utero to effect a delivery: **abortion, n.,** *ăb·ŏr'shŭn* (L. *abortus*), expulsion of the fœtus before its time; miscarriage; in *bot.,* the incomplete or non-formation of a part: **abortive, a.,** *ăb·ŏrt'ĭv,* not come to maturity; treating disease with the view of arresting its further development; barren.

abrachia, n., *ă·brāk'ĭ·ă* (Gr. *a,* without, not; Gr. *brachion,* L. *brachium,* the arm), imperfect development, or entire absence of the arms.

abranchiate, a., *ă·brăng'kĭ·āt* (Gr. *a,* without; Gr. *brangchia,* L. *branchiæ,* the gills of a fish), destitute of gills or branchiæ; without lungs: **abranchiata, n.** plu., *ă·brăng'kĭ·āt'ă,* animals which have no apparent organs of respiration, as the earthworm and leech.

abrasion, n., *ăb·rā'shŭn* (L. *ab,* from; *rāsus,* scraped), a partial rubbing off or tearing of the skin.

abrupt, a., *ăb·rŭpt'* (L. *ab,* from; *ruptus,* broken), appearing as if broken or cut off at the extremity: **abruptly-acuminate, a.,** *-ăk·ūm'ĭn·āt* (L. *acūmen,* a point), having a broad extremity, as a leaf, from which a point arises: **abruptly-pinnate, a.,** *-pĭn'nāt* (L. *pinna,* a feather or fin), having two or more leaflets attached to each side of a central rib without a terminal or odd leaflet.

abscess, n., *ăb'·sĕs* (L. *abscessus*, an abscess—from *abs*, from or away ; *cessum*, to depart), a gathering of humour or pus in some part of the body.

abscission, n., *ăb·sĭsh'·ŭn* (L. *ab*, from ; *scissum*, to cut, *scissus*, cut), in *surg.*, a cutting off or removal of a part ; the premature ending of a malady ; in *bot.*, a cutting off ; the separation of segments or frustules : **abscissa**, n., *ăb·sĭs'·să*, a part of the diameter of a conic section—the plurals are **abscissas**, *ăb·sĭs'·săz*, and **abscissæ**, *ăb·sĭs'·sē*.

absinthe, n., *ăb'·sĭnth* (L. *absinthium*, Gr. *absinthion*, wormwood), a strong alcoholic liqueur, highly flavoured with a tincture of wormwood : **absinthian**, a., *ăb·sĭnth'·ĭ·ăn*, of the nature of wormwood, or pert. to it : **absinthiated**, a., impregnated with wormwood : **absinthate**, n., *ăb·sĭnth'·āt*, combination of absinthic acid with a base : **absinthic**, a., belonging to absinthium ; denoting an acid obtained from it : **absinthin**, n., the bitter principle discovered in absinthium : **absinthium**, n., *ăb·sĭnth'·ĭ·ŭm*, the name, in the pharmacopœia, of the *artemis'ia absinth'ium* : **absinthism**, n., the symptoms produced by the excessive use of the liqueur absinthe.

abstergent, a., n., *ăb·stĕrj'·ĕnt* (L. *abstergens*, wiping dry—from *abs*, from or away ; *tergeo*, I rub off), cleansing; a medicine that cleanses from foulness or sores.

abyssic, a., *ă·bĭs'·ĭk* (Gr. *abussos*, without a bottom—from *a*, without, not ; *bussos*, a bottom), applied to the earths which form the bottoms of ancient seas.

Acacia, n., *ăk ā'·shĭ·ă* (L. *acacia*, a thorn), a genus of Oriental trees ; the Egyptian thorn ; gum arabic, Ord. Leguminosæ, Sub-ord. Mim-

osæ : **Acacia tortilis**, *tŏrt'·ĭl·ĭs* (L. *tortilis*, twined, twisted): A. **Arabica**, *ăr·ăb'·ĭk·ă* (from Arabia) ; **A. vera**, *vēr'·ă* (L. *vērus*, real, genuine); **A. gummifera**, *gŭm·ĭf'·ĕr·ă* (L. *gummi*, gum ; *fero*, I bear) ; **A. albida**, *ăl'·bĭd·ă* (L. *albidus*, whitish), and other species, yield the gums or gummy substances known as gum Arabic, gum Senegal, East Indian gum, etc.: A. **catechu**, *kăt'·ē·kū*, or *kăt'·ē·shōo* (said to be from Japanese *kate*, a tree ; *chu*, juice), an Indian shrub, which furnishes a kind of catechu, is used for tanning, and a powerful astringent : A. **formosus**, *fŏrm·ōz'·ŭs* (L. *formōsus*, finely formed, handsome—from *forma*, shape), a species supplying the Cuban timber called *sabicu*.

Acalephæ, n. plu., *ăk'·ăl·ēf'·ē* (Gr. *akalēphe*, a nettle), a name applied to the jelly-fishes, sea-nettles, and other radiate animals, from their power of stinging : **acalephoid**, n., *ăk'·ăl·ēf'·oyd* (Gr. *eidos*, resemblance), an animal resembling a jelly-fish or sea-nettle, etc. : **acalephous**, a., *ăk·ăl·ēf'·ŭs*, belonging to a nettle ; belonging to the Acalephæ.

Acanthaceæ, n. plu., *ăk'·ănth·ā'·sē·ē* (Gr. *akantha*, a spine), the Acanthus family, an order of herbaceous plants, chiefly tropical : **Acanthus**, n., *ăk·ănth'·ŭs*, a genus of plants ; bear's breech : **Acanthus mollis**, *mŏl'·lĭs* (L. *mollis*, pliant, supple), a species the leaves of which, with their sinuated lobes, are said to have given origin to the capital of the Corinthian pillar **:** **acanthaceous**, *ăk'·ănth·ā'·shŭs*, also **acanthine**, a., armed with prickles: **Acanthocephala**, n. plu., *ăk·ănth'·ō·sĕf'·ăl·ă* (Gr. *kephalē*, the head), a class of parasitic worms in which the head is armed with spines : **Acanthometrina**, n. plu., *ăk·ănth'·ō·mĕt'·rĭn·ă* (Gr. *mētra*, a womb), a family of protozoa characterized by having radiated siliceous spines : **Acanth-opterygii**, n. plu., *ăk'-ănth-ŏp-tĕr-ĭdj'-ĭ-ĭ* (Gr. *pterugion*, a winglet or fin—from *pterux*, a wing), a group of bony fishes with the spinous rays in the front of the dorsal fin.

Acarina, n. plu., *ăk'-ăr-ĭn'-ă* (L. *acărus*, Gr. *akări*, a mite), a division of the Arachnida of which the cheese mite is the type : **acaroid**, a., *ăk'-ăr-ŏyd* (Gr. *eidos*, resemblance), resembling the acarus or louse : **Acarus**, n., *ăk'-ăr-ŭs* (L.), a genus of insects which infest the skin : **Acari**, n. plu., *ăk'-ăr-ĭ;* **Acaridæ**, n. plu., *ăk-ăr'-ĭd-ē*, the systematic name for such insects as the mite, the tick, the water mite, etc.

acaulis, a., *ăk-āwl'-ĭs*, **acaulous**, a., *ăk-āwl'-us*, and **acauline**, a., *ăk-āwl'-ĭn* (Gr. *a*, without; Gr. *kaulos*, L. *caulis*, a stalk), without a stalk; stemless : **acaules-cent**, a., *ăk'-āwl-ĕs'-ĕnt*, having a shortened stem; denoting the non-development of the growing axis : **acaulosia**, n., *ăk'-āwl-ŏz'-ĭ-ă*, the non-development of the stem of a plant.

accelerator, n., *ăk-sĕl'-ĕr-āt'-ŏr* (L. *accelero*, I hasten forward—from *ad*, to; *celero*, I hasten, *celer*, swift), a hastener forward; that which causes to move faster : **accelerator urinæ**, *ŭr-ĭn'-ē* (L. *accelerator*, a hastener; *urina*, urine, *urinæ*, of urine), the accelerator of urine; a muscle of the penis whose action propels the urine; when it does the same for the semen, it is called the *ejacul-ator seminis*.

accessorius, n., *ăk-sĕs-sŏr'-ĭ-ŭs* (L. *accessus*, a coming to, an approach —from *ad*, to ; *cessum*, to go), denoting any muscular appendage which assists the action of a larger muscle ; denoting two nerves of the neck : **ac-cessorius ad sacro-lumbalem**, *ăd sāk'-rō lŭm-băl'-ĕm* (L.

ad, to ; *sacro*, sacred ; *lumbalem*, accus., *lumbalis*, nom., pert. to the *lumbus* or loin), denoting the muscle which acts as an assistant to the *sacro-lumbalis*, consisting of muscular slips which pass from the lower six to the upper six ribs, near their angles : **accessorius obturatorii**, *ŏb'-tūr-ăt-ōr'-ĭ-ĭ* (L. *obturātor*, a stopper up of a cavity, *obturătōrius*, pert. to the stopper up of a cavity, *obturatorii*, gen.— from *obtūro*, I stop up), the accessory or assistant of the obturator—applied to a muscle or nerve which assists, or is an appendage to, the *obturator* : **accessorius pedis**, *pĕd'-ĭs* (L. *pēs*, a foot, *pĕdis*, of a foot), an accessory muscle of the foot, arising from the under surface of the *os calcis* by two heads, and assists to bring the line of traction of the flexor tendons into the centre of the foot.

accouchement, n., *ăk-kōōsh'-mŏng* (F. *accoucher*, to deliver—from *coucher*, to lay down ; L. *ad*, to ; *colloco*, I lay in a place), lying in child-birth ; the act of parturition : **accoucheur**, n., *ăk'-kōōsh-ér'*, a surgeon who attends women in child-birth ; a man-midwife ; an obstetrician : **accoucheuse**, n., *ăk'-kōōsh-āz'*, a female who practises midwifery ; a midwife.

accrescent, a., *ăk-krĕs'-ĕnt* (L. *ad*, to ; *crescens*, growing), denoting plants continuing to grow and increase after flowering.

accrete, a., *ăk-krēt'* (L. *ad*, to; *cret-um*, to grow), grown together : **accretion**, n., *ăk-krēsh'-ŭn*, the act of growing by increase ; growth by external addition to new matter ; in *surg.*, the conjunction of parts naturally separate.

accumbent, a., *ăk-kŭmb'-ĕnt* (L. *accumbens*, lying on—from *ad*, to or on ; *cubo*, I lie down), lying on ; supine ; prostrate ; in *bot.*, applied to the embryo of the Cruciferæ when the cotyledons lie on their edges ; applied to the folded radicle: **accument**, a., *ăk·ŭm'·ĕnt*, in *bot.*, lying against another body.

acephalia, n., *ăs'·ĕ·făl'·ĭ·ă* (Gr. *a*, without ; *kephale*, the head), the condition of a monster without a head: **acephalous**, a., *ăs·ĕf'·ăl·ŭs*, not possessing a distinct head.

Aceraceæ, n. plu., *ăs·ĕr·ā'·sĕ·ē* (L. *ăcer*, a maple tree), the maple family, including the sycamore and Scotch plane tree: **Acer**, n., *ăs'·ĕr*, a genus of trees, for the most part beautiful and of considerable size: **Acer saccharinum**, *săk'·kăr·ĭn'·ŭm* (L. *saccharon*, sweet juice, sugar), the sugar maple of America : **aceric**, a., *ăs·ĕr'·ĭk*, denoting an acid found in its juice.

acerose, a., *ăs'·ĕr·ōz*, and **acerous**, a., *ăs'·ĕr·ŭs* (L. *ăcer*, sharp-pointed, *ăceris*, gen.), having a sharp point; narrow and slender.

acervuli, n. plu., *ăs·ĕrv'·ŭl·ī* (L. *ăscĕrvus*, a heap), in *bot.*, small heaps or clusters.

acetabulum, n., *ăs'·ĕt·ăb'·ŭl·ŭm* (L. *acētăbŭlum*, a sucker, a vinegar cruet, a cup-shaped vessel), the cup-shaped socket of the hip-joint ; the socket of the innominate bone which receives the head of the femur; one of the cup-like sucking discs on the arms of the cuttle-fish: **acetabula**, n. plu., *ăs'·ĕt·ăb'·ŭl·ă*, the sockets of the hip-joints ; the suckers of the cuttle-fishes: **acetabuliform**, a., *ăs'·ĕt·ăb·ŭl'·ĭ·fŏrm* (L. *forma*, shape), in the form of a cup.

acetic, a., *ăs·ĕt'·ĭk* (L. *acētum*, vinegar), denoting an acid ; vinegar : **acetate**, n., *ăs'·ĕt·āt*, the combination of acetic acid with a salifiable base.

acheilary, a., *ă·kīl'·ăr·ĭ* (Gr. *a*, without ; *cheilos*, a lip), in *bot.*, having the labellum undeveloped, as in some orchids.

achene, n., *ăk·ēn'·ĕ*, also **achænium**, n., *ăk·ēn'·ĭ·ŭm;* **achænia**, plu. (Gr. *achanēs*, not gaping, not opening the mouth—from *a*, not; *chainō*, I yawn or crack, as ripe fruit), a monospermal seed-vessel which does not open or crack, whose pericarp does not adhere to the seed: **achenodium**, n., *ăk'·ĕn·ōd'·ĭ·ŭm* (the Latinised suffix, *ode*, signifying 'fulness of'), a fruit composed of many achænia.

Achillis tendo, *ăk·ĭl'·lĭs tĕnd'·ō* (L. *tendo*, a tendon ; *Achillis*, of Achilles), the vulnerable tendon or part in the heel of Achilles ; the united strong tendon of the gastrocnemius and solæus muscles.

achimenes, n., *ăk'·ĭ·mēn'·ēz* (a word of unknown meaning, originally given by Dr. Patrick Browne), an elegant and free-flowering genus of plants, Ord. Gesneraceæ.

achlamydeous, a., *ăk'·lăm·ĭd'·ĕ·ŭs* (Gr. *a*, without; *chlamus*, a loose warm cloak), in *bot.*, having no floral envelope ; denoting naked flowers : **Achlamydeæ**, n. plu., *ăk'·lăm·ĭd'·ĕ·ē*, the class of naked flowers having only the essential organs and no floral envelope.

acholia, n., *ă·kōl'·ĭ·ă* (Gr. *a*, without ; *cholē*, bile), absence or deficiency of bile, occurring in acute atrophy of the liver.

Achras, n., *ăk'·răs* (Gr. *achras*, a species of wild pear tree, its fruit), a genus of trees of the sapotaceæ or sapadilla family, some of which yield edible fruits.

achroma, n., *ăk·rōm'·ă* (Gr. *a*, without ; *chrōma*, colour), a deficiency in the colour of the skin : **achromatic**, a., *ăk'·rŏm·ăt'·ĭk*, free from colour ; applied to lenses which show objects without any prismatic colours: **achromatism**, n., *ăk·rŏm'·ăt·ĭzm*, the state of optical instruments which show objects without prismatic colours : **achromatopsy**, n., *ăk·rŏm'·ăt·ŏps'·ĭ* (Gr. *opsis*, sight), incapacity of distinguishing colours ; colour blindness.

acicular, a., *ăs·ĭk'·ŭl·ăr* (L. *ăcus*, a needle, *acicŭla*, a little needle), in shape like a needle ; having sharp points like needles : **aciculus,** n., *ăs·ĭk'·ŭl·ŭs*, in *bot.*, a strong bristle ; a little spike.

acinaciform, a., *ăs'·ĭn·ăs'·ĭ·fŏrm* (L. *acinăces*, a short sword, a sabre ; *forma*, shape), shaped like a sabre or scimitar : **acinacifolious,** a., *ăs'·ĭn·ăs·ĭ·fŏl'·ĭ·ŭs* (L. *folium*, a leaf), having leaves shaped like a sabre or scimitar.

acinus, n., *ăs'·ĭn·ŭs* (L. *acĭnus*, a berry, a stone or seed in a berry, *acĭni* plu.), in *bot.*, one of the pulpy drupels forming the fruit of the raspberry or bramble ; in *surg.*, small granulations of the liver and similar bodies : **aciniform,** a., *ăs·ĭn'·ĭ·fŏrm* (L. *forma*, shape), having the form or colour of a clustered fruit, as the raspberry.

acne, n., *ăk'·nē* (Gr. *a*, intensive ; *kneō*, I scrape or gnaw), an eruption of hard inflamed tubercles often appearing in youth, generally on the neck, face, shoulders, and breast, sometimes lasting for years—so called in allusion to their appearance : **acne rosacea,** *rōz·ăs'·ē·ă* (L. *rosācĕus*, of or pert. to roses), the ruddy uneven nose of some who indulge in the excessive use of alcoholic liquors.

aconitum napellus, *ăk·ŏn·ĭt'·ŭm năp·ĕl'·lŭs* (Gr. *akoniton*, the poisonous herb monk's-hood, *akontion*, a small dart—from *akōn*, a dart, as darts were dipt in its poisonous juice ; L. *năpellus*, diminutive of *năpus*, a turnip), the plant monk's-hood, friar's-cap, wolf's-bane, or helmet-flower, which contains a narcotic alkaloid, one of the most deadly poisons known : **aconite,** n., *ăk'·ŏn·ĭt*, the common name for aconitum napellus : **aconitine,** or **aconitia,** *ăk·ŏn'·ĭt·ĭn*, or *-ĭsh'·ĭ·ă*, the alkaloid of aconite forming its active principle.

Acontotheci, n. plu., *ăk·ŏn·tŏth'·ĕ·sī* (Gr. *akōn*, unwilling ; *tithēmi*, I put or place), a family of intestinal worms.

Acorus, n., *ăk'·ŏr·ŭs* (L. *acorus*, Gr. *akoros* and *akoron*, the sweet-scented flag), a genus of plants ; the sweet flag, which has an aromatic odour combined with a bitterish acrid taste : **Acorus calamus,** *kăl'·ăm·ŭs* (L. *calamus*, a reed), the systematic name of *acorus* : **Acoreæ,** n. plu., *ăk·ŏr'·ē·ē*, the sweet flag tribe.

acotyledon, n., *ăk'·ŏt·ĭl·ēd'·ŏn* (Gr. *a*, without ; *kotulēdon*, a seed lobe), a plant whose embryos or germs have no seed lobes : **acotyledonous,** a., *ăk'·ŏt·ĭl·ēd'·ŏn·ŭs*, having no seed lobes.

Acrita, n., *ăk'·rĭt·ă* (Gr. *akritos*, confused), the lowest division of the animal kingdom, in which the organs are supposed to be combined confusedly with the other parts ; synonym of *protozoa*.

acrobrya, n. plu., *ăk'·rŏ·brī'·ă* (Gr. *akros*, at the highest point ; *bruo*, I bud forth or germinate), a plant in which the growth is formed by additions in an upward direction ; synonym of *acrogens*.

acrocarpi, n. plu., *ăk'·rŏ·kărp'·ī* (Gr. *akros*, at the highest point ; *karpos*, fruit), mosses having their fructification terminating the axis : **acrocarpous,** a., *ăk'·rŏ·kărp'·ŭs*, having the fructification terminating the axis.

acrogens, n. plu., *ăk'·rŏ·jĕnz* (Gr. *akros*, at the highest point ; *gennaō*, I produce), those plants which increase by growth at the summits or growing points, and whose stems do not materially increase in bulk, as the stems of ferns : **acrogenous,** a., *ăk·rŏdj'·ĕn·ŭs*, increasing by growth at the summit or growing point.

acromium, n., *ăk·rōm'·ĭ·ŭm*, also **acromion,** *ăk·rōm'·ĭ·ŏn* (Gr.

akros, high, extreme ; *omos*, a shoulder), the projecting or outer part or process of the scapula or shoulder : acromial, a., *ăk·rōm'-ĭ·ăl*, of or belonging to the acromium : **acromiales cutanei,** *ăk·rōm'·ĭ·āl'·ēz kū·tān'·ĕ·ī* (L. *acromiales*, plu., pert. to the acromium ; *cutanei*, plu., belonging to the skin), designating those nerves which pass over the acromium, and are thence distributed to the skin : **acromio,** *ăk·rōm'·ĭ·ō*, indicating connection with the acromion : **acromioclavicular,** *klăv·ĭk'·ūl·ăr* (L. *clavis*, a key), denoting the articulation at the shoulder of the outer end of the clavicle in the acromion process of the scapula ; also denoting two ligaments of the scapula and clavicle, named respectively the 'superior' and the 'inferior.'

acropetal, a., *ăk·rŏp'·ĕt·ăl* (Gr. *akros*, at the highest point ; L. *peto*, I seek), in *bot.*, seeking the summit ; applied to the development of lateral shoots from an axis.

acrospire, n., *ăk'·rō·spīr* (Gr. *akros*, the summit ; *speira*, a spiral line), the first shoot or sprout at the end of a germinating seed : **acrospore,** n., *ăk'·rō·spōr* (Gr. *spōra*, seed), a spore borne on the summit of a thread.

actea, n., *ăk·tē'·ă*, or **actæa racemosa,** *ăk·tē'·ă răs'·ēm·ōz'·ă* (Gr. *aktaia*, the elder tree ; L. *racēmōsus*, full of clusters, clustering), in *med.*, the black snake-root, black cohosh or bugbane, a sedative used in the treatment of rheumatism, Ord. Ranunculaceæ.

actinenchyma, n., *ăkt'·ĭn·ĕng'·kĭm·ă* (Gr. *aktin*, a ray ; *engchuma*, an infusion), in *bot.*, cellular tissue having a starlike or stellate form.

actinism, n., *ăkt'·ĭn·ĭzm* (Gr. *aktin*, a ray), the chemical action of sunlight : **actinocarpous,** a., *ăkt'·ĭn·ō·kărp'·ŭs* (Gr. *karpos*, fruit),

having trophosperms radiated like the rays of fruit : **actinograms,** n. plu., *ăct·ĭn'·ō·grăms* (Gr. *gramma*, a letter), the results recorded by the actinograph : **actinograph,** n., *ăkt·ĭn'·ō·grăf* (Gr. *grapho*, I write), an instrument for recording the quantity of actinism present : **actinography,** n., *-răf'·ĭ*, a description of the rays of light : **actinoid,** n., *ăkt'·ĭn·ōyd* (Gr. *eidos*, resemblance), resembling a ray : **actinology,** n., *ăkt'·ĭn·ŏl'·ō·jĭ* (Gr. *logos*, discourse), the doctrine of the rays of light : **actinomeres,** n. plu., *ăkt·ĭn·ŏm'·ĕr·ēz* (Gr. *meros*, a part), in *zool.*, the lobes which are mapped out on the surface of the body of the ctenophora by the ctenophores, or comb-like rows of cilia : **Actinosoma,** n., *ăkt'·ĭn·ō·sōm'·ă* (Gr. *soma*, a body), the entire body of any actinozoön, whether simple as in the sea-anemones, or composed of several zoöids as in most corals : **Actinozoon,** n., *ăkt'·ĭn·ō·zō'·ŏn*, **Actinozoa,** n. plu., *-zō'·ă* (Gr. *zoōn*, an animal), the division of the Cælenterata, of which the sea-anemones are the type.

aculeate, a., *ăk·ūl'·ĕ·āt* (L. *acūleus*, a prickle or thorn—from *ăcus*, a needle), having prickles or sharp points : **aculeiform,** a., *ăk'·ūl·e'·ĭ·fŏrm* (L. *forma*, shape), formed like a prickle or thorn : **aculeus,** n., *ăk·ūl'·ĕ·ŭs*, a prickle forming a process of the bark only, as in the rose : **aculei,** plu., *ăk·ūl'·ĕ·ī*.

acuminate, a., *ăk·ūm'·ĭn·āt* (L. *acūmen*, a point, *acūmĭnis*, of a point), in *bot.*, drawn out into a long point ; tapering : **acuminiferous,** a., *ăk·ūm'·ĭn·ĭf'·ĕr·ŭs* (L. *fero*, I bear), in *zool.*, having pointed tubercles : **acuminulate,** a., *ăk'·ūm·ĭn'·ŭl·āt*, having a very sharp taper-point.

acupressure, n., *ăk·ū·prĕsh'·ūr* (L. *ăcus*, a needle ; *pressum*, to press), the employment of needles instead of ligatures for arresting hæmorrhage from a cut or wounded vessel, by pressing a needle through it and pinning it against an adjacent tissue or bone.

acupuncture, n., *ăk'·ū·pŭngk'·tūr* (L. *ăcus*, a needle ; *punctum*, to prick or puncture), a method of lessening pain in a diseased part, as in neuralgia, by running into it one or more long fine needles.

acute, a., *ăk·ūt'* (L. *acūtus*, sharp, pointed), in *bot.*, terminating gradually in a sharp point.

acyclic, a., *ă·sīk'·lĭk* (Gr. *a*, without ; Eng. *cyclic*), without a cycle or circle.

ad deliquium, *ăd dē·lĭk'·wĭ·ŭm* (L. *ad*, to ; *deliquium*, a want or defect), to fainting—a direction in medicine given for venesection.

adduction, n., *ăd·dŭk'·shŭn* (L. *ad*, to ; *ductus*, led), the act of moving towards ; the movement of a limb towards the median or middle line of the body.

adductor, n., *ăd·dŭct'·ŏr* (L. *adductor*, that which draws towards—from *ad*, to ; *ductus*, led), a muscle that brings one part towards another : **adductor brevis,** *brĕv'·ĭs* (L. *brĕvis*, short), the name of a muscle which brings one part towards another, so called from its shortness : **adductor longus,** *lŏng'·gŭs* (L. *longus*, long), a muscle which brings one part towards another, so called from its length : **adductor magnus,** *măg'·nŭs* (L. *magnus*, great), a muscle, so called from its size, which brings the thigh inwards and upwards : **adductor minimi digiti,** *mĭn'·ĭm·ĭ dĭdj'·ĭt·ĭ* (L. *minimus*, the least, *minimi*, of the least ; *digitus*, a finger, *digiti*, of a finger), the adductor of the little finger ; the deepest of the muscles of the little finger, which arises from the unciform process and the annular ligament : **adductor pollicis**

manus, *pŏl'·lĭs·ĭs măn'·ŭs* (L. *pollex*, the thumb, *pollĭcis*, of the thumb ; *manŭs*, the hand, *manŭs*, of the hand), the adductor of the thumb of the hand ; the muscle that draws inwards the thumb of the hand : **adductor pollicis pedis,** *pĕd'·ĭs* (L. *pēs*, a foot, *pĕdis*, of a foot), the adductor of the great toe or thumb of the foot ; a muscle arising from the bases of the second, third, and fourth metatarsal, which draws the great toe inwards.

Adelarthrosomata, n. plu., *ă·dĕl'·ăr·thrō·sŏm'·ăt·ă* (Gr. *adēlos*, hidden ; *arthros*, a joint ; *soma*, a body), in *zool.*, an order of the Arachnida, comprising the harvest spiders, the book scorpions, etc.

adelphous, a., *ăd·ĕlf'·ŭs* (Gr. *adelphos*, a brother, a blood relation), related ; having an affinity ; in *composition*, a union of filaments.

adesmy, n., *ă·dĕs'·mĭ* (Gr. *a*, without ; *desmos*, a bond or ligament), in *bot.*, a break or division in an organ usually entire.

adherent, a., *ăd·hĕr'·ĕnt* (L. *adhærens*, cleaving or sticking to, *adhærentis*, gen.—from *ad*, to ; *hæreo*, I stick), in *bot.*, denoting the union of parts that are normally separate, and in different verticils, as the calyx when united to the ovary : **adhesion,** n., *ăd·hēzh'·ŭn* (L. *ad*, to ; *hæsum*, to stick), a union of parts of any body by means of cement, glue, growth, etc. ; in *surg.*, the reunion of parts that have been severed.

adiantum, n., *ăd'·ĭ·ănt'·ŭm* (Gr. *adianton*, the herb maidenhair—from *adiantos*, not moistened ; so called from the belief that they will remain dry, though plunged among water), maidenhair, an elegant species of ferns with beautiful leaves : **Adiantum capillus Veneris,** *kăp·ĭl'·lŭs vĕn'·ĕr·ĭs* (L. *capillus*, hair ; *Venus*, goddess of love, *vĕnĕris* of Venus), systematic name for adiantum : **Adiantum pedatum**, *pĕd·ăt'·ŭm* (L. *pedātus*, furnished with feet—from *pes*, a foot, *pedis*, of a foot), this species, and the preceding, furnish the syrup of Capillaire, particularly the latter.

adipose, a., *ăd'·ĭp·ōs* (L. *adipōsus*, fatty—from *adeps*, fat), denoting the fatty tissue which exists more or less throughout the body : **adipocere**, n., *ăd'·ĭp·ŏ·sēr'* (L. *cera*, wax), a fatty substance of a whitish grey colour, into which animal flesh and fat is often changed when buried in moist ground ; grave wax : **adiposis**, n., *ăd'·ĭp·ōz'·ĭs*, great fatness or obesity of the human body : **adiposus panniculus**, *ăd'·ĭp·ōz'·ŭs păn·nĭk'·ŭl·ŭs* (L. *panniculus*, a small piece of cloth—from *pannus*, a cloth or garment), the deep layer of fat in horses and other animals which raises the skin and gives the appearance of roundness and plumpness.

adnate, a., *ăd'·nāt* (L. *ad*, to ; *natus*, born), in *bot.*, applied to an organ united to another throughout its length, as the stipules in the rose ; adhering to other parts.

adnexed, a., *ăd·nĕkst'* (L. *ad*, to ; *nexum*, to bind or tie), in *bot.*, reaching to the stem only, as in the gills of Agarics.

ad pondus omnium, *ăd pŏnd'·ŭs ŏm'·nĭ·ŭm* (L. *ad*, to ; *pondus*, weight ; *omnis*, all, *omnium*, of all), to the weight of the whole ; in *med.*, indicating the proportion of an ingredient in a prescription.

adpressed, a., *ăd·prĕst'* (L. *ad*, to ; *pressum*, to press or squeeze), in *bot.*, closely pressed to a surface, as some hairs ; pressed close to anything ; also spelt **appressed**.

aduncate, a., *ăd·ŭngk'·āt* (L. *aduncus*, hooked), in *bot.*, crooked ; bent in the form of a hook.

adventitious, a., *ăd'·vĕn·tĭsh'·ŭs* (L. *ad*, to ; *ventum*, to come), come to accidentally ; in *bot.*, applied to organs produced in abnormal positions, as in roots arising from aerial stems ; unnatural, accidental, or acquired.

Ægle, n., *ēg'·lē* (L. *Ægle*, a Naiad, daughter of Jupiter—from *aiglē*, brightness), a genus of shrubs producing fragrant flowers, Ord. Aurantiaceæ : **Ægle marmelos**, *măr'·mĕl·ŏs* (Portg. *marmelo*, a quince), a species which yields an excellent fruit, much used in dysentery.

aerophytes, n. plu., *ār'·ō·fīts* (Gr. *aēr*, air ; *phuton*, a plant), plants which grow entirely in the air.

æruginose, a., *ē·rōōdj'·ĭn·ōz* (L. *æruginōsus*, copper-rusted—from *ærūgo*, copper rust), verdigris-green, or copper rust.

Æsculus, n., *ĕs'·kŭl·ŭs* (L. *esca*, food), a genus of large showy trees, Ord. Sapindaceæ : **Æsculus hippocastanum**, *hĭp'·pō·kăst'·ăn·ŭm* (Gr. *hippos*, a horse ; Gr. *kastanon*, L. *castănĕa*, the chestnut-tree), the horse-chestnut, recommended as a febrifuge, seeds sometimes used for coffee : Æ. **ohioensis**, *ō·hī'·ŏ·ĕns'·ĭs* (after *Ohio*, a State of U. States, Amer.), the buck-eye, or American horse-chestnut, leaves and fruit said to be poisonous.

æstival, a., *ĕs·tīv'·ăl* (L. *æstiva*, summer quarters), produced in summer; pert. to summer : **æstivation**, n., *ĕs'·tĭv·ā'·shŭn*, the arrangement of the unexpanded leaves of the flower in the flower-bud which burst in summer; opposed to *vernation*, the arrangement of the leaves of the bud on a branch which burst in spring.

afferent, a., *ăf'·fĕr·ĕnt* (L. *affĕro*, I bring or convey a thing to a place—from *ad*, to ; *fero*, I carry), in *anat.*, conveying from the surface to the centre.

affinity, n., *ăf·fĭn'·ĭt·ĭ* (L. *affĭnis*, bordering on, related to—from *ad*, to ; *finis*, an end), relation ;

agreement ; in *chem.*, combining power of bodies ; in *bot.*, relation in all essential organs.

affusion, n., *ăf·fūzh'·ŭn* (L. *ad*, to ; *fusum*, to pour), the act of pouring a cold or warm liquid on the whole body or a part of it from some elevation, as a remedial measure in many diseases ; in *chem.*, the pouring water on a substance to cleanse it.

agamic, a., *ăg·ăm'·ĭk* (Gr. *a*, without ; *gamos*, marriage), in *zool.*, applied to all forms of reproduction in which the sexes are not directly concerned : **agamous,** a., *ăg'·ăm·ŭs*, in *bot.*, applied to plants without visible organs of fructification; cryptogamous : **agamo-genesis,** n., *ăg'·ăm·ō·jĕn'·ĕs·ĭs* (Gr. *genesis*, generation, origin), the power of non-sexual reproduction.

Agaricaceæ, n. plu., *ăg·ăr'·ĭk·ā'·sĕ·ē* (Gr. *agarikon*, touchwood, a mushroom), the Ord. of plants now called Fungi : **Agarics,** n. plu., *ăg·ăr'·ĭks*, the edible mushrooms of this country : **Agaricus campestris,** *ăg·ăr'·ĭk·ŭs kăm·pĕst'·rĭs* (L. *agaricus*, an agaric ; *campestris*, pert. to a level field), the common mushroom of this country : other edible species are, **A. deliciosus,** *dē·lĭsh'·ĭ·ōz'·ŭs* (L. *deliciōsus*, delightful—from *delĭciæ*, delight) ; **A. Georgii,** *jōrj'·ĭ·ĭ* (L. *Georgius*, George, *Georgii*, of George); **A. procerus,** *prō·sēr'·ŭs* (L. *procērus*, high, tall), eaten abroad, though considered poisonous in Britain ; and **A. prunulus,** *prōōn'·ŭl·ŭs* (L. diminutive of *prunum*, a prune), said to be the finest species of mushroom: **A. oreades,** *ŏr·ē'·ăd·ēz* (L. *Oreades*, mountain nymphs); **A. coccineus,** *kŏk·sĭn'·ĕ·ŭs* (L. *coccinĕus*, of a scarlet colour—from *coccum*, a scarlet colour); and **A. personatus,** *pĕr'·sŏn·ăt'·ŭs* (L. *persōnātus*, provided with a mask, counterfeited), species of Fungi which, being developed in a centrifugal man-

ner, form fairy rings : **A. olearius,** *ŏl'·ē·ār'·ĭ·ŭs* (L. *oleārius*, belonging to oil—from *oleum*, oil); and **A. Gardneri,** *gărd'·nĕr·ĭ* (Latinised proper name, *Gardneri*, of Gardner), these, and other species, give out a sort of phosphorescent light.

agathophyllum aromaticum, *ăg'·ăth·ō·fĭl'·lŭm ăr'·ŏm·ăt'·ĭk·ŭm* (Gr. *agathos*, good, pleasant ; *phullon*, a leaf ; Gr. *aromatikos*, L. *aromaticus*, fragrant), the clove nutmeg of Madagascar ; an ornamental tree.

agave, n., *ăg'·āv* or *ăg'·ăv·ē* (Gr. *agauos*, admirable), the American aloe, from the juice of which the alcoholic liquor pulque is made— the systematic name is **Agave Americana,** *ăg'·ăv·ē ăm·ĕr'·ĭk·ān'·ă*: **Agaveæ,** n. plu., *ăg'·ăv·ē'·ē*, one of Lindley's four tribes into which he divides the Amaryllidaceæ or Amaryllis family.

agglomerate, a., *ăg·glŏm'·ĕr·āt* (L. *agglomero*, I heap up), heaped up ; crowded together.

agrimony, n., *ăg'·rĭ·mŏn·ĭ*, also **agrimonia,** *-mōn'·ĭ·ă* (L. *agrimonia*), a wild British plant of the rose tribe, having bitter astringent properties.

aizoon, n., *ā'·ĭ·zō'·ŏn* (Gr. *aei*, always ; *zōon*, a living creature), one of the fig marigold and ice plant family—so called as the plant lives under almost any treatment.

ala, n., *āl'·ă ;* **alæ,** plu., *āl'·ē* (L. *ala*, a wing), in *anat.*, a part projecting like a wing ; in *bot.*, the lateral petals of a papilionaceous flower ; the membranous appendages of fruit, as in the elm, or of the seed, as in pines : **alary,** a., *āl'·ăr·ĭ*, also **alate,** a., *āl'·āt*, wing-like.

alabastrus, n., *ăl'·ă·bast'·rŭs* (L. *alabaster*, a rose-bud in its green state), in *bot.*, the flower-bud while yet green and before it opens.

Alangiaceæ, n. plu., *ăl·ănj'·ĭ·ā'·sĕ·ē* (from *alangium*, its name in and diuretic: **A. cepa,** *sĕp'.ă* (L. *cœpa,* an onion), the onion : **A. porrum,** *pŏr'.ŭm* (L. *porrum,* a leek), the leek.

allopathy, n., *ăl.lŏp'.ăth.ĭ* (Gr. *allos,* another ; *păthos,* suffering, disease), that mode of medical practice which consists in the use of drugs to produce in the body a condition opposite to the disease to be cured ; opposed to *homœopathy,* which professes to cure diseases by remedies which in a state of health would have produced similar symptoms of disease.

allotropism, n., *ăl.lŏt'.rŏp.ĭsm* (Gr. *allos,* another ; *tropē,* change, conversion), the existence of the same body in more than one usual condition, and with different physical characteristics, as sulphur, which is bright-yellow and brittle in one state, and when melted at a high temperature it becomes dark and extremely tenacious: **allotropy,** n., *ăl.lŏt'.rŏp.ĭ,* same sense as *allotropism*: **allotropic,** a., *ăl'.lŏt.rŏp'.ĭk,* of, or pert. to.

Alnus, n., *ăl'.nŭs* (L. *alnus,* the alder tree), a genus of timber trees thriving best on the banks of rivers and in moist situations : **Alnus glutinosa,** *glŏt'.ĭn.ōz'.ă* (L. *glūtinōsus,* gluey, glutinous—from *gluten,* glue), the common alder, the wood used for underground purposes, and its charcoal in the manufacture of gunpowder: **A. incana,** *ĭn.kăn'.ă* (L. *incānus,* quite grey, hoary), the bark of the species used in Kamtschatka in the preparation of a kind of bread, Ord. Betulaceæ.

Aloe, n., *ăl.ō'.ē* (Gr. and L. *ăloē,* the aloe, bitterness), a genus of plants, Ord. Liliaceæ, various species of which produce the drug called aloes: **aloes,** *ăl'.ōz,* the inspissated juice of various species of the aloë, chiefly the **Aloe Socotrina** (from *Sŏcŏtra*), and **Aloe spic-**

ata, *spĭk.āt'.ă* (L. *spicatus,* furnished with spikes), usually called Socotrine aloes, *sŏk'.ōt.rĭn ăl'.ōz* : **A. dichotoma,** *dĭ.kŏt'.ŏm.ă* (Gr. *dichotomos,* cut into halves—from *dicha,* doubly ; *temno,* I cut), an arborescent species of S. Africa, 30 feet high, called the Quiver tree : **A. vulgaris,** *vŭlg.ār'.ĭs* (L. *vulgaris,* common, vulgar), from E. and W. Indies, and the source of the Barbadoes aloes : **aloetic,** a., *ăl'.ō.ĕt'.ĭk,* also **aloetical,** a., *-ĭk.ăl,* applied to a medicine which contains a large proportion of aloes: **Aloineæ,** *ăl.ō.ĭn'.ĕ.ē,* for **Aloe,** the aloe tribe: aloe, a., *ăl'.ō,* applied to the trees, Ord. Liliaceæ : **aloin,** n., *ăl.ō'.ĭn,* the active principle of aloes.

alopecia, n., *ăl'.ō.pē'.shĭ.ă* (L. *alopecia,* the fox sickness or mange —from Gr. *alōpēx,* a fox), the fox evil or scurf ; any kind of baldness.

Aloysia, n., *ăl'.ŏ.ĭs'.ĭ.ă* (in honour of Maria Louisa, a queen of Spain), a genus of plants, Ord. Verbenaceæ: **Aloysia citriodora,** *sĭt'.rĭ.ōd.ōr'.ă* (*citrus,* the citron tree), the sweet-scented verbena or lemon plant.

Alpinia, n., *ălp.ĭn'.ĭ.ă* (after *Alpini,* an Italian botanist), a genus of plants, Ord. Zingiberaceæ : **Alpinia officinarum,** *ŏf.fĭs'.ĭn.ār'.ŭm* (L. *officina,* a workshop, a laboratory, *officinārum,* of workshops), a Chinese plant, the root stock of which constitutes the Galangal root of commerce, having the same properties as ginger : **A. Galanga,** *găl.ăng'.gă* (a supposed Indian name), supplies a similar rhizome.

alsinaceous, a., *ăl'.sĭn.ā'.shŭs* (Gr. *alsis,* leaping, jumping, in reference to the intervals between the petals ; or *alsos,* a grove or shady place, in reference to usual places of growth ; *alsīnē,* chickweed), denoting a polypetalous corolla, in which there are intervals between the petals, as in chickweed : **Alsineæ,** n. plu., *ăl·sĭn'·ĕ·ē,* a tribe of plants of which the **Alsine media,** *ăl·sĭn'·ē mĕd'·ĭ·ă* (L. *medĭus,* midst, middle), common chickweed, is the type, Ord. Caryophyllaceæ.

Alsodeæ, n. plu., *ăl·sŏd'·ĕ·ē* (Gr. *alsōdes,* blooming, woody), a tribe of the Violaceæ or violet family: **Alsodeia,** n., *ăl'·sŏ·dī'·ă,* a genus of plants, Ord. Violaceæ.

Alstonia, n., *ăl·stōn'·ĭ·ă* (after Dr. *Alston* of Edinburgh), a tribe of plants of the Ord. Apocynaceæ : **Alstonia scholaris,** *skōl·ār'·ĭs* (L. *schola,* a school), a species used in India as a tonic.

Alstromeria, n., *ăl'·strōm·ēr'·ĭ·ă* (after *Alströmer* of Sweden), a tribe of beautiful plants of the Ord. Amaryllidaceæ.

alterative, n., *āwlt'·ĕr·āt'·ĭv* (L. *alter,* another), a medicine which is supposed to alter the condition of the blood and tissues without producing any apparent evacuation : **adj.,** having the power to change or alter without sensible evacuation.

alternate, a., *āwlt·ĕrn'·āt* (L. *altero,* I do everything by turns ; *alternus,* one after another), in *bot.,* arranged at different heights on the same axis, as leaves or branches which stand out singly and by turns with those of the opposite side : **alternate generation,** a mode of reproduction among the lowest animal types, in which the young do not resemble the parent, but the grandparent.

alternipinnate, a., *ălt·ĕrn'·ĭ·pĭn'·nāt* (L. *alternus,* one after another ; *pinna,* a wing), in *bot.,* applied to leaflets or pinnæ which are placed alternately on each side of the midrib, and not directly opposite to each other ; also called 'alternately pinnate.'

Althæa, n., *ăl·thē'·ă* (L. *althœa,* Gr. *althaia,* the wild mallow—

said to be from Gr. *althos,* a cure, a remedy), a genus of tall freeflowering plants, so called from the medicinal qualities of some of the species, Ord. Malvaceæ : **Althæa officinalis,** *ŏf·fĭs'·ĭn·āl·ĭs* (L. *officinālis,* official — from *officĭna,* a workshop), the marsh mallow, employed medicinally as a demulcent and emollient : **A. rosea,** *rōz'·ĕ·ă* (L. *rŏsĕus,* pert. to roses—from *rŏsa,* a rose), the hollyhock, which yields fibres and a blue dye.

alumina, n., *ăl·ŏm'·ĭn·ă* (L. *alūmen,* alum), the clay, loam, or other substance from which alum is obtained ; *pure alumina* is an oxide of the metal *aluminum.*

alveolæ, n. plu., *ăl·vē'·ŏl·ē* (L. *alvĕus,* a hollow, a cavity—from *alvus,* the belly, the abdomen), in *bot.,* regular cavities on a surface, as in the receptacle of the sunflower : **alveolate,** a., *ăl·vē'·ŏl·āt,* deeply pitted so as to resemble a honey-comb ; having little hollows or cavities : **alveoli,** n. plu., *ăl·vē'·ŏl·ī,* the sockets of the teeth : **alveolar,** a., *ăl·vē'·ŏl·ăr,* and **alveolary,** a., *-ăr'·ĭ,* connected with the alveoli or sockets of the teeth : **alveus,** n., *ăl'·vē·ŭs,* in *anat.,* tubes or canals through which a fluid flows, generally the enlarged parts : **alvine,** a., *ăl'·vĭn,* belonging to the belly, stomach, and intestines ; the fæces or dung are termed *alvine* discharges.

amadou, n., *ăm'·ăd·ô* (F.), German tinder ; a substance resembling doeskin leather, prepared from a dry leathery fungus found on old ash and other trees.

amalgam, n., *ăm·ăl'·găm* (Gr. *hama,* together ; *gameo,* I marry), a compound formed of mercury with any other metal.

Amanita muscaris, *ăm·ăn·ĭt'·ă mŭs·kār'·ĭs* (Gr. *amanites,* a mushroom ; *muscarium,* a fly-flap, the hairy parts of plants—from *musca,* a fly), a species of fungus, used as a means of intoxication in Kamtschatka, Ord. Fungi: **amanitin**, n., *ăm'·ăn·ĭt'·ĭn*, the poisonous principle of fungi.

Amaranthaceæ, n. plu., *ăm'·ăr·ănth·ā'·sĕ·ē* (Gr. *amarantos*, L. *amaranthus*, unfading—from Gr. *a*, not; *maraino*, I parch or wither—in reference to the length of time some of them retain their bright colours), the Amaranth family, an Ord. of plants: **Amaranthus**, n., *ăm'·ăr·ănth'·ŭs*, also **Amaranth**, n., *ăm'·ăr·ănth*, a Sub-ord. of the Amaranth family, among which are, 'prince's feather,' 'my-love-lies-bleeding,' and 'cockscomb:' **amaranthaceous**, a., *ăm'·ăr·ănth·ā'·shŭs*, having an arrangement of parts as in the amaranth: **amarantous**, a., *ăm'·ăr·ănt'·ŭs*, undecaying; unfading.

Amaryllidaceæ, n. plu., *ăm'·ăr·ĭl·lĭ·dā'·sĕ·ē* (*Amaryllis*, a country girl celebrated by Virgil, the Latin poet), the Amaryllis family, an Ord. of beautiful bulbous plants: **Amarylleæ**, n. plu., *ăm'·ăr·ĭl'·lĕ·ē*, one of the tribes of the Amaryllis family; the snowdrop, the daffodil, and many other ornamental garden plants belong to this Order.

amaurosis, n., *ăm'·awr·ōz'·ĭs* (Gr. *amaurōsis*, the act of rendering obscure—from *amauros*, obscure), imperfect vision or total blindness, due to paralysis of the optic nerves, arising from various causes: **amaurotic**, a., *ăm'·awr·ŏt'·ĭk*, pert. to the partial blindness or loss of sight produced by paralysis of the optic nerves.

amblyopia, n., *ăm'·blĭ·ŏp'·ĭ·ă* (Gr. *amblus*, blunt, weak; *ōps*, the eye), impaired or weakened vision; obscurity of vision; incomplete amaurosis.

ambrina anthelmintica, *ăm·brĭn'·ă ănth'·ĕl·mĭnt'·ĭk·ă* (not ascertained: Gr. *anti*, against; *helmins*, a tape-worm), a plant of the Ord. Chenopodiaceæ, which yields a volatile oil, used in the cure of worms.

ambulacra, n. plu., *ăm'·bŭl·āk'·ră* (L. *ambulacrum*, a walking-place —from *ambulo*, I walk up and down), the perforated spaces in the crusts or plates of the Echinodermata, through which are protruded the feet, by means of which locomotion is effected by them: **ambulacriform**, a., *-āk'·rĭ·fŏrm* (L. *forma*, shape), having the form and appearance of ambulacra: **ambulatory**, a., *-āt'·ŏr·ĭ*, formed for walking.

ambustio, n., *ăm·bŭst'·ĭ·ō* (L. *ambustio*, a burn), a burn or scald: **ambustial**, a., *ăm·bŭst'·ĭ·ăl*, produced by a burn.

amenorrhœa, n., *ăm·ĕn'·ŏr·rē'·ă* (Gr. *a*, without; *mēn*, a month; *rhēo*, I flow), the absence or retention of the usual flow of the menses.

amentum, n., *ăm·ĕnt'·ŭm* (L. *amentum*, a leathern thong), in *bot.*, a catkin or imperfect flower hanging somewhat like a rope or cat's tail, consisting of an axis covered with bracts in the form of scales: **amenta**, n. plu., *ăm·ĕnt'·ă*: **Amentiferæ**, n. plu., *ăm'·ĕnt·ĭf'·ĕr·ē* (L. *fero*, I bear), a family of fossil plants, bearing amenta: **amentiferous**, a., *-ĕr·ŭs*, denoting plants having amenta or catkins: **amentaceous**, a., *ăm'·ĕnt·ā'·shŭs*, having amenta or catkins.

ametabolic, a., *ă·mĕt'·ă·bŏl'·ĭk* (Gr. *a*, without; *metabolē*, change), applied to insects not possessing wings when perfect, and which, therefore, do not pass through any marked metamorphosis.

Amherstia, n., *ăm·hĕrs'·tĭ·ă* (in honour of the Countess *Amherst*), a tribe of plants of the Sub-ord. Cæsalpineæ, and Ord. Leguminosæ, plants profusely ornamented with pendulous racemes of large vermilion-coloured blossoms.

ammi copticum, *ăm'·mĭ·kŏpt'·ĭk·ŭm* (Gr. *ammos*, sand, in reference to the soil best adapted for its growth), a plant of the Ord. Umbelliferæ ; the Ajowan, or Omam, a condiment of India.

ammonia, n., *ăm·mōn'·ĭ·ă* (*Ammon*, the Libyan Jupiter ; after the place where first found, and where his temple stood), a transparent, pungent gas ; the volatile alkali ; spirits of hartshorn, a substance used in medicine and the arts : **ammoniac,** a., *ăm·mōn'·ĭ·ăk*, also **ammoniacal,** -*ĭ'·ăk·ăl*, and **ammonic,** a., *ăm·mŏn'·ĭk*, pert. to or possessing the properties of ammonia ; pungent : **ammonium,** n., *ăm·mōn'·ĭ·ŭm*, the supposed base of ammonia : **sal-ammoniac,** n., the common name for chloride of ammonium : **ammoniacum,** n., *ăm'·mŏn·ĭ'·ăk·ŭm* (because the tree was supposed to grow chiefly at Ammon), the pharmacopœial name of a gum resin from the north-east of India, exuded from the 'Dorema ammoniacum,' also called **ammoniac,** or **gum-ammoniac**: **ammonio** — with the terminal o which indicates the leading influencing power in a compound.

Ammophila, n., *ăm·mŏf'·ĭl·ă* (Gr. *ammos*, sand ; *phileo*, I love), a genus of grasses which constitute bent and marram of the British shores, Ord. Gramineæ: **ammophila arenaria,** *ăr'·ĕn·ār'·ĭ·ă* (L. *arēnārius*, belonging to sand—from *arēna*, sand), one of the bents of the sea-shore, the roots forming a network among the sand : **ammophilous,** a., *ăm·mŏf'·ĭl·ŭs*, loving sand.

amnesia, n., *ăm·nēzh'·ĭ·ă* (Gr. *a*, without ; *mnesis*, memory), an affection of the brain in which the memory is impaired ; want of memory : **amnesic,** a., *ăm·nēz'·ĭk*, caused by loss of memory.

amnion, n., *ăm'·nĭ·ŏn* (Gr. *amnion*, a vessel for receiving the blood of animals in sacrifice), the internal membrane of the ovum which completely envelopes the embryo, and contains the water surrounding the 'fœtus in utero :' **amnios,** n., *ăm'·nĭ·ŏs*, in *bot.*, the fluid or semi-fluid matter in the embryo-sac : **amniota,** n. plu., *ăm'·nĭ·ōt'·ă*, the group of the vertebrata in which the fœtus is furnished with an amnion, comprising reptiles, birds, and mammals : **amniotic,** a., *ăm'·nĭ·ōt'·ĭk*, pert. to the amnion : **amnitis,** n., *ăm·nīt'·ĭs*, inflammation of the amnion.

amœba, n., *ăm·ēb'·ă* (Gr. *amoibos*, doing in turn, exchanging), in *zool.*, a species of rhizopod, so called from the numerous changes of form which it undergoes : **amœbiform,** a., *ăm·ēb'·ĭ·fŏrm* (L. *forma*, shape), resembling an amœba in shape : **amœboid,** a., *ăm·ēb'·ōyd* (Gr. *eidos*, resemblance), amœbiform ; resembling the movements of amœba.

amomum, n., *ăm·ōm'·ŭm* (Gr. *a*, without ; *mōmos*, a blemish, having a supposed allusion to the qualities of some of the species as counterpoisons), a Sub-ord. of plants, Ord. Zingiberaceæ, the cardamoms of commerce, constituting the seed of several species ; the Grains of Paradise are the seeds of one of the species: **amomeous,** a., *ăm·ōm'·ĕ·ŭs*, having an arrangement of parts as in the amoma : **Amomum cardamomum,** *kărd'·ăm·ōm'·ŭm* (Gr. *kardamōmon*, an aromatic plant ; *kardamon*, a kind of cress), supplies the round cardamoms of Java, Siam, and Sumatra : **A. aromaticum,** *ăr'·ōm·ăt'·ĭk·ŭm* (L. *arōmătĭcus*, aromatic, fragrant), the Bengal cardamom : **A. maximum,** *măks'·ĭm·ŭm* (L. *maxĭmus*, greatest), another Java species furnishing a kind of cardamom : **A. melegueta,** *mĕl·ĕg'·ū·ēt'·ă*, Grains of Paradise, or Melegueta pepper.

amorphous, a., *ăm·ŏrf'·ŭs* (Gr. *a*, without; *morphē*, form or shape), having no regular structure or definite form : **amorphism**, n., *ăm·ŏrf'·ĭzm*, a condition of shapelessness : **amorphophyte**, n., *-ō·fīt* (Gr. *phuton*, a plant), a plant that has irregular or anomalous flowers.

amorphozoa, n. plu., *ăm·ŏrf'·ŏz·ō'·ă* (Gr. *a*, without; *morphē*, shape; *zoōn*, an animal), a name sometimes used to designate the sponges : **amorphozous**, a., *ăm'·ŏrf·ŏz'·ŭs*, designating animals without determinate shape.

ampelideæ, n. plu., *ăm'·pĕl·ĭd'·ĕ·ē* (Gr. *ampĕlos*, the vine plant; *ampĕlis*, a small vine, *ampelĭdos*, of a small vine), the vine family; also called the 'Vitaceæ': **ampelopsis**, n., *ăm'·pĕl·ŏps'·ĭs* (Gr. *opsis*, appearance), certain plants which resemble the vine ; the Virginian creeper, cultivated as a climbing plant.

amphiarthrosis, n., *ăm'·fĭ·ăr·thrŏz'·ĭs* (Gr. *amphi*, about, on both sides ; *arthrosis*, articulation—from *arthron*, a joint), a mixed articulation with obscure and limited motion, of which we have examples in the limited motion of the vertebræ.

amphibia, n. plu., *ăm·fĭb'·ĭ·ă*, also **amphibians**, n. plu., *-ĭ·ănz* (Gr. *amphi*, both, on both sides ; *bios*, life), animals that can live either on land or in water, as frogs, newts, and the like, which have gills when young, but breathe air directly when in the adult state : **amphibial**, a., and **amphibian**, a., pert. to: **amphibious**, a., *-ĭ·ŭs*, able to live on land or in water.

amphicœlous, a., *ăm'·fĭ·sēl'·ŭs* (Gr. *amphi*, on both sides ; *koilos*, hollow), applied to vertebræ which are concave at both ends.

amphicarpous, a., *ăm'·fĭ·kărp'·ŭs* (Gr. *amphi*, both; *karpos*, fruit), possessing two kinds of fruit.

amphidiscs, n. plu., *ăm'·fĭ·dĭsks*

(Gr. *amphi*, on both sides ; *diskos*, L. *discus*, a quoit, a disc), the spicula which surround the gemmules of Spongilla, and resemble two toothed wheels united by an axil.

amphigamæ, n. plu., *ăm·fĭg'·ăm·ē* (Gr. *amphi*, on both sides ; *gamos*, marriage), a Sub-ord. of the Ord. Hepaticæ, plants whose fructification is unknown, and which may therefore be of both sexes: **amphigameous**, a., *ăm·fĭ·gām'·ĕ·ŭs*, designating plants whose fructification has not yet been ascertained.

amphigastria, n. plu., *ăm'·fĭ·găs'·trĭ·ă* (Gr. *amphi*, both ; *găstēr*, the belly), the scale-like stipules of mosses.

amphigenous, a., *ăm·fĭdj'·ĕn·ŭs* (Gr. *amphi*, both ; *gennăō*, I produce), applied to Fungi when the hymenium is not restricted to any particular surface.

amphioxus, n., *ăm·fĭ·ŏks'·ŭs* (Gr. *amphi*, on both sides ; *oxus*, sharp, pointed), the lancelet, a little fish, which itself alone constitutes the Ord. Pharyngobranchii.

amphipneusta, n. plu., *ăm'·fĭp·nŭst'·ă* (Gr. *amphi*, on both sides ; *pneusis*, breathing, respiration), applied to those amphibians which retain their gills along with their lungs, through life : **amphipneustous**, a., *-nŭst'·ŭs*, having both branchiæ and lungs as organs of respiration.

amphipoda, n. plu., *ăm·fĭp'·ŏd·ă* (Gr. *amphi*, on both sides; *pous*, a foot, *podos*, of a foot), an Ord. of Crustaceæ which have feet on both sides, directed partly forwards and partly backwards, as feet for both walking and swimming : **amphipodous**, a., *ăm·fĭp'·ŏd·ŭs*, having feet on both sides.

amphisarca, n., *ăm'·fĭ·sărk'·ă* (Gr. *amphi*, on both sides ; *sarx*, flesh, *sarkos*, of flesh), an indehiscent multilocular fruit with a hard exterior, and pulp round the seeds in the interior, as in the Baobab : **amphisarcous**, a., *-särk'-ŭs*, fleshy or pulpy on all sides.

amphistoma, n., *ăm·fĭs'·tŏm·ă* (Gr. *amphi*, on both sides; *stoma*, a mouth, *stomata*, mouths), a genus of the Entozoa having a cup at each extremity by which they adhere to the intestines : **amphistomous**, a., *ăm·fĭs'·tŏm·ŭs*, belonging to the amphistoma : **amphistomum conicum**, *kŏn'·ĭk·ŭm* (Gr. *kōnĭkos*, belonging to a cone—from *kōnos*, a cone), a parasite met with in the stomachs of the ox and sheep : **A. crumeniferum**, *krŏm'·ĕn·ĭf'·ĕr·ŭm* (L. *crŭmēna*, a bag ; *fero*, I carry), a parasite of the ox : **A. explanatum**, *ĕks'·plăn·āt'·ŭm* (L. *explanātum*, to flatten, to spread out), a parasite of the ox found in the liver : **A. truncatum**, *trŭngk·āt'·ŭm* (L. *truncātus*, cut or lopped off), a parasite of the cat.

amphitropal, a., *ăm·fĭt'·rŏp·ăl* (Gr. *amphi*, on both sides ; *tropē*, a turn, a change), in *bot.*, having an ovule or embryo curved on itself, with the hilum in the middle.

amphora, n., *ăm'·fŏr·ă* (L. and Gr.), an ancient wine vessel of an oblong shape, with a handle on each side of the neck : **amphoric**, a., *ăm·fŏr'·ĭk*, belonging to or shaped like an amphora : **amphoric resonance**, the peculiar clang which may accompany any of the ordinary auscultatory phenomena when resonating within a large cavity.

amplexicaul, a., *ăm·plĕks'·ĭk·awl* (L. *amplector*, I embrace ; *caulis*, the stem), in *bot.*, embracing the stem over a large part of its circumference, as the base of a leaf.

ampulla, n., *ăm·pŏŏl'·lă* (L. *ampulla*, a bottle for liquids, narrow at the neck, and bulging out in the middle), in *anat.*, the trum-

pet - mouthed portions of the semicircular canals of the internal ear ; any part having the same shape ; in *chem.*, a bellied vessel; in *bot.*, a hollow leaf : **ampullaceous**, a., *ăm'·pŏŏl·lā'·shŭs*, like a bellied bottle or inflated bladder.

Amygdaleæ, n. plu., *ăm'·ĭg·dăl'·ĕ·ē*, also called 'Pruneæ' (Gr. *amugdalē*, also *amugdălos*, the almond tree, the nut), a Sub-ord. of the Rosaceæ, chiefly remarkable from the presence of hydrocyanic acid in their kernels, leaves, and flowers : **amygdalæ**, n. plu., *ăm·ĭg'·dăl·ē*, the tonsils, the rounded lobes at the sides of the vallecula on the under surface of the cerebellum : **amygdalate**, n. plu., *ăm·ĭg'·dăl·āt*, made of almonds : **amygdalic**, a., *ăm'·ĭg·dăl'·ĭk*, of or belonging to the almond ; obtained from amygdalin : **amygdalin**, n., a white crystalline substance obtained from bitter almonds ; the peculiar action of Synaptase on amygdalin, produces hydrocyanic acid — also found in bitter almonds : **amygdaloid**, a., *ăm·ĭg'·dăl·ŏyd*, and **amygdaloidal**, a. (Gr. *eidos*, resemblance), almond-shaped : **Amygdalus**, n., *ăm·ĭg'·dăl·ŭs*, the almond tree, the A. **dulcis**, *dŭl'·sĕs* (L. *dulcis*, sweet), yielding the sweet almond ; and the A. **amara**, *ăm·ār'·ă* (L. *amārus*, bitter), the bitter almond.

amylaceous, a., *ăm'·ĭl·ā'·shŭs* (Gr. *amulon*, L. *amylum*, starch), pert. to or resembling starch : **amylene**, n., *ăm'·ĭl·ēn*, a substance obtained from fusel oil distilled with chloride of zinc : **amylic**, a., *ăm·ĭl'·ĭk*, obtained from starch : **amyloid**, a., *ăm'·ĭl·ŏyd* (Gr. *eidos*, resemblance), resembling starch : **amyl**, n., *ăm'·ĭl*, the hypothetical basis of a series of compounds, comprising 'fusel oil': **amyl alcohol**, an oily, colourless liquid, with a peculiar odour, and burning, acrid taste ; fusel oil.

Amyridaceæ, n. plu., *ăm'-ĭr-ĭd-ā'-sĕ-ē* (Gr. *a*, intensive ; *murrhis*, a sweet-scented plant, *murrhidos*, of a sweet-scented plant), an Ord. of plants, now referred to Ord. Burseraceæ, which see ; the balsam trees : **Amyrideæ**, n. plu., *ăm'-ĭr-ĭd'-ĕ-ē*, a Sub-ord. : **Amyris**, n., *ăm'-ĭr-ĭs*, a genus : **Amyris toxifera**, *tŏks-ĭf'-ĕr-ă* (L. *toxicum*, poison ; *fero*, I bear), a species said to be poisonous.

anabasis, n., *ăn-ăb'-ăs-ĭs* (Gr. *anabasis*, an ascent, progress—from *ana*, up ; *baino*, I go, I ascend), in *med.*, the increase of a disease or paroxysm ; **Anabasis ammodendron**, *ăm'-mō-dĕn'-drŏn* (Gr. *ammos*, sand ; *dendron*, a tree), a peculiar leafless shrub of Khiva.

Anacardiaceæ, n. plu., *ăn'-ă-kărd-ĭ-ā'-sĕ-ē* (Gr. *ana*, similar to ; *kardia*, the heart—so called from the form of the nuts of some of them), an Ord. of trees and shrubs ; the cashew nut family, some of which bear edible fruits, as the mango, and many of them furnish gum resins in much request for varnishes and dyes : **Anacardium**, n., *ăn'-ă-kărd'-ĭ-ŭm*, a genus of plants : **Anacard'ium occidentale**, *ŏk'-sĭ-dĕnt-āl'-ĕ* (L. *occidentalis*, western), the tree which furnishes the cashew nut, remarkable for its large succulent peduncle supporting the fruit or nut : **anacardiaceous**, a., *ăn'-ă-kărd-ĭ-ā'-shŭs*, having an arrangement of fruits as in the anacardium.

Anacharis, n., *ăn-ăk'-ăr-ĭs* (Gr. *ana*, without ; *charis*, grace, beauty), an aquatic plant, Ord. Hydrocharidaceæ, which exhibits under the microscope the rotation of protoplasm in its cells.

Anacyclus, n., *ăn'-ă-sĭk'-lŭs* (Gr. *anthos*, a flower ; *kuklos*, a circle—alluding to the rows of ovaries placed around the disc), a genus of hardy annuals, of the Sub-ord.

Corymbiferæ, Ord. Compositæ : **Anacyclus pyrethrum**, *pīr-ĕth'-rŭm* (Gr. *pur*, fire), the pellitory of Spain, an irritant, and promoting the secretion of saliva, properties depending on the presence of a volatile oil.

Anagallis, n., *ăn'-ă-găl'-lĭs* (Gr. and L. *anagallis*, the plant pimpernel—from *ana*, up, through ; *gala*, milk—from its property of coagulating milk), a genus of plants, Ord. Primulaceæ, flowering plants whose flowers are meteoric—that is, open always only during good weather and at particular hours, so as to act as hour glasses and weather glasses : **anagallis arvensis**, *ărv-ĕns'-ĭs* (L. *arvensis*, field inhabiting—from *arvum*, a field), the plant called 'the poor man's weather-glass,' or 'shepherd's weather-glass,' whose flowers open about 8 A.M., but only in fine weather ; the scarlet pimpernel.

anæmia, n., *ăn-ēm'-ĭ-ă* (Gr. *a*, without ; *haima*, blood), diminution in the amount of the blood ; the condition arising from such diminution : **anæmic**, a., *ăn-ēm'-ĭk*, without blood.

anæsthesia, n., *ăn'-ēs-thēz'-ĭ-ă* (Gr. *anaisthesia*, the want or loss of feeling—from *ana*, without ; *aisthesis*, sensation), the loss of feeling or sensation induced by the inhalation of an etherial vapour ; or due to organic or functional disease of the nervous system ; also, in same sense, **anæsthetics**, n. plu., *ăn'-ēs-thĕt'-ĭks* : the agents which take away sensibility from a part, or from the whole system, by acting on the nervous system, are numerous, but those usually employed are such as, opium, ether, chloroform, aconite or aconitia, and belladonna.

anallantoidea, n. plu., *ăn'-ăl-lănt-ōyd'-ĕ-ă* (Gr. *an*, without ; and *allantoidea*, which see), the group of vertebrata in which the embryo is not furnished with an allantois.

analogue, n., *ăn'.ăl.ŏg*(F. *analogue* —from Gr. *ana*, similar to; *logos*, ratio, proportion), an object that has a resemblance to, or correspondence with, another object; a part or organ in one animal which has the same function as another part or organ in a different animal: **analogue** regards similarity of function, **homologue**, identity of parts: **analogy**, n., *ăn.ăl'.ŏ.jĭ*, resemblance between one thing and another in some points; in *anat.*, the relation of parts of a different nature, which, however, perform similar functions: **analogous**, a., *ăn.ăl'.ŏg.ŭs*, applied to parts of a different nature which perform the same or similar functions; in *bot.*, applied to a plant which strikingly resembles one of another genus so as to represent it.

Anamirta cocculus, *ăn'.ăm.ĭrt'.ă kŏk'.ŭl.ŭs* (not ascertained: *coccus*, a berry; *cocculus*, a little berry), a plant of the Ord. Menispermaceæ, whose fruit, called Cocculus Indicus, is extremely bitter; its poisonous seeds were formerly employed to give bitterness to beer and porter.

anamniota, n. plu., *ăn.ăm'.nĭ.ŏt'.ă* (Gr. *an*, without; *amnion*, a vessel for receiving the blood of an animal in sacrifice; see 'amnos'), the group of vertebrata in which the embryo is destitute of an amnion.

Ananassa sativa, *ăn'.ăn.ăs'.să săt.iv'.ă* (said to be from *nanas*, the Guiana name; L. *sativus*, that is sown or planted), a species of the Ord. of trees Bromeliaceæ, producing the well-known Ananas, or Pine-apples.

anantherum, n., *ăn'.ăn.thēr'.ŭm* (Gr. *ana*, without; *anthēros*, flowery, blooming), filaments with anthers.

anarthropoda, n. plu., *ăn'.ăr.thrŏp'.ŏd.ă* (Gr. *an*, without; *arthros*, a joint; *pous*, a foot, *podos*, of a foot), that division of annulose animals in which there are no articulated appendages.

anasarca, n., *ăn'.ă.sărk'.ă* (Gr. *ana*, through; *sarx*, flesh, *sarkos*, of flesh), watery effusion into the cellular tissue; dropsy of the exterior of the body.

anastatica, n., *ăn'.ăs.tăt'.ĭk.ă* (Gr. *anastatikos*, pert. to a recovery—from *anastasis*, a rising up, a recovery), the rose of Jericho, Ord. Cruciferæ, the stalks of which, however curled and dry, will return to their original form when immersed in water.

anastomose, v., *ăn.ăs'.tŏm.ōz* (Gr. *anastomōsis*, the formation of a mouth or aperture—from *ana*, through; *stoma*, a mouth), to unite one vessel to another, as the mouth of a vein to that of another; to inosculate: **anastomosis**, n., *ăn.ăs'.tŏm.ōz'.ĭs*, the union of the branch of a vessel with another from the same trunk, or from other trunks; in *bot.*, union of vessels; union of the final ramifications of the veins of a leaf: **anastomotic**, a., *ăn.ăs'.tŏm.ŏt'.ĭk*, pert. to anastomosis.

anastomotica brachialis, *ăn'.ă.stŏm.ŏt'.ĭk.ă brăk'.ĭ.āl'.ĭs* (Gr. *ana*, by or through; *stoma*, a mouth; L. *brāchiālis*, belonging to the arm—from *brachium*, the arm), in *anat.*, one of the branches of the brachial artery which arises just above the elbow, and runs directly inwards, piercing the internal intermuscular septum, and supplying the parts about the elbow: **anastomotica magna**, *măg'.nă* (L. *magnus*, great), one of the arteries arising from the femoral artery in Hunter's Canal: **anastomotic**, a., see 'anastomose.'

anatropal, a., *ăn.ăt'.rŏp.ăl*, also **anatropous**, a., *ăn.ăt'.rŏp.ŭs* (Gr. *anatropeus*, a subverter—from *ana*, up or through; *trepo*, I turn), in *bot.*, applied to an inverted ovule, the hilum and micropyle being near each other, and the chalaza at the opposite end; having the embryo inverted, so that its base corresponds to the apex of the seed.

anbury, n., also **anberry,** n., *ăn'·bĕr·rĭ* (AS. *ampre* or *ompre*, a crooked swelling vein; Old Eng. *amper*, an inflamed tumour), under these names, and the name *angle-berry*, are included, in veterinary language, both warts and molluscous tumours; in *bot.*, a warty condition or swelling on the roots of such plants as turnips, cabbage, etc., caused by insects.

anceps, a., *ăn'·sĕps* (L. *anceps*, that has two heads, *ancipitis*, of two heads—from *an* for *ambi*, around, round about; *caput*, the head), two-edged; having the sides sharp like a two-edged sword; see 'ancipital' in Dict.

Anchusa, n., *ăng·kūz'·ă* (Gr. *ang-chousa*, the plant alkanet—from *en*, in or on; *cheō* or *cheuō*, I pour, I diffuse), a genus of plants, Ord. Boraginaceæ: **Anchusa tinctoria,** *tĭnk·tōr'·ĭ·ă* (L. *tinctorius*, of or belonging to dyeing—from *tingo*, I dye), a plant which supplies the alkanet root, used as a reddish-brown dye; anciently used for staining the skin: **anchusin,** n., *ăng·kūz'·ĭn*, the red-coloured principle.

anchylosis, n., *ăng'·kĭ·lōz'·ĭs* (Gr. *angkulos*, curved, crooked), the immovable state of a joint resulting from disease, and either osseous or fibrous in character; **anchylosed,** a., *ăng'·kĭ·lōzd*, fixed: **anchylotic,** a., *ăng'·kĭ·lŏt'·ĭk*, pert. to.

Anchylostomum, n., *ăng'·kĭ·lŏs'·tŏm·ŭm* (Gr. *angkulos*, curved, crooked; *stoma*, a mouth), a genus of parasitic worms which infests animal bodies, one of the species of which, the **anchylos-tomum duodenalis,** *dō'·ŏ·dĕn·āl'·ĭs* (L. *duodēni*, twelve each), infests the duodenum of man: **anchy-lostoma,** n. plu.

anconeus, n., *ăng·kōn'·ĕ·ŭs* (L. *ancon*, Gr. *angkon*, the elbow, the curvature of the arm), a triangular muscle situated over the elbow, which assists in extending the forearm: **adj.,** pert. to the elbow: **anconeous,** a., *ăng·kōn'·ĕ·ŭs*, pert. to.

Andira, n., *ănd·ĭr'·ă* (a Brazilian name), a genus of plants, Ord. Leguminosæ, Sub-ord. Papilion-aceæ: **Andira inermes,** *ĭn·ĕrm'·ēz* (L. *inermes*, unarmed, without weapons), the cabbage tree of the West Indies, which acts as a purgative and anthelmintic.

Andreæa, n., *ăn'·drĕ·ē'·ă* (in honour of *Andreæ*, a German professor), a genus differing from all other mosses in having a capsule which splits into four valves, cohering at their ends by means of the persistent lid: **Andreæa,** n., *ăn·drĕ'·ă*, **Andreæ,** plu., *ăn·drĕ'·ē*, also **Andreæaceæ,** n. plu., *ăn'·drĕ·ā'·sĕ·ē*, a Sub-ord. of plants of the Ord. Musci or Bryaceæ, often found in the bleakest places near the limits of perpetual snow, and are usually termed split mosses because the spore cases open by valves without elaters.

andrœcium, n., *ăn·drē'·shĭ·ŭm* (Gr. *anēr*, a man, *andros*, of a man; *oikos*, a house), in *bot.*, the male organs of the flower; the stamens taken collectively.

andrographis, n., *ăn·drŏg'·răf·ĭs* (Gr. *anēr*, a man, *andros*, of a man; *grapho*, I write), a plant of the Ord. Acanthaceæ: **andrographis paniculata,** *păn·ĭk'·ūl·āt'·ă* (L. *panicula*, a tuft, a panicle in plants), a plant of India, employed as a pure bitter tonic by the name of Kariyat or Creyat.

androgynal, a., *ăn·drŏdj'·ĭn·ăl*, also **androgynous,** a., *ăn·drŏdj'·ĭn·ŭs* (Gr. *anēr*, a man, *andros*, of a man; *gunē*, a woman), in *bot.*, having male and female flowers combined on the same peduncle, as in some species of Carex; same as 'Hermaphrodite,' and denoting that the two sexes are united in the same individual: **androgynism**, n., *ăn·drŏdj'·ĭn·ĭzm*, a change from a diœcious to a monœcious condition.

Andromeda, n., *ăn·drŏm'·ĕd·ă* (L. *Andrŏmeda*, Gr. *Andrŏmedē*, a virgin whom, when bound to a rock, Perseus rescued and married), a genus of plants having scaly buds and loculicidal capsules, very ornamental plants, Ord. Ericaceæ.

androphore, n., *ăn·drŏf'·ŏr·ē*, also **androphorum**, n., *ăn·drŏf'·ŏr·ŭm* (Gr. *anēr*, a man, *andros*, of a man; *phoreo*, I bear), a stalk supporting the stamens, often formed by a union of the filaments: **androphores**, plu., *-ŏr·ēz*, the medusiform gonophores of the Hydrozöa which carry the spermatozöa, and differ in form from those in which the ova are developed.

Andropogon, n., *ăn'·drō·pŏg'·ŏn* (Gr. *anēr*, a man, *andros*, of a man; *pōgon*, a beard), a genus of plants, Ord. Gramineæ, having little tufts of hairs on the flowers resembling a man's beard, from some species of which a fragrant oil is procured.

Androsace, n., *ăn·drŏs'·ăs·ē* (Gr. *androsakēs*, among the ancients a plant which has not been identified—said to be from *anēr*, a man, *andros*, of a man; *sakos*, a buckler), a genus of plants, the round hollow leaf resembling an ancient buckler, Ord. Primulaceæ.

androspores, n. plu., *ăn'·drō·spōrz* (Gr. *anēr*, a man, *andros*, of a man; *spora*, a seed), the developed male organs in certain of the Algæ; swarm spores.

Aneimia, n., *ăn·i'·mĭ·ă* (Gr. *aneimōn*, naked), an ornamental genus of ferns having a naked inflorescence, Ord. Filices.

Anemoneæ, n. plu., *ăn'·ĕm·ŏn'·ē·ē* (Gr. *anĕmos*, the wind), a suborder of the Ord. Ranunculaceæ: **Anemone**, n., *ăn·ĕm'·ŏn·ē*, a genus of plants, many of the species of which inhabit elevated windy places: **anemone**, n., *ăn·ĕm'·ŏn·ē*, the wind flower: **anemophilous**, a., *ăn'·ĕm·ŏf'·ĭl·ŭs* (Gr. *philos*, loved), applied to plants fertilized by the agency of wind: **Anemone nemorosa**, *nĕm'·ŏr·ōz'·ă* (L. *nĕmorōsus*, pert. to a grove—from *nĕmus*, a grove), a plant found in woods in Britain.

aner, *ăn'·ĕr*, **andro-**, *ăn'·drō-* (Gr. *anēr*, a man, *andros*, of a man), in *bot.*, terms in composition denoting 'male' or 'stamen.'

Anethum, n., *ăn·ĕth'·ŭm* (L. *anēthum*, Gr. *anēthon*, dill), a genus of plants, the seeds of which are used as flavouring agents by cooks and confectioners, and in *med.* as a carminative: **A. graveolens**, *grăv'·ĕ·ŏl·ĕnz* (L. *gravĕolens*, strong-smelling), common garden dill; see 'fennel' in Dict.

aneurism, n., *ăn'·ūr·ĭzm* (Gr. *aneurusma*, the dilatation of an artery—from *ana*, throughout; *eurus*, broad), a tumour filled with blood, which communicates directly or indirectly with an artery, and arises from a rupture, a wound, an ulceration, or from the simple dilatation of an artery; also applied to enlargement or dilatation of the heart: **aneurismal**, a., *ăn'·ūr·ĭz'·măl*, pert. to.

NOTE.—An *aneurism* is an abnormal dilatation in the course of a vessel due to degeneration of its coats (true *aneurism*); or to the rupture of one or more of these coats (false *aneurism*); or from excessive anastomosis (*aneurism* by anastomosis). When applied to the heart, *aneurism* signifies a circumscribed pouch caused by the breaking down of a limited portion of the heart-wall.

anfractuose, a., *ăn·frăkt'·ū·ōz* (L. *anfractus*, a tortuous or circuitous route), in *bot.*, wavy or sinuous, as the anthers of gourds and cucumbers.

Angelica, n., *ăn·jĕl′·ĭk·ă*(L. *angelus*, an angel, from its supposed angelic virtues), plants whose roots have a fragrant, agreeable smell, and bitterish, pungent taste, Ord. Umbelliferæ.

angienchyma, n., *ăn′·jĭ·ĕng′·kĭm·ă* (Gr. *anggeion*, a vessel; *engchŭma*, an infusion—from *engchuo*, I pour in), in *bot.*, vascular tissue in general : **angiocarpous**, a., *ăn′·jĭ·ō·kărp′·ŭs* (Gr. *karpos*, fruit), applied to lichens having fructification in cavities of the thallus, and opening by a pore; having seed in a vessel : **angiospermous**, a., *ăn′·jĭ·ō·spĕrm′·ŭs* (Gr. *sperma*, seed), having seeds contained in a seed vessel : **angiosperms**, n. plu., *ăn′·jĭ·ō·spĕrms* (Gr. *sperma*, seed), the great mass of flowering plants, so called because the seeds are usually enclosed in a seed vessel or pericarp : **angiosporous**, a., *ăn′·jĭ·ŏs′·pōr·ŭs* (Gr. *spora*, seed), applied to cryptogamic plants having spores or seeds contained in a theca or sporangium, that is, in a spore case or seed vessel : **Angiosporæ**, n. plu., *ăn′·jĭ·ŏs′·pōr·ē* (Gr. *spora*, seed), a sub-class nearly corresponding with the sub - class Acotyledons of the sub-kingdom Cryptogamous plants, having a certain amount of vascular tissue, and sporangia or thecæ containing spores.

angina, n., *ăn·jĭn′·ă* (L. *angina*, quinsy—from *ango*, I choke or strangle), a general term for diseases in which a sense of suffocation is a prominent symptom : **anginal**, a., *ăn·jĭn′·ăl*, also **anginose**, a., *ăn′·jĭn·ōz*, pert. to angina : **angina pectoris**, *pĕkt′·ŏr·ĭs* (L. *pectus*, the breast, *pectŏris*, of the breast), a distressing malady, in which a most excruciating pain

is felt in the chest, with a feeling of strangulation, and a terrible sense of impending death.

angularis faciei, *ăng′·ŭl·ār′·ĭs făs′·ĭ·ĕī* (L. *angŭlāris*, angular—from *angŭlus*, an angle; *facies*, the face), the angular artery of the face, which forms the termination of the trunk of the facial, and ascends to the inner angle of the orbit.

angustiseptæ, n. plu., *ăng′·gŭst·ĭ·sĕpt′·ē* (L. *angustus*, narrow; *septum*, partition), those fruits or seed vessels which have their partition in their narrow diameter : **angustiseptate**, a., *ăng′·gŭst·ĭ·sĕpt′·āt*, having the partition of the fruit or seed vessel very narrow.

anhydride, n., *ăn·hĭd′·rīd* (Gr. *an*, not, without; *hudor*, water), in *chem.*, a body destitute of water : **anhydrous**, a., *ăn·hĭd′·rŭs*, containing no water.

Anigosanthus, n., *ăn′·ĭ·gōz·ănth′·ŭs* (Gr. not ascertained; *anthos*, a flower), a genus of plants so called from their long conspicuous scapes upon which the flowers are raised, Ord. Hæmodoraceæ.

anise, n., *ăn′·ĭs* (L. *anīsum*, Gr. *anizon*, anise), an annual plant whose seeds have an aromatic smell, and pleasant, warm taste, furnishing an aromatic oil : **aniseseed or aniseed**, the seed of the plant ; the Pimpinella anisum.

anisos, *ăn·ĭs′·ŏs* (Gr. *anisos*, unequal), in composition, denoting ‘unequal :’ **anisomerous**, a., *ăn′·ĭ·sŏm′·ĕr·ŭs* (Gr. *meros*, a part), unsymmetrical.

anisostemonous, a., *ăn·ĭs′·ŏs·tĕm′·ŏn·ŭs* (Gr. *anisos*, unequal; *stēma*, the stamen of a plant, the warp of a web), having stamens neither equal in number to the floral envelopes, nor a multiple of them : **anisostemopetalous**, a., *ăn·ĭs′·ŏ·stĕm′·ō·pĕt′·ăl·ŭs* (Gr. *stema*, a stamen; *petalon*, a petal), having stamens unequal in number to the divisions of the corolla: **anisostomous**, a., *ăn'.ĭs·ŏs'.tŏm·ŭs* (Gr. *stoma*, a mouth), having unequal divisions of a calyx or corolla.

Annelida, n. plu., *ăn'.nĕl·ĭd'.ă* or *ăn·nĕl'.ĭd·ă*, also **annelids**, n. plu., *ăn'.nĕl·ĭdz* (L. *annellus*, a little ring; Gr. *eidos*, resemblance), those creatures that have their bodies formed of a great number of small rings, as the earth-worm, forming one of the divisions of the Anarthropoda.

annotinus, n., *ăn·nŏt'.ĭn·ŭs* (L. *annōtĭnus*, a year old—from *annus*, a year), a year old; that produces seed and dies within the same year in which it germinated: also **annual**, a., and **annualis**, in same sense: **annotinous**, a., *ăn·nŏt'.ĭn·ŭs*, showing last year's shoot by a visible point of junction.

annulus, n., *ăn'.nŭl·ŭs* (L. *annŭlus*, a ring), in *bot.*, applied to the elastic rim surrounding the sporangia of some ferns; the cellular rim on the stalk of the mushroom, being the remains of the veil; any circular opening resembling a ring: **annulate**, a., *ăn'.nŭl·āt*, also **annulated**, a., *ăn'.nŭl·āt·ĕd*, composed of a succession of rings: **Annularia**, n. plu., *ăn'.nŭl·ār'.ĭ·ă*, a genus of fossil herbaceous plants, having whorls on the same plane with their stems: **Annuloida**, n. plu., *ăn'.nŭl·ōўd'.ă* (Gr. *eidos*, resemblance), the sub-kingdom comprising Echinodermata and Scolicida: **Annulosa**, n. plu., *ăn'.nŭl·ōz'.ă*, the sub-kingdom comprising the Anarthropoda and the Arthropoda or Articulata; in all, the body is more or less composed of a succession of rings: **annulus ovalis**, *ŏv·āl'.ĭs* (L. *ovālis*, oval), in *anat.*, the prominent oval margin of the *foramen ovale*.

anodyne, n., *ăn'.ŏd·ĭn* (Gr. *an*, without; *odune*, pain), any medicine which relieves pain.

Anomoura, n. plu., *ăn'.ŏm·ôr'.ă*, also **anomura**, n. plu., *ăn'.ŏm·ôr'.ă* (Gr. *anomos*, irregular; *oura*, a tail), a family of crustaceans characterized by their irregular tails, of which the 'hermit crab' is the type.

Anonaceæ, n. plu., *ăn'.ŏn·ā'.sĕ·ē* (from *anona* or *menona*, its native Banda name), the custard apple family, an Order of ornamental trees and shrubs: **Anona**, n., *ăn·ōn'.ă*, a genus of trees, comprising for the most part fruit-bearing plants: **Anona muricata**, *mŭr·ĭk·āt'.ă* (L. *muricātus*, shaped like the murex shell, pointed); **A. squamosa**, *skwăwm·ōz'.ă* (L. *squāmōsus*, scaly—from *squāma*, a scale); and **A. reticulata**, *rē·tĭk'.ŭl·āt'.ă* (L. *reticulātus*, net-like, reticulated—from *rēte*, a net), are the species which furnish the custard apples, the sweet sops, and the sour sops of the East and West Indies: **A. cherimolia**, *kĕr'.ĭ·mōl'.ĭ·ă* (Sp. *chirimoya*, a custard apple), furnishes the cherimoyer, a well-known Peruvian fruit.

Anoplura, n. plu., *ăn'.ō·plôr'.ă* (Gr. *anoplos*, unarmed; *oura*, a tail), an Order of apterous insects.

anorexia, n., *ăn'.ō·rĕks'.ĭ·ă* (Gr. *an*, without; *orexis*, a longing for, eager desire), want of appetite. also **anorexy**, n., *ăn'.ō·rĕks·ĭ*.

Anoura, n., *ăn·ôr'.ă* (Gr. *a*, without; *oura*, a tail), the order of Amphibia, comprising frogs and toads, in which the adult is destitute of a tail; also called Batrachia: **anourous**, a., *ăn·ôr'.ŭs*, tailless.

antacid, n., *ănt·ăs'.ĭd* (Gr. *anti*, against; L. *acidus*, sour), any medicines, as the alkalies and alkaline earths, which counteract the formation of acids in the system.

antenna, n., *ăn·tĕn'.nă*, antennæ, plu., *ăn·tĕn'.nē* (L. *antenna*, a sail-yard), the jointed feelers or horns upon the heads of insects and crustacea : **antennules,** n. plu., *ăn·tĕn'·nŭl·ēz,* the smaller pairs of antennæ in the insects and crustacea.

anterior, a., *ănt·ēr'·ĭ·ėr* (L. *anterior,* former, that which lies before), before; in front: **anterior ligament,** a ligament that lies in front or before another : **anterior superior,** the higher point of two situated anteriorly or in front— see the separate words ; in *bot.,* part of a flower next the bract, or in front ; same as *inferior* when applied to the parts of the flower in their relation to the axis.

anthela, n., *ănth·ēl'·ă* (Gr. *anthēle,* a little blossom—from *anthe,* a blossom), the cymose panicle of the Juncaceæ or Rush family ; a cluster of inflorescence, particularly on rushes, whose branches are widely expanded.

anthelmintic, n., *ănth'·ĕl·mĭnt'·ĭk* (Gr. *anti,* against; *helmins,* a tape-worm, *helminthos,* of a tape-worm), a medicine given for destroying or expelling intestinal worms.

Anthemis, n., *ănth'·ĕm·ĭs* (Gr. *anthemon,* a flower, a blossom— so called from its great production of flowers), a genus of plants of the Sub-ord. Corymbiferæ, Ord. Compositæ: **Anthemis nobilis,** *nŏb'·ĭl·ĭs* (L. *nōbĭlis,* famous, renowned), the chamomile, whose flowers are odoriferous and yield a volatile oil; the flowers are much employed in various ways medicinally : **A. tinctoria,** *tĭngk·tōr'·ĭ·ă* (L. *tinctoria,* dyeing, or belonging to a dyer), a species which supplies a yellow in dyeing.

anther, n., *ănth'·ĕr* (Gr. *anthēros,* flowery, blooming—from *anthos,* a flower), the head part of the stamen of a flower containing the pollen or fertilizing dust.

Anthericeæ, n. plu., *ănth'·ĕr·ĭs'·ĕ·ē* (Gr. *anthĕrix,* a stalk, *antherikos,* of a stalk), a tribe of plants of the Ord. Liliaceæ; the Asphodel

tribe : **Anthericum,** n., *ănth·ĕr'·ĭk·ŭm,* a genus of plants comprising some beautiful species.

antheridium, n., *ănth'·ĕr·ĭd'·ĭ·ŭm,* **antheridia,** plu., *ănth'·ĕr·ĭd'·ĭ·ă* (Gr. *anthēros,* flowery; *eidos,* resemblance), male organs in cryptogamic plants, frequently containing moving filaments analogous to *spermatozŏa* of animals : **antheriferous,** a., *ănth'·ĕr·ĭf'·ĕr·ŭs* (L. *fero,* I bear), bearing anthers.

antherozoa, n. plu., *ănth'·ĕr·ō·zō'·ă* (Gr. *anthēros,* flowery; *zōŏn,* an animal), the spiral filaments or molecules having vibratile appendages discharged from the antheridia; the moving filaments in the antheridium of a flowerless plant: **antherozoids,** n. plu., *ănth·ĕr'·ō·zŏydz* (Gr. *eidos,* resemblance), same sense ; minute bodies which exhibit movements in the antheridium.

anthesis, n., *ănth·ēz'·ĭs* (Gr. *anthos,* a flower), the opening of the flower; the production of flowers.

anthistiria, n. plu., *ănth'·ĭs·tĭr'·ĭ·ă* (Gr. *anthĕstēria,* the feast of the flowers), the kangaroo grass of Australia ; satin grass ; Ord. Gramineæ.

anthocarpous, a., *ănth'·ō·kârp'·ŭs* (Gr. *anthos,* a flower; *karpos,* fruit), formed, as a certain class of fruits, from a number of blossoms united into one body ; applied to multiple, polygynœcial, or confluent fruits, formed by the ovaries of several flowers.

Anthocoroteæ, n. plu., *ănth·ŏs'·ĕr·ŏt'·ĕ·ē* (Gr. *anthos,* a flower; *keras,* a horn—from the horn-like form of the theca), the third of the three sections of the Ord. Hepaticæ : **Anthoceros,** n., *ănth·ŏs'·ĕr·ŏs,* a genus of small frondose plants, so called from the horn-like form of the theca.

anthocyane, n., *ănth'·ō·sĭ'·ăn·ĕ* (Gr. *anthos,* a flower; *kuanos,* dark-blue, sky-coloured), the supposed blue colouring matter in flowers of that hue.

anthodium, n., *ănth·ōd'·ĭ·ŭm* (Gr. *anthōdēs*, flowery—from *anthos*, a flower; *eidos*, resemblance), the common calyx which contains the capitulum or head of flowers of composite plants.

anthophore, n., *ănth'·ō·fōr* (Gr. *anthos*, a flower; *phero*, I carry), a stalk supporting the inner floral envelopes, and separating them from the calyx: **anthophorous**, a., *ănth·ŏf'·ŏr·ŭs*, bearing many flowers.

anthosperm, n., *ănth'·ō·spĕrm* (Gr. *anthos*, a flower; *sperma*, seed), coloured matter in the cells of certain fronds.

anthotaxis, n., *ănth'·ō·tăks'·ĭs* (Gr. *anthos*, a flower; *taxis*, arranging), in *bot.*, the arrangement of the flowers on the axis.

anthoxanthine, n., *ănth'·ŏks·ănth'·ĭn* (Gr. *anthos*, a flower; *xanthos*, yellow), the supposed yellow colouring matter in flowers of that hue.

anthrax, n., *ănth'·răks* (Gr. *anthrax*, a live coal), a carbuncle; a local suppuration which may be idiopathic, or may accompany other diseases as diabetes, or malignant fevers such as the plague, etc.,—common also in the lower animals as well as in man: **anthracoid**, a., *ănth'·răk·ŏyd* (Gr. *eidos*, resemblance), pert. to or resembling an anthrax or carbuncle.

Antiaris, n., *ănt'·ĭ·ār'·ĭs* (*Antiar* or *Antschar*, its Javanese name), a genus of plants of the Sub-ord. Artocarpeæ, Ord. Moraceæ: **Antiaris toxicaria**, *tŏks'·ĭk·ār'·ĭ·ă* (L. *toxicum*, Gr. *toxikon*, poison), the source of the famous poison, called Bohun-Upas or Upas-Antiar by the Javanese: **antiarin**, n., *ănt·ĭ'·ăr·ĭn*, the peculiar principle in the Upas tree to which it is said it owes its deadly properties: A. **saccidora**, *săk'·sĭd·ōr'·ă* (Gr.

sakkos, L. *saccus*, a sack, a bag; Gr. *dōreō*, I give or grant), a gigantic tree, whose fibrous bark is used as sacks.

antibrachium, n., *ănt'·ĭ·brăk'·ĭ·ŭm* (Gr. *anti*, in front of; Gr. *brachiōn*, L. *brachium*, the arm), the fore-arm of the higher vertebrates, composed of the radius and ulna: **antibrach'ial**, a., pert. to.

anticæ, n. plu., *ănt·ĭs'·ē* (L. *antīcus*, that is before or in front), in *bot.*, anthers are so called when they open on the surface next to the centre of the flower, and are called *ănthērœ antīcœ*: **anticus**, a., also **anticous** a., *ănt·ĭk'·ŭs*, placed in front of a flower, as the lip of orchids.

antihelix, n., *ănt'·ĭ·hēl'·ĭks* (Gr. *anti*, opposite to, but here in the sense of 'before'; *helix*, anything twisted or convoluted, the ear), the curved prominence parallel with, and in front of, the helix or external prominent rim of the auricle of the ear.

antiperistaltic, a., *ănt'·ĭ·pĕr·ĭ·stălt'·ĭk* (Gr. *anti*, against; *peristaltikos*, drawing together all round—from *peri*, around; *stello*, I send), applied to the vermicular contraction of the intestinal tube when that takes place in a direction from behind forwards: **antiperistalsis**, n., *-stăls'·ĭs*, the inversion of the peristaltic motion of the intestines.

antiphlogistic, a., *ănt'·ĭ·flŏdj·ĭst'·ĭk* (Gr. *anti*, against; *phlogizo*, I consume or burn), a theoretical term applied to medical treatment intended to subdue inflammation: n., a medicine that checks inflammation.

antipodal, a., *ănt·ĭp'·ŏd·ăl* (Gr. *anti*, opposite; *podes*, feet), having the feet directly opposite; in *bot.*, applied to cells formed by a free-cell formation in phonerogams.

Antirrhineæ, n. plu., *ănt'·ĭr·rĭn'·ĕ·ē* (Gr. *anti*, like, similar; *rhis*,

ANT   27   APH

a nose, *rhinos*, of a nose), the second of the three sections of the Ord. Scrophulariaceæ: **Antirrhinum**, n., *ănt′·ĭr·rĭn′·ŭm*, a genus of plants, the flowers of most of the species bearing a perfect resemblance to the snout of some animal.

antiscorbutic, a., *ănt′·ĭ·skŏr·būt′·ĭk* (Gr. *anti*, against; Eng. *scorbutic*), that is good against scurvy.

antiseptic, n., *ănt′·ĭ·sĕpt′·ĭk* (Gr. *anti*, against; *septos*, putrid), a substance which prevents putrefaction: **adj.**, counteracting putrefaction.

antispasmodic, n., *ănt′·ĭ·spăz·mŏd′·ĭk* (Gr. *anti*, against; Eng. *spasmodic*), any medicine which allays pain, cramp, or spasms in the human body.

antitragus, n., *ănt′·ĭ·trăg′·ŭs* (Gr. *anti*, against; *tragos*, a he-goat), a small tubercle or conical eminence opposite the tragus of the ear, and separated from it by a deep notch; see 'tragus.'

antitropal, a., *ănt·ĭt′·rŏp·ăl* (Gr. *anti*, against; *tropos*, a turn, mode, or manner—from *trepo*, I turn), in *bot.*, applied to an embryo whose radicle is diametrically opposite to the hilum; inverted with respect to the seed, as the radicle: also **antitropous**, a., *ănt·ĭt′·rŏp·ŭs*.

antlia, n., *ănt′·lĭ·ă* (L. *antlia*, a pump), the spiral trunk with which butterflies and other lepidopterous insects suck up the juices of flowers.

antrum Highmori, *ănt′·rŭm hĭ·mŏr′·ĭ* (L. *antrum*, a cave, a hollow; after the English anatomist, *Highmore*, the first describer of it), the maxillary sinus, a large cavity lying above the molar teeth and below the orbital plate: **antrum pylori**, *pĭ·lŏr′·ĭ* (Gr. *pulōros*, a gate-keeper—from *pule*, a gate; *pylorus* is a Latinized form of the Gr. *puloros*; L. *pylori*, of the pylorus), in the

stomach, the lesser pouch near the intestinal opening, which is guarded by a muscular ring called the pylorus.

anus, n., *ān′·ŭs* (L. *anus*, the fundament), the lower orifice of the bowels.

aorta, n., *ā·ŏrt′·ă* (Gr. *aŏrto*, was suspended—from *aeiro*, I raise up), the great trunk artery of the body, which arises from the left side of the heart, and gives origin to all other arteries belonging to the greater or systemic circulation: **aortic**, a., *ā·ŏrt′·ĭk*, pert. to: **aorta abdominalis**, *ăb·dŏm′·ĭn·āl′·ĭs* (L. *abdōmen*, the belly), the abdominal aorta, the direct continuation of the thoracic aorta: **aorta thoracica**, *thŏr·ăs′·ĭk·ă* (L. *thōrax*, the breast, the thorax, *thorācis*, of the breast), the thoracic aorta, the continuation of the arch of the aorta, extending from the lower border of the fifth to the twelfth dorsal vertebra.

aperient, n., *ăp·ēr′·ĭ·ĕnt* (L. *aperiens*, opening), a medicine that opens the bowels: **adj.**, gently purgative.

aperispermic, a., *ăp·ēr′·ĭ·spĕrm′·ĭk* (L. *aperio*, I open; *sperma*, seed), in *bot.*, without separate albumen.

apetalous, a., *ă·pĕt′·ăl·ŭs* (Gr. *a*, without; *petalon*, a leaf), having no petals; monochlamydeous.

Aphaniptera, n., *ăf′·ăn·ĭp′·tĕr·ă* (Gr. *aphanes*, unseen, not apparent—from *a*, not, *phaino*, I show; *pteron*, a wing), an order of insects, comprising fleas, apparently without wings: **aphanipterous**, a., *ăf′·ăn·ĭp′·tĕr·ŭs*, apparently without wings.

aphasia, n., *ă·fā′·zhĭ·ă* (Gr. *aphasia*, inability to speak—from *a*, not, and *phāo*, I speak), amnesic loss of speech from loss of memory of words; ataxic loss of speech from loss of co-ordination of the muscles involved in articulate speech.

Aphelandra, n. plu., *ăf'.ĕl.ănd'.ră* (Gr. *aphĕlēs*, simple, artless; *aner*, a man, *andros*, of a man), a genus of plants, Ord. Acanthaceæ, some of the species of which are cultivated for their showy flowers.

aphonia, n., *ă.fōn'.ĭ.ă* (Gr. *aphōnia*, want of voice—from *a*, without; *phōne*, voice), loss of voice.

aphthæ, n. plu., *ăf'.thē* (Gr. *aphthai*, ulcerations inside the mouth —from *aptō*, I inflame), small white ulcers on the tongue, gums, palate, etc.; thrush: **aphthous,** a., *ăf'.thŭs*, pert. to thrush; having aphthæ or blisters on the skin or mucous membranes: **aphthaphytes,** n. plu., *ăf'.thă.fīts* (Gr. *phuton*, a plant), the mould or fungi that gives rise to aphthæ in the human species: **aphthoid,** a., *ăf'.thŏyd* (Gr. *eidos*, resemblance), resembling aphthæ.

Aphyllantheæ, n. plu., *ăf'.ĭl.ănth'.ĕ.ē* (Gr. *a*, without; *phullon*, a plant; *anthos*, a flower), a tribe of plants, Ord. Liliaceæ; the grass-tree tribe, having a rush-like habit, and membranousimbricated bracts: **Aphyllanthes,** n. plu., *ăf'.ĭl.ănth'.ēz*, a genus of plants, having stems like a rush, and bearing on their summits little tufts of flowers.

aphyllous, a., *ăf.ĭl'.lŭs* or *ăf'.ĭl.lŭs* (G. *a*, without; *phullon*, a leaf), in *bot.*, destitute of leaves: **aphylly,** n., *ăf'.ĭl'.lĭ*, the suppression or want of leaves.

apical, a., *ăp'.ĭk.ăl*, also **apicilar,** a., *ăp.ĭs'.ĭl.ăr* (L. *apex*, a tip or extremity, *apĭcis*, of an extremity), relating to the pointed end of a cone-shaped body; at the apex; in *bot.*, often applied to parts connected with the ovary.

apiculus, n., *ăp.ĭk'.ūl.ŭs*, also **apiculum,** n., *-ūl.ŭm* (L. *apiculus*, a little point—from *apex*, a tip or point), in *bot.*, a terminal soft point springing abruptly: **apiculate,** a., *ăp.ĭk'.ūl.āt*, pert. to an apiculus.

apillary, n., *ăp'.ĭl.lăr.ĭ* (Gr. *a*, without; L. *pĭleus*, Gr. *pilos*, a felt cap), the suppression or want of the upper lip of a flower.

Apios tuberosa, *ăp'.ĭ.ŏs tūb'.ĕr.ōz'.ă* (Gr. *apion*, a pear; *apios*, a pear tree; L. *tūber*, a protuberance, *tūbĕris*, of a protuberance), a plant, of the Sub-ord. Papilionaceæ, and Ord. Leguminosæ, whose roots are used as an article of food in America.

Aplacentalia, n. plu., *ăp'.lăs.ĕnt.āl'.ĭ.ă* (Gr. *a*, without; Eng. *placenta*), the section of the Mammalia, including the Didelphia and Monadelphia, in which the young is not furnished with a placenta: see 'placenta.'

aplanatic, a., *ăp'.lăn.ăt'.ĭk* (Gr. *a*, without; *planăo*, I wander), applied to lenses which entirely correct the aberration of the rays of light.

aplectrum, n., *ă.plĕkt'.rŭm* (Gr. *a*, without; *plēktron*, the point of a spear, the spur of a cock; L. *plectrum*, a little stick or quill for playing on a stringed musical instrument), a curious little plant whose flowers are spurless, and which contains a very glutinous matter, Ord. Orchidiaceæ; in America the plant is called Puttywort.

aploperistomi, n. plu., *ăp'.lō.pĕr.ĭs'.tŏm.ĭ* (Gr. *aplŏŏs*, single; *peri*, round about; *stoma*, a mouth), in *bot.*, a term applied to those mosses which have the mouth of their thecæ naked, or which have a single peristome: **aploperistomatous,** a., *ăp'.lō.pĕr'.ĭ.stŏm'.ăt.ŭs* (Gr. *stoma*, a mouth, *stomătos*, of a mouth), having a single peristome, or composed of only one row of teeth.

aplostemonous, a., *ăp'.lō.stĕm'.ŏn.ŭs* (Gr. *aplŏŏs*, single; Gr. *stēmōn*, L. *stamen*, the upright threads in an ancient loom which stood upright, while the same is now placed horizontally; a warp), in *bot.*, a flower with a single row of stamens.

Aplotaxis, n., *ăp'·lŏ·tăks'·ĭs* (Gr. *aplŏŏs*, single; *taxis*, order), a genus of plants of the Sub-ord. Cynarocephalæ, Ord. Compositæ, found in Cashmere, said to be the ancient Costus, used medicinally and for incense.

apnœa, n., *ăp·nē'·ă* (Gr. *apnoia*, without the power of breathing —from *a*, without; *pneo*, I breathe), absence of respiration; suffocation.

apocarpous, a., *ăp'·ō·kărp'·ŭs* (Gr. *apo*, from; *karpos*, fruit), having the ovary and fruit composed of numerous distinct carpels; applied to fruits when their carpels are either quite separate, or only partially united.

Apocynaceæ, n. plu., *ăp'·ŏs·ĭn·ā'·sĕ·ē* (Gr. *apo*, from; *kuōn*, a dog), the Dog-bane family, an order of plants many of which are poisonous, and not a few bear handsome flowers: **Apocynum,** n., *ăp·ŏs'·ĭn·ŭm*, a genus of plants, so called as believed by the ancients to be fatal to dogs if eaten by them.

Apoda, n. plu., *ăp'·ŏd·ă* (Gr. *a*, without; *pous*, a foot, *podos*, of a foot), applied to those fishes which have no ventral fins; the footless Cæciliæ amongst the Amphibia: **apodal,** a., *ăp'·ŏd·ăl*, also **apodous,** a., *ăp'·ŏd·ŭs*, having no feet; without ventral fins which in fish correspond to legs and feet among animals: **apodia,** n., *ă·pōd'·ĭ·ă*, the absence of feet.

apodema, n. plu., *ăp·ŏd'·ĕm·ă* (Gr. *apo*, from; *dĕma*, a cord, a bond; *demata*, cords or bonds), certain appendages on the bodies of Articulata giving attachment to muscles, or articulating with wings and the like: **apodemata,**

n. plu., *ăp'·ŏd·ĕm'·ăt·ă*, certain chitinous septa which divide the tissues in the Crustacea.

aponeurosis, n., *ăp·ŏn'·ūr·ōz'·ĭs*, **aponeuroses,** plu., *-ōz'·ēz* (Gr. *aponeurosis*, the end of a muscle —from *apo*, from or at; and *neuron*, a nerve, a muscle), the extremity of a muscle where it becomes a tendon; the fibrous sheath of a muscle or investment of a part.

apophyllous, a., *ăp'·ō·fĭl'·lŭs* (Gr. *apo*, from; *phullon*, a leaf), in *bot.*, applied to the parts of a single perianth whorl when they are free leaves.

apophysis, n., *ăp·ŏf'·ĭs·ĭs* (Gr. *apo*, from; *phuo*, I grow), in *anat.*, a process or protuberance on the surface of a bone, generally at the ends; in *bot.*, a swelling at the base of the theca in some mosses; any irregular swelling on the surface: **apophysate,** a., *ăp·ŏf'·ĭs·āt*, having a swelling at the base.

apoplexy, n., *ăp'·ŏ·plĕks'·ĭ* (Gr. *apoplexia*, stupor — from *apo*, from; *plĕsso*, I strike), stupor, or an unconsciousness like that produced by felling an ox : **apoplectic,** a., *ăp'·ŏ·plĕkt'·ĭk*, pert. to.

NOTE.—Many diseases of the brain produce this symptom. As those earliest recognised were accompanied by effusion of blood, the term has been irregularly applied to affections of other organs accompanied by effusion of blood into their tissues, as pulmonary or splenic apoplexy, though these are unaccompanied by stupor.

apostrophe, n., *ăp·ŏs'·trŏf·ē* (Gr. *apo*, from; *strophē*, a turning), in *bot.*, the collection of protoplasm and chlorophyll grains on the walls of cells that are adjacent to other cells.

apothecium, n., *ăp·ŏ·thē'·shĭ·ŭm* (Gr. *apothēkē*, L. *apotheca*, a storehouse — from Gr. *apo*, from; *thēkē*, a box or chest), the rounded shield-like fructification of lichens, forming a receptacle for the reproductive bodies or spores: **apothecia**, n. plu., *ăp·ŏ·thē'·shĭ·ă*.

appendices epiploicæ, *ăp·pĕn'·dĭs·ēz ĕp'·ĭp·lō'·ĭs·ē* (L. *appendix*, an addition, a supplement ; Gr. *epiplŏŏn*, the omentum), the epiploïc appendage ; masses of fat attached by pedicles along the free border of the intestines, which support the intestines : **appendix vermiformis**, *vĕrm'·ĭ·fŏrm'·ĭs* (L. *vermis*, a worm ; *forma*, shape), a small portion of the cæcum which hangs down in a worm-like shape in the centre of the abdomen—remarkable for no known use.

NOTE.—The enormous cæcum of many of the lower animals is, in man, dwindled to a worm-like sac which has received this name.

appendiculate, a., *ăp'·pĕnd·ĭk'·ūl·āt* (L. *appendicula*, a small appendage), having a little appendage, as the scaly appendages of corollas, or found at the base of certain filaments.

applanate, a., *ăp'·plăn·āt* (L. *ad*, to ; *planātus*, made flat—from *plānus*, level, flat), in *bot.*, flattened out ; horizontally expanded.

apposite, a., *ăp'·pŏz·ĭt* (L. *ad*, to ; *positus*, placed or put), in *bot.*, having similar parts similarly placed, as side by side: **appositional**, a., *ăp'·pŏz·ĭsh'·ŭn·ăl*, in algæ, having two branches lying side by side, partly uniting as to appear a compound branch.

appressed, a., *ăp·prĕst'* (L. *ap*, for *ad*, at or to; *pressus*, pressed, kept under), in *bot.*, denoting leaves which are applied to each other, face to face, without being folded or rolled together.

Aptera, n. plu., *ăpt'·ĕr·ă* (Gr. *a*, without ; *pteron*, a wing), a division of insects characterized by the absence of wings in the adult condition : **apterous**, a., *ăpt'·ĕr·*

ŭs, without wings : **apteryx**, n., *ăpt'·ĕr·ĭks* (Gr. *pterux*, a wing), the wingless bird of New Zealand, of the Ord. Cursores.

aqua fortis, *ăk'·wă fŏrt'·ĭs* (L. *aqua*, water ; *fortis*, strong), strong water, the popular name for 'nitric acid:' **aqua regia**, *rēdj'·ĭ·ă* (L. *regius*, royal), a mixture of nitric and hydrochloric acids, so called from its power of dissolving gold, the king of metals.

Aquifoliaceæ, n. plu., *ăk'·wĭ·fōl·ĭ·ās'·ĕ·ē* (L. *aquifolium*, the holly tree ; *aquifolius*, having sharp or pointed leaves — from *acus*, a needle, and *folium*, a leaf), the Holly family, an Order of evergreen trees or shrubs : **Aquifolium**, n., *ăk'·wĭ·fōl'·ĭ·ŭm*, the common holly, indigenous to Britain, forms excellent fences.

Aquilariaceæ, n. plu., *ăk'·wĭl·ăr·ĭ·ās'·ĕ·ē* (L. *aquila*, an eagle—from the genus being called eagle-wood in Malacca), the Aquilaria family: **Aquilaria**, n. plu., *ăk'·wĭl·ăr'·ĭ·ă*, a genus of evergreen shrubs, comprising the eagle-wood, aloeswood, and lign-aloes.

arabin, n., *ăr'·ăb·ĭn* (from *Arabia*, where the gum-producing trees abound), a substance familiarly known as gum-arabic or gumsenegal ; the kind of gum which is soluble in cold water.

Araceæ, n. plu., *ăr·ās'·ĕ·ē* (L. *arum* or *aros*, Gr. *aron*, the plant arum or wakerobin), the Arum family, whose general property is acridity.

Arachis, n., *ăr'·ăk·ĭs* (Gr. *a*, without ; *rhachis*, a backbone or spine), a genus of plants of the Sub-ord. Papilionaceæ, and Ord. Leguminosæ, having only one species, the **Arachis hypogæa**, *hĭp'·ŏ·jē'·ă* (Gr. *hupogaios*, under the earth, subterranean—from *hupo*, under ; *gaia*, the earth), a singular plant that bears no branches, and has the strange power of forcing the fruit or pods as they increase in size into the earth, where they ripen their seeds, usually called the underground kidney-bean or ground-nut; an oil is expressed from their ends, used for cramps in India, and occasionally as a substitute for cod-liver oil in medicine.

Arachnida, n. plu., *ăr·ăk'·nĭd·ă* (Gr. *arachne*, a spider, a spider's web), a class of the Articulata, comprising spiders, scorpions, and ticks: **arachnitis**, n., *ăr'·ăk·nĭt'·ĭs*, inflammation of the arachnoid membrane; sometimes applied to the inflammation of the membranes of the brain: **arachnoid**, a., *ăr·ăk'·noÿd* (Gr. *eidos*, resemblance), applied to a membrane of the brain; in *bot.*, applied to fine hairs so entangled as to resemble a cobweb.

Araliaceæ, n. plu., *ăr·āl'·ĭ·ās'·ĕ·ē* (*aralia*, an American word), the Ivy family: **Aralia**, n. plu., *ăr·āl'·ĭ·ă*, a genus of the above, one species of which has fragrant and aromatic roots which are used in America as a substitute for sarsaparilla: **araliaceous**, a., *ăr·āl'·ĭ·ā'·shŭs*, pert. to the Aralia.

Aranthocephalis, n., *ăr·ănth'·ō·sĕf'·ăl·ĭs* (probably Gr. *arachnē*, a spider; *anthos*, a flower; *kephalē*, the head), an Order of intestinal parasites; the armed worms.

araucaria, n. plu., *ăr'·ăw·kār'·ĭ·ă* (*araucanos*, its name in Chili), the Norfolk Island pine, famed for its size and for its wood: **araucarites**, n. plu., *ăr·ăw'·kăr·ītz*, the fossil wood whose structure is identical with the living araucaria.

arbor vitæ cerebelli, *ărb'·ŏr vĭt'·ē sĕr'·ĕ·bĕl'·lī* (L. *arbor*, a tree; *vitæ*, of life; *cerebelli*, of a small or little brain), the tree of life of the brain; the foliated or arborescent appearance presented by either hemisphere of the cerebellum when a vertical section is made through it: **arbor vitæ uterinus**,

ŭt·ĕr·īn'·ŭs (L. *uterinus*, uterine), the uterine tree of life; the appearance of branches from the stem of a tree presented by the folds on the interior of the 'cervix uteri.'

arborescent, a., *ăr'·bŏr·ĕs'·ĕnt* (L. *arborescens*, growing into a tree—from *arbor*, a tree), branched like a tree.

Arbutus, n., *ăr'·bŭt·ŭs* (L. *arbutus*, the wild strawberry or arbute tree), a genus of plants, Ord. Ericaceæ: **Arbutus unedo**, *ŭn'·ĕd·ō* (L. *unedo*, the arbute or strawberry tree—said to be from *unus*, one, and *edo*, I eat), the strawberry tree, so called from its fruit resembling a strawberry; the fruit is not agreeable, but a wine is prepared from it in Corsica: **arbutean**, a., *ăr·bŭt'·ĕ·ăn* pert. to.

archangelica, n., *ărk'·ăn·jĕl'·ĭk·ă* (Gr. *archos*, chief, and *angelica*, from its supposed virtues), the botanical name for the Angelica plant and root.

archegonium, n., *ărk'·ĭ·gōn'·ĭ·ŭm* (Gr. *archē*, beginning; *gonē*, seed), in *bot.*, the young female cellular organ in cryptogamic plants; the early condition of a spore case.

Archencephala, n. plu., *ărk'·ĕn·sĕf'·ăl·ă* (Gr. *archo*, I command, I rule over; *engkephalos*, the brain), Owen's name for his fourth and highest group of Mammalia, comprising man alone.

archil, n., *ărtsh'·ĭl* (Fr. *orcheil:* Sp. *orchilla*—from Sp. *roca*, a rock), a rich purple colour, obtained from the lichen Roccella tinctoria, found growing on the rocks of the Canaries and other islands.

archisperms, n. plu., *ărk'·ĭ·spĕrmz* (Gr. *archos*, chief; *sperma*, seed), another name for gymnosperms.

arciform, a., *ărs'·ĭ·fŏrm* (L. *arcus*, a bow; *forma*, shape), applied in the medulla oblongata to some of its fibres which emerge at the anterior median fissure, and form a band which curves round the lower border of the olivary body, or which passes transversely across it, and round the sides of the medulla.

Arctium, n., *ărk'·tĭ·ŭm* (Gr. *arktos*, a bear—in reference to its rough, bristly fruit), a genus of plants of the Sub-ord. Cynarocephalæ, Ord. Compositæ: **Arctium lappa**, *lăp'·pă* (L. *lappa*, a bur), the burdock, which is bitterish, and has been used in the form of infusion as a substitute for sarsaparilla.

Arctostaphylos, n., *ărk'·tŏ·stăf'·ĭl·ŏs* (Gr. *arktos*, a bear ; *staphulē*, a grape — in allusion to the rough taste of the fruit), a genus of plants, Ord. Ericaceæ: **Arctostaphylos uva-ursi**, *ŭv'·ă·ĕrs'·ĭ* (L. *ūva*, a grape-berry ; *ursi*, of the bear), the bearberry, whose fruit is used as an astringent : **A. glauca**, *glāwk'·ă* (L. *glaucus*, bluish grey), the manzanita plant, which covers the mountains of California with a thick brushwood.

arcuate, a., *ărk'·ū·āt* (L. *arcus*, a bow), curved in an arched manner like a bow.

arcus senilis, *ărk'·ŭs sĕn·ĭl'·ĭs* (L. *arcus*, a bow, an arch ; *senilis*, aged), the arch of the aged ; a circular, opaque appearance round the margin of the cornea of aged persons, usually affecting both eyes.

ardellæ, n. plu., *ăr·dĕl'·lē* (Gr. *ardālos*, dirty, foul—from *ardō*, I sprinkle), small apothecia of certain lichens, as Arthonia, having the appearance of dust.

Areca, n. plu., *ăr·ēk'·ă* (Indian name), a genus of plants of the Ord. Palmæ: **Areca catechu**, *kăt'·ĕ·shōō* (said to be Japanese *kate*, a tree; *chu*, juice), an elegant palm producing the betel nut, and an extract of an astring-

ent nature like catechu : **Arecineæ**, n. plu., *ăr'·ĕ·sĭn'·ĕ·ē*, the first of the five tribes into which the Ord. Palmæ is divided.

arenaceous, a., *ăr'·ĕn·ā'·shŭs* (L. *arēna*, sand), composed of grains of sand ; having the properties of sand.

areola, n., *ăr·ē'·ŏl·ă* (L. *ārĕŏlă*, a small open place, a small garden bed), the small coloured circle round the nipple, or a pustule : **areolæ**, n. plu., *ăr·ē'·ŏl·ē*, small interstices of cellular or other tissues ; little spaces on the area or surface ; the spaces between the cracks in the lichens : **areolar**, a., *ăr·ē'·ŏl·ăr*, of or like an areola : **areolate**, a., *ăr·ē'·ŏl·āt*, in *bot.*, divided into distinct angular spaces.

Arethusa, n. plu., *ăr'·ĕ·thūz'·ă* (after a nymph of Diana's, who was changed into a fountain), a genus of plants, Ord. Orchidaceæ: **Arethusa bulbosa**, *bŭlb·ōz'·ă* (L. *bŭlbus*, a bulbous root), a plant which has a large fine lilac flower terminating each stem.

arillus, n., *ăr·ĭl'·lŭs*, also **aril**, n., *ăr'·ĭl* (Fr. *arille*, an arillus ; Sp. *arillo*, a small hoop—from *aro*, a hoop; L. *aridus*, dry), the exterior coat of a seed which drying falls off spontaneously : **arillate**, a., *ăr·ĭl'·lāt*, having an aril : **arillode**, n., *ăr'·ĭl·lōd* (Gr. *eidos*, resemblance), an extra covering of the seed ; the 'arillus' proceeds from the placenta, as in the passion-flower, the 'arillode' from the exostome, as in the mace of the nutmeg.

arista, n., *ăr·ĭst'·ă* (L. *arista*, the beard of an ear of corn), a long pointed process, as in barley and many grasses; an awn : **aristate**, a., *ăr·ĭst'·āt*, furnished with beards or spikes, as barley and many grasses; awned: **aristulate**, a., *ăr·ĭst'·ūl·āt*, having a very small arista.

Aristolochiaceæ, n. plu., *ăr·ĭst'·ō·*

lŏk'·ĭ·ā'·sĕ·ē (Gr. *aristos*, best; *locheia*, child - birth, delivery), the Birth-wort family, a small Order of climbing herbaceous plants, bearing mottled and singularly-shaped flowers: **Aristolochia,** n., *ăr·ĭst'·ō·lŏk'·ĭ-ă*, a genus whose flowers have more or less the appearance of a horn; the names Birth-wort and Aristolochias have been given this genus of plants from their supposed action on the uterus: **Aristolochia serpentaria,** *sĕrp'·ĕnt·ăr'·ĭ-ă* (L. *serpens*, a serpent, *serpentis*, of a serpent), the Virginian snake-root, a native of the United States, formerly used as an antidote to snake poison: **arist'olochia'ceous,** a., *-ā'·shŭs*, having an arrangement of parts as in the Aristolochia.

armature, n., *ärm'·ăt·ŭr* (L. *arma*, arms, weapons), in *bot.*, the hairs, prickles, etc. covering an organ; a piece of iron used to connect the poles of magnets.

Armeria, n., *ăr·mēr'·ĭ-ă* (*armeria*, the Latin name of sweet-william), a genus of plants, Ord. Plumbaginaceæ, which, though dwarf, are handsome, and well adapted for ornamenting rock-work: **Armeria maritima,** *măr·ĭt'·ĭm·ă* (L. *maritimus*, belonging to the sea—from *măre*, the sea), thrift or common sea-pink, grows on the sea-shore, and on the top of the highest mountain of Scotland.

arnatto, n., *ăr·năt'·tō*, also spelt **arnotto** and **annotto** (perhaps a corruption of *arnot*, the 'earth-nut,' from a mistaken notion of its origin), a red colour obtained from the reddish pulp which surrounds the seeds of the tree Bixa orellana, used for dyeing cheese and butter, imported into this country in three forms, viz. leaves, eggs, and rolls.

arnica, n., *ăr'·nĭk·ă*, or **arnica montana,** *mŏn·tăn'·ă* (Gr. *arnion*, a little lamb; *montānus*, belonging to a mountain—so called from the resemblance of the leaf to the soft coat of a lamb), mountain tobacco or leopard's bane, the expressed juice of the root of which is used in medicine; Sub-ord. Corymbiferæ, Ord. Compositæ.

Aroideæ, n. plu., *ăr·oȳd'·ĕ·ē* (*arum*, the plant wake-robin; Gr. *eidos*, resemblance), an Order of plants having an arrangement of parts as in the Arum—now called Ord. Araceæ, which see.

Arracacha esculenta, *ăr'·ră·kătsh'·ă ĕs'·kūl·ĕnt'·ă* (*arracacha*, the South American name; L. *esculentus*, fit for eating), a native of Grenada having large and esculent roots, resembling a parsnip in quality, which have been recommended as a substitute for the potato; Ord. Umbelliferæ.

arrack, n., *ăr'·răk* (Arab. *araq*, sweat, juice), a distilled impure spirit, much used in the East, obtained from fermented rice, betel nuts, and the sap and fruit of palms.

Artanthe, n., *ăr·tănth'·ē* (probably *artaō*, I make ready; *anthos*, a flower), a genus of wooded plants with jointed stems, Ord. Piperaceæ: **Artanthe elongata,** *ē'·lŏng·găt'·ă* (L. *elongatus*, made long—from *e*, out; *longus*, long), a shrub of S. America, from which the substance, consisting of the leaves and unripe fruit, called matico or matica is obtained; it possesses aromatic, fragrant, and astringent qualities.

Artemisia, n., *ăr'·tĕm·ĭzh'·ĭ·ă* (from *Artĕmis*, one of the names of Diana, who presided over women in childbed), a genus of plants, Ord. Compositæ, and Sub-ord. Corymbiferæ, the species of which are remarkable for their strong odour and bitter taste: **Artemisia absinthium,** *ăb·sĭnth'·ĭ·ŭm* (L. *absinthium*, wormwood), wormwood, the heads of the flowers of which, as well as other species,

C

under the name of wormseed, are used as anthelmintics and tonics: **A. mutellina** and **spicata**, *mŭt· ĕl·lĭn'·ă spĭk·āt'·ă* (unascertained: L. *spīcātus*, furnished with a point), plants used in the preparation of tincture or distilled spirit, much in use and called in France 'eau' or 'crême d'absinthe' (*ŏ krăm dăb·sāngt*): **A. dracunculus**, *dră·kŭng'·kŭl·ŭs* (L. *dracunculus*, a small serpent, a dragonet), the plant Tarragon, used in pickles and salads, and in the medication of vinegar : **A. abrotanum**, *ăb· rŏt'·ăn·ŭm* (L. *abrŏtŏnum*, Gr. *abrŏtŏnon*, southernwood), the plant southernwood, used on the Continent in the preparation of beer : **A. Indica**, *ĭnd'·ĭk·ă* (L. *Indicus*, Indian), the plant Sikkim-wormseed, grows twelve feet high at elevations of from 2000 to 6000 feet.

arteria centralis retinæ, *ărt·ēr'·ĭ·ă sĕnt·rāl'·ĭs rĕt'·ĭn·ē* (L. *artēria*, an artery ; *centrālis*, central ; *retinæ*, of the retina—from *rēte*, a net), one of the smallest branches of the ophthalmic artery, arising near the optic foramen : **arteriæ propriæ renales**, *ărt·ēr'·ĭ·ē prŏp'· rĭ·ē rĕn·āl'·ēz* (L. *artēriæ*, arteries ; *propriæ*, proper, plu. ; *renales*, renal, plu.—from *rēnēs*, the kidneys), the proper renal arteries which enter the kidney proper in the columns of Bertini : **arteriæ receptaculi**, *rĕs'·ĕp·tăk'·ŭl·ī* (L. *arteriæ*, of an artery; *receptaculi*, receptacles), the receptacles of an artery ; numerous small vessels derived from the internal carotid artery in the cavernous sinus.

arteriolæ rectæ, *ărt·ēr'·ĭ·ŏl·ē rĕkt'·ē* (L. *arteriolæ*, small arteries; *rectæ*, straight, plu.), the straight small or branch arteries ; the second set of arteries which branch off from the 'arteriæ propriæ renales' for the supply of the medullary pyramids, which they enter at their bases.

arteritis, n., *ărt'·ĕr·īt'·ĭs* (L. *artēria*, an artery ; *itis*, denoting inflammation), inflammation of an artery.

artery, n., *ărt'·ĕr·ĭ* (L. *arteria*, an artery—from Gr. *aër*, air, and *tereo*, I preserve, because believed by the ancients to circulate air), one of the vessels that convey the blood from the heart to all parts of the body, having valves only at their origin : **arteriotomy**, n., *ărt·ēr'·ĭ·ŏt'·ŏm·ĭ* (Gr. *tomē*, a cutting), the opening of an artery for the purpose of drawing blood from it.

arthritic, a., *ăr·thrĭt'·ĭk* (Gr. *arthron*, a joint), pert. to the joints or to the gout : **arthritis**, n., *ăr·thrīt'·ĭs*, inflammation of a joint; the gout; a chronic rheumatic disease.

arthrodia, n., *ăr·thrōd'·ĭ·ă* (Gr. *arthron*, a joint ; *arthrodēs*, like joints), that kind of joint which admits of a gliding movement, and is formed by the approximation of plane surfaces, or of one surface slightly concave and the other slightly convex ; the three principal forms of articulation are the Diarthrosis or moveable joints, the Synarthrosis or immoveable joints, the Amphi-arthrosis or mixed joints.

arthrosterigmata, n. plu., *ăr'·thrō· stĕr·ĭg'·măt·ă* (Gr. *arthron*, a joint ; *stērigma*, a joint), jointed Sterigmata, which see.

articular, a., *ărt·ĭk'·ŭl·ĕr* (L. *articulus*, a joint), relating to the joints : **articulation**, n., *ărt·ĭk'· ŭl·ā'·shŭn*, the particular mechanism by which the bones are united to each other in the skeleton : **articular surfaces**, the peculiar gristly surfaces of bone joints : **articularis**, a., *ărt·ĭk'· ŭl·ār'·ĭs*, relating to joints ; applied to the arteries branching off from the popliteal : **Articulata**, n. plu., *ărt·ĭk'·ŭl·āt'·ă*, a division of the Animal kingdom, comprising insects, centipeds, spiders, and crustaceans, which are characterised by the possession of jointed bodies or jointed limbs ; the Arthropoda, which is the term now more usually employed : **articulated**, a., *ărt·ĭk'·ūl·āt·ĕd*, jointed ; having parts separating easily at some point : **articulo mortis**, *ărt·ĭk'·ūl·ō mŏrt'·ĭs* (L. *articulo*, in a joint, in point or moment ; *mors*, death, *mortis*, of death), at the point of death ; about to die.

Artiodactyla, n. plu., *ărt'·ĭ·ō·dăk'·tĭl·ă* (Gr. *artios*, exactly fitted, even ; *daktulos*, a finger or toe), a division of the hoofed quadrupeds, in which each foot has an even number of toes, as two or four.

Artocarpeæ, n. plu., *ărt'·ō·kârp'·ĕ·ē* (Gr. *artos*, bread ; *karpos*, fruit), a sub-order of the Ord. Moraceæ: **Artocarpus**, n., a genus of trees, producing the bread-fruit, and flowers in dense heads: **Artocarpus incisa**, *ĭn·sĭz'·ă* (L. *incīsus*, notched, indented), the well-known bread-fruit tree, which furnishes an abundant supply of food in tropical countries, besides furnishing many other materials for domestic use : **A. integrifolia**, *ĭn·tĕg'·rĭ·fōl'·ĭ·ă* (L. *integrifōlia*, entire leaved—from *intĕger*, entire, undivided ; *folium*, a leaf), the Jack or Jaca tree, the fruit of which attains a large size, weighing sometimes 30 lbs., but is inferior in quality to the bread-fruit—so called from its having entire or undivided leaves.

Arum, n., *ār'·ŭm* (L. *ārum* ; Gr. *āron*, supposed to be an ancient Egyptian word, the plant wake-robin), a genus of plants, Ord. Araceæ : **Arum maculatum**, *măk'·ūl·āt'·ŭm* (L. *maculātum*, stained, spotted), the plant cuckoo-pint or wakerobin ; the species of Arum with spotted leaves, and

poisonous, but yet from the rhizome of which Portland sago is prepared : **A. dracunculus**, *drăk·ŭnk'·ūl·ŭs* (L. *dracŭncŭlus*, a small serpent), the plant dragon's wort, and many-leaved Arum, which is extremely acrimonious : **A. esculentum**, *ĕsk'·ūl·ĕnt'·ŭm* (L. *esculentum*, fit for eating), a species of Arum used as a pot herb in the West Indies.

aryteno, *ăr'·ĭt·ēn'·ō* (Gr. *arutaina*, a pitcher—in animals, the opening of the larynx with the arytenoid cartilages, bearing a resemblance to a pitcher with a spout), denoting connection with the arytenoid cartilages: **aryteno-epiglottidean**, a., *ĕp'·ĭ·glŏt·tĭd'·ē·ăn* (Gr. *epiglōttis*, a little tongue—from *epi*, upon ; *glōttis*, the mouthpiece of a wind instrument, *glōttidos*, of the mouthpiece of a wind instrument — from *glotta*, the tongue), applied to the ligamentous and muscular fibres enclosed by a fold of mucous membrane which are stretched between the sides of the epiglottis and the apex of the arytenoid cartilages : **aryteno-epiglottideus**, *ĕp'·ĭ·glŏt·tĭd'·ē·ŭs*, 'superior' and 'inferior,' designating delicate muscular fasciculi, the former rising from the apex of the arytenoid cartilage, and the latter from the arytenoid cartilage, just above the attachment of the superior vocal cord : **arytenoid**, a., *ăr'·ĭt·ēn'·ōўd* (Gr. *eidos*, resemblance), resembling the mouth of a pitcher: **arytenoid cartilages**, two cartilages, each having a pyramidal form, situated at the upper border of the cricoid cartilage, at the back of the larynx : **arytenoid glands**, the muciparous glands found along the posterior margin of the aryteno-epiglottidean fold, in front of the aryteno-cartilages : **arytenoideus**, n., *ăr'·ĭt·ēn·ōўd'·ē·ŭs*, applied to a single muscle filling up the posterior concave surface of the arytenoid cartilages.

asafœtida, n., *ăs'·ă·fĕt'·ĭd·ă* (L. *asa*, a gum—suggested to be a corruption of the Persian name *anguzeh*; L. *fœtidus*, fetid; Arab. *asâ*, healing), the stinking healer, a fetid gum resin, being the concrete juice of the plant Narthex asafœtida, or Ferula narthex, a plant found in Persia and Affghanistan, and also from Ferula Persica, and Scorodosma fœtidum ; Ord. Umbelliferæ.

Asagræa, n. plu., *ăs'·ă·grĕ'·ă* (in honour of Dr. Asa Gray), a genus of plants, Ord. Melanthaceæ : **Asagræa officinalis,** *ŏf·fĭs'·ĭn·ăl'·ĭs* (L. *officīnālis*, officinal—from *officīna*, a workshop, a laboratory), a plant, a native of Mexico, whose fruit is called Cevadilla, used in the preparation of Veratria, which is employed in cases of neuralgia and rheumatism.

asarabacca, n., *ăs'·ăr·ă·băk'·kă* (from *Asarum*, wild spikenard, but origin unknown), the name given to the powdered leaves of 'Asarum Europæum,' used as an acrid emetic, Ord. Aristolochiaceæ.

Asarum, n., *ăs'·ăr·ŭm* (L. *asarum*, Gr. *asaron*, hazel-wort, wild spikenard), a genus of plants, Ord. Aristolochiaceæ : **Asarum Europæum,** *ūr'·ŏp·ē'·ŭm* (L. *Europæum*, belonging to Europe), a plant whose powdered leaves form an acrid emetic, and whose powdered leaves and roots enter into the composition of cephalic snuffs : **asarin,** n., *ăs'·ăr·ĭn*, an active crystalline substance obtained from the plant : **Asarin Canadense,** *kăn'·ăd·ĕns'·ē* (L. *Canadensis*, belonging to Canada), the wild ginger plant, or Canada snake-root, used as a spice in Canada.

Ascaris, n., *ăsk'·ăr·ĭs* (Gr. *askaris*, a long round worm in the bowels, *askaridos*, of a long round worm), a genus of intestinal worms : **Ascarides,** n. plu., *ăsk·ăr'·ĭd·ēz*, the intestinal thread-worms : **Ascaris lumbricoides,** *lŭm'·brĭk·ōўd'·ēz* (L. *lumbrĭcus*, a mawworm—from *lumbus*, a loin ; Gr. *eidos*, resemblance), the Ascarides, which resemble the earth-worm ; a worm found in the small intestine of man, and probably in the ox : **A. megalocephala,** *mĕg'·ăl·ō·sĕf'·ăl·ă* (Gr. *megalos*, great, large ; *kephalē*, the head), the large-headed Ascarides, the intestinal worms of the horse, ass, mule, etc., found in the small intestine, sometimes in stomach and large intestine : **A. mystax,** *mĭs'·tăks* (Gr. *mustax*, the upper lip, the moustache), the lipped or hairy worms ; the intestinal worms of the cat, lynx, tiger, etc., also of man : **A. marginata,** *mârj'·ĭn·āt'·ă* (L. *marginātus*, furnished with a border), the intestinal worms of the dog, found in the small intestine : **A. suilla,** *sū·ĭl'·lă* (L. *sŭillus*, belonging to a swine—from *sŭs*, a swine), the intestinal worm of the pig.

ascending, a.; *ăs·sĕnd'·ĭng* (L. *ad*, to ; *scandens*, climbing), in *bot.*, applied to a procumbent stem which rises gradually from its base ; applied to ovules attached a little above the base of the ovary ; rising erect from the ground and forming a curve.

asci, n. plu., *ăs'·sī* (Gr. *askos*, L. *ascus*, a cavity or bladder), small membranous cells or bags which contain the sporules of cryptogamic plants : **ascidium,** n., *ăs·sĭd'·ĭ·ŭm*, **ascidia,** n. plu., *ăs·sĭd'·ĭ·ă* (Gr. *askidion*, a little bag), in *bot.*, pitcher leaves ; a form of leaf in which the stalk or petiole is widely and deeply hollowed, and closed by the blade as by a lid ; in *zool.*, an order of shell-less molluscs, having the appearance of small leathern pouches or paps, found in the sea on rocks, old shells, etc., as a pap-like, gelatinous substance : **Ascidioida**, n. plu., *ăs·sĭd′·ĭ·oyd′·ă* (Gr. *eidos*, resemblance), a class of molluscous animals which have often the shape of a two-necked bottle ; synonym of 'Tunicata :' **ascigerous**, a., *ăs·ĭdj′·ĕr·ŭs* (L. *gero*, I bear), producing asci.

ascites, n. plu., *ăs·sīt′·ēz* (Gr. *askos*, a bag, a leathern bottle), dropsy of the abdomen ; a morbid accumulation of serous fluid in the cavity of the peritoneum.

Asclepiadaceæ, n. plu., *ăs·klēp′·ĭ·ăd·ā′·sĕ·ē* (Gr. *Asklēpios*, L. *Æsculāpius*, a celebrated anc. physician), the Asclepias family, an Order of plants : **Asclepias**, n., *ăs·klēp′·ĭ·ăs*, a genus of plants : **Asclepias tuberosa**, *tūb′·ĕr·ōz′·ă* (L. *tuberosus*, having fleshy knobs), the butterfly weed or pleurisy root, a cathartic and diaphoretic : **A. curassavica**, *kŭr′·ăs·săv′·ĭk·ă* (probably L. *cura*, healing, cure—from *curo*, I care for ; *suāvium* or *sāvium*, a mouth), wild ipecacuanha : **A. Syriaca**, *sĭr·ĭ′·ăk·ă* (of or belonging to *Syria*, or connected with it), found in Canada, a very odoriferous plant when in flower—sugar is made from the flowers, and the cotton from its pods is very soft and silky.

ascospore, n., *ăsk′·ō·spōr* (Gr. *askos*, a bag ; *spora*, a seed), spores borne within asci.

asexual, a., *ă·sĕks′·ū·ăl* (Gr. *a*, without ; and *sexual*), applied to modes of reproduction in which the sexes are not concerned ; having no apparent sexual organs.

asiphonate, a., *ă·sĭf′·ŏn·āt* (Gr. *a*, without ; *siphōn*, a siphon), not possessing a respiration tube or siphon ; applied to a division of the lamellibranchiate molluscs.

asparagus, n., *ăs·păr′·ăg·ŭs* (L. *asparagus*, Gr. *asparagos*, the plant asparagus), a well-known plant, whose turios or young shoots, sent up from the underground stem, are cooked and eaten : **asparagine**, n., *ăs·păr′·ă·jĭn*, the active principle of asparagus : **Asparageæ**, n. plu., *ăs·păr·ādj′·ĕ·ē*, the Asparagus tribe of plants, Ord. Liliaceæ.

aspect, n., *ăs′·pĕkt* (L. *ad.* to, at ; *specto*, I look), in *anat.*, look ; appearance.

asperity, n., *ăs·pĕr′·ĭt·ĭ* (L. *asper*, rough), in *bot.*, roughness, as on the leaves of the Ord. Boraginaceæ.

Asperula, n., *ăs·pĕr′·ŭl·ă* (a diminutive of L. *asper*, rough), a genus of plants, Ord. Rubiaceæ : **Asperula odorata**, *ōd′·ŏr·āt′·ă* (L. *odorātus*, that has a smell—from *ŏdor*, smell), woodruff, a plant which gives out a pleasant fragrance when dry.

Asphodeleæ, n. plu., *ăs′·fō·dĕl′·ĕ·ē* (Gr. *asphodĕlos*, asphodel, a plant sacred to Proserpine), a genus of plants, Ord. Liliaceæ, the flowers of which cannot be surpassed : **asphodel**, n., *ăs′·fō·dĕl*, the day-lily, called also king's-spear.

asphyxia, n., *ăs·fĭks′·ĭ·ă* (Gr. *a*, without ; *sphuxis*, the pulse), the temporary or permanent cessation of the motions of the heart and respiration, as in drowning and suffocation ; a curious misnomer for 'suffocation,' in which the pulse never ceases while life lasts : **asphyxiated**, a., *ăs·fĭks′·ĭ·āt·ĕd*, suffocated as by hanging or drowning.

Aspidium, n., *ăs·pĭd′·ĭ·ŭm* (a diminutive from Gr. *aspis*, a shield, *aspĭdos*, of a shield), a genus of ferns, Ord. Filices : **Aspidium filix mas**, *fĭl′·ĭks mās* (L. *filix*, a fern ; *mās*, a male), the male shield-fern, used for tape-worm.

Aspidosperma excelsum, *ăs′·pĭd·ō·spĕrm′·ă ĕk·sĕls′·ŭm* (Gr. *aspis*, a serpent, *aspĭdos*, of a serpent; *sperma*, seed), a Guiana tree, remarkable for the sinuous arrangement of its wood, which gives the stem a deeply-fluted appearance, Ord. Apocynaceæ.

Asplenium, n., *ăs·plēn'·ĭ·ŭm* (Gr. *a*, without; *splēn*, the spleen, from its being believed to remove disorders of that organ), a genus of plants, Ord. Filices; spleenwort.

assurgent, a., *ăs·sėrj'·ĕnt* (L. *assurgens*, rising up—from *ad*, to; *surgo*, I rise), in *bot.*, rising upwards in a curve.

Asteliæ, n. plu., *ă·stēl'·ĭ·ē·ē* (Gr. *a*, without; *stelechos*, the trunk of a tree, a stem), an Order of plants now included in the Ord. Similaceæ: **Astelia**, n., *ă·stēl'·ĭ·ă*, a genus of preceding; the plants have grass-like leaves yielding fibres, natives of New Zealand, Tasmania, and S. Amer.: **Astelia Solandri**, *Sŏl·ănd'·rī* (after *Solandra*, a Swedish botanist), the tree flax of New Zealand.

Asteraceæ, n. plu., *ăst'·ėr·ā'·sĕ·ē* (Gr. *aster*, a star), an Order of plants bearing compound flowers, now included in the vast Ord. Compositæ: **Aster**, n., *ăst'·ėr*, a genus of preceding order, stately and handsome plants, whose flowers have an arrangement resembling little stars.

asteroid, a., *ăst'·ėr·ŏyd* (Gr. *aster*, a star; *eidos*, resemblance), star-shaped; possessing radiating lobes or rays like a star-fish: n., one of the minor planets: **Asteroidea**, n. plu., *ăst'·ėr·ŏyd'·ĕ·ă*, in *zool.*, an Order of the Echinodermata, comprising the star-fishes, which are characterised by their rayed form.

asthenia, n., *ăs·thēn'·ĭ·ă* (Gr. *astheneia*, want of strength, weakness—from *a*, without; *sthenos*, strength), in *med.*, want or loss of strength; debility: **asthenic**, a., *ăs·thēn'·ĭk*, weak; debilitated.

asthma, n., *ăst'·mă* (Gr. *asthma*, shortness of breath—from *āō*, I breathe), a disease of the breathing organs, characterised in its attacks by a gasping for breath.

Astilbe, n., *ă·stĭl'·bē* (Gr. *a*, without; *stilbē*, brilliancy, lustre), a genus of plants, Ord. Saxifrageæ, ornamental, and attaining six feet in height.

astomatous, a., *ă·stŏm'·ăt·ŭs* (Gr. *a*, without; *stoma*, a mouth, *stomăta*, mouths), not possessing a mouth; having no true mouth or aperture.

Astragalus, n., *ăs·trăg'·ăl·ŭs* (Gr. *astragalos*, a die, the ankle joint, the corresponding bones of certain animals, as the sheep, being employed by the ancients as dice), in *anat.*, a bone of the foot which forms part of the ankle joint; in *bot.*, a genus of plants, Ord. Leguminosæ, Subord. Papilionaceæ, so called from the seeds being squeezed into a kind of square form in some of the species: **Astragalus verus**, *vēr'·ŭs* (L. *vērus*, true), A.creticus, *krēt'·ĭk·ŭs* (L. *crētĭcus*, of or from Crete), **A. aristatus**, *ăr'·ĭst·āt'·ŭs* (L. *aristātus*, having an awn—from *arista*, an awn), **A. gummifer**, *gŭm'·mĭf·ėr* (L. *gummi*, gum; *fero*, I bear), and other species, are shrubs which yield gum-tragacanth: **astragaloid**, a.; *ăs·trăg'·ăl·ŏyd* (Gr. *eidos*, resemblance), pert. to or like the astragalus.

astringent, n., *ăs·trĭnj'·ĕnt* (L. *astringens*, drawing or binding tight—from *ad*, to; *stringo*, I bind fast), a medicine which binds or contracts organic textures: **adj.**, binding or contracting as muscular fibre.

Asturian, a., *ăs·tūr'·ĭ·ăn* (*Asturia*, an ancient division of Spain), designating a west Pyrenean flora, confined to the mountainous districts of the west and southwest of Ireland, the nearest Continental parts where they are native being the north of Spain.

atavism, n., *ăt'·ăv·ĭzm* (L. *atăvus*, an ancestor—from *avus*, a grandfather), the disappearance of any peculiarity or disease of a family

during one generation, succeeded
by its reappearance in another;
in *zool.*, the tendency of species
or varieties to revert to an original
type.

ataxia, n., *ă·tăks'·ĭ·ă* (Gr. *a*, not,
without; *taxis*, order—from *tasso*,
I put in order), want of co-
ordination in the movements of a
limb or organ, as 'locomotor
ataxia;' want of co-ordination in
the movements of the arms or
legs, or both, depending upon
fascicular echrosis of the posterior
column of the spinal cord: **ataxic,**
a., inco-ordinate: **ataxic aphasia,**
loss of speech, from want of co-ordination of the muscles employed
in articulate speech.

atheroma, n., *ăth'·ĕr·ōm'·ă* (Gr. or
L. *atheroma*, a tumour filled
with matter; Gr. *athăra*, a pap
made of meal), fatty calcareous degenerations in the body; a curdy
tumour: **atheromatous,** a., *ăth'·
ĕr·ōm'·ăt·ŭs*, containing matter
of the nature of atheroma.

atherosis, n., *ăth'·ĕr·ōz'·ĭs* (a word
formed from Gr. *atherōma*, a
tumour), chronic inflammation of
the internal coat of the arteries.

Atherospermaceæ, n. plu., *ăth'·ĕr·
ō·spėrm·ā'·sĕ·ē* (Gr. *ather*, the
awn or beard of an ear of corn;
sperma, seed—the seeds being
furnished with awns), the plume
nutmeg family, an Order of
plants: **Atherosperma,** n., *ăth'·ĕr·
ō·spėrm'·ă*, a genus of plants of
preceding Order: **Atherosperma
moschatum,** *mŏs·kāt'·ŭm* (mid. L.
moschātus, having a smell like
musk—from *moschus*, musk; Gr.
moschos, a sprout, a shoot), a
native of Australia, the bark of
which resembles sassafras in
flavour.

atlas, n., *ăt'·lăs* (Gr. *Atlas*—from
a, intensive; *tlaō*, I bear, I sus-
tain—in the Greek mythology,
a giant who bore up the earth
upon his shoulders), the top joint
of the neck bones which support

the globe of the head; the first
vertebra of the neck.

atlo-axoid, a., *ăt'·lō-ăks'·oÿd* (Eng.
atlas, the first vertebra of the
neck; Eng. *axis*, the second
vertebra of the neck; Gr. *eidos*,
resemblance), applied to the two
pairs of ligaments which connect
the atlas with the axis of the
vertebræ.

atonic, a., *ă·tŏn'·ĭk* (Gr. *a*, without; *tonos*, a tone), debilitated:
atony, n., *ăt'·ŏn·ĭ*, debility;
muscular weakness.

atrabiliary, a., *ăt'·ră·bĭl'·ĭ·ăr·ĭ* (L.
ater, black; *bilis*, bile), melan-
cholic; hypochondriac.

atractenchyma, n., *ăt'·răk·tĕn'·
kĭm·ă* (Gr. *atraktos*, a spindle, a
distaff; *chumos*, juice, sap), in
bot., tissue composed of spindle-
shaped cells.

Atriplex, n., *ăt'·rĭ·plĕks* (L. *ater*,
black; *plexus*, plaited, twisted),
a genus of plants, Ord. Chenopodiaceæ: **Atriplex hortensis,** *hŏr·
tĕns'·ĭs*, garden Orach or wild
Spinach.

atrium, n., *āt'·rĭ·ŭm* (L. *atrium*,
a front hall), the great chamber
or cloaca into which the intestine
opens in the Tunicata.

atropal, n., *ăt'·rŏp·ăl* (Gr. *a*, without; *tropos*, a turning), in *bot.*,
an ovule in its erect position.

Atropeæ, n. plu., *ăt·rŏp'·ĕ·ē* (Gr.
Atropos, in anc. mythology, one
of the Fates, whose duty it was
to cut short the thread of life), a
Sub-ord. of the Ord. Solonaceæ:
Atropa, n., *ăt'·rŏp·ă*, a genus of
plants: **Atropa belladonna,** *bĕl'·
lă·dŏn'·nă* (see 'belladonna'),
deadly nightshade, a highly
poisonous plant: **atropia,** n.,
ăt·rōp'·ĭ·ă, and **atropin,** n.,
ăt'·rŏp·ĭn, a highly poisonous
alkaloid extracted from the root
of the 'Atropa belladonna,' forming its active principle: **atropism,**
n., *ăt'·rŏp·ĭzm*, the symptoms produced by the frequent medicinal
use of belladonna.

atrophia, n., *ăt·rŏf'·ĭ·ă*, also atrophy, n., *ăt'·rŏf·ĭ* (Gr. *a*, without ; *trophē*, nourishment—from *trepho*, I nourish), a wasting away of the body or of an organ with or without apparent cause, and accompanied by impairment or destruction of functions: atropic, a., *ăt·rŏp'·ĭk*, wasted ; defectively nourished ; in *bot.*, abortion and degeneration of organs.

atropous, a., *ăt'·rŏp·ŭs*, and atropal, a., *ăt'·rŏp·ăl* (Gr. *a*, without ; *tropē*, a turning), in *bot.*, the ovule with foramen opposite to the hilum ; an ovule having its original, erect position ; syn. of 'orthotropous' and 'orthotropal.'

Attalea, n., *ăt·tāl'·ē·ă* (L. *Attalus*, a king of Pergamos renowned for his wealth ; *attalicus*, woven with gold, magnificent), a fine genus of beautiful, ornamental palm trees, attaining a height of from 10 to 70 feet, Ord. Palmæ : Attalea funifera, *fūn·ĭf'·ĕr·ă* (L. *fūnis*, a cord ; *fero*, I bear), a palm whose fruit is known by the name of 'Coquilla nuts,' and the hard pericarps furnish material for making umbrella handles, etc.

attenuation, n., *ăt·tĕn'·ū·ā'·shŭn* (L. *attenuātus*, weak, reduced—from *ad*, to ; *tenuis*, thin), a term employed in homœopathy to denote the dilution of drugs.'

attollens aurem, *ăt·tŏl'·ĕnz ăwr'·ĕm* (L. *attollens*, lifting up on high ; *auris*, the ear, *aurem*, ac.), raising up the ear ; a muscle which raises the ear : attrahens aurem, *ăt'·tră·hĕnz ăwr'·ĕm* (L. *attrahens*, drawing towards ; *aurem*, the ear), drawing towards the ear ; a muscle which draws the ear forwards and upwards : retrahens aurem, *rē'·tră·hĕnz ăwr'·ĕm* (*retrahens*, drawing back ; *aurem*, the ear), drawing the ear back ; a muscle which draws the ear back;—the preceding three small muscles are placed immediately beneath the skin around the external ear, and, though their names express energy, they are rarely active in man.

Aucklandia costus, *ăwk·lănd'·ĭ·ă kŏst'·ŭs* (*Auckland ;* Gr. *kostos*, L. *costum*, an Oriental aromatic plant), another name for ' Aplotaxis lappa,' found in Cashmere, said to be the anc. Costus, the root having been celebrated for its virtues.

Aucuba, n., *ăwk'·ŭb·ă* (name of the shrub in Japan), a genus of plants, Ord. Cornaceæ, fine hardy shrubs: Aucuba Japonica, *jă·pŏn'·ĭk·ă* (*Japŏnĭcus*, of or belonging to Japan), a shrub having beautifully blotched and variegated leaves.

auditory, a., *ăwd'·ĭt·ŏr·ĭ* (L. *audĭtor*, a hearer—from *audio*, I hear), pert. to the sense of hearing.

aura, n., *ăwr'·ă* (Gr. and L. *aura*, the air), a peculiar sensation which sometimes gives warning of a fit of epilepsy.

aural, a., *ăwr'·ăl* (L. *auris*, an ear), pert. to the ear and its diseases.

Aurantiaceæ, n. plu., *ăwr·ăn'·tĭ·ā'·sĕ·ē* (mid. L. *aurantium*, the orange—from *aurum*, gold, in allusion to its colour), the Orange family, many of the species bearing well-known excellent fruit : aurantium, n., *ăwr·ăn'·shĭ·ŭm* (L.), the orange.

aurella, n. plu., *ăwr·ĕl'·lă* (L. *aurellum*, a dimin. from *aurum*, gold), the chrysalides of some Lepidoptera, from their exhibiting a golden lustre.

auricle, n., *ăwr'·ĭ·kl* (L. *aurĭcula*, the ear flap—from *auris*, the ear), the outside ear, which projects as a circular flap from the side of the head ; an ear-like appendage; two muscular cavities of the heart, so called from their resemblance to the ear of a dog, named respectively the *right* and *left* : **auricled**, a., *ãwr'-ĭ-kld*, having ears or ear-like appendages: **auricula**, n., *ãwr-ĭk'-ŭl-ă*, showy garden flowers — see 'Primula:' **auricular**, a., *ãwr-ĭk'-ŭl-ăr*, pert. to the ear ; applied to the ear-shaped cavities of the heart : **auriculate**, a., *ãwr-ĭk'-ŭl-āt*, in *bot.*, having ear-like appendages ; applied to leaves with lobes or leaflets at their base: **auricularis magnus**, *ãwr-ĭk'-ŭl-ār'-ĭs mãg'-nŭs* (L. *auriculāris*, auricular—from *auricula*, the external ear ; *magnus*, great), a name designating the largest nerve of the ascending branches of the cervical plexus : **auriculo-temporalis**, *ãwr-ĭk'-ŭl-ō-tĕmp'-ŏr-āl'-ĭs* (L. *temporālis*, belonging to time—from *tempus*, time), the auriculo-temporal, designating a nerve lying immediately in front of the ear, and close to the temporal artery : **auriculo-ventricular**, *-vĕn-trĭk'-ŭl-ăr* (L. *ventricŭlus*, a little belly, a ventricle of the heart—from *venter*, the belly), of or belonging to the great transverse groove separating the auricles of the heart from the ventricles, or the orifice forming the communication between these chambers. **aurist**, n., *ãwr'-ĭst* (L. *auris*, an ear), one skilled in the cure of diseases of the ear : **auriscope**, n., *ãwr'-ĭ-skōp* (Gr. *skopeo*, I see), an instrument which covers the auricle for ascertaining the condition of the internal ear and its passage. **auscultation**, n., *ãws'-kŭlt-ā'-shŭn* (L. *auscultatio*, a listening to with attention—from Gr. *ous*, L. *auris*, an ear ; L. *cultus*, used or exercised), the method of discovering the extent and seat of any disease by listening with the ear alone (immediate ausc.), or through an instrument called a 'stethoscope' (mediate ausc.). **autonomous**, a., *ãw-tŏn'-ŏm-ŭs* (Gr. *autonomos*, governed by their own

laws—from *autos*, self ; *nomos*, a law), in *bot.*, said of plants which are perfect and complete in themselves. **autophagi**, n. plu., *ãw-tŏf'-ă-jī* (Gr. *autos*, self ; *phago*, I eat), those birds which can run about and obtain food for themselves as soon as they escape from the egg. **autophyllogeny**, n., *ãw-tō-fĭl-ŏdj'-ĕn-ĭ* (Gr. *autos*, self ; *phullon*, a leaf ; *genesis*, birth), in *bot.*, the growth of one leaf upon another. **autopsy**, n., *ãw-tŏps'-ĭ*, also **autopsia**, n., *ãw-tŏps'-ĭ-ă* (Gr. *autos*, self ; *opsis*, sight), seeing a thing one's self ; ocular demonstration ; examination after death. **auxenometer**, n., *ãwks'-ĕn-ŏm'-ĕt-ĕr* (Gr. *auxēsis*, increase ; *metron*, a measure), an instrument for measuring the growth of plants at intervals : **auxospores**, n. plu., *ãwks'-ŏ-spōrz* (Gr. *spora*, a seed), large cells formed as concluding members of a series of smaller cells in Diatomaceæ. **Avena**, n., *ăv-ēn'-ă* (L. *avēna*, the common oats), a genus of plants of the Ord. Gramineæ : **Avena sativa**, *săt-īv'-ă* (L. *satīvus*, fit to be planted—from *sătus*, sown, planted), the cereal oats : **A. farina**, *făr-ĭn'-ă* (L. *farīna*, meal, flour), the farina of oats, the pharmacopœial name for oatmeal: **avenacious**, a., *ăv-ĕn-ā'-shŭs*, pert. to oats, or partaking of the nature of oats. **avenia**, n., *ăv-ēn'-ĭ-ă* (Gr. *a*, without ; *vena*, a vein), without veins or nerves ; in *bot.*, veinless. **Averrhoa**, n., *ăv'-ĕr-rō'-ă* (after *Averrhoes*, a physician of Spain), a genus of trees, Ord. Oxalidaceæ, the fruit of which frequently grows on the trunk itself below the leaves : **Averrhoa bilimbi**, *bĭ-lĭm'-bĭ* (an Indian name), a tree having a green, fleshy, oblong fruit, filled with acid juice, the fruit used as food in the East Indies, and the juice in skin diseases.

aves, n. plu.,*āv'·ēz*(L. *āvis*, a bird), the class of birds.

Avicennia, n., *ăv'·ĭs·ĕn'·nĭ·ă* (after *Avicenna*, a Persian physician), a genus of plants, Ord. Verbenaceæ, which have adventitious roots like the mangrove : **Avicennia tomentosa,** *tŏm·ĕn·tōz'·ă* (L. *tomentum*, a stuffing for cushions, a downy pubescence), a species in great use in Brazil for tanning.

avicularium, n., *ăv·ĭk'·ūl·ār'·ĭ·ŭm* (L. *avicula*, a little bird—from *āvis*, a bird), a singular appendage, frequently shaped like the head of a bird, found in many of the Polyzoa.

awn, n., *āwn* (Icel. *ogn*; Swed. *agn*; Gr. *achne*, chaff), the beard of corn or grass : **awned,** a., *āwnd*, having an awn or beard.

axil, n., *ăks'·ĭl*, also **axilla,** *ăks·ĭl'·lă* (L. *axilla*, the armpit), in *bot.*, the upper angle where the leaf joins the stem : **axilla,** n., *ăks·ĭl'·lă*, in *anat.*, the armpit ; the pyramidal space situated between the upper and lateral part of the chest, and the inner side of the arm ; a part forming a similar angle : **axile,** a., *ăks'·ĭl*, also **axial,** *ăks'·ĭ·ăl*, belonging to the axis : **axillary,** a., *ăks'·ĭl·lăr·ĭ*, in *bot.*, arising from the axis of a leaf ; in *anat.*, designating an artery which commences at the lower border of the first rib, and terminates at the lower border of the tendons of the 'latissimi dorsi' and 'teres major' muscles ; designating parts that belong to the axilla or armpit : **axillary plexus,** in *anat.*, the brachial plexus, formed by the last three cervical and first dorsal nerves : **axial skeleton,** the whole vertebræ of the body, extending in a line from the top of the neck or atlas, to the bottom of the trunk.

axis, n., *ăks'·ĭs*, **axes,** plu., *ăks'·ēz* (L. *axis*, Gr. *axon*, an axle-tree, a pole), in *bot.*, the central portion of the plant from which the plumule and radicle are given off ; the central organ bearing buds ; the common stem or main body of a plant ; in *anat.*, the second cervical vertebra, so called as forming the pivot upon which the atlas and head rotate : **cæliac axis,** the first trunk given off by the abdominal aorta : **thyroid axis,** a short trunk arising from the subclavian artery : **axis cylinder,** the central portion or axis tract of a nerve.

Azalea, n., *ăz·āl'·ĕ·ă* (Gr. *azalĕos*, dry, parched, in allusion to the dry habitat of the plant—from *azō*, I dry or parch), a genus of plants, Ord. Ericaceæ, universally admired for their white, orange, purple, scarlet, and variegated flowers : **Azalea Indica,** *ĭn'·dĭk·ă* (L. *Indicus*, of or from India), a greenhouse plant of great beauty : **A. Pontica,** *pŏnt'·ĭk·ă* (L. *Pontus*, the Black Sea), is supposed to have been the plant whose flowers yielded the poisonous honey noticed by Xenophon in the retreat of the 10,000 : **A. procumbens,** *prŏ·kŭm'·bĕnz* (L. *procumbens*, leaning or bending forwards), grows on the mountains of Scotland and in the Arctic regions.

azote, n., *ăz'·ōt* (Gr. *a*, without ; *zōē*, life), nitrogen gas, so called because it will not support the respiration of animals : **azotic,** a., *ăz·ŏt'·ĭk*, pert. to azote ; fatal to animal life : **azotised,** a., *ăz'·ōt·īzd*, containing nitrogen or azote.

azoturia, n., *ăz·ōt·ūr'·ĭ·ă* (Eng. *azote* ; Gr. *ouron*, L. *urīna*, urine), an excess of urea in the urine ; a disease of animals arising from a too rapid disintegration of tissues, or a defective assimilation of food.

azygos, n., *ăz'·ĭg·ŏs* (Gr. *a*, without; *zugon*, a yoke), a general name applied to muscles, arteries, veins, bones, and other parts that have no fellow or correspondent part—but in *anat.*, the ordinary meaning and application of the term is more or less a misnomer: azygous, a., *ăz'-ĭg-ŭs*, single; without a fellow: azygos processus, *prō-sĕs'-ŭs* (L. *processus*, a going forward, a progression), a process of the sphenoid bone: A. uvulæ, *ŭv'-ūl-ē* (L. *uvula*, a little cluster, a little grape— from *uva*, a cluster, a grape), a muscle of the uvula, but really a pair of muscles: A. vena, *vēn'-ă* (L. *vēna*, a vein), a vein formed by the union of the lower intercostal veins of the left side.

> NOTE.—There are two 'azygous veins,' the greater and the lesser, one on the right side, and the other on the left of the spine, forming a system of communication between the inferior and superior *vena cava.* There are also two 'azygous arteries,' one to each knee-joint. The term is only strictly applicable to the rostrum or central spine of the sphenoid bone, which is a true 'azygous process.'

bacca, n., *băk'-ă* (L. *bacca*, a berry), in *bot.*, a unilocular fruit having a soft outer skin which covers a pulp amongst which the seed is immersed: baccate, a., *băk'-āt*, designating pulpy fruits in general; fleshy: bacciferous, a., *băk-sĭf'-ĕr-ŭs* (L. *fero*, I bear), bearing or producing berries: bacciform, a., *băk'-sĭ-fŏrm* (L. *forma*, shape), having the form or shape of a berry.

bacilli, n. plu., *băs-ĭl'-lī* (L. *băcillum*, a small staff or wand), in *bot.*, the narrow plates of diatoms: bacillar, a., *băs'-ĭl-lăr*, resembling rods; somewhat club-shaped.

bacterium, n., *băk-tēr'-ĭ-ŭm*, bacteria, n. plu., *băk-tēr'-ĭ-ă* (Gr. *baktērion*, a rod, a walking-stick), microscopic, staff-shaped or pointed filaments which are re-

garded as one of the earliest forms of organic life, abounding in animal fluids in a state of decomposition, but their real nature has not yet been ascertained: bacteroid, a., *băk'-tĕr-ŏyd*, resembling the bacteria.

bactridium, n., *băk-trĭd'-ĭ-ŭm* (Gr. *baktron*, a cane, a staff; *eidos*, resemblance), a genus of the Ord. Fungi, found on the horizontal surfaces of old stumps: bacteridia, n. plu., *băk'-tĕr-ĭd'-ĭ-ă*, a term applied to certain straight motionless bodies found in the blood of animals labouring under malignant pustules.

baculiform, a., *băk-ūl'-ĭ-fŏrm* (L. *baculum*, a staff; *forma*, shape), in *bot.*, applied to rod-like bodies in the reproductive organs sphæroplea: baculiferous, a., *băk'-ūl-ĭf'-ĕr-ŭs*, bearing canes or reeds.

Balanidæ, n. plu., *băl-ăn'-ĭ-dē* (Gr. *balanos*, an acorn; and -*idæ*), a family of sessile cirripedes, commonly called 'acorn shells;' balanoid, a., *băl'-ăn-ŏyd* (Gr. *eidos*, resemblance), having the shape of an acorn.

Balanophoraceæ, n. plu., *băl'-ăn-ŏf-ŏr-ā'-sĕ-ē* (Gr. *balanos*, an acorn; *phoreo*, I bear or carry), the Balanophora order, having root-parasites and peculiar fungus-like stems: Balanophora, n., *băl'-ăn-ŏf'-ŏr-ă*, a genus of plants.

balaustia, n., *băl-āws'-tĭ-ă* (Gr. *balaustion*, a pomegranate flower), the fruit of the pomegranate; an indehiscent inferior fruit, with many cells and seeds, the seeds being coated with pulp: balaustine, n., *băl-āws'-tĭn*, the wild pomegranate tree.

baleen, n., *băl-ēn'* (L. *balæna*, a whale), the horny plates which occupy the palate of the true or 'whale-bone' whales.

balm, n., *băm* (Fr. *baume*, balm; Gr. *balsamon*, L. *balsamum*, balsam), a fragrant plant; any ointment that soothes: **balsam**, n., *băl'·săm*, a soothing ointment of an oily nature.

Balsaminaceæ, n. plu., *băl'·săm·ĭn·ā'·sĕ·ē*(Gr. *balsamon*, L. *balsamum*, balsam), the Balsam family, an Order of plants consisting of lofty trees abounding in balsamic juices: **Balsamina**, n., *băl'·săm·ĭn'·ā* (*balassan*, the name given by the Arabs), a genus of above Order: **balsam**, n., *băl'·săm*, a beautiful and popular annual of our gardens, with its white, red, pink, purple, lilac, and finely variegated carnation-like flowers; the juice with alum used by the Japanese to dye their nails red.

Balsamodendron, n., *băl'·săm·ō·děn'·drŏn* (Gr. *balsamon*, balsam; *dendron*, a tree), a genus of plants, Ord. Burseraceæ, which yield a fragrant balsamic and resinous juice, often used as frankincense and in medicine; Elimi is produced by one species: **Balsamodendron myrrha**, *mĭr'·rā* (L. *myrrha*, Gr. *murrha*, myrrh), a shrub of Abyssinia, the source of the officinal myrrh, a bitter aromatic gum resin, anciently used as frankincense: **B. Africanum**, *ăf·rĭk·ān'·ŭm* (L. *Africanus*, belonging to Africa), produces the resin bdellium: **B. Gileadense**, *gĭl'·ĕ·ăd·ĕns'·ĕ* (L. *Gileadensis*, belonging to Gilead), the celebrated balsam called Balm of Gilead.

Bambusa, n., *băm·būz'·ā* (*bambos*, the Indian name; Malay, *bambu*), a genus of plants, Ord. Gramineæ, including the bamboo-cane: **Bambusa arundinacea**, *ăr·ŭnd'·ĭn·ā'·sĕ·ā* (L. *arundinaceus*, pert. to or like a reed—from *arundo*, the reed-cane), the bamboo; a siliceous matter which accumulates in the joints of the stalks is called Tabasheer.

banana, n., *băn·ān'·ā* (Spanish name), a herbaceous plant and its fruit, differing from the plantain in having its stalks marked with dark purple stripes and spots, and the fruit shorter and rounder; the systematic name is Musa sapientum, Ord. Musaceæ.

bangue, n., *băng*; see 'bhang.'

Banisteria, n., *băn'·ĭs·tēr'·ĭ·ā* (after the botanist *Rev. J. Banister*), a genus of plants of beautiful foliage, Ord. Malpighiaceæ.

Banksia, n., *bănk'·sĭ·ā* (in honour of *Sir Joseph Banks*), a genus of plants, Ord. Proteaceæ, so called because they present great diversity of appearance, the clustered cone-like heads of the flowers having a remarkable appearance.

banyan, n., *băn'·yăn* (Sans. *punya*, holy, sacred), the Indian fig tree, Ficus Indicus, which attains to an immense size.

baobab, n., *bā'·ŏb·ăb* (probably from a native name), a tree of Senegal, Monkey-bread, one of the largest known trees—the Adansonia digitata.

Baphia, n., *băf'·ĭ·ā* (Gr. *baphikē*, the art of colouring or dyeing), a genus of plants, Ord. Leguminosæ, Sub-ord. Cæsalpinieæ, which yield ringwood: **Baphia nitida**, *nĭt'·ĭd·ā* (L. *nitidus*, shining, glittering), camwood.

Baptisia, n., *băp·tĭzh'·ĭ·ā* (Gr. *baptizo*, I dip or immerse—from *bapto*, I dye), a genus of ornamental border plants, Ord. Leguminosæ, Sub-ord. Papilionaceæ: **Baptisia tinctoria**, *tĭnk·tōr'·ĭ·ā* (L. *tinctorius*, of or belonging to dyeing), a plant that gives a blue dye; the wild indigo of the United States.

Barbadoes, a., *bărb·ād'·ōz*, of or from Barbadoes, one of the West India islands: **Barbadoes tar**, a mineral tar, a species of naphtha, found naturally in Barbadoes: **Barbadoes aloes**, the inspissated juice of the Aloe vulgaris, the most active form of that drug, imported in gourds from Barbadoes.

barbate, a., *bàrb'·āt* (L. *barba*, a beard), in *bot.*, bearded; having tufts of hair-like pubescence : **barbs**, n., *bàrbs*, hooked hairs: **barbed**, a., *bàrbd*, terminating in the sharp shoulders of a hook or arrow - head : **barbula**, n., *bàrb'·ūl·ă* (L. diminutive, a little beard), the teeth of the peristome of mosses.

barilla, n., *băr·ĭl'·lă* (Sp. *barrilla*, the plant glasswort; *barrillar*, the ashes of the plant), a crude soda extracted from the ashes of the plants Salsola and Salicornia, found growing in salt marshes on the Mediterranean and other shores, Ord. Chenopodiaceæ.

bark, n., *bàrk* (Dan. *bàrk*, Icel. *borkr*, bark), the outer cellular and fibrous covering of the stem, called the Cortex: **bark-bound**, a., having the bark too firm or close.

Barosma, n., *băr·ŏs'·mă* (Gr. *barŭs*, heavy; *osmē*, smell), a genus of plants, so called from the powerful scent of their leaves: **Barosma crenulata**, *krĕn'·ŭl·āt'·ă* (L. *crenulatus*, slightly notched —from *crena*, a notch), as also B. **serratifolia**, *sĕr·răt'·ĭ·fōl'·ĭ·ă* (L. *serratus*, saw-shaped—from *serra*, a saw; *folium*, a leaf), and B. **betulina**, *bĕt'·ŭl·īn'·ă* (L. *betŭla*, the birch), the leaves of these and other species are used in medicine under the name of 'buchu,' and contain a yellowish oil having a powerful odour.

Barringtoniæ, n. plu., *băr'·ĭng·tŏn'·ĭ·ē* (after *Barrington*), a tribe of plants of the Ord. Myrtaceæ, having a fleshy, one-celled fruit: **Barringtonia**, n., *băr'·ĭng·tŏn'·ĭ·ă*, a genus of plants, many of which yield an aromatic, volatile oil.

baryta, n., *băr·īt'·ă*, or **barytes**, n., *băr·īt'·ēz* (Gr. *barutēs*, weight, heaviness—from *barus*, heavy), the heaviest of all the alkaline earths.

basal, a., *băs'·ăl*, also **basilar**, a., *băs'·ĭl·ăr* (L. and Gr. *basis*, the foundation), in *bot.*, attached to the base of an organ—usually the embryo when situated at the bottom of the seed : **basal placenta**, *plă·sĕnt'·ă* (L. *placenta*, a cake), in *bot.*, the placenta at the base of the ovary ; in *anat.*, the placenta at the base of the uterus: **basilar**, a., in *anat.*, at the base, bottom, or foundation of a part ; applied to several bones; also to a process of the occipital bone, and to the artery running over it.

basidium, n., *băs·ĭd'·ĭ·ŭm*, **basidia**, plu., *băs·ĭd'·ĭ·ă* (L. *basidium*, a little pedestal — from *basis*, a pedestal), in some Fungi, a cell bearing on its exterior one or more spores : **basidiospore**, n., *băs·ĭd'·ĭ·ō·spōr* (Gr. *spora*, a spore), a spore borne upon a basidium : **basid'iosporous**, a., -*spōr'·ŭs*, bearing spores upon a basidium.

basilar, see under 'basal:' **basilar aspect**, in *anat.*, that which is towards the base of the head: **basilar artery**, so named from its position at the base of the skull.

basilic, a., *băs·ĭl'·ĭk* (Gr. *basilikos*, royal — from *basileus*, a king), denoting parts supposed to hold a chief place in the animal functions : **basilic vein**, a vein of the upper extremity of considerable size, formed by the coalescence of the anterior and posterior ulnar veins : **basilicon**, n., *băs·ĭl'·ĭk·ŏn*, 'royal ointment,' an old name for old-fashioned remedies for wounds, etc., of three kinds—now restricted to that made of wax, resin, and lard.

basio-glossus, *băz'·ĭ·ŏ·glŏs'·ŭs* (Gr. *basis*, a base; *glossa*, the tongue), the muscle extending from the base of the os hyoïdes to the tongue; one of the three supposed muscles of the hyo-glossus.

basipetal, a., *băs·ĭp'·ĕt·ăl* (Gr.

basis, a base ; *petalon*, a leaf), development of a leaf from apex to base.

basis, n., *bās'ĭs* (Gr. *basis*, a base), in *med.*, the chief ingredient of a prescription.

basis venæ vertebrarum, *bās'ĭs vēn'ē vêrt'ĕb·rār'ŭm* (L. *venæ*, blood-vessels ; *basis*, of a base or body; *vertebrarum*, of the vertebræ), the veins of the body of the vertebræ; the veins contained in large tortuous channels in the substance of the bones of the vertebræ : **basis cordis,** *kŏr'dĭs* (L. *cor*, the heart, *cordis*, of the heart), the base or broad part of the heart.

bass, n., *băs*, also **bast,** n., *băst* (Dut. *bast*, bark, peel; Sw. *basta*, to bind), the inner fibrous bark of dicotyledonous trees, such as the lime tree, from which matting is made.

Bassia, n., *băs'sĭ·ă* (in honour of *Bassi* of Bologna), a genus of handsome, lofty-growing trees, Ord. Sapotaceæ : **Bassia butyracea,** *bŭt'ĕr·ā'sĕ·ă* (L. *būtyrum*, Gr. *boutouron*, butter), a tree which yields a thick, oil-like butter.

bassorin, n., *băs'sŏr·ĭn* (first discovered in *Bassora gum*), a substance obtained by treating gum resin successively with ether, alcohol, and water.

bast, n., see ' bass.'

Batatas, n., *băt·āt'ăs* (Sp. *batata*, the sweet potato), a genus of plants, Ord. Convolvulaceæ : **Batatas edulis,** *ĕd·ūl'ĭs* (L. *edūlis*, eatable), a plant which yields the sweet potato—also called ' Camotas.'

bathymetrical, a., *băth'ĭ·mĕt'rĭk· ăl* (Gr. *bathus*, deep ; *metron*, a measure), applied to the distribution of plants and animals along the sea bottom which they inhabit ; denoting the depths at which plants grow on the sea bottom ; denoting the depth of any tissue or organ.

Batides, n. plu., *băt·ĭd'ēz* (Gr. *batos*, a bramble), the family of the Elasmobranchii, comprising the Rays.

Batrachia, n. plu., *băt·răk'ĭ·ă* (Gr. *batrăchos*, a frog), applied loosely to any of the Amphibia; restricted sometimes to the Amphibians as a class, or to the Anoura : **batrachian,** a., *băt·răk' ĭ·ăn*, relating to frogs, toads, and the like.

Bauhinia, n., *baw·hĭn'ĭ·ă* (in memory of *Bauhin*, a botanist of the 16th cent.), a genus of plants, Ord. Leguminosæ, Sub-ord. Cæsalpinieæ : **Bauhinia tomentosa,** *tŏm'ĕn·tōz'ă* (L. *tomentum*, a stuffing for cushions ; Sp. *tomentoso*, pert. to tow or horsehair), a plant whose dried leaves and young buds are prescribed in dysenteric affections : B. **variegata,** *vār'ĭ·ĕg·āt'ă* (L. *văriĕgātum*, to make of various sorts and colours), a plant, the bark of which is used in tanning leather: B. **racemosa,** *răs'ĕ· mŏz'ă* (L. *racēmōsus*, full of clusters, clustering—from *racēmus*, a cluster of grapes), a plant whose bark is employed in making ropes.

Beaumontia, n., *bō·mŏn'shĭ·ă* (in honour of *Lady Beaumont*), a magnificent Indian climber, having splendid foliage and festoons of enormous funnel-shaped, white flowers.

bebeeru, n., *bĕb·ēr'ŏ*, also **bibiru,** n., *bĭb·ēr'ŏ* (*bebeera*, the greenheart tree, a supposed native name; Latinised name, *bebeerĭna*), the bark of the green-heart, a large tree 60 feet high found in British Guiana, whose wood is imported for shipbuilding : **bebeerin,** n., *bĕb·ēr'ĭn*, a vegetable alkaloid found in bebeerina, possessing tonic and other properties.

begass, n., *bĕ·găs'* (an American word), sugar-cane after being cut and crushed ; called *megass* and *trash* in the West Indies.

Begoniaceæ, n. plu., *bĕ·gōn'·ĭ·ā'·sĕ·ē*
(after *Begon*, a French botanist),
the Begonia family, an Order of
plants : **Begonia**, n., *bĕ·gōn'ĭ·ă*,
a genus of plants, having showy
pink, white, or yellow flowers, and
handsome succulent leaves, great
favourites with cultivators : **Begonia obliqua**, *ŏb·lĭk'·wă* (L.
obliquus, slanting, oblique), a
species said to have purgative
roots, and is sometimes called
wild rhubarb : **B. gemmipara**,
jĕm·ĭp'·ăr·ă (L. *gemma*, a bud ;
pario, I bring forth), a species
from the Himalayas, which has
gemmæ in the axils of the stipules.
belladonna, n., *bĕl·lă·dŏn'·nă* (It.
bella, beautiful ; *donna*, lady—
from its use as a cosmetic by the
ladies of Italy), an extract of the
leaves of the deadly nightshade,
a valuable narcotic in small
doses, but a deadly poison if
exceeded, remarkable for its
power, in certain doses, of dilat-
ing the pupils of the eyes : **Atropa
belladonna**,*ăt'·rŏp·ă*(Gr.*Atropos*,
one of the three Fates, whose
duty it was to cut the thread of
life—in allusion to its deadly
effects), the systematic name for
belladonna, is one of our most
active indigenous poisons, Ord.
Solanaceæ, Sub-ord. Atropeæ.
Bellis, n., *bĕl'·lĭs* (L. *bellus*, pretty,
charming), a genus of plants,
Ord. Compositæ, including the
common daisy : **Bellis perennis**,
pĕr·ĕn'·nĭs (L. *perennis*, that lasts
the whole year through, never-
failing—from *per*, through; *annus*,
a year), the always charming ; the
common wild daisy of our fields
and hills ; in Scotland called the
Gowan : **B. fistulosa**, *fĭst'·ŭl·ōz'·ă*
(L. *fistŭlōsa*, full of holes, porous),
the red daisy of our gardens : **B.
hortensis**, *hŏrt·ĕns'·ĭs* (L. *horten-
sius*, belonging to a garden—from
hortus, a garden), the common
red daisy : **B. prolifera**, *prō·lĭf'·
ĕr·ă* (L. *proles*, offspring ; *fero*,

I bear), the striped daisy, bearing
abnormal buds.
benzoin, n., *bĕn'·zō·ĭn* (said to be
from Ar. *benzoah;* Sp. *benjui*,
benzoin), a concrete, balsamic
exudation obtained by incisions
from a tree of Sumatra and
Borneo—the Styrax benzoin ;
also called **benzoe**, and vulgarly
benjamin : benzoic, a., *bĕn·
zō'·ĭk*, denoting an acid obtained
from benzoin, vulgarly called
benjamin flowers : **benzoinum**,
n., *bĕn'·zō·ĭn'·ŭm*, the pharmacopœial name for ' benzoin.'
Berberidaceæ, n. plu., *bĕr'·bĕr·ĭ·
dā'·sĕ·ē* (L. *berberis*, the barberry ;
Ar. *berberi*, wild), the Barberry
family, an Ord. of plants : **Berberis**, n., *bĕr'·bĕr·ĭs*, a genus of
plants : **Berberis vulgaris**, *vŭlg·
ār'·ĭs* (L. *vulgāris*, general,
common), the common barberry
tree, the bark and stem of which
are astringent, and yield a yellow
dye ; the fruit contains oxalic
acid, and is used as a preserve :
B. lycium, *lĭsh'·ĭ·ŭm* (Gr. *lukion*,
a thorny tree of Thessaly ; *Lycia*
in Asia Minor, where found), a
tree which affords a medicinal
extract in much repute in ancient
times, and still in India, chiefly
for ophthalmia : berberin, n.,
bĕr'·bĕr·ĭn, an alkaline substance
obtained from the root of the
barberry shrub.
Bertholletia, n., *bĕrth'·ŏl·lē'·shĭ·ă*
(in honour of the chemist *Ber-
thollet*), a genus of tall ornamental
trees, Ord. Myrtaceæ : **Bertholletia excelsa**, *ĕk·sĕls'·ă*(L. *excelsus*,
elevated, lofty); or, according to
others, **B. nobilis**, *nōb'·ĭl·ĭs* (L.
nobilis, famous, celebrated), a tree
which produces the well-known
Brazil nuts.
Berzelia, n., *bĕr·zēl'·ĭ·ă* (after the
chemist *Berzelius*), a genus of
pretty flowering plants, Ord.
Braniaceæ.
Beta, n., *bēt'·ă* (L. *bēta*, the beetroot; said to be Celtic *bett*, red), a

genus of plants, Ord. Chenopodiaceæ, many of which are used as
esculent pot-herbs : **Beta vulgaris,** *vŭlg·ār'ĭs* (L. *vulgaris,*
common), the common beetroot
of our gardens and fields ; also
called **B. campestris,** *kăm·pĕst'·rĭs*
(L. *campestris,* belonging to a
field), field beet or mangold-
wurzel.

betel-nut, *bĕt'·l* (F. *betel,* Sp. *betél,*
the fruit of the Areca catechu,
an elegant palm from 40 to 50
feet high ; the powdered nut is
used for tape-worm, and as an
ingredient along with Piper-betle
in the stimulating Eastern masticatory *pan* or *betel.*

Betulaceæ, n. plu., *bĕt'·ŭl·ā'·sĕ·ē*
(as if a Latin word *batula,* a
stroke—from *bātŭo,* 1 strike, I
beat ; *betu,* said to be Celtic name
of the birch), the Birch family,
an Ord. of trees consisting of the
various kinds of birch and alder :
Betula, *bĕt'·ŭl·ă,* a genus of birch
trees, in the sap of which a
saccharine matter exists : **Betula
alba,** *ălb'·ă* (L. *albus,* white), and
B. glutinosa, *glōōt'·ĭn·ōz'·ă* (L.
glutinosus, gluey, glutinous), the
common birch, the oil from the
bark of which gives the peculiar
odour to Russia leather : **B.
papyracea,** *păp'·ĭr·ā'·sĕ·ă* (L.
papyrus, the paper reed), the
canoe birch, whose bark is employed in making boats in North
America : **B. lenta,** *lĕnt'·ă* (L.
lentus, tough, hard), the black
birch of America, called also
'mountain mahogany :' **B. bhaja-
paltra,** *bădj'·ă·pălt'·ră* (an Indian
name), a tree whose bark is used
in India in the manufacture of
paper.

bhang, n., *băng,* and **bangue** or
bang (Sans. *bhangga,* hemp), a
plant, the Cannabis Indica, Indian
hemp, used in India for intoxication, — in some parts, the dried
larger leaves and seeds of fruit ;
in others, the whole plant dried

after flowering, and the tops
and tender parts of the plant
dried.

bi-acuminate, a. (*bis,* twice), two-
pointed with the points diverging.

bi-articulate, a. (*bis,* twice), two-
jointed.

biceps, n., *bi'·sĕps* (L. *biceps,* having
two heads — from *bis,* twice ;
caput, the head ; *bicipitis,* of
having two heads; *bicipites,* plu.),
in *anat.,* a muscle that divides
into two portions, or that has two
distinct origins; applied to a
muscle of the arm and of the thigh:
bicipital, a., *bi·sĭp'·ĭt·ăl,* having
two heads or origins; pert. to the
biceps muscle : **bicipital groove,**
the groove in the bone through
which the biceps muscle passes :
biceps anconeus, *bi'·sĕps ăn'·kŏn·
ē'·ŭs* (L. *biceps,* two-headed ; L.
ancon, Gr. *angkon,* an elbow), the
double-headed muscle at the elbow
which assists in extending the
fore-arm : **b. femoris,** *fĕm'·ŏr·ĭs*
(L. *fĕmur,* the thigh, *fĕmŏris,* of
the thigh), the two-headed muscle
of the thigh ; a large muscle of
considerable length, situated on
the posterior and outer aspect of
the thigh, arising by two heads :
b. flexor cubiti, *flĕks'·ŏr kŭb'·ĭt·ĭ*
(L. *flexor,* that which bends;
cubitus, the elbow, *cubiti,* of the
elbow), the double-headed muscle
that bends the elbow : **b. flexor
cruris,** *krōōr'·ĭs* (L. *crux,* the leg,
cruris, of the leg), the two-headed
muscle which assists in bending
the leg.

bicuspid, a., *bi·kŭsp'·ĭd* (L. *bis,*
twice ; *cuspis,* a spear, a point,
cuspidis, of a spear), having two
points ; applied to teeth that
have two fangs or points, as the
first two molars on each side of
the jaw ; in *bot.,* ending in two
points, as leaves.

bidental, a., *bi·dĕnt'·ăl* (L. *bis,*
twice ; *dens,* a tooth, *dentis,* of a
tooth), having two teeth : **bi-**

dentate, a., *bĭ-dĕnt'-āt*, in *bot.*, having two tooth-like processes.

biennial, a., *bĭ-ĕn'-nĭ-ăl* (L. *biennium*, the space of two years— from *bis*, twice; *annus*, a year), continuing or lasting throughout two years; applied to plants which do not bear flowers and seed till the second year, and then die: n., a plant that stands two years, and then dies.

bifarious, a., *bĭ-fār'-ĭ-ŭs* (L. *bĭfārius*, two-fold, double—from *bis*, twice; *fari*, to speak, to say), in *bot.*, placed in two rows, one on each side of an axis.

bifid, a., *bĭ'-fĭd* or *bĭf'-ĭd* (L. *bĭfĭdus*, cleft or divided into two parts— from *bis*, twice; *fidi*, I cleft or split), forked; cleft in two; opening with a cleft, but not deeply divided.

biflex, a., *bĭ'-flĕks* (L. *bis*, twice; *flexus*, bent, curved), in the sheep, designating a canal between the digits, so called from the peculiar curve which it takes; also called the ' interdigital canal.'

bi-foliate, a., *bĭ-fōl'-ĭ-āt* (L. *bis*, twice ; *foliātus*, leaved), in *bot.*, applied to compound leaves having two leaflets : **bi-follicular**, a., *bĭ-fŏl'-ĭk'-ŭl-ăr* (L. *bis*, twice ; *folliculus*, a small bag or sack), in *bot.*, having a double follicle.

biforine, n., *bĭf'-ŏr-ĭn* (L. *bĭfŏris*, having two doors—from *bis*, twice ; *fŏris*, a door), in *bot.*, an oblong raphidian cell, having an opening at each end.

bifurcate, a., *bĭ-fĕrk'-āt* (L. *bis*, twice, double ; *furca*, a fork), in *bot.*, forked; divided into two as a fork into its two branches : **bifurcation**, n., *bĭ'-fĕrk-ă'-shŭn*, a division into two branches.

bigeminate, a., *bĭ-jĕm'-ĭn-āt* (L. *bis*, twice ; *geminus*, double), in *bot.*, doubly paired, or four in all; twin-forked.

Bignoniaceæ, n. plu., *bĭg'-nōn-ĭ-ā'-sĕ-ē* (in honour of Abbé *Bignon*), the Trumpet-flower family, an

Order having many showy plants, whose flowers are frequently large and trumpet-shaped: **Bignonieæ**, n. plu., *bĭg'-nōn-ĭ'-ĕ-ē*, a Sub-order: **Bignonia**, n., *bĭg-nōn'-ĭ-ă*, a genus whose species are conspicuous objects in tropical forests : **Bignonia chica**, *tshĭk'-ă* (*chica*, Indian name, a beauty, a pretty girl ; *chico*, small), a plant from which the Indians obtain a red ochreous matter for painting their bodies; a fermented liquor among the Indians.

bijugate, a., *bĭ'-jōōg-āt* or *bĭdj'-ōōg-āt* (L. *bĭjŭgus*, yoked two together—from *bis*, twice; *jugum*, a yoke), applied to a compound leaf having two pairs of leaflets.

bikh, n., *'bĭk;* **bish**, n., *bĭsh;* or **nabee**, n., *nă-bē'*, native names for the powerful East Indian poison extracted from the root of Aconitum ferox.

bi-labiate, a., *bĭ-lāb'-ĭ-āt* (L. *bis*, twice; *labium*, a lip), in *bot.*, having the mouth of a tubular organ divided into two parts; two-lipped.

bi-lamellar, a., *bĭ-lăm'-ĕl-lăr* (L. *lamella*, a thin plate), in *bot.*, having two lamellæ or flat divisions; formed of two plates ; also **bi-lamellate**, a., *-lăm'-ĕl-lāt*, in same sense.

bilateral, a., *bĭ-lăt'-ĕr-ăl* (L. *bis*, twice ; *lătus*, a side, *lăteris*, of a side), in *bot.*, arranged on opposite sides ; in *zool.*, having two symmetrical sides.

bile, n., *bĭl* (L. *bilis*, bile), a thick, yellow, bitter liquor separated in the liver, and collected in the gall bladder; the hepatic secretion : **biliary**, a., *bĭl'-yĕr-ĭ*, of or relating to the bile : **bilious**, a., *bĭl'-yŭs*, pert. to or affected by bile : **bilin**, n., *bĭl'-ĭn*, a gummy, pale, yellow mass, said to be the principal constituent of the bile.

bilifulvine, n., *bĭl'-ĭ-fŭlv'-ĭn* (L. *bilis*, bile ; *fulvus*, tawny yellow),

D

the colouring matter of the bile, especially that of the ox.

biliphæin, n., *bĭl'.ĭ.fē'.ĭn* (L. *bilis*, bile; *phaios*, of a brown colour), the brown colouring matter of the bile, and formerly supposed to be its primary form; identical with 'bilifulvine' and 'cholepyrrhine.'

bilirubin, n., *bĭl'.ĭ.rōōb'.ĭn* (L. *bilis*, bile; *rŭbens*, growing red—from *rŭber*, red), a substance identical with the red colouring matter of the blood, from which are obtained, by various degrees of oxidation, a gradation of colours from the green of 'biliverdin' up to pale yellow.

biliverdin, n., *bĭl'.ĭ.vêrd.ĭn'* (Fr. *bile*, bile; *vert*, green colour), the form of pigment into which 'bilirubin' often passes, and into which it may be converted by oxidising agents.

Billardiera, n., *bĭl.lârd'.ĭ.ēr'.ă* (in honour of *Labillardière*, a French botanist), a genus of handsome climbers; Ord. Pittosporaceæ: **Billardiera longiflora**, *lŏnj'.ĭ.flōr'.ă* (L. *longus*, long; *flos*, a flower, *floris*, of a flower), a species producing abundance of flowers and handsome blue berries.

bilobate, a., *bi.lōb'.āt*, also **bilobed**, a., *bĭ'.lōbd* (L. *bis*, twice; Gr. *lobos*, the ear-flap), having two lobes; two-lobed.

bilocular, a., *bi.lŏk'.ūl.ăr* (L. *bis*, twice; *lŏcŭlus*, a little place), in *bot.*, containing two cavities or cells.

bimanous, a., *bi.măn'.ŭs* (L. *bis*, twice; *mănus*, the hand), having two hands, applied to man only: **Bimana**, n. plu., *bi.măn'.ă*, the Order Mammalia, comprising man alone.

binate, a., *bin'.āt* (L. *bini*, two by two), growing in pairs; double; applied to a leaf composed of two leaflets: **binary**, a., *bin'.ăr.ĭ*, in *chem.*, containing two units; in

anat., separating into two, and again into two.

bi-nucleate, a., *bi.nŭk'.lĕ.āt* (L. *bis*, twice; *nuclĕus*, a small nut), having two nuclei.

biogenesis, n., *bi'.ō.jĕn'.ĕs.ĭs* (Gr. *bios*, life; *genĕsis*, origin), a term employed to express the mode by which new species of animal life have been produced; the doctrine that all life springs from antecedent life; in *bot.*, the production of living cells from existing living cells of a similar nature.

biology, n., *bi.ŏl'.ŏ.jĭ* (Gr. *bios*, life; *logos*, discourse), the science which investigates the phenomena of life, both animal and vegetable.

bioplasm, n., *bi'.ō.plăzm* (Gr. *bios*, life; *plasma*, what has been formed, a model), the physical basis of life; the material through which every form of life manifests itself: also **protoplasm**, in same sense.

biparous, a., *bĭp'.ăr.ŭs* (L. *bis*, double; *pario*, I bring forth), having two at a birth; in *bot.*, applied to a cymose inflorescence, in which an axis gives rise to two bracts, from each of which a second axis proceeds, and so on.

bipartite, a., *bĭp'.ărt.ĭt* (L. *bis*, twice; *partītus*, divided), in *bot.*, divided into two parts nearly to the base.

biped, n., *bi'.pĕd* (L. *bis*, twice; *pēs*, a foot, *pĕdis*, of a foot), an animal having two feet: **bipedal**, a., *bĭp'.ĕd.ăl*, having two feet; walking upon two legs.

bipinnate, a., *bi.pĭn'.nāt* (L. *bis*, twice; *pinna* or *penna*, a feather), having a leaf or frond growing from a stem, itself divided into leaflets and ranged in pairs; having leaflets in pairs.

bipinnatifid, a., *bi'.pĭn.năt'.ĭ.fĭd* (L. *bis*, twice; *pinna*, a feather; *findo*, I cleave, *fĭdi*, I cleft), in *bot.*, having pinnatifid leaves, the segments of which are themselves pinnatifid.

bipinnatipartite, a., *bī'·pĭn·năt·ĭ·pärt'·ĭt* (L. *bis*, twice ; *pinna*, a feather ; *partitus*, divided), differing from pinnatifid in having the divisions of a pinnatifid leaf extending to near the midrib.

biplicate, a., *bĭp'·lĭk·āt* (L. *bis*, twice; *plĭcātum*, to fold), in *bot.*, having two folds or plates.

biporose, a., *bī·pōr'·ōz* (L. *bis*, twice ; L. *porus*, Gr. *poros*, a pore), in *bot.*, having two rounded openings.

biramous, a., *bī·rām'·ŭs* (L. *bis*, twice ; *ramus*, a branch), applied to a limb divided into two branches, as in the limbs of the Cirripedes.

bi-septate, a., *bī·sĕpt'·āt* (L. *bis*, twice ; *septum*, a fence ;. an enclosure), having two partitions.

bi-serrate, a., *bī·sĕr'·āt* (L. *bis*, twice ; *serratus*, saw-shaped),. in *bot.*, having serratures which are themselves serrate.

bi-sexual, a., *bī·sĕks'·ū·ăl* (L.. *bis*, twice, and *sexual*),. in *bot.*, male and female organs in the same flowers.

bismuth, n.,. *bĭz'·mŭth* (Ger. *wiszmuth*, bismuth—from *wisz*, white, and *muth*, lively), a hard, brittle, yellowish or reddish-white metal, used in the arts ;. used in medicine in two forms—(1) the subnitrate,. (2) the carbonate of bismuth.

bistort, n.,. *bĭs'·tŏrt* (L. *bis*, twice ; *tortus*, twisted),. the root of the plant Polygonum bistorta, so called on account of its double twist, is a powerful astringent ; snakeweed.

biternate, a., *bī·tĕrn'·āt* (L. *bis*, twice ; *terni*, three by three), in *bot.*, having a leaf divided into three parts, and each division again divided into three parts.

bivalve, n., *bī'·vălv* (L. *bis*, twice ; *valvæ*, folding doors), a shell consisting of two plates or valves, as in the mussel or oyster ; in *bot.*, a seed case or vessel of a similar kind.

biventer cervicis, *bī·vĕnt'·ĕr sĕrv·ĭs'·ĭs* (L. *bis*, twice, double ; *venter*, the belly ; *cervĭcis*, of the neck—from *cervix*, the neck), the double-bellied muscle of the neck ; a muscle of the upper and back part of the neck, formed by a large fasciculus of the 'complexus' or 'trachelo-occipitalis,' remarkable for consisting of two fleshy bellies with an intermediate tendon.

Bixaceæ, n. plu., *bĭks·ā'·sĕ·ē* (*bixa*, the name in S. America), the Arnatto or Anatto· family, an Order of plants, many of which yield edible fruits : **Bixeæ**, n. plu.,. *bĭks'·ĕ·ē*, one of. the four tribes of the Order : **Bixa**, n., *bĭks'·ă*, a genus of plants of the Order : **Bixa orellana**, *ŏr·ĕl'·ăn·ă* (Sp. *orellana*, arnatto or arnotto),. a plant, the reddish pulp surrounding whose seeds yields the red colouring matter known as 'arnatto,' used to give a reddish tinge or colour to butter,. cheese,. etc.

blain, n., *blān* (AS. *blegen*, Dut. or Dan. *blegne*, a boil or pimple), among cattle, a malignant carbuncle in the mouth, and especially on the tongue ; also called. **glossanthrax**,.

blastema, n,, *blăs·tĕm'·ă* (Gr. *blastēma*, a sprout, offspring—from *blastano*, I bud, I germinate), the axis of an embryo ;. the rudimental element of tissues ; an obsolete term for protoplasm.

blastocolla, n.,. *blăst'·ō·kŏl'·lă* (Gr. *blastos*, a bud.; *kolla*, glue), in *bot.*, a gummy substance coating buds.

blastoderm, n.,. *blăst'·ō·dĕrm* (Gr. *blastos*, a bud ; *derma*, skin), the germinal disc or spot which forms on the egg in the early stage of incubation : **blastoderm-**

ic, a., *blăst'·ŏ·dĕrm'·ĭk*, of or belonging to the blastoderm.

Blastoidea, n. plu., *blăst·ōўd'·ĕ·ă* (Gr. *blastos*, a bud; *eidos*, resemblance), an extinct Order of Echinodermata : **blastostyle**, n., *blăst'·ŏ·stīl* (Gr. *stulos*, a column), certain columniform zooids in the Hydrozoa which are destined to bear generative buds.

blebs, see 'bulla.'

bletting, n., *blĕt'·ĭng* (Gr. *blētos*, thrown, wounded ; L. *blŭtĕus*, tasteless, hard), the change that occurs in the pulp of a fruit after being kept for some time, and from which a sour fruit becomes soft, edible, and pleasant.

Blighia, n., *blīg'·ĭ·ă* (after Captain *Bligh*, who carried the breadfruit to the W. Indies), a plant which produces the Akel fruit, whose succulent arillus is used as food, the fruit being as large as a goose's egg, Ord. Sapindaceæ.

Bœhmeria, n., *bĕ·mēr'·ĭ·ă* (in honour of *Bœhmer*, a German botanist), a genus of plants, Ord. Urticaceæ : **Bœhmeria nivea**, *nĭv'·ĕ·ă* (L. *niveus*, snowy—from *nix*, snow), a plant which supplies fibre for Chinese grass cloth, also the Rhea fibre of Assam.

Boldoa, n., *bŏl·dō'·ă* (after *Boldoa*, a Spanish botanist), a genus of plants, Ord. Chonimiaceæ : **Boldoa fragrans**, *frā'·grănz* (L. *fragrans*, emitting a smell), an aromatic tree of Chili whose leaves contain an essential oil.

Boletus, n., *bŏl·ēt'·ŭs* (Gr. *bŏlĭtēs*, L. *bōlētus*, the boletus, the best kind of mushroom—from *bōlos*, a mass or lump, in reference to its massy or globular form), a genus of fungi found in woods, pastures, and on old trees—a curious production whose species are succulent; the Chinese eat fungi largely, and prefer the Boleti to the Agarics : **Boletus granulatus**, *grăn'·ŭl·āt'·ŭs* (L. *grănŭlum*, a little grain—from *grănum*, a grain);

B. subtómentosus, *tōm·ĕn·tōz'·ŭs* (L. *sub*, a less or inferior degree ; *tōmĕntum*, a woolly flocks) ; and **B. edulis**, *ĕd·ūl'·ĭs* (L. *edŭlis*, eatable), are all edible, and the last excellent when cooked.

bolus, n., *bōl'·ŭs* (Gr. *bōlos*, a mass or lump), a medicinal round mass, larger than a pill.

Bombaceæ, n. plu., *bŏm·bā'·sĕ·ē* (L. *bombyx*, cotton, in allusion to the wool in the pods), a tribe of plants of the Ord. Steruliaceæ, having hermaphrodite flowers and palmate or digitate leaves : **Bombax**, n., *bŏm'·băks*, a genus of plants, named 'silk-cotton trees:' **Bombax ceiba**, *sē·ĭb'·ă* (Sp. *céiba*, the silk-cotton tree), the silk-cotton tree ; the cotton, having no cohesion in its fibres, can only be used for stuffing cushions and chairs, and similar domestic purposes.

Boraginaceæ, n. plu., *bŏr·ădj'·ĭn·ā'·sĕ·ē* (Sp. *borrája*, borage; *borago*, a corruption of L. *cor*, the heart, and *ago*, I bring—so called from the nourishing qualities of the plant), the Borage or Bugloss family, an Order of plants which are generally mucilaginous and emollient : **Boragineæ**, n. plu., *bŏr'·ădj·ĭn'·ĕ·ē*, a Sub-order : **Borago**, n., *bŏr·āg'·ŏ*, a genus of plants having succulent stems: **Borago officinalis**, *ŏf·fĭs'·ĭn·āl'·ĭs* (L. *officinal*), borage, which has been used as a remedy in pectoral affections, and otherwise employed.

borax, n., *bōr'·ăks* (Ar. *baurac*, a species of nitre), a salt in appearance like crystals of alum, a compound of boracic acid and soda; used as a domestic remedy for children whose mouths are sore, and for various antiseptic purposes.

Boronia, n., *bŏr·ōn'·ĭ·ă* (after *Boroni*, an Italian), a pretty and interesting genus of New Holland plants, Ord. Rutaceæ, which are remarkable for their peculiar odour.

Boswellia, n., *bŏz·wĕl'·lĭ·ă* (in honour of Dr. *John Boswell*, of Edinburgh), a genus of trees, Ord. Burseraceæ, several of whose species are called Olibanum or frankincense trees, and inhabit the hot and arid regions of Eastern Africa and Southern Arabia, producing fragrant juices and resins which in their dry state are used as frankincense, especially the extract called Olibanum ; the chief species are— **Boswellia Carterii**, *kăr·tĕr'·i·ī* (after *Carter*); **B. Bhau-Dajiana**, *băw-dādj'·ĭ·ăn'·ă* (from native Indian names); **B. Frereana**, *frēr'·ĕ·ān'·ă* (after *Frere*); **B. thurifera**, *thŭr·ĭf'·ĕr·ă* (L. *thŭs*, incense, frankincense, *thŭrĭs*, of incense ; *fero*, I bear).

bot, n., *bŏt*, or **bots**, *bŏtz* (Fr. *bout*, end; Ger. *butt*, a short, thick thing), a worm which infests the intestines of horses, being the larvæ of the horse gadfly—said to be so called, 'bout,' 'bot,' or 'end-worms,' because, after passing through the intestines, they hang for some days upon the margin of the fundament beneath the tail, where they occasion inconvenience and distress, and first attract attention.

botany, n., *bŏt'·ăn·ĭ* (Gr. *bŏtanē*, a herb or plant), the science which treats of plants, their structure, functions, properties, and habits, and their classification and nomenclature.

bothrenchyma, n., *bŏth·rĕng'·kĭm·ă* (Gr. *bothros*, a ditch or furrow; *engchuma*, anything poured in, an infusion), dotted or pitted vessels with depressions inside their walls.

Bothriocephalus, n., *bŏth'·rĭ·ŏ·sĕf'·ăl·ŭs* (Gr. *bothrion*, a little ditch, a little pit; *kephale*, a head), a genus of intestinal worms:

Bothriocephalus latus, *lāt'·ŭs* (L. *lātus*, broad), the pit-headed tape-worm, a Continental form chiefly infesting Switzerland and Russia, the germs of which are conveyed through water : **B. cordatus**, *kŏr·dāt'·ŭs* (L. *cordatus*, heart-shaped), a species infesting the lower animals.

Botrychium, n., *bŏt·rĭk'·ĭ·ŭm* (Gr. *botrus*, a bunch of grapes; *cheo*, I pour out, I scatter), a genus of ferns, so named from the form of their fructification, much like a bunch of grapes, known by the name of 'moon-worts,' Ord. Filices: **Botrychium virginicum**, *vĕr·jĭn'·ĭk·ŭm* (L. *virginicum*, pert. to a virgin), the largest American kind, and named the 'rattlesnake fern' from these reptiles abounding where they grow.

botrytis, n., *bŏt'·rĭt·ĭs* (Gr. *botrus*, a bunch of grapes), a genus of fungi, whose little round seeds or seed vessels resemble a bunch of grapes : **Botrytis bassiana**, *băs'·sĭ·ān'·ă* (after *Bassi*, of Bologna), the fungus which produces the disease in the silkworm called 'muscardine.'

bougie, n., *bŏ'·zhē* or *bŏŏ·zhē'* (Fr. *bougie*, a wax candle or taper), a long slender instrument made of elastic gum, wax, or metal, for assisting in the removal of obstructions in the œsophagus, urethra, rectum, etc., or in the treatment of stricture.

Bovista, n., *bŏ·vĭst'·ă* (a Latinised form of its German name, *bofist*), a genus of fungi, characterised by the enormous size they attain, from 18 to 23 inches in diameter: **Bovista gigantea**, *jĭg'·ănt·ē'·ă* (L. *gĭgantēus*, of or belonging to the giants—from *gigantes*, the giants), a very large and quickly-growing fungus, which has increased from the size of a pea to that of a melon in a single night.

brachia, n. plu., *brăk'·ĭ·ă* (L. arms), two prominent white bands which connect the two pairs of optic lobes on each side with the 'thalamius opticus' and commencement of the optic tracts: **brachialis**, a., *brăk'·ĭ·ăl'·ĭs* (L. *brachialis*, pert. to the arm—from *brachium*, the arm), pert. to the arm : **brachialis internus**, *ĭn·tērn'·ŭs* (L. *internus*, that is, within), an inner muscle of the arm which bends the forearm : **brachial**, a., *brăk'·ĭ·ăl*, of or pert. to the arm: **brachialis anticus**, *ănt·ĭk'·ŭs* (L. *antĭcus*, that is, before or in front—from *ante*, before), a muscle of the arm which arises from the front of the shaft of the humerus in its lower half: **brachio-cephalic**, a., *brăk'·ĭ·ō·sĕ·făl'·ĭk* (Gr. *kephale*, the head), connected with the arm and head; applied to the innominate artery and vein.

brachiate, a., *brăk'·ĭ·āt* (Gr. *brachion*, L. *brachium*, the arm), in *bot.*, having opposite branches on the stem alternately crossing each other at right angles.

Brachiopoda, n. plu., *brăk'·ĭ·ŏp'·ŏd·ă* (Gr. *brachion*, the arm ; *pous*, a foot, *podes*, feet), a class or tribe of the Molluscoida, often called 'lamp-shells,' which possess two fleshy arms continued from the sides of the mouth : **brachiopodous**, a., *brăk'·ĭ·ŏp'·ŏd·ŭs*, having arms in place of feet and legs.

brachium, n., *brăk'·ĭ·ŭm*, brachia, plu. (L. the arm), the upper arm of vertebrates.

Brachychiton, n., *brăk'·ĭ·kīt'·ŏn* (Gr. *brachus*, short; *chaitē*, head of hair, the mane of a horse), a genus of plants having fine flowers and short stellate hairs: **Brachychiton populneum**, *pŏp·ŭl'·nĕ·ŭm* (L. *pŏpulnĕus*, belonging to the poplar—from *pŏpŭlŭs*, the poplar tree), the poplar bottle tree of Australia.

Brachyura, n. plu., *brăk'·ĭ·ūr'·ă* (Gr. *brachus*, short ; *oura*, a tail), a tribe of the decapod Crustaceans having short tails, as the crabs.

bracteæ, n. plu., *brăkt'·ĕ·ē* (L. *bractea*, a thin plate of metal), bracts or floral leaves : **bract**, n., *brăkt*, a floral leaf, a leaf more or less changed in form, from which a flower or flowers proceed : **bracteate**, a., *brăkt'·ĕ·āt*, also **bracteated**, a., *-āt·ĕd*, applied to flowers having bracts: **bracteoles**, n. plu., *brăkt'·ĕ·ōlz* (L. *brăctĕŏla*, a thin leaf of gold, dim. of *bractea*), small or secondary bracts at the base of separate flowers, between the bracts and flowers : **bractlet**, n., *brăkt'·lĕt*, a bracteole.

Bradypodidæ, n. plu., *brăd'·ĭ·pŏd'·ĭd·ē* (Gr. *bradus*, slow ; *podes*, feet), the family of Edentata, comprising the sloths.

branches, n. plu., *brănsh'·ĕs* (It. *branco*, F. *branche*, a branch ; It. *branca*, the claw of a beast), in *bot.*, principal divisions of an axis or stem ; a bough ; in *anat.*, the principal division of an artery or nerve: **branchlets**, n. plu., *brănsh'·lĕts*, little or secondary branches.

branchiæ, n. plu., *brăngk'·ĭ·ē* (Gr. *brangchia*, the gill of a fish), the gills of a fish, respiratory organs adapted to breathe air dissolved in water: **branchiate**, a., *brăngk'·ĭ·āt*, possessing gills or branchiæ: **Branchifera**, n. plu., *brăngk·ĭf'·ĕr·ă* (Gr. *phero*, L. *fero*, I carry), a division of gasteropodous Molluscs having the respiratory organs mostly in the form of distinct gills: **Branchio-gasteropoda**, *brăngk'·ĭ·ō·găst'·ĕr·ŏp'·ŏd·ă*, another name for Branchifera: **branchiopoda**, n. plu., *brăngk'·ĭ·ŏp'·ŏd·ă*, also **branchiopods**, n. plu., *brăngk'·ĭ·ō·pŏds* (Gr. *pous*, a foot, *podes*, feet), crustacea in which the gills are supported by the feet : **branchiopodous**, a., *brăngk'·ĭ·ŏp'·ŏd·ŭs*, gill-footed. **branchiostegal**, a., *brăngk'·ĭ·ŏs'·*

tĕg·ăl (Gr. *brangchia*, a gill; *stĕgō*, I cover, I conceal), among many fishes, having a membrane supported by rays for covering and protecting the gills ; having a gill covering.

brand, a., *brănd* (Icel. *brandr*, Ger. *brand*, a fire-brand), in *bot.*, denoting certain parasitic fungi which produce a scorched or burnt appearance on the living leaves of a tree.

Brassicaceæ, n. plu., *brăs′·sĭ·kā′·sĕ·ē* (L. *brassica*, W. *bresych*, cabbage—said to be in allusion to the bunchy top), an extensive Order of plants, more commonly called Cruciferæ : **Brassica**, n., *brăs′·sĭk·ă*, a genus of the Ord. Cruciferæ ; many of the common culinary vegetables belong to this Order, as cabbage, cauliflower, turnip, radish, cress, etc.: **Brassica oleracea**, *ŏl′·ĕr·ā′·sĕ·ă* (L. *olerāceus*, herb-like — from *ŏlus*, a kitchen herb), the original species, whence all the varieties of cabbage, cauliflower, brocoli, and savoys have been obtained : **B. rapa**, *răp′·ă* (L. *rāpum*, a turnip), the common turnip : **B. campestris**, *kăm·pĕst′·rĭs* (L. *campestris*, belonging to a level field — from *campus*, a field), the source of the Swedish turnip : **B. napus**, *nāp′·ŭs* (L. *nāpus*, a species of turnip), rape or cole-seed, which yields colza and carcel oils : **B. Chinensis**, *tshĭn·ĕns′·ĭs* (mod. L. *Chinensis*, pert. to China), the plant which yields Shanghae oil : **B. nigra**, *nĭg′·ră* (L. *nigra*, fem. black), a plant, the seeds of which furnish table-mustard.

braxy, n., *brăks′·ĭ* (said to be from AS. *breac*, a rheum; AS. *broc*, Icel. *brak*, disease, sickness—may be connected with the root of *brake* and *bracken*, as indicating the nature of the ground where the disease prevails), chronic diarrhœa or dysentery among sheep ; in Scotland, a general term applied to diseases of sheep of the most opposite character.

Brayera anthelmintica, *brā·ēr′·ă ănth′·ĕl·mĭnt′·ĭk·ă* (Gr. *anti*, against ; *helmins*, a worm), the flowers of a tree of Abyssinia which have been found effective in Tænia or tape-worms ; the drug Kousso or Cusso.

bregma, n., *brĕg′·mă* (Gr. *bregma*, the fore-part of the head — from *brecho*, I moisten or wet), the top of the head ; the two spaces in the infant's head where the part of the bone is the longest in hardening.

Brevilinguia, n., *brĕv′·ĭ·lĭng′·gwĭ·ă* (L. *brevis*, short.; *lingua*, the tongue), in *zool.*, a division of the Lacertilia.

Brevipennatæ, n. plu , *brĕv′·ĭ·pĕn·nāt′·ē* (L. *brevis*, short ; *penna*, a wing), a group of the natatorial birds : **brevipennate**, a., *brĕv′·ĭ·pĕn′·nāt*, short-winged.

brevissimus oculi, *brĕv·ĭs′·ĭm·ŭs ŏk′·ŭl·ī* (L. *brevissimus*, very short —from *brevis*, short ; *oculi*, of the eye), the ' obliquus inferior,' from its being the shortest muscle of the eye.

Bromeliaceæ, n. plu., *brŏm·ĕl′·ĭ·ā′·sĕ·ē* (after *Bromel*, a Swedish botanist), the Pine-apple Family, an Order of plants, natives of the warm parts of America : **Bromelia**, *brŏm·ĕl′·ĭ·ă*, a genus of plants, the woody fibres of many of which are used in manufactures: **Bromelia pinguis**, *pĭng′·gwĭs* (L. *pinguis*, fat), a species used as a vermifuge in the W. Indies.

bromine, n., *brōm′·ĭn* (Gr. *brōmos*, a carrion smell, a stench). a red elementary liquid of offensive odour, obtained from sea-water, salt-springs, and sea-weed, used extensively in medicine in the form of bromic acid, and its derivative hydrobromic acid.

Bromus, n., *brōm′·ŭs* (Gr. *bromos*, wild oats), a genus of plants, Ord. Gramineæ : **Bromus purgans**, *pêrg'·ănz* (L. *purgans*, clearing or cleaning out), and **B. catharticus,** *kăth·ărt'·ĭk·ŭs* (Gr. *kathairo*, I clean or purge), grasses which have purgative properties. **bronchus,** n., *brŏngk'·ŭs,* bronchi, plu., *brŏngk'·ī*; also **bronchia,** n., *brŏngk'·ĭ·ă,* bronchiæ, plu., *-ĭ·ē* (Gr. *brongchos,* the windpipe), the two tubes that branch off from the bottom of the trachea or windpipe, by which the air is conveyed to the lungs : **bronchial, a.,** *brŏngk'·ĭ·ăl,* pert. to the bronchi : **bronchiole,** n., *brŏngk'·ĭ·ōl,* a small bronchial tube : **bronchitis,** n., *brŏngk·ī'·ĭs,* the inflammation of the lining membranes of the bronchial tubes : **bronchiocele,** n., *brŏngk'·ĭ·ō·sēl* (Gr. *kēlē,* a tumour), an enlargement of the thyroid glands, known in Alpine regions as goître, and in England as Derbyshire neck : **bronchotomy,** n., *brŏngk·ŏt'·ŏm·ĭ* (Gr. *tome,* a cutting), the operation of making an opening into the air passages ; when the larynx is cut, the operation is termed 'laryngotomy,' and when the trachea, 'tracheotomy :' **bronchial breathing,** a term applied to the sound, resembling that produced by blowing through tubes, which replaces the normal, vesicular, respiratory murmur, when the ear is applied over a solidified portion of lung : **broncophony,** n., *brŏngk·ŏf'·ŏn·ĭ* (Gr. *phōnē,* sound), the peculiarly distant resonance of the voice heard in similar circumstances to preceding ; the muffled and indistinct speech of any one labouring under a bronchial affection.

Brosimum, n., *brŏz'·ĭm·ŭm* (Gr. *brōsimos,* eatable, nutritious—in allusion to their eatable fruit), a genus of plants, Ord. Moraceæ : **Brosimum utile,** *ūt'·ĭl·ĕ* (L. *ūtile,* profitable), the cow tree, whose juice can be employed as a substitute for milk : **B. aubletii,** *ăwb·*

lĕsh'·ĭ·ĭ, the snake-wood or letterwood of Demerara : **B. alicastrum,** *ăl'·ĭk·ăst'·rŭm,* a tree which yields bread-nuts, nutritious and agreeable when boiled or roasted.

Broussonetia, n., *brŏs'·ŏn·ēsh'·ĭ·ă* (after *Broussonet,* a French naturalist), a genus of ornamental and fast-growing trees, Ord. Moraceæ : **Broussonetia papyrifera,** *păp'·ĭr·ĭf'·ĕr·ă* (L. *papyrus,* the paper reed ; *fero,* I bear), the paper mulberry ; the outer bark is used in China and Japan in the manufacture of a kind of paper, the juice as a glue in gilding leather and paper, and the bark produces a fine white cloth.

bruit, n., *brŏ'·ĭ* (F. *bruit,* noise, din), applied to various sounds heard in auscultation in disease of the thorax or its organs.

Bruniaceæ, n. plu., *brŏn'·ĭ·ā'·sĕ·ē* (after *Brun* the traveller), the Brunia family, an Ord. of plants.

Brunoniaceæ, n. plu., *brŏn·ŏn'·ĭ·ā'·sĕ·ē* (after *Dr. Robert Brown*), the Brunonia family, an Order of plants.

bruta, n., *brŏt'·ă* (L. *brūtus,* dull, stupid), used to designate the mammalian order of the Edentata.

Bryaceæ, n. plu., *brī·ā'·sĕ·ē* (Gr. *bruon,* moss, seaweed—from *bruo,* I bud or sprout), another name for the Musci or Moss family, so called because the germination of the seed commences on the plant ; flowerless plants known as ' urn mosses.'

bryology, n., *brī·ŏl'·ŏ·jĭ* (Gr. *bruon,* moss ; *logos,* discourse), the study of the division of mosses ; same as ' muscology.'

Bryonia, n., *brī·ŏn'·ĭ·ă* (L. *bryonia,* Gr. *bruon,* bryonia, a kind of herb—from Gr. *bruo,* I abound, I bud, so named from its abundance), a genus of plants, Ord. Cucurbitaceæ : **Bryonia alba,** *ălb'·ă* (L. *albus,* white), a plant, a powerful purgative, used in medicine ; also **B. dioica,** *dĭ·ŏy'k'·ă*

(Gr. *dis*, twice ; *oikos*, a house, in allusion to the flowers with stamens, and those organs bearing seed growing on different plants), supposed to be the same as **B. alba**; bryony or wild vine; white bryony, applied to the root : **bryonin**, n., *brĭ·ŏn·ĭn*, a yellowish-brown bitter substance obtained from the root.

Bryophillum, n., *brĭ·ŏ·fĭl·ŭm* (Gr. *bruo*, I grow ; *phullon*, a leaf), a genus of curious plants, Ord. Crassulaceæ, so named in allusion to the circumstance that a leaf lying on damp earth emits roots, and throws up stems : **Bryophillum calycinum**, *kăl·ĭ·sĭn·ŭm* (Gr. *kalux*, a flower-cup or calyx, *kalukos*, of a flower-cup), a plant remarkable for producing germinating buds at the edges of its leaves.

bryozoa, n. plu., *brĭ·ŏz·ō·ă* (Gr. *bruon*, moss, seaweed ; *zoōn*, an animal), a synonym of ' polyzoa.'

bubo, n., *bū·bō*, **buboes**, n. plu., *bū·bōz* (Gr. *boubon*, the groin), an inflamed lymphatic gland, common in the groin : **bubonocele**, n., *bū·bŏn·ō·sēl* (Gr. *kēlē*, a tumour), a rupture in which the intestines break down into the groin ; incomplete inguinal hernia.

buccal, a., *bŭk·ăl* (L. *bucca*, the cheek), belonging to the cheek or mouth, as buccal arteries : **buccal membrane**, the lining membrane of the mouth : **buccales**, n. plu., *bŭk·kăl·ēz*, the arteries, veins, nerves, etc. of the cheeks.

buccinator, n., *bŭk·sĭn·āt·ŏr* (L. *buccina*, a kind of trumpet), one of the two broad thin muscles of the cheeks which act during the process of mastication, and in blowing wind instruments.

bulb, n., *bŭlb* (L. *bulbus*, a globular root, an onion), in *anat.*, resembling bulbous roots; in *bot.*, a leaf-bud with fleshy scales, of a globular shape, growing on the soil or partly in it, as the onion : **bulbi**, n. plu., *bŭlb·ī*, bulbs: **bulbil**, n., *bŭlb·ĭl*, also **bulblet**, n., *bŭlb·lĕt*, separable buds in the axil of leaves, as in some lilies : **bulbose**, a., *bŭlb·ōs*, having the structure of a bulb.

bulbus olfactorius, *bŭlb·ŭs ŏl·făkt·ōr·ĭ·ŭs* (L. *bulbus*, a bulb ; *olfactorius*, pert. to smelling), the part of the olfactory nerve which swells into an oval enlargement or bulb : **bulbar**, a., pert. to a bulb, generally used in *med.* to qualify a peculiar form of paralysis due to disease of the medulla oblongata and its immediate surroundings.

bulla, n., *bŭl·lă*, **bullæ**, plu., *bŭl·lē* (L. *bulla*, a water-bubble), the blisters or large vesicles appearing on the body in some forms of skin disease ; blebs : **bullous**, a., *bŭl·lŭs*, pert. to bullæ.

bullate, a., *bŭl·lāt* (L. *bulla*, a bubble), having elevations like blisters ; puckered as in the leaf of a Savoy cabbage.

bunion, n., *bŭn·yŭn* (Fr. *bigne*, a knob rising after a knock ; Icel. *bingh*, a heap ; Gr. *bounion*, a bulbous root—from *bounos*, a mound), a subcutaneous swelling frequently found on the inner side of the ball of the great toe, or it may be elsewhere.

Bunium bulbocastanum, *bŏn·ĭ·ŭm bŭlb·ŏ·kăst·ăn·ŭm* (Gr. *bounion*, a bulbous root—from *bounos*, a mound ; Gr. *bolbos*, L. *bulbus*, a bulb ; Gr. *kastanon*, L. *castănĕa*, a chestnut) ; also **B. flexuosum**, *flĕks·ŭ·ŏz·ŭm* (L. *flexus*, bent), two species of plants of the Ord. Umbelliferæ, the tubers of which are eaten under the name of pig-nuts or earth-nuts.

Burmanniaceæ, n. plu., *bĕr·măn·nĭ·ā·sĕ·ē* (unascertained), the Burmannia family, an Order of tropical weeds allied to the Orchids.

bursa, n., *bĕrs'·ă*, bursæ, plu., *bĕrs'·ē* (Gr. *bursa*, skin, leather); also bursa mucosa, *mū·kŏz'·ă*, bursæ mucosæ, plu., *mū·kŏz'·ē* (L. *mūcōsus*, slimy, mucous), small sacs or cavities enclosing a clear viscid liquid, found interposed between surfaces which move upon each other so as to ensure their free and easy movement: bursa patellæ, *păt·ĕl'·lē* (L. *pătĕlla*, a plate, a knee-pan), the skin of the knee-cap; the subcutaneous synovial sac in front of the knee-cap: bursiform, a., *bĕrs'·ĭ·fŏrm* (L. *forma*, shape), shaped like a purse; subspherical.

Burseraceæ, n. plu., *bĕrs'·ĕr·ā'·sĕ·ē* (after *Burser*, a botanist of Naples), the Myrrh and Frankincense family, an Order of plants: Bursera, n., *bĕrs'·ĕr·ă*, a genus of large trees of considerable value which yield a balsamic and fragrant resinous juice.

bursicule, n., *bĕrs'·ĭk·ūl*, also bursicula, n., *bĕrs·ĭk'·ūl·ă* (L. diminutive of *bursa*, skin), in *bot.*, the part of the rostellum of the Orchids, excavated in the form of a sack: bursiculate, a., *bĕrs·ĭk'·ūl·āt*, purse-like.

Butea, n., *būt'·ĕ·ă* (after *John, Earl of Bute*), a splendid genus of flowering plants, Ord. Leguminosæ: Butea frondoso, *frŏnd·ōz'·ō* (L. *frondōsus*, leafy—from *frons*, a leaf), the Dhak tree of the East Indies, yields a product similar to Kino, and has bright orange-red petals, and a black calyx: B. superba, *sū·pĕrb'·ă* (L. *sŭperbus*, proud), yields with the preceding a beautiful dye, and roots can be made into strong ropes.

Butomaceæ, n. plu., *būt'·ōm·ā'·sĕ·ē* (Gr. *bous*, an ox; *temno*, I cut), the flowering Rush family, an Order of plants: Butomus, n., *būt'·ōm·ŭs*, a genus of aquatic plants which receive their name because they are said to cause the mouths of the cattle to bleed who crop them: Butomus umbellatus, *ŭm'·bĕl·āt'·ŭs* (L. *umbella*, a little shadow—from *umbra*, a shadow), the flowering rush, a beautiful British aquatic plant.

buttocks, n. plu., *bŭt'·tŏks* (Dut. *bout*, the leg or thigh of an animal), the protuberant part of the body behind; the seat.

Buxus, n., *bŭks'·ŭs* (L. *buxus*, the box tree, boxwood), a genus of hardy evergreen shrubs, much valued for its close, hard wood, so useful in the arts, Ord. Euphorbiaceæ: Buxus sempervirens, *sĕmp·ĕr'·vĭr·ĕnz* (L. *semper*, always; *virens*, flourishing), the bark of this species is said to be alterative, and its leaves bitter and purgative.

byssaceous, a., *bĭs·sā'·shŭs* (Gr. *bussos*, L. *byssus*, fine flax), resembling or consisting of fine thread-like filaments: byssiferous, a., *bĭs·sĭf'·ĕr·ŭs* (L. *fero*, I bear), producing or bearing a byssus: byssoid, a., *bĭs'·sŏyd* (Gr. *eidos*, resemblance), resembling very slender threads—like a cobweb: byssus, n., *bĭs'·sŭs*, the silky filaments by which the common mussel and other bivalve mollusca attach themselves to other objects, or to the sea bottom.

Byttneriaceæ, n. plu., *bĭt'·nĕr·ĭ·ā'·sĕ·ē* (after *Buttner*, a German naturalist), the Byttneria and Chocolate family, an Order of shrubs, some bearing showy flowers, others, as the Theobroma cacao, producing the Cocoa of the shops.

cachexia, n., *kă·kĕks'·ĭ·ă* (Gr. *kakos*, bad; *hexis*, habit), a vitiated or deranged condition of the body: cachexia aquosa, *ă·kwŏz'·ă* (L. *aquōsus*, abounding in water—from *aqua*, water), a dropsical disease of sheep.

Cactaceæ, n. plu., *kăk·tā'·sĕ·ē* (Gr. *kaktos*, L. *cactus*, a prickly plant), an Order of succulent herbs, natives of tropical America ; the Cactus or Indian Fig family, many species yielding edible fruits, as the prickly pear : **Cactus**, n., *kăkt'·ŭs*, a genus of plants; the melon thistle.

cadaveric rigidity, *kăd·ăv'·ĕr·ĭk rĭdj·ĭd'·ĭt·ĭ* (L. *cadaver*, a dead body), a term designating the stiffness or rigidity of the body which ensues shortly after death : **cadaveric hyperæmia**, *hī'·pĕr·ēm'·ĭ·ă* (Gr. *huper*, over, above ; *haima*, blood), usually termed 'post-mortem hypostasis,' the livid discoloration on the body after death, caused by the gravitation of the blood.

caducibranchiate, a., *kăd·ŭs'·ĭ·brăngk'·ĭ·āt* (L. *caducus*, falling ; *brangchia*, gills of a fish), having branchiæ or gills which fall off before maturity is reached.

caducous, a., *kăd·ūk'·ŭs* (L. *cadūcus*, falling), in *bot.*, falling off very early, as the calyx of the poppy ; applied to parts of an animal which fall off and are shed during its life.

cæcal, a., *sēk'·ăl* (L. *cœcus*, blind), pert. to the cæcum ; having a blind or closed end : **cæcum**, n., *sēk'·ŭm*, a tube which terminates in a blind or closed end ; a little sac formed in the course of the intestines ; the part of the large intestine situated below the entrance of the ileum—also called **intestinum cæcum**, *ĭn·tĕs·tīn'·ŭm* (L. *intestīnum*, the bowel or gut).

Cæsalpinieæ, n. plu., *sēs·ăl·pĭn·ĭ'·ĕ·ē* (after *Cæsalpīnus*, physician of Pope Clement VIII.), a Sub-order of plants, Ord. Leguminosæ, among which there are many plants which furnish purgative remedies : **Cæsalpinia**, n., *sēs'·ăl·pĭn'·ĭ·ă*, a genus of plants : **Cæsalpinia coriaria**, *kŏr·ĭ·ār'·ĭ·ă* (L. *cŏriārius*, pert. to leather—from *corium*, leather), a species whose curved pods, under the

name of Divi-divi, are used for tanning : **C. brasiliensis**, *brăz·ĭl'·ĭ·ĕns'·ĭs* (L. formative — from Portug. *braza*, glowing embers ; Old-Eng. *brasil*, of a bright red— in allusion to its colour), a tree which yields the Brazil wood of commerce : **C. echinata**, *ĕk'·ĭn·āt'·ă* (L. *echīnātus*, prickly), furnishes Pernambuco wood : **C. sappan**, *săp·păn'*, furnishes the sappan-wood of Scinde.

Cæsarian section or operation, *sēs·ār'·ĭ·ăn* (L. *Cæsărīānus*, of or belonging to Cæsar—said to be so named after *Julius Cæsar*, who was brought into the world in this manner.; probably only an adaptation of L. *cœsus*, cut), the operation of cutting into the womb in order to extract the fœtus.

cæsious, a., *sēzh'·ĭ·ŭs* (L. *cœsius*, bluish-gray, cat-eyed), bluish-gray ; having a fine pale blue bloom.

cæspitellose, a., *sēs'·pĭt·ĕl'·lōz* (L. *cœspes*, turf, sod), a diminutive of cæspitose: **cæspitose**, a., *sēs'·pĭt·ōz*, growing in little tufts; tufted.

Calamus, n., *kăl'·ăm·ŭs* (Gr. *kalamos*, L. *calămus*, a reed, a reed-pen), a genus of plants, Ord. Palmæ, holding the middle place between the grasses and palms : **Calamus draco**, *drāk'·ō* (L. *draco*, Gr. *drakōn*, a species of serpent, a dragon), one of the rattan palms in Sumatra and Borneo, whose resin is one of the substances called 'dragon's blood' : **C. scipionum**, *sĭp·ĭ·ōn'·ŭm* (L. *scīpio*, a staff carried by persons of distinction), a plant whose thinner stems go under the name of 'rattans'; also **C. rotang** : **C. rudentum**, *rŏ·dĕnt'·ŭm* (L. *rŭdens*, a rope, *rudentis*, of a rope, *rudentum*, of ropes), the common or cable cane, a native of the East Indies, etc., growing sometimes to the length of 500 feet.

calamus scriptorius, *kăl'·ăm·ŭs skrĭp·tŏr'·ĭ·ŭs* (Gr. *kalamos*, L. *calamus*, a reed, a reed-pen; *scriptōrius*, for writing), that part of the floor of the fourth ventricle of the brain, the configuration of which resembles the point of a pen.

Calathea, n., *kăl'·ăth·ē'·ă* (Gr. *kalathos*, L. *calăthus*, a wicker basket, a cup), a genus of plants, Ord. Marantaceæ, interesting and ornamental: **Calathea zebrina**, *zĕb·rīn'·ă* (Sp. *zebra*, It. *zebro*, a zebra), a plant, so called from its peculiar striped leaves and velvety aspect.

calathiform, a., *kăl·ăth'·ĭ·fŏrm* (L. *calăthus*, a basket, a cup; *forma*, shape), in *bot.*, hemispherical or concave, like a bowl or cup: **calathium**, *kăl·ăth'·ĭ·ŭm*, in same sense as 'capitulum' and 'anthodium.'

calcaneum, n., *kăl·kān'·ĕ·ŭm* (L. *calcāneum*, the heel—from *calx*, the heel), in *anat.*, the os calcis, or largest bone of the foot, projecting downwards and backwards to form the heel: **calcaneocuboid**, a., *kăl·kān'·ĕ·ō·kŭb'·ŏyd* (Gr. *kubos*, a square; *eidos*, resemblance), an articulation in which the calcaneum is united to the cuboid bone by a synovial joint and ligaments: **calcaneo**, indicates a connection or articulation with the heel.

calcar, n., *kăl'·kăr* (L. *calcar*, a spur), a projecting hollow or solid process from the base of an organ; in *zool.*, the spur of a rasorial bird; the rudiments of hind limbs in certain snakes: **calcarate**, a., *kăl'·kăr·āt*, having a spur or spurs: **calcar avis**, *āv'·ĭs* (L. *avis*, a bird, or 'of a bird'), the spur of the bird, a curved and pointed longitudinal eminence on the inner side of the floor of the cerebrum; also called 'hippocampus minor.'

Calceolaria, n., *kăl'·sĕ·ōl·ār'·ĭ·ă* (L. *calceolus*, a small shoe, in allusion to the form of the corolla), a favourite genus of plants, Ord. Scrophulariaceæ, which contain some very showy species; some of the species used in dyeing: **calceolate**, a., *kăl·sē'·ŏl·āt*, having the form of a slipper, applied to the hollow petals of orchids, and of the calceolaria.

calculus, n., *kăl'·kŭl·ŭs* (L. *calcŭlus*, a small stone), a stony concretion in any gland or organ: **calculi**, n. plu., *kăl'·kŭl·ī*, small stony concretions.

Calendula, n., *kăl·ĕnd'·ūl·ă* (L. *calendæ*, the first days of the Roman month, the calends), a genus of showy plants, flowering almost every month: **Calendula officinalis** (L. *officinālis*, officinal), the common marigold.

calice, n., *kăl'·ĭs·ē* or *kăl'·ĭs* (L. *calix*, a cup; *calĭcis*, of a cup; *calice*, in or with a cup), the little cup in which the polype of a coralligenous zoophyte is contained.

Callitris, n., *kăl'·ĭt·rĭs* (Gr. *kalos*, beautiful; *thrix*, hairy), a beautiful genus of trees, growing 20 or 30 feet high, grows best south of the tropics, Ord. Coniferæ, Sub-ord. Cupressineæ: **Callitris quadrivalvis**, *kwŏd'·rĭ·vălv'·ĭs* (L. *quadrus*, square—from *quatuor*, four; *valvæ*, folding doors), the Arar tree, supplying a solid resin called 'sandarach' or 'pounce,' used instead of blotting-paper to dry the ink by strewing it over MSS.

callosity, n., *kăl·ŏs'·ĭt·ĭ* (L. *callosus*, thick-skinned—from *callus*, hard, thick skin), a horny hardness on the skin; in *bot.*, a leathery or hardened thickening of a part of an organ: **callus**, n., *kăl'·ŭs*, same sense; new bony matter, formed to unite the fractured ends of a bone; in *bot.*, a protuberance on the surface arising from the swelling of cambium cells:

callous, a., *kăl'-ŭs*, hard, indurated.

Calluna, n., *kăl·lŏn'·ă* (Gr. *kalluno*, I make beautiful), a genus of plants, Ord. Ericaceæ : **Calluna vulgaris**, *vŭlg·ār'·ĭs* (L. *vulgāris*, common—from *vulgus*, the multitude), ling or common heather, which has astringent qualities; used commonly for brooms.

calomel, n., *kăl'·ŏ·měl* (Gr. *kalos*, beautiful ; *melas*, black ; rather *meli*, honey), a mild preparation of mercury, chemically known as the sub-chloride of mercury.

Calonyction, n., *kăl'·ŏn·ĭk'·tĭ·ŏn* (Gr. *kalos*, beautiful ; *nux*, night, *nuktos*, of night), a genus of plants, Ord. Convolvulaceæ, so called because they open their flowers at night : **Calonyction speciosum**, *spĕs'·ĭ·ŏz'·ŭm* (L. *spēciōsum*, full of beauty or display—from *species*, look, view, a sort), a plant with large white blossoms, which flowers at night, and has received the name 'moonplant.'

Calophyllum, n., *kăl·ŏ·fĭl'·ŭm* (Gr. *kalos*, beautiful ; *phullon*, a leaf), a genus of large-growing timber trees, Ord. Guttiferæ or Clusiaceæ ; **Calophyllum calaba**, *kăl'·ăb·ă* (*calaba*, native name), the calaba tree, which yields the resin 'tacamahaca:' **C.inophyllum**, *ĭn'·ŏ·fĭl'·ŭm* (Gr. *ĭs*, a fibre, sinew, *ĭnos*, of a fibre; *phullon*, a leaf), a species from the seeds of which a useful oil is obtained.

Calotropis, n., *kăl·ŏt'·rŏp·ĭs* (Gr. *kalos*, beautiful; *tropis*, a keel, the bottom of a vessel), a genus of plants, Ord. Asclepiadaceæ, so called in allusion to the keel of the flowers: **Calotropis procera**, *prŏ·sēr'·ă* (L. *procērus*, high, tall), also **C. gigantea**, *jĭg'·ănt·ē'·ă* (L. *gĭgantēus*, belonging to the giants), the bark of whose roots furnishes the substance called 'mudar'; the essential principle mudarine gelatinises on being heated, and becomes fluid on cooling.

calumba, n., *kăl·ŭm'·bă* (*Colomba*, Ceylon, whence obtained ; *kalumb*, the name in Mozambique), the root of Jateorhiza palmata, from East Africa, an infusion or tincture of which is used as a pure bitter tonic, the bitter crystallisable principle being called 'calumbin.'

calvarium, n., *kăl·vār'·ĭ·ŭm*, **calvaria**, plu. (new L.—from L. *calva*, the scalp without the hair ; *calvus*, bald), the roof of the skull.

Calycanthaceæ, n. plu., *kăl'·ĭk·ănth·ā'·sĕ·ē* (Gr. *kalux*, a calyx ; *anthos*, a flower), the Calycanthus family, a small Order of beautiful early - flowering shrubs, whose flowers are aromatic : **Calycanthus**, n., *kăl'·ĭk·ănth'·ŭs*, a genus of plants, so called in allusion to the colour of the calyx : **Calycanthus floridus**, *flōr'·ĭd·ŭs* (L. *flōridus*, flowery, gay), the Carolina or common American allspice.

calycanthemy, n., *kăl'·ĭk·ănth'·ĕm·ĭ* (Gr. *kalux*, a flower-cup; *anthĕmon*, a flower), the conversion of sepals into petals either wholly or partially; the insertion of the corolla and stamens into the calyx : **cal'ycanth'emous**, a., *·ĕm·ŭs*, having the sepals wholly or partially converted into petals; having the corolla and stamens inserted into the calyx.

Calyceraceæ, n. plu., *kăl·ĭs'·ĕr·ā'·sĕ·ē* (Gr. *kalux*, a flower-cup, *kalukos*, of a flower-cup), a small Order of herbaceous plants inhabiting S. America.

calyces, see 'calyx.'

Calyciflorae, n. plu., *kăl·ĭs'·ĭ·flōr'·ē* (Gr. *kalux*, L. *calyx*, a flower-cup; L. *florālis*, floral—from *Flora*, the goddess of flowers), a sub-class of the Ord. Ochnaceæ: **calycifloral**, a., *kăl·ĭs'·ĭ·flōr'·ăl*, applied to those plants where the petals are separate or united, and the stamens are inserted directly on the calyx.

calycine, a., *kăl'ĭs·ĭn* (L. *calyx*, a flower-cup, *calycis*, of a flower-cup), of or belonging to a calyx or flower-cup: **calycoid,** a., *kăl'ĭk·ŏyd* (Gr. *eidos*, resemblance), resembling a calyx.

Calycophoridæ, n. plu., *kăl'ĭk·ŏ·fŏr'ĭd·ē* (Gr. *kalux*, L. *calyx*, a flower-cup; *phoreo*, I bear or carry), in *zool.*, an Order of the Hydrozoā, so called from possessing bell-shaped swimming organs.

calyculus, n., *kăl·ĭk'ŭl·ŭs* (L. *calyculus*, a little flower-cup, a bud—from *calyx*, a flower-cup), an outer row of leaflets at the base of the calyx giving rise to a double or calyculate calyx; also **calycle,** n., *kăl'ĭ·kl*, in same sense: **calyculate,** a., *kăl·ĭk'ŭl·āt*, applied to flowers which appear as if they possessed a double calyx; a ring or outer covering of bracts appearing to form a distinct whorl of themselves.

calyptoblastic, a., *kăl·ĭp'tŏ·blăst'ĭk* (Gr. *kaluptos*, covered; *blastos*, a sprout or bud), in *zool.*, designating the Hydrozoā in which the nutritive or generative buds possess an external receptacle.

calyptra, n., *kăl·ĭp'tră* (Gr. *kaluptra*, a covering, a veil), the outer covering of the sporangium of mosses; the hood of a moss theca: **calyptrate,** a., *kăl·ĭp'trāt*, hooded.

calyptrimorphous, a., *kăl·ĭp'trĭ·mŏrf'ŭs* (Gr. *kaluptra,* a covering; *morphe*, shape, form), in *bot.*, applied to ascidia or pitchers that have a distinct lid.

calyx, n., *kăl'ĭks*, **calyces,** n. plu., *kăl'ĭs·ēz* (Gr. *kalux*, L. *calyx*, a flower-cup ; Gr. *kalukos*, L. *calycis*, of a flower-cup), in *bot.*, the outer envelope or whorl of a flower, the inner being called the corolla; in *zool.*, the cup-shaped body of the vorticella, or of a crinoid; in *anat.*, **calyces,** short funnel-shaped tubes in the kidneys, into each of which one or more of the papillæ of the renal substance projects.

cambium, n., *kăm'bĭ·ŭm* (new L. *cambium*, nutriment; L. *cambio*, I change), a viscid glutinous substance formed in spring between the bark and the new wood of exogens, the supposed matter for new layers of wood and bark.

Camellia, n., *kăm·ēl'ĭ·ă* (after *Kamel* or *Camellus*, a Moravian Jesuit), a genus of plants whose species are universally prized for their beautiful rose-like flowers, and dark-green shining leaves : **Camellia Japonica,** *jă·pŏn'ĭk·ă* (L. *Japonicus*, of or belonging to Japan), the species from which the cultivated varieties are chiefly obtained: **C. Sasanqua,** *săs·săng'·kwă*, the Sasanqua tea, is cultivated for its flowers, which are used to impart fragrance and flavour to other teas: **C. oleifera,** *ŏl'ē·ĭf'ĕr·ă* (L. *ŏleum*, oil ; *fero*, I carry), yields a valuable oil.

Campanulaceæ, n. plu., *kăm·păn'ŭl·ā'sĕ·ē* (L. *campanula*, a little bell—from *cămpāna*, a bell), the Harebell family, an Order of plants, chiefly herbaceous : **Campanula,** n., *kăm·păn'ŭl·ă*, a genus of plants so called from the resemblance of its corolla to a bell: **Campanula rapuncula,** *răp·ŭngk'ŭl·ă* (dim. of L. *răpum*, a turnip), a species whose roots and young shoots are used for food : **campanulate,** a., *kăm·păn'ŭl·āt*, having the shape of a little bell, as in the flower of the harebell.

Campanularida, n. plu., *kăm·păn'·ŭl·ăr'ĭd·ă* (L. *campanula*, a little bell), an Ord. of hydroid zoophytes.

Camphora officinarum, *kăm'fŏr·ă ŏf·fĭs'ĭn·ār'ŭm* (Sp. *alcanfŏr*, camphor; L. *camphora*, camphor ;

officīna, a workshop, a laboratory, *officīnārum*, of workshops), a tree, Ord. Lauraceæ, which chiefly produces the camphor of the shops, a native of China, Japan, and Cochin-China, obtained from the wood by distillation and sublimation: **camphorated,** a., *kăm'fŏr·ăt·ĕd*, impregnated with camphor.

camptotropal, a., *kămp·tŏt'·rŏp·ăl* (Gr. *kamptos*, flexible, bent; *tropos*, a turn—from *trepo*, I turn), in *bot.*, having curved ovules when the portions on either side of the line of curvation are equal.

campulitropal, see 'campylotropal.'

campylospermæ, n. plu., *kămp'· ĭl·ō·spĕrm'·ē* (Gr. *kampulos*, bent, curved; *sperma*, seed), seeds with the albumen curved at the margins so as to form a longitudinal furrow: **camp'ylosperm'ous,** a., -*spĕrm'·ŭs*, having the albumen of the seed curved at the margin, thus forming a longitudinal furrow.

campylotropal, a., *kămp'·ĭl·ŏt'· rŏp·ăl*, also **camp'ylot'ropous,** a., -*rŏp·ŭs* (Gr. *kampulos*, bent, curved; *tropos*, a turn), in *bot.*, having a curved ovule when the portions on either side of the curvation are unequal; having a curved ovule, with the hilum, micropyle, and chalaza near each other.

canaliculus, n., *kăn·ăl·ĭk'·ŭl·ŭs*, **canaliculi,** n. plu., -*ĭk'·ŭl·ī* (L. *canălĭculus*, a water channel— from *canālis*, a pipe, a channel), a term applied to the minute canals of bone, and to the passages which carry away the tears: **canaliculate,** a., *kăn·ăl· ĭk'·ŭl·āt*, channelled; having longitudinal grooves or furrows.

canalis, n., *kăn·āl'·ĭs* (L. *canālis*, a pipe), a small duct or canal in the human frame for the transmission of nerves, arteries, etc.,

which have received names from their discoverers, as the canals of Fontana, Gärtner, Havers, Nuck, Hunter, Petit, Sylvius, etc., while others are so called from their position, as the following:— **Canalis centralis modioli,** *sĕnt· răl'·ĭs mŏd'·ĭ·ŏl·ī* (L. *centralis*, central, middle; *mŏdĭŏli*, of the nave of a wheel), the central canal of the modiolus; in the ear, one of the many small canals of the modiolus, larger than the rest, running from the base to the centre: **Canalis membranacea,** *mĕm'·brăn·ās'·ĕ·ă* (L. *membrānāceŭs*, belonging to skin or membrane), the membranous canal of the ear: **C. reuniens,** *rĕ·ūn'·ĭ·ĕnz* (L. *re*, again; *ūnĭens*, uniting), a small duct of the ear rendering the cavity of the canalis continuous with that of the saccule: **C. spiralis modioli,** *spĭr·āl'·ĭs* (L. *spirālis*, coiled, twisted — from *spīra*, a coil, a twist), the spiral canal of the modiolus; a small canal of the ear which winds around the modiolus: **C. arteriosus,** *ărt·ēr'·ĭ·ōz'·ŭs* (L. *arteriosus*, pert. to an artery—from *artēria*, an artery), the arterial canal: **C. venosus,** *vēn·ōz'·ŭs* (L. *vēnōsus*, full of veins—from *vēna*, a vein), the venous canal; see 'Ductus arteriosus.'

cancelli, n. plu., *kăn·sĕl'·lī* (L. *cancelli*, a lattice, a grating), the lattice-like texture of the internal bone: **cancellate,** a., *kăn·sĕl'·lāt*, having an appearance like lattice-work; in *bot.*, composed of veins alone, or of lattice-like cells: **cancellous,** a., *kăn·sĕl'·lŭs*, pert. to the net-like tissue, or lattice-work of the inner bone.

cancrum oris, *kănk'·rŭm ōr'·ĭs* (L. *cancrum*, a cancer—from *cancer*, a crab; *ōs*, a mouth, *ōris*, of a mouth), a very rare but dangerous form of gangrenous stomatitis, usually commencing in the cheek, and occurring among children from two to thirteen years of age.

Candollea, n., *kăn·dŏl'lĕ·ă* (after *Decandolle* of Geneva, an eminent botanist), a beautiful genus of plants, Ord. Dilleniaceæ.

Canellaceæ, n. plu., *kăn'·ĕl·ā'·sĕ·ē* (L. *canna,* a reed, a cane), the Canella family, an Order of plants of the West Indies, very aromatic: **Canella,** n., *kăn·ĕl'ă,* a genus of plants, so called from their bark being rolled like cinnamon, valuable and ornamental trees: **Canella alba,** *ălb'ă* (L. *albus,* white), a tall tree yielding the canella bark, or white cinnamon, and likewise several kinds of oil.

canescent, a., *kăn·ĕs'ĕnt* (L. *cānescens,* growing white or hoary), in *bot.,* hoary; somewhat approaching to white.

canine, a., *kăn·īn'* (L. *canis,* a dog), designating the eye-teeth; in mammals, the four teeth which immediately adjoin the incisors, two in each jaw; resembling a dog in qualities or structure : **canine madness,** rabies.

Cannabinaceæ, n. plu., *kăn'·nă·bĭn·ā'·sĕ·ē* (Gr. *kannabis,* L. *cannabis,* hemp), the Hemp and Hop family, an Order of herbaceous plants : **Cannabis,** n., *kăn'·năb·ĭs,* a genus of plants : **Cannabis sativa,** *săt·īv'·ă* (L. *sativus,* fit to be sown or planted), an herbaceous plant yielding the valuable fibre called hemp : **C. Indica,** *ĭn'·dĭk·ă* (L. *Indĭcus,* Indian), a hemp plant used in India to produce intoxication—Bhang is made from the larger leaves and fruit dried ; Gunjah or Ganja consists of the whole plant dried after flowering ; Haschisch or Qinnab, among the Arabs, is made from the tops and tender plants dried : **cannabina,** n., *kăn'·nă·bĭn'·ă,* medicine made from Cannabis Indica.

Cannaceæ, n. plu., *kăn·nā'·sĕ·ē*

(L. *canna,* a reed or cane), the Arrowroot family, an Order of plants, also called Marantaceæ : **Canna,** n., *kăn'·nă,* a genus of plants containing much starch in rhizomes and roots, and producing abundance of bright flowers at all seasons, the seeds of cannas being round and black, and known as 'Indian shot': **Canna coccinea,** *kŏk·sĭn'·ĕ·ă* (L. *coccĭnĕus,* of a scarlet colour — from *coccum,* scar'let); **C. achiras,** *ăk·ĭr'·ăs* (Gr. *a,* without ; *cheir,* the hand ; from W. I.); **C. edulis,** *ĕd·ūl'·ĭs* (L. *edūlis,* eatable), three species of canna from which the arrowroot called 'tous le mois,' *tŏ lĕ mwă,* or St. Kitts' arrowroot, is obtained : **C. iridiflora,** *ĭr'·ĭd·ĭ·flōr'·ă* (L. *ĭris,* the iris or sword lily, *ĭrĭdis,* of the iris; *Flora,* the goddess of flowers), the most splendid flowering plant of the cannas.

cantharis, n., *kănth'·ăr·ĭs,* **cantharides,** n. plu., *kănth·ăr'·ĭd·ēz* (Gr. *kanthăris,* L. *canthăris,* a species of beetle ; Gr. *kantharidos,* L. *cantharidis,* of a beetle), the Spanish fly, so called, collected chiefly in Hungary, and is used in making blistering plasters, etc. ; a weak tincture is occasionally employed as a stimulating wash for promoting the growth of the hair.

canthus, n., also **kanthos,** *kănth'·ŭs,* -*ŏs* (Gr. *kanthos,* the corner of the eye), the angle or corner of the eye, respectively named the outer and inner angles : **canthi,** n. plu., *kănth'·ī,* the corners of the eye.

canula, n., *kăn'·ūl·ă* (L. *canula,* a little reed—from *canna,* a reed), a metallic or elastic tube used for surgical purposes, as for removing a fluid from a tumour.

caoutchouc, n., *kŏŏ'·tshŏŏk* (a native Indian word), india-rubber, the dried juice of various tropical plants, such as Urceola elastica, and Vahea gummifera, Ord. Apocynaceæ; also many o f the Artocarpus tribe, Ord. Moraceæ, furnish caoutchouc.

capillaire, n., *kăp·ĭl·lār'* (F. *capillaire*, capillary, maidenhair), a syrup, prepared from Adiantum pedatum, or Canadian maiden-hair.

capillary, a., *kăp'ĭl·lăr·ĭ* (L. *capillus*, hair), in *bot.*, filiform or thread-like; hair-like.

capillitium, n., *kăp·ĭl·lĭsh'ĭ·ŭm* (L. *capillitium*, the hair collect-ively), the threads or hairs of puff-balls; in prescriptions, the hair of the head—as, **abraditor capillitium**, 'let the head be shaven.'

capitate, a., *kăp'ĭt·āt* (L. *capitatus*, having a head—from *caput*, the head), in *bot.*, having a rounded or pin-like head, as on some hairs; having a globose head: **capitellum**, n., *kăp'ĭt·ĕl'ŭm* (L. *capitellum*, a dim. of *caput*), in *anat.*, a rounded process or knob supported on a narrower, called its neck; in *bot.*, the seed-vessel or head of mosses.

capitulum, n., *kăp·ĭt'·ŭl·ŭm* (L. *capĭtŭlum*, a little head—from *căpŭt*, the head), in *bot.*, a flattened, convex, or slightly concave receptacle covered with flowers, having very short pedicles or none, as in the dandelion, daisy, and other composite flowers; in some lichens, a stalk bearing a round head or knob; in same sense, **Anthodium** and **Calathium**; in *anat.*, a process of bone, same as **capitellum**; a protuberance or round head of bone, fitted into the concavity of another; in *zool.*, the body of a barnacle, from its being supported on a stalk: **capitula**, n. plu., *kăp·ĭt'·ŭl·ă*, small heads or knobs: **capituliform**, a., *kăp'ĭt·ŭl'ĭ·fŏrm* (L. *forma*, shape), having the appearance of a small head or bud.

E

Capparidaceæ, n. plu., *kăp'·ăr·ĭd·ā'·sĕ·ē* (Gr. *kapparis*, L. *cappăris*, the caper tree, the caper; Ar. *algabr*, the caper), the Caper family, an Order of herbaceous plants which have stimulant qualities, also called 'capparids': **Capparis**, n., *kăp'·ăr·ĭs*, a genus of plants, found chiefly in warm countries: **Cappareæ**, n. plu., *kăp·pār'·ĕ·ē*, a Sub-order, having baccate fruit; **Capparis spinosa**, *spīn·ōz'·ă* (L. *spīnōsus*, thorny, prickly—from *spina*, a spine), a species, the flower-buds of which furnish capers.

capreolate, a., *kăp'·rĕ·ŏl·āt* (L. *căprĕŏlus*, the small tendrils of vines—from *căprœa*, a tendril), having tendrils, like vines.

caprification, n., *kăp'·rĭ·fĭk·ā'·shŭn* (L. *caprificare*, to ripen figs by the stinging of the gall-fly—from *căper*, a he-goat, and *fīcus*, a fig), a process of accelerating the ripening of fruit by punctur-ing or by insects, particularly of the cultivated fig.

Caprifoliaceæ, n. plu., *kăp'·rĭ·fŏl·ĭ·ā'·sĕ·ē* (L. *căprœa*, a tendril; *folium*, a leaf), the Honeysuckle family, an Order of plants, many of which, as the elder and honey-suckle, have odoriferous flowers: **Caprifolium**, n., *kăp'·rĭ·fŏl'·ĭ·ŭm*, a genus of favourite climbing plants, including the common honey-suckle—so called from the climb-ing and twining habit of the plant.

Capsicum, n., *kăps'·ĭk·ŭm* (L. *capsa*, a chest or case for fruit, in allusion to the fruit being con-tained in the pods), a genus of plants, Sub-ord. Solaneæ, and Ord. Solanaceæ, the different species of which furnish Cayenne pepper and chillies, natives of hot climates: **Capsicum annuum**, *ăn'·nū·ŭm* (L. *annŭŭs*, annual—from *annus*, a year), the plant from which capsicum and Cayenne or Guinea pepper are obtained:

C. fastigiatum, *făs·tĭdj'ĭ·āt'·ŭm* (L. *fastigiatus*, pointed at the top—from *fastigĭum*, a projecting point), name for same plant and its products as preceding : C. frutescens, *frŏt·ĕs'·ĕnz* (L. *frŭtex*, a shrub or bush, *frŭticis*, of a shrub), a shrubby plant which along with preceding supplies bird-pepper : C. baccatum, *băk·āt'·ŭm* (L. *baccatus*, furnished with berries — from *bacca*, a berry), yields a globular fruit, furnishing cherry or berry capsicum.

capsula circumscissa, *kăps'·ŭl·ă sĕrk'·ŭm·sĭs'·ă* (L. *capsŭla*, a little chest—from *capsa*, a chest or box ; *circumscissus*, torn or cut off around), in *bot.*, a capsule opening with a lid ; a pyxidium.

capsule, n., *kăps'·ŭl* (L. *capsula*, a little chest), in *anat.*, a membranous bag enclosing an organ ; in *bot.*, a dry seed-vessel opening by valves, teeth, or pores ; in *chem.*, a small shallow cup : capsular, a., *kăps'·ŭl·ẽr*, relating to a capsule ; hollow ; full of cells : capsular ligament, *lĭg'·ă·mĕnt*, a little loose bag at a joint which contains the peculiar liquid for its lubrication : capsuliferous, a., *kăps'·ŭl·ĭf'·ẽr·ŭs* (L. *fero*, I bear), bearing capsules.

caput, n., *kăp'·ŭt* (L. *caput*, the head), in *anat.*, a rounded process supported on a narrower part called its neck : caput cæcum coli, *kăp'·ŭt sēk'·ŭm kōl'·ĭ* (L. *cœcus*, blind ; Eng. *colon*, the great gut), a blind sac about two inches and a half in length, situated at the upper extremity of the great gut.

caramel, n., *kăr'·ăm·ĕl* (F.), burnt sugar, chiefly used for colouring wines and brandies; a black porous substance obtained by heating sugar to about 400°.

carapace, n., *kăr'·ă·pās* (Gr. *kara-bos*, a crustaceous animal like the crab), the crustaceous and horny coverings of certain classes of animals, as crabs and lobsters, the tortoise, etc.; the protective shield or case of certain of the Infusoria.

carbon, n., *kårb'·ŏn* (L. *carbo*, a coal, *carbonis*, of a coal), pure charcoal, exists pure only in the diamond : carbonate, n., *kårb'·ŏn·āt*, a compound formed by the union of carbonic acid with a base.

carbuncle, n., *kårb·ŭngk'·l* (L. *carbuncŭlus*, a little coal—from *carbo*, coal), a gem ; an intense inflammation occupying the whole thickness of the skin within a limited area : carbuncular angina, see 'angina'; a disease of pigs, chiefly characterized by difficulty of breathing and painful inflammatory swellings around the pharynx and larynx.

carcerule, n., *kår'·sĕr·ŭl*, also carcerulus, n., *kår·sĕr'·ŭl·ŭs* (L. dim. from *carcer*, a prison, a gaol), in *bot.*, a dry, indehiscent fruit, with the carpels adhering around a common axis, as in a mallow.

carcinoma, n., *kår'·sĭn·ōm'·ă* (Gr. *karkinoma*, cancer, *karkinomatos*, of a cancer—from *karkinos*, a crab-fish), cancer: carcinomatous, a., *kår'·sĭn·ŏm'·ăt·ŭs*, pert. to cancers.

cardamoms, n. plu., *kård'·ăm·ŏmz* (Gr. *kårdămōmŏn*, an aromatic plant, spice), oval trivalvular capsules containing seeds, furnished by various species of Amomum, Elettaria, and Renealmia ; much used in giving colour and pleasant flavour to medicines.

cardia, n., *kård'·ĭ·ă* (Gr. *kardia*, the heart), the opening which admits the food into the stomach: cardiac, a., *kård'·ĭ·ăk*, pert. to the heart ; invigorating the heart, as by stimulants : n., a medicine or cordial which animates the spirits : **cardiac polypus**, *pŏl'-ĭp-ŭs*, a pre-mortem coagulation of the blood within the heart : **cardialgia**, n., *kărd'-ĭ-ălj'-ĭ-ă* (Gr. *algos*, pain), pain in the stomach; heart-burn : **carditis**, n., *kărd-īt'-ĭs*, inflammation of the tissues of the heart.

Carduus, n., *kărd'-ū-ŭs* (L. *cardŭus*, a thistle), a genus of plants, Ord. Compositæ, Sub-Ord. Cynaro-cephalæ, which includes the various species of thistle : **Carduus benedictus**, *bĕn'-ĕ-dĭkt'-ŭs* (L. *benĕdĭctus*, commended, praised), the blessed thistle, formerly used as a stomachic.

carex, n., *kār'-ĕks*, **carices**, n. plu., *kār'-ĭs-ēz* (L. *cārex*, reed-grass), a genus of plants, Ord. Cyperaceæ : **Carex arenaria**, *ăr'-ĕn-ār'-ĭ-ă* (L. *arēnāria*, a sand-pit—from *arēna*, sand), **C. disticha**, *dĭst'-ĭk-ă* (L. *distichus*, consisting of two rows), **C. hirta**, *hĕrt'-ă* (L. *hirtus*, rough, hairy), have been used under the name German sarsaparilla ; some of the carices, having creeping stems, bind together the loose moving sand of the sea-shore.

Cariceæ, n. plu., *kăr-ĭs'-ĕ-ē* (origin unknown—said to be from *Cāria* in Asia Minor, where cultivated), a tribe of plants, Ord. Papayaceæ: **Carica**, n., *kăr'-ĭk-ă*, a genus of plants: **Carica papaya**, *păp-ā'-yă*, the Papaw tree, which yields an acrid milky juice, and an edible fruit.

carices, n. plu., see 'carex.'

caries, n., *kăr'-ĭ-ēz* (L. *cărĭes*, rottenness), ulceration or rotten-ness of a bone, caries having the same relation to bone which ulceration has to soft parts, as in a decaying tooth : **carious**, a., *kăr'-ĭ-ŭs*, affected with caries.

carina, n., *kăr-ĭn'-ă* (L. *cărīna*, the bottom of a ship, the keel), the two partially united lower petals of a papilionaceous flower, as in the lower petals of pea-flowers,

which have a keel-like shape : **carinal**, a., *kăr-īn'-ăl*, said of the æstivation when the carina includes the other part oi the flower : **carinate**, a., *kăr-ĭn'-āt*, keeled.

cariopsis, see 'caryopsis.'

Carludovica, n., *kăr'-lŏ-dŏv'-ĭk-ă* (in honour of *Charles* IV. of Spain and his queen), a genus of plants, Ord. Pandanaceæ : **Carludovica palmata**, *păl-māt'-ă* (L. *palmātus*, marked with the palm of the hand), a plant from whose leaves Panama hats are made, a valuable industry.

carminative, n., *kăr-mĭn'-ăt-ĭv* (It. *carminare*, to card wool, to make gross humours fine and thin by medicine—from *carmen*, a card for wool), remedies which relieve flatulence and alleviate colicky pains, as on the supposed old medical theory of humours.

Carnivora, n. plu., *kăr-nĭv'-ŏr-ă* (L. *căro*, flesh, *carnis*, of flesh ; *vŏro*, I devour), the flesh-eating animals, an Order of the Mammalia : **carnivorous**, a., *kăr-nĭv'-ŏr-ŭs*, feeding upon flesh.

carnose, a., *kăr-nōz'* (L. *carnōsus*, fleshy—from *căro*, flesh), fleshy ; having a consistence resembling flesh : **carnosity**, n., *kăr-nŏs'-ĭt-ĭ*, a small fleshy excrescence.

carotid, n. or a., *kăr-ŏt'-ĭd* (car-*.otīdēs*, plu., a modern L. or Gr. formative — from Gr. *karoō*, I stupefy, from the idea of the ancients that by these arteries an increased flow of blood pro-duced sleep or stupor ; said also to be a Latinised formation from Gr. *kara*, the head ; *ous*, the ear, *ōtos*, of the ear, from the con-nection of the arteries with the face and ear ; more likely, from Gr. *karos*, deep sleep, because compression of them was sup-posed to produce sleep, hence they were also called 'arteriæ soporiferæ'), one of the two large arteries of the neck, subdivided into the 'external carotid,' supplying the face and head, and 'internal carotid,' which divides into the interior and middle cerebral arteries, supplying the anterior and part of the middle lobes of the brain.

carpel, n., *kårp'.ĕl* (Gr. *karpos,* fruit), the name of one or more modified leaves forming the pistil of a plant—when formed of a single leaf, then pistil and carpel are identical; one of the parts which compose the innermost of the four sets of floral whorls into which the complete flower is separable ; also called ' carpidium': **carpellary,** a., *kårp·ĕl'. ĕr-ĭ,* belonging to a carpel.

carpoclonium, n., *kårp'·ō·klŏn'·ĭ· ŭm* (Gr. *karpos,* fruit; *klōnion,* a small branch or shoot), in *bot.,* a free spore case in certain Algæ.

carpogonium, n., *kårp'·ō·gōn'·ĭ·ŭm* (Gr. *karpos,* fruit; *goneus,* a parent, or *gune,* a woman), in *bot.,* in certain Fungi, the twisted end of a branch of Mycelium, forming the female organs.

carpology, n., *kårp·ŏl'·ŏ·jĭ* (Gr. *karpos,* fruit; *logos,* discourse), the part of botany which treats of the structure of fruits and seeds.

carpophaga, n. plu., *kårp·ŏf'·åg·å* (Gr. *karpos,* fruit; *phago,* I eat), fruit-eating animals, a section of the Marsupialia : **carpophagous,** a., *kårp·ŏf'·åg·ŭs,* living on fruits.

carpophore, n., *kårp'·ō·fōr* (Gr. *karpos,* fruit; *phoreo,* I carry or bear), in *bot.,* a stalk raising the pistil above the whorl of the stamens.

carpos, n., *kårp'·ŏs* (Gr. *karpos,* fruit), fruit ; in composition, assumes the form **carpo.**

carpus, n., *kårp'·ŭs* (Gr. *karpos,* Latinised form *carpus,* the wrist), the wrist : **carpal,** a., *kårp'·ål,* belonging to the wrist : **carpometacarpal,** a., *kårp'·ō - mĕt'·å·*

kårp'·ål, pert. to the hand and wrist, excluding the fingers.

Carthamus tinctorius, *kårth'·åm·ŭs tĭnk·tōr'·ĭ·ŭs* (said to be a corruption of the Latinised Arabic name *quortum,* to paint ; L. *tinctōrĭus,* belonging to dyeing), a species of plant, Ord. Compositæ, whose dried flowers constitute safflower or bastard saffron, which yields a pink dye.

cartilage, n., *kårt'·ĭl·ādj* (L. *cartilāgo,* gristle), gristle ; a whitish elastic substance, such as is attached to bones, but softer than bone : **cartilaginous,** a., *kårt'·ĭl· ådj'·ĭn·ŭs,* consisting of gristle instead of bone ; hard and tough : **cartilagines alarum nasi,** *kårt·ĭl· ådj'·ĭn·ēz ål·år'·ŭm nāz'·ĭ* (L. *cartilāgo,* gristle ; *ålå,* a wing ; *nāsus,* a nose), the cartilages of the wings of the nose ; the lower lateral cartilages of the nose, having a peculiar curved form : **c. laterales nasi,** *låt'·ĕr·āl'·ēz* (L. *laterāles,* adj. plu., lateral), the lateral cartilages of the nose ; the upper lateral cartilages of the nose, situated in the upper part of the projecting portion of the nose : **c. minores nasi,** *mĭn·ōr'·ēz* (L. *minōres,* adj. plu., lesser), the lesser cartilages of the nose ; two or three cartilaginous nodules connected with the ascending process of the upper maxilla— also called **c. sesamoideæ nasi** (Gr. *sēsamon,* fruit of the sesame ; *eidos,* resemblance—in allusion to their shape of seeds or nodules): **cartilago nictitans,** *kårt·ĭl·āg'·ō nĭkt'·ĭt·ånz* (L. *nictitans,* winking), a small cartilage contained in the membrana nictitans, which see : **c. triticea,** *trĭt·ĭs'·ĕ·å* (L. *trĭtĭcĕus,* wheaten—from *trĭtĭcum,* wheat), the wheat-shaped cartilage ; a small oblong cartilaginous nodule connected with the lateral thyro-hyoid ligaments.

Carum, n., *kår'·ŭm* (said to be so called as coming from *Caria,* in

Asia Minor; It., F., Scot. *Carvi*; Span. *alcaravéa*, caraway seed), a genus of plants, Ord. Umbelliferæ, sometimes called Apiaceæ: **Carum carui**, *kār'·ū·ī* (the Latinised form of Gr. *karuon*, a nut, signifying 'of a nut'), the species which produces the seeds or fruit known as 'caraway seeds,' which furnish a volatile oil, and are carminative and aromatic.

caruncula, n., *kăr·ŭngk'·ŭl·ă*, also **caruncle**, n., *kăr'·ŭngk·l* (L. *căruncŭla*, a little piece of flesh —from *caro*, flesh), a small fleshy excrescence, diseased or natural, as the comb of a cock; in *bot.*, a fleshy or thickened appendage of the seed: **carunculate**, a., *kăr·ŭngk'·ŭl·āt*, having a fleshy excrescence or protuberance: **caruncula lachrymalia**, *lăk'·rĭm·āl'·ĭs* (L. *lăchrymālis*, belonging to the tears—from *lachryma*, a tear), the lachrymal fleshy excrescence; a spongy-looking reddish elevation, formed by a group of glandular follicles, situated in the internal cavity of each eye: **carunculæ myrtiformes**, *kăr·ŭngk'·ŭl·ē mērt'·ĭ·fŏrm'·ēz*, plu. (L. *myrtus*, a myrtle; *forma*, shape — the myrtle being sacred to Venus), the myrtle-shaped fleshy excrescences; small rounded elevations near the vaginal orifice.

Carya, n., *kār'·ĭ·ă* (Gr. *karuon*, a nut; *karua*, a walnut), a genus of plants, Ord. Inglandaceæ, yielding edible oily nuts: **Carya alba**, *ălb'·ă* (L. *albus*, white), a species which yields the American hickory nut.

Caryocar, n., *kăr·ĭ'·ŏk·ăr* (Gr. *karuon*, a nut), a genus of fruit-bearing trees—so called because the fruit of the species contains edible nuts, Ord. Ternstrœmiaceæ: **Caryocar butyrosum** (L. *bŭtyrōsum*, pert. to butter—from *bŭtyrum*, butter), a tree which yields the Sonari or butter-nuts.

Caryophyllaceæ, n. plu., *kăr'·ĭ·ŏ·*

fĭl·ā'·sĕ·ē (Gr. *karuon*, a nut; *phullon*, a leaf), the Chickweed and Clovewort family, an Order of plants, including the clove-pink or carnation and its numerous varieties: **Caryophylla'ceous**, a., *-ă'·shŭs*, belonging to the clove tribe; having a corolla in which there are five petals with long, narrow, tapering claws, as in many pinks.

Caryophyllus, n., *kār'·ĭ·ŏ·fĭl'·ŭs* (Gr. *karuon*, a nut; *phullon*, a leaf), a genus of plants, Ord. Myrtaceæ—so called from the flower-bud being round like a nut: **Caryophyllus aromaticus**, *ăr'·ōm·ăt'·ĭk·ŭs* (L. *arōmătĭcus*, aromatic, fragrant), a tree originally of the Moluccas, whose dried flower-buds in the form of nails constitute the cloves of commerce.

caryopsis, n., *kār'·ĭ·ŏps'·ĭs* (Gr. *karuon*, a nut; *opsis*, sight, appearance), a dry, one-seeded, indehiscent fruit, having the endocarp adhering to the spermoderm ; a seed having the pericarp so incorporated with itself as to be inseparable from it, as in grains of wheat, maize, and other grasses.

Caryota, n., *kār'·ĭ·ōt'·ă* (Gr. *karuōtoi*, dates of the palm), a genus of palm-trees, Ord. Palmæ: **Caryota urens**, *ŭr'·ĕnz* (L. *ŭrens*, parched, dried up), a species of palm from which sago, as well as sugar and a kind of wine, are procured.

cascarilla, n., *kăsk'·ăr·ĭl'·ă* (Sp. *cascarilla*, thin bark—from *cascara*, bark), the bark of several species of Croton, as 'Croton eleuteria,' 'C. cascarilla,' and 'C. eleutheria,' used in *med.* as a tonic and stimulant.

Casearia, n., *kăs'·ĕ·ār'·ĭ·ă* (after *Casearius*, a botanist), a genus of plants, Ord. Samydaceæ, some of which are bitter and astringent.

casein, n., *kās'·ĕ·ĭn* (F. *caseine*, casein—from L. *cāsĕus*, cheese), the cheesy portion obtained from the curd of milk ; a substance procured from milk, animal flesh, or vegetables.

cashew, n., *kăsh'.ū* (F. *acajou*—from the native name), the nut of the Anacardium occidentale, remarkable for the large succulent peduncle which supports the fruit or nut.

Cassia, n., *kăsh'.ĭ.ă* (Gr. *kassia*, L. *cassia*, a tree with an aromatic bark), a genus of plants, Ord. Leguminosæ, Sub-ord. Cæsalpinieæ, whose species furnish purgative remedies : **Cassia lanceolata,** *lăn'.sĕ.ŏl.āt'.ă* (L. *lansĕŏlātus*, lance-shaped—from *lancĕa*, a lance or spear); **C. acutifolia,** *ăk.ūt'.ĭ.fŏl'.ĭ.ă* (L. *acūtus*, sharp-pointed ; *folium*, a leaf); **C. elongata,** *ē.lŏng.āt'.ă* (L. *elongātus*, lengthened out—from *e*, out ; *longus*, long); **C. obtusata,** *ŏb'.tūz.āt'.ă* (L. *obtūsātus*, blunted); **C. obovata,** *ŏb'.ō.vāt'.ă* (L. *obōvātus*, egg-shaped, but inversely —from *ob*, opposite, and *ovum*, an egg), supply the various kinds of senna, while other species have purgative leaves : **cassia-bark,** believed to be obtained chiefly from the Cinnamomum cassia, Ord. Lauraceæ; yields also an oil, and both oil and bark are used as aromatic stimulants ; the flower-buds are used in confectionery.

cassideous, a., *kăs.sĭd'.ĕ.ŭs* (L. *cassis*, a helmet, *cassĭdis*, of a helmet), helmet-shaped ; having a large helmet-shaped petal, as the aconite.

cassowary, n., *kăs'.sŏ.wăr'.ĭ* (Hind. *kassuwaris*, a large bird), a tree yielding excellent timber ; see ' Casuarinaceæ.'

Cassytheæ, n. plu., *kăs.sĭth'.ĕ.ē* (unascertained), Dodder Laurels, a tribe of plants, Ord. Lauraceæ : **Cassytha,** n., *kăs.sĭth'.ă*, a genus of plants which are generally aromatic and fragrant.

castoreum, n., *kăst.ōr'.ĕ.ŭm* (L. *castŏrĕum*, a secretion of the

beaver—from *castor*, a castor or beaver), a peculiar concrete substance obtained from the follicles of the prepuce of the castor or beaver.

Casuarinaceæ, n. plu., *kăs'.ū.ăr.ĭn.ā'.sĕ.ē* (*kassuwaris*, native name —in allusion to the resemblance of the leaves to the feathers of the cassowary), the Beef-wood family, an Order of Australian trees or shrubs with filiform branches and toothed sheaths in place of leaves : **casuarina,** n., *kăs'.ū.ăr.ĭn.ă*, the Cassowary tree, yielding excellent timber, having somewhat the colour of raw beef, whence the name Beef-wood.

cataclysm, n., *kăt'.ă.klĭzm* (Gr. *kataklusmos*, inundation—from *kata*, down, and *kluzein*, to wash), a deluge.

catacorolla, n., *kăt'.ă.kŏr.ŏl'.ă* (Gr. *kata*, down, upon), in *bot.*, another corolla, formed inside or outside the first one.

catalepsy, n., *kăt'.ă.lĕps.ĭ* (Gr. *katalēpsis*, a seizing or grasping—from *kata*, down ; *lepsis*, a seizing), a peculiar disease in which motion and sensation seem to be suspended ; a trance : **cataleptic,** a., *kăt'.ă.lĕpt'.ĭk*, pert. to catalepsy.

catalysis, n., *kăt.ăl'.ĭs.ĭs* (Gr. *katalusis*, disbandment, destruction—from *kata*, down ; *lŭsis*, a loosening), in *chem.*, the influence which induces changes in the composition of substances by their mere contact with another body or power.

catamenia, n., *kăt.ă.mēn'.ĭ.ă* (Gr. *katamēnios*, monthly—from *kata*, down ; *mēn*, a month), the monthly discharges of females.

cataphyllary, a., *kăt'.ă.fĭl'.ĕr.ĭ* (Gr. *kata*, down ; *phullon*, a leaf), applied to the leaves of a plant when they are mere scales; having the leaves enclosed in buds by perules, or on a root-stock by scales.

cataplasm, n., *kăt'-ă-plăzm* (Gr. *kataplasma*, L. *cataplasma*, a poultice—from Gr. *kata*, down; *plasso*, I form), a poultice or plaster.

cataract, n., *kăt'-ăr-ăkt* (Gr. *kataraktēs*, L. *cataracta*, a waterfall —from Gr. *kata*, down; *rhaktos*, a precipice), a large body of water rushing and falling over rocks; in *med.*, a disease of the eyes in which the vision becomes impaired or destroyed, due to opacity of the crystalline lens.

Catarhina, n., *kăt'-ăr-in'-ă* (Gr. *kata*, down; *rhines*, nostrils), in *zool.*, a group of the Quadrumana, characterised by twisted or curved nostrils placed at the end of the snout: **catarhine,** a., *kăt'-ăr-in*, of or belonging to. -

catarrh, n., *kăt-ăr'* (L. *catarrhus*, a catarrh—from Gr. *katarrheō*, I flow down—from *kata*, down; *rheo*, I flow), a nasal catarrh, a disease well known by its producing a running or flow of mucus from the nostrils, caused by exposure to sudden alternations of temperature;—catarrhal affections may implicate either the skin or mucous membranes; they are mainly characterised by their superficial and spreading character, but do not necessarily augment the secretions of the part affected: **catarrhus sinuum frontalium,** *kăt-ăr'-ŭs sin'-ŭ-ŭm frŏnt-ăl'-ĭ-ŭm* (L. *sinuum*, of curves—from *sinus*, a curve, a hollow; *frontalium*, gen. plu. of *frontālis*, pert. to the front), the catarrh of the frontal sinuses, the sinuses being the hollow spaces in the bones which communicate with the nostrils.

catechu, n., *kăt'-ĕ-shoō*, also **cutch,** n., *koŏtsh* (said to be from Japanese *cate*, a tree; *chu*, juice), the heart wood of the Acacia catechu, an Indian shrub which contains much tannin, and is a powerful astringent.

catenulate, a., *kăt-ĕn'-ŭl-āt* (L. *catēna*, a chain), put together like the links of a chain.

Catha, n., *kăth'-ă* (a native Arabian name), a genus of plants, Ord. Celastraceæ: **Catha edulis,** *ĕd-ūl'-ĭs* (L. *edūlis*, eatable—from *edo*, I eat), a species, the young shoots of which furnish the Arabian drug called 'kât,' used as a stimulant.

cathartic, n., *kăth-ărt'-ĭk* (Gr. *kathairo*, I clean or purge), a medicine which purges, as senna, castor-oil, etc.: adj. purgative.

Cathartocarpus fistula, n., *kăth-ărt'-ō-kărp'-ŭs fist'-ŭl-ă* (Gr. *kathairo*, I clean or purge; *karpos*, fruit; *fistŭla*, a hollow reed, a stalk), a species of Sub-ord. Cæsalpinieæ, whose indehiscent pod contains a laxative pulp.

catheter, n., *kăth'-ĕt-ĕr* (Gr. *kathĕtēr*, that which is let down, a probe — from *kathiĕmi*, I let down), a curved tube of silver, india-rubber, or gum-elastic, employed for drawing off the urine from the bladder: **catheterism,** n., *kăth-ĕt'-ĕr-ĭzm*, the art or operation of introducing a catheter.

catkin, n., *kăt'-kin* (after the domestic *cat*, and *kin*, little), a kind of flower, long and slender, resembling a cat's tail, as in the willow or hazel, the birch, etc.; same as Amentum.

cauda equina, *kăwd'-ă ĕ-kwin'-ă* (L. *cauda*, a tail; *equinus*, belonging to a horse—from *equus*, a horse), the horse-tail; the bundle or brush of nervous cords terminating the spinal marrow in man; the corresponding part in lower animals.

caudal, a., *kăwd'-ăl* (L. *cauda*, a tail), pert. to a tail, or a tail-like appendage: **caudate,** a., *kăwd'-āt*, having a tail or feathery appendage: **caudicle,** n., *kăwd'-ĭk-l*, also **caudicula,** n., *kăwd-ĭk'-ŭl-ă*, a small membranous process supporting a pollen mass in orchids.

caudex, n., *kăwd'ĕks* (L. *caudex*, the trunk or stem of a tree, *caudĭcis*, of a trunk), the axis of a plant ; the stem of a palm or of a tree-fern : **caudex ascendens**, *ăs·sĕnd'·ĕnz* (L. *ascendens*, ascending), the trunk or stem above ground : **c. descendens**, *dē·sĕnd'·ĕnz* (L. *descendens*, descending), the root, being the stem below ground.

caul, n., *kăwl* (AS. *cawl*, F. *câle*, a kind of little cap), the membrane which sometimes covers the head and face or greater part of the body of a child when born, and consisting of the amniotic membranes ; a netted membrane covering the lower intestines; the omentum.

caulescent, a., *kăwl·ĕs'·ĕnt* (L. *caulis*, a stalk or stem), growing up into a stem ; having an evident stem.

caulicle, n., *kăwl'·ĭk·l*, also **cauliculus**, n., *kăwl·ĭk'·ŭl·ŭs* (L. *caulĭcŭlus*, a small stalk—from *caulis*, a stalk), a stalk connecting the axis of the embryo and the cotyledons ; the part of the axis which intervenes between the collar and cotyledons.

Caulinia fragilis, *kăwl·ĭn'·ĭ·ă frădj'·ĭl·ĭs* (L. *caulis*, a stem ; *frăgĭlis*, easily broken, brittle), one of the plants in which protoplasmic rotation has been observed, Ord. Naiadaceæ or Potameæ.

caulis, n., *kăwl'·ĭs* (L. *caulis*, a stalk), the stalk or stem of a plant ; an aerial stem : **cauline**, a., *kăwl'·ĭn*, belonging to a stem or growing immediately upon it : **cauline bundles**, fibro-vascular bundles on a stalk which do not pass into leaves : **caulinary**, a., *kăwl'·ĭn·ĕr·ĭ*, belonging to the stem or growing immediately from it—same as 'cauline.'·

caustic, n., *kăwst'·ĭk* (Gr. *kaustikos*, having the power to burn), a substance which possesses the property of corroding any part of a living body by its chemical action —one of the mildest cauteries is the nitrate of silver or lunar caustic :

cautery, n., *kăwt'·ĕr·ĭ*, any substance or agent employed for firing and searing any superficial part of the living body; **potential cautery** designates the various forms of caustic applications ; **actual cautery** consists in a rod or knob of iron heated to a dull red or white heat according to the effect desired ; **galvanic cautery** is applied by means of wires heated by a galvanic battery : **cauterisation**, n., *kăwt'·ĕr·ĭz·ā'·shŭn*, the act or effect of burning or searing a living part.

cavernous respiration, a peculiar hollow sound, as that produced by blowing into a wide-mouthed glass vessel, heard by auscultation over a large dry cavity in a lung: **cavernous tissue**, in *bot.*, any tissue consisting of layers or groups of cells with cavities between them.

cavicornia, n. pln., *kăv'·ĭ·kŏrn'·ĭ·ă* (L. *cavus*, hollow ; *cornu*, a horn), ruminants whose horns consist of a central bony core surrounded by a horny sheath.

Ceanothus, n., *sē·ăn·ōth'·ŭs* (said to be from Gr. *keanōthos*, a kind of thorn; *keanthos*, a kind of thistle), a genus of plants, often spiny, Ord. Rhamnaceæ : **Ceanothus Americanus**, *ăm·ĕr'·ĭk·ān'·ŭs*, a plant whose leaves have been sometimes used in America as a substitute for tea, the roots used as an astringent.

Cecropia, n., *sē·krōp'·ĭ·ă* (after *Cecropia*, the citadel of Athens, named in honour of *Cecrops*, king of Attica, whose legs were fabled to have been serpents), a genus of trees, Ord. Moraceæ, having peltate leaves, and attaining a height of 30 feet: **Cecropia peltata,** *pĕlt·āt'·ă* (L. *peltāta,* armed with a small peltata, or small half-moon-shaped shield), the Trumpet-wood, so called from the hollowness of its stem and branches, which are used for wind instruments, the fibrous bark being used as cordage.

Cedrelaceæ, n. plu., *sĕd'·rĕl·ā'·sĕ·ē* (Gr. *kedros,* L. *cedrus,* a cedar tree), the Mahogany family, an Order of trees having an aromatic fragrance: **Cedrela,** n., *sĕd·rēl'·ă,* a genus of trees: **Cedrela febrifuga,** *fĕb·rĭf'·ŭg·ă* (L. *fĕbris,* a fever; *fŭgo,* I drive away), a species whose bark is used for the cure of intermittent fevers, and the wood is sometimes called 'bastard cedar.'

Cedrus, n., *sĕd'·rŭs* (Gr. *kĕdros,* L. *cedrus,* the cedar tree), a genus of cedar trees found on the Cedron, Judæa, whence it is said the name, Ord. Coniferæ, very valuable for their timber: **Cedrus Libani,** *lĭb'·ăn·ī* (L. *Libănus,* Lebanon, a mountain of Syria), the Cedar of Lebanon: **C. deodara,** *dē'·ŏd·ār'·ă* (said to be from Hind. *deva,* a deity; *dara,* timber; Sans. *div,* heaven), the Deodar or Himalayan Cedar.

Celastraceæ, n. plu., *sĕl'·ăs·trā'·sĕ·ē* (Gr. *kĕlas,* a winter's day, the fruit remaining on the tree all winter), the Spindle-tree family, an Order of small trees or shrubs, having sub-acrid properties, and the seeds of some yielding a useful oil: **Celastreæ,** n. plu., *sĕl·ăs'·trĕ·ē,* a tribe or Sub-order: **Celastrus,** n., *sĕl·ăs'·trŭs,* a genus: **Celastrus nutans** or **paniculatus,** *nūt'·ănz* or *păn·ĭk'·ūl·āt'·ŭs* (L. *nūtans,* nodding, tottering; *pănĭculātus,* tufted), two species which are said to be of a stimulating nature: **C. venenatus,** *vĕn'·ĕn·āt'·ŭs* (L. *venēnātus,* furnished with poison—from *vĕnĕnum,* poison), this, as well as other species, are said to be poisonous.

cell, n., *sĕl* (L. *cella,* a storeroom), in *bot.,* one of the minute globules or vesicles composing cellular tissue; a small cavity or hollow part: **cellular,** a., *sĕl'·ūl·ăr,* composed or made up of cells: **cellule,** n., *sĕl'·ūl,* the very minute cells or vesicles composing the leaves of mosses and other plants: **cellulose,** n., *sĕl'·ūl·ōz,* the substance of which cell walls are composed, constituting the material for the structure and growth of plants; a similar material in animal tissue: **cellular tissue,** tissue formed by the union of minute globules or bladders, named 'cells,' 'cellules,' 'vesicles,' or 'utricles.'

Cellulares, n. plu., *sĕl'·ūl·ār'·ēz* (L. *cellula,* a small storeroom—from *cella,* a hiding-place), a Sub-class of the Ord. Hepaticæ, plants which are acotyledons, and entirely composed of cellular tissue, having no distinct axis, and their leaves no stomata; also called 'cryptogamous,' and 'acotyledonous' plants.

Celosia, n., *sēl·ōz'·ĭ·ă* (Gr. *kēlos,* dry, burnt), a genus of plants, Ord. Amaranthaceæ, some of which appear as if they were singed: **Celosia cristata,** *krĭst·āt'·ă* (L. *cristātus,* crested—from *crista,* a tuft or crest), the plant cockscomb.

celotomy, n., *sēl·ŏt'·ŏm·ĭ* (Gr. *kēlē,* a tumour; *tomē,* a cutting, a section), the operation for removing the stricture in strangulated hernia.

Celtideæ, n. plu., *sĕlt·ĭd'·ĕ·ē* (said to be from *celtis,* an old name of the lotus), a Sub-order of plants, Ord. Ulmaceæ: **Celtis,** n., *sĕlt'·ĭs,* a genus: **Celtis occidentalis,** *ŏk·sĭd·ĕnt·āl'·ĭs* (L. *occĭdentālis,* western—*occĭdens,* the west), the nettle-tree or sugar-berry, which has a sweet drupaceous fruit.

cenanthy, n., *sĕn·ănth'ĭ* (Gr. *kĕnos*, void, empty ; *anthos*, a flower), the absence of stamens and pistils in flowers.

centrifugal, a., *sĕnt·rĭf'ŭg·ăl* (L. *centrum*, the centre ; *fugio*, I flee), tending to go away from the centre ; in *bot.*, applied to the inflorescence in which the flowering commences first at the centre: centripetal, a., *sĕnt·rĭp'ĕt·ăl* (L. *peto*, I seek), tending to the centre ; in *bot.*, applied to that inflorescence in which the flowering commences first at the circumference or base.

centrum ovale cerebri, *sĕnt'·rŭm ŏv·āl'ĕ sĕr'·ĕb·ri* (L. *centrum*, the centre ; *ovāle*, oval—from *ovum*, an egg ; *cerebri*, of the brain), the oval centre of the brain, called respectively, minus, *mīn'·ŭs*, little, and majus, *mādj'·ŭs*, great, being an oval central mass of white cerebral matter of the hemisphere of the brain.

Cephaelis, n., *sĕf'·ă·ēl'·ĭs* (Gr. *kephalē*, the head, the flowers being disposed in heads), a genus of plants, Ord. Rubiaceæ, which furnish important articles to the Materia Medica : Cephaelis ipecacuanha, *ĭp'·ĕ·kăk·ū·ăn'·ă* (a native Brazilian word), the plant whose roots yield the ipecacuanha, extensively employed in medicine.

cephalagra, n., *sĕf·ăl'·ăg·ră* (Gr. *kephalē*, the head; *agra*, seizure), acute pain in the head, either from gout or rheumatism : cephalalgia, n., *sĕf'·ăl·ălj'·ĭ·ă* (Gr. *algos*, pain), headache; continuous pain in the head.

cephalic, a., *sĕf·ăl'·ĭk* (Gr. *kephalē*, the head), pert. to the head ; pert. to a disease or affection of the head : cephalo, *sĕf'·ăl·ō*, denoting attachment to or connection with the head : cephalobranchiate, a., *brăngk'·ĭ·āt* (Gr. *brangchia*, a gill), carrying gills upon the head ; applied to a

section of the Annelida : cephaloid, a., *sĕf'·ăl·ōyd* (Gr. *eidos*, resemblance), in *bot.*, capitate or head-shaped.

cephalophora, n. plu., *sĕf'·ăl·ŏf'·ŏr·ă* (Gr. *kephalē*, the head ; *phoreo*, I bear, I carry), a name for those Mollusca which have a distinct head ; more usual term is 'encephala.'

Cephalopoda, n. plu., *sĕf'·ăl·ŏp'·ŏd·ă* (Gr. *kephalē*, the head ; *podes*, feet), a class of the Mollusca in which there is a series of arms around the head, as in the cuttle-fishes : cephalopodous, a., *sĕf'·ăl·ŏp'·ŏd·ŭs*, pert. to those animals which have the feet or arms arranged around the head, or the head between the body and the feet, as in cuttle-fishes.

Cephalotaxus, n., *sĕf'·ăl·ō·tăks'·ŭs* (Gr. *kephale*, the head; L. *taxus*, a yew), a genus of plants, Ord. Coniferæ, Sub-ord. Taxineæ, handsome coniferous shrubs and trees, which have the habit of the yew.

Cephaloteæ, n. plu., *sĕf'·ăl·ōt'·ĕ·ē* (Gr. *kephalōtos*, having a head or top), an Order of plants according to some, and by others included under the Ord. Saxifragaceæ: Cephalotus, n., *sĕf'·ăl·ōt'·ŭs*, an anomalous apetalous genus of the Ord. Saxifragaceæ, of which there is only one species : Cephalotus follicularis, *fŏl·ĭk'·ŭl·ār'·ĭs* (L. *follĭcŭlus*, a little bag inflated with air, a little bag—from *follis*, an air-ball), a native of S.W. Australia, having leaves arranged as a rosette at the top of the rhizome, one kind having the true ascidia or pitchers.

cephalo-thorax, n., *sĕf'·ăl·ō·thŏr'·ăks* (Gr. *kephalē*, the head ; *thorax*, the chest), the anterior division of the body, composed of the coalesced head and chest, in many Crustaceæ and Arachnida.

cephalotomy, n., *sĕf'·ăl·ŏt'·ŏm·ĭ* (Gr. *kephalē*, the head ; *tomē*, a cutting), the art or operation of dissecting or opening the head.

Ceradia, n., *sĕr·ād′·ĭ·ă* (Gr. *keras*, a horn—so called from the horn-like appearance of the branches); called also **Ceradia furcata**, *fĕrk·āt′·ă* (L. *furcātus*, forked—from *furca*, a fork), a peculiar plant, having the appearance of a shrub of coral spreading its short leather-coated branches upwards like a candelabra, a native of dry, sterile places in the S. and W. of Africa, yields a resinoid substance called sometimes African bdellium.

ceramidium, n., *sĕr′·ăm·ĭd′·ĭ·ŭm* (Gr. *keramis*, a tile, a copying stone, *keramidis*, of a tile; *keramion*, an earthen vessel, a jar), a pear-shaped capsule or pitcher with a terminal opening, and a tuft of spores arising from the base, as seen in some Algæ: **ceramium**, n., *sĕr·ăm′·ĭ·ŭm*, an extensive genus of sea-weeds, so called from their numerous pear-shaped capsules, Ord. Algæ or Hydrophyta.

cerasin, n., *sĕr′·ăs·ĭn* (Gr. *kerasos*, L. *cerasus*, the cherry tree—so called from *Cerasus* of Pontus, in Asia), that part of the gum of the cherry, the plum, and almond trees, insoluble in cold water: **Cerasus**, n., *sĕr′·ăs·ŭs*, a valuable genus of fruit trees, Ord. Rosaceæ: **Cerasus laurocerasus**, *lāwr′·ŏ·sĕr′·ăs·ŭs* (L. *laurus*, a laurel tree; *cerasus*, the cherry tree), the cherry laurel, the common bay laurel—an oil in large quantities exists in the young leaves, giving to the water distilled from them poisonous qualities: **C. avium**, *āv′·ĭ·ŭm* (L. *ăvium*, of birds—from *ăvis*, a bird), the cherry of the birds, the common cherry, used in the manufacture of kirschenwasser, cherry-brandy, literally cherry-water: **C. Occidentalis**, *ŏk′·sĭ·dĕnt·ăl′·ĭs* (L. *Occĭdentālis*, western — from *Occĭdens*, the

West), used for flavouring Noyau; the kernels of the cerasus give flavour to Ratafia, Cherry-brandy, and Maraschino.

cerate, n., *sĕr′·āt* (L. *ceratum*, overlaid with wax—from *cera*, wax), an ointment or unguent in which wax forms a chief ingredient : **cerated**, a., *sĕr·āt′·ĕd*, covered with wax.

ceratiasis, n., *sĕr′·ăt·ĭ′·ăs·ĭs* (Gr. *keras*, horn), the growth of hard horny tumours : **ceratitis**, n., *sĕr′·ăt·ĭt′·ĭs*, inflammation of the cornea.

ceratium, n., *sĕr·ā′·shĭ·ŭm* (Gr. *keration*, a little horn), in *bot.*, a long one-celled pericarp with two valves, containing many seeds attached to two placentæ, which are alternate with the lobes of the stigma, as in Glaucium and Corydalis; a genus of minute Fungi, so called from the plants resembling small horns, found on dead wood.

cerato, *sĕr′·āt·ō* (Gr. *keras*, a horn), in composition, expressing a connection with the cornua of the hyoid bone, or with the cornea : **cerato-genesis**, *jĕn′·ĕs·ĭs* (Gr. *gennaö*, I beget), the formation or production of horn.

Ceratonia, n., *sĕr′·ă·tōn′·ĭ·ă* (Gr. *keratia* and *keratonia*, the carob tree—from *keration*, a little horn, a pod), a genus of plants, Ord. Leguminosæ, Sub-ord. Cæsalpinieæ : **Ceratonia siliqua**, *sĭl′·ĭ·kwă* (L. *sĭlĭqua*, a pod of leguminous plants, the carob), a pod known as the Algaroba bean; the carob tree, locust tree, or St. John's bread—so called from the mistaken idea that the pods were John the Baptist's food in the wilderness.

Ceratophyllaceæ, n. plu., *sĕr′·ăt·ō·fĭl·lā′·sĕ·ē* (Gr. *keras*, a horn ; *phullon*, a leaf), the Hornwort family, an Order of plants : **Ceratophyllum**, n., *sĕr′·ăt·ō·fĭl′·lŭm*, a genus of plants, so named from the petals being so cut as to resemble stags' horns : **ceratophyllous,** a., *sĕr'·ăt·ŏ·fĭl'·lŭs,* horn-leaved.

Cerbera, n., *sĕr'·bĕr·ă* (L. *Cĕrbĕrus,* the three-headed dog of Pluto, whose bite was poisonous), a genus of plants, Ord. Apocynaceæ, generally poisonous : **Cerbera Ahouai,** *ă·hŏ'·ĭ* (*Ahouai,* an Indian name), a plant whose fruit, contained in a nut, is deadly poison.

Cercariæ, n. plu., *sĕr·kār'·ĭ·ē* (Gr. *kerkos,* a tail), a genus of infusory animalcules ; a remarkable genus of intestinal parasites, so called because in one stage of their existence they have a rudder tail: **cercæ,** n. plu., *sĕr'·sē,* the feelers which project from behind in some insects.

cercidium, n., *sĕr·sĭd'·ĭ·ŭm* (Gr. *kerkos,* a tail), in *bot.,* tail-like roots of some Fungi.

cercomonas, n., *sĕrk·ŏm'·ŏn·ăs* (Gr. *kerkos,* a tail ; *mŏnos,* single, solitary), a minute animalcule having a tail-like prolongation : **cercomonas urinarius,** *ūr'·ĭn·ār'·ĭ·ŭs* (L. *urinārius,* urinary—from *ūrina,* urine), a minute intestinal parasite in the urine of animals, frequently in the fresh urine of the horse.

cere, n., *sēr* (L. *cera,* wax), the naked space found at the base of the bills of some birds.

cerealia, n. plu., *sĕr'·ĕ·āl'·ĭ·ă* (L. *cĕrēālis,* pert. to Ceres, or to grain —from *Cĕrēs,* the goddess of corn and fruits), the different grains used for food ; also called **cereals,** n. plu., *sēr'·ĭ·ălz* : **cerealin,** n., *sĕr'·ĕ·ăl·ĭn,* the nutritious or fleshforming principle in flour.

cerebellum, n., *sĕr'·ĕb·ĕl'·lŭm* (L. *cerebellum,* a small brain—from *cĕrĕbrum,* the brain), the hinder or lower part of the brain : **cerebral,** a., *sĕr'·ĕb·răl,* pert. to the brain : **cerebria,** n. plu., *sĕr·ĕb'·rĭ·ă,* mental derangement : **cerebriform,** a., *sĕr·ĕb'·rĭ·fŏrm* (L.

forma, shape), having an appearance like brain matter ; designating a form of cancer : **cerebric,** a., *sĕr·ĕb'·rĭk,* denoting one of the peculiar acids found in the fatty matter of the brain.

cerebro, *sĕr'·ĕb·rō* (L. *cĕrĕbrum,* the brain), a prefix indicating a connection with the cerebrum or brain : **cerebro-spinal,** an adjective indicating connection or association with the brain and spine, as 'cerebro-spinal' axis : **cerebrum,** n., *sĕr'·ĕb·rŭm,* the brain proper.

Cereus, n., *sēr'·ĕ·ŭs* (L. *cērĕus,* waxen, pliant, soft), a very beautiful genus of plants, Ord. Cetaceæ, many of which show a tendency to spiral development : **Cereus flagelliformis,** *flăj·ĕl'·lĭ·fŏrm'·ĭs* (L. *flăgellum,* a whip, a vine-shoot ; *forma,* shape), one of the species in which setæ, spines, and hairs have a tendency to arrange themselves spirally : **Cereus grandiflorus,** *grănd'·ĭ·flōr'·ŭs* (L. *grăndis,* great, grand ; *flōrus,* having flowers—from *flos,* a flower, *flōris,* of a flower), one of the plants remarkable for only flowering at night, expanding its flower about 10 P.M., and lasting only for the night ; other two night-flowering plants are **C. M'Donaldiæ,** *măk'·dŏn·ăld'·ĭ·ē,* the Cereus of M'Donald, and **C. nycticalus,** *nĭk·tĭk'·ăl·ŭs* (Gr. *nux,* night, *nuktis,* of night ; *kaleo,* I call or summon), the plant that summons in the night ; plants which flower only at night.

ceriferous, a., *sĕr·ĭf'·ĕr·ŭs* (L. *cera,* wax ; *fero,* I produce), in *bot.,* bearing or producing wax : **cereous,** a., *sēr'·ĕ·ŭs,* like wax ; waxen.

cernuous, a., *sĕrn'·ū·ŭs* (L. *cernŭus,* bending or stooping with one's head to the ground), in *bot.,* hanging down the head ; nodding, pendulous.

Ceroxylon, n., *sĕr·ŏks'·ĭl·ŏn* (L. *cera,* wax ; Gr. *xulon,* wood, timber), a palm tree which yields wax, forming a coating over its trunk, Ord. Palmæ.

cerumen, n., *sĕr·ūm'ĕn* (L. *cera*, wax), the wax of the ear secreted by ceruminous glands: **ceruminous,** a., *sĕr·ūm'ĭn·ŭs*, of or belonging to the cerumen.

cervical, a., *sĕrv·ĭk'·ăl* or *sĕrv'ĭk·ăl* (L. *cervix*, the neck, *cervīcĭs*, of the neck), connected with the region of the neck: **cervical vertebræ,** n. plu., *vĕrt'ĕb·rē* (L. *vertebra*, a joint), the seven bones of the spine of the neck: **cervico,** *sĕrv·ĭk'·ō* or *sĕrv'ĭk·ō*, denoting connection or association with the region of the neck proper, or simply with a neck: **cervix,** n., *sĕrv'·ĭks*, the neck: **cervix cornu posterioris,** *kŏrn'ū pŏst·ĕr'·ĭ·ōr'·ĭs* (L. *cornū*, a horn; *postĕriōris*, of posterior), the neck of the posterior horn; a part of the grey substance of the spinal cord: **cervix femoris,** *fĕm'·ŏr·ĭs* (L. *fĕmur*, the thigh, *fĕmŏris*, of the thigh), the neck of the thigh bone between the head and the trochantes: **c. uteri,** *ūt'·ĕr·ī* (L. *ūtĕrus*, the womb, the matrix), the neck of the womb.

cervicalis ascendens, *sĕrv'·ĭk·āl'·ĭs ăs·ĕnd'·ĕnz* (L. *cervīcālis*, belonging to the neck—from *cervix*, the neck; *ascĕndens*, ascending), the muscle which forms the continuation of the accessorius upwards into the neck.

cestoidea, n. plu., *sĕst·ōўd'·ĕ·ă* (Gr. *kestos*, a girdle; *eidos*, appearance), an old name for Tæniada, a class of intestinal worms with flat bodies like tape; tapeworms: **cestoid,** a., *sĕst'·ōўd*, pert. to the cestoidea or tapeworms.

Cestraphori, n. plu., *cĕst·răf'·ŏr·ī* (Gr. *kestra*, a military weapon; *phoreo*, I bear), a group of Elasmobranchii, represented by the Port Jackson shark.

Cetacea, n. plu., *sēt·ā'·shĭ·ă* (Gr.

ketos, L. *cetus*, a whale), the Order of the Mammals comprising the whales and dolphins: **cetaceous,** a., *sēt·ā'·shŭs*, pert. to the whale kind.

Cetraria, n. plu., *sĕt·rār'·ĭ·ă* (L. *cētra*, a short shield or buckler), a genus of lichens: **Cetraria Islandica,** *ĭs·lănd'·ĭk·ă* (L. *Islăndĭca*, of or belonging to Iceland), Iceland moss, a lichen which contains a nutritious matter called lichen-starch or lichenin: **cetrarin,** n., *sĕt'·răr·ĭn*, the bitter principle existing in Iceland moss.

Chærophyllum, n., *kēr'·ŏ·fĭl'·lŭm* (Gr. *chairō*, I am glad, I rejoice; *phullon*, a leaf), a genus of plants so called from the pleasant smell of the leaves: **Chærophyllum bulbosum,** *bŭlb·ŏz'·ŭm* (L. *bulbōsus*, bulbous—from *bulbus*, a bulb), bulbous chervil, which is used like carrots.

chætognatha, n. plu., *kēt'·ŏg·nāth'·ă* (Gr. *chaitē*, horse-hair; *gnathos*, the cheek or jaw bone), an Order of the Anarthropoda, having only one genus, the oceanic Sagitta.

chaffy, a., *tshăf'·ĭ* (AS. *ceaf*, Ger. *kaff*, chaff), in *bot.*, covered with minute membranous scales.

Chailletiaceæ, n. plu., *kĭl·lē'·shĭ·ā'·sĕ·ē* (after *Chaillet*, a Swiss botanist), the Chailletia family, a small Order of trees and shrubs: **Chailletia,** n., *kĭl·lē'·shĭ·ă*, a genus: **Chailletia toxicaria,** *tŏks'·ĭ·kār'·ĭ·ă* (Gr. *toxikon*, L. *toxicum*, poison in which arrows were dipped), a species whose fruit is poisonous, known in Sierra Leone as ratsbane.

chalaza, n., *kăl·āz'·ă* (Gr. *chalaza*, hail, a small tubercle resembling a hailstone), in *bot.*, the disc-like scar where the nourishing vessels enter the nucleus of the ovule: **chalazæ,** n. plu., *kăl·āz'·ē*, in an egg, two spirally twisted bands having a pyramidal slope, one at each end, the apex adhering to the yolk, and the base to the white or glair : **chalazion, n.**, *kăl·āz'·ĭ·ŏn*, a little tumour on the edge of the eyelid, so called from its supposed resemblance to a hailstone.

Chamælauciæ, n. plu., *kăm'·ē·lăw'·sĭ·ē* (Gr. *chamai*, upon the ground; and said to be *lauchis*, a poplar), fringe myrtles, a tribe of the Ord. Myrtaceæ, heath-like plants with fragrant foliage, and opposite dotted leaves : **Chamæ·laucium, n.**, *kăm'·ē·lăw'·sĭ·ŭm*, a genus of plants.

Chamærops, n., *kăm·ē'·rŏps* (Gr. *chamai*, upon the ground ; *rhŏps*, a thicket, a twig), a handsome genus of palms,- so called from their lower growth : **Chamærops humilis**, *hŭm'·ĭl·ĭs* (L. *hŭmĭlis*, lowly, small—from *hŭmus*, the earth, the ground), the only European species of palm.

chancre, n., *shăng'·kẽr* (Fr. *chancre*, a sore), a venereal ulcer or sore : **chancroid, n.**, *shăng'·krŏyd* (Gr. *eidos*, resemblance), a venereal ulcer having a soft base.

channelled, a., *tshăn'·nĕld* (L. *canālĭs*, a pipe for water), hollowed out like a gutter.

Characeæ, n. plu., *kăr·ā'·sĕ·ē* (Gr. *chairo*, I am glad), the Chara family, an Order of curious water-plants : **Charas, n.** plu., *kăr'·ăz*, also **Charæ, n.** plu., *kăr'·ē*, a genus of water-plants which grow in stagnant water ; some of them have their stems encrusted with carbonate of lime, and are used for polishing plate ; in others not so encrusted, the movement of rotation in the protoplasmic matter of the tubes is well seen.

charpie, n., *shărp'·ē* (Fr. *charpie*, lint compress), the fine flock obtained by scraping linen rags or lint ; a coarse kind of lint or tow, used for absorbing blood, matter, and the like.

Chavica, n., *shăv·ĭk'·ă* (native name), a genus of plants, Ord. Piperaceæ, natives of the hottest parts of the world : **Chavica Roxburghii**, *rŏks·bẽrg'·ĭ·ī* (*Roxburgh*, a county of Scotland), a plant which supplies long pepper : **C. betle**, *bēt'·l* (Sp. *betle*, the betel-nut), the leaf of betel pepper, which is chewed with the areca nut in the East, as a means of intoxication : 'Piper' is the common systematic name for 'chavica.'

Cheiroptera, n. plu., *kīr·ŏp'·tẽr·ă* (Gr. *cheir*, the hand ; *pteron*, a wing), the Order of Mammals comprising the bats and the bat kind: **cheiropterous, a.**, *kīr·ŏp'·tẽr·ŭs*, pert. to the bat kind.

Cheirostemon, n., *kīr'·ŏ·stēm'·ŏn* (Gr. *cheir*, the hand ; *stēmōn*, a stamen), a genus of plants, Ord. Sterculiaceæ, so called from having five stamens, and the filaments united at the base : **Cheirostemon platanoides**, *plăt'·ăn·ŏyd'·ēz* (L. *platus*, broad, wide ; Gr. *eidos*, appearance), the hand - plant of Mexico, so called from its five peculiarly-curved anthers, which resemble a claw or the human hand.

chelæ, n. plu., *kēl'·ē* (Gr. *kēlē*, a claw), the bifid claws or pincers terminating some of the limbs in such Crustacea as the crab, lobster, etc.: **chelate, a.**, *kēl'·āt*, having chelæ or two cleft claws.

cheliceræ, n. plu., *kēl·ĭs'·ẽr·ē* (Gr. *kēlē*, a claw; *kēras*, a horn), the prehensile claws of the scorpion.

Chelidonium, n., *kēl'·ĭ·dōn'·ĭ·ŭm* (Gr. *chelidonion*, the celandine—from *chelidōn*, a swallow), a genus of plants, Ord. Papaveraceæ, possessing narcotic properties ; an orange-coloured juice : **Chelidonium majus**, *mādj'·ŭs* (L. *mājus*, great), celandine, which yields an orange-coloured juice, and is said to have acrid properties.

Chelonia, n. plu., *kēl·ōn'·ĭ·ă* (Gr.

chelōnē, a shell, a tortoise), an Order of reptiles which comprise the tortoises and turtles : **chelonian**, a., *kĕl·ōn'·ĭ·ăn*, pert. to animals of the tortoise kind : **chelonobatrachia**, n. plu., *kĕl·ŏn'·ō·bă·trăk'·ĭ·ă* (Gr. *batrăchos*, a frog), sometimes applied to the Anoura, comprising frogs and toads. **Chenopodiaceæ**, n. plu., *kĕn'·ō·pŏd·ĭ·ā'·sĕ·ē*, also **Chenopods**, n. plu., *kĕn'·ō·pŏdz* (Gr. *chĕn*, a goose, *chĕnos*, of a goose ; *pous*, a foot, *podes*, feet), the Goosefoot family, an Order of plants, so called in allusion to many of the species having leaves resembling the webbed feet of the goose : **Chenopodium**, n., *kĕn'·ō·pōd'·ĭ·ŭm*, a genus comprising several culinary herbs : **Chenopodium bonus Henricus**, *bōn'·ŭs hĕn·rĭk'·ŭs* (L. *bŏnus*, good ; *Henrĭcus*, Henry), English mercury, the seeds of which are used in the manufacture of shagreen : **C. quinoa**, *kwĭn'·ō·ă* (unascertained), a plant which grows at a great elevation, whose seeds are used in Peru as food by the name of 'petty rice,' the leaves as spinach, and which contains much starch and oil : **C. erosum**, *ĕ·rōz'·ŭm* (L. *erosum*, to eat away, to corrode), the Australian spinach : **C. tomentosum**, *tōm'·ĕn·tōz'·ŭm* (L. *tomentōsum*, covered with a whitish, down-like wool—from *tŏmentum*, a woolly pubescence), the tea plant of Tristan d'Acunha and Inaccessible Island : some of the Chenopodiums emit a very fetid odour.

Chiasma, n., *kĭ·ăz'·mă* (Gr. *chiasmos*, a marking with the Gr. letter χ, a cut crosswise), in *anat.*, the central body of nervous matter formed by the junction and the crossing of the inner fibres of the optic nerves, which go to opposite eyes, the outer fibres proceeding direct to the eye on the same side.

chigoe, n., *tshĭg'·ō* (of Peruvian origin ; Sp. *chico*, small), a painful sore beneath the epidermis of the toes or part of the feet in warm countries, caused by the entrance of flea - like insects of the same name—the systematic names being 'pulex penetrans,' and 'pulex irritans': **chigger**, n., *tshĭg'·gėr*, another spelling of 'chigoe.'

Chilognatha, n. plu., *kĭl'·ŏg·năth'·ă* (Gr. *cheilos*, the lip, the snout of an animal ; *gnathos*, a jaw), an Order of the Myriopoda : **Chilopoda**, n. plu., *kĭl·ŏp'·ŏd·ă* (Gr. *podes*, feet), an Order of the Myriopoda.

Chimaphila, n., *kĭm·ăf'·ĭl·ă* (Gr. *cheima*, a storm, frost ; *phileo*, I love), a genus of plants, Ord. Ericaceæ, plants which are green in winter, and are ornamental and medicinal : **Chimaphila umbellata**, *ŭm'·bĕl·lăt'·ă* (L. *umbellătus*, bearing umbels—from *umbella*, a sunshade), a North American plant, the winter - green, the only bitter tonic which is also diuretic.

Chimonanthus, n., *kĭm'·ŏn·ănth'·ŭs* (Gr. *cheimōn*, winter ; *anthos*, a flower), a genus of plants, Ord. Calycanthaceæ, which flower in the winter-time, and the flowers have a delightful fragrance.

China, *kĭn'·ă*, or **China nova**, *nŏv'·ă* (It. *china*, Sp. *quina*, china; Swed. *kinabark;* L. *novus*, new), the German name for Peruvian or Jesuits' bark ; various kinds of cinchona bark.

chiragra, n., *kĭr·āg'·ră* (L. *chĭrāgra*, Gr. *cheiragra*, gout in the hand—from Gr. *cheir*, the hand ; *agra*, a seizure), gout in the hand.

chiretta, n., *kĭr·ĕt'·tă* (a corruption of the systematic name *chirayta* — from Tamil, *shayraet*), a name for the whole plant, including the flowers and roots, of Agathotes chirayta, found in

Northern India, very bitter, and is an esteemed and slightly laxative tonic : **Chironia**, n., *kĭr·ŏn'·ĭ·ă*, a genus of plants, Ord. Gentianaceæ.

chiropodist, n., *kĭr·ŏp'·ŏd·ĭst* (Gr. *keiro*, I clip or pare ; *podes*, the feet), one who extracts corns and removes bunions ; a corn and wart doctor.

chitine, n., *kĭt'·ĭn* (Gr. *chitōn*, a coat of mail), the peculiar chemical substance, nearly allied to horn, which forms the covering of many of the crustacea, insecta, etc.: **chitinous**, a., *kĭt'·ĭn·ŭs*, consisting or having the nature of chitine.

Chlænaceæ, n. plu., *klē·nā'·sĕ·ē* (Gr. *chlaina*, a gown or cloak), a small Order of trees or shrubs found in Madagascar.

chlamys, n., *klăm'·ĭs* (Gr. *chlamus*, L. *chlamys*, a coat, an upper garment, *chlamydis*, of a coat), in *bot.*, a covering, the floral envelope : **chlamydeous**, a., *klăm·ĭd'·ĕ·ŭs*, pert. to.

Chloranthaceæ, n. plu., *klōr'·ănth·ā'·sĕ·ē* (Gr. *chloros*, green ; *anthos*, a flower), the Chloranthus family, an Order of plants esteemed in tropical countries for medicinal properties: **Chloranthus**, n., *klōr·ănth'·ŭs*, a genus of curious plants : **Chloranthus officinalis**, *ŏf·fĭs'·ĭn·āl'·ĭs* (L. *officina*, a workshop), a species which is aromatic and fragrant : **chloranthia**, n., *klōr·ănth'·ĭ·ă*, also **chloranthy**, n., *klōr·ănth'·ĭ*, a vegetable luxuriance consisting of a bunch of leaves into which the floral organs of a flower have been converted.

chlorine, n., *klōr'·ĭn* (Gr. *chlōros*, grass-green), a greenish - yellow gas, possessing great power as a bleacher : **chloride**, n., *klōr'·ĭd*.

chloroform, n., *klōr'·ō·fŏrm* (so called because it cousists of one atom of *formyle* and three atoms of *chlorine*), a clear, transparent,

watery-looking liquid, produced in the crude state by distilling rectified spirit from off chlorinated lime, usually called chloride of lime, remarkable for its property of producing insensibility to pain when inhaled ; also called the 'perchloride of formyle.'

chlorofucine, n., *klōr'·ō·fūs'·ĭn* (Gr. *chloros*, grass-green; Gr. *phukos*, L. *fucus*, the plant alkanet, the red colour from the same), a clear, yellow-green colouring matter.

chlorophyll, n., *klōr'·ō·fĭl* (Gr. *chlōros*, grass-green ; *phullon*, a leaf), the green colouring matter of plants.

chloros, n., *klōr'·ŏs* (Gr. *chlōros*, grass-green), in *composition*, chloro- : **chlorosis**, n., *klōr·ōz'·ĭs*, a loss of colour ; a diseased state in which the skin assumes a sallow tint, its most prominent phenomenon being a spanæmic condition of the blood, with diminution of the red corpuscles : **chlorotic**, a., *klōr·ŏt'·ĭk*, pert. to or affected with chlorosis.

Chlorosporeæ, n. plu., *klōr'·ō·spŏr'·ĕ·ē* (Gr. *chlōros*, grass-green ; *spora*, a spore), a Sub-order of the Algæ, plants growing in damp situations, and usually of a grass-green colour.

Chloroxylon, n., *klōr·ŏks'·ĭl·ŏn* (Gr. *chlōros*, grass-green ; *xulon*, wood), a genus of fine timber trees, Ord. Cedrelaceæ, so named from the deep yellow colour of the wood : **Chloroxylon Swietenia**, *swēt·ēn'·ĭ·ă* (after *Swieten*, a Dutch botanist), a species which produces satin-wood, and a kind of oil.

cholagogue, n., *kŏl'·ă·gŏg* (Gr. *cholē*, bile ; *agōgos*, a leader), a medicine which acts on the liver, and increases the flow of bile.

choledochus, a., *kŏl·ĕd'·ŏk·ŭs* (Gr. *cholē*, bile ; *dechomai*, I receive), denoting the common bile duct, conveying bile both from the liver and the gall-bladder into the duodenum.

cholepyrrhine, n., *kŏl′·ĕ·pĭr′·rĭn* (Gr. *cholē*, bile; *purrhos*, red), a yellow substance in the bile.

cholera, n., *kŏl′·ĕr·ă* (Gr. *cholera*, a water gutter from the roof of a house; L. *cholera*, the gall, bile — from *cholē*, bile; *rheo*, I flow; or Gr. *cholas*, the bowels), a disease characterised in its severer forms by rice - water vomiting and purging, — of the two kinds, British and Asiatic, the latter is terribly fatal : **cholera morbus,** *mŏrb′·ŭs* (L. *morbus*, sickness, disease), British cholera, a vomiting and purging, rarely fatal to adults : **cholera maligna,** *măl·ĭg′·nă* (L. *malignus*, malignant), Asiatic cholera.

cholesteatoma, n., *kŏl·ĕst′·ē·ăt·ōm′·ă* (Gr. *cholē*, bile ; *steatōma*, tallow, a swelling resembling fat—from *stear*, fat), an encysted tumour consisting almost entirely of cholesterin packed in spherical masses, and surrounded by a somewhat dense capsule : **cholesteatomatous,** a., *kŏl·ĕst′·ĕ·ăt·ōm′·ăt·ŭs*, pert. to or consisting of an encysted fatty tumour.

cholesterin, n., *kŏl·ĕst′·ĕr·ĭn* (Gr. *cholē*, bile ; *stear*, fat, *steatos*, of fat ; or *stereos*, hard, solid), a white fatty matter found in the blood, brain, and bile, but chiefly in the bile.

cholicele, n., *kŏl′·ĭ·sēl* (Gr. *cholē*, bile; *kēlē*, a tumour, a swelling), the gall-bladder when unnaturally distended with bile.

cholic, a., *kŏl′·ĭk* (Gr. *cholē*, bile), of or belonging to bile ; an acid obtained from bile ; also **choleic,** a., *kŏl·ē′·ĭk* : **choloidic,** a., *kŏl·ŏȳd′·ĭk* (Gr. *eidos*, resemblance), denoting an acid obtained from bile.

cholochrome, n., *kŏl′·ō·krōm* (Gr. *cholē*, bile ; *chrōma*, colour), the colouring matter of bile ; biliphæin.

F

Chondodendron, n., *kŏn′·dō·dĕn·drŏn* (unascertained ; Gr. *dendron*, a tree), a genus of plants, Ord. Monospermaceæ : **Chondodendron tomentosum,** *tōm′·ĕn′·tōz′·ŭm* (L. *tomentōsum*, woolly, downy—from *tomentum*, a woolly pubescence), a species found in Peru and Brazil, whose stem and root furnish 'Pareira brava,' used in chronic inflammation of the bladder.

chondrin, n., *kŏn′·drĭn* (Gr. *chondros*, a grain, a clot, cartilage), a substance, a kind of animal gelatine, found in cartilages, fungous bone, and the cornea : **chondroglossus,** *kŏn′·drō·glŏs′·sŭs* (Gr. *glossa*, the tongue), a muscle, being simply one of the three fibres of the hyo-glossus muscle running to the tongue : **chondroma,** n., *kŏn·drōm′·ă*, a growth of cartilage from bones; a cartilaginous tumour : **chondrosis,** n., *kŏn·drōz′·ĭs*, a diseased condition or formation of cartilage.

chondrus crispus, *kŏn′·drŭs krĭsp′·ŭs* (Gr. *chondros*, a clot, cartilage ; L. *crispus*, curled, wrinkled), a name frequently given to carrageen or Irish moss; its systematic name in America.

chorda, n., *kŏrd′·ă* (Gr. *chordē*, L. *chorda*, a gut, a string, a chord), a cord; a tendon ; a collection of fibres : **chorda dorsalis,** *dŏr·săl′·ĭs* (L. *dorsālis*, pert. to the back—from *dorsum*, the back), the linear condensed structure which appears in the fœtal development immediately below the cerebro-spinal groove : **c. tympani,** *tĭm′·păn·ī* (L. *tympănum*, a drum, a tambourine, *tympăni*, of a drum), the chord of the tympanum, a branch of the facial nerve which crosses the tympanum to join the gustatory nerve: **chordæ tendineæ,** plurals, *kŏrd′·ē tĕnd·ĭn′·ĕ·ē* (L. *tendo*, a tendon, *tendĭnis*, of a tendon, *tendĭnĕus*, belonging to a tendon

—from *tendo*, I stretch out), the tendinous chords of the heart which connect the carneæ columnæ to the valves guarding the auricular orifice.

chorea, n., *kŏr·ē'ǎ* (Gr. *choreia*, a dance), St. Vitus's dance; a disease attended with irregular and involuntary movements of the voluntary muscles, except when asleep, occurring mostly in the young.

chorion, n., *kōr'·ĭ·ŏn* (Gr. *chōrion*, skin or leather), in *anat.*, the external membrane investing the fœtus in the womb; in *bot.*, a fluid pulp composing the nucleus of the ovule in the earliest stage:

choroid, a., *kŏr'·ŏyd* (Gr. *eidos*, resemblance), resembling the chorion; denoting a highly vascular membrane: n., the membrane of the eye, situated between the sclerotica and the retina.

chorisis, n., *kŏr'·ĭs·ĭs* (Gr. *chorizo*, I separate), in *bot.*, separation of a lamina from one part of an organ, so as to form a scale or a doubling of the organ: **chorisation**, n., *kŏr'·ĭz·ā'·shŭn*, in same sense.

Choristosporei, n. plu., *kŏr·ĭst'·ō·spōr'·ē·ī* (Gr. *chōristos*, separate, distinct; *spora*, a seed), a Sub-order of Algæ, consisting of rose or purple-coloured sea weeds, with fronds formed of a single row of articulated cells.

chroma, n., *krōm'·ǎ* (Gr. *chrōma*, colour; in *composition*, **chromo-**: **chromatism**, n., *krōm'·ǎt·ĭzm*, also **chromism**, n., *krōm'·ĭzm*, in *bot.*, an unnatural colouring of plants and leaves.

chromatometer, n., *krōm'·ǎt·ŏm'·ĕt·ĕr* (Gr. *chroma*, colour, *chrōmǎtos*, of colour; *metron*, a measure), a measurer of colours, especially as applied to plants.

chromatophores, n. plu., *krōm·ǎt'·ō·fōrz* (Gr. *chrōma*, colour, *chrōmǎtos*, of colour; *phoreo*, I carry), little sacs containing pig-

ment-granules, found in the integument of cuttle-fishes: **chromatophorous**, a., *krōm'·ǎt·ŏy'·ŏr·ŭs*, containing or secreting colouring matter.

chromatosis, n., *krōm'·ǎt·ōz'·ĭs* (Gr. *chrōma*, colour), constitutional discoloration.

chromogen, n., *krōm'·ō·jĕn* (Gr. *chroma*, colour; *gennaō*, I produce), a vegetable colouring matter, acted upon by acids and alkalies to produce red, yellow, or green tints: **chromule**, n., *krōm'·ŭl* (a diminutive of Gr. *chrōma*, colour), the colouring matter of flowers; the colouring matter of plants except green.

chrysalis, n., *krĭs'·ǎl·ĭs* (Gr. *chrusallus*, L. *chrysalis*, the gold-coloured chrysalis of the butterfly—from Gr. *chrusos*, gold), the second stage in the state of such insects as the butterfly, the moth, etc., so named as sometimes exhibiting a golden lustre; some spell **chrysalid**, *krĭs'·ǎl·ĭd*.

Chrysanthemum, n., *krĭs·ǎnth'·ĕm·ŭm* (Gr. *chrusos*, gold; *anthemion*, a flower, a blossom), a genus of plants, Ord. Compositæ, Sub-ord. Corymbiferæ, so called alluding to some of the flowers being yellow; the numerous species are exceedingly beautiful: **Chrysanthemum carneum**, *kăr'·nĕ·ŭm* (L. *carnĕus*, fleshy—from *cǎro*, flesh, *cǎrnis*, of flesh), a species, the flowers of which are said to destroy fleas.

Chrysobaleneæ, n. plu., *krĭs'·ō·bǎl·ǎn'·ĕ·ē* (Gr. *chrusos*, gold; *balanos*, an acorn), a Sub-order of the Order of plants Rosaceæ, this Sub-order being chiefly natives of tropical parts of Africa and America: **Chrysobalanus**, n., *krĭs'·ō·bǎl'·ǎn·ŭs*, a genus of plants, the species bearing the common edible fruits, raspberries, strawberries, brambles, apples, pears, plums, cherries, quinces, almonds, peaches, etc.

Chrysophyll, n., *krĭs'·ŏ·fĭl* (Gr. *chrusos*, gold; *phullon*, a leaf), the golden - yellow colouring matter in many plants and their flowers : **Chrysophyllum**, n., *krĭs'·ŏ·fĭl'·lŭm*, a fruit - bearing genus of plants, Ord. Sapotaceæ, the under surface of the leaves having dense hairs of a bright yellow colour : **Chrysophyllum Cainito**, *kĭn·ĭt'·ŏ* (a native name), a species which yields the fruit star-apple,

chrysops cæcutiens, *krĭs'·ŏps sē·kū'·shĭ·ĕnz* (Gr. *chrusōps*, gold-coloured—from *chrusos*, gold, *ōps*, the eye; L. *cæcūtiens*, blinding—from *cæcus*, blind), an African fly which attacks horses' eyes and blinds them.

churrus, n., *kŭr'·rŭs* or *tshŭr'·rŭs* (native name), the Indian variety of the hemp plant, having a marked resinous varnish on its leaves ; a resinous extract from the Indian hemp or 'canna-bis.'

chylaqueous, a., *kĭl·āk'·wĕ·ŭs* (Gr. *chulos*, juice, humour ; L. *aqua*, water), in *zool.*, applied to a fluid consisting partly of water taken in from the exterior, and partly of the products of digestion which occupy the body cavity in many Invertebrates ; applied also to the special canal sometimes existing for its conduction.

chyle, n., *kĭl* (Gr. *chulos*, juice), a white or milky fluid separated from the substances digested in the stomach, and conveyed into the circulation of the blood by the lacteal vessels : **chylific**, a., *kĭl·ĭf'·ĭk* (L. *facio*, I make), pro-ducing chyle ; designating a part of the digestive apparatus of insects ; applied to one of the stomachs, where more than one is present : **chylous**, a., *kĭl'·ŭs*, pert. to or full of chyle : **chylification**, n., *kĭl'·ĭf·ĭk·ā'·shŭn*, the process of making chyle from food : **chyliferous**,

a., *kĭl·ĭf'·ĕr·ŭs* (L. *fero*, I bear), bearing or carrying chyle.

chyli receptaculum, *kĭl'·ĭ rēs'·ĕp·tăk'·ŭl·ŭm* (L. formative, *chyli*, of chyle ; L. *rĕceptăculum*, a magazine, a receptacle), the re-ceptacle or reservoir of the chyle, a triangular dilatation of the thoracic duct, commencing in the abdomen.

chylopoiesis, n., *kĭl'·ō·pŏy·ēz'·ĭs* (Gr. *chulos*, juice ; *poiĕo*, I make; *poiēsis*, a making or forming), the process of making chyle from food : **chylopoietic**, a., *kĭl'·ō·pŏy·ĕt'·ĭk*, making or producing chyle ; belonging to the stomach and intestines;—same meaning as 'chylification' and 'chylific,' but more correct in their formation.

chyme, n., *kĭm* (Gr. *chumos*, juice, moisture), the pulpy mass of digested food before being changed into chyle: **chyme mass**, the central semi-fluid sarcode in the interior of the Infusoria : **chymif-erous**, a., *kĭm·ĭf'·ĕr·ŭs* (L. *fero*, I bear), containing or bearing chyme: **chymification**, n., *kĭm·ĭf'·ĭk·ā'·shŭn*, the process of changing into chyme.

Cibotium, n., *sĭb·ō'·shĭ·ŭm* (Gr. *kibōtos*, a chest, a casket), a genus of ferns, Ord. Filices, so named in reference to the form of the indusium : **Cibotium barom-etz**, *băr'·ŏm·ĕtz* (a Russian name), a fern called the Scythian or Tartarean lamb, because, prepared in a particular way, it resembles a lamb.

cicatricula, n., *sĭk'·ăt·rĭk'·ŭl·ă* (L. dim. of *cĭcātrix*, a mark or scar), the scar left after the falling of a leaf ; the hilum or base of the seed ; the point in the ovum or egg in which life first shows itself : **cicatricose**, a., *sĭk·ăt'·rĭ·kōz*, marked with scars or cicat-rices: **cicatrix**, n., *sĭk·ăt·rĭks*, the scar or seam that remains on the skin after a wound has skinned over and healed.

Cichoraceæ, n. plu., *sĭk'·ŏr·ā'·sĕ·ē* (Gr. *kichōriŏn*, L. *cĭchŏrium*, succory or endive), a Sub-order of the Ord. Compositæ, most of the plants of which yield a milky juice, and are bitter and astringent : **Cichorium**, n., *sĭk·ōr'·ĭ·ŭm*, a genus of plants : **Cichorium endivia**, *ĕn·dĭv'·ĭ·ă* (F. *endive*, a salad), a species, the blanched leaves of which constitute endive : **C. intybus**, *ĭn'·tĭb·ŭs* (said to be from L. *in*, in ; *tŭbus*, a tube—from the hollow form of its stem), the succory or chicory, cultivated for the sake of its root, used for mixing with coffee when roasted and ground, or used alone as coffee : **cichoriaceous**, a., *sĭk·ōr'·ĭ·ā'·shŭs*, having the qualities of chicory or wild endive.

Cicuta, n., *sĭk·ūt'·ă* (L. *cĭcūta*, the plant hemlock), a genus of plants, Ord. Umbelliferæ : **Cicuta virosa**, *vĭr·ōz'·ă* (L. *virōsus*, slimy, poisonous — from *vīrus*, slime, poison), water-hemlock or cowbane.

cilia, n. plu., *sĭl'·ĭ·ă* (L. *cilium*, an eyelid with the hairs growing on it ; *cilia*, eyelids), the hairs on the edge of the eyelids ; hairs on the margin of any body ; thin hair-like projections from an animal membrane which have a quick, vibratory motion — in insects only microscopic ; in *bot.*, short stiff hairs fringing the margin of a leaf : **ciliary**, a., *sĭl'·ĭ·ĕr·ĭ*, belonging to the eyelids or cilia : **ciliate**, a., *sĭl'·ĭ·āt*, also **ciliated**, a., *sĭl'·ĭ·āt·ĕd*, provided with cilia ; fringed.

ciliograda, n. plu., *sĭl'·ĭ·ō·grād'·ă* (L. *cilium*, an eyelid with the hairs on its margin ; *grădĭor*, I walk, *grădus*, a step), animals that swim by means of cilia—same as 'Ctenophora :' **ciliograde**, a., *sĭl'·ĭ·ō·grād*, swimming by the vibratory motion of cilia.

Cinchoneæ, n. plu., *sĭn·kōn'·ĕ·ē* (after the wife of the *Conde del*

Cinchon, a viceroy of Peru, who was cured of a fever by the Peruvian bark, 1638), a Sub-order of the Ord. Rubiaceæ : **Cinchona**, n., *sĭn·kōn'·ă*, a genus of trees and shrubs, various species of which furnish Peruvian or Jesuit's bark, growing abundantly in Upper Peru : **Cinchona Condaminea**, *kŏn'·dă·mĭn'·ĕ·ă* (after *De la Condamin*, a celebrated navigator); **C. calisaya**, *kăl'·ĭs·ā'·yă* ; **C. succirubra**, *sŭk'·sĭ·rŏb'·ră* (L. *succus*, juice, moisture ; *rŭber* or *rŭbra*, red), are the three species which furnish the pharmaceutical bark ; about twelve species furnish the commercial bark, and for the manufacture of quinine, which the pharmacopœia, however, directs to be prepared from the yellow bark, the C. calisaya, and C. lancifolia : **cinchonin**, *sĭn'·kŏn·ĭn*, also **cinchonia**, n., *sĭn·kōn'·ĭ·ă*, an alkaloid obtained from cinchona bark : **cinchonism**, n., *sĭn'·kŏn·ĭzm*, a disturbed condition of the general health by overdoses and too frequent use of quinine.

cincinnus, n., *sĭn·sĭn'·ŭs*, or **cicinus**, n., *sĭs·ĭn'·ŭs* (Gr. *kikīnŏs* or *kikinnos*, a lock of hair, a curled lock), applied to the hair on the temples ; in *bot.*, an inflorescence ; a scorpioid cyme.

cinclides, n. plu., *sĭn·klĭd'·ēz* (Gr. *kingklis*, a lattice, a grating), apertures in the column walls of some sea anemones, which probably serve for the emission of the cord-like craspeda.

cinenchyma, n., *sĭn·ĕng'·kĭm·ă* (Gr. *kineo*, I move ; *engchuma*, an infusion), in *bot.*, laticiferous tissue formed by anastomising vessels ; applied to laticiferous vessels of plants on account of the granules contained in the 'latex' exhibiting certain movements under the microscope : **cinenchymatous**, a., *sĭn'·ĕng·kĭm'·ăt·ŭs*, having laticiferous tissue.

cinereous, a., *sĭn·ēr'·ĕ·ŭs*, also **cineritious**, a., *sĭn'·ēr·ĭsh'·ŭs* (L. *cĭnĕrācĕus*, and *cĭnĕrĕus*, resembling ashes, ash-coloured—from *cĭnĭs*, ashes, *cĭnĕris*, of ashes), resembling ashes in colour, appearance, or consistence ; in *anat.*, applied to the outer or cortical substance of the brain, which has a grey colour.

Cinnamodendron, n., *sĭn'·năm·ŏ·dĕn'·drŏn* (Gr. *kinnamōmŏn*, Ar. *kinamon*, cinnamon; Gr. *dendron*, a tree), a genus of trees, Ord. Canellaceæ : **Cinnamodendron corticosum**, *kŏrt'·ĭk·ōz'·ŭm* (L. *cortĭcōsus*, full of bark—from *cortex*, bark, *cortĭcis*, of bark), a tree of the West Indies which yields an aromatic bark : **Cinnamomum**, n., *sĭn'·năm·ōm'·ŭm*, a genus of plants, Ord. Lauraceæ : **Cinnamomum Zeylanicum**, *zĭ·lăn'·ĭk·ŭm* (from *Zeylan*, Ceylon), the true cinnamon tree of commerce, cultivated in Ceylon : **C. cassia**, *kăsh'·ĭ·ă* (see 'cassia'), the chief source of Cassia lignea, or cassia bark of commerce.

circinate, a., *sĕr'·sĭn·āt* (L. *circĭno*, I turn round; *circĭnātum*, to turn round—from *circĭnus*, a pair of compasses), in *bot.*, rolled inwards from the summit towards the base like a crosier, as the young fronds of ferns : **circinal**, a., *sĕr'·sĭn·ăl*, rolled in spirally with the summit in the centre.

circulus articuli vasculosus, *sĕrk'·ŭl·ŭs ărt·ĭk'·ŭl·ī văsk'·ŭl·ōz'·ŭs* (L. *circŭlus*, a circle ; *artĭcŭlus*, a joint, *artĭcŭlī*, of a joint; *văscŭl·ōsus*, full of vessels, as veins and arteries), the vascular circle of a joint ; a narrow vascular border around an articular cartilage.

circulus major, *sĕrk'·ŭl·ŭs mădj'·ŏr* (L. *circŭlus*, a circle ; *mājor*, greater), the greater circle ; a vascular ring in the ciliary muscle of the iris : **circulus minor**, *mĭn'·ŏr* (L. *mĭnor*, less or lesser), a second and lesser circle of anas-

tomosis ending in small veins : **c. tonsillaris**, *tŏn'·sĭl·lār'·ĭs* (L. *tonsillāris*, belonging to the tonsils—from *tonsĭlis*, shorn, cut, or clipped), the tonsillar circle ; a kind of plexus formed by some branches of the glosso-pharyngeal nerve around the tonsil : **c. venosus**, *vĕn·ōz'·ŭs* (L. *vēnōsus*, full of veins—from *vēna*, a vein), an anastomatic venous circle surrounding the base of the nipple.

circumduction, n., *sĕrk'·ŭm·dŭk'·shŭn* (L. *circum*, around ; *ductum*, to lead), a slight circular motion which the head of a long bone describes in its socket, caused by the movement of the extremity of a limb describing a large circle on a plane—said of the movements of the shoulder and hip-joints.

circumferential, a., *sĕrk'·ŭm·fēr·ĕn'·shăl* (L. *circum*, around ; *ferens*, carrying, *ferentis*, of carrying), pert. to the circumference ; n., a marginal fibro-cartilage attached around the lip of the cotyloid cavity as seen in the hip-joint.

circumflexus, n., *sĕrk'·ŭm·flĕks'·ŭs* (L. *circum*, around; *flĕxus*, bent), bent circularly ; circumflex ; applied to certain vessels and nerves from their course : **circumflexus palati**, *păl·āt'·ī* (L. *palātŭs*, the palate, *palāti*, of the palate), a broad, thin, ribbon-like muscle of the palate.

circumscissile, a., *sĕrk'·ŭm·sĭs'·ĭl* (L. *circum*, around ; *scissum*, to cut), cut round in a circular manner, as in seed vessels opening by a lid.

circumscription, n., *sĕrk'·ŭm·skrĭp'·shŭn* (L. *circum*, around ; *scriptus*, written), limitation ; the periphery or margin of a leaf or other organ.

Cirrhipedia or **Cirripedia**, n. plu., *sĭr'·rĭ·pēd'·ĭ·ă* (L. *cirrus*, F. *cirrhe*, a lock, a curl ; *pĕdēs*, feet), a Sub-class of Crustacea, having curled, jointed feet: also, in same sense, **Cirrhopoda** or **Cirropoda**, *sīr·rŏp'·ŏd·ā* (Gr. *pous*, a foot, *podes*, feet): **cirropodous**, a., *sīr·rŏp'·ŏd·ŭs*, having filaments or cirri arranged in pairs on the abdomen, forming a sort of feet or fins.

cirrhose, a., *sīr'·rōz*, also **cirrhous**, *sīr'·ŭs* (F. *cirrhe*, L. *cirrus*, a lock, a curl), having or giving off tendrils : **cirrhus**, n., *sīr'·ŭs*, also **cirrus**, n., *sīr'·ŭs*, a tendril ; a modified leaf in the form of a twining process : **cirrhiform**, a., *sīr'·rī·fŏrm* (L. *forma*, shape), having a tendril-like shape: **cirrhi** or **cirri**, n. plu., *sīr'·ī*, in *bot.*, tendrils ; in *zool.*, tendril-like appendages, such as the feet of barnacles and acorn shells ; the lateral processes on the arms of the Brachiopoda: **cirrif'erous**, a., (L. *fero*, I carry), also **cirrigerous**, a., *sīr·ĭdj'·ĕr·ŭs* (L. *gero*, I carry), carrying cirri.

cirrhosis, n., *sīr·rōz'·ĭs* (Gr. *kirrhos*, tawny-coloured), a pathological condition consisting of an excessive formation of fibrous connective tissue, which conduces to various secondary changes ; a diseased state of the liver, in which it becomes smaller and firmer than usual, known commonly as 'hob-nailed' or 'gin-drinker's liver.'

Cissampelos, n., *sĭs·ăm'·pĕl·ŏs* (Gr. *kissos*, ivy ; *ampelos*, a vine), a beautiful genus of stove climbers, Ord. Menispermaceæ : **Cissampelos ovalifolia**, *ŏv·ăl'·ĭ·fōl'·ĭ·ā* (L. *ovālis*, oval ; *fŏlium*, a leaf, *folia*, leaves); also **C. Mauritiana**, *măw'·rĭsh·ĭ·ān'·ā* (after Prince *Maurice* of Nassau), species which are tonic and diuretic.

Cissus, n., *sĭs'·sŭs* (Gr. *kissos*, ivy), a genus of climbers, Ord. Ampelideæ or Vitaceæ : **Cissus cordata**, *kŏrd·āt'·ā* (L. *cordātus*, heart-shaped), and **C. setosa**, *sĕt·ōs'·ā* (L. *sētōsus*, full of coarse hairs or bristles—from *sēta*, a bristle), species the leaves of which are

said to possess acrid properties : **C. tinctoria**, *tĭnk·tōr'·ĭ·ā* (L. *tinctōrius*, belonging to dyeing), a species whose leaves and fruit abound in a green colouring matter, which on exposure becomes blue, used as a dye for cotton fabrics.

Cistaceæ, n. plu., *sĭst·ā'·sĕ·ē* (Gr. *kistos*, the cistus or rock rose), the Rock Rose family, an Order of shrubs or herbaceous plants : **Cistus**, n., *sĭst'·ŭs*, a genus of plants, many of which yield a resinous balsamic juice : **Cistus Creticus**, *krēt'·ĭk·ŭs* (L. *Crēticus*, of or from Crete, in the Levant), the principal species which produces the resinous matter called 'ladanum' or 'labdanum.'

cistella, n., *sĭst·ĕl'·lā* (L. *cistella*, a small basket—from *cista*, a basket of wicker-work), in *bot.*, a capsular shield of some lichens.

cistolith, n., *sĭst'·ō·lĭth* (L. *cista*, a basket of wicker-work ; Gr. *lithos*, a stone), in *bot.*, an agglomeration of raphides suspended in a sac by a tube, as in Ficus elastica.

cistome, n., *sĭst'·ŏm·ē* (Gr. *kistē*, a small box or chest, or L. *cista*, a basket of wicker-work ; Gr. *stoma*, a mouth), in *bot.*, a funnel-shaped prolongation of the cuticle into the openings of the stomata.

citrate, n., *sĭt'·rāt* (L. *citrus*, a lemon, or the tree), a salt of citric acid, a common form of giving many remedies : **citric acid**, *sĭt'·rĭk*, the substance which gives the pleasant acid flavour to oranges, lemons, and most other fruits: **citron**, n., *sĭt'·rŏn*, the fruit of the citron tree : **citrine**, a., *sĭt'·rĭn*, like a citron; yellow-green.

cladanthi, n. plu., *klăd·ănth'·ī* (Gr. *klados*, a tender branch, a twig ; *anthos*, a flower), in *bot.*, flowers which terminate a lateral branch in mosses.

cladenchyma, n. plu., *klăd·ĕng'·kĭm·ā* (Gr. *klados*, a tender branch; *engchuma*, an infusion), tissue composed of branching cells, as in some hairs.

cladocarpi, n. plu., *klăd'·ō·kârp'·ĭ* (Gr. *klados*, a tender branch; *karpos*, fruit), in *bot.*, mosses which produce sporangia on short lateral branches.

cladocera, n. plu., *klăd·ŏs'·ĕr·ă* (Gr. *klădos*, a branch, a twig; *kĕras*, a horn), an Order of Crustacea having branched antennæ.

cladodium, n., *klăd·ōd'·ĭ·ŭm* (Gr. *klados*, a tender branch), in *bot.*, a plant that has flattened out branches, as in the butcher's broom and some cacti.

Cladonia, n., *klăd·ōn'·ĭ·ă* (Gr. *klados*, a tender branch), a genus of lichens: **Cladonia rangiferina**, *rānj·ĭf'·ĕr·īn'ă* (Lap. and Finn. *raingo*, the reindeer; *ferīnus*, of or belonging to a wild beast), the lichen upon which the reindeer feeds.

cladoptosis, n., *klăd'·ŏp·tōz'·ĭs* (Gr. *klados*, a branch; *ptōsis*, a fall), in *bot.*, the fall of branches, as in Thuja, Taxodium, etc.

Cladosporium, n., *klăd'·ō·spōr'·ĭ·ŭm* (Gr. *klados*, a branch; *spora*, seed), a genus of minute fungi, having the sporules attached to the branches, mostly found on old decaying wood: **Cladosporium herbarum**, *hĕrb·ār'·ŭm* (L. *herba*, grass, an herb, *herbārum*, of herbs), the minute fungi which cause the disease in silkworms called 'gattine,' which is a corruption of ' catkin,' from its appearance.

clathrate, a., *klăth'·rāt* (L. *clāthrī*, a trellis, a lattice), in *bot.*, latticed like a grating.

claustrum, n., *klăwst'·rŭm* (L. *claustrum*, that which shuts off, a lock, a bar), a thin lamelliform deposit of grey matter in the cerebrum.

clavate, a., *klăv'·āt* (L. *clāvātus*, club - shaped — from *clāvus*, a cudgel, a club), club-shaped; becoming gradually thicker towards the top: **claviform**, a., *klăv'·ĭ·*

fŏrm (L. *forma*, shape), same sense as preceding: **clavellose**, a., *klăv'·ĕl·lōz*, having club-like processes.

Claviceps purpurea, *klāv'·ĭ·sĕps pĕr·pūr'·ĕ·ă* (L. *clāvĭceps*, club-headed — from *clāvus*, a club; *căpŭt*, a head; *purpŭrĕus*, purple-coloured), a species of fungi producing the disease called ' ergot,' which attacks rye and other grasses.

clavicle, n., *klăv'·ĭk·l* (L. *clāvicula*, a small key — from *clāvis*, a key), the collar-bone, so called from its supposed resemblance to an ancient key.

clavula, n., *klăv'·ŭl·ă* (L. *clăvula*, a little nail — from *clāvus*, a nail), in *bot.*, the receptacle of certain fungi.

clavus, n., *klāv'·ŭs* (L. *clāvus*, a nail), a corn or callosity: **clavus hystericus**, *hĭs·tĕr'·ĭk·ŭs* (Gr. *husterikos*, L. *hystĕrĭcus*, pert. to the womb, hysterical — from Gr. *hustera*, the womb), an acute pain in the head, having the feeling as if a nail were being driven into the part, occurring in hysterical persons.

claw, n., *klāw* (Dut. *klauwe*, a ball or claw; F. *clou*, a nail), in *bot.*, the narrow end or base of some petals.

Claytonia, n., *klā·tōn'·ĭ·ă* (after *Clayton*, an American botanist), a genus of very pretty plants, Ord. Portulacaceæ: **Claytonia tuberosa**, *tŭb'·ĕr·ōz'·ă* (L. *tuberōsus*, having fleshy knobs — from *tŭber*, a bump, a knob), a species of plants whose roots are eaten in Siberia.

cleido-mastoid, a., *klīd'·ō·măst'·ŏyd* (Gr. *kleis*, a key, or the clavicle, *kleidos*, of a key; Eng. *mastoid*, nipple-like, as on the breast), one of two muscles which are attached inferiorly to the anterior surface of the sternum, and the inner third of the clavicle.

Clematideæ, n. plu., *klĕm'·ăt·ĭd'·ĕ·ē*

(Gr. *klēma*, a vine branch, *klēmătos*, of a vine branch ; L. *clēmătis*, the clematis, *clēmătĭdis*, of the clematis), a Sub-order of plants, Ord. Ranunculaceæ: **Clematis**, n., *klĕm'-ăt-ĭs*, a genus of highly ornamental, and for the most part, climbing plants, so called because most of the species climb like the vine : **Clematis recta**, *rĕkt'-ă* (L. *rectus*, straight, upright) ; **C. flammula**, *flăm'-ŭl-ă* (L. *flammŭla*, a little flame—from *flamma*, a flame), two species, the leaves of which have been used as vesicants.

Cleomeæ, n. plu., *klē-ōm'-ĕ-ē* (Gr. *kleiō* or *klēō*, I close or shut), a Sub-order of plants, Ord. Capparidaceæ : **Cleome**, n., *klē-ōm'-ē*, a genus of very pretty free-flowering plants, so called alluding to the parts of the flower ; some species are very pungent, and are used as substitutes for mustard : **Cleome dodecandra**, *dŏd'-ĕk-ănd'-ră* (Gr. *dōdeka*, twelve ; *anēr*, a man, *andros*, of a man), a species whose root is used as an anthelmintic.

Clerodendron, n., *klēr'-ō-dĕn'-drŏn* (Gr. *klēros*, a share, a lot ; *dendron*, a tree), a beautiful genus of plants, Ord. Verbenaceæ, so named from the uncertain medicinal properties of the species ; the leaves when bruised are employed to kill vermin on cattle in India : **Clerodendron Thomsonæ**, *tŏm'-sŏn-ē*, and its variety **C. Balfourianum**, *băl-fōōr'-ĭ-ăn'-ŭm* (*Thomson, Balfour*), are beautiful climbing plants, from the contrast between their scarlet flowers and white calyx.

clestines, n. plu., *klĕs'-tĭn-ēz* (Gr. *klēstos* or *kleistos*, shut or closed), in *bot.*, cells containing raphides.

Clianthus, n., *klī-ănth'-ŭs* (Gr. *kleos* or *kleios*, glory, renown ; *anthos*, a flower), a genus of plants so called in allusion to the noble appearance of the species, Ord.

Leguminosæ, Sub-ord. Papilionaceæ.

clinandrium, n., *klĭn-ănd'-rĭ-ŭm* (Gr. *klinē*, a bed ; *anēr*, a man, *andros*, of a man), in *bot.*, the part of the column of orchideous plants in which the anther lies : **clinanthium**, n., *klĭn-ănth'-ĭ-ŭm* (Gr. *anthos*, a flower), a common receptacle, assuming a flattened, convex, or concave form, bearing numerous flowers, as in the head of the daisy.

clinical, a., *klĭn'-ĭk-ăl*, sometimes **clinic**, a., *klĭn'-ĭk* (Gr. *klinē*, a bed), pert. to a bed ; applied to the instruction of a teacher to students of medicine at the bedside of the patient, or from notes taken by a teacher at the bedside : **clinoid**, a., *klĭn'-oÿd* (Gr. *eidos*, resemblance), resembling a bed or parts of a bed ; applied to processes of bone of the sphenoid bone bearing a resemblance to the knobs of a bed.

clitoris, n., *klĭt'-ŏr-ĭs* (Gr. *kleitoris*, the clitoris—from *kleio*, I shut), a small elongated body in the female, corresponding in conformation and structure to a diminutive penis : **clitoritis**, n., *klĭt'-ŏr-īt'-ĭs*, inflammation of the clitoris.

cloaca, n., *klō-āk'-ă* (L. *cloāca*, a common sewer), the common cavity into which the intestinal canal and the ducts of the generative and urinary organs open, and from which they discharge their contents, as in some Invertebrates, as among insects, and in many vertebrates, as among domestic fowls.

clonic, a., *klŏn'-ĭk* (Gr. *klŏnos*, tumult), denoting a convulsion with alternate contraction and relaxation.

Clusiaceæ, n. plu., *klōōz'-ĭ-ā'-sĕ-ē* (after *Charles de l'Ecluse*, a botanist, 1609), an Order of beautiful trees and shrubs, yielding resinous juices, known also as Guttiferæ or Guttifers, or the

Gamboge family: **Clusia**, n., *klōōzh'ĭ·ă*, a very ornamental genus of trees, remarkable for the mode in which they send out adventitious roots: **Clusia flava**, *flăv'ă* (L. *flăvus*, golden - yellow), a species whose fruit, called also wild mango or balsam tree, yields a yellow juice like gamboge.

clypeate, a., *klĭp'ĕ·āt* (L. *clypeatus*, furnished with a shield—from *clypeus*, a shield), in *bot.*, having the shape of a shield: **clypeiform**, a., *klĭp'ĕ·ĭ·fŏrm* (L. *forma*, shape), shield-shaped, as the carapace of the king-crab: **clypeus rugulose**, *klĭp'ĕ·ŭs rŭg'ŭl·ōz* (L. *clypeus*, a shield; a dim. of L. *ruga*, a plait or wrinkle), a shield or horny covering full of wrinkles.

clyster, n., *klĭst'ĕr* (Gr. *klustēr*, a clyster—from *kluzō*, I wash), an injection into the bowels by the anus.

cnidæ, n. plu., *nĭd'ē* (Gr. *knidē*, a nettle, because it stings—from *knaō*, I excite itching), the urticating cells, or thread cells, which give many cœlenterate animals power to sting.

coagulum, n., *kō·ăg'ŭl·ŭm*, **coagula**, n. plu., *kō·ăg'ŭl·ă* (L. *coăgŭlum*, curdled milk), clot of blood; the curd of milk; a thickened or fixed mass of a liquid.

coarctate, a., *kō·ărk'tāt* (L. *cŏarctātum*, to press together), in *bot.*, closely pressed; enclosed in a case or covering in such a manner as to give no indication of what is within, as in the transformation of insects: **coarctation**, n., *kō'ărk·ta'shŭn*, the act of straitening or pressing together, as in strictures of the intestine or urethra.

cocci, n. plu., *kŏk'sī*, see 'coccus.'

coccidium, n., *kŏk·sĭd'ĭ·ŭm* (Gr. *kokkos*, a seed, a kernel; *eidos*, resemblance), in Algæ, a round conceptacle without a pore and containing a tuft of spores: **coc-**

codes, n. plu., *kŏk'kŏd·ēz*, round protuberances like peas.

coccoliths, n. plu., *kŏk'kŏ·lĭths* (Gr. *kokkos*, a berry; *lĭthos*, a stone), minute oval or rounded bodies, found either free or attached to the surface of coccospheres, probably of vegetable origin.

Coccoloba, n., *kŏk'kŏ·lŏb'ă* (Gr. *kokkos*, a berry; *lobos*, a lobe), a genus of plants, Ord. Polygonaceæ: **Coccoloba uvifera**, *ūv·ĭf'ĕr·ă* (L. *ūvifĕra*, bearing grapes — from *uva*, a grape; *fero*, I bear), the sea-side grape, so called from the appearance of the fruit, which yields an astringent substance called Jamaica kino.

coccospheres, n. plu., *kŏk'kŏ·sfērs* (Gr. *kokkos*, a berry; *sphaira*, a sphere), spherical masses of sarcode, bearing coccoliths upon their external surface.

cocculum, n., *kŏk'ūl·ŭm* (*coccus*, a L. formative from Gr. *kokkos*, a berry, a seed, a scarlet colour; L. *coccum*, the berry of the scarlet oak), in *bot.*, a seed cell which opens with elasticity: **Cocculus**, n., *kŏk'ūl·ŭs*, a genus of plants, Ord. Menispermaceæ, remarkable for their medicinal virtues, so named because most of the species bear scarlet berries: **Cocculus Indicus**, *ĭn'dĭk·ŭs* (L. *Indicus*, belonging to India), the fruit of the Anamirta cocculus, which is extremely bitter, and the seeds contain a poisonous narcotic principle, called Picrotoxin; the pericarp yields a non-poisonous substance called Menispermin: **coccus**, n., *kŏk'ŭs*, and **coccum**, n., *kŏk'ŭm*, **cocci**, n. plu., *kŏk'sī*, portions of the dry elastic fruit of many of the Euphorbiaceæ, which separate with great force and elasticity in order to project their seeds: **Coccus cacti**, *kŏk'ŭs kăk'tī* (*coccus*, the scarlet-colour; L. *cacti*, of the cactus), the name of the cochineal insect, which feeds upon cactuses;

the female insect when dried constitutes the cochineal of commerce.

coccyx, n., *kŏk'·sĭks* (Gr. *kokkux*, the cuckoo, imitation of its cry, a crest, *kokkūgos*, of the cuckoo; L. *coccyx*, the cuckoo, *coccygis*, of the cuckoo), the terminal portion of the spinal column in man, commonly consisting of four rudimentary vertebræ, so called from its resemblance to a cuckoo's beak or bill : **coccygeal**, a., *kŏk·sĭdj'·ĕ·ăl*, connected with the coccyx : **coccygeus**, a., *kŏk·sĭdj'·ĕ·ŭs*, applied to a muscle consisting of a thin, flat, and triangular sheet of fleshy and tendinous fibres connected with the coccyx.

cochlea, n., *kŏk'·lĕ·ă* (L. *cochlea*, a snail; Gr. *kochlias*, a cockle, a snail with a spiral shell), in *anat.*, the most interior division of the internal ear, consisting externally of a tapering spiral tube: **cochleate**, a., *kŏk'·lĕ·āt*, twisted like a snail shell.

cochlear, n., *kŏk'·lĕ·ăr* (L. *cochlĕar*, a spoon, *cochlĕāris*, of a spoon—from *cochlea*, a snail shell), in *bot.*, a kind of æstivation, in which a helmet-shaped part covers all the others in the bud : **cochleariform**, a., *kŏk'·lĕ·ăr'·ĭ·fŏrm* (L. *forma*, shape), shaped somewhat like a spoon.

Cochlearia, n. plu., *kŏk'·lĕ·ār'·ĭ·ă* (L. *cochlĕar*, a spoon), a genus of plants, the leaves of which are hollowed like the bowl of a spoon, Ord. Cruciferæ : **Cochlearia officinalis**, *ŏf·fĭs'·ĭn·āl'·ĭs* (L. *officina*, a workshop), the common scurvy-grass, used as a stimulant: **Ō. Armoracia**, *ăr'·mōr·ā'·shĭ·ă* (*Armorica*, Brittany, the district of France from which first brought), the horse-radish, which has irritant and also vesicant properties.

Cocoineæ, n. plu., *kŏk'·ō·ĭn'·ĕ·ē* (Prtg. *coco*, an ugly mask to frighten children, so named from the monkey-like face at the base of the nut), the Cocoa-nut tribe, a Sub-order of trees, Ord. Palmæ, which consist of the oil-bearing palms : **Cocos**, n., *kŏk'·ŏs*, a genus of palm trees, including the cocoa-nut tree: **Cocos nucifera**, *nū·sĭf'·ĕr·ă* (L. *nux*, a nut, *nŭcis*, of a nut; *fĕro*, I bear), the coco or cocoa-nut palm, the most useful in the world for its various products: **cocoa**, n., *kŏk'·ō*, the very large nut of the cocos palm ; also the name given to the fruit of the Theobroma cacao, which is of the size of a kidney-bean, and when dried and ground into powder, and variously prepared, is sold under the names cocoa and chocolate.

codeia, n., *kŏd·ī'·ă* or *kŏd·ē'·yă*, also **kodein**, n., *kŏd·ē'·ĭn* (Gr. *kōdeia*, a poppy head), an alkaloid, one of the active medicinal principles of opium.

codonostoma, n. plu., *kŏd'·ŏn·ŏs'·tŏm·ă* (Gr. *kodon*, a bell ; *stoma*, a mouth), the aperture or mouth of the disc of a medusa, or of the bell of a medusiform gonophore.

Cœlenterata, n. plu., *sēl·ĕn·tēr·āt'·ă* (Gr. *koilos*, hollow ; *enteron*, a bowel or gut, *entera*, entrails), in *zool.*, the Sub-kingdom comprising the Hydrozoa and Actinozoa, used instead of the old term Radiata.

cœliac, a., *sēl'·ĭ·ăk* (Gr. *koilia*, belly), pert. to the cavity of the belly : **cœliac passion** (*passion*, suffering), another name for colic: **cœlitis**, *sēl·īt'·ĭs*, abdominal inflammation.

cœlosperm, n., *sēl'·ŏ·spĕrm*, **cœlospermæ**, n. plu., *sēl'·ŏ·spĕrm'·ē* (Gr. *koilos*, hollow ; *koilia*, the belly ; *sperma*, seed), seeds with the albumen curved at the ends.

cœnenchyma, n., *sēn·ĕng'·kĭm·ă* (Gr. *koinos*, common ; *engchuma*, an infusion, tissue), the calcareous tissue which unites together the various corallites of a compound corallum.

cœnœcium, n., *sēn·ē'·shĭ·ŭm* (Gr. *koinos*, common ; *oikos*, a house), in *zool.*, the plant-like structure or dermal system of any polyzoön; another name for ' polyzoary ' or ' polypidom.'

cœnosarc, n., *sēn'·ŏ·särk* (Gr. *koinos*, common ; *sarx*, flesh), the common organized medium by which the separate polypites of compound hydrozoa are connected together.

cœnurus, n., *sēn·ūr'·ŭs*, cœnuri, n. plu., *sēn·ūr'·ī* (Gr. *koinos*, common; *oura*, tail), intestinal worms, consisting of cystose bladders, each of which contains several animals grouped together, and adhering to its sides : cœnurus cerebralis, *sēr'·ĕb·rāl'·ĭs* (L. *cerebrālis*, belonging to the brain—from *cerebrum*, the brain), the brain cœnurus ; the disease sturdy in sheep, caused by cœnuri.

cœrulescent, a., *sēr'·ŭl·ĕs'·ĕnt* (L. *cœrŭlĕus*, dark-blue, sky-coloured), of a blue or sky-blue colour.

Coffea, n., *kŏf·fē'·ă* (Ar. *kawah*, Sp. *cafe*, coffee), the coffee trees, a genus of very ornamental trees, Sub-ord. Coffeæ, *kŏf·fē'·ē*, which furnish important articles of materia medica, Ord. Rubiaceæ : Coffea Arabica, *ăr·ăb'·ĭk·ă* (*Arabica*, from Arabia), the tree and its varieties which furnish the coffee of commerce, said to be a native of Caffa in Arabia : caffein, n., *kăf·fē'·ĭn*, the bitter principle of coffee, identical with Theine, obtained from tea.

Colchiceæ, n. plu., *kŏl·tshĭs'·ē·ē* (said to be after *Colchis*, its original habitat), a Sub-order of the Ord. Melanthaceæ, which have, in general, poisonous properties : Colchicum, n., *kŏl'·tshĭk·ŭm*, a genus of bulbous plants having important medicinal properties, acrid, purgative, emetic, and narcotic : Colchicum autumnale, *aw'·tŭm·nāl'·ĕ* (L. *autumnālis*, autumnal — from *autumnus*, autumn), meadow saffron, or autumn crocus, used in medicine as an extract, vinegar, or tincture : colchicin, n., *kŏl'·tshĭs·ĭn*, an alkaloid obtained from the corn and seeds of the preceding species: Colchicum variegatum, *vār'·ĭ·ĕ·gāt'·ŭm* (L. *variĕgātum*, to make of various sorts or colours), a species used for diseases of the joints by the ancient physicians under the name ' hermodactyle.'

Coleoptera, n. plu., *kŏl'·ĕ·ŏp'·tĕr·ă* (Gr. *koleos*, a sheath ; *pteron*, a wing), the Beetle family, an Order of insects which have horny outer cases or sheaths for the protection of their membranous wings : coleopterous, a., *kŏl'·ĕ·ŏp'·tĕr·ŭs*, having horny sheaths or coverings for their wings, as the Beetle family.

coleorhiza, n., *kŏl'·ĕ·ŏ·rīz'·ă* (Gr. *koleos*, a sheath ; *rhiza*, a root), a sheath which covers the young rootlets of monocotyledonous plants.

colesule, n., *kŏl'·ĕs·ūl* (diminutive of *cōles*, old name for the penis —from Gr. *kōlē*, the penis), in *bot.*, a cellular ring surrounding the pistillidia in Jungermanniæ.

colic, n., *kŏl'·ĭk* (Gr. *kolikos*, L. *colicus*, pert. to the colic—from Gr. *kōlon*, the largest intestine), severe twisting pain in the bowels, especially near the navel: colica pictonum, *kŏl'·ĭk·ă pĭkt'·ŏn·ŭm* (L. *Pictŏnes*, the Pictavians or inhabitants of Poitou, France, where endemic, *Pictŏnum*, of the Pictavians), lead colic : colitis, n., *kŏl·īt'·ĭs*, inflammation of the colon.

colica dextra, *kŏl'·ĭk·ă dĕks'·tră* (L. *colĭcus*, pert. to the colic— from Gr. *kōlon*, the great gut ; L. *dextra*, right), the right colic, an artery which arises about the middle of the mesenteric artery : c. media, *mēd'·ĭ·ă* (L. *mĕdius*, middle), the middle colic, an artery which arises from the upper part of the mesenteric artery.

collateral, a., *kŏl·lăt'·ĕr·ăl* (L. *con*, together ; *lătus*, a side, *lătĕris*, of a side), in *bot.*, placed by the side of another, as in some ovules.

Collemaceæ, n. plu., *kŏl'·lĕ·mā'·sĕ·ē* (Gr. *kollē*, glue), a small Order of flowerless plants, intermediate between the Algæ and Lichens, bearing the thallus of an alga and the fruit of a lichen : **Collema**, n., *kŏl·lēm'·ă*, a genus of Lichens, all the species of which are gelatinous.

collenchyma, n., *kŏl·lĕng'·kĭm·ă* (Gr. *kollē*, glue ; *engchuma*, an infusion, tissue), in *bot.*, the substance lying between and uniting cells.

colleters, n. plu., *kŏl·lēt'·ĕrz* (Gr. *kollētos*, glued or cemented together—from *kollē*, glue), in *bot.*, glandular hairs on the leaves of a bud producing 'blastocolla.'

colletic, a., *kŏl·lĕt'·ĭk* (Gr. *kollēt-ĭkos*, L. *collētĭcus*, sticky, gluey —from Gr. *kollē*, glue), of the nature of glue ; gluey.

collodion, n., *kŏl·lōd'·ĭ·ŏn* (Gr. *kollē*, glue ; *eidos*, resemblance), a solution of gun-cotton in ether and spirit : **colloid**, n., *kŏl'·lŏyd*, in *chem.*, an inorganic compound having a gelatinous appearance ; a substance which cannot diffuse through organic membranes ; opposed to a 'crystalloid,' which does readily diffuse itself so : **colloid cancer**, a soft, jelly-like form of cancer.

Collomia, n., *kŏl·lōm'·ĭ·ă* (Gr. *kollē*, glue, referring to the glutinous seeds), a genus of plants, Ord. Polemoniaceæ.

collum, n., *kŏl'·lŭm* (L. *collum*, a neck), in *bot.*, the part where the plumule and radicle, or root and stem, unite.

collyrium, n., *kŏl·lĭr'·ĭ·ŭm* (Gr. *kollurion*, an eye-salve), a lotion or wash for any part of the body, latterly applied chiefly to a wash for the eyes ; an eye water.

Colocasia, n., *kŏl'·ō·kāz'·ĭ·ă* (Gr. *kolokasia*, the root of the Egyptian bean), a genus of plants, Ord. Araceæ : **Colocasia esculenta**, *ĕsk'·ūl·ĕnt'·ă* (L. *esculentus*, fit for eating—from *esca*, food), a species which has edible corms or bulbs, which are called Eddoes and Cocoes in the W. Indies.

colocynth, n., *kŏl'·ō·sĭnth* (Gr. *kolokunthis*, the wild or purging gourd), the pulp of a kind of gourd, common in many parts of Asia ; the bitter apple, which is a powerful purgative.

colon, n., *kōl'·ŏn* (Gr. *kōlon*, L. *colon*, the great gut), the large intestine, from the cæcum to the rectum.

coloquintida, n., *kŏl'·ō·kwĭnt'·ĭd·ă* (the Latinised form of the Fr. *coloquinte*, colocynth), the bitter globular fruit, the pulp of which constitutes the medicinal colocynth ; see 'colocynth' ; the Cucumis colocynthis, Ord. Cucurbitaceæ.

colostrum, n., *kŏl·ŏst'·rŭm* (L. *colostrum*, the first milk of animals after delivery), the milk first secreted in the breasts after childbirth.

colotomy, n., *kŏl·ŏt'·ŏm·ĭ* (Gr. *kōlon*, the colon or great gut ; *tomē*, a cutting), an operation for opening the bowel in the left loin, to remove an obstruction in the lower part of the intestine.

colpenchyma, n., *kŏl·pĕng'·kĭm·ă* (Gr. *kolpos*, the fold of a garment; *engchuma*, an infusion, tissue), in *bot.*, tissue composed of wavy or sinuous cells.

Colubrina, n. plu., *kŏl'·ū·brīn'·ă* (L. *coluber*, a snake), in *zool.*, a division of the Ophidia : **colubrine**, a., *kŏl'·ū·brĭn*, pert. to serpents ; having the appearance of a serpent.

columba, n., *kŏl·ŭm'·bă*, or **calumba**, n., *kăl·ŭm'·bă* (from *Col-*

omba, in Ceylon), the root of the plant Cocculus palmatus, or Menispermum palmatum, an excellent tonic.

Columbacei, n. plu., *kŏl'·ŭm·bā'·sĕ·ī* (L. *columba*, a dove), the division of rasorial birds which include doves and pigeons.

Columbine, n., *kŏl'·ŭm·bīn* (L. *columbīnus*, dove-like—from *columba*, a dove ; may be only *column*, and *bine*—from AS. *bindan*, Icel. *binda*, to bind, as in woodbine), the common climbing plant Aquilegia vulgaris, Ord. Ranunculaceæ.

columella, n., *kŏl'·ŭm·ĕl'·lă* (L. *columella*, a small column or pillar—from *columna*, a column), the central column, as in the sporangia of mosses ; an axis which has carpels arranged around it ; the central axis round which the whorls of a spiral univalve are wound ; the central pillar found in the thecæ of many corals : **column**, n., *kŏl'·ŭm*, the solid body formed by the union of the styles and filaments in some plants ; the cylindrical body of a sea anemone.

columella cochleæ, *kŏl'·ŭm·ĕl'·lă kŏk'·lē·ē* (L. *columella*, a small column ; *cochlĕa*, a spiral shell, *cochlĕæ*, of a spiral shell), the central pillar round which turns the spiral tube of the ear.

Columelliaceæ, n. plu., *kŏl'·ŭm·ĕl'·lĭ·ā'·sĕ·ē* (after *Columella*, a Spaniard), a small Order of evergreen shrubs and trees: **Columellia**, n., *kŏl'·ŭm·ĕl'·lĭ·ă*, a genus of evergreens.

columnæ carneæ, *kŏl·ŭm'·nē kâr'·nĕ·ē* (L. *columnæ*, columns ; *carneæ*, fleshy—from *caro*, flesh, *carnis*, of flesh), the fleshy columns or pillars ; the irregular rounded muscular bands on a great part of the inner surfaces of the ventricles of the heart : **columnæ recti**, *rĕkt'·ī* (L. *rectum*, the rectum, *recti*, of the rectum—

from *rectus*, straight), the larger folds of the rectum.

Colutea, n., *kŏl·ūt'·ĕ·ă* (Gr. *koloutea*, a kind of tree which dies if mutilated ; L. *colūtĕa*, a pod-like kind of fruit), a genus of plants, Ord. Leguminosæ, Sub-ord. Papilionaceæ : **Colutea arborescens**, *âr'·bŏr·ĕs'·ĕns* (L. *arborescens*, growing into a tree—from *arbor*, a tree), the bastard or bladder senna, whose leaves are used abroad to adulterate the true senna.

coma, n., *kōm'·ă* (Gr. *kōma*, a deep sleep), a kind of stupor, deep sleep, or insensibility ; a deep lethargic sleep from which the person cannot be awakened : **comatose**, a., *kŏm'·ăt·ōz*, excessively drowsy ; lethargic.

coma, n., *kōm'·ă* (L. *coma*, Gr. *komē*, the hair of the head), tufts of hairs terminating certain seeds ; bracts or tufts as at the summit of a pine - apple : **comose**, a., *kōm·ōz'*, furnished with hairs, as the seeds of the willow.

Combretaceæ, n. plu., *kŏm'·brĕt·ā'·sĕ·ē* (*combretum*, said to have been a Latin name for a climbing plant), an Order of climbing plants: **Combreteæ**, n. plu., *kŏm·brēt'·ĕ·ē*, a Sub-order: **Combretum**, n., *kŏm·brēt'·ŭm*, a genus.

comes nervi ischiadici, *kōm'·ēz nĕrv'·ī ĭs'·kĭ·ăd'·ĭs·ī* (L. *comes*, a companion ; *nervus*, a nerve, *nervi*, of a nerve ; *ischiadicus*, that has hip-gout—from *ischias*, hip-gout), the companion of the ischiadic nerve ; a branch of the sciatic artery : **comes nervi phrenici**, *frĕn'·ĭs·ī* (L. *phrenĭcus*, belonging to the diaphragm—from Gr. *phren*, the diaphragm), the companion of the phrenic nerve ; a very slender but long branch of the phrenic artery which accompanies the phrenic nerve.

Commelynaceæ, n. plu., *kŏm·mĕl'·ĭn·ā'·sĕ·ē* (after *Commelin*, a

Dutch botanist), the Spider-wort family, an Order of plants, some of which have fleshy rhizomes which are used for food : **Commelyna**, n., *kŏm'·mĕl·ĭn'·ă*, a genus of very handsome plants.

commissure, n., *kŏm·mĭsh'·ōōr* (L. *commissura*, a knot, a joint— from *con*, together; *missus*, sent), the place where two parts meet and unite ; the point of union between two parts that meet closely : **commissural**, a., *kŏm·mĭsh'·ōōr·ăl*, connecting together; applied to nerve-fibres which unite different ganglia.

complanate, a., *kŏm'·plăn·āt* (L. *complāno*, I make level—from *con*, together ; *plānus*, level), in *bot.*, flattened.

complicate, a., *kŏm'·plĭk·āt* (L. *complicātŭm*, to fold together—from *con*, together; *plico*, I fold), in *bot.*, folded up upon itself.

Compositæ, n. plu., *kŏm·pŏz'·ĭt·ē* (L. *compositus*, put together, compounded ; *compōno*, I compound—from *con*, together; *pŏno*, I put or place), the Composite family, one of the largest and most important Orders in the vegetable kingdom, and distributed over all quarters of the world.

compress, n., *kŏm'·prĕs* (L. *compressus*, pressed together—from *con*, together ; *pressus*, pressed, kept under), folds of soft linen cloth, used to cover the dressings of wounds, etc. : **compressed**, a., *kŏm·prĕst'*, in *bot.*, flattened laterally : **compression**, n., *kŏm·prĕsh'·ŭn*, in *anat.*, pressure upon the brain caused by some severe injury : **compressor**, n., *kŏm·prĕs'·ŏr*, a muscle which compresses the parts on which it acts: **compressorium**, n., *kŏm'·prĕs·ŏr'·ĭ·ŭm*, that which compresses or fixes ; a compressor.

Conanthereæ, n. plu., *kŏn'·ănth·ĕr'·ĕ·ē* (Gr. *kōnos*, a cone ; Eng.

anther—from Gr. *anthĕros*, L. *anthēra*, flowery), a Sub-order of plants, Ord. Liliaceæ, so called because their anthers are united into a cone ; the stemless herbs of Peru and Chili : **Conanthera**, n., *kŏn'·ănth·ĕr'·ă*, a genus.

conarium, n., *kŏn·ār'·ĭ·ŭm* (Gr. *kōnos*, L. *cōnus*, the fruit of the fir), in *anat.*, a small reddish body, about the size of a small cherry-stone, in the cerebrum, called also the 'pineal body' or 'gland.'

concatenate, a., *kŏn·kăt'·ĕn·āt* (L. *con*, together ; *catēnātus*, chained —from *catēna*, a chain), chained together.

concentric, a., *kŏn·sĕnt'·rĭk* (L. *con*, together ; *centrum*, the middle point), in *bot.*, having a common centre.

conceptacle, n., *kŏn·sĕpt'·ă·kl* (L. *conceptaculum*, a receptacle), in *bot.*, a hollow sac containing a tuft or cluster of spores ; the thecæ of ferns.

concha, n., *kŏngk'·ă* (Gr. *kongchē*, L. *concha*, a shell), the external ear, by which sounds are collected and transmitted through the modiolus to the internal ear.

Conchifera, n. plu., *kŏngk·ĭf'·ĕr·ă*, also **Conchifers**, n. plu., *kŏngk'·ĭf·ĕrs* (L. *concha*, a shell ; *fero*, I bear or carry), an extensive class of bivalve shell-fish, including the oyster, mussel, cockle, and scallop ; a synonym for 'lamelli-branchiate :' **conchiferous**, a., *kŏngk·ĭf'·ĕr·ŭs*, producing or having shells : **conchiform**, a., *kŏngk'·ĭ·fŏrm* (L. *forma*, shape), having the shape of a shell.

concolorate, a., *kŏn·kŏl'·ŏr·āt*, also **concolorous**, a., *kŏn·kŏl'·ŏr·ŭs* (L. *con*, together ; *color*, colour), similar in colour.

concrete, a., *kŏng'·krēt* (L. *con*, together ; *cretum*, to grow), united in growth ; growing together : **concretion**, n., *kŏn·krēsh'·ŭn*, a mass formed by the union of various parts adhering to each other.

concussion, n., *kŏn·kŭsh'·ŭn* (L. *concussio*, a shaking—from *con*, together; *quassum*, to shake), in *med.*, a severe shattering or injury of some internal organ in consequence of a fall, or heavy blow.

conduplicate, a., *kŏn·dū'·plĭk·āt* (L. *con*, together; *duplico*, I double—from *duo*, two; *plico*, I fold), doubled; folded upon itself.

condyle, n., *kŏn'·dĭl* (Gr. *kondulos*, a knuckle, a knob), in *anat.*, a protuberance having a flattened articular surface; **condyles,** n. plu., the articular surfaces by which the skull articulates with the vertebral column: **condyloid,** a., *kŏn'·dĭl·ŏyd* (Gr. *eidos*, resemblance), resembling a condyle.

cone, n., *kōn* (Gr. *kōnos*, L. *cōnus*, a cone), the scaly fruit of the fir, pine, etc.

conenchyma, n., *kŏn·ĕng'·kĭm·ă* (Gr. *kōnos*, the cone of the pine; *engchuma*, an infusion, tissue), tissue composed of conical cells, as in the form of hairs.

Conferva, n., *kŏn·fĕrv'·ă*, **Confervæ,** n. plu., *kŏn·fĕrv'·ē* (L. *conferveo*, I grow together), a very extensive and interesting genus of Algæ, having branched cellular expansions, and nutritive and reproductive cells, often distinct and separate—so called on account of their coherence in a branched linear or lateral expansion: **Confervaceæ,** n. plu., *kŏn'·fĕrv·ā'·sĕ·ē*, a Sub-order of flowerless water-plants of the simplest structure, of various colours—green, olive, violet, and red: **Conferva crispa,** *krĭsp'·ă* (L. *crĭspus*, crisp, curled), the water-plant called Water - flannel, forming beds of entangled filaments which enclose pentagonal and hexagonal spaces: **confervoid,** a., *kŏn·fĕrv'·ŏyd* (Gr. *eidos*, resemblance), formed of single rows of cells, as in the Confervæ; having thread-like articulations.

confluent, a., *kŏn'·flŏō·ĕnt* (L. *con*, together; *fluens*, flowing), in *bot.*, gradually uniting in the progress of growth.

congenital, a., *kŏn·jĕn'·ĭt·ăl* (L. *congenĭtus*, born together—from *con*, together; *genitus*, brought forth, produced), existing from birth, as a disease or some deformity.

congested, a., *kŏn·jĕst'·ĕd* (L. *congestus*, pressed together — from *con*, together; *gestum*, to carry), in *bot.*, heaped together; in *med.*, having an unnatural accumulation of blood: **congestion,** n., *kŏn·jĕst'·yŭn*, an unnatural collection of blood in any part or organ of a body.

conglobate, a., *kŏn'·glŏb·āt* (L. *conglobātum*, to gather into a ball—from *con*, together; *globus*, a ball), in the shape of a ball or sphere.

conglomerate, a., *kŏn·glŏm'·ĕr·āt* (L. *conglomeratum*, to roll together — from *con*, together; *glomĕro*, I wind into a ball or heap), in *bot.*, clustered together; applied to a gland composed of various glands or lobules with a common excretory duct; denoting a stony mass composed of sandstone and various pebbles.

conglutinate, a., *kŏn·glŏōt'·ĭn·āt* (L. *conglutinātus*, glued or cemented), glued together in heaps; united together as by a tenacious substance.

conia, n. plu., *kŏn'·ĭ·ă* (Gr. *kōneion*, L. *conium*, hemlock), the active principle of hemlock, consisting of a volatile oleaginous alkali, which acts as an energetic poison: **Conium,** n., *kŏn'·ĭ·ŭm*, a genus of plants, Ord. Umbelliferæ: **Conium maculatum,** *măk'·ŭl·āt'·ŭm* (L. *maculātum*, to spot, to stain—from *macŭla*, a stain), the plant hemlock, probably the 'koneion' of the Greeks, and 'cicuta' of the Romans.

conidia, n., *kŏn·ĭd'·ĭ·ă* (Gr. *konis*, a nit, the egg of a louse, flea, or bug, *konidos*, of a nit), in *bot.*, the peculiar spores in fungi which resemble buds : **conidiiferous**, a., *kŏn·ĭd'·ĭ·ĭf'·ĕr·ŭs* (L. *fero*, I bear or carry), producing or bearing conidia.

Coniferæ, n. plu., *kōn·ĭf'·ĕr·ē*, also **Conifers**, n. plu., *kŏn'·ĭf·ĕrs* (L. *cōnus*, a cone ; *fĕro*, I bear), the cone-bearing family, a very extensive Order of trees having four Sub-orders : **coniferous**, a., *kŏn·ĭf'·ĕr·ŭs* (L. *fĕro*, I bear), producing or bearing cones.

coniocyst, n., *kŏn'·ĭ·ō·sĭst* (Gr. *konis*, dust, a nit ; *kustis*, a bladder), in *bot.*, spore cases resembling tubercles.

Coniomycetes, n., *kŏn'·ĭ·ō·mī·sēt'·ēz* (Gr. *kŏnis*, a nit, dust ; *mukēs*, a fungus, a mushroom, *mukētos*, of a fungus), a Sub-order of Fungi, in which the flocci of the fruit are obsolete or mere peduncles.

coniothalameæ, n. plu., *kŏn'·ĭ·ŏ·thăl·ăm'·ē·ē* (Gr. *kŏnis*, a nit, dust ; *thalămos*, a bedchamber, a nest), a Sub-order of lichens ; pulverulent lichens.

Conirosters, n. plu., *kŏn'·ĭ·rŏst'·ĕrs* (L. *cōnus*, a cone ; *rostrum*, a beak), in *zool.*, the division of perching birds with conical beaks.

conium, *kōn'·ĭ·ŭm*, see ' conia.'

conjugate, a., *kŏn'·jŏŏg·āt* (L. *conjugatum*, to unite—from *con*, together ; *jugum*, a yoke or bond), paired ; joined by pairs : **conjugation**, n., *kŏn'·jŏŏg·ā'·shŭn*, the union of two cells in such a way as to develop a spore : **conjugate spirals**, in *bot.*, whorled leaves so arranged as to give two or more generating spirals running parallel to each other.

conjunctiva, n., *kŏn'·jŭnkt·īv'·ă* (L. *conjunctīvus*, fastening together—from *con*, together ; *jungo*,

I join), the fine sensitive membrane which covers the front of the eyeball, and lines the eyelids.

Connaraceæ, n. plu., *kŏn'·năr·ā'·sĕ·ē* (Gr. *konnăros*, a species of tree), the Connarus family, a small Order of tropical trees and shrubs, some bearing handsome flowers, and others edible fruits : **Connarus**, n., *kŏn'·năr·ŭs*, a genus of trees and shrubs.

connate, a., *kŏn·nāt'* (L. *con*, together ; *nātus*, born), in *bot.*, having two leaves with their bases united ; having parts united in any stage of development, which are normally distinct.

connective, n., *kŏn·nĕkt'·ĭv* (L. *con*, together ; *necto*, I tie), in *bot.*, the fleshy part which connects the lobes of an anther.

connivent, a., *kŏn·nĭv'·ĕnt* (L. *connivens*, winking or blinking), in *bot.*, having two organs arching over so as to meet above, as petals ; converging.

conoid, a., *kŏn·ōyd'*, also **conoidal**, a., *kŏn·ōyd'·ăl* (Gr. *kōnos*, a cone ; *eidos*, resemblance), shaped like a cone.

constipation, n., *kŏn'·stĭp·ā'·shŭn* (L. *constipātum*, to press closely together — from *con*, together ; *stipo*, I press together), sluggish action of the bowels ; difficult expulsion of the hardened fæces.

constricted, a., *kŏn·strĭkt'·ĕd* (L. *constrictus*, drawn or bound together), in *bot.*, tightened or contracted in width, as if tied with a cord : **constrictor**, n., *kŏn·strĭkt'·ŏr*, in *anat.*, a muscle which draws together or contracts an opening of the body, as the pharynx.

consumption, n., *kŏn·sŭm'·shŭn* (L. *con*, together ; *sumptum*, to take) a gradual and general wasting of the body from diseased lungs, or other cause ; phthisis.

contagion, n., *kŏn·tādj'·ŭn* (L. *contagio*, contact, touch), the communication of disease by contact or touch.

continuous, a., *kŏn·tĭn′·ŭ·ŭs* (L. *continŭus*, uninterrupted), in *bot.*, without joints or articulations.

contorted, a., *kŏn·tŏrt′·ĕd* (L. *con*, together; *tortus*, twisted), in *bot.*, twisted regularly in one direction —applied to a form of æstivation: contortive, a., *kŏn·tŏrt′·ĭv*, applied to the parts of a single whorl placed in a circle, each exhibiting a torsion of its axis.

contortuplicate, a., *kŏn′·tŏr·tŭp′·lĭk·āt* (L. *contortus*, twisted; *plĭcātum*, to fold), in *bot.*, twisted and folded in plaits.

Contrayerva or Contrajerva, n., *kŏn′·tră·yĕrv′·ă* (Indian - Spanish, *contrayerva*—from *contra*, against; *yerva*, poison, as supposed to be good against poison), the plant whose root yields the Contrayerva root of commerce, Ord. Moraceæ.

contusion, n., *kŏn·tū′·zhŭn* (L. *contūsum*, a bruise—from *con*, together; *tūsus*, beaten), a form of injury caused by heavy pressure or a sharp blow without any external wound; a bruise: contused wound, when, in addition to the injury of the soft parts, the skin is broken.

conus arteriosus, *kōn′·ŭs ărt·ēr′·ĭ·ōz′·ŭs* (L. *conus*, a cone; *artēria*, an artery), the arterial cone; a smooth, conical prolongation of the left ventricle upwards, from which the pulmonary artery arises.

convergent, a., *kŏn·vĕrj′·ĕnt* (L. *con*, together; *vergens*, bending, turning), in *bot.*, applied to ribs of leaves running from base to apex in a curved manner.

convolute, a., *kŏn′·vŏl·ōt*, also convolutive, a., *kŏn′·vŏl·ōt′·ĭv* (L. *con*, together; *volūtus*, rolled), in *bot.*, having a leaf in a bud rolled upon itself; rolled up laterally so as partially to embrace each other.

Convolvulaceæ, n. plu., *kŏn·vŏlv′·ūl·ā′·sĕ·ē* (L. *convolvŭlus*, the plant bindweed—from *con*, together; *volvo*, I roll), the Convolvulus or Bindweed family, an extensive Order of twining shrubs and herbaceous plants, having generally an acrid juice in the roots, which is purgative: Convolvulus, n., *kŏn·vŏlv′·ūl·ŭs*, a genus of plants so called from their twining or winding habit: Convolvulus scammonia, *skăm·mōn′·ĭ·ă* (Gr. *skammōnia*, L. *scammōnia*, the plant scammony), a species the root of which yields a gummy resinous exudation: C. sepium, *sĕp′·ĭ·ŭm* (L. *sēpium*, the internal shell of the cuttle-fish), a species which yields a spurious kind of scammony: C. batatas, *băt·āt′·ăs* (a Spanish or Mexican word), a species which yields the sweet potato or yam, used as food in tropical countries: C. scoparius, *skōp·ār′·ĭ·ŭs* (L. *scopārius*, of or belonging to a broom; *scopæ*, a bunch of twigs, a broom), yields the oil called Rhodium.

convulsions, n. plu., *kŏn·vŭl′·shŭns* (L. *convulsĭō*, a convulsion— from *con*, together; *vulsum*, to pluck or tear away), violent and involuntary contractions of certain muscles of the body, as in fits.

copaiba, n., *kō·pāb′·ă*, or copaiva, n., *kō·pāv′·ă* (Sp. and Portg.), an oleo-resin or turpentine, obtained from various parts of S. America; a balsam.

Copaifera, n., *kō·pāf′·ėr·ă* (*copaiba*, and L. *fero*, I bear), a genus of plants, Ord. Leguminosæ, Subord. Cæsalpinieæ: Copaifera Jacquinii, *jăk·wĭn′·ĭ·ī* (after *Jacquinia*, a botanist of Vienna), the copaiba of Jacquinia: C. Langsdorfii, *lăngs·dŏrf′·ĭ·ī* (of *Langsdorff*); C. bijuga, *bĭdj′·ūg·ă* (L. *bĭjŭgus*, yoked two together— from *bis*, twice; *jŭgum*, a yoke); C. multijugus, *mŭlt·ĭdj′·ūg·ŭs* (L. *multus*, many; *jŭgum*, a yoke); C. Martii, *mărt′·shĭ·ī* (L. of the month of *March*); C. Guianensis, *gwī′·ăn·ĕns′·ĭs* (of or from *Guiana*); C. coriacea, *kŏr′·ĕ·ā′·sĕ·ă* (L. *cori-*

G

āceus, leathery — from *corium*, skin, hide), are species which yield the balsam of copaiba.

Copepoda, n., *kŏp·ĕp'·ŏd·ă* (Gr. *kōpe*, an oar ; *podes*, feet), oar-footed animals, an Order of Crustacea.

coprolite, n., *kŏp'·rŏ·līt* (Gr. *kopros*, dung ; *lithos*, a stone), the petrified dung of animals, chiefly of saurians.

Coptis, n., *kŏp'·tĭs* (Gr. *kopto*, I cut), a genus of plants, so called in allusion to the division of the leaves, Ord. Ranunculaceæ : **Coptis teetæ,** *tē'·tē* (a native name), a pretty species, the rhizome being used in India as a bitter tonic.

coracoid, a., *kŏr'·ăk·ōyd* (Gr. *koraks*, a crow, *korakos*, of a crow, crow's; *eidos*, shape, likeness), applied to a process of bone of the shoulder-blade, so named from its resemblance to a crow's beak ; one of the bones of the pectoral arch in birds and reptiles : **coraco brachialis,** *kŏr'·ăk·ō brăk'·ĭ·āl'·ĭs* (Gr. *koraks*, a crow ; L. *brachĭālis*, pert. to an arm—from *brāchium*, an arm), the muscle of the arm connected with the coracoid process ; a muscle connected with the coracoid process and inserted into the humerus or arm bone.

coralliform, a., *kŏr·ăl'·lĭ·fŏrm* (Gr. *korallion*, L. *corālium*, red coral ; L. *forma*, shape), also **coralloid,** a., *kŏr'·ăl·lōyd* (Gr. *eidos*, resemblance), resembling coral : **corallum,** n., *kŏr·ăl'·lŭm*, the hard structure deposited in the structures of the Actinozoa, called coral : **corallite,** n., *kŏr'·ăl·līt*, the portion of a corallum secreted by a single polype: **coralliginous,** a., *kŏr'·ăl·lĭdj'·ĕn·ŭs*, producing a corallum: **coralline,** a., *kŏr'·ăl·lĭn* like or containing coral: **Corallina,** n., *kŏr'·ăl·lĭn'·ă*, a genus of plants resembling a coral, Ord. Algæ : **Corallina officinalis,** *ŏf·fĭs'·ĭn·āl'·ĭs* (L. *officinālis*, officinal), one of numerous species considered

vegetable, which are reckoned animal by many.

Corchorus, n., *kŏrk'·ŏr·ŭs* (Gr. *korē*, the pupil of the eye ; *koreō*, I cleanse, I purge), a genus of plants, Ord. Tiliaceæ : **Corchorus capsularis,** *kăps'·ūl·ār'·ĭs* (L. *capsulāris*, pert. to a capsule—from *capsŭla*, a little chest), a species which in India furnishes the jute used in making coarse carpets and gunny bags : **C. olitorius,** *ŏl'·ĭt·ōr'·ĭ·ŭs* (L. *olitōrius*, belonging to vegetables—from *olĭtor*, a market gardener), Jew's mallow, the leaves of which are used as a culinary vegetable: **C. pyriformis,** *pĭr'·ĭ·fŏrm'·ĭs* (L. *pirum*, a pear ; *forma*, shape), a species in Japan which furnishes fibres.

corculum, n., *kŏrk'·ŭl·ŭm* (L. *corcŭlum*, a little heart—from *cor*, the heart), in *bot.*, a name for the embryo.

cord, n., *kŏrd* (L. *chorda*, Gr. *chordē*, a string, a gut), the string or process which attaches the seed or embryo to the placenta.

cordate, a., *kŏrd'·āt* (L. *cor*, the heart, *cordis*, of the heart), heart-shaped ; having the broad, heart-shaped part next the stalk or stem : **cordiform,** a., *kŏrd'·ĭ·fŏrm* (L. *forma*, shape), a solid body having the shape of a heart : **cordate-hastate,** a., *-hăst'·āt* (L. *hasta*, a spear), of a shape between a heart and a spear: **cordate-sagittate,** a., *-sădj·ĭt'·āt* (L. *sagitta*, an arrow), of a shape between a heart and an arrow-head : **cordate-ovate,** a., *-ōv'·āt* (L. *ovātus*, shaped like an egg), of a shape between a heart and an egg.

Cordiaceæ, n. plu., *kŏrd'·ĭ·ā'·sē·ē* (after *Cordus*, a German botanist), the Cordia family, an Order of trees, some of which yield edible fruits : **Cordia,** n., *kŏrd'·ĭ·ă*, a genus of plants : **Cordia myxa,** *mĭks'·ă* (Gr. *muxa*, mucus, mucilage); also **C. latifolia,** *lāt'·ĭ·fōl'·ĭ·ă*

(L. *lātus*, broad ; *folium*, a leaf), are species whose succulent, mucilaginous fruits are known by the name of Sebesten plums.

Cordyline, n., *kŏrd'·ĭl·ĭn'·ē* (Gr. *kordulē*, a club, a bump), a genus of ornamental shrubs, Ord. Liliaceæ : **Cordyline Australis**, *aws·trāl'·ĭs* (from *Australasia*), the plant Ti of New Zealand ; also **C. Banksii**, *bănks'·ĭ·ī* (of *Banks*— after *Sir Joseph Banks*), are species which yield fibres.

coriaceous, a., *kōr'·ĭ·ā'·shŭs* (L. *corium*, skin, hide), consisting of or resembling leather ; tough ; leathery.

Coriandrum sativum, *kŏr'·ĭ·ănd'·rŭm săt·ĭv'·ŭm* (L. *coriandrum,* Gr. *koriannon*, the coriander— from Gr. *koris*, a bug, alluding to the smell of the seed ; L. *sativus*, fit to be planted), a plant yielding seeds which are a warm and agreeable aromatic, Ord. Umbelliferæ.

Coriariaceæ, n. plu., *kŏr'·ĭ·ăr·ĭ·ā'·sĕ·ē* (L. *corium*, skin, hide), the Coriaria family, an Order of plants : **Coriaria**, n., *kŏr'·ĭ·ār'·ĭ·ă*, a genus of plants : **Coriaria myrtifolia**, *mĕrt'·ĭ·fōl'·ĭ·ă* (Gr. *murtos*, the myrtle tree ; L. *folium*, a leaf), a species whose leaves have been employed on the Continent to adulterate senna; used for dyeing black and tanning, and with sulphate of iron makes a dark blue: **C. ruscifolia**, *rŭs'·ĭ·fōl'·ĭ·ă* (L. *ruscus*, a probable adaptation of *Russo-colore*, from its colour), the Toot or Tutu plant of New Zealand—the seeds and young shoots are poisonous.

corium, n., *kōr'·ĭ·ŭm* (L. *corium*, skin), the cutis vera or true skin, consisting of a fibro-vascular layer ; called also the 'derma,' and is covered by the epidermis or scarf skin.

corm, n., *kŏrm* (Gr. *kormos*, a trunk, a log), the thickened or bulb-like solid base of the stems of plants, such as in the Colchicum and Arum : **cormogenous**, a., *kŏr·mŏdj'·ĕn·ŭs* (Gr. *gennaō*, I produce), having a corm or stem: **cormus**, n., *kŏrm'·ŭs*, same as ' corm.'

Cornaceæ, n. plu., *kŏrn·ā'·sĕ·ē* (L. *cornĕus*, belonging to cornelwood—from *cornu*, a horn, as the wood is thought to be hard and durable as horn ; *cornus*, the cornel cherry tree), the Cornel family, an Order of trees, shrubs, and herbs : **Cornus**, n., *kŏrn'·ŭs*, an ornamental genus of plants ; the cornel tree : **Cornus florida**, *flŏr'·ĭd·ă* (L. *florĭdus*, flowery) ; and **C. sericea**, *sĕr·ĭs'·ĕ·ă* (L. *serĭcĕus*, silky—from *sĕrĭca*, silks), species used in America as tonics and febrifuges: **C. mascula**, *măsk'·ūl·ă* (L. *mascŭlus*, male), a species whose fruit is used for food; the red-wood of Turkey, from which the Turks obtain the dye for their red fezes: **C. sanguinea**, *săng·gwĭn'·ĕ·ă* (L. *sanguinĕus*, of blood, bloody—from *sanguis*, blood), a species whose seeds furnish oil : **C. Suecica**, *sū·ĕs'·ĭk·ă* or *swĕs'·ĭk·ă* (L. *Suecia*, Sweden), a Scotch species whose fruit is said to be tonic: **cornel**, n., *kŏrn'·ĕl*, a tree yielding small edible cherries; the dog - wood tree.

cornea, n., *kŏrn'·ĕ·ă* (L. *corneus*, horny—from *cornu,* a horn), a horny transparent membrane forming the front part of the eyeball— also called the **cornea pellucida**, *pĕl·lōs'·ĭd·ă* (L. *pellucĭdus,* transparent): **c. opaca**, *ō·pāk'·ă* (L. *opācus,* shady, dark), the hinder part of the eyeball, which is opaque and densely fibrous ; also called the 'sclerotic coat.'

corneous, a., *kŏrn'·ĕ·ŭs* (L. *cornĕus*, horny—from *cornu*, a horn), in *bot.*, having the consistence of horn ; horny : **corniculate**, a., *kŏrn·ĭk'·ūl·āt* (L. *cornĭcŭla*, a little horn), having a horn-like appendage.

cornicula laryngis, *kŏrn·ĭk'·ŭl·ă lăr·ĭnj'·ĭs* (L. *corniculum*, a little horn—from *cornu*, a horn ; Gr. *larunx*, the upper part of the windpipe, *larunggos*, of the windpipe), the little horns of the larynx ; two small cartilaginous nodules of a somewhat conical shape at the summits of the arytenoid cartilages.

cornu, n., *kŏrn'·ū* (L. *cornu*, a horn), a horn : **cornua**, n. plu., *kŏrn'·ū·ă*, horns : **cornus**, see under 'cornaceæ :' **cornute**, a., *kŏrn'·ūt*, horn - shaped : **cornu Ammonis**, *ăm·mōn'·ĭs* (of *Ammon*), the horn of Ammon, a long white eminence on the brain, indented or notched so as to present some resemblance to the paw of an animal ; called also 'pes hippoeampi.'

corolla, n., *kŏr·ŏl'·lă* (L. *corolla*, a small wreath or crown), in *bot.*, the second whorl of leaves in a flower, commonly the most brilliantly coloured, the separate pieces of which are called 'petals': **corollifloral**, a., *kŏr·ŏl'·lĭ·flŏr'·ăl* (L. *flos*, a flower, *floris*, of a flower), applied to those plants that have the united petals placed under the ovary, and the stamens either borne by the petals, or inserted independently into the torus : **corollifloræ**, n. plu., *kŏr·ŏl'·lĭ·flŏr'·ē*, all plants that have the calyx and corolla present, the corolla gamopetalous, hypogynous, usually bearing the stamens: **corolline**, a., *kŏr'·ŏl·lĭn*, of or belonging to a corolla.

corona, n., *kŏr·ōn'·ă* (L. *corōna*, Gr. *korōnē*, a garland, a wreath), in *bot.*, a corolline appendage between the corolla and stamens; cup-like or in rays, as the crown of the Daffodil ; in *anat.*, the upper surface of the molar teeth.

corona glandis, *kŏr·ōn'·ă glănd'·ĭs*

(L. *corōna*, a crown ; *glans*, an acorn, a gland, *glandis*, of an acorn), the crown of the gland ; the elevated margin or bean of the glan penis.

coronal, a., *kŏr·ōn'·ăl* (L. *corōna*, a crown, a wreath), in *anat.*, pert. to the crown of the head : **coronary**, a., *kŏr'·ŏn·ēr·ĭ*, encircling like a crown : **coronate**, a., *kŏr'·ŏn·āt*, having little crown - like eminences : **coronet**, n., *kŏr'·ŏn·ĕt*, a little or inferior crown : **corona radiata**, *răd'·ĭ·āt'·ă* (L. *radiātus*, furnished with rays—from *rădius*, a staff, a ray), an assemblage of radiating fibres in each hemisphere of the cerebrum that may be compared to a fan: **coronula**, n. plu., *kŏr·ŏn'·ŭl·ă*, little crowns ; small calyx-like bodies ; borders surrounding the seeds of certain flowers.

Coronilla, n., *kŏr'·ŏn·ĭl'·lă* (L. *corona*, a crown, referring to the arrangement of the flowers), an interesting genus of plants, Ord. Leguminosæ, Sub-ord. Papilionaceæ : **Coronilla varia**, *văr'·ĭ·ă* (L. *varius*, changing, varying), a species which acts as a narcotic poison: **C. emerus**, *ēm'·ĕr·ŭs* (Gr. *ēmeros*, not wild, cultivated), the leaves of this and preceding are sometimes used to adulterate senna.

coronoid, a., *kŏr'·ŏn·ōyd* (Gr. *corōnē*, a crow, the beak of a crow ; *eidos*, resemblance), in *anat.*, applied to a process of the large bone of the fore-arm, so named from its being shaped like the beak of a crow : **coronoid fossa**, *fŏs'·să* (L. *fossa*, a ditch), a depression on the head of the ulna.

corpus, n., *kŏrp'·ŭs* (L.), a body : **corpora**, n. plu., *kŏrp'·ŏr·ă* (L.), bodies : **corpus albicans**, *ălb'·ĭk·ănz* (L. sing.), **corpora albicantia**, *ălb'·ĭk·ăn'·shĭ·ă* (L. *albicans*, being white—from *albus*, white), two round white eminences, situated behind the tuber cinereum, and between the crura cerebri: **corpora amylacea,** *ăm'·ŭl·ā'·sē·ă* (L. *amyl-āceus,* belonging to starch—from *amylum,* starch), minute, mostly microscopical, bodies, formerly believed to consist of starch, but of unknown chemical composition, found in various organs: **c. cavernosa,** *kăv'·ĕr·nōz'·ă* (L. *cav-ernōsus,* full of cavities; *cāvus,* hollow), the principal part of the body of the penis, consisting of two cylindrical bodies placed side by side, closely united and in part blended together : **c. geniculata,** *jĕn·ĭk'·ŭl·āt'·ă* (L. *genicŭlātus,* having knots—from *genĭcŭlum,* a little knot), two little masses of grey matter, about the size and shape of coffee beans, placed on each side of the genu of the optic tract of the cerebrum, and named respectively the 'ex-ternum' and 'internum' : **c. quadrigemina,** *kwŏd'·rĭ·jĕm'·ĭn·ă* (L. *quădrĭgæ,* a set of four—from *quatuor,* four ; *gemĭnus,* twin or twin-born), four rounded emin-ences separated by a crucial depression, and placed two on each side of the middle line of the cerebrum : **c. striata,** *strĭ·āt'·ă* (L. *strĭāta,* a fluted shell), two large ovoid masses of grey matter, situated in front and to the outer side of the optic thalami, in the cerebrum.

corpus callosum, *kŏrp'·ŭs kăl·lōz'·ŭm* (L. *corpus,* a body ; *callosus,* thick-skinned — from *callum,* hard thick skin), a thick layer of medullary fibres passing transversely between the two hemispheres of the brain, and forming their great commissure : **corpus dentatum,** *dĕnt·āt'·ŭm* (L. *dentātus,* toothed, having teeth), an open bag or capsule of grey matter, the section of which pre-sents a dentated outline, in the cerebellum ; also called 'ganglion of the cerebellum': **c. fimbriatum,**

fĭm'·brĭ·āt'·ŭm (L. *fimbrĭātus,* fibrous, fringed—from *fimbriæ,* fibres, threads), a narrow, white, tape-like band situated immedi-ately behind the choroid plexus in the cerebrum.

corpuscle, n., *kŏrp·ŭsk'·l* (L. *corp-usculum,* a little body—from *corpus,* a body), a small body ; a particle: **corpuscula tactus,** *kŏrp·ŭsk'·ŭl·ă tăkt'·ŭs* (L. *corpusc-ula,* corpuscles ; *tactŭs,* touch, *tactŭs,* of touch), the touch bodies, or tactile corpuscles; min-ute bodies having the appearance of a miniature fir-cone, with great tactile sensibility, found in the skin of the hand and foot, and some other parts: **corpusculated,** a., *kŏrp·ŭsk'·ŭl·āt'·ĕd,* applied to fluids which, like the blood, con-tain floating solid particles or corpuscles.

Correa, n., *kŏr·rē'·ă* (after *Corræa,* a Portuguese botanist), a genus of shrubs, Ord. Rutaceæ, re-markable for their gamopetalous corollas.

corrosive, a., *kŏr·rōz'·iv* (L. *con,* together; *rosum,* to gnaw), con-suming; wearing away : **corrosive sublimate,** *sŭb'·lĭm·āt* (L. *sublim-ātum,* to lift up on high), a very poisonous preparation of mercury; the perchloride of mercury.

corrugate, a., *kŏr'·rŏŏg·āt,* and **corrugated,** a., *kŏr'·rŏŏg·āt'·ĕd* (L. *corrugātum,* to make full of wrinkles—from *con,* together ; *ruga,* a wrinkle), in *bot.,* crumpled; wrinkled.

cortex, n., *kŏrt'·ĕks* (L. *cortex,* bark, *cortĭcis,* of bark), the bark of trees: **cortical,** a., *kŏrt'·ĭk·ăl,* belonging to the bark: **corticate,** a., *kŏrt'·ĭk·āt,* covered with a layer of bark; resembling bark : **cortical layer,** in *zool.,* the layer of sarcode enclosing the chyme mass, and surrounded by the cuticle, in the Infusoria: **cortic-olous,** a., *kŏrt·ĭk'·ŏl·ŭs* (L. *cŏlo,* I dwell, I abide), growing on bark; also applied to that portion of the kidney which lies between the cones and the surface of the organ.

cortina, n., *kŏrt·ĭn'·ă* (L. *cortĭna*, the tripod of Apollo, a veil), in *bot.*, the remains of the veil which continue attached to the edges of the pileus in Agarics: **cortinate,** a., *kŏrt'·ĭn·āt*, like a cobweb in texture.

Corydalis, n., *kŏr'·ĭ·dăl'·ĭs* (Gr. *korudăllĭs*, the bulbous fumitory; *korudălos*, the lark), a beautiful genus of plants, Ord. Fumariaceæ, so called because the spurs of the flowers resemble the spurs of the lark: **Corydalis bulbosa,** *bŭlb·ōz'·ă* (L. *bulbōsus*, full of bulbs—from *bulbus*, a bulb), a species whose tubes have been used as a substitute for Birthworts in expelling intestinal worms: **corydaline,** a., *kŏr'·ĭ·dāl'·ĭn*, resembling the flower of the corydalis.

Corylaceæ, n. plu., *kŏr'·ĭ·lā'·sĕ·ē* (L. *corylus*, Gr. *korŭlos*, a hazel or filbert tree; said also to be derived from Gr. *korus*, a helmet, *korŭthos*, of a helmet, in reference to the calyx enwrapping the fruit), the Nut family or Mastworts, an Order containing such timber trees as the oak, beech, and chestnut; the Order also called the **Cupaliferæ: Corylus,** n., *kŏr'·ĭl·ŭs*, a genus of trees, chiefly cultivated for the sake of their fruit: **Corylus Avellana,** *ăv'·ĕl·lăn'·ă* (L. *Avellanus*, belonging to Avella, a town of Campania near which hazel trees were numerous), the species which produces the hazel-nut, with its involucral appendage.

corymb, n., *kŏr'·ĭm* (Gr. *korumbos*, L. *cŏrymbus*, the top, a cluster), in *bot.*, an inflorescence in which the lower stalks are longest, and all the flowers come nearly to the same level: **Corymbiferæ,** n. plu., *kŏr'·ĭm·bĭf'·ĕr·ē* (L. *fĕro*, I bear), the second of the three sections into which Jussieu divides the Compositæ, included under the section Tubulifloræ of De Candolle: **corymbiferous,** a., *kŏr'·ĭm·bĭf'·ĕr·ŭs*, bearing a cluster of flowers in the form of a corymb —also in same sense, **corymbose,** a., *kŏr'·ĭm·bōz*.

Coryphineæ, n. plu., *kŏr'·ĭf·ĭn'·ē·ē* (Gr. *korŭphē*, the top, the summit), a Sub-order or tribe of palms, Ord. Palmæ; the talipot and date palms: **Corypha,** n., *kŏr'·ĭf·ă*, a beautiful genus of palms, from 15 ft. to 150 ft. high.

coryza, n., *kŏr·ĭz'·ă* (Gr. *korŭza*, mucus of the nose), an inflammatory affection of the mucous membrane lining the nose, resulting in an increased defluxion of mucus: **coryza gangrenosa,** *găng'·grĕn·ōz'·ă* (L. *gangrœna*, a cancerous ulcer), malignant catarrh, in which there is a discharge of ichor mixed with blood, and accumulations of pus in the nasal sinuses.

Coscinium, n., *kŏs·sĭn'·ĭ·ŭm* (Gr. *koskinon*, a sieve), a genus of climbing plants, Ord. Menispermaceæ, so called in allusion to the cotyledons being perforated: **Coscinium fenestratum,** *fĕn'·ĕs·trāt'·ŭm* (L. *fenestrātum*, to furnish with openings), a species which supplies a false calumba-root containing much berberine.

costa, n., *kŏst'·ă* (L. *costa*, a rib, a side), a rib; the mid-rib: **costæ,** n. plu., *kŏst'·ē*, in *bot.*, the prominent bundles of vessels in the leaves; in *zool.*, the rows of plates which succeed the inferior or basal portion of the cup among Crinoidea; vertical ridges on the outer surface of theca among corals: **costal,** a., *kŏst'·ăl*, connected with the ribs: **costate,** a., *kŏst'·āt*, provided with ribs; having longitudinal ridges.

costo, *kŏst'·ō* (L. *costa*, a rib, a side), denoting muscles which arise from the ribs: **costo-clavicular**, denoting a ligament attached inferiorly to the cartilage of the first rib near its sternal end: **costo-sternal**, applied to the ribs united to the sternum: **costo-vertebræ**, the ribs arising from the vertebræ or backbone.

cotunnius, n., *kŏt·ŭn'·nĭ·ŭs*, also **liquor cotunnii** (after first describer), the perilymph, a limpid fluid secreted by the lining membranes of the osseous labyrinth of the ear.

cotyledon, n., *kŏt'·ĭl·ēd'·ŏn* (Gr. *kotulēdōn*, the socket of the hip joint, a cup-like hollow—from *kotulē*, a hollow), in *bot.*, the temporary leaf, leaves, or lobes of the embryo of a plant which first appear above ground; in *anat.*, applied to the portions of which the placentæ of some animals are formed: **cotyloid**, a., *kŏt'·ĭl·ōyd* (Gr. *eidos*, resemblance), a deep cup-shaped cavity in the os innominatum; the acetabulum.

couch, v., *kowtsh* (F. *coucher*, to lay down), in *med.*, to depress or remove the film called a cataract which overspreads the pupil of the eye: **couching**, n., *kowtsh'·ing*, the operation itself.

coup-de-soleil, n., *kōō'·dĕ·sŏl·ĕl'* (F. *coup*, stroke; *de*, of; *soleil*, the sun), sunstroke; a dangerous disease produced by exposure of the head to the rays of the sun, almost wholly confined to tropical countries.

crampons, n. plu., *krăm'·pŏnz*, (Dut. *krampe*, F. *crampon*, hooks, claspers), in *bot.*, the adventitious roots which serve as fulcra or supports, as in the ivy.

cranial, a., *krăn'·ĭ·ăl* (Gr. *krānion*, Mid. L. *crānium*, the skull), of or pert. to the cranium or skull: **craniology**, n., *krăn'·ĭ·ŏl'·ō·jĭ* (Gr. *logos*, discourse), a treatise on the skull: **craniotomy**, n., *krăn'·ĭ·ŏt'·ŏm·ĭ* (Gr. *tomē*, a cutting), the operation of opening the skull,

sometimes rendered necessary in effecting delivery: **cranium**, n., *krăn'·ĭ·ŭm*, the bony or cartilaginous case containing the brain.

craspeda, n. plu., *krăsp'·ĕd·ă* (Gr. *kraspedon*, a border, a tassel), the long cords, containing thread cells, which are attached to the free margins of the mesenteries of a sea anemone.

Crassulaceæ, n. plu., *krăs'·ūl·ā'·sĕ·ē* (diminutive of L. *crassus*, solid, thick, in allusion to the fleshy leaves and stems), the Houseleek or Stonecrop family, an Order of succulent herbs and shrubs of considerable beauty, found in the driest and most arid situations, where not a blade of grass or a tuft of moss could live: **Crassula**, n., *krăs'·ūl·ă*, a genus of succulent plants: **Crassula profusa**, *prō·fūz'·ă* (L. *profūsus*, spread out, extended); C. **lactea**, *lăkt'·ĕ·ă* (L. *lactĕus*, containing milk, milky —from *lac*, milk); and C. **marginata**, *mârj'·ĭn·āt'·ă* (L. *marginātum*, to furnish with a border), are species in the leaves of which there are two kinds of stomata, one scattered over the leaves, and the other, very minute, raised on discs arranged in a row within the margin of the leaf.

cratera, n., *krăt·ēr'·ă* (L. *crātēra*, a wine-cup—from Gr. *kratēr*, a cup), in *bot.*, a cup-shaped receptacle: **crateriform**, a., *krăt·ēr'·ĭ·fŏrm* (L. *forma*, shape), cup-shaped; concave: **Craterina**, n., *krăt'·ĕr·ĭn'·ă*, a genus of parasitic insects: **Craterina hirundinis**, *hĭr·ŭnd'·ĭn·ĭs* (L. *hirundo*, a swallow, *hirundĭnis*, of a swallow), a species found upon swallows.

crease, v., *krēs* (Breton or Prov. F. *krĭz*, a wrinkle, a tuck in a garment), in *far.*, to groove around the outer circumference of the iron shoe of a horse: **creasing**, n., *krēs'·ing*, the art of grooving an iron horse-shoe which marks the line where the nails are to be placed; also called 'ful-lering.'

creasote, n., *krē'·ăs·ŏt*, also cre-osote, n., *krē'·ŏs·ŏt* (Gr. *kreas*, flesh; *zōzō*, I preserve), an oily, colourless liquid, with a charac-teristic smell, obtained from wood or coal tar.

creatine or creatin, n., *krē'·ăt·ĭn* (Gr. *kreas*, flesh, *kreătos*, of flesh), a substance in the form of colour-less transparent crystals, obtained from flesh of different animals, as sheep, oxen, fowls, fish : creat-inin, n., *krē·ăt'·ĭn·ĭn*, an alkaline substance in the form of prism-atic crystals, procured chiefly from the urine.

cremaster, n., *krĕm·ăst'·ĕr* (Gr. *kremaō*, I suspend ; *kremămai,* I am suspended, I hang), a muscle which draws up or suspends the testis in males: cremasteric, a., *krĕm'·ăst·ĕr'·ĭk*, pert. to the cremaster muscle ; applied to a fascia.

cremocarp, n., *krĕm'·ō·kărp* (Gr. *kremaō*, I suspend ; *karpos*, fruit), the fruit of the Umbellif-eræ, consisting of two one-seeded carpels, completely invested by the tube of the calyx.

crenate, a., *krēn'·āt* (Mid. L. *crena*, a notch), in *bot.*, having a series of rounded marginal prominences; having convex teeth, as on the margin of a leaf : crenature, n., *krĕn'·ăt·ŭr*, a di-vision or notch of the margin of a crenate leaf; a notch in a leaf or style : crenulate, a., *krĕn'·ŭl·āt* (dim. of *crena*), having the edge slightly or minutely notched.

crepitant, a., *krĕp'·ĭt·ănt* (L. *crepitans,* creaking or crackling —gen. *crepĭtantis*), crackling: crepitant rale, *răl* (F. *râle*, a rattling in the throat), a fine crackling sound heard in respira-tion, caused by the passage of the air through mucus in the bron-chial tubes : crepitation, n., *krĕp'·ĭt·ā'·shŭn*, a small, sharp,

crackling noise, as of salt when thrown on the fire ; a rubbing of hair, a similar sound heard in inspiration at the commencement of pneumonia.

crepuscular, a., *krĕp·ŭsk'·ŭl·ĕr* (L. *crepusculum*, twilight, dusk), applied to animals which are active in the dusk or twilight.

Crescentieæ, n. plu., *krĕs'·sĕn·tĭ'·ĕ·ē* (after *Crescenti* of Bologna), a Sub-order of trees, whose fruit is woody and melon-shaped, Ord. Bignoniaceæ : Crescentia, n., *krĕs·sĕn'·shĭ·ă*, a genus of hand-some trees, some of them having edible fruits : Crescentia Cujete, *kūdj·ēt'·ĕ* (from Jamaica), also called C. cuneifolia, *kūn'·ĕ·ĭ·fōl'·ĭ·ă* (L. *cunĕus*, a wedge ; *folium,* a leaf), the calabash tree of tropical America, the hard pericarps of whose melon-like fruit are used as cups and bottles.

crest, n., *krĕst* (L. *crista*, the tuft or plume on the head of birds), an appendage to fruits or seeds having the form of a crest.

cretaceous, a., *krēt·ā'·shŭs* (L. *creta*, chalk), composed of chalk ; chalky ; in *bot.*, chalky-white: creta preparata, *krēt'·ă prĕp'·ăr·āt'·ă* (L. *preparātus*, prepared), a medical preparation of chalk.

cretinism, n., *krēt'·ĭn·ĭzm* (F. *Crétin*, one of certain inhabitants of the Alps and other mountains, remarkable for their stupid and languid appearance), a peculiar kind of idiocy, attended with goître and other malformations, with arrest of development, especially of the skull, which prevails in districts about the Alps and other mountains : Cretin, n., *krēt'·ĭn*, one of the deformed idiots of the Alpine and other mountainous regions : in the Pyrenees they are called 'Cagots.'

cribriform, a., *krĭb'·rĭ·fŏrm* (L. *cribrum*, a sieve ; *forma*, shape), pierced with small holes like a sieve ; perforated : **cribrose, a.,** *krĭb'·rōz*, same sense.

cricoid, a., *krĭk'·ŏÿd* (Gr. *krĭk'·ŏs,* a ring ; *eidos,*. resemblance), shaped like a ring : **crico-arytenoid,** *krĭk'·ō-ăr·ĭt'·ĕn·ŏÿd,* applied to articulations which are surrounded by a series of thin capsular fibres, and connected with the bases· of the arytenoïd cartilages.

crinite, a., *krĭn'·ĭt* (L. *crĭnis,* the hair), in *bot.,* having the appearance of a tuft of hair ; bearded.

crinoids, n. plu., *krĭn'·ŏÿds,* also **crinoidea, n. plu.,** *krĭn·ŏÿd'·ĕ·ă* (Gr. *krĭnon,* a lily ; *eidos,* resemblance), in *zool.,* an Order of Echinodermata, including forms which are usually stalked, and which sometimes resemble lilies.

Crinum, n., *krĭn'·ŭm* (Gr. *krĭnon,* a lily), a fine genus of bulbous plants, Ord. Amaryllidaceæ, having many of the species very beautiful, and producing delightfully fragrant flowers in large umbels.

crisis, n., *krĭs'·ĭs,* **crises, n. plu.,** *krĭs'·ēz* (L. *crisis,* Gr. *krisis,* a decision), that important stage of some diseases which may eventuate in recovery or death.

crisp, a., *krĭsp* (L. *crispus,* Old F. *crespe,* curled), in *bot.,* having an undulated margin : **crispate, a.,** *krĭsp'·āt,* irregularly curled or twisted : **crispation, n.,** *krĭsp·ā'·shŭn,* in *anat.,* the permanent shrinking of a tissue.

cristate, a., *krĭst'·āt* (L. *crista,* a crest), in *bot.,* crested ; tufted : **crista frontalis,** *krĭst'·ă frŏnt·āl'·ĭs* (L. *frontālis,* belonging to the forehead—from *frons,* the forehead), the frontal crest ;· a ridge of bone on· the inner surface of the frontal bone of the skull : **c. galli,** *găl'·lī* (L. *gallus,* a cock, *galli,* of a cock), the crest of the cock, or cock's crest ; a ridge of bone in the skull which rises into a thick process of the ethmoid bone.

Crithmum, n., *krĭth'·mŭm* (Gr. *krithē,* barley), a genus of plants, Ord. Umbelliferæ, so called in allusion to the singularity of the seeds : **Crithmum maritimum,** *măr·ĭt'·ĭm·ŭm* (L. *maritimus,* belonging. to the sea—from *mărĕ,* the sea), the samphire, found growing abundantly on the rocks near the sea,—is used as a pickle.

Crocodilia, n. plu., *krŏk'·ŏd·ĭl'·ĭ·ă* (L. *crocodilus,*. Gr. *krokodeilos,* a crocodile),. a well-known Order of reptiles.

Crocus, n., *krŏk'·ŭs* (L. *crocus,* Gr. *krokos,* Gael. *croch,* red), a well-known and much admired genus of early spring plants, Ord. Iridaceæ : **Crocus sativus,** *săt·ĭv'·ŭs* (L. *satīvus,* that is·fit to be planted), a species which furnishes the colouring material called saffron : **C. autumnalis,** *āw'·tŭm·nāl'·ĭs* (L. *autumnālis,*. autumnal —from *autumnus,* autumn); and **C. odorus,** *ŏd·ōr'·ŭs* (L. *odōrus,* sweet-smelling—from *ŏdor,* scent, smell), are species also supplying saffron.

crotaphyte, n., *krŏt'·ă·fīt* (Gr. *krotaphos,* a temple of the head), the temporal·muscle, which fills the temporal fossa and extends itself over a considerable part of the side of the head.

Croton, n., *krŏt'·ŏn* (Gr. *krŏtōn,* a tick, with reference to the resemblance of the seeds), a genus of plants,. Ord. Euphorbiaceæ : **Croton tiglium,** *tĭg'·lĭ·ŭm* (an Indian word), an Indian and Asiatic shrub, from whose seed croton-oil is expressed,— internally the oil acts as. an irritant purgative, externally it produces pustules : **C. Pavana,** *păv·ăn'·ă* (unascertained);·also **C. Roxburgii,** *rŏks·bĕrg'·ĭ·ī* (L. *Roxburgii,* of Roxburgh in Scotland), species which yield purgative oils : **C.**

Malambo, *măl·ămb′ō* (unascertained), yields a tonic bark : C. **eleuteria,** *ĕl′·ū·tēr′·ĭ·ă* (L. and Gr. *eleutheria*, freedom, liberty), produces cascarilla bark, also called sea-side balsam or sweet wood, used as a tonic and stimulant : **C. pseudo-china** or **niveum,** *sūd′·ō·kīn′·ă* or *nĭv′·ĕ·ŭm* (Gr. *pseudēs*, false, lying ; *china*, Ger. name for Peruvian bark ; L. *nivĕus*, snowy), produces copalchi bark, used as a tonic.

croup, n., *krōp* (Icel. *kropa*, Scot. *roup*, to cry), an inflammatory disease of the trachea, chiefly occurring in early childhood, attended by very noisy breathing : **croupous,** a., *krōp′·ŭs*, pert. to croup ; fibrinous.

crown, n., *krown* (W. *crwn*, round, circular ; Gael. *crùn*, a boss, a garland ; L. *corona*, a crown), in *bot.*, the short stem at the upper part of the root of perennial herbs.

Crozophora, n. plu., *krōz·ŏf′·ŏr·ă* (Gr. *krossos*, a pitcher, a pail ; *phoreo*, I bear), a genus of plants, Ord. Euphorbiaceæ : **Crozophora tinctoria,** *tĭnkt·ōr′·ĭ·ă* (L. *tinctorius*, belonging to dyeing—from *tinctus*, dyed), furnishes a purple dye called turnsole, which becomes blue on the addition of ammonia.

Cruciferæ, n. plu., *krōs·ĭf′·ér·ē* (L. *crux*, a cross ; *fero*, I bear), the cruciferous or Cresswort family, an Order of herbaceous plants, comprising many of the common culinary vegetables, as cabbages, turnips, radish, cress—so called from having the four petals of the flowers arranged in the form of a cross : **cruciferous,** a., *krōs·ĭf′·ér·ŭs*, having flowers arranged in the form of a cross.

cruciform, a., *krōs′·ĭ·fŏrm* (L. *crux*, a cross, *crucis*, of a cross ; *forma*, shape), in *bot.*, like the parts of a cross, as in flowers of Cruciferæ ; arranged in the form

of a cross ; also **cruciate,** a., *krōsh′·ĭ·āt*, same sense ; said of a flower when four petals are placed opposite each other and at right angles: **crucial,** a., *krōsh′·ĭ·ăl*, applied to certain ligaments of the knee which cross or intersect each other somewhat like the letter x.

cruor, n., *krō′·ŏr* (L. *crŭor*, blood, *cruōris*, of blood), the soluble coloured ingredient of blood, separable into two substances, globulin and hæmatin.

crura, n. plu., *krōr′·ă* (L. *crūs*, the leg, *crūris*, of a leg, *crūra*, legs), the legs ; parts of the body which resemble legs ; in *bot.*, divisions of a forked tooth : **crus cerebelli,** *krŭs sĕr′·ĕb·ĕl′·ĭ* (L. *cerebelli*, of the cerebellum); the leg of the cerebellum: **crura cerebelli,** the legs of the cerebellum : **crura cerebri,** *sĕr′·ĕb·rī* (L. *cerebrum*, the cerebrum), the legs or pillars of the brain ; terms denoting parts or divisions of the brain, so called from their appearance : **crural,** a., *krōr′·ăl*, pert. to the legs or lower limbs : **cruræus,** a., *krōr·ē′·ŭs*, in *anat.*, applied to one of the extensor muscles of the leg, arising from the thigh-bone and inserted into the knee-pan.

crusta, n., *krŭst′·ă* (L. *crusta*, skin, bark), in *bot.*, the frosted appearance on the fronds of some lichens.

Crustaceæ, n. plu., *krŭst·ā′·sĕ·ē* also **Crustaceans,** n. plu., *krŭst·ā′·sē·ăns* (L. *crusta*, skin, bark), the articulate animals, comprising lobsters, crabs, etc., which have a hard shell or crust, which they cast periodically : **crustaceous,** a., also **crustose,** a., *krŭst′·ōz*, pert. to the Crustacea ; of the nature of crust or shell ; in *bot.*, hard, thin, and brittle ; applied to lichens hard and expanded like a crust ; having the appearance of hoar-frost.

crusta petrosa, *krŭst'·ă pĕt·rōz'·ă*
(L. *crusta*, skin, bark; *petrōsus*,
very rocky—from Gr. and L.
petra, a rock), a rocky crust; the
layer of true bone which invests
the root part of the teeth, or the
part not protected by enamel.

Cryptocarya, n., *krĭpt'ō·kār'·ĭ·ă*
(Gr. *kruptos*, hidden; *karŭa*,
a walnut—alluding to the fruit
being covered), a genus of shrubs,
Ord. Lauraceæ: **Cryptocarya
moschata**, *mŏs·kăt'·ă* (Mod. L.
moschātŭs, having a smell like
musk—from Arab.*mosch* or *mesk*,
musk), a species which produces
the Brazilian nutmegs.

cryptogamia, n., *krĭpt'ō·găm'·ĭ·ă*
(Gr. *kruptos*, hidden, concealed;
gamos, marriage), a general name
applied to all the lower orders of
plants which have no apparent
or true flowers, such as mushrooms, lichens, mosses, seaweeds,
and ferns: **cryptogamous**, a.,
krĭpt·ŏg'·ăm·ŭs, having their
organs of reproduction obscure:
cryptogamic, a., *krĭpt'·ō·găm'·ĭk*,
in same sense as cryptogamous.

cryptorchismus, n., *krĭpt'·ŏr·kĭz'·
mŭs* (Gr. *kruptos*, concealed;
orchis, the testicles), the retention of the testes in the abdomen.

cryptos, a., *krĭpt'·ŏs* (Gr. *kruptos*,
concealed, hidden), concealed;
not readily observed; in *composition*, crypto-.

cryptostomata, n. plu., *krĭpt'·ō·
stŏm'·ăt·ă* (Gr. *kruptos*, hidden;
stoma, a mouth, *stomăta*,
mouths), circular nuclei on the
surfaces of some algæ.

crystalloid, a., *krĭst'·ăl·ōyd* (Gr.
krustallos, L. *crystallum*, ice,
rock crystal; Gr. *eidos*, resemblance), crystalline; resembling
crystal; in *bot.*, applied to a
portion of the protoplasmic substance of cells having a crystalline form.

ctenocyst, n., *tĕn'·ō·sĭst* (Gr. *kteis*,
a comb, *ktenos*, of a comb;
kustis, a bag, a bladder), the

sense organ in the ctenophora:
ctenoid, a., *tĕn'·ōyd* (Gr. *eidos*,
resemblance), having the appearance of a comb; applied to those
scales of fishes the hinder margins of which are fringed with
spines having the appearance of
a comb: **Ctenophora**, n. plu.,
tĕn·ŏf'·ŏr·ă (Gr. *phoreo*, I bear),
an Order of Actinozoa, including
sea creatures which swim by
means of bands of cilia arranged
in comb-like plates; also called
ctenophores, n. plu., *tĕn'·ō·fōrz*.

Cubeba, n., *kū·bēb'·ă* (Arab. *cubabah*), a small genus of plants,
Ord. Piperaceæ: **Cubeba officin-
alis**, *ŏf·fĭs'·ĭn·āl'·ĭs* (L. *officinalis*,
officinal—from *officīna*, a workshop), the fruit of a climbing
plant of Java and other Indian
islands, used, under the name
'cubeb-pepper,' or 'cubebs,' in
arresting discharges from mucous
membranes; also called 'Piper
cubeba.'

cubit, n., *kūb'·ĭt*, also cubitus, n.,
kūb'·ĭt·ŭs (L. *cubitum*, Gr. *kubiton*,
the elbow or bending of the arm),
the fore-arm.

cuboid, a., *kūb'·ōyd* (Gr. *kubos*, a
cube; *eidos*, resemblance), applied to a bone situated at the
outer side of the foot.

cucullaris, a., *kū'·kŭl·lār'·ĭs* (L.
cŭcullus, a hood), the trapezius
muscle: **cucullate**, a., *kū·kŭl'·lāt*,
in *bot.*, hooded; formed like a
hood.

Cucumis, n., *kūk'·ŭm·ĭs* (L.
cŭcŭmis, a cucumber, *cŭcŭmeris*,
of a cucumber), an extensively
cultivated genus of plants for
culinary purposes, Ord. Cucurbitaceæ: **Cucumis melo**, *mēl'·ō* (Gr.
mēlon, L. *mēlo*, a melon), the
common melon: **C. sativus**, *săt·
īv'·ŭs* (L. *sativus*, that is fit to
be planted), the cucumber: **C.
colocynthis**, *kŏl'·ō·sĭnth'·ĭs* (Gr.
kolokunthis, the wild gourd), a
species which yields the globular
fruit 'coloquintida' or 'bitter-

apple,' the pulp of which is the colocynth of medicine.

Cucurbitaceæ, n. plu., *kū·kẽrb'·ĭt·ā'·sẽ·ē* (L. *cŭcurbĭta*, a gourd), the Cucumber family, an Order of plants, many of which are drastic purgatives : **Cucurbita**, n., *kū·kẽrb'·ĭt·ă*, a genus of plants : **Cucurbita citrullus**, *sĭt'·rŭl'·lŭs* (new L. *citrullus*, the Sicilian citrul or water-melon plant), the water-melon, prized for its cool, refreshing juice : C. **pepo**, *pẽp'·ō* (L. *pẽpo*, a large melon, a pumpkin ; *pẽponis*, of a pumpkin), the white gourd : C. **maxima**, *măks'·ĭm·ă* (L. *maxĭmus*, greatest), the pumpkin or red gourd : C. **melo-pepo**, *mēl'·ō-pẽp'·ō* (L. *mēlo*, a melon ; *pẽpo*, a pumpkin), the squash : C. **ovifera**, *ŏv·ĭf'·ẽr·ă* (L. *ovum*, an egg ; *fero*, I bear), the egg gourd, or vegetable marrow : **cucurbitaceous**, a., *kū·kẽrb'·ĭt·ā'·shŭs*, resembling a gourd or cucumber.

cudbear, n., *kŭd'·bār* (after *Sir Cuthbert Gordon*), a purple or violet colouring matter obtained from a lichen Lecanora tartarea.

culm, n., *kŭlm* (L. *culmus*, a stalk, a stem), the stalk or stem of corn or grasses, usually hollow and jointed : **culmicolous**, a., *kŭl·mĭk'·ŏl·ŭs* (L. *colo*, I clothe or dress), growing on the culm of grasses.

Cuminum Cyminum, *kūm·ĭn'·ŭm sĭm·ĭn'·ŭm* (L. *cumĭnum*, Gr. *cumĭnon*, cumin—the systematic name being made up of a repetition of the same word in its L. and Gr. forms), cumin, a plant, Ord. Umbelliferæ, whose seeds have a very peculiar odour and bitter aromatic taste.

cuneate, a., *kūn'·ē·āt* (L. *cunĕātus*, pointed like a wedge—from *cunĕŭs*, a wedge), shaped like a wedge standing upon its point ; wedge-shaped : **cuneiform**, a., *kūn'·ē·ĭ·fŏrm* (L. *forma*, shape), same sense as 'cuneate.'

cuniculate, a., *kŭn·ĭk'·ŭl·āt* (L. *cuniculus*, a rabbit burrow), having a long pierced passage.

Cunonieæ, n. plu., *kŭn'·ōn·ĭ'·ĕ·ē* (after *M. Cuno* of Amsterdam), a Sub-order of trees and shrubs of the Southern Hemisphere, Ord. Saxifragaceæ.

Cupania, n., *kū·pān'·ĭ·ă* (after *Father Cupani* of Italy), a genus of plants, Ord. Sapindaceæ : **Cupania sapida**, *săp'·ĭd·ă* (L. *sapĭdus*, tasting, savouring—from *săpĭo*, I taste), yields the Akel fruit, whose succulent arillus is used as food.

cupel, n., *kūp'·ĕl* (L. *cupella*, a little cup), a very porous cup-like vessel used in refining metals.

cupola, n., *kūp'·ŏl·ă* (It. *cupola*, a round vaulted chapel behind the chancel), an arched or spherical vault on the top of an edifice; in *anat.*, an arched and closed extremity of the ear, forming the apex of the cochlea.

cupping, n., *kūp'·ĭng* (It. *coppa*, a head ; L. *cupa*, a cask), a method of local blood-letting by means of a bell-shaped glass and a scarificator.

cupreous, a., *kūp'·rĕ·ŭs* (L. *cuprĕus*, of copper—from *cuprum*, copper), consisting of or resembling copper; coppery.

Cupressineæ, n. plu., *kūp'·rĕs·sĭn'·ĕ·ē* (L. *cupressus*, the cypress tree), a Sub-order of trees, comprising the cypress and juniper, Ord. Coniferæ: **Cupressus**, n., *kūp·rĕs'·sŭs*, a genus of handsome evergreen trees : **Cupressus sempervirens**, *sĕm·pẽr'·vĭr·ĕnz* (L. *semper*, always ; *virens*, verdant), the common cypress tree, which yields a durable wood, supposed to be the gopher wood of the Scriptures.

cupula, n., *kūp'·ūl·ă* (L. *cūpulus*, a little cup—from *cupa*, a cask, a cup), the cup of the acorn, formed by an aggregation of bracts: cupule, n., *kŭp′·ŭl*, a part of a fruit surrounding its lower part like a cup, as an acorn; a cupula: **cupuliform,** a., *kŭp·ŭl′·i·fŏrm* (L. *forma*, shape), shaped like the cup of an acorn.

Cupuliferæ, n. plu., *kŭp′·ŭl·if′·ér·ē* (L. *cupula*, a little cup, a cupule; *fero*, I bear), the Nut family, an Order of trees, including the hazel and the oak; also named **Ccrylaceæ,** n. plu., *kŏr′·il·ā′·sĕ·ē:* **cupuliferous,** a., *kŭp′·ŭl·if′·ér·ŭs*, having or bearing cupules.

Curculigo, n., *kér·kŭl′·ig·ō* (L. *curcŭlio*, a corn worm, a weevil), a genus of pretty herbaceous plants, so called from the seeds having a process resembling the beak of the weevil, Ord. Bromeliaceæ.

Curcuma, n., *kérk·ūm′·ŭ* (Ar. *kurkum*), a genus of plants, Ord. Zingiberaceæ : **Curcuma longa,** *lŏng′·gŭ* (L. *longus*, long), a species which furnishes Turmeric, a yellow lemon powder used as a dyestuff, employed medicinally as a carminative, and enters into the composition of curry powder : **curcumin,** n., *kérk·ūm′·in*, the yellow colouring matter of turmeric: **Curcuma angustifolia,** *ăng·gŭst′·i·fōl′·i·ŭ* (L. *angustus*, narrow, contracted; *folium*, a leaf), a species which furnishes the East Indian arrowroot : **C. zerumbet,** *zér·ŭm′·bĕt* (from the East Indies), and **C. leucorhiza,** *lō′·kŏr·iz′·ŭ* (Gr. *leukos*, white; *rhiza,* a root), yield a starch similar in kind to East Indian arrowroot.

Cursores, n. plu., *kérs·ōr′·ēz* (L. *cursor*, a runner—from *curro*, I run), an Order of birds formed for running swiftly, and destitute of the power of flight, as the ostrich and emu.

curvembryeæ, n. plu., *kérv′·ĕm·brī′·ĕ·ē* (L. *curvus*, bent, crooked ; Gr. *embruon*, an infant in the womb), plants which have their embryos curved: **curvembryonic,** a., *kérv·ĕm′·brī·ŏn′·ik*, having the embryo curved.

Cuscuteæ, n. plu., *kŭs·kūt′·ĕ·ē* (said to be a corruption of Gr. *kadutas,* a Syrian parasitical plant ; Arab. *chessuth* or *chasuth*), a Sub-order of the Ord. Convolvulaceæ : **Cuscuta,** n., *kŭs·kūt′·ŭ*, a genus of curious parasitical plants producing abundance of sweet-scented flowers in autumn: **Cuscuta epithymum,** *ē·pĭth′·ĭm·ŭm* (Gr. *epithumon*, L. *epithymon*, the flower of a species of thyme—from Gr. *epi*, upon ; *thumon*, the herb thyme), the dodder or scold weed, a parasitic plant ; this and other species have acrid purgative properties.

cusparia, n., *kŭs·pār′·i·ŭ* (a native name), a name given to the bark of the Galepea cusparia, Ord. Rutaceæ, which is used as a tonic and febrifuge ; called also 'Angostura bark.'

cuspidate, a., *kŭsp′·id·āt* (L. *cusp-is*, point of a spear, *cŭspĭdis*, of the point of a spear), in *bot.*, gradually tapering to a sharp stiff point ; in *zool.*, furnished with small pointed eminences or cusps : **cuspis,** n., *kŭsp′·is*, also **cusp,** n., *kŭsp*, a spike, somewhat thick at the base, and tapering gradually to a point.

cuticle, n., *kūt′·ĭk·l* (L. *cutīcŭla*, a little skin—from *cŭtis*, the skin), the thin exterior coat of the skin; in *zool.*, the pellicle which forms the outer layer of the body amongst the Infusoria ; in *bot.*, the thin layer that covers the epidermis : **cuticular,** a., *kūt·ĭk′·ŭl·ăr*, belonging to the cuticle or outer surface : **cutis,** n., *kūt′·is*, the true skin ; the inferior layer of the integument called skin ; in *bot.*, the peridium of some fungi: **cutis vera,** *kūt′·is vēr′·ŭ* (L. *vērus*, true), the true skin, the sentient and vascular texture, which is covered and defended by the insensible and non-vascular 'cuticle'; also called 'derma' or 'corium.'

cyanic, a., *sī·ăn'·ĭk* (Gr. *kuanos*, dark-blue), denoting the blue appearance which a patient has in certain diseases ; in *bot.*, denoting a series of colours in plants of which blue is the type, and which do not pass into yellow, comprising 'greenish-blue, blue, violet-blue, violet, violet-red, and red': cyanide, n., *sī'·ăn·ĭd*, a salt which is a compound of cyanogen with an elementary substance : cyanogen, n., *sī·ăn'·ŏ·jĕn* (Gr. *gennaō*, I produce), a gas with an odour like crushed peach leaves, which burns with a rich purple flame, and is an essential ingredient in Prussian blue : cyanosis, n., *sī'·ăn·ōz'·ĭs*, a diseased condition arising from a defect or malformation in the heart, characterised by blueness of the skin, markedly in the hands, tip of nose, and ears ; also called 'morbus cœruleus.'

Cyathea, n., *sī'·ăth·ē'·ă* (Gr. *kuathos*, L. *cyathus*, a cup), a genus of fine tropical ferns, Ord. Filices, so named from the cup-shaped form of the indusium : Cyathea medullaris, *mĕd'·ŭl·lār'·ĭs* (L. *medullāris*, belonging to or having the nature of marrow—from *medulla*, the marrow), the ponga of New Zealand, which furnishes a gum used as a vermifuge.

cyathiform, a., *sī·ăth'·ĭ·fŏrm* (L. *cyathus*, a cup ; *forma*, shape), shaped like a cup ; resembling a cup : cyathoid, a., *sī'·ăth·oyd* (Gr. *eidos*, resemblance), having the appearance of a cup ; cyathiform : cyathus, n., *sī'·ăth·ŭs*, the cup-like bodies in marchantia or liverworts.

Cycadaceæ, n. plu., *sĭk'·ăd·ā'·sĕ·ē* (Gr. *kukas*, a kind of palm ; new L. *cycas*, a kind of palm, *cycados*, of a palm), the Cycas family, an Order of trees and shrubs with cylindrical trunks : Cycas, n.,

sĭk'·ăs, a genus of trees and shrubs which yield much starchy matter along with mucilage : Cycas revoluta, *rĕv'·ōl·ōōt'·ă* (L. *revolūtum*, rolled back—from *re*, back or again ; *volvo*, I roll); also C. circinalis, *sĕrs'·in·āl'·ĭs* (L. *circinālis*, encircled, whorled—from *circĭnus*, a pair of compasses), are species from whose stems a kind of sago is made, and a clear transparent gum exudes from them : cycadaceous, a., *sĭk'·ăd·ā'·shŭs*, pert. to the Cycads or Cycadaceæ.

Cyclamen, n., *sĭk'·lăm·ĕn* (L. *cyclaminos*, Gr. *kuklaminon*, the plant sow-bread—from Gr. *kuklos*, a circle), a pretty bulbous genus of plants, Ord. Primulaceæ, having round leaves, the principal food of the wild boars of Sicily, hence its common name.

Cyclantheæ, n. plu., *sĭk·lănth'·ĕ·ē* (Gr. *kuklos*, a circle ; *anthos*, a flower), a Sub-order of the Ord. Pandanaceæ, having fan-shaped or pinnate leaves : Cyclanthus, n., *sĭk·lănth'·ŭs*, a genus of plants.

cycle, n., *sĭk'·l* (Gr. *kuklikos*, L. *cyclĭcus*, circular—from Gr. *kuklos*, a circle), in *bot.*, the turn of a spiral as of a leaf so arranged : cyclic, a., *sĭk'·lĭk*, also cyclical, a., *sĭk'·lĭk·ăl*, arranged in a circle round an axis ; coiled up.

cyclogens, n. plu., *sĭk'·lō·jĕns* (Gr. *kuklos*, a circle ; *gennaō*, I produce), a name given to exogenous plants, in consequence of exhibiting concentric circles in their stems : cyclogenous, a., *sĭk·lŏdj'·ĕn·ŭs*, having concentric woody circles.

cycloid, a., *sĭk'·loyd* (Gr. *kuklos*, a circle ; *eidos*, resemblance), in *zool.*, applied to those scales of fishes which have a regularly circular or elliptical outline with an even margin.

cyclosis, n., *sĭk·lōz'·ĭs* (Gr. *kuklosis*, a surrounding, a circulation—

from *kuklos*, a circle), the partial circulation observable in the milky juice of certain plants.

cyclostomi, n. plu., *sĭk·lŏs'·tŏm·ī* (Gr. *kuklos*, a circle ; *stoma*, a mouth), a name applied to the hag-fishes and lampreys, forming the Ord. Marsipobranchii, so called from their circular mouths: **cyclostomous,** a., *sĭk·lŏs'·tŏm·ŭs*, having a circular mouth or aperture for sucking, among certain fishes, as the lamprey.

Cydonia, n., *sĭd·ōn'·ĭ·ă* (so called as from *Kydon*, in Crete), a genus of fruit trees, Ord. Rosaceæ : **Cydonia Japonica,** *jă·pŏn'·ĭk·ă* (*Japonicus*, belonging to Japan), a handsome hardy shrub, producing beautiful scarlet or white flowers in great abundance : C. **vulgaris,** *vŭlg·ār'·ĭs* (L. *vulgāris*, common, ordinary), the quince, the seeds or pips of which, when boiled in water, yield a mucilaginous decoction.

cylindrenchyma, n., *sĭl'·ĭn·drĕng'·kĭm·ă* (Gr. *kulindros; engchuma,* an infusion—from *chumos,* juice), in *bot.,* tissue composed of cylindrical cells.

cymbellæ, n. plu., *sĭm·bĕl'·lē* (L. *cymbŭla,* a little boat — from *cymba,* a boat), in Algæ, reproductive locomotive bodies : **cymbiform,** a., *sĭm'·bĭ·fŏrm* (L. *forma,* shape), having the shape of a boat.

cyme, n., *sīm* (Gr. *kuma,* L. *cyma,* the young sprout of a cabbage), in *bot.,* a mode of inflorescence resembling a flattened panicle, as that of the elder tree : **cymose,** a., *sīm·ōz',* having an inflorescence in the form of a cyme.

cynanche, n., *sĭ·năng'·kē* (Gr. *kuōn,* a dog ; *angcho,* I strangle), a disease of the windpipe, attended with inflammation, so called from the dog-like bark by which it is sometimes accompanied : **cynanche maligna**

carbuncularis, *mă·lĭg'·nă kărb·ŭngk'·ŭl·ār'·ĭs* (L. *maligna,* malignant ; *carbuncŭlus,* a small coal, a carbuncle), a malignant carbuncular cynanche; malignant sore throat : **cynanche tonsillaris,** *tŏns'·ĭl·lār'·ĭs* (L. *tonsillæ,* the tonsils in the throat), quinsy, a troublesome affection, consisting of inflammation of the tonsils and adjacent parts of the fauces : c. **trachealis,** *trăk'·ĕ·āl'·ĭs* (L. *trachēdlis,* belonging to the trachea or windpipe), croup : c. **parotidea,** *păr·ŏt'·ĭd·ē'·ă* (Gr. *para,* about ; *ous,* the ear, *otos,* of the ear), mumps ; another name for ' parotitis.'

Cynanchum, n., *sĭ·năng'·kŭm* (Gr. *kuon,* a dog ; *angchō,* I strangle, in allusion to its poisonous qualities), a genus of plants, Ord. Asclepiadaceæ : **Cynanchum monspeliacum,** *mŏns'·pĕl·ĭ'·ăk·ŭm* (after *Montpellier,* in France), a species which furnishes Montpellier scammony.

cynarocephalæ, n. plu., *sĭn·ār'·ō·sĕf'·ăl·ē* (Gr. *kuōn,* a dog ; *kephalē,* a head), a Sub-order of plants, Ord. Compositæ, which are usually tonic and stimulant : **Cynara,** n., *sĭn'·ăr·ă* (Gr. *kuōn,* a dog), a genus of plants, so called in allusion to the spines of the involucrum: **Cynara cardunculus,** *kărd·ŭngk'·ŭl·ŭs* (L. *carduncŭlus,* a diminutive of *cardŭus,* a thistle), the cardoon, a species resembling the artichoke, whose blanched stems and stalks are eaten : C. **scolymus,** *skŏl'·ĭm·ŭs* (Gr. *skolumos,* L. *scolymos,* an edible kind of thistle), the artichoke, the root of which the Arabians considered an aperient.

cynarrhodon, n., *sĭn·ăr'·rŏd·ŏn* (Gr. *kuōn,* a dog ; *rhodon,* a rose), applied to the hips or fruit of dog-roses, and roses in general.

Cynodon, n., *sĭn'·ŏd·ŏn* (Gr. *kuōn,* a dog ; *odous,* a tooth, *odontos,* of a tooth), a genus of grasses, Ord. Gramineæ: **Cynodon dactulon**, *dăkt'·ŭl·ŏn* (Gr. *daktulos*, L. *dactylus*, a finger, a sort of muscle), a species from whose roots a cooling drink is made in India ; is used in mucous discharges from the bladder.

Cynoglossum, n., *sĭn'·ō·glŏs'·sŭm* (Gr. *kuōn*, a dog ; *glossa*, a tongue), a genus of plants, Ord. Boraginaceæ, so called from their leaves resembling dogs' tongues, hence the common name 'hound's tongue' ; the species are pretty border plants.

Cynomorium, n., *sĭn'·ō·mōr'·ĭ·ŭm* (Gr. *kuōn*, a dog ; L. *morion*, a narcotic plant, nightshade), a genus of plants, Ord. Balanophoraceæ, which are root-parasites and tropical : **Cynomorium coccineum**, *kŏk·sĭn'·ĕ·ŭm* (L. *coccĭnĕus*, of a scarlet colour—from *coccum*, the berry of the scarlet oak), grows in Malta and Sardinia, and was long celebrated for arresting hæmorrhage ; usually known under the name of Fungus Melitensis, after Melita, the old name of Malta.

Cynosurus, n., *sĭn'·ŏs·ūr'·ŭs* (Gr. *kuōn*, a dog ; *oura*, a tail), a genus of grasses, Ord. Gramineæ, so called from its resemblance to a dog's tail, hence the common name, 'dog's-tail grass' : **Cynosurus cristatus**, *krĭst·āt'·ŭs* (L. *cristātus*, crested, tufted—from *crista*, a tuft on the head), esteemed one of the best fodder grasses in Europe.

Cyperaceæ, n. plu., *sĭp'·ĕr·ā'·sĕ·ē* (Gr. *kupeiros*, a kind of rush, the water-flag), the Sedge family, an Order of grass-like herbs, which do not supply nutriment to cattle : **Cyperus**, n., *sĭp·ēr'·ŭs*, a genus of sedges growing in water or in moist situations : **Cyperus papyrus**. *păp·ir'·ŭs* (L. *papyrus*, Gr. *papuros*, the paper-reed), the Papyrus of the Nile, the

cellular tissue of which was used in the manufacture of paper : **C. Syriacus**, *sĭr·i'·ăk·ŭs* (from *Syria*), differs from the C. papyrus in having the leaves and floral clusters drooping : **C. longus**, *lŏng'·gŭs* (L. *longus*, long), a species whose roots have been used as bitter and tonic remedies : **C. odoratus**, *ŏd'·ōr·āt'·ŭs* (L. *odorātus*, sweet-smelling—from *ŏdor*, a smell, scent), a species whose roots are aromatic : **C. esculentus**, *ĕsk'·ŭl·ĕnt'·ŭs* (L. *esculentus*, edible—from *esca*, food), supposed to be the flag of the Bible.

cyphellæ, n. plu., *sĭf·ĕl'·lē* (Gr. *kuphella*, things which are hollow, cups—from *kuphos*, crooked, bent), in *bot.*, urn-shaped soredia on the under surface of the thallus of some lichens : **cyphellate**, a., *sĭf·ĕl'·lāt*, having minute sunken cup-like spots, as the under surface of the thallus of Sticta.

cypsela, n., *sĭp'·sĕl·ă* (Gr. *kupselē*, a hollow, a chest), the inferior, monospermal, indehiscent fruit of Compositæ ; an achænium.

Cyrtandreæ, n. plu., *sĕr·tănd'·rĕ·ē* (Gr. *kurtos*, crooked ; *anēr*, a man, *andros*, of a man), a Sub-order of the Ord. Bignoniaceæ, having their fruit succulent or capsular, or siliquose and two-valved : **Cyrtandra**, n., *sĕr·tănd'·ră*, a genus of plants.

cyst, n., *sĭst*, also **cystis**, n., *sĭst'·ĭs* (Gr. *kustis*, a bladder), in animal bodies, a bag containing morbid matter ; a sac or vesicle ; in *bot.*, a sub-globose cell or cavity : **cystalgia**, n., *sĭst·ăl'·jĭ·ă* (Gr. *algos*, pain, grief), pain in the bladder : **cystic**, a., *sĭst'·ĭk*, pert. to or contained in a cyst ; pert. to the gall-bladder: **cystica**, n. plu., *sĭst'·ĭk·ă*, the embryonic forms of certain intestinal worms, as tapeworms.

Cysticercus, n., *sĭst'·ĭ·sĕrk'·ŭs* (Gr. *kustis*, a bladder ; *kerkos*, a tail), an embryo tapeworm which, in this stage, inhabits the flesh of various animals according to its species, and which consists of a head and neck like those of a tapeworm, from which, however, a vesicular appendage hangs down; the tailed bladder-worm : **Cysticercus cellulosæ**, *sĕl'·ŭl·ŏz'·ē* (L. *cellulōsus*, having cells), a species of tailed bladder-worm found in the substance of the heart of the pig ; the embryo of the tænia solium : **C. pisiformis**, *pīs'·ĭ·fŏrm'·ĭs* (L. *pĭsum*, a pea ; *forma*, shape), a species found in the rabbit : **C. tenuicollis**, *tĕn'·ŭ·ĭ·kŏl'·lĭs* (L. *tĕnŭis*, thin; *collum*, the neck), a species about an inch long with a very small head : **C. cucumerinus**, *kŭk·ūm'·ĕr·īn'·ŭs* (L. *cŭcŭmis*, a cucumber, *cŭcŭmĕris*, of a cucumber), a species of bladder-worms found in the rabbit : **C. fasciolaris**, *făs'·sĭ·ŏl·ār'·ĭs* (L. *fascĭŏla*, a small baudage—from *fascia*, a bandage), a species of bladder-worms found in the rat and mouse : all the preceding, and many others, are the embryos of the different species of tapeworm.

cystidium, n., *sĭst·ĭd'·ĭ·ŭm*, **cystidia**, n. plu., *sĭst·ĭd'·ĭ·ă* (dim. of Gr. *kustis*, a bladder), in *bot.*, sacs containing spores ; a kind of fructification in fungi.

cystitis, n., *sĭst·īt'·ĭs* (Gr. *kustis*, a bladder), inflammation of the bladder : **cystirrhœa**, n., *sĭst'·ĭr·rē'·ă* (Gr. *rheo*, I flow), a discharge of mucus from the bladder.

cystocarp, n., *sĭst'·ō·kărp* (Gr. *kustis*, a bladder; *karpos*, fruit), the receptacle in which the spores are ultimately formed in Florideæ, a Sub-order of seaweeds.

cystocele, n., *sĭst'·ō·sēl* (Gr. *kustis*, a bladder ; *kele*, a tumour), hernia of the bladder: **cystodynia**, n., *sĭst'·ō·dīn'·ĭ·ă* (Gr. *odunē*, pain), pain in the bladder.

cystocestoid, a., *sĭst'·ō·sĕst'·ōўd* (Gr. *kustis*, a bladder; *kestos*, a girdle;

eidos, resemblance), applied to intestinal, cystose parasites having flat bodies.

cystoid, a., *sĭst'·ōўd* (Gr. *kustis*, a bladder ; *eidos*, resemblance), resembling a cyst: **cystose**, a., *sĭst'·ōz*, containing cysts.

cystolith, n., *sĭst'·ō·lĭth* (Gr. *kustis*, a bladder ; *lithos*, a stone), a cell containing numerous crystals, usually lying loose, as in the leaf of Ficus.

Cytisus, n., *sĭt'·ĭs·ŭs* (Gr. *kutisos*, the bean trefoil tree), a very ornamental genus of trees and shrubs, Ord. Leguminosæ, Sub-ord. Papilionaceæ : **Cytisus scoparius**, *skŏp·ār'·ĭ·ŭs* (L. *scopārius*, a broom to sweep with), a species whose broom-tops are used as a diuretic: **C. laburnum**, *lăb·ĕrn'·ŭm* (L. *laburnum*, the laburnum tree or shrub), the laburnum tree, the seeds and bark of which are narcotic.

cytoblast, n., *sĭt'·ō·blăst* (Gr. *kutos*, a vessel, a cell ; *blastano*, I bud), the nucleus of animal and vegetable cells : **cytoblastema**, n., *sĭt'·ō·blăst·ēm'·ă*, the viscous fluid, or formative material, in which animal and vegetable cells are produced, and by which they are held together ; protoplasm.

cytogenesis, n., *sĭt'·ō·jĕn'·ĕs·ĭs* (Gr. *kutos*, a vessel, a cell ; *genesis*, origin), the development of cells in animal and vegetable structures: **cytogenetic**, a., *sĭt'·ō·jĕn·ĕt'·ĭk*, pert. to cell formation : **cytogenous**, a., *sĭt·ŏdj'·ĕn·ŭs*, having connective tissue : **cytogeny**, n., *sĭt·ŏdj'·ĕn·ĭ*, cell formation.

Cyttaria Darwinii, *sĭt·tār'·ĭ·ă dăr·wĭn'·ĭ·ī* (Gr. *kuttăros*, the cell in the comb of bees), a species of fungi named in honour of Darwin, found on the bark of the beech, globular, and of a bright yellow colour, and which the natives of Tierra del Fuego eat.

Dactylis, n., *dăkt'·ĭl·ĭs* (Gr. *daktulos*,

H

L. *dactylis*, a finger), the cock's-foot grass, a genus of grasses, Ord. Gramineæ, so called from the fancied resemblance of the head to fingers.

dædalenchyma, n., *dēd'·ăl·ĕng'·kĭm·ă* (Gr. *daidaleos*, skilfully wrought, variegated ; *engchuma*, an infusion, tissue), tissue composed of entangled cells, as in some fungi.

Dalbergieæ, n. plu., *dăl'·bérj·ĭ'·ĕ·ē* (after *Dalberg*, a Swedish botanist), a tribe of plants, Sub-ord. Papilionaceæ, Ord. Leguminosæ : Dalbergia, n., *dăl·bérj'·ĭ·ă*, a genus of plants : Dalbergia sissoo, *sĭs'·sŏ·ō* (a native name), an Indian forest tree, valued for its wood.

daltonism, n., *dăwlt'·ŏn·ĭzm*, colour blindness ; a condition of the eye in which the individual cannot distinguish one colour from another, so called from Dr. Dalton, the chemist, who suffered under this defect.

Dammar, n., *dăm'·ăr* (a native name), a very handsome genus of trees, Ord. Coniferæ, from a species of which, it is said, liquid storax is obtained : Dammar Australis, *ăws·trăl'·ĭs* (L. *Australis*, Southern—from *Auster*, the South), the Kawri pine of New Zealand, which yields a hard resin : D. Orientalis, *ŏr'·ĭ·ĕnt·ăl'·ĭs* (L. *Orientalis*, Eastern—from *Oriens*, the East), a species called the Amboyna pitch tree, also yielding resin : white Dammar, Indian copal or gum animi, used in India as a varnish, is obtained from Vateria Indica.

dandelion, n., *dăn'·dĕ·lĭ'·ŏn* (F. *dent*, tooth ; *de*, of ; *lion*, lion), a well-known plant having a yellow flower on a naked stem, and deeply-notched leaves ; yields a milky juice, used medicinally ; the Taraxacum Dens Leonis, Ord. Compositæ.

dandriff, n., *dăn'·drĭf*, also dan-

druff, n., *dăn'·drŭf* (Bret. *tañ*, F. *teigne*, scurf ; W. *drwg*, bad, evil), a disease of the scalp, characterised by quantities of little scales on the skin ; pityriasis.

Daphnæ, n. plu., *dăf'·nē* (Gr. and L. *Daphnē*, the daughter of the river god Peneus, changed into a laurel tree), a Sub-order of the Ord. Thymelæaceæ, having hermaphrodite or nearly unisexual flowers : Daphne, n., *dăf'·nē*, a genus of handsome dwarf shrubs, mostly evergreens : Daphne mezereum or mezereon, *mĕz'·ĕr·ĕ'·ŭm* or -*ĕ'·ŏn* (said to be Pers. *madzaryoun*), the bark of the root and branches used in decoction as a diaphoretic in cutaneous and syphilitic affections, in large doses acts as an irritant poison, and the succulent fruit is poisonous : D. gnidium, *nĭd'·ĭ·ŭm* (*gnidia*, ancient name of the laurel), the spurge flax or flax-leaved daphne, bark has been used in medicine : D. Alpina, *ălp·ĭn'·ă* (*Alpinus*, from the Alps), a dwarf olive tree, said to be purgative : D. cneorum, *nē·ōr'·ŭm* (Gr. *kneōron*, a kind of nettle, a species of daphne), a beautiful species, having similar properties to the D. mezereum : D. Pontica, *pŏnt'·ĭk·ă* (L. *Ponticus*, pert. to the Black Sea—from *Pontus*, the Black Sea), a spurge laurel, having diaphoretic qualities : D. laureola, *lăwr·ē'·ŏl·ă* (L. *laurĕŏlus*, a small laurel—from *laurĕa*, a laurel tree), the spurge laurel, bark used in medicine, the berries are poisonous to all animals except birds : daphnein, n., *dăf'·nē·ĭn*, the neutral crystalline principle contained in the D. mezereum.

Darlingtonia, n., *dăr'·lĭng·tōn'·ĭ·ă* (after *Dr. Darlington*, of America), a genus of the Ord. Sarraceniaceæ, pretty plants from the Rocky Mountains.

darnel, n., *dărn'·ĕl* (Prov. F. *dar*

nelle; Lith. *durnas*, foolish as in intoxication), a weed among corn, supposed to induce intoxication ; the grass **Lolium temulentum**, said to be poisonous, but erroneously.

dartos, n., *dȧrt'ŏs* (Gr. *dartos*, the fleshy coat covering the testes—from *děro*, I flay), a thin layer of loose, reddish, musculo-cutaneous tissue, forming the tunic of the scrotum : **dartoid**, a., *dȧrt'oyd*, resembling the dartos.

Datiscaceæ, n. plu., *dăt'ĭs·kā'sě·ē* (origin unknown), the Datisca family, an Order of herbaceous plants : **Datisca**, n., *dăt·ĭsk'ă*, a genus of hardy herbaceous plants, some of which are bitter : **Datisca cannabina**, *kăn'ă·bĭn'ă* (Gr. *kannăbis*, L. *cannăbis*, the hemp), a species said to have purgative qualities.

Datura, n., *dăt·ūr'ă* (a corruption of Arabic name *tatorah*), a genus of plants, Ord. Solanaceæ, Subord. Atropeæ, many of the species of which are powerfully narcotic : **Datura stramonium**, *stră·mōn'ĭ·ŭm* (L. *strāmen*, straw, *strāminis*, of straw, so called from its fibrous roots), the thorn apple, so called from its prickly capsule ; the leaves and seeds are used as narcotics, and in the form of powder and tincture as anodynes and antispasmodics, the leaves are smoked for asthma : **D. Tatula**, *tăt'ūl·ă* (N. American word) ; **D. Metel**, *mět'ĕl* (from Asia) ; **D. sanguinea**, *săng·gwĭn'ĕ·ă* (L. *sanguĭnĕus*, of blood —from *sanguis*, blood), the red thorn apple ; **D. ferox**, *fēr'ŏks* (L. *ferox*, wild, fierce) : **D. fastuosa**, *făst'ū·ōz'ă* (new L. *fastuōsus*, pert. to pride — from *fastus*, arrogance, pride), are species which have properties similar to D. stramonium : **D. alba**, *ălb'ă* (L. *albus*, white), the white-flowered Datura, whose leaves

and seeds are used in India as sedative and narcotic.

Daucus, n., *dȧwk'ŭs* (Gr. *daukōn*, a kind of wild carrot), a genus of plants, Ord. Umbelliferæ : **Daucus carota**, *kăr·ōt'ă* (mid. L. *carōta*, a carrot), a species producing the esculent root, the common carrot ; from this and other species is made the Ajowan or Omam, a condiment of India.

Davallia, n., *dăv·ăl'lĭ·ă* (in honour of *Davall*, a Swiss botanist), a genus of very beautiful ferns : **Davallia Canariensis**, *kăn·ăr'ĭ·ĕns'ĭs* (new L. *Canariensis*, of or from the Canary Islands), a beautiful species of fern, whose root-stock, covered with coarse brown hair, very much resembles a hare's foot, hence the name ' hare's-foot ' fern.

decandrous, a., *děk·ănd'rŭs* (Gr. *děka*, ten ; *anēr*, a male, *andros*, of a male), in *bot.*, applied to a flower that has ten stamens.

decapoda, n. plu., *děk·ăp'ŏd·ă* (Gr. *děka*, ten ; *podes*, feet), a section of the Crustaceæ which have ten ambulatory feet ; the family of cuttle-fishes, having ten arms or cephalic processes.

deciduous, a., *dě·sĭd'ū·ŭs* (L. *deciduus*, that falls down or off— from *de*, down ; *cado*, I fall), not perennial or permanent ; applied to parts which fall off or are shed during the life of the animal ; in *bot.*, falling off after performing its functions for a limited time ; applied to trees which lose their leaves annually.

declinate, a., *děk'lĭn·āt* (L. *declino*, I turn aside, I bend—from *de*, down ; *clino*, I lean), in *bot.*, directed downwards from its base ; bent downwards or on one side.

decoction, n., *dě·kŏk'shŭn* (L. *decoctus*, a boiling down—from *de*, down ; *coctus*, boiled or baked), the extraction of the virtues of any substance by boiling it in water; the fluid in which the substance has been boiled.

decollated, a., *dĕ·kŏl'·lāt·ĕd* (L. *dēcollātum*, to behead—from *de*, down; *collum*, the neck), applied to univalve shells, the apex of which falls off in the course of growth: **decollation,** n., *dĕ·kŏl·lā'·shŭn*, the separation of the head from the trunk.

decompound, a., *dē'·kŏm·po̅w̅nd* (L. *de*, down, from; and Eng. *compound*), in *bot.*, applied to a leaf cut into numerous compound divisions.

decorticate, a., *dĕ·kŏrt'·ĭk·āt*, also **decorticated,** a., *-āt'·ĕd* (L. *decorticātum*, to deprive of the bark —from *de*, down, from; *cortex*, bark), deprived of the bark or cortical layer: **decortication,** n., *dĕ·kŏrt'·ĭk·ā'·shŭn*, the operation of stripping off bark.

decumbent, a., *dĕ·kŭmb'·ĕnt* (L. *decumbens*, lying down—from *de*, down; *cumbo*, I lie), in *bot.*, lying flat along the ground, but rising from it at the extremity.

decurrent, a., *dĕ·kŭr'·rĕnt* (L. *decurrens*, running down from a higher point—from *de*, down; *currens*, running), in *bot.*, applied to leaves which adhere to the stem beyond their point of attachment, forming a sort of winged or leafy appendage, as in thistles.

decussate, a., *dĕ·kŭs'·sāt* (L. *decussātum*, to divide crosswise, as in the form of an x), in *bot.*, applied to opposite leaves crossing each other in pairs at right angles; v., to cross or intersect in the form of x: **decussation,** n., *dĕk'·ŭs·sā'·shŭn*, the intersection or crossing of lines, etc. in the form of an x; union in the shape of an x or cross: **decussative,** a., *dē·kŭs'·sāt·ĭv*, formed in the shape of a cross.

dedoublement, n., *dē·dŭb'·l·mĕnt* (L. *de*, down; Eng. *double*); also **deduplication,** n., *dē·dŭp'·lĭk·ā'·*

shŭn (L. *de*, down; Eng. *duplication*), the act of doubling down; in *bot.*, the separation of a layer from the inner side of a petal, either presenting a peculiar form, or resembling the part from which it is derived; chorisis—which see.

defecation, n., *dĕf'·ĕk·ā'·shŭn* (L. *defæcātum*, to cleanse from dregs, to refine—from *de*, down, from; *fæx*, dregs or refuse matter, *fæcis*, of dregs), the act of discharging the fæces from the bowels; the removal of the lees or sediment of a liquid.

defervescence, n., *dĕf'·ĕrv·ĕs'·ĕns* (L. *defervescens*, ceasing to boil, cooling down—from *de*, down; *fervesco*, I become hot), in *med.*, the fall in the temperature in a patient, when convalescent from an acute disease.

definite, a., *dĕf'·ĭn·ĭt* (L. *definītum*, to limit—from *de*, down, from; *fīnis*, an end), in *bot.*, applied to inflorescence, when it ends in a single flower, and the expansion of the flower is centrifugal; having the number of the parts of an organ limited and not exceeding twenty.

deflexed, a., *dĕ·flĕkst'* (L. *deflexum*, to bend or turn aside—from *de*, down; *flexum*, to bend, to curve), in *bot.*, bent in a continuous curve.

defoliation, n., *dĕ·fōl'·ĭ·ā'·shŭn* (mid. L. *defoliātum*, to shed leaves—from *de*, down; *folium*, a leaf), the fall or shedding of the leaves of plants.

degeneration, n., *dĕ·jĕn'·ĕr·ā'·shŭn* (L. *degenerātum*, to depart from its race or kind—from *de*, down; *genus*, race, kind, *gĕnĕris*, of a kind), a gradual deterioration in a part of a living body, in the whole living body, or in a race; in *bot.*, a deterioration of growth or development in a part, as when scales take the place of leaves.

deglutition, n., *dĕg'·lŏ·tĭsh'·ŭn* (L. *de*, down; *glūtio*, I swallow), the act of swallowing food after mastication.

dehisce, v., *dē·hĭs'* (L. *dehisco*, I split open, I part asunder—from *de*, down, from ; *hisco*, I open, I gape), in *bot.*, to open or part asunder, as the seed - pods of plants: **dehiscence**, n., *dē·hĭs'ĕns*, the mode of opening an organ, as of the seed vessel and anther, generally along a determinate line : **dehiscing**, a., *dē·hĭs'ĭng*, splitting into regular parts.

Delesseria, n., *dĕl'ĕs·sēr'ĭ·ă* (in honour of *Delessert*, a French botanist), a beautiful genus of mostly deep-green Algæ, found in the ocean and on the sea-shore.

Delima, n., *dĕl·ĭm'ă* (L. *delīmo*, I file or shave off—from *de*, down ; *līmo*, I file), a genus of very fine climbing plants, Ord. Dilleniaceæ, so called from their leaves being used for polishing.

deliquesce, v., *dĕl'ĭ·kwĕs'* (L. *deliquesco*, I dissolve—from *de*, down; *liqueo*, I am fluid), to melt or become liquid by attracting moisture from the air : **deliquescent**, a., *dĕl'ĭ·kwĕs'ĕnt*, liquefying by contact with the air : **deliquescence**, n., *dĕl'ĭ·kwĕs'ĕns*, the melting by absorbing moisture from the air, as certain substances do.

delirium, n., *dĕ·lĭr'ĭ·ŭm* (L. *delirium*, madness), that condition of the mind in acute disease in which the mind wanders, resulting in incoherent speech : **delirium tremens**, *trĕm'ĕnz* (L. *trĕmens*, shaking, quivering), temporary insanity accompanied with a tremulous condition of the body and limbs, a disease of habitual drinkers : **d. traumaticum**, *trăwm·ăt'ĭk·ŭm* (Gr. *traumătĭkos*, L. *traumătĭcus*, fit for healing wounds), a similar disease which may follow serious accidents or surgical operations.

delitescence, n., *dĕl'ĭt·ĕs'ĕns* (L. *delitescens*, lying hid—from *de*,

down ; *latescens*, hiding oneself), in *med.*, the period during which morbid poisons, as smallpox, lie hid in the system ; the sudden termination of an inflammation.

Delphinium, n., *dĕl·fĭn'ĭ·ŭm* (L. *delphin*, a dolphin—from a supposed resemblance in the nectary of the plant to the imaginary figure of the dolphin), a genus of showy plants, Ord. Ranunculaceæ: **Delphinium staphysagria**, *stăf'ĭ·săg'rĭ·ă* (Gr. *staphis*, a dried grape ; *agria*, belonging to the country, rustic), the plant Stavesacre, whose seeds are irritant and narcotic, used for destroying vermin : **D. glaciale**, *glăs'ĭ·āl'ĕ* (L. *glăciālis*, icy, frozen), a species which grows at the height of 16,000 feet on the Himalayas : **D. ajacis**, *ădj·ās'ĭs* (from Switzerland); and **D. consolida**, *kŏn·sŏl'ĭd·ă* (L. *consŏlĭdo*, I make very solid or firm—from *con*, together; *sŏlĭdus*, whole, complete), two species universally grown among border annuals; the latter is regarded as a simple astringent.

deltoid, a., *dĕlt'·oÿd* (the Gr. letter Δ, called delta ; *eidos*, resemblance), shaped somewhat like a delta ; triangular in the outline or section ; denoting a large, thick, triangular muscle, which forms the convexity of the shoulder, and pulls the arm directly outwards and upwards ; **deltoids**, n. plu., also **deltoides**, n. plu., *dĕlt·oÿd'·ēz* : **deltoid ligament**, the internal lateral ligament of the ankle joint, consisting of a triangular layer of fibres.

dementia, n., *dĕ·mĕn'shĭ·ă* (L. *dēmentĭa*, the being out of one's mind—from *de*, down ; *mens*, the mind), that form of insanity in which the powers of the mind gradually fade away, or become a perfect blank : **dementia senilis**, *sĕn·ĭl'ĭs* (L. *senĭlis*, aged, senile), the loss of intellect in old age.

demodex folliculorum, *dĕm'·ŏd·ĕks fŏl·lĭk'·ŭl·ōr'·ŭm* (Gr. *dēmos,* fat; *dēx,* a worm that devours wood; L. *follĭcŭlus,* skin, follicle, *folliculōrum,* of skins), the worm-like parasite found in the hair follicles of the human skin, especially those on the side of the nose.

demulcent, a., *dĕ·mŭls'·ĕnt* (L. *demulcens,* stroking down—from *de,* down; *mulcens,* soothing gently), softening; mollifying: n., a medicine which softens or mollifies.

dendriform, .a., *dĕnd'·rĭ·fŏrm* (Gr. *dendron,* a tree; L. *forma,* shape), also **dendroid,** a., *dĕnd'·rŏyd* (Gr. *dendron,* a tree; *eidos,* resemblance), and **dendritic,** *dĕnd·rĭt'·ĭk,* branched like a tree; arborescent.

Dendrobium, n., *dĕnd·rōb'·ĭ·ŭm* (Gr. *dendron,* a tree; *bios,* life), a splendid genus of orchidaceous plants, Ord. Orchidaceæ, the species being generally found upon trees in the places of their natural growth: **Dendrobium nobile,** *nŏb'·ĭl·ĕ* (L. *nōbĭlis,* famous); **D. chrysanthum,** *krĭs·ănth'·ŭm* (Gr. *chrusos,* gold; *anthos,* a flower); **D. Gibsoni,** *gĭb·sŏn'·ī* (*Gibsōni,* of Gibson); **D. fimbriatum,** *fĭm'·brĭ·āt'·ŭm* (L. *fimbriātum,* fringed —from *fimbriæ,* fibres, threads); **D. densiflorum,** *dĕns'·ĭ·flōr'·ŭm* (L. *densus,* thick, dense; *flōrum,* shining, bright), are a few species unsurpassed in the beauty of their flowers.

dengue, n., *dĕng'·gā* (in the British West Indian Islands, this disease was called *dandy,* in reference to the stiffness and restraint it gave to the limbs, afterwards translated by the Spaniards into their *dangue,* meaning prudery, fastidiousness, from its similarity of sound), a violent and singular form of fever and rheumatism which is an occasional epidemic in tropical regions.

dens prolifer, *dĕns prŏl'·ĭf·ĕr* (L.

dens, a tooth; *prōles,* offspring; *fero,* I bear), a tooth growing apparently on a parent tooth: **dens sapientiæ,** *săp'·ĭ·ĕn'·shĭ·ē* (L. *sapientia,* wisdom, *sapientiæ,* of wisdom), the tooth of wisdom, or the wisdom tooth, the last molar in each range of teeth, so called from its late appearance through the gums.

dentate, a., *dĕnt'·āt* (L. *dentātus,* having teeth—from *dens,* a tooth), in *bot.,* toothed; having short triangular divisions of the margin: **denticulate,** a., *dĕnt·ĭk'·ŭl·āt* (L. *denticulātus,* furnished with small teeth), having very small tooth-like projections along the margin: **denticulations,** n. plu., *dĕnt·ĭk'·ŭl·ā'·shŭns,* very small teeth.

dentine, n., *dĕnt'·ĭn* (L. *dens,* a tooth, *dentis,* of a tooth), the principal mass or foundation of the body and root of a tooth, resembling very compact bone, though not identical with it in structure.

dentirosters, n. plu., *dĕnt'·ĭ·rŏst'·ĕrs,* or **dentirostres,** n. plu., *-rŏst'·rēz* (L. *dens,* a tooth, *dentis,* of a tooth; *rostrum,* a beak), the group of perching birds in which the upper mandible of the beak has its lower margin toothed · **dentirostrate,** a., *dĕnt'·ĭ·rŏst'·rāt,* having the beak like a tooth.

denudate, a., *dĕn'·ŭd·āt* (L. *denudātum,* to lay bare, to make naked —from *de,* down; *nŭdus,* naked), in *bot.,* having a downy or hairy surface made naked: **denudation,** n., *dĕn'·ŭd·ā'·shŭn,* the act or state of being laid bare or made naked.

deobstruent, n., *dē·ŏb'·strōō·ĕnt* (L. *de,* down; *obstruens,* building anything for the purpose of stopping the way, gen. *obstruentis*), any medicine supposed to be able to remove an obstruction in a part of the body, such as enlargements, tumours, etc.

deodorant, n., *dē·ŏd'·ŏr·ănt* (L.

de, down; *odorans*, giving a smell to—from *odor*, a smell, good or bad), a substance which purifies the air and removes noxious vapours or gases which may be injurious to human life: **deodorise**, v., *dē-ŏd'-ŏr-īz*, to disinfect : **deodorisation**, n., *dē-ŏd'-ŏr-īz-ā'-shŭn*, the art or act of depriving of odour or smell.

depilation, n., *dĕp'-il-ā'-shŭn* (L. *depilātum*, to pull out the hair—from *de*, down; *pĭlus*, a hair), loss of hair, naturally or by art : **depilatory**, a., *dĕ-pĭl'-ăt-ŏr-ĭ*, having the quality or power of removing hair: **n.**, any ointment or lotion to take off hair without injuring the skin.

deplanate, a., *dĕp-lăn'-āt* (L. *de*, down; *plānātum*, to make level), in *bot.*, flattened.

deplete, v., *dĕ-plēt'* (L. *deplētum*, to empty out—from *de*, down; *pleo*, I fill), to reduce in quantity by taking away : **depletion**, n., *dĕ-plē'-shŭn*, the act of emptying; the act diminishing the quantity: **depletives**, n. plu., *dĕ-plēt'-ĭvs*, substances calculated to diminish fulness of habit.

depressed, a., *dĕ-prĕst'* (L. *depressum*, to press or weigh down—from *de*, down; *pressum*, to press), in *bot.*, applied to a solid organ having the appearance of being flattened from above downwards : **depression**, n., *dĕ-prĕsh'-ŭn*, a hollow; the hollow formed by the fractured portion of the cranial bone.

depressor, n., *dĕ-prĕs'-sŏr* (L. *depressum*, to press or weigh down), applied to certain muscles which draw down the parts on which they act : **depressor alæ nasi**, *dĕ-prĕs'-sŏr ăl'-ē nāz'-ī* (L. *depressor*, that which depresses; *āla*, a wing, *alœ*, of a wing ; *nāsus*, the nose, *nāsi*, of the nose), a short radiated muscle whose fibres are inserted into the septum and back part of the ala of the nose :

depressor anguli oris, *ăng'-gŭl-ī ŏr'-ĭs* (L. *angŭlus*, an angle, *angŭli*, of an angle ; *ōs*, the mouth, *ōris*, of the mouth), a triangular muscle arising from its broad base from the external oblique line of the lower jaw, and passing upwards into the angle of the mouth.

depurant, n., *dĕp'-ŭr-ănt* (L. *depūrans*, purifying or cleansing—from *de*, down ; *pūrus*, clean, pure), a medicine supposed to be capable of purifying the blood : **depuration**, n., *dĕp'-ŭr-ā'-shŭn*, the act or process of freeing from impurities ; the cleansing of a wound.

derma, n., *dĕrm'-ă* (Gr. *derma*, a skin, *dermatos*, of a skin), the true skin; see 'cutis vera': **dermal**, a., *dĕrm'-ăl*, belonging to or consisting of the true skin : **dermatoid**, a., *dĕrm'-ăt-ōyd* (Gr. *eidos*, resemblance), resembling the skin : **dermatitis**, n., *dĕrm'-ăt-īt'-ĭs*, inflammation of the skin : **dermatogen**, n., *dĕrm-ăt'-ŏ-jĕn* (Gr. *gennao*, I beget, I produce), the outermost layer or covering of the skin in plants which becomes the epidermis.

Dermatodectes, n. plu., *dĕrm'-ăt-ō-dĕk'-tēz* (Gr. *derma*, skin, *dermatos*, of a skin; *dēktēs*, a biter), a genus of parasites of the horse, ox, and sheep, so called because they simply bite and hold on to the skin.

dermoid, a., *dĕrm'-ōyd* (Gr. *derma*, skin ; *eidos*, resemblance), resembling skin ; dermatoid : **dermoid papilla**, *păp-ĭl'-lă* (L. *papilla*, a nipple), in *surg.*, a small eminence, covered with a skin-like substance.

dermosclerites, n. plu., *dĕrm-ŏs'-klĕr-īt'-ēz* (Gr. *derma*, skin ; *sklĕros*, hard), masses of spicules found in the tissues of some of the Alcyonidæ.

dermo-skeleton, n., *dĕrm'-ō-skĕl'-ĕt-ŏn* (Gr. *derma*, skin ; Eng.

skeleton), the hard integument which covers many animals, and affords protection to them, making its appearance as a leathery membrane, or as shell, crust, scales, or scutes.

descendens abdominis, *dĕ·sĕnd'-ĕns ăb·dŏm'·ĭn·ĭs* (L. *descendens*, descending ; *abdōmen*, the belly, *abdŏminis*, of the belly), the muscle that supports and compresses the abdomen : **descendens noni,** *nŏn'·ĭ* (L. *nonus*, ninth, *nōni*, of ninth), applied to a branch of the ninth pair of nerves of the neck.

desiccation, n., *dĕs'·ĭk·ā'·shŭn* (L. *dēsiccātum*, to dry up—from *de*, down ; *siccus*, dry), the act of making quite dry ; the state of being dried : **desiccant,** a., *dĕs'·ĭk·ănt*, drying : n., a medicine that dries a sore : also **desiccative,** a., *dĕs·ĭk'·āt·ĭv*, in same sense.

Desmidieæ, n. plu., *dĕs'·mĭd·ĭ'·ĕ·ē* (said to be from Gr. *desmos*, a bond, from the parts cohering when in a state of dissolution), a Sub-order of plants, Ord. Conjugatæ, of the great combined Ord. Algæ or Hydrophyta : **Desmidium,** n., *dĕs·mĭd'·ĭ·ŭm*, a genus of minute green Algæ, found in summer in still waters : **Desmidiæ,** n. plu., *dĕs·mĭd'·ĭ·ē*, minute fresh-water plants of a green colour, without a siliceous epidermis.

Desmodium, n., *dĕs·mōd'·ĭ·ŭm* (Gr. *desmos*, a bond, having reference to the stamens being joined), a genus of plants, Subord. Papilionaceæ, Ord. Leguminosæ : **Desmodium gyrans,** *jïr'·ăns* (L. *gyrans*, turning round in a circle), the Gorachand of Bengal, a sensitive plant, whose compound leaves are in constant movement, in jerks, oscillatory movements, or movements upwards and downwards, and which also exhibit a remarkable irritability : **D. gyroides,** *jïr·ŏyd'·ēz*

(L. *gyrus*, Gr. *guros*, a circular course ; *eidos*, resemblance) ; **D. vespertilionis,** *vĕsp'·ĕr·tĭl·ĭ·ŏn'·ĭs* (L. *vespertīlīo*, a bat, *vespertīliōnis*, of a bat—from *vesper*, the evening), are species which exhibit similar movements : **D. diffusum,** *dĭf·fūz'·ŭm* (L. *diffūsum*, to pour or spread out—from *dif*, asunder ; *fundo*, I pour), a species affording a fodder plant.

desquamation, n., *dĕs'·kwăm·ā'·shŭn* (L. *dēsquāmātum*, to scale or peel off—from *de*, down ; *squama*, a scale), the act of throwing off in scales, as the skin ; the separation of the scurf-skin in the form of scales, layers, or patches.

desudation, n., *dĕs'·ūd·ā'·shŭn* (L. *desudo*, I sweat greatly—from *de*, down ; *sudo*, I sweat), a profuse sweating.

detergent, n., *dĕ·tĕrj'·ĕnt* (L. *detergens*, wiping off—from *de*, down ; *tergĕo*, I wipe clean), cleansing : n., a medicine that cleans wounds, ulcers, etc.

determinate, a., *dĕ·tĕrm'·ĭn·āt* (L. *determinātum*, to border off, to bound—from *de*, down ; *terminus*, a boundary), in *bot.*, having a definite or cymose inflorescence ; the opposite of 'effuse.'

detrusor urinæ, *dĕ·trŏz'·ŏr ūr·ĭn'·ē* (L. *detrūsor*, that which forces away—from *de*, down ; *trūdo*, I thrust ; *urīna*, the urine, *urinæ*, of the urine), the external muscular coat of the bladder, which expels the urine.

deuterozooids, n. plu., *dūt'·ĕr·ō·zō'·ōyds* (Gr. *deuteros*, second ; *zoön*, an animal ; *eidos*, resemblance), those zooids which are produced by germination from zooids.

Deutzia, n., *dūtz'·ĭ·ă* (after *Deutz* of Amsterdam), a genus of very ornamental plants, Ord. Philadelphaceæ : **Deutzia scabra,** *skăb'·ră* (L. *scăber* or *scăbra*, rough, scabby), a species which has a scurfy matter on its leaves, which are used for polishing in Japan.

development, n., *dĕ·vĕl'·ŏp·mĕnt* (F. *développer*, to unfold), the progressive changes taking place in living bodies until maturity is reached ; increase ; growth : **theory of development,** the progressive advancement of life from its lowest types as they first appeared on the earth, or are supposed to have first existed, up to those highest forms of life now existing on the earth, as contradistinguished from acts of direct creation ; evolution.

devitalise, v., *dĕ·vit'·ăl·īz* (L. *de*, down ; Eng. *vitalise*), to deprive of vitality or life, as the part of an animal body.

dewlap, n., *dū'·lăp* (Dan. *dog·lœp*, dew-sweeping ; Prov. Sw. *dogg*, Dut. *douw*, dew ; Dan. *lœp*, a flap), the loose skin which hangs down from the neck of an ox.

dextral, a., *dĕks'·trăl* (L. *dextra*, the right hand), right hand, or to the right hand ; denoting the direction of the spiral in the greater number of univalve shells.

dextrine, n., *dĕks'·trĭn* (L. *dexter*, right, on the right hand), a gummy matter into which the interior substance of starch globules is convertible by diastase, and by certain acids, so called from turning the plane in polarized light to the right hand.

dextrorse, a., *dĕks·trŏrs'* (L. *dexter*, to the right ; *versus*, turned), directed towards the right.

diabetes, n., *dī'·ă·bēt'·ēz* (Gr. *diabetes*, a siphon—from *dia*, through; *baino*, I go), a disease characterised by passing an immoderate quantity of urine, with great thirst and general debility : **diabetes insipidus,** *ĭn·sĭp'·ĭd·ŭs* (L. *insĭpidus*, unsavoury—from *in*, not ; *săpĭo*, I taste), diabetes in which the urine is limpid and

devoid of sugar : **d. mellitus,** *mĕl·līt'·ŭs* (L. *mellītus*, tasting like honey—from *mel*, honey), diabetes in which the urine is sweet, and contains sugar.

diacetate, n., *dī·ăs'·ĕt·āt* (Gr. *dis*, twice ; Eng. *acetate*), an 'acetate' is a combination of acetic acid with a salifiable base; a 'diacetate' is an 'acid acetate,' or a combination of two parts of acetic acid with a salifiable base : **diacetate of lead,** sugar of lead ; commonly called 'acetate of lead.'

diachænium, n., *dī'·ă·kēn'·ĭ·ŭm* (Gr. *dis*, twice; Eng. *achænium*), fruit composed of two achænia united by a commissure to a common axis ; same as 'cremocarp.'

diachylon, n., *dī·ăk'·ĭl·ŏn* (Gr. *dia*, through, by means of ; *chulos*, juice), an adhesive plaster, formerly made from expressed juices, now made of an oxide of lead and oil ; litharge plaster.

diachyma, n., *dī·ăk'·ĭm·ă* (Gr. *dia*, through ; *chumos*, a fluid, juice), the cellular tissue of leaves occupying the space between their two surfaces.

diadelphous, a., *dī'·ă·dĕlf'·ŭs* (Gr. *dis*, twice ; *adelphos*, a brother), having stamens in two bundles united by their filaments.

diagnosis, n., *dī'·ăg·nōz'·ĭs* (Gr. *diagnōsis*, a judging power or faculty—from *dia*, through ; *gignōskō*, I know), the art of distinguishing one disease from another.

dialycarpous, a., *dī'·ăl·ĭ·kărp'·ŭs* (Gr. *dialūo*, I part asunder ; *karpos*, fruit), in *bot.*, having a pistil or fruit composed of distinct carpels : **dialypetalous,** a., *dī'·ăl·ĭ·pĕt'·ăl·ŭs* (Gr. *petalon*, a leaf), having corollas composed of several petals : **dialysepalous,** a., *dī'·ăl·ĭ·sĕp'·ăl·ŭs* (Eng. *sepal*), having a calyx composed of separate sepals ; also **diaphyllous,** a., *dī·ăf'·ĭl·lŭs* (Gr. *phullon*, a leaf), in same sense.

dialysis, n., *dī·ăl'·ĭs·ĭs* (Gr. *dialusis*,

a dissolving or dissolution—from *dia*, through ; *luo*, I loose), in *chem.*, a process of analysis of a liquid by diffusion through organic membranes, or such artificial septa of organic matter as parchment - paper ; the separation of crystallisable from uncrystallisable substances, a septum allowing the passage of the former and not of the latter ; in *bot.*, the separation of parts usually joined.

diandrous, a., *dĭ-ănd′.rŭs* (Gr. *dis*, twice ; *anēr*, a male, *andros*, of a male), having two stamens, as a flower.

Dianthus, n., *dĭ-ănth′.ŭs* (Gr. *dios*, divine ; *anthos*, a flower—having allusion to the fragrance and beautiful arrangement of the flowers), a very beautiful and ornamental genus of plants, Ord. Caryophylleæ, containing some of the most prized flowers we possess, such as clove-pink and carnation : **Dianthus caryophyllus**, *kăr′.ĭ-ō-fil′.lŭs* (Gr. *karuon*, a nut ; *phullon*, a leaf—in reference to the shape of the flower-buds), the clove - pink, or clove gilly-flower ; sometimes used in making a syrup.

Diapensieæ, n. plu., *dĭ′.ă-pĕns-ĭ′.ĕ-ē* (Gr. *dia*, through ; *pente*, five—alluding to flowers being five cleft), a Sub-order or tribe of plants, Ord. Hydrophyllaceæ.

diaphanous, a., *dĭ-ăf′.ăn-ŭs* (Gr. *dia*, through ; *phaino*, I show), allowing light to pass through ; nearly transparent.

diaphoresis, n., *dĭ′.ă-fŏr-ēz′.ĭs* (Gr. *diaphoresis*, a carrying through, perspiration—from *dia*, through ; *phoreo*, I carry), an increase of perspiration : **diaphoretic**, n., *dĭ′.ă-fŏr-ĕt′.ĭk*, a medicine which increases perspiration.

diaphragm, n., *dĭ′.ă-frăm* (Gr. *diaphragma*, a partition wall—from *dia*, through ; *phrasso*, I hedge or fence in), the midriff ; the large muscle which forms the

partition between the abdominal and thoracic cavities, also called the 'phren' or 'septum transversum'; in *bot.*, a dividing membrane or partition : **diaphragmitis**, n., *dĭ′.ă-frăm-ĭt′.ĭs*, inflammation of the diaphragm.

diaphysis, n., *dĭ-ăf′.ĭs-ĭs* (Gr. *diaphŭsis*, the state of growing between or through—from *dia*, through ; *phuo*, I produce), the central point of ossification for the shaft in the long bones ; in *bot.*, the prolongation of the inflorescence.

diapophysis, n., *dĭ′.ă-pŏf′.ĭs-ĭs*, **diapophyses**, n. plu., *-pŏf′.ĭs-ēz* (Gr. *dia*, through ; *apophuo*, I send out shoots, I sprout), in *anat.*, the upper transverse process of a vertebra, as the dorsal transverse processes, and the posterior parts of the cervical transverse processes.

diarrhæmia, n., *dĭ′.ăr-rēm′.ĭ-ă* (Gr. *dia*, through; *rheō*, I flow; *haima*, blood), among cattle, a disease characterised by breaking up of the blood, ecchymoses, and secretions tinged with blood.

diarrhœa, n., *dĭ′.ăr-rē′.ă* (Gr. *diarrhoia*, a violent purging—from *dia*, through ; *rheo*, I flow), a purging or flux ; frequent loose evacuations from the bowels.

diarthrosis, n., *dĭ′.ăr-thrōs′.ĭs* (Gr. *diarthrōsis*, a separation or division by joints—from *dia*, through ; *arthron*, a joint), in *anat.*, a connection of two joints admitting of motion between them, which includes the greater proportion of the joints of the body : **diarthrodial**, a., *dĭ′.ăr-thrōd′.ĭ-ăl*, of or belonging to diarthrosis.

diastase, n., *dĭ′.ăs-tās* (Gr. *diastăsis*, a standing apart, separation—from *dia*, through, asunder ; *histĕmi*, I cause to stand), a peculiar azotised principle which has the property of converting starch into sugar : **diastema**, n., *dĭ′.ăs-*

tēm'ă, a gap or interval, especially between teeth.

diastole, n., *dĭ-ăs'tŏl-ē* (Gr. *diastolē*, separation — from *dia*, through, separation ; *stellō*, I set or place), the dilatation or opening of the heart after contraction; the contraction is the 'systole.'

diathesis, n., *dĭ-ăth'ĕs-ĭs* (Gr. *diathesis*, a disposing or putting in order — from *dia*, through ; *tithēmi*, I put or place), a peculiar state or condition of body, which predisposes an individual `to a disease or a group of diseases.

Diatomaceæ, n. plu., *dĭ'-ă-tŏm-ā'-sĕ-ē*, also **Diatoms**, n. plu., *dĭ'-ăt-ŏms* (Gr. *diatomē*, dissection, division — from *dia*, through, asunder ; *tomē*, a cutting — the filaments being divided into joints), an Order or tribe of Algæ, which are provided with siliceous envelopes, the fronds consisting of frustula or fragments united by a gelatinous substance, and which inhabit still waters and moist places : **Diatoma**, n., *dĭ-ăt'-ŏm-ă*, very minute species of Algæ, found in the sea, and ditches, at all seasons : **diatomine**, n., *dĭ-ăt'-ŏm-ĭn*, a buff-coloured substance found in diatoms, which conceals the green colour of the chlorophyl.

Dibothria, n. plu., *dĭ-bŏth'-rĭ-ă* (Gr. *dis*, twice ; *bothros*, a hole, a pit), a genus of tape-worms : **dibothrium decipiens**, *dĭ-bŏth'-rĭ-ŭm dē-sĭp'-ĭ-ĕns* (L. *decipiens*, ensnaring), a parasitic worm infesting the small intestine of the cat: **d. serratum**, *sĕr-rāt'-ŭm* (L. *serratus*, saw-shaped), a parasitic worm infesting the small intestine both of the dog and fox.

Dibranchiata, n. plu., *dĭ-brăngk'-ĭ-āt'-ă* (Gr. *dis*, twice, double ; *brangchia*, the gills of a fish), the Order of Cephalopoda in which only two gills are present, as in the cuttle-fishes : **dibranch-**iate, a., *dĭ-brăngk'-ĭ-āt*, having two gills.

dichasium, n., *dĭ-kāz'-ĭ-ŭm* (Gr. *dichasō*, I divide into two), in *bot.*, a form of definite inflorescence in which each primary axis produces a pair of opposite lateral axes, each of which produces a similar pair.

dichlamydeous, a., *dĭk'-lăm-ĭd'-ĕ-ŭs* (Gr. *dis*, twice, double ; *chlamus*, a cloak, a garment), in *bot.*, having a calyx and corolla ; having two whorls in the flowers.

dichogamous, a., *dĭ-kŏg'-ăm-ŭs* (Gr. *dicha*, in two parts ; *gameo*, I marry), applied to plants in which the stamens and stigmas of the same flower do not reach maturity at the same time.

Dichopetalum, n., *dĭk'-ō-pĕt'-ăl-ŭm* (Gr. *dicha*, in two parts; *petalon*, a leaf), a genus of plants, Ord. Umbelliferæ, one of whose species in Victoria has five petaloid sepals.

dichotomous, a., *dĭk-ŏt'-ŏm-ŭs* (Gr. *dichotomos*, cut in two — from *dicha*, in two parts ; *tomē*, a cutting), in *bot.*, having the divisions of a stem always in pairs; furcate or forked : **dichotomous cyme**, a definite inflorescence in which the secondary axes are produced in pairs, each one ending in a single flower : **dichotomy**, n., *dĭk-ŏt'-ŏm-ĭ*, a mode of branching by constant forking.

diclesium, n., *dĭ-klēz'-ĭ-ŭm* (Gr. *diklis*, twice-shutting—from *dis*, twice ; *kleiō*, I shut), a small, dry, indehiscent pericarp, having the indurated perianth adherent to the carpel, and forming part of the shell ; a fruit composed of an indehiscent, one-seeded pericarp, invested by a persistent and indurated perianth, as in 'mirabilis.'

diclinous, a., *dĭ-klīn'-ŭs* (Gr. *dis*, twice ; *klinē*, a couch), in *bot.*, having the male and female organs in separate flowers ; unisexual.

dicoccous, a., *dĭ·kŏk'·kŭs* (Gr. *dis*, twice, double ; *kokkos*, a berry, a kernel), having two capsules united, one cell in each ; split into two cocci.

dicotyledonous, a., *dĭ'·kŏt·ĭl·ēd'·ŏn·ŭs* (Gr. *dis*, twice ; Eng. *cotyledonous*), in *bot.*, having two lobes, seed-leaves, or kotyledons: **dicotyledon**, n., *dĭ'·kŏt·ĭl·ēd'·ŏn*, a plant whose seed consists of two lobes.

Dicranum, n., *dĭ·krān'·ŭm* (Gr. *dikranos*, having two heads, cloven — from *krānion*, the skull, having reference to the divisions of the teeth of the capsule), a fine genus of mosses, Ord. Musci or Bryaceæ, many of whose species form broad masses of turfy vegetation.

Dictamnus, n., *dĭk·tăm'·nŭs* (Gr. *diktamnos*, L. *dictamnum*, the plant dittany of Crete), a genus of very ornamental plants, Ord. Rutaceæ, which emit a strong odour : **Dictamnus fraxinella**, *frăks'·ĭn·ĕl'·lă* (L. *fraxĭnus*, the ash tree, alluding to its leaves resembling those of the ash), the false dittany, whose leaves, when rubbed, emit a fine odour, somewhat resembling that of lemon peel ; this and other species abound so much in volatile oil, that, it is said, the atmosphere around them becomes inflammatory in hot, dry, and calm weather.

dictyogens, n. plu., *dĭk·tĭ'·ŏ·jĕns* (Gr. *diktuon*, a net ; *gennaō*, I produce), a great class of plants which have a cellular system, the latter consisting partly of elastic spiral vessels : **dictyogenous**, a., *dĭk'·tĭ·ŏdj'·ĕn·ŭs*, applied to monocotyledons which have netted veins.

Didelphia, n. plu., *dĭ·dĕlf'·ĭ·ă* (Gr. *dis*, twice ; *delphus*, the womb), the subdivision of Mammals comprising the Marsupials.

Didymocarpeæ, n. plu., *dĭd'·ĭm·ō·kărp'·ĕ·ē* (Gr. *didumos*, two-fold, twin ; *karpos*, fruit—in allusion to the twin capsules), a Sub-order of plants, Ord. Bignoniaceæ, having succulent or capsular fruit : **Didymocarpus**, n., *dĭd'·ĭm·ō·kărp'·ŭs*, a genus of pretty plants : **didymosis**, n., *dĭd'·ĭm·ōz'·ĭs*, in *bot.*, two united ; union of two similar organs.

didynamous, a., *dĭd·ĭn'·ăm·ŭs* (Gr. *dis*, twice, double ; *dunamis*, power), in *bot.*, having two long and two short stamens.

Dielytra, n., *dĭ·ĕl'·ĭt·ră* (Gr. *dis*, twice, double ; *elutron*, a case, a sheath), a genus of herbaceous plants, Ord. Fumariaceæ, the base of whose flowers is furnished with two sheath-like spurs.

diencephalon, n., *dĭ·ĕn·sĕf'·ăl·ŏn* (Gr. *dia*, through, between ; *engkephalon*, the brain), the second of the divisions of the anterior primary vesicle of the brain.

Diervilla, n., *dĭ'·ĕr·vĭl'·lă* (after *Dierville*, a French surgeon), a genus of plants, Ord. Caprifoliaceæ.

dietary, n., *dĭ'·ĕt·ăr·ĭ* (Gr. *diaita*, L. *diæta*, mode or place of life, means of life ; F. *diète*), a systematic course or order of diet with the view of maintaining the body in perfect health : adj., relating to diet : **dietetics**, n. plu., *dĭ'·ĕt·ĕt'·ĭks*, that branch of medicine which relates to the regulation of diets in sickness and health.

diffluent, n., *dĭf'·flŏŏ·ĕnt* (L. *diffluens*, dissolving—from *dis*, asunder ; *fluo*, I flow), in *bot.*, dissolving ; having the power to dissolve.

diffract, a., *dĭf·frăkt'* (L. *diffractus*, broken in pieces, shattered—from *dis*, apart, asunder ; *fractus*, broken), in *bot.*, broken into distinct areolæ separated by chinks.

diffuse, a., *dĭf·fūs'* (L. *diffusus*, spread abroad—from *dis*, asunder ; *fusus*, poured or spread), widely spread ; in *bot.*, spreading irregularly : **diffusion**, n., *dĭf·fūzh'·ŭn*, in *chem.*, the property of becoming uniformly mixed.

digastric, a., *dĭ·găst'·rĭk* (Gr. *dis*, twice ; *gastēr*, the belly), having a double belly—applied to a muscle of the lower jaw.

digestion, n., *dĭ·jĕst'·yŭn* (L. *digestio*, the dissolving of food, *digestiōnis*, of the dissolving of food), the changing of the food in the stomach into a substance called chyme, preparatory to its being fitted for circulation and nourishment.

digit, n., *dĭdj'·ĭt* (L. *digitus*, a finger), a finger or toe : **digital**, a., *dĭdj'·ĭt·ăl*, pert. to or resembling a finger: **digitate**, a., *dĭdj'·ĭt·āt* (L. *digitātus*, having fingers), branched like fingers ; in *bot.*, having a compound leaf composed of several leaflets attached to one point: **digitate-pinnate**, applied to a digitate leaf with pinnate leaflets.

Digitalis, n., *dĭdj'·ĭt·āl'·ĭs* (L. *digitālis*, of or belonging to the finger—from *dĭgĭtus*, a finger, in reference to the flower having some resemblance to a finger), a genus of plants, Ord. Scrophulariaceæ, the most of whose species are showy flowers : **Digitalis purpurea**, *pér·pŭr'·ĕ·ă* (L. *purpŭreus*, purple-coloured—from *purpŭra*, a purple colour), foxglove, the most important medicinal plant of the Order, the seeds and leaves of which are employed in the form of powder, tincture, and infusion : **digitalin**, n., *dĭdj'·ĭt·āl'·ĭn*, a crystalline principle which contains the active properties of digitalis : **Digitalis levigata**, *lĕv'·ĭg·āt'·ă* (L. *lēvigātus*, softened, macerated well) ; **D. grandiflora**, *grănd'·ĭ·flōr'·ă* (L. *grandis*, high, grand ; *flos*, a flower, *flōris*, of a flower); **D. lutea**, *lōōt'·ĕ·ă* (L. *lūtĕus*, yellow, of the colour of the plant lutum); **D. tomentosa**, *tōm'·ĕn·tōz'·ă* (L.

tomentōsus, downy—from *tōmentum*, a stuffing for cushions, a flock of wool), are other species which have similar properties: **digitaliform**, a., *dĭdj'·ĭt·āl'·ĭ·fŏrm* (L. *forma*, shape), having a shape like the corolla of digitalis.

Digitigrada, n., *dĭdj'·ĭt·ĭ·grād'·ă* (L. *digitus*, a finger ; *gradior*, I walk), a subdivision of the Carnivora : **digitigrade**, a., *dĭdj'·ĭt·ĭ·grăd*, walking upon the tips of the toes, and not upon the soles of the feet, as the cat, the weasel, and the lion.

digitipartite, a., *dĭdj'·ĭt·ĭ·pàrt'·īt* (L. *digitus*, a finger ; *partītus*, divided—in allusion to the five fingers of the hand), in *bot.*, applied to a leaf with five divisions extending to near the base ; also called ' quinquepartite.'

digynous, a., *dĭdj'·ĭn·ŭs* (Gr. *dis*, twice ; *gunē*, a woman), having two styles or pistils.

dilamination, n., *dĭ·lăm'·ĭn·ā'·shŭn* (L. *dis*, asunder ; *lamina*, a blade), in *bot.*, the separation of a layer from the inner side of a petal, either presenting a peculiar form, or resembling the part from which it is derived ; also called ' deduplication ' and ' chorisis.'

dilatation, n., *dĭl'·ăt·ā'·shŭn* (L. *dilatātus*, enlarged, amplified—from *dis*, asunder ; *lātus*, wide), a spreading or extending in all directions: **dilatator**, n., *dĭl'·ăt·āt'·ŏr*, a muscle that dilates or expands a part: **dilatator naris**, *năr'·ĭs* (L. *nāris*, the nose, of the nose), one of two muscles which expand the nose, or widen the nostrils.

dill, n., *dĭl* (AS. *dile*, anise ; Swed. *dill*, Prov. Dan. *dull*, still, quiet), the seeds of an aromatic plant, the Anethum graveolens, belonging to the Hemlock family, whose distilled oil or prepared water is used as a soothing medicine in maladies accompanied with flatulence.

Dilleniaceæ, n. plu., *dĭl-lĕn'ĭ-ā'-sĕ-ē* (after *Professor Dillenius,* of Oxford), the Dillenia family, an Order of trees and shrubs of considerable beauty, some yielding fruit, others producing fine timber: **Dillenia,** n., *dĭl-lĕn'ĭ-ă,* a genus of very elegant shrubs when in flower.

diluents, n., *dĭl'ū-ĕnts* (L. *dilūtus,* washed away, weakened — from *dis,* asunder ; *lutus,* washed), in *med.,* remedies made use of to quench thirst, or to make the blood thinner and cooler, such as toast-and-water, barley-water, etc.

Dimerosomata, n. plu., *dĭm'ĕr-ō-sŏm'ăt-ă* (Gr. *dis,* twice ; *mĕros,* a part ; *sōma,* a body, *sōmătos,* of a body), an Order of Arachnida, comprising spiders, so called from the marked division of the body into two parts, viz. the cephalothorax and abdomen.

dimerous, a., *dĭm-ĕr'ŭs* (Gr. *dis,* twice ; *mĕros,* a part), in *bot.,* composed of two pieces ; having parts arranged in twos.

dimidiate, a., *dĭm-ĭd'ĭ-āt* (L. *dimidiātus,* divided into halves— from *dimidium,* the half), applied to an organ when the one half is smaller than the other half ; split into two on one side, as the calyptra of some mosses ; applied to the gills of Argarics when they proceed only half-way to the stem.

dimorphic, a., *dī-mŏrf'ĭk* (Gr. *dis,* twice ; *morphē,* shape), having two forms of flowers, differing in the size and development of the stamens and pistils, as in Primula and Linum : **dimorphous,** a., *dī-mŏrf'ŭs,* assuming different forms in similar parts of a plant : **dimorphism,** n., *dī-mŏrf'ĭzm,* the occurrence of the same species of plant in two or three different states.

dimyary, a., *dī-mĭ'ăr-ĭ* (Gr. *dis,* twice ; *muŏn,* a muscle of the body), closed by two muscles ; applied to those bivalve molluscs which have their shells closed by two adductor muscles.

diœcious, a., *dī-ē'shŭs* (Gr. *dis,* twice ; *oikos,* a house), in *zool.,* having the sexes distinct, applied to species which consist of male and female individuals ; in *bot.,* having staminiferous flowers on one plant, and pistiliferous flowers on another plant : **diœcia,** n. plu., *dī-ē'shĭ-ă,* a class of plants having male flowers on one plant, and female on another : **diœciously-hermaphrodite,** a., having hermaphrodite flowers, but only one of the essential organs perfect in a flower.

Dion, n., *dī'ŏn* (Gr. *dis,* twice ; *oŏn,* an egg—from each scale bearing two ovules), a remarkable genus of Mexican plants, Ord. Cycadaceæ : **Dion edule,** *ĕd-ūl'ĕ* (L. *edūlis,* eatable), a species which yields a kind of arrowroot in Mexico.

Dionæa, n. plu., *dī'ŏn-ē'ă* (*Diōnæa,* Venus, being a patronymic from *Diōne,* the mother of Venus ; *Diōne,* a name of Venus herself), a genus of curious plants, Ord. Droseraceæ : **Dionæa muscipula,** *mŭs-sĭp'ŭl-ă* (L. *mŭscipŭla,* a mouse-trap—from *mŭs,* a mouse ; *căpio,* I take), Venus's fly-trap, a North American plant, having the laminæ of the leaves in two lobes, the irritable hairs on which being touched cause the folding of the lobes and thus entrap flies.

Dioscoreaceæ, n. plu., *dī'ŏs-kōr'ĕ-ā'-sĕ-ē* (after *Dioscorĭdes,* a famous Greek physician), the Yam tribe, an Order of twining shrubs, natives of tropical countries : **Diascorea,** n., *dī'ăs-kōr'ĕ-ă,* a genus of climbing plants cultivated in tropical climates for the sake of its roots, which are called yams, and are used in the same way as potatoes : **Dioscorea alata,** *ăl-āt'ă* (L. *ālātus,* furnished with wings) ; **D. sativa,** *săt-īv'ă* (L. *satīvus,* fit to be sown or planted) ;

D. aculeata, *ăk·ūl'·ĕ·āt'·ă* (L. *acūlĕātus*, thorny, prickly—from *acūlĕus*, the sting of a bee), are the species which produce the tubers called Yams, used as potatoes.

Diosma, n., *dĭ·ŏs'·mă* (Gr. *diosmos*, transmitting odours—from *dios*, godlike ; *osmē*, smell), a genus of very beautiful heath-like shrubs, Ord. Rutaceæ.

Diospyros, n., *dī·ŏs'·pĭr·ŏs* (Gr. *diospŭros*, the fruit that caused oblivion — from *dios*, godlike ; *pŭros*, wheat, fruit ; or *pŭren*, a kernel, a berry), a genus of ornamental and very valuable timber trees, Ord. Ebenaceæ, remarkable for the hardness and durability of their wood : **Diospyros lotus,** *lŏt'·ŭs* (Gr. *lōtos*, L. *lōtus*, the water-lily of the Nile), a species which is said to have produced the fruit which caused oblivion : **D. reticulata,** *rĕ·tĭk'·ūl·āt'·ă* (L. *reticŭlātus*, made like a net—from *rēte*, a net) ; **D. ebenum,** *ĕb'·ĕn·ŭm* (L. *ebĕnus*, the ebon-tree), along with other species furnish ebony, which is the black duramen of the tree : **D. virginiana,** *vĕrj·ĭn'·ĭ·ān'·ă* (L. *virgĭnĕus*, belonging to a virgin—from *virgo*, a virgin), the persimon tree, yields a fruit, sometimes called the date-plum, which is sweet and eatable when ripe, especially after frost, and the bark has been employed as a febrifuge : **D. kaki,** *kăk'·ĭ* (a native name), the Keg fig of Japan, the fruit resembling a plum : **D. embryopteris,** *ĕm'·brĭ·ŏp'·tĕr·ĭs* (Gr. *embrŭo*, an embryo ; *pteris*, a fern —from *pteron*, a wing), yields a succulent fruit, the pulp of which is astringent and very glutinous: **D. quæsitus,** *kwĕs·ĭt'·ŭs* (L. *quæsītus*, sought out, select), a species which supplies the Coromandel wood of Ceylon.

dipetalous, a., *dī·pĕt'·ăl·ŭs* (Gr. *dis*, twice *petalon*, a petal), having two petals.

diphtheria, n., *dĭf·thĕr'·ĭ·ă* (Gr. *dĭphthera*, skin, leather), a disease characterised by the forming of a leathery, false membrane on a diseased surface; a disease of the pharynx and tonsils, so named, having a croupous, false membrane : **diphtheritis,** n., *dĭf'·thĕr·ĭt'·ĭs* (*itis*, inflammation), same sense as 'diphtheria' : **diphtheritic,** a., *dĭf'·thĕr·ĭt'·ĭk*, pert. to diphtheria.

diphyodont, n., *dĭf·ĭ'·ŏ·dŏnt* (Gr. *dis*, twice ; *phuo*, I generate ; *odous*, a tooth, *odontos*, of a tooth), one of the Mammals which have two sets of teeth.

diphyozooids, n. plu., *dĭf·ĭ'·ŏ·zō'·ŏўds* (Gr. *dis*, twice ; *phuo*, I generate ; *zoon*, an animal ; *eidos*, resemblance), detached reproductive portions of Calycophoridæ, an Order of ocean Hydrozoa.

diplecolobeæ, n. plu., *dĭp'·lĕ·kō·lŏb'·ĕ·ē* (Gr. *dis*, twice ; *plĕkō*, I twine, I plant ; *lobos*, the lobe of the ear), in *bot.*, cotyledons twice folded transversely.

diploe, n., *dĭp'·lō·ē* (Gr. *diploē*, a fold), in *bot.*, the cellular tissue surrounding the vessels of the leaf, and enclosed within the epidermis—sometimes called the 'diachyma' and 'mesophyllum'; in *anat.*, the network of bone tissue which fills up the interval between the two compact plates in the bones of the skull: **diploic,** a., *dĭp·lō'·ĭk*, of or pert. to the diploë.

diploperistomi, n. plu., *dĭp'·lō·pĕr·ĭs'·tŏm·ī* (Gr. *diploös*, double ; *peri*, about ; *stoma*, a mouth), mosses which have a double peristome : **diploperistomous,** a., *dĭp'·lō·pĕr·ĭs'·tŏm·ŭs*, having a double peristome.

diplostemonous, a., *dĭp'·lō·stĕm'·ŏn·ŭs* (Gr. *diploös*, double; *stĕmōn*, the thread called the warp, *stĕmŏnos*, of the warp—from *histĕmi*, I cause to stand, the ancient looms being upright), in *bot.*, having a double row of stamens, often double the number of the petals or sepals.

diplotegia, n., *dĭp'·lō·tēdj'·ĭ·ă* (Gr. *diploos*, double ; *tĕgos*, a covering), in *bot.*, an inferior, dry seed vessel, usually opening by valves or by pores, as in Campanula ; sometimes applied to a double covering, as a calyx and an epicalyx.

Diplozygiæ, n. plu., *dĭp'·lō·zĭdj'·ĭ·ē* (Gr. *diploos*, double ; *zugia*, the hornbeam, a tree having a smooth grey bark, ridged trunk, and very hard, white wood—from *zugon*, a yoke, the wood being fit for the yokes of cattle), a section or Sub-order of trees, Ord. Umbelliferæ; also the name of a genus.

Dipnoi, n. plu., *dĭp'·nō·ī* (Gr. *dis*, twice ; *pnoē*, breath), an Order of fishes represented by the Lepidosiren, which has twofold respiratory organs, both gills and true lungs.

Dipsacaceæ, n. plu., *dĭps'·ăk·ā'·sĕ·ē* (Gr. *dipsakos*, the fuller's thistle—said to be from *dipsa*, thirst, their hollow leaves holding water to satisfy thirst), the Teazel family, an Order of plants : **Dipsacus**, n., *dĭps'·ăk·ŭs*, a curious genus of plants : **Dipsacus sylvestris**, *sĭl·vĕst'·rĭs* (L. *sylvestris*, woody—from *sylva*, a wood), the plant Venus's bath, so called from the water contained in the hollow leaves being considered good for bleared eyes ; some of the species are considered febrifugal : **D. fullonum**, *fŭl·lōn'·ŭm* (L. *fŭllo*, a fuller, *fŭllōnis*, of a fuller, *fŭllōnum*, of fullers), a species the heads of which are called fuller's teazel, from their spiny bracts being used in dressing cloth : **D. pilosus**, *pĭl·ōz'·ŭs* (L. *pĭlōsus*, hairy, shaggy—from *pĭlus*, a hair), a very pretty flowering species.

dipsomania, n., *dĭps'·ŏ·mān'·ĭ·ă* (Gr. *dipsa*, thirst ; *mānia*, mad-

ness), the irresistible longing for alcoholic liquors, either developed or innate in some men and women.

Diptera, n. plu., *dĭp'·tĕr·ă* (Gr. *dis*, twice ; *pteron*, a wing), an Order of insects having two wings: **dipterous**, a., *dĭp'·tĕr·ŭs*, having two wings, or two wing-like appendages: **Dipteraceæ**, n. plu., *dĭp'·tĕr·ā'·sĕ·ē*, an old term for Dipterocarpaceæ, which see.

Dipterix, n., *dĭp'·tĕr·ĭks* (Gr. *dis*, twice ; *pterux*, a wing, the two upper lobes of the calyx, appearing as wings), a genus of ornamental trees, Sub-ord. Papilionaceæ, Ord. Leguminosæ : **Dipterix odorata**, *ŏd'·ŏr·āt'·ă* (L. *odorātus*, sweet-smelling ; *ŏdor*, scent, smell), a species, the fragrant seeds of which are known as Tonka or Tonquin beans, used in giving a pleasant scent to snuff.

Dipterocarpaceæ, n. plu., *dĭp'·tĕr·ō·kărp·ā'·sĕ·ē* (Gr. *dis*, twice ; *pterux*, a wing ; *karpos*, fruit), the Sumatra camphor family, an Order of handsome ornamental trees abounding in resinous juice: **Dipterocarpus**, n., *dĭp'·tĕr·ō·kărp'·ŭs*, a genus of trees, various species of which yield a substance like balsam of copaiva : **Dipterocarpus lævis**, *lēv'·ĭs* (L. *lævis* or *lĕvis*, light, not heavy) ; **D. angustifolius**, *ăng·gŭst'·ĭ·fōl'·ĭ·ŭs* (L. *angustus*, narrow ; *fŏlium*, a leaf); **D. turbinatus**, *tĕrb'·ĭn·āt'·ŭs* (L. *turbinātus*, cone-shaped — from *turbo*, a whipping-top); **D. hispidus**, *hĭsp'·ĭd·ŭs* (L. *hispĭdus*, shaggy, hairy); **D. Zeylanicus**, *zī·lăn'·ĭk·ŭs* (*Zeylan*, Ceylon), are species which yield wood oil.

Dirca, n., *dĕrk'·ă* (Gr. *Dirka*, a fountain, in reference to the natural habitat of the plant), a genus of little shrubs growing in the marshes of N. America, Ord. Thymelæaceæ: **Dirca palustris**, păl·ŭst'·rĭs (L. *pălustris*, marshy —from *pălŭs*, a swamp or marsh), the N. American leather wood, a species whose bark is used for cordage, and the twigs are made into ropes and baskets,—the fruit is said to be narcotic.

diremption, n., dĭr·ĕm'·shŭn (L. *diremptus*, separation or division), in *bot.*, the occasional separation or displacement of leaves.

Disa, n., dīz'·ă (origin of name unknown), a genus of interesting tuberous-rooted plants, Ord. Orchidiaceæ: **Disa grandiflora,** gründ'·ĭ·flŏr'·ă (L. *grandis*, large, great; *flos*, a flower, *flōris*, of a flower), a species found on Table Mountain in marshes: **D. ferruginea,** fĕr'·ōō·jĭn'·ĕ·ă (L. *ferrŭginĕus*, of an iron-rust colour—from *ferrŭgo*, iron-rust); also **D. tenuifolia,** tĕn'·ū·ĭ·fōl'·ĭ·ă (L. *tenŭis*, thin; *folium*, a leaf), are species found in same place at an elevation of 3582 feet.

Dischidia, n., dĭs·kĭd'·ĭ·ă (Gr. *dis*, twice; *schizō*, I split), a genus of ornamental plants, Ord. Asclepiadaceæ: **Dischidia Rafflesiana,** răf·flĕs'·ĭ·ān'·ă (after *Sir Stamford Rafflen*), an Indian climber whose pitchers are formed by the lamina of the leaf, and have an open orifice into which the rootlets at the upper part of the plant enter, thus probably furnishing a fluid for the nourishment of the upper branches.

disciform, a., dĭs'·ĭ·fŏrm (L. *discus*, a quoit; *forma*, shape), in the form of a disc; flat and circular.

discocarp, n., dĭsk'·ō·kârp (Gr. *diskos*, a disc; *karpos*, a fruit), applied to a collection of fruits in a somewhat globose receptacle.

discoid, a., dĭsk'·oўd, also **discoidal,** a., dĭsk·oўd'·ăl (Gr. *diskos*, a disc; *eidos*, resemblance), in the form of a disc; disciform; round, or having a convex face; applied to the flosculous or tubular flowers of Compositæ.

Discomycetes, n., dĭsk'·ō·mĭs'·ĕt·ēz (Gr. *diskos*, a disc; *mukēs*, a fungus, *mukētos*, of a fungus), a section or Sub-order of the Fungi, including Morels and Truffles.

Discophora, n., dĭsk·ŏf'·ŏr·ă (Gr. *diskos*, a quoit; *phoreo*, I bear), the Medusæ or jelly-fishes, so called from their shape; applied sometimes to the leeches, Hirudinea, from their suctorial discs.

discrete, a., dĭs·krēt' (L. *discretus*, separated—from *dis*, asunder; *cretus*, separated), separated from each other; distinct; not continuous or confluent.

discus proligerus, dĭsk'·ŭs prō·lĭdj'·ĕr·ŭs (L. *discus*, a quoit, a disc; *prōles*, offspring; *gero*, I bear), in *anat.*, a small flattened heap of granular cells, in the centre of which is embedded the ovum or germinal vesicle.

disgorgement, n., dĭs·gŏrj'·mĕnt (L. *dis*, asunder; F. *gorge*, the throat), the discharge of a certain quantity of fluid or semi-fluid matter by the mouth.

disinfectant, n., dĭs'·ĭn·fĕkt'·ănt (L. *dis*, asunder; Eng. *infect*), a substance or fluid which destroys the evil effects of foul or infectious matter.

disintegration, n., dĭs'·ĭn·tĕg·rā'·shŭn (L. *dis*, asunder; *integer*, whole, entire), the breaking into numerous large and small pieces of any solid body; the wearing down or away from atmospheric influences.

dislocation, n., dĭs'·lōk·ā'·shŭn (L. *dis*, asunder; *locatus*, put or placed), in *surg.*, the displacement of one or more bones.

dispermous, a., dĭ·spĕrm'·ŭs (Gr. *dis*, double; *sperma*, seed), having two seeds.

dissect, v., dĭs·sĕkt' (L. *dissectus*, cut asunder—from *dis*, asunder; *sectus*, cut), to cut and separate parts of a body in order to examine minutely its structure: **dissected,** a., dĭs·sĕkt'·ĕd, in *bot.*,

I

cut into a number of narrow divisions : **dissection**, n., *dĭs·sĕk'·shŭn*, the cutting or separating parts of a body with the view of examining minutely its structure and arrangement of parts.

dissepiment, n., *dĭs·sĕp'·ĭ·mĕnt* (L. *dissēpto*, I separate or divide—from *dis*, asunder; *sepes*, a hedge), in *bot.*, a partition in an ovary or fruit; used sometimes to designate certain imperfect transverse partitions found growing from the septa of many corals.

dissilient, a., *dĭs·sĭl'·ĭ·ĕnt* (L. *dissiliens*, leaping asunder, flying apart—from *dis*, asunder; *saliens*, leaping), in *bot.*, bursting and opening with an elastic force.

distal, a., *dĭst'·ăl* (L. *disto*, I stand apart; a probable corruption of Eng. *distant*), in *anat.*, remote from the place of attachment, as the 'distal' extremity of a bone ; farthest from the heart or trunk; in *zool.*, applied to the quickly growing end of the hydrosoma of a hydrozoon by which the organism is fixed, when attached at all ; the opposite end is called the 'proximal.'

distant, a., *dĭst'·ănt* (L. *distantia*, remoteness—from *dis*, asunder ; *stans*, standing), in *bot.*, applied to the gills of Agarics when widely separated.

distemper, n., *dĭs·tĕmp'·ĕr* (L. *dis*, not; Eng. *temper*, the condition of the animal body in all its parts in health), a disease of some animals, chiefly the dog, whose leading symptoms are a running from the nose and eyes, and a loss of strength and spirits.

distichous, a., *dĭs'·tĭk·ŭs* (Gr. *dis*, twice ; *stichos*, a row), in *bot.*, disposed in two rows on the opposite sides of a stem, as the grains in an ear of barley.

Distoma, n., *dĭs'·tŏm·ă* (Gr. *dis*, twice ; *stŏma*, a mouth, *stŏmăta*, mouths), a genus of the Entozoa, having two pores or suckers :

Distoma hepaticum, *hĕ·păt'·ĭk·ŭm* (Gr. *hēpătikos*, L. *hēpătĭcus*, one diseased in the liver), a small, flat, flounder-like worm found in the livers of sheep in a perfect condition, and in the bile ducts of sheep and oxen ; it also attacks the horse, the ass, the pig, and other animals, sometimes even man ; the 'Fasciola hepatica,' which see: **D. lanceolatum**, *lăns'·ē·ō·lāt'·ŭm* (L. *lanceŏla*, a little spear—from *lancĕa*, a spear), a species of intestinal worm which attacks the pig, cat, rabbit, etc., but finds its most frequent habitat in the liver of the ox : **distomidæ**, n. plu., *dĭs·tŏm'·ĭd·ē*, a family of the Entozoa or fluke-worms, comprising several genera, of which Distoma is one ; see 'Cobbold.'

distractile, a., *dĭs·trăkt'·ĭl* (L. *distractus*, divided, perplexed—from *dis*, asunder ; *tractus*, drawn or dragged), in *bot.*, separating two parts to a distance from each other ; torn asunder.

dithecal, a., *dĭ·thēk'·ăl* (Gr. *dis*, twice ; *thēkē*, a receptacle, a chest), in *bot.*, having two loculaments or cavities,—said of an anther.

dittany, n., *dĭt'·ăn·ĭ* (Gr. *diktamnos*, L. *dictamnus*, the plant dittany—from *Dictē*, the mountain in Crete where found), an aromatic plant whose leaves resemble lemon thyme in smell ; wild or bastard dittany is **Dictamnus fraxinella**, Ord. Rutaceæ, which abounds in a volatile oil ; the dittany of Crete is **Origanum dictamnus**, Ord. Labiatæ.

diuresis, n., *dī'·ūr·ēz'·ĭs* (Gr. *dioureō*, I void by urine—from *dia*, through; *ouron*, urine), an increased or excessive flow of urine: **diuretic**, a., *dī'·ūr·ĕt'·ĭk* (Gr. *diouretikos*), having the power of provoking urine : n., a medicine which increases the discharge of urine.

divaricate, a., *dĭ·văr'·ĭk·āt* (L. *divāricātus*, spread asunder—from

dis, asunder; *vārĭcus*, with feet spread apart), in *bot.*, having branches coming off from the stem at a very wide or obtuse angle; spreading irregularly and widely.

divergent, a., *dĭ·vérj'ĕnt* (L. *dis*, asunder; *vergens*, inclining), in *bot.*, radiating or spreading outwards from a common centre.

diverticulum, n., *dĭ·vért·ĭk'·ūl·ŭm*, **diverticula**, n. plu., *-ūl·ă* (L. *diverticulum*, a bye-way—from *diverto*, I turn aside), in *anat.*, a cul-de-sac, or blind lateral tube given off from the main tube.

Dochmius, n., *dŏk'·mĭ·ŭs* (Gr. *dochmios*, L. *dochmius*, an ancient poetic foot), a genus of intestinal worms: **Dochmius hypostomus**, *hĭ·pŏs'tŏm·ŭs* (Gr. *hupo*, under; *stŏma*, a mouth), a parasite of the sheep, goat, and other ruminants, found in the intestines: **D. trigonocephalus**, *trĭg'·ŏn·ō·sĕf'·ăl·ŭs* (Gr. *trigōnon*, a triangle; *kephălē*, the head), a species of parasites which infest the stomach and intestines of the dog: **D. tubæformis**, *tūb'·ē·fŏrm'·ĭs* (L. *tuba*, a trumpet; *forma*, shape), a species found in the duodenum of the cat; see 'Gamgee.'

Dodder, n., *dŏd'·dĕr* (Ger. *dotter*, the dodder; Irish, *dodd*, a bunch), curious leafless parasitical plants, whose slender, entangled, threadlike stems run over other plants and often smother them; the genus is Cuscuta, Ord. Convolvulaceæ: **Cuscuta Europæa**, attacks thistles, oats, etc.: **C. spithymum**, found on heath, furze, etc.: **C. epilinum**, attacks flax: **C. trifolii**, is the pest of clover fields.

dodecagynous, a., *dŏd'·ĕk·ădj'·ĭn·ŭs* (Gr. *dōdĕka*, twelve; *gŭne*, a woman), having twelve pistils: **dodecandrous**, a., *dŏd'·ĕk·ănd'·rŭs* (Gr. *anēr*, a man, a male, *andros*, of a man), having twelve stamens.

dolabriform, a., *dō·lăb'·rĭ·fŏrm* (L. *dŏlābra*, an axe; *forma*, shape), in *bot.*, shaped like an axe.

dolichocephali, n. plu., *dŏl'·ĭk·ō·sĕf'·ăl·ī*, also **dolichocephalia**, n. plu., *dŏl'·ĭk·ō·sĕf·ăl'·ĭ·ă* (Gr. *dolichos*, long; *kephalē*, the head), in *anat.*, a monstrosity in which the head is unnaturally long, in a direction from before backwards; a term applied to a long-headed race of cave-dwellers who inhabited Britain in prehistoric times; **dolicocephalic**, a., *dŏl'·ĭk·ō·sĕf·ăl'·ĭk*, long-headed or long-skulled.

Dorema, n., *dŏr·ēm'·ă* (Gr. *dōrēma*, a gift), a genus of plants, Ord. Umbelliferæ, which produce gum ammoniac, natives of Persia; **Dorema ammoniacum**, *ăm'·ŏn·ĭ'·ăk·ŭm* (Gr. *Ammōn*, Egyptian name of Jupiter, whose temple was in the sandy deserts of Libya, where the tree grew), a tree which yields ammoniac, a fetid gum resin; the tree yields resin, gum, and volatile oil, all used medicinally.

dorsal, a., *dŏrs'·ăl* (L. *dorsum*, the back), pert. to the back, as the *dorsal* fin of a fish; in *bot.*, applied to the suture of the carpel which is farthest from the axis; fixed upon the back: **dorsiferous**, a., *dŏrs·ĭf'·ĕr·ŭs* (L. *fero*, I bear), applied to ferns which bear fructification on the back of their fronds: **dorsum**, n., *dŏrs'·ŭm*, the part of the carpel farthest from the axis: **dorsal surface**, in *anat.*, the back or posterior, as distinguished from the ventral or anterior surface: **dorsal vertebræ**, the bones in the spine of the back, twelve in number.

dorsales pollicis, *dŏrs·ăl'·ēz pŏl'·lĭs·ĭs* (new L. *dorsālis*, dorsal—from L. *dorsum*, the back; *pollex*, a thumb, *pollĭcis*, of a thumb), in *anat.*, the dorsal arteries of the thumb; two small arteries which run along the sides of the dorsal aspect of the thumb : **dorsalis indicis**, *dŏrs·āl'·ĭs ĭn'·dĭs·ĭs* (L. *index*, anything that points out, the forefinger, *indĭcĭs*, of the forefinger), a small branch artery which runs along the radial side of the back of the index finger : **dorsalis hallucis**, *hăl'·ūs·ĭs* (new L. *hallux*, the great toe, *hallŭcis*, of the great toe—from L. *hallex*, the great toe, said to be from Gr. *hallŏmai*, I leap, as being chiefly employed in leaping), an artery along the outer border of the first metatarsal bone, and at the cleft between the first and second toes: **dorsalis pedis**, *pĕd'·ĭs* (L. *pēs*, a foot, *pĕdĭs*, of a foot), the dorsal artery of the foot.

dorsibranchiate, a., *dŏrs'·ĭ·brăngk'·ĭ·āt* (L. *dorsum*, the back ; Gr. *brangchia*, gills of a fish), in *zool.*, having external gills attached to the back.

dorsi-lumbar, a., *dŏrs'·ĭ·lŭmb'·ăr* (L. *dorsum*, the back, *dorsi*, of the back ; *lumbus*, a loin), a small off-set from the lumbar plexus nerve.

Dorstenia, n., *dŏr·stĕn'·ĭ·ă* (after *Dorsten*, a German botanist), a genus of very curious plants, Ord. Moraceæ, having a slightly concave, broad receptacle, bearing numerous flowers : **Dorstenia contrayerva**, *kŏn'·trā·yĕrv'·ă* (L. *contra*, against ; *yerba*, the native name for maté or Paraguayan tea, so called as esteemed good against poison); **D. Houstoni**, *hŏws'·tŏn·ĭ* (after *Houston*); **D. Brasiliensis**, *brăz·ĭl'·ĭ·ĕns'·ĭs* (of or from *Brazil*), are species which furnish the contrayerva root of commerce, used as a stimulant, tonic, and diaphoretic.

dossil, n., *dŏs'·sĭl* (F. *dousil*, a peg or tap to draw off liquor from a cask ; Ger. *docke*, a bunch), a small portion of lint made round, or in the form of a date, to be laid on a sore.

douche, n., *dŏsh* (F. *douche*, a

shower bath), a bath given by a jet or stream of water poured from above on some part of the body.

Dracæna, n., *dră·sēn'·ă* (Gr. *drakaina*, a she-dragon), a genus of trees, Ord. Liliaceæ, whose inspissated juice is said to become a powder like dragon's blood ; they often branch in a dichotomous manner, and attain large dimensions : **Dracæna draco**, *drăk'·ō* (Gr. *drakŏn*, L. *drăco*, a species of serpent), a species which, with others, yields an astringent resin called dragon's blood : **D. terminalis**, *tèrm'·ĭn·āl'·ĭs* (L. *terminālis*, terminal — from *termĭnus*, a boundary, so called because planted in India to make boundaries), a species which in Java is considered valuable in dysenteric affections : **Dracontium**, n., *dră·kŏn'·shĭ·ŭm*, a genus of plants, so called because the stems are spotted like the skin of a snake, or from the appearance of its root: **Dracontium fœtidum**, *fĕt'·ĭd·ŭm* (L. *fœtĭdus*, fetid, stinking), the skunk cabbage, which exhales a very fetid odour, and the powdered root used as an antispasmodic : **D. pertusum**, *vèr·tūz'·ŭm* (L. *pertūsus*, perforated — from *per*, through, thoroughly ; *tūsus*, beaten), a very acrimonious plant, the fresh leaves used by the Indians over dropsical parts to produce vesications : **D. polyphyllum**, *pŏl·ĭ·fĭl'·lŭm* (Gr. *polus*, many ; *phullon*, a leaf), a species whose prepared root in India is supposed to possess antispasmodic virtues, and to be a remedy in asthma.

dracunculus, n., *dră·kŭnk'·ūl·ŭs* (a diminutive of Gr. *drakŏn*, a serpent), the Guinea-worm, the adult female of a nematode parasite, a worm which burrows beneath the skin of the legs and feet of human beings in certain limited intertropical districts of Asia and Africa.

drastic, n., *drăst'ĭk* (Gr. *drastikos,* active, vigorous—from *draō,* I do or act), a purgative whose action is somewhat rapid and violent: **adj.,** acting violently.

dropsy, n., *drŏps'ĭ* (L. *hydrops,* Gr. *hudrōps,* the dropsy ; Gr. *hudōr,* water ; *ōps,* the eye—the word formerly spelt *hydropsy*), an unnatural accumulation of fluid in the cellular tissues, or in other cavities of the body.

Droseraceæ, n. plu., *drŏs'ĕr·ā'·sĕ·ē* (Gr. *droseros,* dewy—from *drosos,* dew), the Sundew family, an Order of herbaceous plants growing in damp places : **Drosera, n.,** *drŏs'ĕr·ă,* a genus of herbaceous plants, having acid taste combined with slight acridity, and the leaves furnished with red glandular hairs, discharging from their ends drops of a viscid acrid juice in sunshine — hence the name Sundew or 'Ros solis,' *rŏs sōl'ĭs,* dew of the sun ; some Droseras yield a dye, and their leaves fold upon insects that touch the hairs : **Drosophyllum, n.,** *drŏs'·ō·fĭl'·lŭm* (Gr. *phullon,* a leaf), another genus of the same family.

drug, n., *drŭg* (F. *drogue,* a drug; Dut. *droog,* dry), a general name for all medicinal substances.

Drupaceæ, n. plu., *drŏ·pā'·sĕ·ē* (L. *drupa,* Gr. *druppa,* an over-ripe wrinkled olive), the almond-worts, an Order of trees and shrubs, now included under the Sub-ord. Amygdaleæ or Pruneæ, of the Ord. Rosaceæ, which bear such stone fruits as the cherry, plum, peach, etc.: **drupe, n.,** *drŏp,* a fleshy or purple fruit without valves, and containing a hard stony kernel; a stone fruit : **drupaceous, a.,** *drŏp·ā'·shŭs,* consisting of or producing drupes : **drupel, n.,** *drŏp'ĕl* (a diminutive of *drupa*), a small drupe; a fleshy or purple fruit containing many small stony seeds, as the raspberry and blackberry.

Dryandra, n., *drī·ănd'·ră* (after *Dryander,* a Swedish botanist), a genus of splendid plants nearly allied to Banksia, Ord. Proteaceæ.

Drymis, n., *drĭm'·ĭs* (Gr. *drumos,* a forest, a grove), a genus of plants, Ord. Magnoliaceæ: **Drymis Winteri,** *wĭnt'·ĕr·ī* (after Captain *Winter*), also called **D. aromatica,** *ăr'·ōm·ăt'·ĭk·ă* (L. *arōmătĭcus,* Gr. *arōmatikos,* aromatic, fragrant), a species brought by Captain Winter from the Straits of Magellan, 1578 ; yields Winter's bark ; has been employed as an aromatic stimulant.

Dryobalanops, n., *drī'·ō·băl'·ăn·ŏps* (Gr. *drus,* an oak tree ; *bălănos,* an acorn), a genus of trees, Ord. Dipterocarpaceæ : **Dryobalanops camphora,** *kămf'·ŏr·ă* (F. *camphre,* Ar. *kafur,* Gr. *kaphoura,* camphor), also called **D. aromatica,** *ăr'·ōm·ăt'·ĭk·ă* (L. *arōmătĭcus,* aromatic, fragrant), a tree which furnishes camphor oil, while solid camphor is found in the cavities of the wood, but only after the tree attains a considerable age.

ductus ad nasum, *dŭkt'·ŭs ăd nāz'·ŭm* (L. *ductus,* a leading or conducting ; *ad,* to ; *nāsus,* the nose), a duct to the nose; the nasal duct descending to the fore part of the lower meatus of the nose : **ductus arteriosus,** *ărt·ēr'·ĭ·ōz'·ŭs* (L. *artēriōsus,* full of arteries—from *artēria,* an artery), a short tube about half an inch in length at birth which unites the pulmonary artery with the aorta, but becomes obliterated after birth: **d. communis choledochus,** *kŏm·mūn'·ĭs kŏl·ĕd'·ŏk·ŭs* (L. *communis,* common ; Gr. *cholē,* bile ; *dochos,* holding or containing—from *dechomai,* I receive), the common bile duct, the largest of the ducts, conveying the bile both from the liver and the gall-bladder into the duodenum : **d. cysticus,** *sĭst'·ĭk·ŭs* (Gr. *kustis,* a bladder, a purse), the cystic or excretory duct which leads from the neck of the gall-bladder to join the hepatic : **d. hepaticus,** *hē·păt'ĭk·ŭs* (Gr. *hēpatĭkos,* affecting the liver—from *hēpar,* the liver), the hepatic duct, formed by the union of the biliary pores, and proceeds from the liver to the duodenum : **d. lachrymalis,** *lăk'rĭ·māl'ĭs* (L. *lachrymālis,* lachrymal — from *lāchryma,* a tear), the lachrymal duct ; the excretory ducts of the lachrymal gland : **d. thoracicus,** *thŏr·ăs'ĭk·ŭs* (Gr. *thōrax,* the breast, *thōrăkos,* of the breast; L. *thōrax, thōrācis*), the great trunk formed by the junction of the absorbent vessels.

dulcamara, n., *dŭlk'ăm·ār'ă* (L. *dulcis,* sweet ; *amarus,* bitter), a common British hedge-plant, called 'bitter-sweet' or 'woody nightshade,' from the root when chewed first tasting bitter, and then sweet ; the Solanum dulcamara, Ord. Solanaceæ : **dulcamarine, n.,** *dŭlk'ă·mār'ĭn,* an extract from the plant.

dumose, a., *dūm·ōz'* (L. *dūmōsus,* covered with bushes—from *dūmus,* a thorn-bush), full of bushes ; having a low, shrubby aspect.

duodenum, n., *dū'ŏ·dēn'ŭm* (L. *duodēni,* twelve each), the first portion of the small intestines immediately succeeding the stomach, which in man is about eight or ten inches in length : **duodenal, a.,** *dū'ŏ·dēn'ăl,* connected with or relating to the duodenum.

Dura-Mater, n., *dūr'ă·māt'ėr* (L. *durus,* hard ; *māter,* a mother), the semi-transparent outer membrane which invests and protects the brain and spinal cord.

duramen, n., *dūr·ām'ĕn* (L. *dūrāmen,* hardness — from *dūrus,* hard), the inner or heart wood of a tree.

Durio, n., *dūr'ĭ·ō* (from *duryon,* the native Malay name for the fruit), a genus of trees, Ord.

Sterculiaceæ : **Durio zibethinus,** *zĭb'ĕth·ĭn'ŭs* (said to be from Arab. *zobeth,* civet), the tree which produces the fruit called durian, or civet durian, in the Indian Archipelago ; the fruit is about the size of a man's head, and considered the most delicious of Indian fruits, though of a very fetid odour.

Durvillea, n., *dŭr·vĭl'lē·ă* (after *D'Urville*), a genus of sea-plants, Ord. Algæ : **Durvillea utilis,** *ŭt'ĭl·ĭs* (L. *utĭlis,* useful), one of the large-stemmed species of Algæ.

dynamics, n. plu., *dĭn·ăm'ĭks* (Gr. *dunamis,* power), that branch of mechanics which investigates the effects of forces not in equilibrium but producing motion : **dynamometer, n.,** *dĭn'ăm·ŏm'ĕt·ėr* (Gr. *metron,* a measure), an instrument for measuring the muscular power of men and animals.

dyscrasia, n., *dĭs·krāz'ĭ·ă* (Gr. *duscrāsĭa,* a bad mixture—from *dus,* an inseparable particle, denoting 'with pain,' 'with difficulty,' 'badly'; *krasis,* a mixture), a morbid or bad state of the vital fluids.

dysentery, n., *dĭs'ĕnt·ĕr·ĭ* (Gr. *dusĕntĕrĭa,* L. *dysentĕria,* a flux, dysentery — from *dus,* badly ; *entĕra,* the bowels), a flux or looseness of the bowels, with a discharge of blood and mucus, and griping pains.

dysmenorrhœa, n., *dĭs'mĕn·ŏr·rē'ă* (Gr. *dus,* badly ; *mēnes,* the menstrual discharges ; *rhēo,* I flow), difficult menstruation.

dyspepsia, n., *dĭs·pĕps'ĭ·ă* (Gr. *duspepsia,* difficulty of digestion — from *dus,* badly ; *pepto,* I digest), bad or difficult digestion.

dysphagia, n., *dĭs·fādj'ĭ·ă* (Gr. *dus,* badly ; *phago,* I eat), difficulty of swallowing.

dyspnœa, n., *dĭsp·nē'ă* (Gr. *duspnoia,* L. *dyspnœa,* difficulty of breathing—from *dus,* badly ; *pnĕo,*

I breathe), a difficulty of breathing.

dysuria, n., *dĭs·ūr'·ĭ·ă* (Gr. *dus*, badly ; *ouron*, urine), difficulty in making urine.

Ebenaceæ, n. plu., *ĕb'·ĕn·ā'·sĕ·ē* (Gr. *ebĕnos*, L. *ebĕnus*, the ebon tree, ebony), the Ebony family, an Order of trees remarkable for the durability and hardness of its wood, and some bear edible fruits: **ebony**, n., *ĕb'·ŏn·ĭ*, the black duramen of the species Diospyros reticulata and ebonum.

ebracteate, a., *ĕ·brăk'·tĕ·āt* (L. *e*, from ; *bractĕa*, a thin layer of wood), in *bot.*, without a bract or floral leaf.

eburnation, n., *ēb'·ĕr·nā'·shŭn* (L. *ebur*, ivory), an ivory-like condition of bone arising from disease, chiefly in connection with rheumatoid arthritis.

Ecballium agreste, *ĕk·băl'·lĭ·ŭm ăg·rĕst'·ĕ* (Gr. *ekballo*, I cast out, I expel ; L. *agrestis*, belonging to the fields), or **Ecballium officinarum**, *ŏf·fĭs'·ĭn·ār'·ŭm* (L. *officĭna*, the shop, *officĭnārum*, of the shops), the wild or squirting cucumber ; the latter is the officinal name of the Momordica elaterium, Ord. Cucurbitaceæ.

ecchymosis, n., *ĕk'·ĭ·mōz'·ĭs* (Gr. *ek*, out of ; *chumos*, juice), livid spots or blotches on the skin arising from an escape of blood into the connective tissues of the skin, as may be caused by a fall or blow, or resulting from disease ; a bruise.

Eccremocarpus, n., *ĕk'·krĕ·mō·kărp'·ŭs* (Gr. *ekkrĕmēs*, hanging down ; *karpos*, fruit), a genus of ornamental climbing plants, Ord. Bignoniaceæ, so called from the pendant character of its fruit : **Eccremocarpus scaber**, *skăb'·ĕr* (L. *scăber*, rough), a commonly cultivated species.

ecderon, n., *ĕk'·dĕr·ŏn* (Gr. *ek*, out; *deros*, skin, hide), in *zool.*, the

outer of the two layers of that part of the skin called 'ectoderm,' corresponding to the 'epidermis' in man, into which it shows a tendency to break up.

ecdysis, n., *ĕk'·dĭs·ĭs* (Gr. *ekdusis*, the act of stripping, an emerging), a shedding or moulting of the skin.

echinate, a., *ĕk·īn'·āt* or *ĕk'·ĭn·āt* (L. *echīnātus*, prickly—from Gr. *echīnos*, L. *echīnus*, a sea-urchin, a hedgehog), covered with prickles like a hedgehog; prickly: **echinus**, n., *ĕk·īn'·ŭs*, a sea-hedgehog; the prickly head or top of a plant.

Echinocactus, n., *ĕk·īn'·ō·kăk'·tŭs* (L. *echīnus*, a hedgehog ; *cactus*, the cactus), a genus of spiny plants, Ord. Cactaceæ, of great beauty and interest: **Echinocactus viznaga**, *vĭz·nāg'·ă* (*viznăga*, a carrot-like ammi), a species which attains large dimensions.

Echinococcus, n., *ĕk·īn'·ō·kŏk'·kŭs*, **Echinococci**, n. plu., *ĕk·īn'·ō·kŏk'·sī* (Gr. *echīnos*, a hedgehog ; *kokkos*, a berry), the larval form of a minute tapeworm of the dog, the Tænia echinococcus commonly called 'hydatid'; known by many other names, as **Echinococcus hominis**, *hŏm'·ĭn·ĭs* (L. *homo*, man, *homĭnis*, of man), a species which infests man ; and **E. veterinorum**, *vĕt'·ĕr·ĭn·ōr'·ŭm* (L. *veterĭnōrum*, of beasts of burden), a species which infests cattle, etc.

Echinodermata, n. plu., *ĕk·īn'·ō·dĕrm'·ăt·ă* (Gr. *echīnos*, a sea-hedgehog ; *derma*, skin), a class of animals comprising sea-urchins, star-fishes, etc., most of which have spiny skins : **Echinoidea**, n., *ĕk'·īn·oÿd'·ĕ·ă* (Gr. *eidos*, resemblance), an Order of animals which comprises sea-urchins.

Echinorhynchus, n., *ĕk·īn'·ō·rĭngk'·ŭs* (Gr. *echīnos*, a hedgehog ; *rungchos*, a snout, a beak), a genus of intestinal worms : **Echinorhynchus gigas**, *jīg'·ăs* (L.

gigas, a giant), a parasite which infests the intestines of the pig.

echinulate, a., *ĕk·ĭn'·ŭl·āt* (dim. of L. *echīnus*, a hedgehog), possessed of small spines or prickles.

Echites, n. plu., *ĕk·ĭt'·ēz* (Gr. *echis*, a viper, from its smooth, twining shoots), a beautiful genus of ever-green twiners, Ord. Apocynaceæ : **Echites scholaris**, *skōl·ār'·ĭs* (L. *scholāris*, scholarly—from *schōla*, a school), a species used in India as a tonic : **E. antidysenterica**, *ănt'·ĭ·dĭs·ĕn·tĕr'·ĭk·ă* (Gr. *anti*, against; *dusenterĭkos*, one who has the dysentery), a species said to be astringent and febrifugal.

Echium, n., *ĕk'·ĭ·ŭm* (Gr. *echis*, a viper), a pretty genus of shrubs, Ord. Boraginaceæ, whose seeds are said to resemble the head of the viper.

eclampsia, n., *ĕk·lămps'·ĭ·ă* (Gr. *eklampsis*, a shining forth—from *ek*, forth; *lampein*, to shine), a convulsive attack, so termed from its suddenness.

ecraseur, n., *ĕk'·răz·ār'* (F. from *écraser*, to crush, to grind), a surgical instrument for removing tumours by a combined process of crushing and tearing, attended by much less bleeding than cutting out.

ecstasy, n., *ĕk'·stăs·ĭ* (Gr. *ekstasis*, change of state—from *ek*, out; *stasis*, standing, state), intense nervous and emotional excitement, in which the functions of the senses are suspended, and which is frequently accompanied by rigid immobility of one or more series of muscles.

ectasis, n., *ĕk'·tăs·ĭs* (Gr. *ektăsis*, extension), the dilated condition of an artery, as in aneurisms, or of a vein, as in varices ; usually applied to the dilatation of small blood-vessels.

ecthyma, n., *ĕk·thĭm'·ă* (Gr. *ek-thūma*, an eruption), a skin disease consisting of large, circular, raised pustules, sur-rounded by livid, purplish zones.

Ectocarpus, n., *ĕk'·tō·kărp'·ŭs* (Gr. *ektos*, outside ; *karpos*, fruit), a genus of dark - green marine plants, Ord. Algæ, whose thecæ are not enclosed, hence the name.

ectocyst, n., *ĕk'·tō·sĭst* (Gr. *ektos*, outside ; *kustis*, a bladder), in *zool.*, the external investment of the cœnœcium of a polyzoön.

ectoderm, n., *ĕk'·tō·dĕrm* (Gr. *ektos*, outside ; *derma*, skin), in *zool.*, the external integumentary layer of the Cœlenterata, corre-sponding to the epidermis in man ; the outer or upper layer of cells into which the blastoderm is divided after the completion of the segmenting process.

ectopia, n., *ĕk·tōp'·ĭ·ă* (Gr. *ek*, out of ; *topos*, place), the displacement of a part : ectopia cordis, *kŏrd'·ĭs* (L. *cor*, the heart, *cordis*, of the heart), the displacement of the heart, in which the heart is situ-ated outside the chest at birth : **e. vesicæ**, *vĕs·ĭ'·sē* (L. *vēsīca*, the bladder, *vēsīcæ*, of the bladder), a deficiency in the abdominal wall of the bladder, in which the bladder appears as a red surface on which the ureters open.

ectosarc, n., *ĕk'·tō·sărk* (Gr. *ektos*, outside ; *sarx*, flesh, *sarkos*, of flesh), in *zool.*, the outer trans-parent sarcode-layer of certain rhizopods, such as the Amœba.

ectozoon, n., *ĕk'·tō·zō'·ŏn*, ectozoa, n. plu., *ĕk'·tō·zō'·ă* (Gr. *ektos*, out-side ; *zoön*, an animal, *zoä*, animals), animal parasites which attach themselves to the skin of the human body, as 'the itch insect,' 'the louse,' 'the chegoe,' and 'the Guinea worm.'

ectropion, n., *ĕk·trōp'·ĭ·ŏn*, also **ectropium**, n., *-ĭ·ŭm* (Gr. *ek*, out ; *trepo*, I turn), a disease in which the eyelids are everted.

ecyphellate, a., *ē·sĭf'·ĕl·lāt* (Gr. *e*, for *ex* or *ek*, without ; Eng.

cyphellate), in *bot.*, not having minute sunken cup-like spots.

eczema, n., *ĕk'zĕm·ă* (Gr. *ekzĕsis*, an eruption on the skin—from *ek*, out; *zeo*, I boil), a catarrhal affection of the skin, which may be an erythema, a vesicle, a pustule, a fissure, etc., and has received various names accordingly, as **eczema chronicum**, *krŏn'ĭk·ŭm* (Gr. *chronos*, time), chronic eczema; also psoriasis; a chronic inflammation of the skin, associated with some thickening, and the formation of cracks and fissures; popularly, the disease in horses is called 'rat tails,' from the elevated patches of scabs on the back part of the limbs: **e. impetiginodes**, *ĭm'pĕt·ĭdj'ĭn·ōd'ēz* (L. *impĕtīgo*, a skin disease, *impĕtīgĭnes*, skin diseases), the eruption in dogs suffering from red mange; grocer's itch: **e. rubrum**, *rōōb'rŭm* (L. *rŭbrum*, red), the common red mange of smooth terriers and greyhounds; the eruption of vesicles occurring on an inflamed skin: **e. simplex**, *sĭm'plĕks* (L. *simplex*, simple, unmixed), one of the mangy affections of dogs; 'humid tetter' in man: **e. solare**, *sōl·ār'ē* (L. *sōlāris*, belonging to the sun—from *sol*, the sun), an eruption on the skin from the effects of the sun or heated air in summer; heat spots: **eczematous**, a., *ĕk·zĕm'ăt·ŭs*, of or belonging to the disease eczema.

Edentata, n. plu., *ē'dĕnt·āt'ă* (L. *e*, without; *dens*, a tooth, *dentes*, teeth), an Order of Mammalia, so called because destitute of front or incisive teeth: **edentate**, a., *ē·dĕnt'·āt*, without front teeth; deprived of teeth: **edentulous**, a., *ē·dĕnt'·ūl·ŭs*, toothless; applied to the mouth of an animal without dental apparatus; applied to the hinge of the bivalve molluscs.

Edriophthalmata, n. plu., *ĕd'rĭ·*

ŏf·thăl'·măt·ă (Gr. *hedraios*, sitting, sedentary—from *hedzō*, I sit; *ophthalmos*, an eye), the division of the Crustacea in which the eyes are not supported upon stalks: **edriophthalmous**, a., *-thăl'·mŭs*, having immovable sessile eyes.

efferent, a., *ĕf'·fĕr·ĕnt* (L. *ef* for *ex*, out; *fĕro*, I bear or carry), conveying from or outwards; carrying from the centre to the periphery: n., a vessel which carries outwards, distinguished from *afferent*, which means 'conveying into or towards.'

effervescence, n., *ĕf'·fĕr·vĕs'·sĕns* (L. *effervesco*, I boil up or over), the frothing or bubbling up of liquids from the generation and escape of gas.

efflorescence, n., *ĕf'·flŏr·ĕs'·sĕns* (L. *effloresco*, I blow or bloom as a flower), a mealy-like substance which covers certain minerals when exposed to the influence of the atmosphere; the conversion of a solid substance into a powder.

effluvium, n., *ĕf·flŏv'·ĭ·ŭm* (L. *effluvium*, a flowing out—from *ex*, out; *fluo*, I flow), the invisible vapour arising from putrefying matter or from diseased bodies.

effusion, n., *ĕf·fūzh'·ŭn* (L. *effusus*, poured out or forth—from *ex*, out; *fusus*, poured), the act of pouring a liquid into or over; what is poured out.

egranulose, a., *ē·grăn'·ūl·ōz* (L. *e*, without; Eng. *granulose*), in *bot.*, without granules.

Ehretiaceæ, n. plu., *ĕr·ēsh'·ĭ·ā'·sē·ē* (after *Ehret*, a German botanical draughtsman), a Sub-order of plants, Ord. Boraginaceæ: **Ehretia**, n., *ĕr·ēsh'·ĭ·ă*, a genus of plants of much beauty.

ejaculator, n., *ē·jăk'·ūl·āt'·ŏr* (L. *ejaculātus*, cast or thrown out), name of one or two muscles: **ejaculatores**, n. plu., *ē·jăk'·ūl·āt·ŏr'·ēz*, the two muscles which surround the bulb of the urethra.

Elæagnaceæ, n. plu., *ĕl'ē·ăg·nā'sĕ·ē* (Gr. *elaios*, the wild olive; *agnos*, the 'agnus castus' or chaste tree), the Oleaster family, an Order of trees and shrubs usually covered with silvery stellate hairs : **Elæagnus**, n., *ĕl'ē·ăg'nŭs*, a genus, several species of which bear edible fruit : **Elæagnus arborea**, *ăr·bōr'ē·ă* (L. *arbŏrĕus*, tree-like—from *arbor*, a tree) ; **E. conferta**, *kŏn·fĕrt'ă* (L. *confertus*, thick, dense); and **E. Orientalis**, *ōr'ĭ·ĕnt·āl'ĭs* (L. *Orĭentālis*, Eastern—from *orĭens*, the rising sun), species which yield eatable fruit, the latter a dessert fruit called 'zinzeya': **E. parvifolia**, *părv'ĭ·fōl'ĭ·ă* (L. *parvus*, little ; *folium*, a leaf), yields an edible fruit, has highly fragrant flowers, and abounds in honey.

Elæocarpeæ, n. plu., *ĕl'ē·ō·kărp'ē·ē* (Gr. *elaios*, a wild olive ; *karpos*, fruit), a Sub-order of plants, Ord. Tiliaceæ, whose fruit has been compared to an olive : **Elæocarpus**, n., *ĕl'ē·ō·kărp'ŭs*, a very beautiful genus of plants,—the bark is used as a tonic.

Elæodendron, n., *ĕl'ē·ō·dĕnd'rŏn* (Gr. *elaios*, a wild olive ; *dendron*, a tree), an ornamental genus of plants, Ord. Celastraceæ.

Elais, n., *ĕl·ā'ĭs* (Gr. *elaia*, an olive tree), a genus of palm trees, Ord. Palmæ, from the fruit of which the natives of Guinea express an oil as the Greeks do from the olive, hence the name: **Elais Guineensis**, *gĭn'ē·ĕns'ĭs* (from *Guinea*, in Africa); and **E. melanococca**, *mĕl'ăn·ō·kŏk'kă* (Gr. *melan*, black ; *kokkos*, a seed, a berry), species of palms from whose fruit the palm-oil imported from the W. Coast of Africa is obtained.

Elaphrium, n., *ĕl·ăf'rĭ·ŭm* (Gr. *elaphros*, light, of no value), a genus of ornamental trees, Ord. Burseraceæ, whose wood is of no value : **Elaphrium tomentosum**, *tŏm'ĕnt·ōz'ŭm* (L. *tōmentum*, a stuffing for cushions), yields the Indian Tacamahac, a balsamic bitter resin.

Elasmobranchii, n. plu., *ĕl·ăs'mō·brăngk'ĭ·ī* (Gr. *elasma*, a plate of metal; *brangchia*, the gills of fish), an Order of fishes, including the sharks and rays.

elaterium, n., *ĕl·ăt·ēr'ĭ·ŭm* (L. *elatērĭum*, Gr. *elatērion*, the juice of the wild cucumber—from Gr. *elatēr*, a driver), the sediment from the expressed juice of the squirting gourd or wild cucumber, which is a powerful drastic purgative : **elaterin**, n., *ĕl·ăt'ĕr·ĭn*, the active principle of elaterium : **elaters**, n. plu., *ĕl'ăt·ĕrs*, elastic, spirally-twisted filaments for dispersing spores, found with spores in liverworts, etc.

Elatinaceæ, n. plu., *ĕl'ăt·ĭn·ā'sĕ·ē* (Gr. *elatē*, a pine tree, from the supposed resemblance of the leaves of some of them to those of the pine), the Water-pepper family, an Order of marsh plants found in all parts of the world : **Elatine**, n., *ĕl·ăt'ĭn·ē*, a genus of curious little aquatic plants.

elecampane, n., *ĕl'ē·kăm·pān'* (F. *énule-campane;* L. *inula helenium* —from Gr. *helenion*, a plant said to have sprung from Helen's tears), the common name of Inula Helenium, whose root has stimulant and aromatic qualities.

electrode, n., *ē·lĕk'trōd* (Gr. *ēlektron*, amber ; *hodos*, a way), the direction of an electric current ; the extremities of the conductors through which the electric current enters or quits a body.

electuary, n., *ē·lĕk'tū·ĕr·ĭ* (mid. L. *electŭārium*, a confection— from Gr. *ek*, out ; *leicho*, I lick), a medicine made up as a confection with honey or sugar.

elemi, n., *ĕl'ĕm·ĭ* (F. *elemi*, but probably a native word), a resinous substance from several species of trees, brought from Ethiopia in masses of a yellowish colour, from species of Canarium commune and balsamiferum, Ord. Burseraceæ.

elephantiasis, n., *ĕl'·ĕ·făn·tĭ'·ăs·ĭs* (Gr. *elephas*, an elephant, *elephantis*, of an elephant), a disease of the skin, in which it becomes thick and rugose ; the disease chiefly affects the lower limbs, and depends on different causes.

Elettaria, n., *ĕl'·ĕt·ār'ĭ·ă* (*elettāri*, a Malabar word for the lesser cardamom), a genus of plants, Ord. Zingiberaceæ : **Elettaria cardamomum**, *kărd'·ăm·ōm'·ŭm* (Gr. *kardămŏn*, a kind of cress), the species which yields the Malabar cardamoms, the fruit being ovoid and three-sided : E. **major**, *mādj'·ŏr* (L. *major*, greater), a variety, formerly so called, growing in Ceylon.

eleutheropetalous, a., *ĕl·ōōth'·ĕr·ŏ·pĕt'·ăl·ŭs* (Gr. *eleutheros*, free ; *petalon*, an unfolded leaf), in *bot.*, polypetalous: **eleutherosepalous**, a., *-sĕp'·ăl·ŭs* (a simple arbitrary conversion of *petalon* into *sepalon*), polysepalous.

elixir, n., *ĕ·lĭks'·ĭr* (Ar. *el iksir*, the philosopher's stone), a refined spirit ; a medicine supposed to be particularly efficacious.

ellipsoidal, a., *ĕl'·lĭps·ŏȳd'·ăl* (L. *ellipsis*, Gr. *elleipsis*, an ellipsis, an omission ; Gr. *eidos*, resemblance), nearly oval in shape.

Elodea, n., *ĕl·ōd'·ē·ă* (Gr. *elōdēs*, marshy, boggy), a genus of aquatic plants, Ord. Hypericaceæ.

elutriation, n., *ĕ·lōt'·rĭ·ā'·shŭn* (L. *elutriātus*, washed out—from *e*, out of ; *lutus*, washed), a process of washing for separating the finer particles of a powder from the coarser ; also for separating the lighter earthy parts of metallic ores.

Elymus, n., *ĕl'·ĭm·ŭs* (Gr. *eluō*, I cover or wrap up), a genus of plants, Ord. Gramineæ : **Elymus condensatus**, *kŏn'·dĕns·āt'·ŭs* (L.

condensātus, made very dense— from *con*, together ; *densus*, dense, close), the bunch-grass of California, an early fodder-grass in Britain : E. **arenarius**, *ăr'·ĕn·ār'·ĭ·ŭs* (L. *ărēnārĭa*, a sand-pit), this species, and Ammophila arenaria, form the 'bent' and 'marram' of our own shores.

elytrum, n., *ĕl'·ĭt·rŭm*, **elytra**, n. plu., *ĕl'·ĭt·ră* (Gr. *elutron*, a covering or sheath), the hard wing-sheaths of beetles ; scales or plates on the back of the sea-mouse, Aphrodite : **elytriform**, a., *ĕl'·ĭt'·rĭ·fŏrm* (L. *forma*, shape), in the form of a wing-sheath : **elytrine**, n., *ĕl'·ĭt·rĭn*, the substance of the coriaceous wing-sheaths of such insects as beetles.

emarginate, a., *ĕ·mārj'·ĭn·āt* (L. *emarginātus*, deprived of its edge —from *e*, out of ; *margo*, the extremity or margin), in *bot.*, having a notch at the end or summit, as if a piece had been cut out.

embolism, n., *ĕm'·bŏl·ĭzm* (Gr. *embolisma*, a patch ; *embŏlos*, what is thrust or put in—from *en*, in ; *ballō*, I throw or cast), the plugging or blocking of an artery by any migratory foreign body, as an air bubble, an oil globule, a blood clot, or a granule of fibrine ; also called **embole**, *ĕm'·bŏl·ē*: **embolon**, n., *ĕm'·bŏl·ŏn*, the clot or other matter which, carried into the circulation of the blood, produces an embolism.

embrocation, n., *ĕm'·brŏk·ā'·shŭn* (Gr. *embrochē*, a steeping, an embrocation), the act of bathing and rubbing a diseased part with a liquid medicine ; the mixture so employed.

embryo, n., *ĕm'·brĭ·ŏ* (Gr. *embruon*, an infant in the womb—from *en*, in ; *bruō*, I shoot or bud), the first rudiments of an animal or plant ; in *bot.*, the young plant contained in the seed : **embryo-**

buds, nodules in the bark of the beech and other trees : **embryogeny**, n., *ĕm'·brĭ·ŏdj'·ĕn·ĭ* (Gr. *gennao*, I produce), in *bot.*, the development of the embryo in the ovule : **embryogenic**, a., *-jĕn'·ĭk*, of or belonging to : **embryo-sac**, same as **embryonary sac**, which see.

embryology, n., *ĕm'·brĭ·ŏl'·ŏ·jĭ* (Gr. *embruon*, an infant in the womb ; *logos*, discourse), the study of the formation of the embryo; the anatomy which traces the development of the creature from the impregnated ovum.

embryonary, a., *ĕm·brĭ'·ŏn·ĕr·ĭ* (Gr. *embruon*, an infant in the womb), relating to the embryo ; rudimentary : **embryonal**, a., *ĕm·brĭ'·ŏn·ăl*, same sense : **embryonary sac**, in *bot.*, the cellular bag in which the embryo is formed.

embryotega, n., *ĕm'·brĭ·ŏt'·ĕg·ă* (Gr. *embruon*, an infant in the womb ; *tĕgos*, a covering), in *bot.*, a process or callosity raised from the spermoderm by the embryo of some seeds during germination, as in the bean.

emergent, a., *ĕ·mĕrj'·ĕnt* (L. *emergo*, I rise up, I come forth—from *e*, out of ; *mergo*, I plunge or dip), rising out of ; in *bot.*, protruding through the cortical layer.

emersed, a., *ē·mĕrst'* (L. *e*, out of ; *mersus*, plunged or dipped), in *bot.*, protruded upwards.

emesia, n., *ĕm·ēs'·ĭ·ă*, also **emesis**, n., *ĕm'·ĕs·ĭs* (Gr. *emesia*, an inclination to vomit ; *emesis*, the act of vomiting), the act of vomiting.

emetic, n., *ĕ·mĕt'·ĭk* (Gr. *emetikos*, that causes vomiting—from *emeo*, I vomit ; L. *emetica*, an emetic), a medicine or other agent which produces vomiting : **adj.**, that causes vomiting : **emetin**, n., *ĕm'·ĕt·ĭn*, the active principle of ipecacuanha.

emiction, n., *ĕ·mĭk'·shŭn* (L. *e*, out of ; *mictus*, made water), the discharging of urine ; what is voided by the urinary passages.

eminentia collateralis, *ĕm'·ĭn·ĕn'·shĭ·ă kŏl·lăt'·ĕr·āl'·ĭs* (L. *ēmĭnentia*, a prominence ; *collaterālis*, collateral — from *con*, together ; *lātus*, a side), a smooth eminence between the middle and posterior horns of the cerebrum.

emmenagogue, n., *ĕm·mĕn'·ă·gŏg* (Gr. *emmēna*, the menses—from *en*, in ; *mēn*, a month ; *agō*, I lead, I bring), a remedy supposed to promote the menstrual discharges.

emollient, n., *ē·mŏl'·lĭ·ĕnt* (L. *emolliens*, making soft—from *e*, out of ; *mollis*, soft), a liquid remedy meant to soothe a part and diminish irritation, when applied externally.

Empetraceæ, n. plu., *ĕm'·pĕt·rā'·sĕ·ē* (Gr. *ĕmpĕtros*, growing among rocks—from *en*, in, among; *petra*, a rock), the Crowberry family, an Order of heath-like shrubs, bearing small sub-acid berries : **Empetrum**, n., *ĕm·pĕt'·rŭm*, a genus of heath-like shrubs, so called from the character of their place of growth : **Empetrum nigrum**, *nĭg'·rŭm* (L. *nigrum*, black), the black crowberry, common in the mountainous parts of Northern Europe.

emphysema, n., *ĕm'·fĭs·ēm'·ă* (Gr. *emphusēma*, a puffing up, inflation—from *en*, in ; *phusaō*, I blow), the distension of a tissue with air ; a disease of the lungs in which the air cells become unduly distended, and ultimately ruptured : **emphysematous**, a., *ĕm'·fĭs·ēm'·ăt·ŭs*, characterised by an abnormal distension of the air in the lungs, or by the presence of air as the result of injury or decomposition in a tissue.

empiricism, n., *ĕm·pĭr'·ĭs·ĭzm* (L. *empirici*, Gr. *empeirikoi*, ancient physicians who followed a system based on practical experience alone), practice in a profession founded on experience alone, as opposed to experience based on scientific knowledge; the practice of medicine without a medical education; quackery.

emprosthotonos, n., *ĕm'·prŏs·thŏt'·ŏn·ŏs* (Gr. *emprosthen*, in front; *teinō*, I bend), a form of tonic convulsion in which the patient is thrown forwards, as occurs in some cases of tetanus.

empyema, n., *ĕm'·pī·ēm'·ă* (Gr. *empuēma*, a purulent discharge —from *en*, in; *puon*, pus), a collection of purulent matter in the pleural cavity.

emulsin, n., *ē·mŭls'·ĭn* (L. *ĕmŭlsus*, milked out, drained out—from *e*, out of; *mulgeo*, I milk), a nitrogenous compound found in certain oily seeds, as in almonds: **emulsion,** n., *ē·mŭl'·shŭn*, a smooth liquid for softening; a cough mixture; a bland fluid medicine having a milky appearance, produced chiefly by the combination of an oily substance with water and an alkali.

emunctory, n., *ē·mŭngk'·tēr·ĭ,* **emunctories,** n. plu., *-tēr·ĭz* (L. *emunctus*, wiped or blown, as one's nose), a part of the body where anything excrementitious is collected or separated in readiness for ejectment.

enarthrosis, n., *ĕn'·ăr·thrōs'·ĭs* (Gr. *enarthros*, jointed—from *en*, in; *arthron*, a joint), a ball-and-socket joint, like the shoulder and hip, allowing motion in every direction.

enation, n., *ē·nā'·shŭn* (L. *enātus*, grown or sprung up—from *e*, out of; *nātus*, born), the changes produced by excessive development in various organs of plants; the growth of adventitious lobes.

Encephalartos, n., *ĕn'·sĕf·ăl·ărt'·ŏs* (Gr. *engkephalos*, that which is in the head—from *en*, in; *kephalē*, the head; *artos*, bread), a genus of trees, Ord. Cycadaceæ, whose various species are known by the

Hottentots under the general name 'bread-tree.'

encephalitis, n., *ĕn'·sĕf·ăl·īt'·ĭs* or *ĕng'·kĕf·ăl·īt'·ĭs* (Gr. *engkephalos*, that which is in the head, the brain—from *en*, in; *kephalē*, the head), inflammation of the brain:

encephaloid, a., *ĕn·sĕf'·ăl·ŏyd* or *ĕng·kĕf'* (Gr. *eidos*, resemblance), resembling the materials of the brain.

encephalon, n., *ĕn·sĕf'·ăl·ŏn* or *ĕng·kĕf'·ăl·ŏn* (Gr. *engkephalos*, the brain—from *en*, in; *kephalē*, the head), the whole contents of the cranium; the brain: **encephalous,** a., *ĕn·sĕf'·ăl·ŭs* or *ĕng·kĕf'*, possessing a distinct head, applied to certain of the molluscs: **encephalocele,** n., *ĕn'·sĕf·ăl'·ō·sēl* or *ĕng'·kĕf·ăl'* (Gr. *kēlē*, a tumour), a congenital condition in which, owing to a deficiency in the cranial walls, a portion of the brain and its membranes are protruded; also called 'hernia cerebri.'

enchondroma, n., *ĕn'·kŏn·drōm'·ă* (Gr. *en*, in; *chondros*, cartilage), a tumour somewhat smooth on its surface, essentially consisting of cartilaginous structure.

encipient, n., *ĕn·sĭp'·ĭ·ĕnt* (L. *en*, in; *capio*, I take), a palatable vehicle in which cattle may take a medical preparation, such as bruised coriander seeds.

encysted, a., *ĕn·sĭst'·ĕd* (Gr. *en*, in; *kustis*, a bladder), enclosed in a bag, sac, or cyst; consisting of cysts: **encystation,** n., *ĕn'·sĭst·ā'·shŭn*, the transformation undergone by certain of the Protozoa, when they become motionless, and surround themselves with a thick coating or cyst.

endecagynian, a., *ĕn'·dĕk·ă·jĭn'·ĭ·ăn*, also **endecagynous,** a., *ĕn'·dĕk·ădj'·ĭn·ŭs* (Gr. *hendeka*, eleven; *gunē*, a woman), in *bot.*, having eleven pistils.

endemic, a., *ĕn·dĕm'·ĭk* (Gr. *en*, in; *demos*, a people), peculiar to a district or to a certain class of persons; applied to a prevalent disease arising from local causes, as bad air or water: **n.**, a disease prevailing in a particular locality, or among a particular class of persons: **epidemic** is an infectious or contagious disease attacking many persons at the same time, but of a temporary character; while an **endemic** is due to local conditions, and is always more or less permanent in a district.

endermic, a., *ĕn·dĕrm'ĭk* (Gr. *en*, in; *derma*, skin), applied to the method of using certain medicines by injecting them under the skin.

enderon, n., *ĕn'dĕr·ŏn* (Gr. *en*, in; *deros*, skin), in *zool.*, the inner of the two layers of that part of the skin called 'ectoderm' or 'epidermis·'; see 'ecderon.'

endocardium, n., *ĕn'dō·kärd'ĭ·ŭm* (Gr. *endon*, within; *kardia*, the heart), the membrane lining the interior of the heart: **endocarditis**, n., *ĕn'dō·kärd·ĭt'ĭs* (L. *itis*, inflammation), the inflammation of the membrane lining the interior of the heart.

endocarp, n., *ĕn'dō·kärp* (Gr. *endon*, within; *karpos*, fruit), in *bot.*, the membrane which lines the cavity containing the seeds, as in the apple; the stone or shell which encloses the seed or embryo, as in the plum.

endochrome, n., *ĕn'dō·krōm* (Gr. *endon*, within; *chroma*, colour), the colouring matter of cellular plants, exclusive of the green; the cell contents of Algæ.

endocyst, n., *ĕn'dō·sĭst* (Gr. *endon*, within; *kustis*, a bag or cyst), in *zool.*, the inner membrane or integumentary layer of a polyzoön.

endoderm, n., *ĕn'dō·dĕrm* (Gr. *endon*, within; *derma*, skin), in *zool.*, the inner or lower of the two layers of cells into which the blastoderm is divided after the completion of the segmenting

process: **endodermic**, a., *ĕn'dō·dĕrm'ĭk*, of or belonging to the endoderm.

endogenæ, n. plu., *ĕn·dŏdj'ĕn·ē*, also **endogens**, n. plu., *ĕn'dō·jĕns* (Gr. *endon*, within; *gennaō*, I produce), that division of the vegetable kingdom, as palms, grasses, rushes, and the like, whose growth takes· place from within, and not by external concentric layers, as in the 'exogens'; also called Monocotyledons: **endogenous**, a., *ĕn·dŏdj'ĕn·ŭs*, increasing by internal growth.

endolymph, n., *ĕn'dō·lĭmf* (Gr. *endon*, within; L. *lympha*, a water-nymph, water), the liquid contained within the membranous labyrinth of the ear.

endometritis, n., *ĕn·dŏm'ĕt·rĭt'ĭs* (Gr. *endon*, within; *mētra*, the womb),· inflammation of the lining membrane of the uterus.

endophlœum, n., *ĕn'dō·flē'ŭm* (Gr. *endon*, within; *phloios*, the bark of trees), the inner layer of the bark of trees;· the liber.

endopleura, n., *ĕn'dō·plŏr'ă* (Gr. *endon*, within; *pleura*, a side), in *bot.*, the inner covering of the seed immediately investing the embryo and albumen.

endopodite, n., *ĕn·dŏp'ŏd·ĭt* (Gr. *endon*, within; *pous*, a foot, *podes*, feet), in *zool.*, the inner of the two secondary joints into which the typical limb of a crustacean is divided.

endorhizal, a., *ĕn'dŏ·rĭz'ăl* (Gr. *endon*, within; *rhiza*, a root), having a root within,—applied to monocotyledonous plants, whose young root or radicle, when piercing the lower part of the axis, appears covered with a cellular sheath; the sheath is denominated the 'coleorhiza.'

endosarc, n., *ĕn'dō·särk* (Gr. *endon*, within; *sarx*, flesh), the inner molecular layer of sarcode in the Amœba.

endo-skeleton, n., *ĕn'·dō-skĕl'·ĕt·ŏn* (Gr. *endon*, within ; Eng. *skeleton*), the internal hard structures, such as bones, which serve for the attachment of muscles, or the protection of organs, as opposed to the external hard covering of shell.

endosmometer, n., *ĕn'·dŏs·mŏm'·ĕt·ĕr* (Eng. *endosmosis;* Gr. *metron*, a measure), an instr. to show Endosmose and Exosmose, consisting of a bladder of syrup attached to a tube and plunged into a vessel of water.

endosmose, n., *ĕn'·dŏs·mōz*, also **endosmosis**, n., *ĕn'·dŏs·mōz'·ĭs* (Gr. *endon*, within ; *ōsmos*, a thrusting, impulsion), that property of membranous tissue by which fluids of unequal densities, when placed on opposite sides of it, are enabled to pass through and intermix.

endosperm, n., *ĕn'·dō·spĕrm* (Gr. *endon*, within ; *sperma*, seed), in *bot.*, albumen formed within the embryo-sac : **endospermic**, a., *ĕn'·dō·spĕrm'·ĭk*, of or belonging to endosperm.

endospore, n., *ĕn'·dō·spōr* (Gr. *endon*, within ; *spora*, seed), the inner integument of spores: **endosporous**, a., *ĕn'·dō·spōr'·ŭs*, applied to Fungi which have their spores contained in a case.

endosteum, n., *ĕn·dŏs'·tē·ŭm* (Gr. *endon*, within ; *osteon*, a bone), the medullary membrane, a fine layer of highly vascular, areolar tissue within the bones.

endostome, n., *ĕn'·dŏ·stŏm* (Gr. *endon*, within ; *stŏma*, mouth), in *bot.*, the passage through the inner integument of an ovule.

endothecium, n., *ĕn'·dō·thē'·shĭ·ŭm* (Gr. *endon*, within ; *thēkē*, a box), in *bot.*, the inner lining of the anther cells.

enema, n., *ĕn·ēm'·ă*, **enemata**, n. plu., *ĕn·ĕm'·ăt·ă* (Gr. *eniēmi*, I cast or throw in), a medicine or preparation of food thrown into the lower bowel ; injections ; clysters.

enervation, n., *ĕn'·ĕrv·ā'·shŭn* (L. *enervatus*, having the nerves and sinews taken out from—from *e*, out of ; *nervus*, a nerve), a weak state of body or nervous debility arising from nervous disorders ; the state of being weakened.

enervis, n., *ē·nĕrv'·ĭs* (L. *enervis*, nerveless—from *en*, out of ; *nervus*, a nerve), in *bot.*, without nerves or veins.

enneagynian, a., *ĕn'·nĕ·ă·jĭn'·ĭ·ăn*, also **enneagynous**, a., *ĕn'·nĕ·ădj'·ĭn·ŭs* (Gr. *ennea*, nine ; *gunē*, a woman), in *bot.*, having nine pistils.

enneandrous, a., *ĕn'·nĕ·ănd'·rŭs* (Gr. *ennea*, nine ; *anēr*, a male, a man, *andros*, of a male), in *bot.*, having nine stamens.

enostosis, n., *ĕn'·ŏs·tōz'·ĭs* (Gr. *en*, in ; *osteon*, a bone), a bony tumour growing inward into the medullary canal of a bone ; see 'exostosis.'

ensiform, a., *ĕns'·ĭ·fŏrm* (L. *ensis*, a sword ; *forma*, a shape), in the form of a sword, as the leaves of Iris ; sword-shaped.

enteric, a., *ĕn·tĕr'·ĭk* (Gr. *entĕron*, an intestine), belonging to the intestines : **enteritis**, n., *ĕn'·tĕr·ĭt'·ĭs*, inflammation of the intestines, especially of the small intestine : **enterocele**, n., *ĕn·tĕr'·ō·sēl* (Gr. *kēle*, a tumour), a hernial tumour containing intestine.

enterorrhœa, n., *ĕn'·tĕr·ō·rē'·ă* (Gr. *enteron*, an intestine ; *rheo*, I flow), an abnormal increase of the secretions of the mucous glands of the intestines.

enterotomy, n., *ĕn'·tĕr·ŏt'·ŏm·ĭ* (Gr. *enteron*, an intestine ; *tomē*, a cutting), an operation on, or dissection of, the intestines.

enterozoa, n. plu., *ĕn'·tĕr·ō·zō'·ă* (Gr. *enteron*, an intestine ; *zoön*, an animal), a general name for the intestinal parasites which infest the bodies of animals.

enthelmins, n., *ĕn·thĕl'·mĭns* (Gr. *entos*, within ; *helmins*, a worm), an intestinal worm.

entire, a., *ĕn·tīr'* (F. *entier*, whole, complete ; L. *integer*, whole), in *bot.*, having no lobes or marginal divisions.

entomic, a., *ĕn·tŏm'·ĭk* (Gr. *entoma*, insects), pert. to insects : **entomoid**, a., *ĕn'·tŏm·ōyd* (Gr. *eidos*, resemblance), resembling an insect : **entomology**, n., *ĕn'·tŏm·ŏl'·ŏ·jĭ* (Gr. *logos*, discourse), the history and habits of insects: **entomophaga**, n. plu., *ĕn'·tŏm·ŏf'·ăg·ă* (Gr. *phago*, I eat), the section of the Marsupials which live chiefly on insects : **entomophagous**, a., *ĕn'·tŏm·ŏf'·ăg·ŭs*, chiefly subsisting on insects.

entomophilous, a., *ĕn'·tŏm·ŏf'·ĭl·ŭs* (Gr. *entoma*, insects ; *philo*, I love), in *bot.*, applied to flowers in which pollination is effected by insects.

entomostraca, n. plu., *ĕn'·tŏm·ŏs'·trăk·ă* (Gr. *entoma*, insects ; *ostrakon*, a shell), in *zool.*, a division of the Crustacea covered with a delicate membranaceous shell, of which the water-flea may be looked on as the type— they are chiefly fresh-water, and usually microscopic : **entomostracous**, a., *ĕn'·tŏm·ŏs'·trăk·ŭs*, enclosed in an integument, as an insect.

entophyte, n., *ĕn'·tō·fĭt*, **entophyta**, n. plu., *ĕn·tŏf'·ĭt·ă* (Gr. *entos*, within ; *phuton*, a plant), vegetable parasites which exist within the body, found in some diseases of the mucous membranes of the mouth and alimentary canal ; plants growing within others : **entophytic**, a., *ĕn'·tō·fĭt'·ĭk*, developing in the interior of plants and afterwards appearing on the surface, as fungi.

entozoon, n., *ĕn'·tō·zō'·ŏn*, **entozoa**, n. plu., *ĕn'·tō·zō'·ă* (Gr. *entos*,

within ; *zoŏn*, an animal), animal parasites which infest the interior of the bodies of other animals : **entozoology**, n., *ĕn'·tō·zō·ŏl'·ŏ·jĭ* (Gr. *logos*, discourse), a discourse or treatise on internal parasites.

entropion, n., *ĕn·trōp'·ĭ·ŏn* (Gr. *en*, in ; *trōpē*, a turning), the inversion or turning in of the eyelashes : **entropy**, n., *ĕn'·trōp·ĭ*, dissipation of energy.

enuresis, n., *ĕn'·ūr·ēz'·ĭs* (Gr. *enoureo*, I make water—from *en*, in ; *ouron*, urine), incontinence or involuntary escape of the urine.

envelope, n., *ĕn'·vĕl·ŏp* (F. *envelopper*, to fold up), a wrapper ; an investing integument : floral **envelopes**, in *bot.*, the calyx and corolla.

enzootic, a., *ĕn'·zō·ŏt'·ĭk* (Gr. *en*, in ; *zŏŏtŏkos*, bringing forth living animals—from *zoŏn*, an animal ; *tiktō*, I bring forth), applied to diseases peculiar to a district among the lower animals: **enzootic hæmaturia**, an endemic disease causing bloody urine among animals.

Epacridaceæ, n. plu., *ĕp·ăk'·rĭd·ās'·ē·ē* (Gr. *epi*, upon ; *akros*, the top, from the species found on hill-tops), the Epacris family, an Order of small shrubs and trees, allied to Ericaceæ, which represent the heaths in Australia: **Epacreæ**, n. plu., *ĕp·ăk'·rē·ē*, a tribe or Suborder : **Epacris**, n., *ĕp'·ăk·rĭs*, a genus of very elegant greenhouse plants.

epanody, n., *ĕp·ăn'·ŏd·ĭ* (Gr. *epanodos*, a return—from *epi*, upon ; *ana*, up ; *hodos*, a way), in *bot.*, the return of an irregular flower to a regular form.

epencephalon, n., *ĕp'·ĕn·sĕf'·ăl·ŏn* (Gr. *epi*, upon ; *egkephalos*, what is in the head, the brain), one of the five primary divisions of the brain, including the cerebellum, pons varolii, and the anterior part of the fourth ventricle : **epencephalic**, a., *ĕp·ĕn·sĕf·ăl'·ĭk*, situated over the contents of the head, or the brain.

epenchyma, n., ĕp-ĕng'kĭm-ă (Gr. *epi*, upon; *chumos*, juice), in *bot.*, the fibro-vascular tissues.

ependyma, n., ĕp-ĕn'dĭm-ă (Gr. *ependuma*, an outer or upper tunic—from *epi*, upon; *enduma*, clothing), the delicate epitheliated structure which lines the canal of the spinal cord and the cerebral ventricles : **ependyma ventriculorum**, vĕn·trĭk·ŭl·ŏr'ŭm (L. *ventrĭcŭlus*, the belly), the ependyma of the ventricles, the epithelial membranes lining the ventricles.

Ephedra, n., ĕf'·ĕd·ră (Gr. *ephedra*, a sitting, the plant horse-tail), a genus of curious plants, Sub-ord. Gnetaceæ, Ord. Coniferæ, whose berries are eaten in Russia, and by the wandering tribes of Great Tartary.

ephelis, n., ĕf·ēl'·ĭs (Gr. *epi*, upon; *helios*, the sun), sun - burn ; freckles.

ephemera, n., ĕf·ĕm'ĕr·ă (Gr. *epi*, upon; *hēmera*, a day), a fever which runs its course in a day : **ephemeral**, a., ĕf·ĕm'ĕr·ăl, applied to flowers which open and decay in a day.

ephippium, n., ĕf·ĭp'·pĭ·ŭm (Gr. *ephippeion*, a saddle—from *epi*, upon; *hippos*, a horse), the deep pit in the middle of the superior surface of the sphenoid bone, so called from its shape.

epiblast, n., ĕp'·ĭ·blăst (Gr. *epi*, upon ; *blastos*, a shoot), an abortive organ in the oat, supposed to be the rudiment of a second cotyledon.

epiblema, n., ĕp'·ĭ·blĕm'·ă (Gr. *epi*, upon; *blĕma*, a wound), an imperfectly formed epidermis covering the newly formed extremities of roots, etc., being, as it were, the tissue which first covers wounds.

epicalyx, n., ĕp'·ĭ·kăl'·ĭks (Gr. *epi*, upon ; Eng. *calyx*), the outer

calyx, consisting either of sepals or bracts, as in mallows.

epicarp, n., ĕp'·ĭ·kărp (Gr. *epi*, upon ; *karpos*, fruit), in *bot.*, the outer coat or covering of the fruit.

epichilium, n., ĕp'·ĭ·kĭl'·ĭ·ŭm (Gr. *epi*, upon or above; *cheilos*, a lip), in *bot.*, the label or terminal portion of the articulated lip of orchids.

epicline, n., ĕp'·ĭ·klīn (Gr. *epi*, upon ; *klĭnē*, a bed), in *bot.*, the nectary when placed on the receptacle of the flower : **epiclinal**, a., ĕp'·ĭ·klĭn'·ăl, seated on the disc or receptacle.

epicondyle, n., ĕp'·ĭ·kŏn'·dĭl (Gr. *epi*, upon ; *kondulos*, the elbow-joint), the protuberance on the external side of the distal end of the os humeri or shoulder-bone.

epicorolline, n., ĕp'·ĭ·kŏr'·ŏl·lĭn (Gr. *epi*, upon ; Eng. *corolline*), in *bot.*, inserted upon the corolla.

epicranium, n., ĕp'·ĭ·krān'·ĭ·ŭm (Gr. *epi*, upon ; *krānion*, the skull), the scalp or integuments lying over the cranium : **epicranial**, a., ĕp'·ĭ·krān'·ĭ·ăl, applied to the muscle which extends over the upper surface of the cranium uniformly from side to side, without division.

epidemic, a., ĕp'·ĭ·dĕm'·ĭk (Gr. *epi*, upon; *demos*, the people), prevailing generally ; affecting great numbers : n., a disease universally prevalent in a district or country.

Epidendrum, n., ĕp'·ĭ·dĕnd'·rŭm (Gr. *epi*, upon; *dendron*, a tree —as usually found growing on branches of trees), a very extensive genus of 'epiphytes,' Ord. Orchidaceæ, many of which are deserving of culture for the beauty and delicious fragrance of their flowers : **Epidendrum frigidum**, frĭdj'·ĭd·ŭm (L. *frĭgĭdus*, cold), a species in Columbia, at an elevation of 12,000 or 13,000 feet, covered with a sort of varnish.

epidermis, n., ĕp'·ĭ·dĕrm'·ĭs (Gr.

K

epi, upon; *derma*, skin), the scarf or outermost layer of the skin; in *bot.*, the cellular layer covering the external surface of plants, the true skin of plants: **epidermoid**, a., *ĕp'·ĭ·dẽrm'·ŏўd* (Gr. *eidos*, resemblance), like the epidermis: **epidermic**, a., *ĕp'·ĭ·dẽrm'·ĭk*, pert. to the epidermis.

epididymis, n., *ĕp'·ĭ·dĭd'·ĭm·ĭs* (Gr. *epi*, upon; *didumos*, a testicle), a long, narrow, flattened body lying upon the outer edge of the posterior border of the testis.

epigæous, a., *ĕp'·ĭdj·ē'·ŭs*, or **epigæal**, a., *ĕp'·ĭdj·ē'·ăl* (Gr. *epi*, upon; *gĕä* or *gē*, the earth), in *bot.*, growing on the ground or close to it.

epigastric, a., *ĕp'·ĭ·găst'·rĭk* (Gr. *epi*, upon; *gastēr*, the belly, the stomach), pert. to the upper part of the abdomen: **epigastrium**, n., *ĕp'·ĭ·găst'·ri·ŭm*, the upper and middle part of the abdomen, nearly coinciding with the pit of the stomach.

epigeal, a., *ĕp'·ĭ·jē'·ăl* (Gr. *epi*, upon; *gē*, the earth), in *bot.*, above ground, applied to cotyledons; synonym of **epigæal** and **epigæous**, which see.

epiglottis, n., *ĕp'·ĭ·glŏt'·tĭs* (Gr. *epi*, upon; *glottis*, the mouth of the windpipe — from *glotta*, the tongue), the valve or cartilage that covers the upper part of the windpipe when food or drink is passing into the stomach: **epiglottitis**, n., *ĕp'·ĭ·glŏt·tīt'·ĭs*, inflammation of the epiglottis.

epigone, n., *ĕ·pĭg'·ŏn·ē* (Gr. *epi*, upon; *gonē*, seed, offspring), in *bot.*, the cellular layer which covers the young seed-case in mosses and the liverworts: **epigonium**, n., *ĕp'·ĭ·gŏn'·ĭ·ŭm*, in same sense.

epigynous, a., *ĕp'·ĭdj'·ĭn·ŭs* (Gr. *epi*, upon; *gunē*, a female, a woman), in *bot.*, above the ovary and attached to it.

epihyal, a., *ĕp'·ĭ·hī'·ăl* (Gr. *epi*, upon; Eng. *hyoid*, which see), applied to a considerable portion of the stylo-hyoid ligament, which is sometimes converted into bone in the human subject, and is in animals naturally osseous.

epilepsy, n., *ĕp'·ĭ·lĕps'·ĭ* (Gr. *epilepsia*, a seizure, the falling sickness—from *epi*, upon; *lambanō*, I seize), a disease characterised by a sudden loss of consciousness, and convulsions of greater or less severity: **epileptic**, a., *ĕp'·ĭ·lĕpt'·ĭk*, affected with falling sickness: **epileptoid**, a., *ĕp'·ĭ·lĕpt'·ŏўd* (Gr. *eidos*, resemblance), resembling epilepsy.

Epilobium, n., *ĕp'·ĭ·lōb'·ĭ·ŭm* (Gr. *epi*, upon; *lobos*, a lobe), a genus of plants, Ord. Onagraceæ, so called from the flowers having the appearance of being seated on the top of the pod; many of the species are very ornamental.

Epimedium, n., *ĕp'·ĭ·mēd'·ĭ·ŭm* (Gr. *epi*, upon; *Media*, an anc. country), a genus of elegant little plants, Ord. Berberidaceæ, which were said to grow in Media.

epimera, n. plu., *ĕp'·ĭ·mēr'·ă* (Gr. *epi*, upon; *mēros*, the upper part of the thigh), the parts lying immediately above the joints of the limb, as the 'epimera' or side segments of the lobster: **epimeral**, a., *ĕp'·ĭ·mēr'·ăl*, applied to that part of the segment of an articulate animal which lies immediately above the joint of the limb.

epinasty, n., *ĕp'·ĭ·năst'·ĭ* (Gr. *epi*, upon; *nastos*, pressed together, stuffed), in *bot.*, the nutation of bilateral, appendicular organs, when the growth is most rapid on the inner or upper side.

epipetalous, a., *ĕp'·ĭ·pĕt'·ăl·ŭs* (Gr. *epi*, upon; *petalon*, a leaf), inserted upon the petals, or growing upon them.

Epiphegus, n., *ĕp·ĭf'·ĕg·ŭs* (Gr. *epi*, upon; *phēgos* or *phagos*, a beech tree—from *phagō*, I eat), a genus of herbaceous parasitical plants, Ord. Orobanchaceæ, which are, in general, astringent and bitter : **Epiphegus Virginiana**, *vĕr·jĭn'·ĭ-ān'·ă* (after *Virginia*, an American State—from *virgo*, a virgin), a species called beech-drops, has been used in powder in cancerous sores.

epiphlœum, n., *ĕp'·ĭ-flē'·ŭm* (Gr. *epi*, upon, on the outside; *phloios*, bark), an external layer of bark : **epiphlœodal**, a., *ĕp'·ĭ-flē·ōd'·ăl*, existing superficially in the epidermis of bark.

epiphora, n., *ĕp·ĭf'·ŏr·ă* (Gr. *epi-phora*, a bringing to or upon—from *epi*, upon ; *phero*, I bring), watery eye, a derangement of the tear duct which allows the tears to flow down the cheeks.

epiphragm, n., *ĕp'·ĭ-frăm* (Gr. *epi*, upon ; *phragma*, a division), in *bot.*, the membrane closing the orifice of the thecæ in the Urn mosses.

epiphyllous, a., *ĕp'·ĭ-fŭl'·lŭs* (Gr. *epi*, upon ; *phullon*, a leaf), inserted or growing upon a leaf.

epiphysis, n., *ĕp·ĭf'·ĭs-ĭs*, **epiphyses**, n. plu., *-ĭs-ēz* (Gr. *epiphusis*, a growing upon, an additional growth—from *epi*, upon ; *phuo*, I grow), part of a bone separated from the shaft in early life by gristle, which finally becomes ossified to the main bone.

epiphyta, n. plu., *ĕp'·ĭ-fīt'·ă*, also **epiphyte**, n., *ĕp'·ĭ-fīt*, **epiphytes**, n. plu., *-fīts* (Gr. *epi*, upon ; *phuton*, a plant), vegetable parasites found on the skin of the human body, forming very troublesome skin affections ; plants attached to other plants, and growing suspended in the air ; a plant which grows on another plant, but not nourished by it : **epiphytal**, a., *ĕp'·ĭ-fīt'·ăl*, growing upon another plant.

epiploon, n., *ĕ-pĭp'·lŏ-ŏn* (Gr. *epiploos*, the caul — from *epi*, upon ; *pleō*, I swim), the omentum or caul ; a portion of the peritoneum or lining membrane of the abdomen, which covers in front, and as it were floats or sails on the intestines : epiploic, a., *ĕp'·ĭp-lō'·ĭk*, also epiploical, a., *-lō'·ĭk·ăl*, of or pert. to the epiploon or caul.

epipodia, n. plu., *ĕp'·ĭp-ōd'·ĭ-ă* (Gr. *epi*, upon ; *pous*, the foot, *podos*, of the foot), the muscular lobes developed from the lateral and upper surfaces of the 'foot' of some Molluscs : **epipodite**, n., *ĕ-pĭp'·ōd·ĭt*, a process developed upon the basal joint of some of the limbs of certain Crustacea : **epipodium**, n., *ĕp'·ĭ-pōd'·ĭ·ŭm*, a disc formed of several knobs or glands.

epirreology, n., *ĕp·ĭr'·rĕ·ŏl'·ŏ·jĭ* (Gr. *epirrheō*, I flow upon or over—from *epi*, upon ; *rhĕō*, I flow ; and *logos*, speech), that branch of natural history which treats of the influence of external agents on living plants.

episepalous, a., *ĕp'·ĭ-sĕp'·ăl·ŭs* (Gr. *epi*, upon ; Eng. *sepal*), in *bot.*, growing upon the sepals.

epispadias, n., *ĕp'·ĭ-spād'·ĭ·ăs* (Gr. *epi*, upon ; *spaō*, I draw), a term applied to a malformation of the wall of the bladder and adjacent parts ; one whose urethral orifice is on the upper part of the penis.

epispastic, a., *ĕp'·ĭ-spăst'·ĭk* (Gr. *epi*, upon ; *spaō*, I draw), applied to substances, 'epispastics,' which excite the skin and cause blisters, such as Spanish flies.

episperm, n., *ĕp'·ĭ-spĕrm* (Gr. *epi*, upon; *sperma*, seed), the external covering of the seed.

episporangium, n., *ĕp'·ĭ-spōr·ănj'·ĭ·ŭm* (Gr. *epi*, upon ; *spora*, a seed ; *anggos*, a vessel), an indusium overlying the spore cases of certain ferns, as Aspidium.

epispore, n., *ĕp'·ĭ-spōr* (Gr. *epi*, upon ; *spora*, a seed), the outer covering of some spores.

epistaxis, n., *ĕp'-ĭs-tăks'-ĭs* (Gr. *epistazō*, I cause to drop or trickle down, *epistaxō*, I shall cause to drop down—from *epi*, upon; *stazo*, I drop), hæmorrhage or bleeding from the nose.

episterna, n., *ĕp'-ĭ-stĕrn'-ă* (Gr. *epi*, upon; *sternon*, the breastbone), the lateral pieces of the dorsal arc of the somite of a Crustacean : **episternal**, a., *ĕp'-ĭ-stĕrn'-ăl*, situated on or above the sternum or breast-bone.

epistome, n., *ĕp-ĭs'-tŏm-ē* (Gr. *epi*, upon; *stoma*, a mouth), a valve-like organ which arches over the mouth in certain of the Polyzoa.

epistrophy, n., *ĕp-ĭs'-trŏf-ĭ* (Gr. *epistrophē*, a turning about, conversion—from *epi*, upon; *strophē*, a turning), in *bot.*, the reversion of a monstrous or variegated form to a normal one ; a mode of distribution of protoplasm and chlorophyll granules on free cell-walls under the action of light.

epithallus, n., *ĕp'-ĭ-thăl'-lŭs* (Gr. *epi*, upon; *thallos*, L. *thallus*, a young shoot or branch), the cortical layer of Lichens: **epithalline**, a., *ĕp'-ĭ-thăl'-lĭn*, growing on the thallus.

epitheca, n., *ĕp'-ĭ-thēk'-ă* (Gr. *epi*, upon; *thēkē*, a sheath, a box), a continuous layer surrounding the thecæ in some corals externally : **epithecium**, n., *ĕp'-ĭ-thē'-shĭ-ŭm*, the surface of the fructifying disc in certain Fungi and Lichens.

epithelioma, n., *ĕp'-ĭ-thēl'-ĭ-ōm'-ă* (formed from *epithelium*, which see), epithelial cancer, occurring on tegumentary or mucous surfaces, the lips and cheeks being the parts most commonly affected by it.

epithelium, n., *ĕp'-ĭ-thēl'-ĭ-ŭm* (Gr. *epi*, upon; *thēlē*, the nipple, or *thallō*, I grow), the layer of cells forming the surface of all the internal membranes of the body—of the same nature as epidermis, but much finer ;

in plants, a finer epidermis having thin cells filled with colourless fluid, and lining the ovary, etc. : **epithelial**, a., *ĕp'-ĭ-thēl'-ĭ-ăl*, pert. to or formed of epithelium : **epitheliated**, a., *ĕp'-ĭ-thēl'-ĭ-āt-ĕd*, covered with the delicate lining called epithelium, as a serous cavity, a membrane, etc.

epitrochlea, n., *ĕp'-ĭ-trŏk'-lĕ-ă* (Gr. *epi*, upon; *trochilia*, L. *trochlĕa*, a pulley, a roller), in *anat.*, the inner condyle of the humerus.

epizoon, n., *ĕp'-ĭ-zō'-ŏn*, **epizoa**, n. plu., *ĕp'-ĭ-zō'-ă* (Gr. *epi*, upon; *zoŏn*, an animal), animals which are parasitic upon other animals, infesting the surface of the body; a division of the Crustacea which are parasitic upon fishes; opposed to 'entozoon' and 'entozoa.'

epizootic, a., *ĕp'-ĭ-zō-ŏt'-ĭk* (Gr. *epi*, upon; *zoŏn*, an animal), applied to diseases prevailing among animals, corresponding to 'epidemic' diseases among men.

epulis, n., *ĕp'-ūl-ĭs* (Gr. *epi*, upon; *oulon*, gum), a tumour of the gum, often connected with a carious tooth.

equinia, n., *ĕ-kwĭn'-ĭ-ă*, also **equina**, n., *ĕ-kwĭn'-ă* (L. *equinus*, of or belonging to a horse—from *equus*, a horse), glanders and farcy, a contagious disease peculiar to the horse and mule, but capable of transmission to man.

Equisetaceæ, n. plu., *ĕk'-wĭ-sē-tā'-sĕ-ē* (L. *equisētis*, the plant horse-tail—from *equus*, a horse; *seta*, hair), the Horse-tail family, an Order of plants found in ditches, lakes, rivers, and damp places, so called in allusion to the fine hair-like branches : **Equisetum**, n., *ĕk'-wĭ-sēt'-ŭm*, the only known genus of the Order ; from the quantity of silicic acid contained in them, some of the species are used in polishing mahogany : **Equisetum hyemale**, *hĭ'-ĕm-āl'-ĕ* (L. *hĭĕmālis*, of or belonging to winter—from *hĭĕms*, winter), a species, often called Dutch rushes.

equitant, a., *ĕk'·wĭ·tănt* (L. *equitans*, riding), in *bot.*, having leaves folded longitudinally, and overlapping each other without any involution.

erect, a., *ĕ·rĕkt'* (L. *erectus*, raised or set up), in *bot.*, having an ovule rising from the base of the ovary; having innate anthers, that is, anthers attached to the top of the filament: **erectile tissue,** *ĕ·rĕkt'·ĭl tĭsh'·ū*, in *anat.*, a peculiar structure forming the principal part of certain organs which are capable of being rendered turgid or erected by distension with blood: **erector,** n., *ĕ·rĕkt'·ŏr*, a muscle which causes a part to erect or set up.

ergot, n., *ĕr'·gŏt* (F. *ergot*, cock's-spur), a diseased state in the grains of rye caused by the fungus Claviceps purpurea, appearing as a black-looking protuberance or spur from the ear, hence the name 'spurred rye'; in *anat.*, a name given to a curved and pointed longitudinal eminence on the inner side of the floor of the cerebrum; also called 'calcar avis,' the bird's spur: **ergotin,** n., *ĕr'·gŏt·ĭn*, the active principle of ergot, principally used for hypodermic injection to arrest hæmorrhage: **ergotism,** n., *ĕr'·gŏt·ĭzm*, the effect sometimes produced in the individual who eats rye bread containing ergot.

Ericaceæ, n. plu., *ĕr'·ĭ·kā'·sĕ·ē* (L. *ĕricæus*, of heath or broom—from L. *ĕrīce*, Gr. *ereikē*, heath, broom), the Heath family, an Order of shrubs or herbaceous plants: **Ericeæ,** n. plu., *ĕr·ĭs'·ĕ·ē*, a Sub-order, including the true heaths with naked buds, and the rhododendrons with scaly conical buds: **Erica,** n., *ĕr·ĭk'·ă*, a genus comprising a large number of very beautiful and interesting plants, mostly natives of the Cape of Good Hope: **Erica cinerea,** *sĭn·ĕr'·ĕ·ă* (L. *cĭnĕrĕus*, ash-coloured — from *cĭnis*, ashes), and **E. tetralix,** *tĕt'·răl·ĭks* (L. and Gr. *tĕtralix*, the heath plant), are common in Britain: **E. Mackaiana,** *măk'·ĭ·ān'·ă* (after the discoverer), and **E. Mediterranea,** *mĕd'·ĭ·tĕr·rān'·ĕ·ă* (after the sea so called), are peculiar to Ireland: **E. ciliaris,** *sĭl'·ĭ·ār'·ĭs* (L. *cĭlĭāris*, ciliary—from *cĭllum*, an eyelid), and **E. vagans,** *văg'·ănz* (L. *vagans*, wandering about), are two species common to England and Ireland.

Eriobotrya, n., *ĕr'·ĭ·ō·bŏt'·rĭ·ă* (Gr. *erion*, wool; *botrus*, a bunch of grapes), a genus of plants, Ord. Rosaceæ, Sub-ord. Pomeæ, whose racemes are very woolly: **Eriobotrya Japonica,** *jă·pŏn'·ĭk·ă* (*Japonĭcus*, of or from Japan), yields the Japanese fruit loquat.

Eriocaulon, n., *ĕr'·ĭ·ō·kāwl'·ŏn* (Gr. *erion*, wool; *kaulos*, a stem or stalk), a genus of very interesting plants having woolly stems, Ord. Restiaceæ: **Eriocaulon septangulare,** *sĕpt·ăng'·gŭl·ār'·ĕ* (L. *septangūlāris*, seven-angled — from *septem*, seven; *angulus*, an angle), a native of Britain and Ireland.

Eriogonum, n., *ĕr'·ĭ·ŏg'·ŏn·ŭm* (Gr. *erion*, wool; *gonu*, the knee), a genus of pretty plants having their stems woolly at the joints, Ord. Polygonaceæ, Sub-ord. or Tribe **Eriogoneæ,** n. plu., *ĕr'·ĭ·ō·gŏn'·ĕ·ē*.

Eriophorum, n., *ĕr'·ĭ·ŏf'·ŏr·ŭm* (Gr. *erion*, wool; *phoreo*, I bear), a genus of interesting plants, Ord. Cyperaceæ, whose seeds are covered with a woolly substance, found in boggy situations; the species are called 'cotton-grass.'

Eriospermeæ, n. plu., *ĕr'·ĭ·ō·spĕrm'·ĕ·ē* (Gr. *erion*, wool; *sperma*, seed), a tribe of plants, Ord. Liliaceæ, the stemless plants of S. Africa whose seeds are

covered with long silky hairs :
Eriospermum, n., *ěr'-ĭ-ō-spérm'-*
ŭm, a genus of Cape bulbs, ornamental when in flower.

eroded, a., *ĕr-ōd'-ĕd*, also **erose**,
a., *ĕr-ōz'* (L. *erodo*, I consume or
eat away ; *erosus*, consumed or
eaten away), in *bot.*, irregularly
toothed as if gnawed.

Errantia, n. plu., *ĕr-răn'-shĭ-ă* (L.
errans, wandering, *errantis*, of
wandering), an Order of Annelida,
distinguished by their great locomotive powers.

erratic, a., *ĕr-răt'-ĭk* (L. *errāticus*,
wandering about—from *erro*, I
wander), in *med.*, showing or
having a tendency to spread.

eructation, n., *ĕ'-rŭk-tā'-shŭn* (L.
eructatus, belched out—from *e*,
out of ; *ructatus*, belched), the
act of belching wind or foul air
from the stomach, often a sign of
indigestion.

erumpent, a., *ĕ-rŭmp'-ĕnt* (L. *e*,
out of ; *rumpens*, breaking,
rumpentis, of breaking), in *bot.*,
showing prominence, as if bursting through the epidermis.

Eryngium, n., *ĕr-ĭnj'-ĭ-ŭm* (L.
eryngion, a species of thistle ; Gr.
ēruggion, the plant eryngian—
said to be from *erengo*, I belch),
an extensive genus of extremely
ornamental and beautiful plants,
Ord. Umbelliferæ, some species
of which are said to be good
against flatulence : **Eryngium**
campestre, *kăm-pĕst'-rĕ* (L. *cam-*
pestris, belonging to a field—from
campus, a field), and **E. marit-**
imum, *măr-ĭt'-ĭm-ŭm* (L. *marīt-*
ĭmus, belonging to the sea—from
măre, the sea), are species whose
roots are sweet, aromatic, tonic,
and diuretic : **Eryngo**, n., *ĕr-*
ĭng'-ō, the sea holly, growing
abundantly on almost every sea-
coast ; a name for either of preceding, particularly the latter.

erysipelas, n., *ĕr'-ĭ-sĭp'-ĕl-ăs* (Gr.
erusipelas, a red eruption on the
skin—from *eruthros*, red ; *pella*,

skin), an acute, diffuse, and
specific inflammation of the
skin, which frequently involves
the subcutaneous cellular tissue ;
the Rose ; St. Anthony's fire :
erysipelatous, a., *ĕr'-ĭ-sĭp-ĕl'-ăt-ŭs*,
eruptive ; of or resembling erysipelas : **erysipelacea**, n. plu.,
ĕr'-ĭ-sĭp-ĕl-ā'-sē-ă, a class of diseases, such as erysipelas, small-
pox, measles, and scarlet fever.

erythema, n., *ĕr'-ĭth-ēm'-ă* (Gr.
eruthema, redness—from *eruth-*
aino, I make red), a superficial
redness of the skin ; a form of
eczema : **erythematous**, a., *ĕr'-ĭ-*
thĕm'-ăt-ŭs, having a superficial
redness of some portion of the
skin : **erythema nodosum**, *nŏd-*
ōz'-ŭm (L. *nodōsum*, full of knots
—from *nōdus*, a knot), a form of
erythema attended by an eruption of red oval patches, chiefly
on the lower limbs, most
common in young women : **e.**
intertrigo, *ĭn'-tĕr-trĭg'-ō* (L. *inter-*
trīgo, a chafing or galling—from
inter, between ; *tero*, I rub),
applied to those inflammations of
the cutaneous surface arising from
the friction of one part of the skin
against another ; irritation of the
skin from discharges flowing over
it : **e. paratrimma**, *păr'-ă-trĭm'-mă*
(Gr. *para*, near to, side by side ;
trimma, what has been rubbed or
bruised — from *tribō*, I rub or
bruise), a form of erythematous
inflammation due to pressure or
rubbing, such as arises in horses
from saddles or collars.

Erythræa, n., *ĕr'-ĭ-thrē'-ă* (Gr.
eruthros, red—from the colour of
the flowers), a genus of pretty
plants, Ord. Gentianaceæ :
Erythræa centaurium, *sĕnt-ăwr'-*
ĭ-ŭm (Gr. *kentaurion*, L. *centaur-*
ēum, the plant centaury), the
common centaury, whose flowering
cymes are used as a substitute for
gentian.

Erythrina, n., *ĕr'-ĭ-thrīn'-ă* (Gr.
eruthros, red, from the colour of

the flowers), the coral flower; a genus of splendid plants, with fine large leaves, and brilliant scarlet or red flowers, Ord. Leguminosæ, Sub-ord. Papilionaceæ: **Erythrina monosperma**, *mŏn'·ō·spĕrm'·ă* (Gr. *monos*, alone; *sperma*, seed), a species which yields gum lac: **erythrine, n.**, *ĕr'·ĭ·thrĭn*, in great part or wholly red.

Erythronium, n., *ĕr'·ĭ·thrōn'·ĭ·ŭm* (Gr. *eruthros*, red—from the colour of the leaves and flowers), a genus of handsome, dwarf - growing plants, Ord. Liliaceæ: **Erythronium Americanum**, *ăm·ĕr'·ĭk·ān'·ŭm* (from *America*), a species whose root is used as an emetic: **E. dens caninus**, *dĕnz kăn·īn'·ŭs* (L. *dens*, a tooth; *caninus*, belonging to a dog—from *cănis*, a dog), the dog-tooth violets, whose roots have been used in colic and epilepsy.

erythrophyll, n., *ĕr'·ĭ·thrō·fĭl* (Gr. *eruthros*, red; *phullon*, a leaf), the red colouring matter of leaves, indicating change and low vitality in them.

Erythroxylaceæ, n. plu., *ĕr'·ĭ·thrŏks·ĭl·ā'·sĕ·ē* (Gr. *eruthros*, red; *xulon*, wood), the Erythroxylon family, an Order of shrubs and trees, chiefly from W. Indies and S. America, whose species have tonic, purgative, and narcotic properties: **Erythroxylon, n.**, *ĕr'·ĭ·thrŏks·ĭl·ŏn*, a genus of trees whose wood is of a bright red colour, and yields a dye: **Erythroxylon coca**, *kōk'·ă* (a Spanish name; Gr. *kokkos*, a seed, a kernel), a plant whose leaves are used by the miners of Peru as a stimulant, and which are chewed with a small mixture of finely-powdered chalk; the common name for the prepared leaves is 'coca' or 'ipadu.'

Escalloniæ, n. plu., *ĕsk'·ăl·lōn·ĭ'·ĕ·ē* (in honour of *Escallon*, a Spanish traveller in S. Amer.), a Sub-ord.

of the Ord. Saxifragaceæ: **Escallonia, n.**, *ĕsk'·ăl·lōn'·ĭ·ă*, a genus of plants whose species are very fine evergreen greenhouse shrubs: **Escallonia macrantha**, *măk·rănth'·ă* (Gr. *makros*, of great extent, high; *anthos*, a flower), and **E. rubra**, *rōōb'·ră* (L. *ruber*, red), are grown in the milder parts of Britain.

eschar, n., *ĕsk'·ăr* (Gr. *eschăra*, a hearth, a scab), a crust or scab on a part, produced by burning or caustic: **escharotic, a.**, *ĕsk'·ăr·ŏt'·ĭk*, having the power to sear or burn the flesh: **n.**, any powerful chemical substance which, when applied to the body, destroys the vitality of a portion of it.

Eschscholtzia, n., *ĕsh·shŏltz'·ĭ·ă* (after *Eschscholtz*, a botanist), a genus of plants, Ord. Papaveraceæ, natives of California, etc., some species of which produce beautiful yellow flowers; the dilated apex of the peduncle resembles the extinguisher of a candle.

Esculapian, a., *ĕsk'·ūl·āp'·ĭ·ăn* (L. *Æsculapius*, Father of medicine), pert. to the healing art; medical.

esculent, a., *ĕsk'·ūl·ĕnt* (L. *esculentus*, fit for eating — from *esca*, food), good as food for man: **n.**, something that can be eaten, and good for food.

esparcet, n., *ĕs·pârs'·ĕt* (F. *esparcet*, Sp. *esparceta*), a green crop, something like the leguminous plant sainfoin.

essence, n., *ĕs'·sĕns* (L. *essentia*, the being of anything; F. *essence*), the concentrated odour of a plant, occurring in any part, procured by distillation with water.

estivation, n., *ĕs'·tĭv·ā'·shŭn* (L. *æstiva*, summer quarters), in *bot.*, the disposition of the parts of the perianth in the flower-bud; the arrangement of the unexpanded leaves of the flower-bud which burst in summer—as opposed to vernation, the arrangement of the leaves of the bud on a branch which burst in spring.

etærio, n., ĕ·tēr′·ĭ·ō (Gr. *etairia*, fellowship, society), the aggregate drupes which form the fruit of such as the strawberry and bramble.

ether, n., ēth′·ĕr (L. *œther*, Gr. *aither*, the upper or pure air), a very light, volatile, and inflammable liquid, obtained from alcohol and an acid by distillation.

ethmoid, a., ĕth′·mŏӯd, also ethmoidal, a., ĕth·mŏӯd′·ăl (Gr. *ēthmos*, a sieve ; *eidos*, resemblance), in *anat.*, a sieve-like bone which projects downwards from between the orbital plates of the frontal bone, and enters into the formation of the cranium; the bone of the nose which is perforated like a sieve for the passage of the olfactory nerves.

ethnography, n., ĕth·nŏg′·răf·ĭ (Gr. *ethnos*, a race, a nation; *grapho*, I write), an account of the origin, dispersion, connection, and characteristics of the various races of mankind.

etiolation, n., ēt′·ĭ·ōl·ā′·shŭn (mid. L. *etiolātus*, blanched, deprived of colour: F. *étioler*, to grow up long-shanked and colourless, as a plant), in *bot.*, the process of blanching plants by excluding the action of light; absence of green colour : etiolated, a., ēt′·ĭ·ōl·āt′·ĕd, blanched; deprived of colour.

etiology, n., ēt·ĭ·ŏl′·ŏ·jĭ (Gr. *aitia*, a cause; *logos*, discourse), in *med.*, the doctrine of causes, particularly with reference to diseases.

Eucalyptus, n., ūk′·ăl·ĭp′·tŭs (Gr. *eu*, well; *kalupto*, I cover), a genus of tall, handsome, fast-growing plants, Ord. Myrtaceæ, so called from the limb of the calyx covering the flower before expansion, which afterwards falls off in the shape of a lid or cover;

the species yield an astringent matter used for tanning; some of the species constitute the gigantic gum trees of Australia : Eucalyptus amygdalinus, ăm·ĭg′·dăl·ĭn′·ŭs (L. *amygdălĭnus*, of or made from almonds—from *amygdăla*, an almond), an Australian gum tree which attains the height of 400 feet : E. mannifera, măn·nĭf′·ĕr·ă (Eng. *manna;* L. *fero*, I bear), a species which furnishes a saccharine exudation resembling manna : E. globus, glōb′·ŭs (L. *glŏbus*, a ball, a sphere), the blue gum tree, or fever gum tree, furnishes good timber, an astringent bark, and a fragrant oil : E. dumosa, dŭm·ōz′·ă (L. *dūmōsus*, bushy—from *dūmus*, a thorn bush), a species on whose leaves is found a saccharine substance mixed with cellular hairs, produced by the attacks of a species of insect : E. perfoliata, pĕr′·fōl·ĭ·āt′·ă (L. *per*, through; *folĭātus*, leaved—from *folĭum*, a leaf), a handsome species, having hoary, bluish foliage, and a neat growth of the branches : E. resinifera, rĕz′·ĭn·ĭf′·ĕr·ă (L. *resina*, resin; *fĕro*, I produce), the brown gum tree of New Holland, yields an astringent, resinous-like substance, called 'kino,' which exudes from incisions in the bark as a red juice, a single tree producing as much as sixty gallons : Eucalypti, n. plu., ūk′·ăl·ĭp′·tī, a general name for all the species of Eucalyptus.

Eugenia, n., ū·jēn′·ĭ·ă (in honour of *Prince Eugene* of Savoy), a very ornamental and highly useful genus of plants, Ord. Myrtaceæ : Eugenia caryophyllata, kăr′·ĭ·ō·fīl·āt′·ă (Gr. *karuon*, a nut; *phullon*, a leaf), yields the cloves of commerce ; also called Caryophyllus aromaticus : E. pimenta, pĭm·ĕnt′·ă (Sp. *pimiento*, Indian pepper), a tree of the W.

Indies and Mexico, producing pimento, allspice, or Jamaica pepper; also called Pimenta officinalis, which see: **E. acris**, *āk'·rĭs* (L. *ācer* or *ācris*, sharp, pointed), a species used for pimento: **E. jambos**, *jăm'·bŏs* (corrupted from *schambu*, the Malay name for one of the species); and **E. Malaccensis**, *măl'·ăk·sĕns'·ĭs* (from *Malacca*), are species which produce the rose apples: **E. cauliflora**, *kāwl'·ĭ·flōr'·ă* (L. *caulis*, a stem; *flōrus*, shining, bright—from *flos*, a flower); and **E. ugni**, *ŭg'·nī* (*ugni*, a probable Chili name), species which produce fruits, the former eaten in Brazil, the latter in Peru.

Eulophia, n., *ū·lōf'·ĭ·ă* (Gr. *eulophos*, having a splendid crest), a very pretty genus of tuberous-rooted plants, Ord. Orchidiaceæ, so called in allusion to the labellum bearing elevated lines or ridges: **Eulophia herbacea**, *hèrb·ā'·sĕ·ă* (L. *herbācěus*, grassy —from *herba*, grass); and **E. campestris**, *kăm·pĕst'·rĭs* (L. *campester* or *campestris*, of or belonging to a level field—from *campus*, a plain), are species producing from their tuberous roots, in common with other orchidaceous plants, a substance called salep, which forms an article of diet for convalescents.

Euonymus, n., *ū·ŏn'·ĭm·ŭs* (Gr. *euōnŭmos*, of good name, but, by a euphemism, unlucky, hurtful—from *eu*, well; *onŏma*, a name), spindle-tree, a genus of ornamental shrubs, Ord. Celastraceæ; some species present a very showy appearance when the fruit is ripe: **Euonymus tingens**, *tĭnj'·ĕnz* (L. *tingens*, dyeing, colouring), a species whose bark yields a yellow dye, used in marking the *tika* on the forehead of Hindoos: **E. Europæus**, *ūr'·ŏp·ē'·ŭs* (L. *Europæus*, belonging to Europe), the young shoots, when

charred, are used to form a particular kind of drawing pencil; its fruit and inner bark are said to be purgative and emetic.

Eupatorium, n., *ūp'·ăt·ōr'·ĭ·ŭm* (Gr. *eupatŏricn*, agrimony—from *eupătōr*, well-born), a genus of plants, Ord. Compositæ: **Eupatorium ayapana**, *ā'·yă·pān'·ă* (a Brazilian native name), a powerful sodorific, has been used to cure the bites of snakes: **E. perfoliatum**, *pèr·fōl'·ĭ·āt'·ŭm* (L. *per*, through; *foliātus*, leafy—from *folium*, a leaf), a species yielding a tonic stimulant, used as a substitute for Peruvian bark.

Euphorbiaceæ, n. plu., *ū·fŏrb'·ĭ·ā'·sĕ·ē* (in honour of *Euphorbus*, an ancient physician), the Spurge family, an Order of trees and shrubs, often abounding in acrid milk; many species are poisons, others medicinal, some contain starch for food, some dyes, and others furnish wood highly useful in the arts, as boxwood: **Euphorbia**, n., *ū·fŏrb'·ĭ·ă*, an extensive genus of plants, many of which abound in a milky, caustic juice, and others irritant resins: **Euphorbia ipecacuanha**, *ĭp'·ĕ·kăk·ū·ăn'·ă* (in S. America a word denoting simply a vomiting root), a species whose root has been employed as a substitute for ipecacuan: **E. antiquorum**, *ănt'·ĭk·wōr'·ŭm* (L. *antiquorum*, of the ancients —from *antiquus*, ancient); and **E. Canariensis**, *kăn·ār'·ĭ·ĕns'·ĭs* (of or from the *Canary* Islands), with some other fleshy species, produce the drug euphorbium: **E. nereifolia**, *nĕr'·ē·ĭ·fōl'·ĭ·ă* (L. *Nērēus*, a sea-god; *folium*, a leaf), the juice of the leaves used in India as a purge and deobstruent: **E. thymifolia**, *tĭm'·ĭ·fōl'·ĭ·ă* (L. *thymum*, thyme; *folium*, a leaf), leaves and seed used in India for intestine worms: **E. tirucalli**, *tĭr'·ū·kăl'·lī* (an Indian name), a species whose fresh acrid juice is used in India as a vesicatory : **E. heptagona,** *hĕpt'·ă·gŏn'·ă* (Gr. *hepta,* seven ; *gōnia,* an angle), is said to furnish the Ethiopians with a deadly poison for their arrows : **euphorbium,** n., *ū·fŏrb'·ĭ·ŭm,* the inspissated milky juice of certain species of euphorbia, a violent irritant, whether applied internally or externally : **Euphorbium lathyris,** *lăth'·ĭr·ĭs* (Gr. *lathŭris,* a plant of the spurge kind), caper spurge, has cathartic properties : **E. pilosa,** *pĭl·ōz'·ă* (L. *pĭlōsus,* hairy, shaggy) ; and **E. palustris,** *păl·ŭst'·rĭs* (L. *păluster* and *pălustris,* marshy, swampy—from *pălŭs,* a marsh), species, the roots of which are used as purgatives, and are said to be useful in hydrophobia : **E. phosphorea,** *fŏs·fōr'·ĕ·ă* (Gr. *phŏsphŏros,* L. *phŏsphŏrus,* the light-bringer, the morning star—from Gr. *phŏs,* light ; *phoreo,* I bear), a species whose milky sap is said to emit a peculiar phosphorescent light.

Euphrasia, n., *ū·frāz'·ĭ·ă* (Gr. *euphrăsia,* gladness, joy), a genus of interesting plants, Ord. Scrophulariaceæ : **Euphrasia officinalis,** *ŏf·fĭs'·ĭn·āl'·ĭs* (L. *officinālis,* officinal—from *officina,* a workshop), eye-bright or euphrasy, so called because formerly used in ophthalmia.

Euryale, n., *ūr·ĭ'·ăl·ē* (Gr. *Euruălē,* one of the Gorgons), a genus of handsome water-plants, whose fine large leaves float on the surface of the water, Ord. Nymphæaceæ.

Euryangium, n., *ūr'·ĭ·ănj'·ĭ·ŭm* (Gr. *eurus,* broad ; *anggos,* a vessel), a genus of plants, Ord. Umbelliferæ : **Euryangium sumbul,** *sŭm'·bŭl* (an Eastern name), the sumbul root, brought to this country in large pieces like huge bungs, a nervin stimulant said to be employed in Germany and Russia with success against cholera.

Eustachian tube, *ūs·tāk'·ĭ·ăn* (after *Eustachius,* its discoverer), in *anat.,* a tube or canal extending from behind the soft palate to the tympanum of the ear, to which it conveys the air : **Eustachian valve,** a fold of the lining membrane of the right auricle of the heart, supposed to assist in the proper direction of the fœtal blood current.

Eustrongylus, n., *ūs·trŏng'·gĭl·ŭs* (Gr. *eu,* well ; *stronggulos,* round, globular), a genus of intestinal worms : **Eustrongylus gigas,** *jīg'·ăs* (L. *gĭgas,* a giant), a species found in the kidneys, bladder, and other parts of the horse, ox, dog, etc.

Eutassa, n., *ū·tăs'·să* (Gr. *eu,* well ; *tassō,* I set in order), a genus of trees, Ord. Coniferæ : **Eutassa excelsa,** *ĕk·sĕls'·ă* (L. *excelsus,* lofty, high), the Norfolk Island pine, famed for its size and for its wood.

Euterpe, n., *ū·tĕrp'·ē* (Gr. *euterpēs,* pleasing, charming— from *Euterpe,* one of the Muses), a fine genus of palms, some attaining a height of 40 feet, Ord. Palmæ : **Euterpe montana,** *mŏn·tān'·ă* (L. *montānus,* mountain—from *mons,* a mountain), the cabbage palm, the terminal buds of which are used as culinary vegetables : **E. oleracea,** *ŏl'·ĕr·ā'·sē·ă* (L. *olerācĕus,* resembling herbs—from *ŏlus,* a herb), the tallest of American palms, the white hearts of the green tops of which are eaten.

Eutoca, n., *ū'·tŏk·ă* (Gr. *eutokos,* prolific), a genus of very pretty flowering plants, Ord. Hydrophyllaceæ.

evacuant, n., *ĕ·văk'·ū·ănt* (L. *evacuātus,* emptied out—from *e,* out ; *vacuus,* empty), a medicine used for producing evacuation.

eversion, n., *ē·vér'·shŭn* (L. *ĕversio,* a turning out or expulsion—from *e,* out ; *versus,* turned), the protrusion of organs from a cavity ;

the state of being turned back or outward.

evolution, n., *ĕv'·ŏl·ū'·shŭn* (L. *evŏlūtus*, rolled out—from *e*, out; *volvo*, I roll), the theory which maintains that the first created animals contained the germs of all future possible successors, successively included one within the other, and that generation is merely the act of unfolding the germ ; the theory of the gradual development, at various periods of the world's history, of animals and of man from simpler forms and lower types to their present more complex structures.

exacerbation, n., *ĕks·ăs'·ĕr·bā'·shŭn* (L. *exacerbātus*, provoked—from *ex*, out ; *acerbus*, bitter, harsh), the increase of violence in the symptoms of a disease.

exalbuminous, a., *ĕks'·ăl·būm'·ĭn·ŭs* (L. *ex*, out of ; Eng. *albumen*), in *bot.*, without a separate store of albumen.

exania, n., *ĕks·ān'·ĭ·ă* (L. *ex*, out of ; *ānŭs*, the fundament), a falling down of the anus ; prolapsus ani.

exannulate, a., *ĕks·ăn'·nŭl·āt* (L. *ex*, out of ; *annulus*, a ring), not having a thecal ring, applied to some ferns.

exanthema, n., *ĕks'·ănth·ēm'·ă* (Gr. *exanthēma*, a blossom—from *ex*, out of ; *anthos*, a flower), an eruption ; applied to contagious febrile diseases terminating in an eruption on the skin, such as scarlet fever, measles, etc. : **exanthemata**, n. plu., *ĕks'·ănth·ĕm'·ăt·ă* : **exanthematous**, a., *-ĕm'·ăt·ŭs*, of or pert. to.

exasperate, a., *ĕgz·ăsp'·ĕr·āt* (L. *exasperātus*, made rough, sharpened—from *ex*, out of ; *asper*, rough), in *bot.*, covered with hard, stiff, short points.

excentric, a., *ĕks·sĕnt'·rĭk* (L. *ex*, out of ; *centrum*, the centre), out of the centre ; removed from the centre or axis.

exciple, n., *ĕks'·sĭp·l*, also **excipulum**, n., *ĕks·sĭp'·ŭl·ŭm* (L. *excipulum*, a receptacle—from *excipio*, I catch, I receive), the external investment of the thalamium in the apothecia of lichens: **excipulus**, n., *ĕks·sĭp'·ŭl·ŭs*, a receptacle containing fructification in lichens ; a minute black fungus upon dead raspberry stems.

excision, n., *ĕk·sĭzh'·ŭn* (L. *excisus*, cut out or off—from *ex*, out of ; *cæsus*, cut), in *surg.*, the removal by operation of a part of the body, but short of amputation.

excoriation, n., *ĕks'·kōr·ĭ·ā'·shŭn* (L. *ex*, out of ; *corium*, skin, hide), a slight wound which only abrades the skin.

excrement, n., *ĕks'·krē·mĕnt* (L. *excrementum*, that which passes from the body—from *ex*, out of ; *crētus*, separated), the matter discharged from animal bodies after digestion ; fæcal evacuation : **excrementitious**, a., *ĕks'·krē·mĕnt·ĭsh'·ŭs*, consisting of fæcal matter evacuated from an animal body.

excrescence, n., *ĕks·krĕs'·ĕns* (L. *excrescentia*, morbid excrescences on the body—from *ex*, out of ; *crescens*, growing), a preternatural growth on any part of the body ; in *bot.*, a gnarr or wart on the stem of a tree.

excreta, n. plu., *ĕks·krēt'·ă* (L. *excretus*, carried off or discharged from the body by stool or urine —from *ex*, out of ; *crētus*, separated), the natural secretions or discharges which are thrown off from the body, as from the bowels, the bladder, or by perspiration ; also **excretions**, n. plu., *ĕks·krē'·shŭns*, in same sense.

excurrent, a., *ĕks·kŭr'·rĕnt* (L. *ex*, out of ; *currens*, running), in *bot.*, running out beyond the edge or point ; central, as the stem of a fir with branches disposed regularly around it.

exfoliation, n., *ĕks·fōl'·ĭ·ā'·shŭn* (L. *ex*, out of ; *folium*, a leaf), the separation of a scale or dead portion of bone from the living.

exindusiate, a., *ĕks'·ĭn·dūz'·ĭ·āt* (L. *ex*, out of; *indusium*, a shirt), in *bot.*, not having an indusium.

exintine, n., *ĕks·ĭn'·tĭn* (L. *ex*, from; *intus*, within), in *bot.*, one of the inner coverings of the pollen grain.

exogens, n., *ĕks'·ō·jĕnz* (Gr. *exo*, without; *gennaō*, I produce), that division of the vegetable kingdom in which the plants grow by additions to the outside of the wood in the form of annual concentric layers, as in the oak, ash, elm, etc.—the 'endogens' being those whose growth is from within outwards: **exogenous,** a., *ĕks·ŏdj'·ĕn·ŭs*, growing or increasing in size by annual additions to the outside: **exogenæ,** n. plu., *ĕks·ŏdj'·ĕn·ē*, another name for exogens.

Exogonium, n., *ĕks'·ō·gōn'·ĭ·ŭm* (Gr. *exo*, without; *gonu*, the knee), a genus of plants, Ord. Convolvulaceæ : **Exogonium purga,** *pérg'·ă* (L. *purgo*, I clear or clean out), the jalap plant, a native of the Mexican Andes, whose roots, in size and shape from a walnut to a moderately-sized turnip, form the officinal part; used in the form of powder and tincture as an active irritant cathartic.

exomphalos, n., *ĕks·ŏmf'·ăl·ŏs* (Gr. *exo*, without; *omphalos*, a navel), umbilical hernia; the protrusion of the intestine through the umbilicus.

exophthalmos, n., *ĕks·ŏf·thăl'·mŏs*, also **exophthalmia,** n., *-thăl'·mĭ·ă* (Gr. *exo*, without, outside; *ophthalmos*, the eye), great prominence of the eyes, in which the individual has a marked and peculiar stare : **exophthalmic,** a., *-thăl'·mĭk*, of or pert. to exophthalmia.

exopodite, n., *ĕks·ŏp'·ŏd·ĭt* (Gr. *exo*, outside; *pous*, a foot, *podos*,

of a foot), in *zool.*, the outer o the two secondary joints int which the typical limb of a Crust acean is divided.

exorhizal, a., *ĕks'·ō·rĭz'·ăl* (Gr. *exo*, outside; *rhiza*, a root), in *bot.*, applied to those plants whose roots in germination proceed at once from the radicular extremity of the embryo.

exoskeleton, n., *ĕks'·ō·skĕl'·ĕt·ŏn* (Gr. *exo*, outside; *skeleton*, a dry body or skeleton), the external skeleton, constituted by a hardening of the integument; also called dermo-skeleton.

exosmose, n., *ĕks'·ŏs·mōz*, also **exosmosis,** n., *ĕks'·ŏs·mōz'·ĭs* (Gr. *exo*, outside; *ōsmos*, a thrusting, an impulsion), the passing outwards of a fluid through a membrane from the inside; the passing inwards from the outside being called **endosmose.**

exospore, n., *ĕks'·ō·spōr* (Gr. *exo*, outside; *spora*, a seed), the outer covering of a spore: **exosporous,** a., *ĕks·ŏs'·pōr·ŭs*, having naked spores, as in fungi.

Exostemma, n., *ĕks'·ō·stĕm'·mă* (Gr. *exo*, outside; *stemma*, a crown—alluding to the exserted stamens), a genus of plants, Ord. Rubiaceæ, whose species yield various kinds of false cinchona bark, which do not contain the cinchona alkalies : **Exostemma floribundum,** *flōr'·ĭ·bŭnd'·ŭm* (L. *flos*, a flower, *floris*, of a flower; *abundus*, abundant), a species described as a timber tree.

exostome, n., *ĕks'·ō·stōm* (Gr. *exo*, outside; *stŏma*, a mouth), in *bot.*, the outer opening of the foramen of the ovule.

exostosis, n., *ĕks'·ŏs·tōz'·ĭs* (Gr. *exostōsis*, a bony excrescence—from *exo*, outside; *ostĕon*, a bone), an unnatural projection or growth from a bone; a wart-like excrescence often seen on the roots of leguminous plants.

exothecium, n., *ĕks'·ō·thē'·shĭ·ŭm*

(Gr. *exo*, without; *thēkē*, a case), in *bot.*, the outer coat of the anther.

exotic, a., *ĕgz·ŏt'·ĭk* (Gr. *exōtikos*, foreign, strange—from *exō*, outside; L. *exōtĭcus*, foreign), foreign; not native: n., a shrub or tree introduced from a foreign country;—*indigenous*, the opposite of *exotic*, means belonging naturally to a country; native.

expectorant, n., *ĕks·pĕkt'·ŏr·ănt* (L. *expectorātus*, driven from the breast—from *ex*, out of; *pectus*, the breast), any medicine supposed capable of promoting the expulsion of fluid or phlegm from the air-passages or lungs: **expectoration**, n., *-ŏr·ā'·shŭn*, the act of discharging matter from the air-passages or lungs.

exserted, a., *ĕks·sèrt'·ĕd* (L. *exsertus*, thrust forth), in *bot.*, projecting beyond something else, as stamens beyond the corolla; proceeding from a common base: **exsertile**, a., *ĕks·sèrt'·ĭl*, capable of being thrust out or excluded.

exsiccation, n., *ĕks'·sĭk·kā'·shŭn* (L. *exsiccātus*, made quite dry—from *ex*, out of; *siccus*, dried up), the expulsion of moisture from solid bodies by heat, pressure, or by any other means.

exstipulate, a., *ĕks·stĭp'·ŭl·āt* (L. *ex*, without; *stipula*, a stalk or stem), in *bot.*, having no stipules.

extension, n., *ĕks·tĕn'·shŭn* (L. *extensus*, stretched out), the pulling strongly a fractured or dislocated limb in order to reduce it: **extensor**, n., *ĕks·tĕns'·ŏr*, a muscle which extends or stretches out a part.

extensor carpi radialis brevior, *ĕks·tĕns'·ŏr kârp'·ī rād'·ĭ·āl'·ĭs brēv'·ĭ·ŏr* (L. *extensor*, that which stretches out; *carpus*, the wrist; *radius*, the small bone of the fore-arm; *brevior*, shorter), the shorter radial extensor of the wrist; the muscle at the wrist

which extends and brings the hand backwards: **e. carpi radialis longior**, *lŏn'·jĭ·ŏr* (L. *longior*, longer), the longer radial extensor of the wrist; the muscle which assists in extending and bringing the hand backwards: **e. carpi ulnaris**, *ŭl·nār'·ĭs* (L. *ulna*, the large bone of the fore-arm), the ulnary extensor of the wrist; the muscle at the wrist: **e. communis digitorum manus**, *kŏm·mūn'·ĭs dŭlj'·ĭt·ōr'·ŭm mān'·ŭs* (L. *communis*, common; *digitus*, a finger or toe; *manus*, the hand, *manus*, of the hand), the common extensor of the fingers of the hand; the muscle which extends all the joints of the finger: **e. longus digitorum pedis**, *lŏng'·ŭs dĭdj'·ĭt·ōr'·ŭm pēd'·ĭs* (L. *longus*, long; *digitus*, a finger or toe, *digitōrum*, of the fingers; *pes*, a foot, *pedis*, of a foot), the long extensor of the toes of the foot; the muscle which extends all the joints of the four small toes: **e. proprius pollicis pedis**, *prŏp'·rĭ·ŭs pŏl'·lĭs·ĭs pēd'·ĭs* (L. *proprius*, proper; *pollex*, the great toe of the foot, *pollicis*, of the great toe; *pes*, a foot, *pedis*, of the foot), the proper extensor of the great toe of the foot; the muscle which extends the great toe.

extine, n., *ĕks'·tĭn* (L. *exter*, on the outside), in *bot.*, the outer covering of the pollen-grain.

extra-axillary, a., *ĕks'·tră·ăks·ĭl'·ăr·ĭ* (L. *extra*, on the outside; *axilla*, arm-pit), removed from the axil of the leaf, as some buds; growing from above or below the axils.

extract, n., *ĕks'·trăkt* (L. *extractus*, drawn out or forth—from *ex*, out of; *tractus*, drawn), an infusion, decoction, or tincture of a medicine evaporated to a paste.

extravasation, n., *ĕks·trăv'·ăs·ā'·shŭn* (L. *extra*, without; *vasa*, vessels of any kind), in *med.*, the unnatural escape of a fluid from its vessel or its channel, and infiltration into surrounding tissues, as the blood after the rupture of a vessel.

extrorse, a., *ĕks·trŏrs'* (L. *extra*, on the outside ; *orsus*, beginning, commencement), in *bot.*, applied to anthers in which the slit through which the pollen escapes is towards the outside of the flower, and not, as usual, towards the pistil ; turned outwards.

exudation, n., *ĕks'·ŭd·a'·shŭn* (L. *exsudo*, or *exudo*, I sweat out—from *ex*, out of ; *sudo*, I sweat), the discharge of moisture or juices from animal bodies or from plants ; the abnormal escape of the blood-plasma occurring in inflammation of certain tissues, as the lungs.

exutive, a., *ĕks·ūt'·ĭv* (L. *exūtus*, drawn out or off), in *bot.*, applied to seeds wanting the usual integumentary covering.

exuviæ, n. plu., *ĕks·ūv'·ĭ·ē* (L. *exuviæ*, that which is laid aside or taken off from the body), the cast off parts of animals or plants, as skins, shells, etc. : **exuviation,** n., *ĕks·ūv'·ĭ·a'·shŭn*, the process by which animals and plants throw off their old coverings or shells and assume new ones, as serpents their skins, and crustacea their shells.

Fabaceæ, n. plu., *făb·a'·sĕ·ē* (L. *făba*, a bean ; connected with Sansc. *bhac*, Gr. *phagein*, to eat, to devour), Lindley's Order of the Pea and Bean tribe, now called Leguminosæ : **fabaceous,** a., *făb·a'·shŭs*, of or like a bean.

facial, a., *fā'·shĭ·ăl* (L. *făcĭes*, the face), the parts of the face, as opposed to the cranial parts of the head.

fæces, n. plu., *fēs'·ēz* (L. *fæx*, dregs or sediment, *fæcis*, of sediment), the excrement or contents of the bowels ; sediment or settlings :

fæcal, a., *fēk'·ăl*, relating to excrement.

Fagopyrum, n., *făg'·ō·pīr'·ŭm* (Gr. *fēgos*, L. *fāgus*, the beech-tree—from *phāgo*, I eat ; Gr. *pūros*, wheat), the genus of buckwheat plants, so called from the seeds being three-cornered like beech-nuts, Ord. Polygonaceæ : **Fagopyrum esculentum,** *ĕsk'·ŭl·ĕnt'·ŭm* (L. *escŭlentus*, full of food—from *esca*, food), and **F. tataricum,** *tăt·ăr'·ĭk·ŭm* (probably, of or from *Tartary*), species of buckwheat whose seeds are used as food.

Fagus, n., *făg'·ŭs* (L. *fāgus*, Gr. *fēgos*, the beech tree—from Gr. *phāgo*, I eat), a genus of handsome ornamental timber-trees, Ord. Cupuliferæ or Corylaceæ, so called as the nuts of the beech tree were used in early times as food : **Fagus sylvatica,** *sĭlv·ăt'·ĭk·ă* (L. *sylvătĭcus*, living in the woods—from *sylva*, a wood), is the common beech tree : **F. Forsteri** (after Forster, a botanist), the evergreen beech of S. America : **F. antarctica,** *ănt·ărk'·tĭk·ă* (new L. *antarcticus*, southern —from Gr. *anti*, opposite ; *arktikos*, northern), a species of beech found in the Antarctic regions.

fairy-rings, n. plu., *fār'·ĭ-rĭngs*, scorched-like circles, or circles of greener grass, found frequently in pasture-lands in Britain, produced by a peculiar mode in the growth of several species of Agarics.

falcate, a., *fălk'·āt* (L. *falcatus*, scythe-shaped — from *falx*, a reaping-hook), in *bot.*, bent or shaped like a reaping-hook ; crescent-shaped : **falciform,** a., *făls'·ĭ·fŏrm* (L. *forma*, shape), in same sense ; shaped like a reaping-hook.

Fallopian tubes, *făl·lōp'·ĭ·ăn tūbz* (after *Fallopius*, their discoverer), hollow canals forming appendages to the womb and ducts of the ovaries.

falx cerebelli, *fălks sĕr'·ĕ·bĕl'·lĭ* (L. *falx*, a reaping-hook; *cerebellum*, a little brain), in *anat.*, a small triangular process of the dura mater received into the indentation between the two lateral lobes of the cerebellum behind : **falx cerebri,** *sĕr'·ĕb·rī* (L. *cĕrĕbrum*, the brain, *cĕrĕbri*, of the brain), a strong arched process of the dura mater, which descends vertically in the longitudinal fissure between the two hemispheres of the brain—so named from its sickle-like form.

Family, n., *făm'·ĭl·ĭ* (L. *familia*, a family or household; F. *famille*), the systematic name for the group above a Genus and below an Order; often used in a loose and general sense for Kind, Tribe, or Order.

farcy, n., *fărs'·ĭ* (It. *farcina*, F. *farcin*), a disease allied to glanders.

farina, n., *făr·īn'·ă* (L. *farina*, meal—from *far*, grain), meal or flour ; the dust or pollen of plants: **farinaceous,** a., *făr'·ĭn·ā'·shŭs*, mealy ; having the texture or consistence of flour ; chaffy : **farinose,** a., *făr'·ĭn·ōz*, in *bot.*, covered with a white, mealy powder.

fascia, n., *făs'·sĭ·ă*, **fasciæ,** n. plu., *făs'·sĭ·ē* (L. *fascia*, a bandage, a swathe), a surgical bandage ; a membranous lamina of a variable extent and thickness, investing and protecting as a sheath a delicate organ of the body : **fascia dentata,** *dĕnt·āt'·ă* (L. *dentātus*, toothed — from *dens*, a tooth, *dentis*, of a tooth), a serrated band of grey substance in the lower boundary or floor of the middle or descending cornu of the cerebrum: **f. lata,** *lāt'·ă* (L. *lātus*, broad), a broad, dense, fibrous aponeurosis, which forms a uniform investment for the upper part of the thigh: **f. obturator,** *ŏb'·tŭr·āt'·ŏr* (L. *obtūro*, I stop up;

obturātor, that which stops up), a fascia which descends and covers the obturator internus muscle : **f. palmar,** *pălm'·ăr* (L. *palmāris*, a hand's-breadth—from *palma*, the palm, the hand), a strong aponeurosis on the palm of the hand : **f. plantar,** *plănt'·ăr* (L. *plantāris*, belonging to the sole of the foot—from *planta*, the sole), a fibrous membrane on the sole of the foot, stronger and thicker than any other.

fasciated, a., *făs'·sĭ·āt·ĕd* (L. *fascia*, a bandage), bound with a bandage ; in *bot.*, having a stem flattened out ; having several leaf-buds united in growth so as to produce a branch presenting a flattened appearance: **fasciation,** n., *făs'·sĭ·ā'·shŭn*, the act or manner of binding up diseased parts ; in *bot.*, the union of branches or stems in growth presenting a flattened appearance.

fascicle, n., *făs'·ĭk·l*, also **fasciculus,** n., *făs·ĭk'·ūl·ŭs* (L. *fascicŭlus*, a small bundle—from *fascia*, a bandage), a little bunch ; a cluster ; in *anat.*, a bundle of muscular fibre: **fascicled,** a., *făs'·ĭk·ld*, also **fasciculated,** a., *făs·ĭk'·ūl·āt·ĕd*, the bunches or bundles proceeding from a common point ; arranged in bundles: **fasciculate,** a., *făs·ĭk'·ūl·āt*, and **fascicular,** a., *făs·ĭk'·ūl·ăr*, in same sense as fascicled : **fasciculus,** n., *făs·ĭk'·ūl·ŭs*, in *bot.*, a small collection of nearly sessile flowers, forming a dense, flat-topped bunch, as in the sweet-william ; same sense as fascicle: **fascicular tissue,** in *bot.*, a tissue lying inside another tissue, in which the growth proceeds at one or both ends, so as greatly to elongate it.

fasciculi graciles, *făs·ĭk'·ūl·ī grăs'·ĭl·ēz* (L. *fascicŭlus*, a small bundle, *fascicŭli*, small bundles ; *gracĭlis*, sing., *gracĭles*, plu., small, slender), two narrow white cords placed one on each side of the posterior median fissure of the medulla oblongata: f. teretes, *těr'·ět·ēz* (L. *těres*, rounded off, *těrětis*, gen., *těrětes*, plu.), two bundles of white fibres mixed with much grey matter in the medulla oblongata: **fasciculus cuneatus**, *kūn'·t̄·āt'·ŭs* (L. *cŭněātus*, shaped like a wedge—from *cŭněus*, a wedge), the part of the posterior column of the cord which belongs to the restiform body of the medulla· f. uncinatus, *ŭn'·sĭn·āt'·ŭs* (L. *uncīnātus*, furnished with hooks—from *uncus*, a hook), a white bundle of fibres seen on the lower aspect of the hemisphere, the more superficial being curved.

fastigiate, a., *făs·tĭdj'·ĭ·āt* (L. *fastīgĭum*, a projecting ridge—from *fastīgo*, I slope up to a point), in *bot.*, having a pyramidal form, from the branches being parallel and erect, as the Lombardy poplar; nearly parallel and pointing upwards.

Fatsia, n., *făts'·ĭ·ă* (a native Chinese name), a genus of plants, Ord. Araliaceæ: **Fatsia papyrifera**, *păp'·ĭr·ĭf'·ěr·ă* (L. *papyrus*, the paper-reed, paper; *fěro*, I bear), a species from whose pith the Chinese prepare the celebrated rice paper.

fauces, n., *fāws'·ēz* (L. *fauces*, the upper part of the throat; *faux*, sing., the throat), the upper part of the throat, from the root of the tongue to the entrance of the gullet: **faux,** n., *fāwks*, in *bot.*, the throat or constricted part of a flower.

fauna, n., *fāwn'·ă* (L. *Faunus*, one of the gods of the fields or woods), all the animals peculiar to a country, area, or period; ' flora ' denotes all the plants.

favella, n., *făv·ěl'·lă*, **favellæ,** n. plu., *făv·ěl'·lē* (L. *favilla*, hot cinders or ashes), in *bot.*, a kind

of conceptacle among Algæ; conceptacular fruit in certain Algæ: **favellidium,** n., *făv'·ěl·lĭd'·ĭ·ŭm*, **favellidia,** n. plu., *-lĭd'·ĭ·ă*, spherical masses of spores contained in capsules; a favella immersed in the frond of Algæ.

favus, n., *făv'·ŭs* (L. *făvus*, a honeycomb), a parasitic disease of the skin produced by the Achorion Schönleinii; a form of ringworm: **favose,** a., *făv·ōz'*, and **faveolate,** a., *făv·ē'·ōl·āt*, honeycombed.

feather-veined, a., *fěth'·ěr·vānd*, in *bot.*, applied to a leaf having the veins running from the midrib to the margin, at a more or less acute angle: **feathery,** a., *fěth'·ěr·ĭ*, having hairs which are themselves hairy.

febrile, a., *fěb'·rĭl* (L. *fēbris*, a fever; F. *febrile*, pert. to a fever), pert. to a fever; indicating fever; feverish: **febricula,** n., *fěb·rĭk'·ŭl·ă* (L. *febrĭcŭla*, a slight fever), a fever characterised by its short duration and mildness of symptoms: **febrifuge,** n., *fěb'·rĭ·fŭdj* (L. *fugo*, I drive away), any medicine which mitigates or removes a fever, as quinia, bark, and arsenic.

fecula, n., also **fæcula,** n., *fěk'·ŭl·ă* (L. *fæcula*, salt of tartar deposited from urine—from *fœx*, dregs or sediment), a powder obtained from plants and their seeds, etc., by crushing and washing them and allowing the matter to settle: **feculence,** n., *fěk'·ŭl·ěns*, in *phar.*, any substance settling from turbid fluids: **feculent,** a., *fěk'·ŭl·ěnt*, abounding with sediment or excrementitious.

fecundation, n., *fěk'·ŭnd·ā'·shŭn* (L. *fēcundus*, fruitful, fertile), the act of making fruitful; state of being impregnated: **fecundity,** n., *fěk·ŭnd'·ĭt·ĭ*, fruitfulness; the power of producing or bringing forth young.

felo de se, *fēl'·ō dě sē* (mid. L., a

felon upon himself), a suicide; in *law*, one who commits a felony by suicide.

female flower, *fĕm'·āl flōw'·r*, in *bot.*, a flower producing pistils only.

femur, n., *fĕm'·ŭr* (L. *fĕmur*, the thigh, *fĕmoris*), one of the thigh-bones; one of the two largest and longest bones of the body: **femoral**, a., *fĕm'·ŏr·ăl*, pert. to the thigh: **femoral condyles**, the rounded eminences at each end of the thigh-bones; **f. region**, the thighs — 'region' simply denoting any artificial division of the body, as 'chest,' 'abdomen,' etc.

fenestra, n., *fĕn·ĕst'·ră* (L. *fenestra*, a window), in *anat.*, applied to two small openings in the petrous portion of the temporal bone: **fenestra ovalis**, *ōv·āl'·ĭs* (L. *ovālis*, of or belonging to an egg—from *ōvum*, an egg), in the ear, a reniform opening, leading from the tympanum into the vestibule: **f. rotunda**, *rōt·ŭnd'·ă* (L. *rŏtundus*, wheel-shaped—from *rŏta*, a wheel), an oval aperture placed at the bottom of a funnel-shaped depression leading into the cochlea: **fenestrate**, a., *fĕn·ĕst'·rāt* (L. *fenestrātus*, furnished with openings or windows), in *bot.*, having openings like a window; having small perforations.

fennel, n., *fĕn'·nĕl* (L. *fĕnĭcŭlum*, fennel; AS. *feonel*), an umbelliferous plant of various species; one is cultivated as a pot herb, and for its seeds and an essential oil; systematic name is **Fœniculum vulgare**, *fĕn·ĭk'·ŭl·ŭm vŭlg·ār'·ĕ* (L. *vulgāris*, general, ordinary—from *vulgus*, the people); also **F. dulce**, *dŭls'·ĕ* (L. *dulcis*, sweet), sweet fennel, Ord. Umbelliferæ.

Ferns, n., *fĕrnz* (AS. *fearn*, fern; Swed. *fara*, to go—applied to events produced by diabolic art),

a family of cryptogamic plants, usually with broad feathery leaves or fronds, Ord. Filices; the fern or male shield-fern (Asplenium filix mas) is a remedy of very great value in the treatment of tape-worms.

Feronia, n., *fĕr·ōn'·ĭ·ă* (L. *Ferōnia*, an old Italian goddess of plants), a genus of fruit-bearing plants, Ord. Aurantiaceæ: **Feronia elephantum**, *ĕl'·ĕ·fănt'·ŭm* (L. *elephantus*, an elephant), a species from which is procured a gum, like gum-arabic; a genus of ticks infesting the horse and the ass, etc.

ferruginous, a., *fĕr·rŏdj'·ĭn·ŭs* (L. *ferrŭgĭnĕus*, of the colour of iron rust—from *ferrum*, iron), impregnated or coated with oxide of iron; chalybeate; applied to medicines having iron for their active principle; in *bot.*, rust-coloured.

Ferula, n., *fĕr'·ŭl·ă* (L. *fĕrŭla*, the plant fennel-giant, a rod for punishment), the giant-fennels, a genus of plants, Ord. Umbelliferæ: **Ferula galbaniflua**, *găl'·băn·ĭ·flŏ'·ă* (L. *galbănum*, the resinous sap of an umbelliferous plant in Syria; *fluo*, I flow); also **F. rubricaulis**, *rŏŏb'·rĭ·kāwl'·ĭs* (L. *rŭber*, red; *caulis*, a stem), are species which produce the gum-resin galbanum, consisting of resin, gum, and a volatile oil, used as an antispasmodic and emenagogue: **F. Persica**, *Pĕrs'·ĭk·ă* (L. *Persicus*, of or from Persia), a plant whose leaves are very much divided, yields an inferior sort of asafœtida, consisting of a resinous and gummy matter with a sulphur oil, used as a stimulant, antispasmodic, and anthelmintic.

Festuca, n., *fĕst·ŭk'·ă* (old F. *festu*, a straw; L. *festūca*, the young shoot or stalk of a tree), a genus of plants, Ord. Gramineæ: **fescue**, n., *fĕsk'·ŭ*, a sharp-pointed

L

kind of grass : **Festuca flabel-loides,** *flăb'·ĕl·lŏyd'·ēz* (L. *flābel-lum,* a fly-flap ; Gr. *eidos,* resem-blance), the Tussac grass of the Falkland Islands, which, though tender enough for animal food, attains a height of five or six feet : **festucine,** a., *fĕst·ū'·sĭn,* of a straw colour.

fetlock, n., *fĕt'·lŏk* (Dut. *vitlok,* Swiss, *fiesloch,* the pastern of a horse ; Ger. *fitze,* a bundle of threads), the tuft of hair growing a little above the back part of the hoof of a horse ; the joint on which such hair grows.

Feverfew, n., *fēv'·ĕr·fū* (F. *fièvre,* L. *febris,* a fever ; *fugāre,* to put to flight), a herb good against fevers ; the plant Pyrethrum parthenium, which is aromatic and stimulant.

fibra primitiva, *fīb'·ră prĭm'·ĭt·iv'·ă* (L. *fibra,* a fibre, a band ; *primitivus,* that which is first or original — from *primus,* first), the primitive band in the nervous system : **fibræ vel processus ar-ciformes,** *vĕl prō·sĕs'·sŏŏs ärs·ĭ·fŏrm'·ēz* (L. *vel,* or ; *processûs,* processes ; *arciformes,* a. plu., shaped like a bow—from *arcus,* a bow ; *forma,* shape), the arciform fibres or processes, a set of superficial white fibres on the forepart and sides of the medulla oblongata.

fibril, n., *fīb'·rĭl* (a dimin. of L. *fibra,* a fibre), a very minute or ultimate fibre : **fibrillæ,** n. plu., *fīb·rĭl'·lē,* in *bot.,* the thread-like divisions of roots : **fibrillation,** n., *fīb'·rĭl·lā'·shŭn,* the state or condition of becoming fibrils, or in appearance like fibrils : **fibril-lose,** a., *fīb'·rĭl·lōz,* in *bot.,* covered with little strings or fibres : **fibrillous,** a., *fīb·rĭl'·lŭs,* consisting of or formed of small fibres : **fibrin,** n., *fīb'·rĭn,* a peculiar substance found in animals and vegetables, which forms fibres and muscular flesh ; a substance formed in the act of coagulation of the blood by the union of fibrinogen, a body peculiar to intercellular fluid, with a fibrinoplastic substance termed paraglobulin, derived from the cellular structures of the body : **fibrinogen,** n., *fīb·rĭn'·ō·jĕn* (Gr. *gennăo,* I pro-duce), one of the two substances which produce fibrin, the coag-ulum in hydrocele fluid, in serous fluids, and in blood, the other substance being named 'globulin': **fibrinogenous,** a., *fīb'·rĭn·ŏdj'·ĕn·ŭs,* denoting a substance found in a hydrocele fluid, etc. ; pro-ducing fibrin : **fibrinoplastic,** a., *fīb'·rĭn·ō·plăst'·ĭk* (Eng. *plastic*), denoting one of the ingredients which produce fibrin ; also de-noting globulin : **fibrinoplastin,** n., *-plăst'·ĭn,* another name for glob-ulin ; a substance supplied from the blood : **fibroid,** a., *fīb'·rŏyd* (Gr. *eidos,* resemblance), resem-bling simple fibre in structure ; denoting a tumour in which the cell elements have assumed the appearance of fibres : **fibro-cellul-ar,** in *bot.,* tissue composed of spiral cells : **fibrous,** a., *fīb'·rŭs,* composed of numerous fibres : **fibro-vascular tissue,** a tissue composed of mixed vessels, con-taining spiral and other fibres.

fibula, n., *fīb'·ūl·ă* (L. *fibŭla,* a buckle), the outer and smaller bone of the leg, so named as being opposite the part where the knee-buckle was placed when these were worn ; the part cor-responding to the ulna in the fore-arm.

Ficoideæ, n. plu., *fĭk·ŏyd'·ĕ·ē* (L. *fĭcus,* a fig), the Fig-marigold and Ice-plant family, an Order of plants, the greater part found at the Cape of Good Hope—some are used as food, others yield soda : **ficoidean,** a., *fĭk·ŏyd'·ĕ·ăn,* having an arrangement of parts as in the fig plant.

Ficus, n., *fĭk'ŭs* (L. *fĭcus*, the fig tree and its fruit), an extensive genus of plants, Ord. Moraceæ: **Ficus Carica**, *kăr'ĭk·ă* (L. *Cărĭcus*, of or from Caria; *Cărĭcı*, a Carian-dried fig), the common fig mentioned in the Old and New Testaments, consisting of a succulent hollow receptacle, enclosing numerous single-seeded carpels, called a 'syconus': **F. Indica**, *ĭn'dĭk·ă* (L. *Indĭcus*, of or from India), the banyan tree of India, whose juice is sometimes used in toothache, and bark as a tonic: **F. australis**, *āws·trāl'ĭs* (L. *austrālis*, southern), a species which can live suspended in the air for a long time: **F. religiosa**, *rĕ·lĭdj'ĭ·ōz'ă* (L. *relĭgiōsus*, pious, religious—from *relĭgio*, religion), the pippul tree, or sacred fig of India: **F. elastica**, *ē·lăst'ĭk·ă* (mid. L. *elastĭcus*, It. *elastĭco*, elastic), a species which produces a large amount of caoutchouc, as also a few others: **F. sycomorus**, *sĭk'ō·mōr'ŭs* (L. *sycomōros*, a mulberry tree), probably the sycamore of the Bible, whose wood is said to be very durable: **F. racemosa**, *răs'ĕ·mōz'ă* (L. *răcemōsus*, full of clusters), a species which is slightly astringent, and the juice of the root a powerful tonic.

filament, n., *fĭl'ă·mĕnt* (L. *fĭlum*, a thread), a thread; a fibre; in *bot.*, the stalk supporting the anther; a thread-like substance formed of cells placed end on end: **filamentous**, a., *fĭl'ă·mĕnt'ŭs*, denoting a string of cells placed end to end; thread-like; bearing filaments: filiform, a., *fĭl'ĭ·fŏrm*, slender like a thread.

Filaria, n., *fĭl·ār'ĭ·ă* (L. *fĭlum*, a thread), a genus of parasitic worms: **Filaria bronchialis**, *brŏng'kĭ·ăl'ĭs* (new L. *bronchialis*, bronchial—from Gr. *brongchos*, the windpipe), a species once

found in diseased bronchial glands: **F. immitis**, *ĭm·mĭt'ĭs*, (L. *immĭtis*, not soft, rough, fierce), a parasite of the dog, found in the heart: **f. lachrymalis**, *lăk'rĭm·āl'ĭs* (L. *lăchryma*, a tear), a parasite of the horse and ox: **F. Medinensis**, *mĕd'ĭn·ĕns'ĭs* (*Medĭna*, in Arabia, where frequently met with), the Guinea worm, a parasite met with chiefly on some parts of the shores of Africa, which penetrates the skin of the feet and legs, causing painful symptoms: **F. oculi humani**, *ŏk'ŭl·ī hŭm·ān'ī* (L. *oculus*, the eye; *humānus*, human), the filaria of the human eye, a species discovered in the surrounding fluid and in the crystalline lens: **F. papillosa**, *păp'ĭl·lōz'ă* (L. *papillōsus*, having many small nipples—from *păpilla*, a nipple), a parasite of the horse, ox, and ass, found in the globe of the eye: **F. sanguinis hominis**, *săng'gwĭn·ĭs hŏm'ĭn·ĭs* (L. *sanguis*, blood, *sanguĭnis*, of blood; *homo*, man, *homĭnis*, of man), the filaria of the blood of man; a parasite found in the blood of man, usually in connection with elephantiasis of the skin, and a milky state of the urine (chylous urine): **F. tripinnulosa**, *trĭ·pĭn'ŭl·ōz'ă* (L. *tris*, three; *pinnŭla*, a little wing), a parasite of the dog, found in capsule of crystalline lens.

Filices, n. plu., *fĭl'ĭs·ēz* (L. *fĭlix*, a fern, *fĭlĭces*, ferns), the Fern family, elegant, leafy, herbaceous plants, which in tropical and mild climates become large trees.

filum terminale, *fĭl'ŭm tĕrm'ĭn·āl'ĕ* (L. *terminālis*, terminal—from *termĭnus*, a bound, a limit), the terminal thread or ligament; the central ligament of the spinal cord.

fimbria, n., *fĭm'brĭ·ă*, fimbriæ, n. plu., *fĭm'brĭ·ē* (L. *fimbriæ*, threads, fringe), in *anat.*, any structure resembling a fringe : **fimbriated**, a., *fĭm'·brĭ·āt'·ĕd*, fringed at the margin.

finger and toe, a diseased form of turnip growth, in which the bulbs are divided into two or more forks.

first intention, the healing up of a cut or wound without suppuration.

fissile, a., *fĭs'·sĭl* (L. *fissĭlis*, that may be split—from *fissus*, cleft or split), having a tendency to become split or divided.

fissilinguia, n., *fĭs'·sĭ·lĭng'·gwĭ·ă* (L. *fissus*, cleft ; *lingua*, the tongue), a division of Lacertilia having bifid tongues.

fission, n., *fĭsh'·ŭn* (L. *fissus*, cleft or split), in *zool.*, multiplication by means of a process of self-division ; in *bot.*, the division of an organ which is usually entire : **fissiparous**, a., *fĭs·sĭp'·ăr·ŭs* (L. *pario*, I produce), applied to the multiplying or increasing certain animal forms by the self-division of the individual into two or more parts, each of which becomes a perfect creature, similar to the parent original ; in *bot.*, propagating by a division of cells ; dividing spontaneously into two parts by means of a septum : **fissipara**, n. plu. *fĭs·sĭp'·ăr·ă*, a name applied to those creatures which propagate by spontaneous fission : **fissiparation**, n., *fĭs·sĭ·păr·ā'·shŭn*, the act or process of propagating by spontaneous fission, as among the Infusoria and Polyps, etc.

Fissirostres, n. plu., *fĭs·sĭ·rŏs'·trēz* (L. *fissus*, cleft ; *rostrum*, a beak), a Sub-order of the perching birds.

fissura palpebrarum, *fĭs·sūr'·ă pălp'·ĕ·brār'·ŭm* (L. *fissūra*, a cleft, a chink ; *palpĕbra*, the eyelid, *palpebrārum*, of the eyelids), the fissure of the eyelids ; the interval between the angles of the eyelids.

fissure, n., *fĭsh'·ūr* (L. *fissura*, a cleft or slit; F. *fissure*), a straight slit in an organ for the discharge of its contents ; a slit or cleft.

fistula, n., *fĭst'·ŭl·ă* (L. *fistŭla*, a hollow reed), a narrow channel or tube leading to a cavity containing matter or dead bone, or communicating with the intestinal canal or other cavity, and lined with a membrane which secretes a puriform fluid ; a deep narrow ulcer or sore ; **fistular**, a., *fĭst'·ŭl·ăr*, also **fistulous**, a., *fĭst'·ŭl·ŭs*, having the nature of a fistula ; in *bot.*, hollow like the stem of grasses.

flabelliform, a., *flăb·ĕl'·lĭ·fŏrm* (L. *flabellum*, a fan ; *forma*, shape), in *bot.*, shaped like a fan; plaited like a fan.

flaccid, a., *flăk'·sĭd* (L. *flaccidus*, flabby, withered), soft and weak; wanting in stiffness.

Flacourtieæ, n. plu., *flă'·kōōr·tĭ·ĕ·ē* (in honour of *Flacourt*, a French botanist), a tribe of shrubs or small trees, Ord. Bixaceæ : **Flacourtia**, *flă·kōōr'·shĭ·ă*, a genus of ornamental fruit trees or shrubs, some bearing edible fruits, and others useful in medicine : **F. ramontchi**, *ră·mŏn'·tshĭ* (a native name), a species from Madagascar, bearing leaves and fruit similar to those of a plum.

flagellum, n., *flă·jĕl'·lŭm* (L. *flăgellum*, a whip, a scourge), in *bot.*, a runner ; a creeping stem, bearing rooting buds at different points, as in the strawberry ; the lash-like appendage exhibited by many Infusoria: **flagelliform**, a., *flă·jĕl'·lĭ·fŏrm* (L. *forma*, shape), tapering and supple like a whip : **flagellate**, a., *flă·jĕl'·lāt*, in same sense ; having a long lash-like appendage.

flavescent, a., *flăv·ĕs'·ĕnt* (L. *flavesco*, I become golden-yellow), in *bot.*, growing yellow : **flavicant**, a., *flăv'·ĭk·ănt*, yellow.

fleam, n., *flēm* (F. *flamme*, a lancet ; Dut. *vlieme*, a sharp-pointed thing, a lancet), an instrument used for bleeding horses and cattle.

flex, v., *flĕks* (L. *flexus*, bent), to extend the leg upon the thigh or upon the pelvis: **flexing,** bending: **flexed,** *flĕkst*, bent: **flexes,** it bends.

flexion, n., *flĕk'·shŭn* (L. *flexus*, bent), the bending of a limb, as opposed to 'extension,' the stretching out of a limb: **flexor,** n., *flĕks'·ŏr*, a muscle which bends or contracts a part of the body, and is opposed to 'extensor,' a muscle which extends a part.

flexor carpi radialis, *flĕks'·ŏr kârp'·ī rād'·ĭ·āl'·ĭs* (L. *flexor*, that which bends or contracts; *carpus*, a wrist, *carpi*, of a wrist; *radius*, the rotatory bone of the fore-arm), the radial flexor of the wrist; the muscle which bends the hand and assists to turn its palm towards the ground: **f. carpi ulnaris,** *ŭl·nār'·ĭs* (L. *ulna*, the large bone of the fore-arm), the ulnary flexor of the wrist; the muscle which assists in bending the arm: **f. longus digitorum manus,** *lŏng'·gŭs dĭdj'·ĭt·ōr'·ŭm mān'·ŭs* (L. *longus*, long; *digitus*, a finger or toe, *digitorum*, of fingers or toes; *mănŭs*, a hand, *mănūs*, of a hand), the long flexor of the fingers of the hand; the muscle which bends the joint or phalanx of the fingers: **f. sublimis perforatus,** *sŭb·lĭm'·ĭs pĕrf'·ŏr·āt'·ŭs* (L. *sublimis*, high; *perforātus*, bored through), the high perforated muscle, so named from its being perforated by the tendon of another flexor, the 'flexor profundis'; the muscle which bends the second joint or phalanx of the fingers.

flexuose, a., *flĕks'·ū·ōz*, or **flexuous,** a., *flĕks'·ū·ŭs* (L. *flexuōsus*, full of windings, tortuous—from *flexus*, a turning, a winding), in *bot.*, having alternate curvatures in opposite directions; bent in a zigzag manner.

flexure, n., *flĕks'·ūr* (L. *flexūra*, a bending or winding), a joint; the part bent; a curvature.

flocculent, a., *flŏk'·kūl·ĕnt* (L. *flocculus*, a small lock of wool—from *floccus*, a lock of wool), having the appearance of flocks or flakes; adhering in flocks or flakes: **flocculence,** n., *flŏk'·kūl·ĕns*, the state of being in flocks or flakes; adhesion in flocks: **flocculose,** a., *flŏk'·kūl·ōz*, woolly; like wool: **flocculus,** n., *flŏk'·kūl·ŭs*, in *anat.*, a prominent tuft or lobule, situated behind and below the middle peduncle of the cerebellum.

floccus, n., *flŏk'·kŭs*, **flocci,** n. plu., *flŏk'·sī* (L. *floccus*, a lock of wool), a tuft of hair terminating in a tail; woolly hairs or threads; woolly filaments with sporules in Fungi and Algæ: **floccose,** a., *flŏk·kōz'*, covered with wool-like tufts.

flora, n., *flōr'·ă* (L. *Flōra*, the goddess of flowers—from *flos*, a flower, *flōris*, of a flower), plants peculiar to a country, or to a geological era; the opposite, 'fauna,' signifies the animals peculiar to a district: **floral,** a., *flōr'·ăl*, pert. to flowers; in *bot.*, seated near the flower, and about the flower-stalk: **floral envelopes,** the calyx and corolla: **florets,** n. plu., *flōr'·ĕts*, the little flowers collected into a head in composite plants.

Florideæ, n. plu., *flōr·ĭd'·ĕ·ē* (L. *flōridus*, flowery—from *flos*, a flower), a Sub-order of Algæ; rose or purple coloured sea-weeds, with fronds formed of a single row of articulated cells, or of several rows; also called 'Rhodosporeæ,' or 'Choristosporei.'

flosculous, a., *flŏsk'·ūl·ŭs*, and **floscular,** a., *flŏsk'·ūl·ăr* (L. *flosc·ūlus*, a little flower—from *flos*, a flower), in *bot.*, applied to the tubular florets of Compositæ : **floscule**, n., *flŏsk'·ŭl*, the partial or lesser floret.

floss, n., *flŏs* (It. *floscio*, Pied. *flos*, drooping, flaccid ; F. *flosche*, weak, soft ; Bav. *floss*, loose, not fast), a downy or silky substance found in the husks of certain plants.

fluctuation, n., *flŭk'·tū·ā'·shŭn* (L. *fluctuātus*, moved like a wave— from *fluctus*, a wave), the wave-like movement, when there is any accumulation of fluid in a part, felt by manipulation.

fluorine, n., *flŏ'·ŏr·ĭn* (L. *fluo*, I flow ; *flŭor*, a flowing—so called from being used as a flux), an elementary substance first found in fluor-spar.

flux, n., *flŭks* (L. *fluxus*, a flow, a flux), an abnormal discharge of fluid matter from the bowels or other part.

fœtor, n., *fēt'·ŏr* (L. *fœtor*, a stench), a strong offensive smell.

fœtus, n., *fēt'·ŭs* (L. *fœtus*, filled with young, pregnant), the young of animals in the womb, or in the egg after assuming a perfect form : **fœtal**, a., *fēt'·ăl*, pert. to the fœtus : **fœtation**, n., *fēt·ā'·shŭn*, the formation of a fœtus : **fœticide**, n., *fēt'·ĭ·sīd* (L. *cædo*, I kill), the killer of a fœtus.

foliaceous, a., *fōl'·ĭ·ā'·shŭs* (L. *foliācĕus*, like leaves — from *fŏlĭum*, a leaf), leafy ; having the form or texture of a leaf : **foliar**, a., *fōl'·ĭ·ăr*, pert. to or growing upon leaves : **foliation**, n., *fōl'·ĭ·ā'·shŭn*, the leafing of plants ; the manner in which the young leaves of plants are arranged in the leaf-bud.

foliicolous, a., *fōl'·ĭ·ĭk'·ŏl·ŭs* (L. *folium*, a leaf ; *cŏlo*, I dwell), growing on leaves : **foliiferous**, a., *fōl'·ĭ·ĭf'·ĕr·ŭs* (L. *fĕro*, I bear), leaf-bearing ; also spelt **foliferous**, a., *fōl·ĭf'·ĕr·ŭs*.

foliola, n., *fōl·ĭ'·ŏl·ă*, and **foliole**, n., *fōl'·ĭ·ŏl* (new L. *folĭŏlum*, a

little leaf—from *folĭum*, a leaf), a leaflet : **foliolose**, a., *fōl'·ĭ·ŏl·ōz*, consisting of minute leaf-like scales.

follicle, n., *fŏl'·lĭk·l* (L. *folliculus*, a small bag or sac inflated with air—from *follis*, a bag or bellows), a little bag ; a cavity ; in *bot.*, a seed-vessel opening along the side, to which the seeds are attached, as in the pea : **follicular**, a., *fŏl·lĭk'·ŭl·ăr*, also **folliculous**, a., *fŏl·lĭk'·ŭl·ŭs*, having follicles, or producing follicles.

fomes, n., *fōm'·ēz*, **fomites**, n. plu., *fōm'·ĭt·ēz* (L. *fōmes*, touch-wood, fuel, *fomites*, touchwoods), porous substances capable of absorbing and retaining contagious matter (probably germs); woollen cloth and wood are said to be excellent 'fomites.'

fontanelles, n. plu., *fŏn'·tăn·ĕlz'* (F. *fontanelle*, the meeting of the seams of the skull—from F. *fontaine*, L. *fons*, a fountain—so called from the pulsations of the brain, perceptible at the anterior fontanelle, as of a rising of water in a fountain), four spaces in the skull, opposite the angles of the parietal bones, which remain unoccupied by bone after the osseous wall has been formed elsewhere.

NOTE.—The smaller spaces at the inferior angles of the parietal bones are of little consequence, and are filled in soon after birth. The anterior fontanelle between the anterior and superior angles of the parietal bones, and the superior angles of the ununited frontal segments, is of great importance to the accoucheur in determining the positions of the child during labour. The anterior and superior fontanelles are generally not wholly filled in till the second year.

foramen, n., *fŏr·ām'·ĕn*, **foramina**, n. plu., *fŏr·ăm'·ĭn·ă* (L. *forāmen*, an aperture or opening—from *foro*, I bore), in *anat.*, a small opening such as may be made into a substance by boring ; in *bot.*, the opening in the coverings of the ovule : **foramen obturator**,

ŏb'·tūr·āt'·ŏr (L. forāmen, an
aperture or opening ; obturātor,
that which stops or closes up),
an oval opening in both sides of
the large bone that ends or closes
up the trunk ; the large oval
interval between the ischium and
the pubes : f. thyroid, thīr'·ŏyd
(Gr. thureos, a shield ; eidos,
resemblance), one of the two
openings of the shield-like bones
which terminate the trunk; same
as 'foramen obturator': f. cæcum,
sēk'·ŭm (L. cæcus, blind), a small
opening which terminates below
the frontal crest of the skull: f.
commune anterius, kŏm·mūn'·ē
ănt·ēr'·ĭ·ŭs (L. commūnis, common; antērior, that which is
placed before—from ante, before),
the anterior common foramen,
an opening under the arch of the
fornix: f. commune posterius,
pŏst·ēr'·ĭ·ŭs (L. postērior, that
which is placed behind—from
post, behind), the posterior
common foramen, an opening between the middle and the posterior commissure of the brain :
f. incisivum, ĭn'·sĭs·īv'·ŭm (L.
incīsīvus, of or belonging to the
incisor teeth — from incīsus,
notched, indented), the incisor
foramen ; the opening immediately behind the incisor teeth :
f. magnum occipitis, măg'·nŭm
ŏk·sĭp'·ĭt·ĭs (L. magnus, great;
occiput, the back part of the
head, occĭpĭtis, of the back part
of the head), the great foramen
of the occiput ; the great opening
at the under and fore part of the
occipital bone: f. ovale, ŏv·āl'·ē
(L. ovālis, an oval), the oval
foramen or aperture between the
auricles of the fœtal heart; an
oval aperture between the tympanum and the vestibule of the
ear: f. rotundus, rōt·ŭnd'·ŭs (L.
rotundus, round, circular), the
round or triangular aperture of
the internal ear.
Foraminifera, n. plu., fŏr·ăm'·ĭn·

ĭf'·ĕr·ă (L. foramen, an aperture ;
fero, I carry), an Order of Protozoa having shells perforated by
numerous pseudopodial apertures;
many-celled organisms : foramin-
iferous, a., fŏr·ăm'·ĭn·ĭf'·ĕr·ŭs,
having many chambers or holes.
forceps, n. plu., fŏr'·sĕps (L. forceps, a pair of tongs, as if ferrĭceps
—from ferrum, iron ; capio, I
take), a kind of tongs of various
sizes and shapes, used by surgeons, and by anatomists and
accoucheurs : **forcipate**, a., fŏr'·
sĭp·āt, in bot., forked like
pincers.
formication, n., fŏrm'·ĭk·ā'·shŭn
(L. formĭca, an ant), a sensation
resembling that caused by ants
creeping on the skin.
Fornasinia, n., fŏr'·năs·ĭn'·ĭ·ă (not
ascertained), a genus of plants,
Ord. Leguminosæ, Sub - ord.
Papilionaceæ: **Fornasinia ebenif-
era**, ĕb'·ĕn·ĭf'·ĕr·ă (L. ĕbĕnus, the
eben tree, ebony ; fero, I bear),
produces a kind of ebony, a
papilionaceous plant found in
Caffraria.
fornix, n., fŏrn'·ĭks, **fornices**, n.
plu., fŏrn'·ĭs·ēz (L. fornix, an
arch, fornĭcis, of an arch), an
arched sheet of white longitudinal
fibres, which appears partly in
the floor of both lateral ventricles
of the brain, situated beneath
the corpus callosum ; in bot.,
arched scales in the orifice of
some flowers : **fornicate**, a., fŏrn'·
ĭk·āt (L. fornicātus, arched),
arched.
fossa, n., fŏs'·să, **fossæ**, n. plu.,
fŏs'·sē (L. fossa, a ditch ; fossus,
dug), in anat., a little cavity or
depression in a bone ; any depression in the human body : **fossa
cystis felleæ**, sĭst'·ĭs fĕl'·lĕ·ē (Gr.
kustis, a bladder, a pouch; L.
fellĕus, pert. to the gall—from
fel, the gall-bladder), the fossa
of the gall-bladder, a shallow
oblong cavity on the under surface of the right lobe of the liver

for the lodgment of the gall-bladder: **f. hyaloidea,** *hī'ȧl·ōȳd'ĕ·ȧ* (Gr. *hualos,* glass; *eidos,* resemblance), a cup-like depression on the anterior surface of the vitreous humour containing the crystalline lens: **f. innominata,** *ĭn·nŏm'ĭn·āt'ȧ* (L. *in,* not; *nominātus,* named), in the external ear, a narrow curved groove between the helix and antihelix: **f. lachrymalis,** *lăk'rĭm·āl'ĭs* (L. *lachryma,* a tear), a depression in the frontal bone of the cranium for the reception of the lachrymal-gland: **f. navicularis,** *năv·ĭk'ŭl·ār'ĭs* (L. *nāvĭcula,* a boat—from *nāvis,* a ship), a depression separating the two roots of the antihelix; a depression on the floor of the urethra; a small cavity within the fourchette; **f. ovalis,** *ōv·āl'ĭs* (L. *ovālis,* oval), an oval depression situated above the orifice of the inferior vena cava: **f. scaphoides,** *skăf·ōȳd'ēz* (Gr. *skaphē,* a little boat; *eidos,* resemblance), another name for 'fossa navicularis.'

fossil, n., *fŏs'sĭl* (L. *fossus,* dug), any remains of plants or animals dug out of the earth's crust changed into a stony consistence: adj., dug out of the earth: **fossiliferous,** a., *fŏs·sĭl·ĭf'ĕr·ŭs* (L. *fero,* I bear).

Fothergilla, n., *fŏth'ĕr·gĭl'lȧ* (after *Dr. Fothergill,* of London), a genus of beautiful shrubs, whether in leaf or flower, bearing pretty, sweet-scented flowers, Ord. Hamamelidaceæ.

founder, n., *fownd'ĕr* (L. *fundus,* F. *fond,* the ground or bottom), a diseased state of the plantar region of the foot in the horse, generally both fore-feet, producing lameness.

fourchette, n., *fŏŏr·shĕt'* (F. *fourchette,* a fork), a small transverse fold, just within the posterior commissure of the vulva; the bone in birds formed by the junction of the clavicles; the wishing-bone in a fowl.

fovea, n., *fōv'ĕ·ȧ,* **foveæ,** n. plu., *fōv'ĕ·ē* (L. *fovea,* a pit, a depression), in *anat.,* a slight depression; in *bot.,* a depression in front of a leaf of some Lycopodiaceæ, containing the sporangium: **foveate,** a., *fōv'ĕ·āt,* also **foveolate,** a., *fōv·ĕ'ŏl·āt,* having pits or depressions called foveæ or foveolæ: **foveola,** n., *fōv·ĕ'ŏl·ȧ* (L., a little pit), in *bot.,* little pits or regular depressions.

fovilla, n., *fōv·ĭl'lȧ* (L. *foveo,* I nourish), in *bot.,* the matter contained in the grains of pollen, consisting of minute granules floating in a liquid.

fracture, n., *frăkt'ŭr* (L. *fractūra,* a breach, a fracture—from *fractus,* broken), a broken bone: **simple fracture,** the breaking of a bone without the injury of the skin or adjacent soft parts: **compound fracture,** a broken bone with a wound through the skin and muscles, and exposure of bone: **comminuted fracture,** a bone broken into several small fragments: **compound comminuted fracture,** a bone broken into several small fragments, together with injury of the soft parts, whereby the bones are visibly exposed, or are accessible to the probe.

frænum, n., *frēn'ŭm,* **fræna,** n. plu., *frēn'ȧ* (L. *frœnum,* a bit, a curb), in *anat.,* a part which checks or curbs; a membranous fold which keeps an organ in position: **frænulum,** n., *frēn'ŭl·ŭm* (dim. of *frœnum*), used in same sense: **frænum linguæ,** *lĭng'gwē* (L. *lingua,* a tongue), the curb of the tongue; a fold at the under surface of the tongue—when short or too far forward in infants, they are said to be tongue-tied.

Francoaceæ, n. plu., *frăngk'ō·ā'sĕ·ē,* also **Francoads,** n. plu., *frăngk'·ō·ăds* (after *Franco*, a botanist of the 16th century), an Order of herbaceous plants without stems, natives of Chili; the species regarded in Chili as cooling and sedative, and their roots are used to dye black ; the Order is sometimes included under the Ord. Saxifragaceæ : **Francoa**, n., *frăngk·ō'·ă*, a genus of plants beautiful when in flower.

Frankeniaceæ, n. plu., *frăngk·ēn'·ĭ·ā'·sĕ·ē*, or **Frankeniads**, n. plu., *frăngk·ēn'·ĭ·ăds* (after *Frankenius*, of Upsal, 1661), the Frankeniad family, an Order of herbaceous plants : **Frankenia**, n., *frăngk·ēn'·ĭ·ă*, a genus of beautiful evergreen shrubs or herbs, said to be mucilaginous and slightly aromatic.

Frasera, n., *frāz·ēr'·ă* (after *Fraser*, a collector of N. American plants), a genus of curious little plants, Ord. Gentianaceæ : **Frasera Walteri**, *wălt'·ĕr·ī* (*Walter*, a proper name; *Walteri*, of Walter, a Latinised spelling), a species sometimes called the American calumba, found in the morasses of N. America; the root is said to furnish an excellent bitter.

Fraxinella, see 'Dictamnus.'

Fraxinus, n., *frăks'·ĭn·ŭs* (L. *fraxinus*, an ash tree), a genus of trees, Ord. Oleaceæ : **Fraxinus excelsior**, *ĕk·sĕls'·ĭ·ŏr* (L. *excelsior*, loftier—from *excelsus*, lofty), the common ash, whose tough, elastic wood is much used by coachmakers, wheelwrights, and implement makers ; the 'weeping ash' is a pendulous variety ; the wood of the roots is beautifully veined ; for 'manna ash,' see 'Ornus.'

freckle, n., *frĕk'·l* (Icel. *frekna*, Norse *flukr*, freckles; Ger. *fleck*, a spot), congenital pigmentation of the rete mucosum, the spots being the size of split peas or less, occurring on the skin beneath the clothing, as well as on the skin when exposed to light; lentigo ; also minute coloured specks often seen on the skin, generally the face ; any small discoloured spot ; ephelis.

Freycinetia, n., *frā'·sĭn·ĕ'·shĭ·ă* (after *Captain Freycinet*, a French circumnavigator), a genus of ornamental tree-like plants, some having a climbing habit, Ord. Pandanaceæ : **Freycinetia Banksii**, *bănks'·ĭ·ī* (after *Sir Joseph Banks*), the kie-kie or screw pine of New Zealand, whose fleshy bracts, called 'tawhara,' are eaten by the natives, and made into a luscious jelly by the colonists.

Freziera, n., *frĕz'·ĭ·ēr'·ă* (after *Frezier*, a French traveller), a genus of tall ornamental trees, Ord. Ternstrœmiaceæ : **Freziera theoides**, *thĕ·oyd'·ēz* (*thea*, the tea-plant; Chin. *tshă*, Russ. *tshai*, tea ; Gr. *eidos*, resemblance), a species whose leaves are used as tea in Panama.

Fritillaria, n., *frĭt'·ĭl·lār'·ĭ·ă* (L. *fritillus*, a dice-box, a chess-board, alluding to the chequered sepals of the flowers), a genus of plants, Ord. Liliaceæ, having singular and showy flowers : **Fritillary**, n., *frĭt'·ĭl·ăr·ĭ*, the name of a common showy garden flower.

frond, n., *frŏnd* (L. *frons*, a leaf, *frondis*, of a leaf), the peculiar leafing of palms and ferns ; the union of a leaf and branch: **frondescence**, n., *frŏnd·ĕs'·sĕns* (L. *frondesco*, I shoot forth leaves), the time or season of putting forth leaves ; the conversion of petals or other organs into leaves : **frondlet**, n., *frŏnd'·lĕt*, a little frond : **frondose**, a., *frŏnd·ōz'*, having a foliaceous or leaf-like expansion.

frugivorous, a., *frŏ·jĭv'·ŏr·ŭs* (L. *frux*, fruit, *frugis*, of fruit ; *voro*, I devour), applied to animals that feed upon fruits.

frustules, n. plu., *frŭst'·ūlz*, alsc **frustula**, n. plu., *frŭst'·ūl·ă* (dim.

of L. *frustum*, a fragment), in *bot.*, the parts or fragments into which certain sea - weeds, the diatoms, separate : **frustulose,** a., *frŭst'·ūl·ōz*, consisting of fragments.

frutex, n., *frŏt'·ĕks* (L. *frŭtex*, a shrub, *frŭtĭcis*, of a shrub), in *bot.*, a shrub : **fruticose,** a., *frŏt'·ĭk·ōz*, shrub-like : **fruticulose,** a., *frŏt·ĭk'·ūl·ōz*, a dim. of fruticose ; somewhat shrub-like ; slightly shrubby : **fruticulus,** n., *frŏt·ĭk'·ūl·ŭs*, an under shrub not exceeding the length of the arm.

Fucaceæ, n. plu., *fū·kā'·sĕ·ē* (L. *fucus*, the rock - lichen ; Gr. *phukos*, the plant alkanet, seaweed), a Sub-order of Algæ, brown or olive coloured plants, growing chiefly in salt water, consisting of cells which unite so as to form various kinds of thalli ; the brown sea-weeds or sea-wracks, some of which are eatable, and others possess medicinal properties : **Fucus,** n., *fūk'·ŭs*, a genus of sea-weeds : Fucus bacciferus, *băk·sĭf'·ĕr·ŭs* (L. *bacca*, a berry ; *fero*, I bear), the Gulf-weed, eaten as a raw salad, and pickled : F. **digitatus,** *dĭdj'·ĭt·āt'·ŭs* (L. *digitatus*, having fingers or toes—from *digĭtus*, a finger), the sea-girdle and hangers, growing on stones and rocks in the sea near the shore : F. **edulis,** *ĕd·ūl'·ĭs* (L. *edūlis*, eatable—from *edo*, I eat), the red dulse, eaten raw or broiled : F. **esculentus,** *ĕsk'·ūl·ĕnt'·ŭs* (L. *escŭlentus*, fit for eating — from *esca*, food) ; and F. **fimbriatus,** *fĭm'·brĭ·āt'·ŭs* (L. *fimbrĭātus*, fibrous, fringed—from *fimbriæ*, fibres, threads), edible sea-weeds or daber locks : F. **natans,** *nāt'·ăns* (L. *natans*, swimming, floating), the sea lentil, said to be useful in dysuria : F. **palmatus,** *pălm·āt'·ŭs* (L. *palmātus*, marked like the palm of a hand—from *palma*, the palm of the hand), the handed

fucus or dulse, eaten raw or cooked : F. **pinnatifidus,** *pĭn'·năt·ĭf'·ĭd·ŭs* (L. *pinna*, a feather, a wing ; *findo*, I cleave, *fĭdi*, I have cleft), the pepper dulse, eaten as a salad, is warm like cresses : F. **saccharinus,** *săk'·kăr·ĭn'·ŭs* (L. *saccharum*, Gr. *sakcharon*, sugar), the sweet fucus or seabelts growing on stones and rocks ; leaves sweet, which exude a sugary substance when dry : F. **vesiculosus,** *vĕs·ĭk'·ūl·ōz'·ŭs* (L. *vesicŭla*, a little blister, a vesicle), the plant bladder-fucus, sea-oak, or sea-wrack.

Fuchsia, n., *fū'·shĭ·ă* (after *Fuchs*, a German botanist), a very beautiful and well-known genus of shrubs, of numerous species ; some of the garden varieties are exceedingly beautiful, Ord. Onagraceæ.

fucoxanthine, n., *fūk'·ō·zănth'·ĭn* (Gr. *phukos*, the plant alkanet, sea-weed ; *xanthos*, yellow), a colouring matter of the Xanthophyll group, found in Melanosporeæ.

fugacious, a., *fūg·ā'·shŭs* (L. *fugax*, swift, *fugācis*, of swift — from *fugĭo*, I fly), in *bot.*, falling off early, as the petals of Cistus ; evanescent.

fuliginous, a., *fūl·ĭdj'·ĭn·ŭs* (L. *fulīgo*, soot, *fulĭgĭnis*, of soot), sooty ; in *bot.*, smoke-coloured, or brownish-black.

fulvous, a., *fŭlv'·ŭs* (L. *fulvus*, of a deep yellow), tawny yellow ; of a saffron colour.

Fumariaceæ, n. plu., *fūm·ār'·ĭ·ā'·sĕ·ē* (L. *fūmus*, smoke, from the smell of some of the species, or from the effect of the juice upon the eyes being the same as smoke), the Fumeworts or Fumitory family, an Order of herbaceous plants, said to be bitter and diaphoretic in their properties : **Fumaria,** n., *fūm·ār'·ĭ·ă*, a genus of plants : **Fumitory,** n., *fūm'·ĭt·ŏr·ĭ* (*fumeterre,* a French name for the genus—from L. *fumus,* smoke ; *terræ,* of the earth), the English name for the genus.

fundament, n., *fŭnd'·ă·mĕnt* (L. *fundāmentum,* groundwork, basis —from *fundo,* I lay the foundation), the lower part of the rectum ; the anus ; the seat of the body : **fundus,** n., *fŭnd'·ŭs* (L. the bottom of a thing), the base or lower part of an organ which has a neck or external opening : **fundal,** a., *fŭnd'·ăl,* pert. to the fundus.

fungous, a., *fŭng'·ŭs,* having the character or consistence of Fungi: **fungosity,** n., *fŭng·ŏs'·ĭ·tĭ,* a soft excrescence: **fungiform,** a., *fŭnj'·ĭ·fŏrm* (L. *forma,* shape), having the shape of a fungus ; like a fungus : **fungoid,** a., *fŭng'·ŏyd* (Gr. *eidos,* resemblance), like a fungus ; fungiform : **Fungus hæmatodes,** *hēm'·ăt·ŏd'·ēz* (Gr. *haima,* blood; *eidos,* resemblance), a variety of soft cancer in which the tumour is large and of rapid growth, composed of soft cancerous tissue mixed with large clots of blood : **F. vinosus,** *vĭn·ōz'·ŭs* (L. *vīnōsus,* having the taste of wine —from *vīnum,* wine), a dark-coloured fungus which vegetates in dry cellars where wine, ale, porter, etc. are kept.

Fungus, n., *fŭng'·ŭs,* **Fungi,** n. plu., *fŭnj'·ĭ* (L. *fungus,* a mushroom or toadstool), a mushroom or toadstool : the Mushroom family, an Order of plants ; in *surg.,* the unnatural formation of flesh about an ulcer, commonly called 'proud flesh.'

funiculus, n., *fŭn·ĭk'·ŭl·ŭs,* also **funicle,** n., *fŭn'·ĭk·l* (L. *funĭcŭlus,* a slender rope—from *fūnis,* a cord), the umbilical cord connecting the hilum of the ovule to the placenta ; a cord-like appendage by which, in many cases, the seeds are attached ; in *anat.,* a number of nerve-fibres enclosed in a tubular sheath forming a slender round cord of no determinate size.

fur, n., *fĕr* (Goth. *fôdr,* a sheath ; Icel. *fôthr,* Sp. *forro,* sheath, lining), soft, thick hair on certain animals ; a layer of morbid matter, resembling fur, indicating a diseased state.

furcate, a., *fĕrk'·āt* (L. *fŭrca,* a fork), branching like the prongs of a fork : **furcation,** n., *fĕrk·ā'·shŭn,* the branching like a fork : **furculum,** n., *fĕrk'·ŭl·ŭm,* also **furcula,** n., *fĕrk'·ŭl·ă* (L. *furcŭla,* a forked prop, a dim. of *furca,* a fork), the v-shaped bone of birds, formed by the united clavicles ; the merry-thought ; in *anat.,* the middle one of the three deep notches of the manubrium, or thickest part of the sternum.

furfur, n., *fĕr'·fĕr* (L. *furfur,* bran), scales like bran ; dandriff : **furfuraceous,** a., *fĕr'·fĕr·ā'·shŭs,* scurfy or scaly ; covered with a meal-like powder : **furfuration,** n., *fĕr'·fĕr·ā'·shŭn,* the state of suffering from scurf or scaliness of the skin.

furunculus, n., *fūr·ŭngk'·ŭl·ŭs* (L. *fūruncŭlus,* a petty thief, a boil —from *fūr,* a thief), a boil or small tumour having a central core, and suppurating imperfectly: **furuncular,** a., *fūr·ŭngk'·ŭl·ăr,* of or belonging to a furunculus.

fuscous, a., *fŭsk'·ŭs* (L. *fuscus,* dark, dusky), in *bot.,* blackish-brown, or darkish-brown : **fuscescent,** a., *fŭs·sĕs'·ĕnt,* tending to a darkish brown.

fusel oil, n., *fūz'·ĕl* (Ger. *fusel,* bad or poor brandy), an alcohol or volatile oil of a nauseous and irritating odour, contained in a greater or less quantity in all forms of crude spirits, and to which substance bad spirits owe their noxious qualities ; found only in minute quantity in fine wine spirits ; it is also spelt **fousel.**

fusiform, a., *fūz'·ĭ·fŏrm* (L. *fūsus*, a spindle ; *forma*, shape), shaped like a spindle ; tapering at both ends.

Galactodendron, n., *găl·ăkt'·ō·dĕnd'·rŏn* (Gr. *gala*, milk, *galaktos*, of milk ; *dendron*, a tree), a lofty-growing tree, called the cow-tree because its milky juice is used as a substitute for milk ; also called Brosmium utile, Ord. Moraceæ.

galactophorous, a., *găl'·ăkt·ŏf'·ŏr·ŭs* (Gr. *gala*, milk, *galaktos*, of milk ; *phoreo*, I carry), conveying milk or white juice ; applied to certain ducts or canals in the mamma which convey the milk to the summit of the mammilla.

galangal, a., *găl'·ăn·găl* (native name ; Sp. *galanga*, a species of the arrowroot), applied to a dried root brought from China, having an aromatic smell and a pungent bitter taste, formerly used in medicine ; the root-stock of Alpinia officinarum, Ord. Zingiberaceæ.

Galanthus, n., *găl·ănth'·ŭs* (Gr. *gala*, milk ; *anthos*, a flower, alluding to the milk-white flowers), a genus of plants, Ord. Amaryllidaceæ : **Galanthus nivalis**, *nĭv·āl'·ĭs* (L. *nĭvālis*, of or belonging to snow—from *nix*, snow), the common snowdrop : **G. plicatus**, *plĭk·āt'·ŭs* (L. *plĭcātus*, folded), a larger and finer species of snowdrop, native of the Crimea.

galbanum, n., *gălb'·ăn·ŭm* (L. *galbănum*, Gr. *chalbanē*, the resinous sap of a Syrian plant), the resinous gum of an umbelliferous plant imported from India and the Levant.

galbulus, n., *gălb'·ŭl·ŭs* (L. *galbulus*, Sp. *galbulo*, the nut or little round ball of the cypress tree), in *bot.*, a modification of the cone, where the apex of each carpellary scale is much enlarged or fleshy, so that collectively they form a round, compact fruit.

gale, n., *gāl* or *gāl'·ĕ* (probably Norse *galen*, angry, mad; or Icel. *gala*, to sing—from its supposed medical qualities ; F. *galé*), the Myrica gale ; the gale, Scotch myrtle, or bog myrtle, a native fragrant bush, common to marshy grounds and damp heaths in Britain, Ord. Myricaceæ.

galea, n., *găl'·ĕ·ă* (L. *gălĕa*, a helmet, a headpiece), in *bot.*, a sepal or petal shaped like a helmet: **galeate**, a., *găl'·ĕ·āt*, shaped in a hollow vaulted manner like a helmet.

Galieæ, n. plu., *găl·ĭ'·ĕ·ē* (Gr. *gala*, milk—from the flowers of one of the species being used for curdling milk), one of the three series or Sub-orders into which the Ord. Rubiaceæ has been divided ; also named 'Stellatæ,' because they have verticillate leaves : **Galium**, n., *găl'·ĭ·ŭm*, a genus of plants, common weeds.

galipea, n., *găl'·ĭ·pē'·ă* (a native name; Sp. *galipot*, white frankincense), a genus of plants found in Venezuela, Ord. Rutaceæ : **Galipea cusparia**, *kŭs·pār'·ĭ·ă* (L. *cuspis*, a spear; but not ascertained); also **G. officinalis**, *ŏf·fĭs'·ĭn·āl'·ĭs* (L. *officinālis*, official — from *officina*, a workshop), plants which supply the Angostura bark, used as a tonic and febrifuge.

gall, n., *gāwl* (AS. *gealla*, gall ; *gealew*, yellow ; Ger. *galle*, gall ; *gelb*, yellow), a bitter, yellowish-green fluid secreted by the liver ; bile : **gall-bladder**, n., a small pear-shaped sac which receives the bile from the liver : **gall-stone**, n., a concretion formed from the gall.

gall, n., *gāwl*, also **gall-nut** (L. *galla*, F. *galle*, the oak-apple; It. *galla*, a bubble, an oak-gall), hard, round excrescences on the Quercus infectoria, caused by the punctures, and deposited eggs, of the Diplolepis Gallæ tinctoriæ, etc. : **gallic**, a., *găl'·lĭk*, denoting an acid obtained from gall-nuts.

gall, v., *găwl* (F. *galler*, to fret, to itch ; It. *galla*, scab ; Icel. *galli*, a fault or imperfection), to injure or break the skin by rubbing : n., a wound in the skin produced by rubbing.

Gallinacei, n. plu., *găl'·lĭn·ā'·sĕ·ī* (L. *gallīna*, a hen, a fowl), that section of the Order of Rasorial birds of which the common fowl is the type ; sometimes applied to the whole Order.

galvanism, n., *gălv'·ăn·ĭzm* (after *Galvani*, the discoverer), electricity developed from the chemical action which takes place from certain bodies placed in contact, as different metals ; often applied to the body as a remedial agent, especially in the case of nervous diseases.

Gamassia, n., *găm·ăs'·sĭ·ă* (a native name), the Gamass or Squamash, a genus of plants, Ord. Liliaceæ : **Gamassia esculenta**, *ĕsk'·ūl·ĕnt'·ă* (L. *esculentus*, eatable, esculent — from *esca*, food), a plant whose root bulb is used as food, and is called by the Indians of N. America 'biscuit-root.'

Gambier, n., *găm'·bĭ·ĕr*, or **Gambir**, n., *găm'·bĭr* (from *Gambier*, East Indies), an astringent drug, and used as a substitute for catechu, obtained from the Uncaria gambier, Ord. Rubiaceæ.

gamboge, n., *găm·bŏdj'* (from *Cambodia*, in Asia), a yellow or greenish kind of resin, used as a pigment, and in medicine as a powerful purgative.

gamogastrous, a., *găm'·ō·găs'·trŭs* (Gr. *gamos*, marriage, union ; *gastēr*, the belly, an ovary), in *bot.*, applied to a pistil formed by a union of the ovaries more or less complete, while the styles and stigmata remain free : **gam-**

opetalous, a., *găm'·ō·pĕt'·ăl·ŭs* (Gr. *petalon*, a leaf), having a corolla formed by the union or grafting together of several petals, so as to form a tube ; monopetalous : **gamosepalous**, a., *găm'·ō·sĕp'·ăl·ŭs* (*sepalon*, an adapted word, formed from Gr. *petalon*), having a calyx formed by the union of several petals ; monosepalous : **gamophyllous**, a., *găm'·ō·fŭl'·lŭs* (Gr. *phullon*, a leaf), having one leaf or membrane ; monophyllous.

ganglion, n., *găng'·glĭ·ŏn*, **ganglions**, n. plu., *-ŏnz*, or **ganglia**, n. plu., *-lĭ·ă* (Gr. *gangglion*, a little tumour under the skin near the sinews), in *surg.*, a tumour in the sheath of a tendon ; a mass of nervous matter containing nerve cells and giving origin to nerve fibres; a nerve centre: **ganglionic**, a., *găng'·glĭ·ŏn'·ĭk*, relating to ganglia ; applied to collections of nucleated nerve cells which are centres of nervous power to the fibres connected with them ; in *bot.*, a swelling in the mycelium of some fungi : **gangliated**, a., *găng'·glĭ·āt·ĕd*, having ganglions; intertwined: **gangliform**, a., *găng'·glĭ·fŏrm* (L. *forma*, shape), having the form of a ganglion : **ganglioma**, n., *găng'·glĭ·ōm'·ă*, a glandular or ganglionic tumour : **ganglion intercaroticum**, *ĭn'·tĕr·kăr·ŏt'·ĭk·ŭm* (L. *inter*, between, amidst; new L. *carōticus*, carotid — from Gr. *karoō*, I throw into a deep sleep), a large ganglionic body placed on the inner side of the angle of division of the common carotid artery : **g. thyroideum**, *thŭr·ōŷd'·ĕ·ŭm* (new L. *thyroīdeus*, resembling the shape of an oblong shield—from Gr. *thureos*, a shield ; *eidos*, resemblance), the smallest of the cervical ganglia, placed on or near the inferior thyroid artery.

gangrene, n., *găng'·grēn* (L. *gangræna*, Gr. *ganggraina*, a gangrene

—from Gr. *graino*, I eat or gnaw), a condition of some soft part of a living body causing mortification and death of the part: **gangrenous**, a., *găng'·grēn·ŭs*, showing a tendency to gangrene; having the character of gangrene: **gangrena senilis**, *găng·grēn'·ă sĕn·il'·ĭs* (L. *senilis*, aged, senile), the gangrene which occurs in aged people from imperfect nutrition of a part, due to a diseased condition of the supplying bloodvessels.

ganoid, a., *găn'·ŏyd* (Gr. *ganos*, splendour; *eidos*, resemblance), applied to an Order of fishes, living and extinct, having angular scales, composed of horny or bony plates covered with a shining enamel: **Ganoidei**, n. plu., *găn·ŏyd'·ē·ī*, an Order of fishes.

gapes, n. plu., *gāpz* (AS. *geap*, wide; AS. *geapan*, Icel. *gapa*, to gape; Gael. *gab*, a mouth), a fatal disease among poultry and birds, in which they open their mouths wide and gasp for breath, caused by the presence of the parasite sclerostoma syngamus in large numbers in the trachea, or partially developed in the lungs.

Garcinia, n., *găr·sĭn'·ĭ·ă* (in honour of *Dr. Laurent Garcin*, a traveller), a valuable genus of fruit-bearing trees, Ord. Guttiferæ or Clusiaceæ: **Garcinia morella** or **pedicellata**, *mŏr·ĕl'·lă* or *pĕd'·ĭ·sĕl·lāt'·ă* (It. *morello*, dark, blackish; F. *morelle*, the night-shade; L. *pĕdĭcellus*, a foot-stalk), a diœcious tree, with laurel-like foliage and small yellow flowers, found in Camboja, Siam, etc., produces gamboge: **G. pictoria**, *pĭk·tōr'·ĭ·ă* (L. *pictōrĭus*, pictorial — from *pictor*, a painter); and **G. Travancorica**, *trăv'·ăn·kŏr'·ĭk·ă* (from *Travancore*, India), also furnish gamboge, which in commerce is received in the form of pipe, roll,

lump, or cake gamboge: **G. elliptica**, *ĕl·lĭp'·tĭk·ă* (L. *ellĭptĭcus*, oval—from *ellipsis*, an oval), a species producing a kind of gamboge, called 'coorg': **G. mangostana**, *măn'·gŏs·tān'·ă* (Malay *mangusta*), a tree which bears the mangosteen, an E. Indian fruit, one of the finest known, resembling a middle-sized orange, filled with a sweet and highly-flavoured pulp.

Gardenia, n., *găr·dēn'·ĭ·ă* (after *Dr. Garden* of Charleston, America), a splendid genus of plants, producing sweet-scented flowers of various colours, Ord. Rubiaceæ.

gargle, n., *gărg'·l* (F. *gargouiller*, to gargle, a word imitative of the sound produced; *gargareōn*, the throat), a liquid medicinal preparation, used for washing the mouth and throat: v., to wash the mouth and throat by gargling the liquid up and down in them.

Garryaceæ, n. plu., *găr'·rĭ·ā'·sĕ·ē* (after *Nicholas Garry*, of Hudson's Bay Company), a small Order, or rather Sub-order, of shrubs, with opposite leaves and pendulous amentaceous racemes of flowers, included in the Ord. Cornaceæ: **Garrya**, n., *găr'·rĭ·ă*, a genus of ornamental shrubs, similar in appearance to Viburnum, and a great botanical curiosity: **Garrya elliptica**, *ĕl·lĭp'·tĭk·ă* (L. *ellĭptĭcus*, oval — from *ellipsis*, an oval), a species which has unisexual flowers, and is prized for its peculiar silky catkins.

Gasteromycetes, n., *găst'·ĕr·ō·mĭ·sēt'·ēz* (Gr. *gastēr*, the belly; *mukēs*, a fungus), a division of the Fungi in which the hymenium is enclosed in a membrane, the spores being scattered over it in sets of four, as seen in puff-balls.

Gasteropoda, n. plu., *găst'·ĕr·ŏp'·ŏd·ă* (Gr. *gastēr*, the belly; *podes*, feet), an Order of molluscous animals which have their feet along the belly, or a ventral muscular disc adapted for creeping, as in the periwinkle.

Gasterothalameæ, n., *găst'·ĕr·ō·thăl·ām'·ĕ·ē* (Gr. *gastēr*, the belly; *thalămos*, a bed-chamber, a receptacle), a section of the Lichens having the shields always closed, or opening by bursting through the cortical layer of the thallus, the nucleus containing the deliquescing or shrivelled sporangia.

gastric, a., *găst'·rĭk* (Gr. *gastēr*, the belly or stomach), relating to the belly or stomach in man; popularly applied to certain forms of fever: **gastric juice**, the fluid in the stomach which acts as the principal agent in digestion: **gastritis**, n., *găst·rīt'·ĭs*, inflammation of the stomach: **gastro**, *găst'·rō*, signifying, related to, or connected with the stomach.

gastricolæ, n. plu., *găst·rĭk'·ŏl·ē* (Gr. *gastēr*, the belly; *colo*, I inhabit), intestinal parasites, being the larvæ of certain flies, found in the intestines of various animals.

gastro-cephalitis, *găst'·rō·sĕf'·ăl·īt'·ĭs* (Gr. *gastēr*, the belly; *kephalē*, the head), inflammation of the stomach with excitement of the brain and head.

gastrocnemius, n., *găst'·rŏk·nēm'·ĭ·ŭs* (Gr. *gastēr*, the belly; *knēmē*, the leg), the muscle or muscles which principally form the calf of the leg, and whose office it is to extend the foot; this muscle is also called 'gemellus superior.'

gastrodynia, n., *găst'·rō·dĭn'·ĭ·ă* (Gr. *gastēr*, the belly; *odunē*, pain), a painful affection of the stomach.

gastro - enteritis, *găst'·rō·ĕn'·tĕr·īt'·ĭs* (Gr. *gastēr*, the belly; *enteron*, an intestine), inflammation of the stomach and small intestines.

gastromalacia, n., *găst'·rō·măl·ā'·shĭ·ă* (Gr. *gastēr*, the belly; L.

malăcus, Gr. *malakos*, soft to the touch, tender), a softening of the stomach, due to the action of the gastric juice on the coats of the stomach after death.

gastrorrhœa, n., *găst'·rŏr·rē'·ă* (Gr. *gastēr*, the belly; *rheo*, I flow), catarrh of the stomach, attended with the discharge of abundant mucus.

gastro-splenic, a., *găst'·rō·splēn'·ĭk* (Gr. *gastēr*, the belly; *splēn*, the spleen), pert. to the stomach and spleen.

gastrula, n., *găst·rŏl'·ă* (a dim. formed from Gr. *gastēr*, the belly), a name applied to the developmental stage in various animals, in which the embryo consists of two fundamental membranes, an outer and an inner, enclosing a central cavity.

gattine, n., *găt'·tĭn* (a corruption of *catkin*; It. *gattino*, a kitten), a disease in silkworms caused by the fungus Cladosporium herbarum, so called from the dead caterpillars presenting the appearance of a kind of pastille, as the disease 'muscardine' has that of a little cake.

Gaultheria, n., *găŵl·thēr'·ĭ·ă* (after *Gaulthier*, a botanist of Canada), a genus of ornamental shrubs, Ord. Ericaceæ: **Gaultheria procumbens**, *prō·kŭm'·bĕnz* (L. *procumbens*, leaning forward, bending down); and G. **shallon**, *shăl'·lŏn* (name in Amer.), furnish succulent and grateful berries which yield a volatile oil.

gelatine, n., *jĕl'·ăt·ĭn* (F. *gélatine*, It. and Sp. *gelatina*, gelatine; L. *gelătum*, to congeal), the principle of jelly; animal jelly: **gelatinous**, a., *jĕl·ăt'·ĭn·ŭs*, resembling or consisting of jelly; having the consistence of jelly.

gelatio, n., *jĕl·ā'·shĭ·ŏ* (L. *gelătio*, frost), frostbite as it affects man: **gelation**, n., *jĕl·ā'·shŭn*, the rigid state of the body in catalepsy, as if frozen.

gemelli, n. plu., *jĕm·ĕl'.lĭ* (a dim. of L. *gemini*, twins), the names of two muscles, named respectively the 'gemellus superior' and 'gemellus inferior,' see 'gastrocnemius.'

geminate, a., *jĕm'.ĭn·āt* (L. *gemini*, twins ; *geminātus*, doubled), in *bot.*, growing in pairs ; same as 'binate.'

gemma, n., *jĕm'.mă*, gemmæ, n. plu., *jĕm'.mē* (L. *gemma*, a bud ; *gemmæ*, buds), in *bot.*, a bud ; leaf buds as distinguished from flower buds ; reproductive buds found in liverworts ; the buds produced by any animal, whether detached or not : gemmate, a., *jĕm'.māt*, having buds : gemmation, n., *jĕm·mā'-shŭn*, the development of leaf buds ; in *zool.*, the process of producing new structures by budding : gemmiferous, a., *jĕm·mĭf'·ĕr·ŭs* (L. *fero*, I bear), bearing buds : gemmiform, a., *jĕm'·mĭ·fŏrm* (L. *forma*, shape), shaped liked a bud : gemmiparous, a., *jĕm·mĭp'·ăr·ŭs* (L. *pario*, I produce), in *bot.*, reproducing by buds ; in *zool.*, giving origin to new structures by a process of budding.

gemmule, n., *jĕm'.mūl* (L. *gemmŭla*, a little bud—from *gemma*, a bud), in *bot.*, the first bud of the embryo ; same as 'plumule' ; in *zool.*, the ciliated embryos of many Cœlenterata ; the seed-like reproductive bodies or spores of Spongilla.

genera, see 'genus.'

generation, n., *jĕn'.ĕr·ā'·shŭn* (L. *generātum*, to beget, to engender —from *genus*, a race or kind), production ; formation : generative, a., *jĕn'.ĕr·āt·ĭv*, that generates or produces.

generic, pert. to a 'genus.'

genestade, n., *jĕn'.ĕs·tād* (a corruption of L. *genista* or *genesta*, the broom plant), an enzootic disease of cattle, sthenic hæmaturia, due principally to the astringent principle of the plants and young trees which animals eat ; so called in France as due to animals eating the plant Genistica Hispanica, *hĭs·păn'·ĭk·ă* (of or from *Hispania* or Spain) : Genista, n., *jĕn·ĭst'.ă* (called in F. *plantegenêt*), a plant from which the Plantagenets took their name: Genista tinctoria, *tĭngk·tōr'·ĭ·ă* (L. *tinctōrius*, belonging to dyeing —from *tingo*, I dye), a dye-plant, formerly known as 'dyer's greenweed' or 'dyer's broom.'

genial, a., *jĕn'.ĭ·ăl* (Gr. *geneion*, the chin), of or belonging to the chin : genio, *jĕn'.ĭ·ō*, signifying in compounds a connection with the jaw : genio-hyo-glossus, *hĭ'.ō-glŏs'·sŭs* (Gr. letter *v*, upsilon, or Eng. *u* ; *glossa*, the tongue), a thin, flat, triangular muscle, so named from its triple attachment to the jaw, the hyoid bone, and tongue : genio-hyoid, *hĭ'.ōyd*, (Gr. *eidos*, resemblance), a narrow, slender muscle situated immediately beneath the inner border of the mylo-hyoid.

genitals, n. plu., *jĕn'.ĭt·ăls* (L. *genitālis*, serving to beget—from *gigno*, I beget), the parts of an animal which are the immediate instruments of generation.

genito-crural, a., *jĕn'.ĭt·ō-krŏr'·ăl* (L. *genitālis*, serving to beget— from *gigno*, I beget), applied to a nerve which belongs partly to the external genital organs, and partly to the thigh : genito indicates connection with the genital organs.

Gentianaceæ, n. plu., *jĕn'.shĭ·ăn·ā'·sĕ·ē* (after *Gentius*, the anc. king of Illyria who first proved its virtues), the Gentian family, a well-known Order of plants, principally herbaceous, distributed over nearly every part of the world, prized for their beauty ; many exhibit great variety of colours, have a bitterness in their roots, leaves, and flowers, used as tonics : **Gentianeæ**, n. plu., *jĕn'-shĭ-ăn'-ĕ-ē*, a tribe of the Order : **Gentiana**, n., *jĕn'-shĭ-ān'-ă*, an extremely beautiful genus of plants : **Gentiana lutea**, *lōōt'-ĕ-ă* (L. *lūtum*, a plant used in dyeing yellow), a species whose root is principally used in medicine, found at a high elevation on Pyrenees and Alps, produces showy yellow flowers, and the root yellow internally : **G. punctata**, *pŭngk-tāt'-ă* (L. *punctātus*, punctured — from *punctum*, a puncture, a sting) ; **G. purpurea**, *pĕr-pūr'-ĕ-ă* (L. *purpŭreus*, purple-coloured—from *purpŭra*, a purple colour) ; and **G. Pannonica**, *păn-nŏn'-ĭk-ă* (L. *Pannŏnĭa*, an anc. country of Turkey), are species whose roots are often mixed with the root of G. lutea : **G. kurroo**, *kŭr-rŏ'* (a native name), a species of the Himalayas having similar properties : **G. campestris**, *kăm-pĕst'-rĭs* (L. *campestris*, pert. to a level field—from *campus*, a flat field) ; and **G. amarella**, *ăm'-ăr-ĕl'-lă* (a dim. of L. *amārus*, bitter), British species which have also been used as bitter tonics.

genu, n., *jĕn'-ū* (L. *gĕnū*, the knee), the knee or bend of the corpus callosum.

genus, n., *jĕn'-ŭs*, **genera**, n. plu., *jĕn'-ĕr-ă* (L. *gĕnŭs*, birth, race, *gĕnĕris*, of a race), that which has several species under it ; a group next lower to an Order ; a Species is one of the group called a Genus, while accidental differences in species give rise to Varieties ;—we have accordingly in natural history, the Order, the Genus, the Species, the Variety, while to indicate minor differences we have often groups called Suborders, Sub-genera, Sub-species, and Sub-varieties ; see ' Species.'

Geoffroya, n., *jĕf-frŏy'-ă* (after M. *Geoffroy*, author of *Materia Med-*

M

ica, died 1731), a genus of trees, Ord. Leguminosæ, Sub-ord. Papilionaceæ : **Geoffroya superba**, *sŭ-pĕrb'-ă* (L. *superbus*, excellent, splendid), a species whose fruit, called Umari, is much used by the inhabitants of Brazil, etc.

geotropism, n., *jē-ŏt'-rŏp-ĭzm* (Gr. *gĕă*, the earth ; *tropē*, a turning, a change), in *bot.*, the influence of gravitation on growth.

Gephyrea, n. plu., *jĕf-ĭr'-ĕ-ă* (Gr. *gephūra*, a mound or dyke), a class of the Anarthropoda, comprising the spoon-worms and their allies.

Geraniaceæ, n. plu., *jĕr-ăn'-ĭ-ā'-sĕ-ē* (Gr. *gerănos*, a crane, in allusion to the long beak-like prolongation of the axis), the Cranesbill family, an Order of plants which are astringent and aromatic : **Geranium**, n., *jĕr-ăn'-ĭ-ŭm*, a genus, some of whose species produce very handsome flowers: **Geranium maculatum**, *măk'-ūl-āt'-ŭm* (L. *maculātus*, spotted, speckled— from *măcŭla*, a spot), a species whose root is called ' alum root,' from its being a very powerful astringent : **G. oblongatum**, *ŏb'-lŏng-gāt'-ŭm* (L. *oblongus*, rather long, oblong), the yellow geranium, whose root-stock is used by the natives of Namaqualand, S. Africa, as an article of food ; **G. Robertianum**, *rŏb-ĕrt'-ĭ-ăn'-ŭm* (from *Robert*, proper name), a species used in N. Wales in nephritic complaints.

germ, n., *jĕrm* (L. *germen*, the bud of a tree, a young twig), that from which anything springs; the rudiment of an undeveloped new being: **germ-cells**, the cells or nuclei which contain active germinal matter or protoplasm ; **germ-mass**, the germinal matter or protoplasm ; the materials prepared for the future formation of the embryo : **germen**, n., *jĕrm'-ĕn*, in *bot.*, a name for the ovary : **germinal**, a., *jĕrm'-*

ĭn·ăl, pert. to a germ : **germinal vesicle**, in *bot.* and *zool.*, a cell contained in the embryo sac from which the embryo is developed ; the small vesicular body within the ovum or the yolk of the egg : **germination**, n., *jĕrm'ĭn·ā'shŭn*, the first beginning of vegetation in seed ; the first act of growth.

Gesneraceæ, n. plu., *gĕs'·nĕr·ā'sĕ·ē* (after the botanist *Gesner*, of Zürich), the Gesnera family, an extensive Order of little, soft-wooded herbs or shrubs, generally possessing considerable beauty, natives chiefly of the warmer regions of America : **Gesnera**, n., *gĕs'·nĕr·ŏ*, a genus whose species are very handsome plants.

gestation, n., *gĕst·ā'shŭn* (L. *gest-ātio*, a bearing or carrying—from *gesto*, I bear or carry), the period during which females carry the embryo in the womb from conception to delivery ; the state of pregnancy.

Geum, *gē'ŭm* (Gr. *geuo*, I give to taste, I entertain), a genus of ornamental plants, Ord. Rosaceæ, distinguished by astringent and tonic qualities : **Geum urbanum**, *ĕrb·ān'ŭm* (L. *urbānus*, belonging to the city or town—from *urbs*, a city), the common and water avens; and **G. rivale**, *rĭv·āl'ĕ* (L. *rīvālis*, belonging to a brook—from *rīvus*, a small stream), have been employed as tonics and astringents, and for efficacy have been compared to Cinchona : **G. coccineum**, *kŏk·sĭn'ĕ·ŭm* (L. *coccīnĕus*, of a scarlet colour—from *coccum*, the berry of the scarlet oak), an extremely handsome species.

gibber, n., *gĭb'bĕr* (L. *gibber*, crook-backed, hunch-backed; *gibbus*, hunched, humped), in *bot.*, a pouch at the base of a floral envelope : **gibbosity**, n., *gĭb·bŏs'ĭ·tĭ*, a round or swelling prominence ; in *bot.*, a swelling at the base of an organ : **gibbous**, a.,

gĭb'bŭs, swollen at the base ; having a distinct swelling at some part of the surface.

gid, n., *gĭd* (a corruption of *giddy*, unsteady, alluding to their tottering gait ; Norse *gidda*, to shake, to tremble), the disease called 'sturdy' among sheep, caused by parasites on the brain, viz. the Cœnurus cerebralis.

Gilliesiaceæ, n. plu., *gĭl·lĭz'ĭ·ā'sĕ·ē* (after *Dr. Gillies*, of Chili), the Gilliesia family, an Order of herbs with tunicated bulbs, grass-like leaves, and umbellate, spathaceous flowers : **Gilliesia**, n., *gĭl·lĭz'ĭ·ă*, a genus of the Order.

gills, n. plu., *gĭlz* (AS. *geaflas*, the chaps, the jaws ; Swed. *gel*, a jaw, the gill of a fish), the organs of breathing in fishes, forming reddish fibrous flaps, or fringe-like processes, placed on both sides of the head ; in *bot.*, the thin vertical plates on the under side of the cap of certain Fungi.

ginger, n., *jĭnj'ĕr* (F. *gingembre*, L. *zingiber*, ginger), the underground stem or rhizome of the Indian plant Zingiber officinale, also named Amomum zingiber, Ord. Zingiberaceæ, used as an aromatic stimulant.

gingivæ, n. plu., *jĭn·jĭv'ē* (L. *gingiva*, a gum), the gums ; a dense fibrous tissue, very closely connected with the periosteum of the alveolar processes, and covered by a red mucous membrane : **gingivitis**, n., *jĭn'jĭv·ĭt'ĭs*, inflammation of the gums.

ginglymus, n., *gĭng'glĭm·ŭs* (Gr. *gingglumos*, the hinge of a door, a joint), in *anat.*, a joint which allows motion in two directions only, as the joint of the elbow and the lower jaw : **ginglymoid**, a., *gĭng'glĭm·ŏyd* (Gr. *eidos*, resemblance), resembling a hinge.

gizzard, n., *gĭz'zĕrd* (F. *gésier*, a gizzard; Prov. F. *grezie*, a gizzard—from *gres*, gravel), the strong muscular division of the stomach in fowls, birds, and insects; the crop.

glabella, n., *glăb·ĕl'·lă* (L. *glăbellus*, without hair, smooth—dim. from *glăber*, smooth), in *anat.*, the triangular space between the eyebrows; the nasal eminence lying between the superciliary ridges.

glabrous, a., *glăb'·rŭs* (L. *glăber*, without hair, smooth), in *bot.*, smooth; devoid of hair.

glacial, a., *glā'·shi·ăl* (L. *glacies*, ice), consisting of ice; frozen: **glacial acetic acid,** the strongest acetic acid, so named from its crystallizing in ice-like leaflets at the ordinary temperature of 55°.

Gladiolus, n., *glăd·i'·ŏl·ŭs* (L. *glădiŏlus*, a small sword—from *glădĭus*, a sword), an extensive genus of plants, consisting chiefly of beautifully flowering bulbs from the Cape of Good Hope, Ord. Iridaceæ; in *anat.*, the second piece of the sternum, considerably longer, narrower, and thinner than the first piece.

gladius, n., *glăd'·i·ŭs* (L. *glădĭus*, a sword), the horny endoskeleton or pen of certain cuttle-fishes: **gladiate,** a., *glăd'·i·āt*, in *bot.*, shaped like a short, straight sword.

glair, n., *glār* (F. *glaire*, white of an egg; Scot. *glair* or *glaur*, mud or slime), the white of a raw egg; any viscous transparent substance resembling it: **glairy,** a., *glār'·i*, like glair; slimy.

gland, n., *glănd* (L. *glans*, an acorn, *glandis*, of an acorn), organs of manifold forms and structure which perform the functions of secretion, or when ductless are believed to modify the composition of the blood, found in all parts of the body; a similar combination of ducts or vessels in plants; an organ of secretion in plants consisting of cells, generally on the epidermis; wart-like swelling on

plants: **glans,** n., *glănz*, in *bot.*, the acorn or hazel nut, and such like, which are enclosed in bracts; the nut-like extremity of the penis.

glanders, n. plu., *glănd'·ĕrz* (old F. *glandre*, a swelling of the glands; L. *glans*, an acorn), a malignant contagious disease of equine animals capable of being conveyed to man, which primarily affects the mucous membranes of the nose, and is accompanied by a starchy or gluey (fibrinous) discharge.

glandule, n., *glănd'·ŭl*, also **glandula,** n., *glănd'·ŭl·ă* (L. *glandulæ*, the glands of the throat, dim. of *glans*, an acorn), a small gland or secreting vessel: **glandular,** a., *glănd'·ŭl·ăr*, consisting of or pert. to glands; in *bot.*, applied to hairs in plants having glands on their tips: **glandulæ ceruminosæ,** *glănd'·ŭl·ē sĕr·ŏm'·ĭn·ōz'·ē* (new L. *cerūmen*, the wax secreted by the ear—from *cēra*, wax), the ceruminous glands; the numerous small glands or follicles which secrete the ear-wax.

glaucium, n., *glăws'·i·ŭm* (Gr. *glaukos*, a colour between green and blue), a genus of very pretty plants, Ord. Papaveraceæ.

glaucoma, n., *glăwk·ōm'·ă* (Gr. *glaukōma*, a certain disease of the eye—from *glaukos*, blue-grey or sea-green; L. *glaucōma*, an obscuration of the crystalline lens—from *glaucus*, bluish-grey), a disease of the eye giving to it a bluish or greenish tinge: **glaucous,** a., *glăwk'·ŭs*, of a sea-green colour; in *bot.*, covered or frosted with a pale-green bloom: **glaucescent,** a., *glăws·ĕs'·sĕnt*, having a bluish-green or sea-green appearance.

Glaux, n., *glăwks* (Gr. *glaukos*, blue-grey or sea-green), a very pretty genus of plants, so called in allusion to the colour of the leaves, Ord. Primulaceæ: **Glaux**

maritima, *măr·ĭt'ĭm·ă* (L. *marĭt-ĭmus*, maritime—from *măre*, the sea), a species having the corolla abortive, and the calyx coloured.

gleba, n., *glēb'ă* (L. *glēba*, a lump of earth, a clod ; *glēbula*, a small clod), the spore-forming apparatus of Phalloideæ : **glebulæ**, n. plu., *glĕb'ul·ē*, crumb-like masses.

gleet, n., *glēt* (F. *glette*, the froth of an egg ; low Ger. *glett*, slippery), a slimy or glairy discharge from a wound ; the thin humour as the result of gonorrhœal disease.

Gleichenieæ, n. plu., *glīk'ĕn·ĭ'ĕ·ē* (after *Baron Gleichen*, a German botanist), a Sub-order of the Ord. Filices, having the sori dorsal, and the sporangia opening vertically : **Gleichenia**, n., *glīk·ĕn'ĭ·ă*, a genus of pretty ferns.

glenoid, a., *glĕn'ŏyd* (Gr. *glēnē*, the pupil of the eye, a socket ; *eidos*, resemblance), in *anat.*, applied to a part having a shallow cavity, as the socket of the shoulder joint : **glenoid fossa**, *fŏs'să* (L. *fossa*, a ditch), the socket of the shoulder joint : **glene**, n., *glēn'ē*, the hollow part of a bone ; a socket.

glioma, n., *glī·ōm'ă*, **gliomata**, n. plu., *glī·ōm'ăt·ă* (Gr. *glia*, glue), a tumour peculiar to the brain and similar nervous structures, generally the former ; a tumour very nearly allied to the sarcoma, consisting of primitive cells resembling those of the insterstitial substance of nervous structure.

globate, a., *glōb'āt* (L. *globātus*, made round—from *globus*, a ball), globe-shaped : **globoids**, n. plu., *glōb'ŏyds* (Gr. *eidos*, resemblance), non-crystalline, clustered granules enclosed in grains of aleuron : **globose**, a., *glōb·ōz'* (L. *globōsus*, round as a ball), having the form of a ball ; spherical : **globosity**, n., *glōb·ŏs'ĭt·ĭ*, the quality of being round.

Globularia, n., *glōb'·ul·ăr'·ĭ·ă* (L.

globulus, a little ball—from *glŏbus*, a ball), a very handsome genus of plants, Ord. Verbenaceæ, so named from the production of the flowers in globose heads.

globule, n., *glŏb'·ul* (L. *glŏbŭlus*, a little globe—from *glŏbus*, a globe), a very minute particle of matter in a round form ; in *bot.*, the male organ of the Chara : **globulin**, n., *glŏb'·ul·ĭn*, the albuminous matter which forms the principal part of the blood corpuscles ; in *bot.*, the round, transparent granules formed in the cellular tissue, which constitute fæcula : **globulus**, n., *glŏb'·ul·ŭs*, the round deciduous shield of some lichens.

globus hystericus, *glŏb'·ŭs hĭst·ĕr'·ĭk·ŭs* (L. *glŏbus*, a ball ; *hystĕr-ĭcus*, Gr. *husterikos*, pert. to the womb, hysterical—from *hustĕra*, the womb), in hysteria, the sensation of a ball rising up in the chest and throat ; the hysterical ball.

glochidiate, a., *glō·kĭd'·ĭ·āt* (Gr. *glōchis*, the angular end of anything, as of an arrow—from *glōx*, the awn or beard of grain), in *bot.*, applied to hairs on plants, the divisions of which are barbed like a fish hook.

glomerate, a., *glŏm'·ĕr·āt* (L. *glomerātus*, gathered into a round heap—from *glomus*, a ball or clew of thread), gathered into round heaps or heads.

glomerule, n., *glŏm'·ĕr·ul*, also **glomerulus**, n., *glŏm·ĕr'·ul·ŭs* (dim. of L. *glomus*, a ball or clew of thread), a head or dense cluster of flowers ; the powdering leaf lying on the thallus of lichens : **glomerulus**, n., **glomeruli**, n. plu., *glŏm·ĕr'·ul·ī*, granulous cells, being the result of the transformation of other cells, either of normal or pathological formation, as in the case of mucus or pus corpuscles ; in *bot.*, powdery masses on the surfaces of some lichens: **glomeruliferous**, a., *glŏm·ĕr′·ŭl·ĭf′·ĕr·ŭs* (L. *fero*, I bear), in *bot.*, bearing clusters of minutely - branched, coral - like excrescences.

glossanthrax, n., *glŏs′·săn·thrăks* (Gr. *glossa*, the tongue; *anthrax*, burning coal), among cattle, a disease characterised by a development of malignant carbuncle in the mouth, especially on the tongue.

glosso, *glŏs′·sō* (Gr. *glossa*, the tongue), a prefix in compounds denoting 'attachment to or connection with the tongue': **glossopharyngeal**, *făr′·ĭn·jē′·ăl* (Gr. *pharungx*, the gullet), a nerve connected with the tongue and pharynx : **glossitis**, n., *glŏs·sīt′·ĭs*, inflammation of the tongue : **glossoid**, a., *glŏs′·ōȳd* (Gr. *eidos*, resemblance), resembling the tongue : **glossology**, n., *glŏs·sŏl′·ŏ·jĭ* (Gr. *logos*, discourse), the explanation of the special terms used in any science : **glottis**, n., *glŏt′·tĭs* (Gr. *glotta* or *glossa*, the tongue), the narrow opening at the upper part of the windpipe at the back of the tongue : **glottitis**, n., *glŏt·tīt′·ĭs*, inflammation of the glottis.

Gloxinia, n., *glŏks·ĭn′·ĭ·ă* (after the botanist *Gloxin*, of Colmar), a splendid genus of plants, worthy of extensive cultivation, Ord. Gesneraceæ.

glucose, n., *glō·kōz′* (Gr. *glukus*, sweet), grape sugar ; the peculiar form of sugar which exists in grapes and other fruits ; also found in animals, as in the blood; also excreted by the urine in Diabetes mellitus.

glume, n., *glŏm* (L. *gluma*, the husk of corn), the husk of corn or grasses formed of flaps or valves embracing the seed : **glumaceous**, a., *glŏm·ā′·shŭs*, resembling the dry, scale-like glumes of grasses : **glumiferous**,

a., *glŏm·ĭf′·ĕr·ŭs* (L. *fero*, I bear), bearing or producing glumes : **glumelle**, n., *glŏm·ĕl′*, or glumellule, n., *glŏm·ĕl′·ŭl*, the inner husk of the flowers of grasses ; the palea or fertile glume of a grass : **glumellæ**, n. plu., *glŏm·ĕl′·lē*, a plural used to denote the paleæ or fertile glumes of grasses.

gluten, n., *glŏt′·ĕn*, also **glutin**, n., *glŏt′·ĭn* (L. *gluten*, paste or glue), a tough substance obtained from wheat and other grains : **glutenoid**, a., *glŏt′·ĕn·ōȳd* (Gr. *eidos*, resemblance), resembling gluten or allied to it.

gluteus, n., *glŏt·ē′·ŭs* (Gr. *gloutos*, the buttock or hip), one of the three large muscles which form the seat: they are named respectively **gluteus maximus**, *măks′·ĭm·ŭs* (L. *maximus*, the greatest), which extends the thigh, and is the largest ; **g. medius**, *mēd′·ĭ·ŭs* (L. *medius*, the middle), which acts when we stand, and is the second in size ; and **g. minimus**, *mĭn′·ĭm·ŭs* (L. *minimus*, the least), which assists the other two, and is the third in size : **gluteal**, a., *glŏt·ē′·ăl*, pert. to the buttocks.

glycerine, n., *glĭs′·ĕr·ĭn* (Gr. *glukus*, sweet), the sweet principle of oils and fats : **glyceric acid**, *glĭs·ĕr′·ĭk*, an acid produced by the action of nitric acid on glycerine : **nitro - glycerine**, *nĭt′·rō·glĭs′·ĕr·ĭn*, a powerful blasting oil, and very dangerous explosive agent, prepared by the action of nitric and sulphuric acids on glycerine: **glycogen**, n., *glĭk′·ō·jĕn* (Gr. *gennao*, I produce), a substance formed by the liver, and capable of being converted into grape sugar, or into glucose.

glycocholic, a., *glīk′·ō·kŏl′·ĭk* (Gr. *glukus*, sweet ; *cholē*, bile), denoting an acid obtained from the bile of the ox and other animals.

Glycyrrhiza, n., *glĭs′·ĕr·rīz′·ă* (Gr. *glŭkus*, sweet ; *rhiza*, a root), a genus of plants, Ord. Leguminosæ, Sub-ord. Papilionaceæ, the sweet, sub-acrid, and mucilaginous juice of whose roots is used as a pectoral : **Glycyrrhiza glabra**, *glăb'·ră* (L. *glăber*, without hair, smooth), the plant which yields liquorice root, used medicinally as a demulcent : G. **echinata**, *ĕk'·ĭn·ăt'·ă* (L. *echinātus*, set with prickles, prickly) ; and G. **glandulifera**, *glănd'·ŭl·ĭf'·ĕr·ă* (L. *glandulæ*, glands ; *fero*, I bear), also possess a sweetness in their roots and leaves : **glycyrrhizin**, n., *glĭs'·ĕr·rĭz'·ĭn*, or **glycion**, n., *glĭs'·ĭ·ŏn*, the peculiar sweet principle in the roots and leaves of the Glycyrrhiza, and other papilionaceous plants ; liquorice-sugar.

gnathic, a., *năth'·ĭk* (Gr. *gnathos*, the cheek or jaw bone), belonging to the cheek or superior maxilla : **gnathites**, n. plu., *năth·ĭt'·ēz*, in *zool.*, the masticatory organs of the Crustacea : **gnathitis**, n., *năth·ĭt'·ĭs*, inflammation of the jaw : **gnatho**, *năth'·ō*, a prefix in compounds indicating connection with the jaw.

gnaurs, n. plu., *năwrs*, better spelling **gnar** or **gnarr**, *năr* (Dut. *knarren*, to growl; Swed. *knarra*, to creak; *knorla*, to twist or curl), excrescences or warts on the stem of a tree.

Gnetaceæ, n. plu., *nĕt·ā'·sĕ·ē* (from *gnemon*, a native name), the Joint-firs, an Order of small trees or shrubs, some species bearing eatable, fleshy fruit : **Gnetum**, n., *nēt'·ŭm*, a genus whose seeds in India are cooked and eaten, and the green leaves are used as spinach.

gnomonicus, n., *nŏm·ŏn'·ĭk·ŭs* (Gr. *gnōmon*, the pin or style of a dial), in *bot.*, applied to a stalk which is bent at right angles.

goitre, n., *goÿt'·ĕr* (F. *goitre*, a wen), a large tumour or swelling on the fore part of the neck,

prevalent chiefly in Alpine districts.

Gomphocarpus, n., *gŏm'·fō·kârp'·ŭs* (Gr. *gomphos*, a peg, a club ; *karpos*, fruit), a pretty Cape genus of plants, Ord. Asclepiadaceæ : **Gomphocarpus fruticosus**, *frŏt'·ĭ·kōz'·ŭs* (L. *frŭtĭcōsus*, shrubby—from *frutex*, a shrub), the silk plant of Madeira.

Gompholobium, n., *gŏm'·fō·lōb'·ĭ·ŭm* (Gr. *gomphos*, a peg, a club ; *lobos*, a pod), a splendid genus of New Holland plants, Ord. Leguminosæ, Sub-ord. Papilionaceæ, having club- or wedge-shaped pods : **Gompholobium uncinatum**, *ŭn'·sĭn·āt'·ŭm* (L. *uncĭnātus*, furnished with hooks — from *uncus*, a hook), a species which has poisoned sheep in Swan River Colony.

gomphosis, n., *gŏm·fōz'·ĭs* (Gr. *gomphos*, a peg, a wedge), in *anat.*, a form of joint in which a conical body is fastened into a socket, as the teeth in the jaw.

Gomphrena, n., *gŏm·frēn'·ă* (Gr. *gomphos*, a club, from the shape of the flowers), a genus of plants having round heads of purple and white flowers, Ord. Amaranthaceæ : **Gomphrena globosa**, *glōb·ōz'·ă* (L. *glŏbōsus*, round—from *glŏbus*, a ball, a globe), the Globe amaranth.

gonangium, n., *gŏn·ăn'·jĭ·ŭm* (Gr. *gonos*, offspring ; *anggeion*, a vessel), the chitinous receptacle in which the reproductive buds of certain of the Hydrozoa are produced.

gongylus, n., *gŏng'·jĭl·ŭs*, **gongyli**, n. plu., *gŏng'·jĭl·ī* (Gr. *gonggulos*, round), in *bot.*, round, hard bodies produced on certain Algæ, which become ultimately detached, and germinate ; same as 'gonidia.'

gonidia, n. plu., *gŏn·ĭd'·ĭ·ă* (Gr. *gonos*, offspring, seed ; *eidos*, resemblance), green germinating cells in the thallus of lichens immediately beneath the surface.

gonoblastidia, n. plu., *gŏn′·ō·blăst·ĭd′·ĭ·ă* (Gr. *gonos*, offspring ; *blastidion*, a dim. of *blastos*, a bud), the processes which carry the reproductive receptacles or 'gonophores,' in many of the Hydrozoa.

gonocalyx, n., *gŏn′·ō·kăl′·ĭks* (Gr. *gonos*, offspring ; *kalux*, a cup), the swimming-bell in a medusiform gonophore; the same structure in a gonophore which is not detached.

gonophore, n., *gŏn′·ō·fōr* (Gr. *gonos*, offspring ; *phoreo*, I bear, I carry), in *bot.*, an elevated or elongated receptacle bearing the stamens and carpels in a prominent and conspicuous manner ; in *zool.*, the generative buds or receptacles of the reproductive elements in the Hydrozoa, whether these become detached or not.

gonorrhea, n., *gŏn′·ŏr·rē′·ă* (Gr. *gonorrhoia*, a gonorrhea—from *gonē*, semen ; *rheo*, I flow), the discharge of a purulent or mucopurulent fluid from the inflamed mucous membranes of the generative organs, the result of infection, and highly contagious ; urethritis : **gonorrheal**, a., *gŏn′·ŏr·rē′·ăl*, pert. to : **gonorrheal ophthalmia**, inflammation of the eye from the contact of gonorrheal matter.

gonosome, n., *gŏn′·ō·sōm* (Gr. *gonos*, offspring ; *sōma*, body), a term applied to the reproductive zooids of a hydrozoön.

gonotheca, n., *gŏn′·ō·thēk′·ă* (Gr. *gonos*, offspring ; *thēkē*, a chest, a case), the chitinous receptacle within which the gonophores of certain of the Hydrozoa are produced.

gonus, *gŏn′·ŭs* (Gr. *gŏnu*, the knee), and **gonum**, *gŏn′·ŭm* (Gr. *gōnia*, a corner or angle), in *bot.*, words in composition signifying either 'kneed' or 'angled,'—the *o* short when the former, and long when the latter ;—*polygŏnum*, many-kneed ; *tetragōnum*, four-angled.

Goodeniaceæ, n. plu., *gōōd′·ĕn·ĭ·ā′·sĕ·ē* (in honour of *Dr. Goodenough*, Bishop of Carlisle), the Goodenia family, an Order mostly of herbaceous plants, of Australia and S. Sea Islands : **Goodenia**, n., *gōōd·ĕn′·ĭ·ă*, a very elegant genus of plants : **Goodenieæ**, n. plu., *gōōd′·ĕn·ĭ′·ĕ·ē*, a Sub-order.

Gossypium, n., *gŏs·sĭp′·ĭ·ŭm* (L. *gossypion*, the cotton tree—said to be from Ar. *goz* or *gothn*, a soft substance), a highly valuable genus of plants comprising the various species of cotton plants, cotton being nothing more than the collection of hairs which surround the seeds, Ord. Malvaceæ : **Gossypium Barbadense**, *bărb′·ă·dĕns′·ĕ* (of or from *Barbadoes*), the species which yields the best cotton, the Sea Island, New Orleans, and Georgian cotton : **G. Peruvianum**, *pĕr·ŏv′·ĭ·ān′·ŭm* (new L. *Peruviănus*, of or from Peru); and **G. acuminatum**, *ăk·ūm′·ĭn·āt′·ŭm* (L. *acūmĭnātum*, pointed, sharpened—from *acūmen*, a point), species which furnish the S. American cotton: **G. herbaceum**, *hĕrb·ā′·sĕ·ŭm* (L. *herbāceus*, grassy—from *herba*, grass), the common cotton of India; the Chinese Nankin cotton : **G. arboreum**, *ăr·bŏr′·ĕ·ŭm* (L. *arbŏreus*, pert. to a tree — from *arbor*, a tree), the Indian tree cotton.

gout, n., *gŏwt* (L. *gutta*, a drop—from the old medical theory which attributed all disorders to the settling of a drop of morbid humour upon the part affected ; F. *goutte*, a drop, the gout), a well-known painful disease of the joints.

gracilis, n., *grăs′·ĭl·ĭs* (L. *gracilis*, slender), the name of a long, thin, flat muscle of the thigh, which assists the 'sartorius.'

grain, n., *grān* (L. *granum*, grain of corn ; F. *grain*), the fruit of cereal grasses ; the smallest weight, so named as supposed to be of equal weight with a grain of corn : **grains of Paradise**, the seeds of 'Amomum malegueta.'

Grallatores, n. plu., *grăl·lă·tŏr'·ēz* (L. *grallator*, he that goes on stilts—from *grallæ*, stilts), the Order of the long-legged wading birds.

Gramineæ, n. plu., *grăm·ĭn'·ĕ·ē*, also **Graminaceæ**, *grăm'·ĭn·ā'·sĕ·ē* (L. *grāmĭnĕus*, pert. to grass—from *grāmen*, grass), the Grass family, an Order of plants forming the most important in the vegetable kingdom, as furnishing the chief supply of food for man in the cereals, etc., and herbage for animals : **graminaceous**, a., *grăm'·ĭn·ā'·shŭs*, pert. to grass ; like grass : **graminivorous**, a., *grăm'·ĭn·ĭv'·ŏr·ŭs* (L. *voro*, I devour), feeding or subsisting on grass.

Granatum, n., *grăn·āt'·ŭm* (L. *granum*, a grain), the pomegranate, so called because full of seed; see 'Punica granatum.'

granivorous, a., *grăn·ĭv'·ŏr·ŭs* (L. *granum*, grain ; *voro*, I devour), living upon grains or other seeds.

granule, n., *grăn'·ūl*, **granules**, n. plu., *grăn'·ūls* (dim. from L. *grānum*, a grain), minute particles of matter, either organic or inorganic ; in *bot.*, minute bodies varying greatly in size, having distinct, external, shadowed rings or margins, the external edges of which are abrupt : **granular**, a., *grăn'·ūl·ėr*, also **granulose**, a., *grăn'·ūl·ōz*, consisting of grains or granules; resembling granules: **granulated**, a., *grăn'·ūl·āt·ĕd*, roughish on the surface ; composed of granules : **granulations**, n., *grăn'·ūl·ā'·shŭns*, the small,

soft nodules of a florid red colour which appear on the surface of healthy healing wounds or ulcers: **granula-gonima**, *grăn'·ŭl·ă·gŏn'·ĭm·ă* (L. *granula*, a little grain ; Gr. *gonimos*, having the power of generating), clusters of spherical cells filled with green granular matter, seated beneath the cortical layer in lichens.

Gratiola, n., *grăt·ĭ'·ŏl·ă* (L. *grātia*, grace, favour, the grace of God—from their supposed medicinal virtues), a genus of pretty free - flowering plants, Ord. Scrophulariaceæ : **Gratiola officinalis**, *ŏf·fĭs'·ĭn·āl'·ĭs* (L. *officinalis*, officinal), the plant hedgehyssop, bitter and acrid, formerly called 'Gratia Dei,' the grace of God, from its efficacy as a medicine.

gravel, n., *grăv'·ĕl* (It. *gravella*, F. *gravelle*, sand), small stony concretions formed in the kidneys, which, when passed, form a gravelly kind of sediment in the urine ; the disease thus caused.

gravid, a., *grăv'·ĭd* (L. *gravidus*, pregnant—from *grăvis*, heavy), pregnant; heavy or great with child.

grease, n., *grēs* (F. *graisse*, It. *grascia*, grease), a disease in horses, consisting of inflammation of the skin at the back of the fetlock and heels, on which pustules form, yielding a fetid, purulent discharge.

Gregarina, n., *grĕg'·ăr·ĭn'·ă* (L. *gregarius*, belonging to a herd or flock—from *grex*, a flock), one of the **Gregarinidæ**, *grĕg'·ăr·ĭn'·ĭd·ē*, a class of the Protozoä.

Grevillea, n., *grĕv·ĭl'·lĕ·ă* (after *Greville*, a patron of botany), a handsome genus of New Holland plants, Ord. Proteaceæ: **Grevillea robusta**, *rō·bŭst'·ă* (L. *robustus*, oaken—from *rōbur*, an oak tree), the silver oak.

Grewia, n., *grō'·ĭ·ă* (in honour of *Dr. Grew*, the botanist), a genus of plants, Ord. Filiaceæ, having elm-looking leaves : **Grewia** mic-rocos, *mĭk'·rōk·ŏs* (Gr. *mikros*, little ; *kosmos*, a world); and **G. Asiatica,** *āzh'·ĭ·ăt'·ĭk·ă* (from *Asia*), species whose fruits are agreeable, and are largely employed in making sherbet in N. W. India : **G. oppositifolia,** *ŏp·pŏz'·ĭt·ĭ·fōl'·ĭ·ă* (L. *opposĭtus,* placed before or opposite; *folium,* a leaf), a species whose fibres are used in making paper.

groin, n., *grŏÿn* (F. *groin,* snout of a hog), in the human body, the depressed part between the belly and the thigh.

grossification, n., *grŏs'·sĭ·fĭk·ā'·shŭn* (L. *grossus,* thick ; *facio,* I make), in *bot.,* the swelling of the ovary after impregnation.

Grossulariaceæ, n. plu., *grŏs'·ū·lăr·ĭ·ă'·sĕ·ē* (mid. L. *grossŭla,* a gooseberry ; *grossŭlus,* a small unripe fig—from *grossus,* an unripe fig), the Gooseberry and Currant family, natives of temperate regions, and many yield edible fruits.

grumous, a., *grŏm'·ŭs* (L. *grumus,* a little heap or hillock), clotted ; knotted ; in *bot.,* collected into granule masses.

Guaiacum, n., *gwā'·yă·kŭm* or *gwă·yā'·kŭm* (Sp. *guayaco,* S. Amer. *guaiac,* the name of the tree), a genus of lofty ornamental trees, Ord. Zygophyllaceæ : **Guaiacum offic-inale,** *ŏf·fĭs'·ĭn·āl'·ē* (L. *officinālis,* officinal), a beautiful W. Indian tree, whose wood, lignum vitæ, is prized for its hardness, yields the resinous substance known as **guaiac** or **gum-guaiac,** *gwĭ'·ăk* or *gwā'·yăk,* the gum and wood used medicinally as a stimulant and diaphoretic : **G. sanctum,** *săngk'·tŭm* (L. *sanctus,* holy), a species which also yields gum-guaiac.

guano, n., *gŏŏ·ăn'·ō* or *gwăn'·ō* (Sp. *guano* or *huano*—from Peruvian *huanu,* dung), the vast accumulations of the droppings of sea-fowls found on islands on the coast of S. America, much used as a manure.

guaranine, n., *gwăr'·ăn·ĭn* (after a tribe of American Indians so named), a bitter crystalline substance obtained from the Guarana bread or Brazilian cocoa, identical with caffeine.

guard, a., n., *gărd* (F. *garder,* to keep ; It. *guardare,* to guard), in *bot.,* applied to sister cells bounding a stoma, formed by bipartition of a mother cell ; in *zool.,* the cylindrical fibrous sheath with which the internal chambered shell of a Belemnite is protected.

Guatteria, n., *gwăt·tēr'·ĭ·ă* (after *Guatteri,* an Italian botanist), a splendid genus of plants, Ord. Anonaceæ : **Guatteria virgata,** *vérg·āt'·ă* (L. *virgātus,* made of twigs or osiers—from *virga,* a twig), a species yielding the lance-wood of commerce.

gubernaculum, n., *gŏŏb'·ér·năk'·ŭl·ŭm* (L. *gŭbernāculum,* a helm, a rudder—from *gŭberno,* I steer), a conical-shaped cord, attached above to the lower end of the epididymis, and below to the bottom of the scrotum.

Guilandina, n., *gĭl'·ăn·dīn'·ă* (after the Prussian traveller and botanist, *Guilandina*), a genus of pretty shrubs, Ord. Leguminosæ, Sub-ord. Cæsalpinieæ : **Guilandina bonducella,** *bŏnd'·ū·sĕl'·lă* (unascertained), the nicker tree, yields a bitter and tonic, and its seeds are said to be emetic.

Guinea-worm, n., *gĭn'·ĕ·wérm* (of or from *Guinea,* in Africa), a worm which infests the skin of man in certain warm countries ; the Dracunculus, which see.

gullet, n., *gŭl'·lĕt* (F. *goulet,* the gullet—from *goule,* the mouth ; L. *gula,* the windpipe), the passage in the neck of an animal down which food and drink pass into the stomach ; the œsophagus.

gum, n., *gŭm* (L. *gummi*, F. *gomme*, gum), a vegetable mucilage found thickened on the surface of certain trees : **gum-resin**, an exudation from certain trees and shrubs partaking of the nature of a gum and a resin : **gum-arabic**, a gum procured from several species of acacia in Africa and S. Asia : **gum-lac**, a resinous substance exuded from the bodies of certain insects, chiefly found upon the banyan tree.

gumma, n., *gŭm'·mă*, **gummata**, n. plu., *gŭm'·măt·ă* (L. *gummi*, gum ; *gummātus*, containing gum, gummy), a species of new growths produced in various organs of the body, arising from constitutional syphilis ; also called 'syphiloma.'

gustatory, a., *gŭst'·ăt·ĕr·ĭ* (L. *gustus*, taste, flavour), pert. to the taste ; applied to a nerve of the sense of taste which supplies the papillæ and mucous membrane of the tongue.

gutta percha, n., *gŭt'·tă pĕr'·tshă* (Malay, ragged gum), a kind of caoutchouc which softens at a moderate temperature, used for soles of shoes, straps, and numerous articles of domestic use, the produce of Isonandria gutta, Ord. Sapotaceæ, obtained from Borneo and Singapore.

NOTE.—Said also to be from the name of the tree, or the name of the island from which first imported, viz. Pulo-Percha.

Guttiferæ, n. plu., *gŭt·tĭf'·ĕr·ē* (L. *gutta*, a drop ; *fero*, I bear), the Gamboge family, an Order of beautiful trees and shrubs yielding a resinous juice of a yellow colour, acrid and purgative ; Order also named Clusiaceæ: **guttiferous**, a., *gŭt·tĭf'·ĕr·ŭs*, yielding gum or resinous substances.

guttulate, a., *gŭt'·tŭl·āt* (L. *guttula*, a little drop), in *bot.*, in the form of small drops ; composed of small round vesicles.

guttural, a., *gŭt'·tĕr·ăl* (L. *guttur*, the throat), formed in the throat ; pert. to the throat.

gymnaxony, n., *jĭm·năks'·ŏn·ĭ* (Gr. *gumnos*, naked ; *axon*, an axletree), in *bot.*, a state in which the placenta protrudes through the ovary and alters its position.

gymnoblastic, a., *jĭm'·nō·blăst'·ĭk* (Gr. *gumnos*, naked ; *blastos*, a bud), applied to the Hydrozoa in which the nutritive and reproductive buds are not protected by horny receptacles.

gymnocarpous, a., *jĭm'·nō·kărp'·ŭs* (Gr. *gumnos*, naked ; *karpos*, fruit), in *bot.*, applied to naked fruit, that is, fruit having no pubescence or floral envelope around it ; applied to lichens having fructifications in the form of a scutellate, cup-shaped, or linear thallus.

gymnogen, n., *jĭm'·nō·jĕn* (Gr. *gumnos*, naked ; *gennao*, I produce), a plant with naked seeds, that is, seeds which are not enclosed in an ovary ; a gymnospermous plant.

Gymnolæmata, n. plu., *jĭm'·nō·lēm'·ăt·ă* (Gr. *gumnos*, naked ; *laimos*, the neck or throat), an Order of the Polyzoa, having the mouth devoid of the valvular structure known as the epistome.

Gymnophiona, n. plu., *jĭm·nŏf'·ĭ·ōn'·ă* (Gr. *gumnos*, naked ; *ophis*, a serpent, *ophios*, of a serpent), the Order of the Amphibia comprising the snake-like Cæciliæ.

Gymnosomata, n. plu., *jĭm'·nō·sŏm'·ăt·ă* (Gr. *gumnos*, naked ; *sōma*, a body, *sōmătos*, of a body), the Order of Pteropoda which have not the body protected by a shell.

gymnospermous, a., *jĭm'·nō·spĕrm'·ŭs* (Gr. *gumnos*, naked ; *sperma*, seed), having naked seeds, or seeds not enclosed in a true ovary, as Conifers: **gymnospermæ**, n. plu., *jĭm'·nō·spĕrm'·ē*, also **gymnosperms**, n. plu., *·spĕrmz*, monochlamydeous or achlamydeous plants differing from Exogens in having naked ovules ; ovules developed without the usual integuments.

gymnospore, n., *jĭm'.nō.spŏr* (Gr. *gumnos*, naked ; *spora*, seed), a naked spore: **gymnosporous**, a., *-spŏr'.ŭs*, pert. to: **gymnosporæ**, n. plu., *jĭm'.nō.spŏr'.ē*, the class of plants having naked spores.

gymnostomi, n. plu., *jĭm-nŏs'.tŏm.ī* (Gr. *gumnos*, naked ; *stoma*, a mouth), mosses without a peristome, or naked mouthed: **gymnos'tomous**, a., *-tŏm.ŭs*, naked mouthed; without a peristome: **Gymnos'tomum**, n., *-tŏm.ŭm*, a numerous genus of plants, growing in tufts and patches of various colours, found in almost every situation, Ord. Musci or Bryaceæ, so called in allusion to the open orifice of the theca.

gynandrophore, n., *gĭn.ănd'.rō.fŏr* (Gr. *gunē*, a female ; *anēr*, a male, *andros*, of a male ; *phoreo*, I bear), in *bot.*, a column bearing stamens and pistil.

gynandrous, a., *gĭn.ănd'.rŭs* (Gr. *gunē*, a female ; *anēr*, a male, *andros*, of a male), having stamens and pistil in a common column, as in orchids.

gynantherous, a., *gĭn.ănth'.ĕr.ŭs* (Gr. *gunē*, a female ; *anthēros*, belonging to a flower—from *anthos*, a flower), having the stamens converted into pistils.

Gynerium, n., *gĭn.ēr'.ĭ.ŭm* (Gr. *gunē*, a female ; *erion*, wool), a genus of plants, Ord. Gramineæ ; the pampas-grass, covering the vast plains of S. America, very ornamental in the flower-garden: **Gynerium argenteum**, *ăr.jĕnt'.ĕ.ŭm* (L. *argentĕus*, silvery), the pampas-grass of the Cordilleras.

gynizus, n., *gĭn.iz'.ŭs* (Gr. *gunē*, a female, a pistil ; *hizō*, I cause to sit, I seat), the position of the stigma on the columns of orchids.

gynobase, n., *gĭn'.ō.bāz* (Gr. *gunē*, a female ; *basis*, a base), in *bot.*, a central axis, to the base of which the carpels are attached ; a fleshy receptacle bearing separate fruits : **gynobasic**, a., *gĭn'.ō.bāz'.ĭk*, having a gynobase.

Gynocardia, n., *gĭn'.ō.kărd'.ĭ.ă* (Gr. *gunē*, a female ; *kardia*, the heart), a genus of plants, Ord. Bixaceæ : **Gynocardia odorata**, *ŏd'.ōr.āt'.ă* (L. *odorātus*, scented —from *odor*, smell, scent), a species, called Chalmugra seeds, from whose seeds an oil is expressed, used in India for the cure of leprosy and various cutaneous diseases.

gynœcium, n., *gĭn.ē'.shĭ.ŭm* (Gr. *gunē*, a female ; *oikos*, a house), in *bot.*, the female organ of the flowers ; the pistil.

gynophore, n., *gĭn'.ō.fŏr* (Gr. *gunē*, a female ; *phoreo*, I bear), in *bot.*, a stalk supporting the ovary; in *zool.*, the generative buds or gonophores of Hydrozoa containing ova alone, and differing in form from those which contain spermatozoa.

gynostegium, n., *gĭn'.ō.stĕdj'.ĭ.ŭm* (Gr. *gunē*, a female, a pistil ; *stego*, I cover), the staminal crown of Asclepias.

gynostemium, n., *gĭn'.ō.stĕm'.ĭ.ŭm* (Gr. *gunē*, a female, a pistil ; *stēmōn*, a thread, a stamen), a column in orchids bearing the organs of reproduction ; the united stamens and pistil of orchids.

gyrate, a., *jĭr'.āt* (Gr. *guros*, L. *gyrus*, a ring, a circle), winding or going round as in a circle ; turning in a circular manner : **gyration**, n., *jĭr.ā'.shŭn*, a turning or whirling round ; rotation as in cells : **gyri**, n. plu., *jĭr'.ī*, in the cerebrum, the numerous smooth and tortuous eminences into which the grey matter of the surface of the hemispheres is moulded.

Gyrencephala, n. plu., *jĭr'.ĕn.sĕf'.*

ăl·ă (Gr. *gurŏ̄*, I curve or bend ; *engkephalos*, the brain), a section of the Mammalia, in which the cerebral hemispheres are abundantly convoluted.

Gyrocarpeæ, n. plu., *jĭr'·ō·kårp'·ĕ·ē* (L. *gyro*, I turn round in a circle ; *karpos*, fruit), a Suborder or tribe of the Ord. Combretaceæ, so called in allusion to the fruit moving in the air : **Gyrocarpus**, n., *jĭr'·ō·kårp'·ŭs*, a genus of very ornamental plants.

gyroma, n., *jĭr·ōm'·ă* (Gr. *gurōmă*, a circle—from *gūros*, round), the annulus or ring around the sporecase of ferns.

gyrophora, n., *jĭr·ŏf'·ŏr·ă* (Gr. *guros*, a circle ; *phoreo*, I bear— in allusion to the disc of the shield), a very interesting genus of plants of the Lichen family, found growing chiefly upon exposed rocks, Ord. Lichenes ; several species of Gyrophora constitute the Tripe-de-roche, on which Franklin and his companions existed for some time.

gyrose, a., *jĭr·ōz'* (Gr. *guros*, a circle), in *bot.*, turned round like a crook ; folded and waved.

habit, n., *hăb'·ĭt* (L. *habĭtus*, state of the body, dress), in *bot.*, the general external appearance of a plant : **habitat**, n., *hăb'·ĭt·ăt* (L. *habĭtat*, it inhabits), the natural locality of an animal or plant ; the situation, district, or country inhabited by an animal or plant in its wild state.

Habrothamnus, n., *hăb'·rō·thăm'·nŭs* (Gr. *habros*, graceful, elegant; *thamnos*, a shrub, a thicket), a genus of elegant greenhouse plants, bearing pannicles of flowers in profusion, Ord. Solanaceæ.

hæmal, a., *hĕm'·ăl* (Gr. *haima*, blood), connected with the blood or blood vessels ; applied to the arch under the vertebral column

which encloses and protects the organs of circulation.

Hæmanthus, n., *hĕm·ănth'·ŭs* (Gr. *haima*, blood ; *anthos*, a flower— in allusion to the colour of the flowers), a genus of fine bulbous plants, Ord. Amaryllidaceæ : **Hæmanthus toxicarius**, *tŏks'·ĭk·ār'·ĭ·ŭs* (Gr. *toxikon*, poison), a species whose root is poisonous.

hæmapoiesis, n., *hĕm'·ă·pŏy·ēz'·ĭs* (Gr. *haima*, blood ; *poiēsis*, the making or forming of a thing), the production or formation of blood : **hæmapoietic**, a., *hĕm'·ă·pŏy·ĕt'·ĭk* (Gr. *poiētikos*, making, effecting), making or producing blood.

hæmapophyses, n. plu., *hĕm'·ă·pŏf'·ĭs·ēz* (Gr. *haima*, blood; Eng. *apophysis*), in *anat.*, the parts projecting from a vertebra which form the hæmal arch.

hæmatemesis, n., *hĕm'·ă·tĕm'·ĕs·ĭs* (Gr. *haima*, blood, *haimătos*, of blood ; *emeō*, I vomit), a vomiting of blood.

hæmatin, n., *hĕm'·ăt·ĭn* (Gr. *haima*, blood, *haimatos*, of blood), the colouring matter resulting from the decomposition of hæmoglobin by heat : **hæmatine**, n., *hĕm'·ăt·ĭn*, the colouring matter of logwood : **hæmatoidin**, n., *hĕm'·ăt·ŏyd'·ĭn* (Gr. *eidos*, resemblance), the blood crystals found as a pathological production in old extravasations of blood : **hæmatitis**, n., *hĕm'·ăt·ĭt'·ĭs*, inflammation of the blood : **hæmatinuria**, n., *hĕm'·ăt·ĭn·ŭr'·ĭ·ă* (Gr. *ouron*, urine), a condition of the urine in which it contains hæmatin.

hæmatocele, n., *hĕm·ăt'·ō·sēl* (Gr. *haima*, blood, *haimătos*, of blood; *kēlē*, a tumour), a tumour formed by an effusion of blood from the vessels of the testis or its coverings, or of the sprematic cord ; any tumour consisting principally of blood,—*e.g.*, 'pelvic hæmatocele.'

hæmatocrya, n. plu., *hĕm'·ăt·ŏk'·rĭ·ă* (Gr. *haima*, blood ; *kruos*, cold), applied by Owen to the cold - blooded Vertebrates, viz. the Fishes, Amphibia, and Reptiles : hæmatocryal, a., *hĕm'·ăt·ŏk'·rĭ·ăl*, cold blooded.

hæmatoidin, see ' hæmatin.'

hæmatoma, n., *hĕm'·ăt·ōm'·ă*, hæmatomata, n. plu., *hĕm'·ăt·ŏm'·ăt·ă* (Gr. *haima*, blood, *haimătos*, of blood), a kind of tumour formed from an effused blood mass resulting from a hæmorrhage.

hæmatometra, n., *hĕm'·ăt·ō·mĕt'·ră* (Gr. *haima*, blood, *haimătos*, of blood ; *mētra*, womb), an accumulation of menstrual blood in the uterus, which becomes thick, black, and tarry, and often causes great dilatation.

Hæmatopinus, n., *hĕm'·ăt·ŏp'·ĭn·ŭs* (L. *hæmătŏpus*, Gr. *haimătŏpous*, a blood foot—from Gr. *haima*, blood ; *pous*, a foot, *podos*, of a foot), a genus of animal parasites : Hæmatopinus asini, *ăs'·ĭn·ī* (L. *asĭnus*, an ass), the louse of the ass, sometimes found on the horse : H. eurysternus, *ŭr'·ĭs·tĕrn'·ŭs* (Gr. *eurusternos*, having a broad breast — from *eurus*, broad ; *sternon*, the breast), the louse of the ox : H. piliferus, *pĭl·ĭf'·ĕr·ŭs* (L. *pĭlus*, hair ; *fĕro*, I carry.), the louse of the dog, but not common, also found on ferrets : H. stenopsis, *stĕn·ŏps'·ĭs* (Gr. *stĕnos*, narrow ; *ŏpsis*, sight), the louse of the goat : H. suis, *sū'·ĭs* (L. *sūs*, a swine, *sŭĭs*, of a swine), the louse of the swine, occurring on it in great numbers : H. vituli, *vĭt'·ŭl·ī* (L. *vĭtŭlus*, a calf), the louse of the calf.

hæmatotherma, n. plu., *hĕm'·ă·tō·thĕrm'·ă* (Gr. *haima*, blood ; *thermos*, warm), the warm-blooded Vertebrates, viz. Birds, and Mammals : hæmatothermal, a., *-thĕrm'·ăl*, hot blooded.

Hæmatoxylon, n., *hĕm'·ăt·ŏks'·ĭl·ŏn* (Gr. *haima*, blood, *haimătos*, of blood ; *xulon*, wood), a genus of trees, Ord. Leguminosæ, which furnish dyes : Hæmatoxylon Campechianum, *kăm·pĕtsh'·ĭ·ān'·ŭm* (from Bay of *Campeachy*, where largely obtained ; Sp. *campéche*, logwood), the logwood tree, or Campeachy wood of commerce, the inner wood of which is used both as a dye and an astringent : hæmatoxylin, n., *hĕm'·ăt·ŏks'·ĭl·ĭn*, the colouring principle of logwood, chiefly used for staining preparations for microscopic purposes.

hæmaturia, n., *hĕm'·ăt·ŭr'·ĭ·ă* (Gr. *haima*, blood ; *ouron*, urine), a discharge of urine containing blood.

hæmin, n., *hĕm'·ĭn*, also called ' hydrochlorate of hæmatin ' (Gr. *haima*, blood), a crystalline derivative from hæmoglobin, which forms a most delicate medico-legal test of the presence of blood.

Hæmodoraceæ, n. plu., *hĕm'·ō·dŏr·ā'·sĕ·ē* (Gr. *haima*, blood ; *dōron*, a gift), the Blood-root family, an Order of plants, so called from the red colour of their roots, used for dyeing : Hæmodorum, n., *hĕm'·ō·dŏr'·ŭm*, a genus of ornamental plants.

hæmoglobin, n., *hĕm'·ō·glōb'·ĭn* (Gr. *haima*, blood ; L. *globus*, a ball), a red colouring matter which infiltrates the stroma of the blood corpuscles, and which may be decomposed into an albuminous substance called ' globulin,' and a colouring matter called ' hæmatin ' ; also hæmatoglobulin, n., *hĕm'·ăt·ŏ·glŏb'·ŭl·ĭn*.

hæmoptysis, n., *hĕm·ŏp'·tĭs·ĭs* (Gr. *haima*, blood ; *ptuo*, I spit), a coughing up or expectoration of blood.

hæmorrhage, n., *hĕm'·ŏr·rădj* (Gr. *haimorrhagia*, a flowing of blood —from *haima*, blood ; *rhegnumi*,

I burst forth), a discharge of blood from the lungs, nose, or intestines, or an effusion of blood into the brain, arising from the rupture of one or more blood vessels; any bleeding: **hæmorrhagic**, a., *hĕm'·ŏr·rădj'·ĭk*, pert. to or consisting of hæmorrhage.

hæmorrhoids, n. plu., *hĕm'·ŏr·rōy̆ds* (Gr. *haima*, blood; *rheo*, I flow; *eidos*, resemblance), piles, consisting of tumours, situated at or near the anus, varying in size from a pea to a pigeon's egg, and consisting essentially in a dilated and varicose condition of the hæmorrhoidal veins: **hæmorrhoidal**, a., *hĕm'·ŏr·rōy̆d'·ăl*, pert. to piles.

hæmothorax, *hĕm'·ō·thōr'·ăks* (Gr. *haima*, blood; *thōrax*, the trunk of the body), applied to the pleural sac filled with blood, or with a fluid of a sanguineous character, which undergoes various secondary changes and degenerations in which the surrounding tissues are also involved.

Halesia, n., *hăl·ēsh'·ĭ·ă* (after *Dr. Hales*, a vegetable physiologist), a genus of plants, Ord. Styracaceæ; the snowdrop trees of California, whose species are beautiful and valuable from their flowering so early in the season.

Halimocnemis, n., *hăl'·ĭ·mŏk'·nĕm·ĭs* (Gr. *halimos*, brackish, marine—from *hals*, salt; *nĕmos*, a grove), a genus of plants, Ord. Chenopodiaceæ, a species of which, growing in salt marshes, yields soda.

halitus, n., *hăl'·ĭt·ŭs* (L. *halitus*, breath—from *halo*, I breathe), a breathing; the vapour arising from new-drawn blood.

hallux, n., *hăl'·lŭks* (L. *hallex*, the thumb or great toe), the great toe in man; the innermost of the five digits which normally compose the hind foot of a vertebrate animal.

halophytes, n. plu., *hăl'·ō·fĭtz* (Gr. *hals*, the sea; *phuton*, a plant), plants of salt marshes, containing salts of soda in their composition.

Haloregeaceæ, n. plu., *hăl'·ō·rādj'·ĕ·ā'·sĕ·ē* (Gr. *hals*, the sea; *rhax*, a berry, a bush, *rhăgos*, of a berry), the Mare's-tail family, an Order of herbs or under shrubs, often aquatic, having whorled leaves and sessile flowers: **Haloragis**, n., *hăl'·ŏr·ādj'·ĭs*, a genus of rather curious plants.

Halteres, n. plu., *hălt·ēr'·ēz* (Gr. *haltēres*, masses of lead held in the hands to balance leapers), the rudimentary filaments or balancers which represent the posterior pair of wings in the Order of insects called the Diptera.

Hamamelidaceæ, n. plu., *hăm'·ă·mĕl·ĭ·dā'·sĕ·ē* (Gr. *hama*, together, with; *melon*, an apple, in allusion to the fruit accompanying the flower), the Witch-hazel family, an Order of small trees and shrubs: **Hamamelis**, n., *hăm'·ă·mĕl'·ĭs*, a genus of plants whose species are ornamental trees, producing a fruit somewhat like a nut: **Hamamelis virginica**, *vĕr·jĭn'·ĭk·ă* (of or from *Virginia*, Amer.; L. *virgo*, a maid, a virgin, *virgĭnis*, of a virgin), a species whose seeds are used as food, while its leaves and bark are astringent and acrid.

hamose, a., *hăm·ōz'*, and **hamous**, a., *hăm'·ŭs* (L. *hamus*, a hook), in *bot.*, having the end hooked or curved.

hamular, a., *hăm'·ūl·ăr* (L. *hamulus*, a little hook—from *hāmus*, a hook), in *anat.*, having a hook-like appearance; having small hooks: **hamulose**, a., *hăm'·ūl·ōz'*, in *bot.*, covered with little hooks: **hamulus**, n., *hăm'·ūl·ŭs*, in *bot.*, a kind of hooked bristle; in *anat.*, a hook-like process: **hamulus lachrymalis**, *lăk'·rĭ·māl'·ĭs* (L. *lachrymālis*, lachrymal — from *lachryma*, a tear), the lachrymal hook-like process.

harmonia, n., *hăr·mōn'·ĭ·ă* (Gr.

harmŏnia, a joining together—from *harmozō*, I fit together), in *anat.*, a form of articulation in which there is neither serration of the edges of the bones nor interposed cartilage, and in which of course there is no movement.

hastate, a., *hăst'·āt* (L. *hasta*, a spear), shaped like a halbert, applied to leaves; applied to a leaf with two portions of the base projecting more or less completely at right angles to the blade.

haulm, n., also **halm**, n., *hăwm* or *hăm* (Ger. *halm*, F. *chaulme*, straw), the stem or stalk of grain; the dead stems of herbs, as of the potato.

haunch, n., *hăwnsh* (F. *hanche*, the hip; old H. Ger. *hlancha*, the flank), the hip; that part of a man or quadruped which lies between the last rib and the thigh; a joint of mutton or venison.

haustellate, a., *hăws·tĕl'·lāt* (L. *haustellum*, a sucker — from *haurio*, I draw water), provided with suckers, applied to the mouths of certain Crustacea and Insecta: **haustorium**, n., *hăws·tōr'·ĭ·ŭm* (L. *haustor*, a drawer), the sucker at the extremity of the parasitic root of the dodder; the root-like sucker of the ivy, etc.

haw, n., *hăw* (AS. *haga*, Ger. *hag*, a hedge, an enclosure), the berry of the hawthorn; the membrana nictitans, or third eyelid of birds and quadrupeds.

heart, n., *hărt* (AS. *heorte;* Goth. *hairto;* Sans. *hardi;* Gr. *kardia*, the heart), the central organ of the circulation, which, by alternate contracting and expanding, sends the blood through the arteries, to be again received by it from the veins.

hectic, a., *hĕk'·tĭk* (Gr. *hektikos*, pert. to habit of body—from *hexis*, habit of body), constitutional; habitual: **hectic fever**, a peculiar form of remittent fever, the result of exhausting disease.

hectocotylus, n., *hĕk'·tō·kŏt'·ĭl·ŭs* (Gr. *hekatōn*, a hundred; *kotulos*, a cup), the metamorphosed reproductive arm of certain of tho male cuttle-fishes.

Hedera, n., *hĕd'·ĕr·ă* (L. *hĕdĕra*, the plant ivy), a genus of ivy plants, Ord. Araliaceæ: **Hedera helix**, *hĕl'·ĭks* (Gr. *hĕlix*, anything twisted, a fold; L. *hĕlix*, a kind of ivy), a species of ivy whose succulent fruit is emetic and purgative: **hederaceous**, a., *hĕd'·ĕr·ā'·shŭs*, of or pert. to ivy.

Hedysarum, n., *hĕd'·ĭs·ār'·ŭm* (Gr. *hēdus*, sweet), a genus of very handsome flowering plants, producing racemes of beautiful pea-flowers, Ord. Leguminosæ, Subord. Papilionaceæ: **Hedysarum gyrans**, *jīr'·ănz* (L. *gyrans*, turning round in a circle), a species which exhibits a remarkable irritability in its leaves; the Gorachand of Bengal.

Heimia, n., *hīm'·ĭ·ă* (after *Dr. Heim*, a celebrated physician of Berlin), a genus of plants very pretty when in blossom, Ord. Lythraceæ: **Heimia salicifolia**, *săl·ĭs'·ĭ·fōl'·ĭ·ă* (L. *salix*, a willow tree, *salĭcis*, of a willow tree; *folĭum*, a leaf), a species said to have diaphoretic properties, and by the Mexicans considered a potent remedy in venereal diseases.

Hekistotherms, n. plu., *hē·kĭs'·tō·thĕrmz* (Gr. *hēkistos*, very little; *thĕrmē*, heat), plants of the arctic and antarctic regions, and the higher regions of mountains in temperate climates, such as Mosses, Lichens, Coniferæ, etc., which can bear darkness under snow, and require a small amount of heat.

Helianthemum, n., *hĕl'·ĭ·ănth'·ĕm·ŭm* (Gr. *hēlios*, the sun; *anthĕm·on*, a flower—in allusion to the yellow flowers), a genus of showy.

free-flowering plants, including some of the prettiest little shrubs in cultivation for rock-work, Ord. Cistaceæ.

Helianthus, n., *hēl'-ĭ-ănth'-ŭs* (Gr. *helios*, the sun; *anthos*, a flower—so called from the brilliant colour of the flowers, or from the erroneous belief that the flowers always turned towards the sun), a highly ornamental and extensive genus of plants, producing large heads of beautiful flowers, Ord. Compositæ, Sub-ord. Corymbiferæ : **Helianthus annuus**, *ăn'-nū-ŭs* (L. *annŭus*, that lasts a year—from *annus*, a year), the common sunflower, whose seeds contain a bland oil, and when roasted have been used as a substitute for coffee : **H. tuberosus**, *tūb'-ĕr-ōz'-ŭs* (L. *tuberōsus*, full of humps or swellings—from *tuber*, a hump), the Jerusalem or Girasole artichoke, whose roots are used as substitutes for potatoes.

helicine, a., *hĕl'-ĭs-ĭn* (Gr. *hĕlix*, anything twisted, a fold, *helĭkos*, of a twisted thing), in *anat.*, applied to certain arteries connected with the penis which assume a convoluted or tendril-like appearance ; winding ; spiral.

helicis major, *hĕl'-ĭs-ĭs mādj'-ŏr* (L. *hĕlix*, a fold, ivy, *helĭcis*, of a twisted thing ; *major*, greater), the greater (muscle) of the helix ; applied to a narrow, vertical band of muscular fibres on the anterior margin of the helix : **h. minor**, *mĭn'-ŏr* (L. *minor*, less or lesser), the lesser (muscle) of the helix ; applied to an oblique fasciculus attached to the part of the helix commencing from the bottom of the concha; see 'helix.'

helicoid, a., *hĕl'-ĭk-ŏyd*, also **helicoidal**, a., *hĕl'-ĭk-ŏyd'-ăl* (Gr. *helix*, a twisted thing ; *eidos*, resemblance), twisted like a snail shell, applied to inflorescence : **helicoid cyme**, a cyme in which the flowers are arranged in a continuous spiral round a false axis.

helicotrema, n., *hĕl'-ĭk-ō-trēm'-ă* (Gr. *hĕlix*, anything twisted ; *trēma*, an opening, a hole), in *anat.*, a small opening placed at the apex of the cochlea of the ear.

Helictereœ, n. plu., *hĕl'-ĭk-tēr'-ĕ-ē* (Gr. *hĕlix*, a spiral, a screw), a Tribe or Sub-order of the Ord. Sterculiaceæ : **Helecteres**, n. plu., *hĕl'-ĕk-tēr'-ēz*, the screw trees, a genus of free-flowering shrubs, so named in reference to the carpels being twisted.

Heliotropieœ, n. plu., *hĕl'-ĭ-ō-trōp-ĭ'-ĕ-ē* (Gr. *hēlios*, the sun ; *tropē*, a turning), a Sub-order of the Ord. Boraginaceæ, so called from their flowers being said to turn towards the sun : **Heliotrope**, n., *hĕl'-ĭ-ō-trōp*, also **Heliotropium**, n., *hĕl'-ĭ-ō-trōp'-ĭ-ŭm*, a genus of plants, some of whose species are highly valued from the fragrance of their flowers : **heliotropism**, n., *hĕl'-ĭ-ŏt'-rŏp-ĭzm*, that property by which certain plants constantly turn their leaves and flowers towards the sun ; the bending of a plant either from or towards light.

helix, n., *hĕl'-ĭks*, **helices**, n. plu., *hĕl'-ĭs-ēz* (Gr. *helix*, the twisted thing), something that is spiral ; in *anat.*, the curved rim of the external body of the ear; the snail shell.

Helleboreœ, n. plu., *hĕl'-lĕ-bōr'-ĕ-ē* (Gr. *hellĕboros*, L. *hellĕbŏrus*, hellebore—from Gr. *hĕlein*, to kill or overcome ; *bŏra*, food), a Sub-order of the Ord. Ranunculaceæ, so called in reference to the poisonous qualities of the plants: **Helleborus**, n., *hĕl'-lĕb'-ŏr-ŭs*, a genus of plants having poisonous qualities : **Helleborus officinalis**, *ŏf'-fĭs'-ĭn-āl'-ĭs* (L. *officinālis*, officinal); **H. niger**, *nĭdj'-ĕr* (L. *nĭger*, black), the Christmas rose; **H. fœtidus**, *fĕt'id·ŭs* (L. *fœtidus*, stinking, fetid); **H. viridis**, *vĭr'id·ĭs* (L. *vĭrĭdis*, green), are species which act as drastic purgatives; powerful cardiac sedatives; some of them were used in ancient times in cases of mania : Hellebore, n., *hĕl'lĕ·bōr*, the common name of several of the species; the Christmas rose or flower; still employed in medicine.

helminthoid, a., *hĕl'minth·ōyd* (Gr. *helmins*, an intestinal worm; *eidos*, resemblance), worm-shaped; vermiform.

hemelytra, n. plu., *hĕm·ĕl'ĭt·ră* (Gr. *hēmi*, half; *elŭtron*, a sheath), among certain insects, wings which have the apex membranous, while the inner portion is chitinous, and resembles the elytron of a beetle.

hemeralopia, n., *hĕm'ĕr·ă·lōp'ĭ·ă* (Gr. *hēmera*, day; the latter part of doubtful formation, usually referred to Gr. *ops*, the eye, or *opsomai*, I see; the l may be introduced for the sake of euphony), day vision only; night blindness; intermittent amaurosis, in which the person is able to see only in daylight : **hemeralops**, n., *hĕm'ĕr·ă·lŏps*, one afflicted with night blindness.

Hemerocallideæ, n. plu., *hĕm'ĕr·ō·kăl·lĭd'ĕ·ē* (Gr. *hēmera*, a day; *kallos*, beauty), a Sub-order of the Ord. Liliaceæ, the Day lily tribe : **Hemerocallis**, n., *hĕm'ĕr·ō·kăl'lĭs*, an ornamental genus of flowering plants, whose beautiful flowers last a day; the day lily.

hemicarp, n., *hĕm'ĭ·kărp* (Gr. *hemi*, half; *karpos*, fruit), in *bot.*, one portion of a fruit which spontaneously divides into halves.

hemicrania, n., *hĕm'ĭ·krān'ĭ·ă* (Gr. *hēmi*, half; *kranion*, the skull), pain confined to one side of the head; brow ague.

hemicyclic, a., *hĕm'ĭ·sĭk'lĭk* (Gr. *hēmi*, half; Eng. *cycle*), in *bot.*, applied to the transition from one floral whorl to another when it coincides with a definite number of turns of the spiral.

Hemidesmus, n., *hĕm'ĭ·dĕz'·mŭs* (Gr. *hēmisus*, a half; *dĕsmos*, a bond, a tie, in allusion to its filaments), a genus of pretty climbing plants, Ord. Asclepiadaceæ : **Hemidesmus Indicus**, *ĭn'dĭk·ŭs* (L. *Indicus*, of or belonging to India), a species whose fragrant roots are used in Madras as a substitute for sarsaparilla, under the name ' Country Sarza.'

hemimetabolic, a., *hĕm'ĭ·mĕt·ă·bŏl'ĭk* (Gr. *hēmi*, half; *metabolē*, change), applied to insects which undergo an incomplete metamorphosis.

hemiplegia, n., *hĕm'ĭ·plēdj'ĭ·ă* (Gr. *hēmi*, half; *plēgē*, a blow, a stroke), a paralysis of one lateral half of the body.

Hemiptera, n. plu., *hĕm·ĭp'tĕr·ă* (Gr. *hēmi*, half; *pteron*, a wing), an Order of insects which have sometimes the anterior wings hemelytra : **hemipteral**, a., *hĕm·ĭp'tĕr·ăl*, also **hemipterous**, a., *hĕm·ĭp'tĕr·ŭs*, having the upper wings partly coriaceous and partly membranous.

hemisphere, n., *hĕm'ĭ·sfēr* (Gr. *hēmi*, half; *sphaira*, a globe), in *anat.*, applied to each lateral half of the brain.

hemlock, n., *hĕm'lŏk* (AS. *hemleac*), an indigenous plant which possesses sedative properties, and is employed both internally and externally; the Conium maculatum, Ord. Umbelliferæ.

hemp, n., *hĕmp* (Dut. *hennip*, Ger. *hanf*, Icel. *hanpr*, hemp), a plant which yields the valuable fibres or threads of the same name; the Cannabis sativa, Ord. Cannabinaceæ, a species of hemp used in India under various names as a narcotic and intoxicant.

henbane, n., *hĕn'·bān* (Eng. *hen*, and *bane*), a poisonous wild British herb, possessing narcotic properties, and used in medicine, so called from its being supposed to be poisonous to domestic fowls ; the Hyoscyamus niger, Ord. Solanaceæ.

henna, n., *hĕn'·nă*, also called **alhenna** (Ar. *hinna*), a tropical shrub, whose powdered leaves made into a paste are used in Asia and Egypt in dyeing the nails, etc. an orange hue ; the Lawsonia inermis, Ord. Lythraceæ.

hepatic, a., *hĕp·ăt'·ĭk* (Gr. *hepatikos*, affecting the liver—from *hĕpar*, the liver), belonging to the liver; applied to a duct conveying the bile from the liver ; having a liver-like colour and consistency : **hepatitis**, n., *hĕp'·ăt·ī'·tĭs*, inflammation of the liver : **hepatisation**, n., *hĕp'·ăt·īz·ā'·shŭn*, a diseased part having the appearance of liver; the second stage of pneumonia.

Hepaticæ, n. plu., *hĕp·ăt'·ĭs·ē* (Gr. *hepatikos*, belonging to the liver —from *hepar*, the liver), the Liverwort family, an Order of plants the lobes of whose leaves have been compared to the lobes of the liver : **Hepatica**, n., *hĕp·ăt'·ĭk·ă*, a genus of pretty plants, producing abundant flowers, Linnæan Ord. Ranunculaceæ.

hepato-cystic, a., *hĕp·ăt'·ō·sĭst'·ĭk* (Gr. *hepar*, the liver ; *kustis*, a bladder), applied to small ducts passing from the liver to the gall-bladder ; pert. to the liver and gall-bladder.

heptagynous, a., *hĕp·tădj'·ĭn·ŭs* (Gr. *hepta*, seven ; *gunē*, female), in *bot.*, having seven styles.

heptandrous, a., *hĕp·tănd'·rŭs* (Gr. *hepta*, seven ; *anēr*, a male, man), in *bot.*, having seven stamens.

herb, n., *hĕrb* (L. *herba*, F. *herbe*, grass, vegetation), a plant with an annual stem, as opposed to one with a woody fibre : **herbaceous**, a., *hĕrb·ā'·shŭs*, applied to green succulent plants which die down to the ground in winter ; having annual shoots ; applied to green-coloured cellular parts : **herbarium**, n., *hĕrb·ār'·ĭ·ŭm*, a prepared collection of dried plants.

hermaphrodite, n., *hĕr·măf'·rŏd·īt* (Gr. *Hermēs*, the god Mercury; *Aphrodite*, the goddess Venus), a living creature which is neither perfect male nor female ; in *bot.*, a plant which has the male and female organs, that is, stamens and pistil, in the same flower.

hermodactyle, n., *hĕrm'·ō·dăk'·tĭl* (Gr. *Hermēs*, Mercury ; *daktulos*, a finger—that is, the finger of Mercury), a species of colchicum, famous among the ancients for diseases of the joints ; probably a species of Colchicum Illyricum, or according to others of C. variegatum, Ord. Melanthaceæ.

Hernandieæ, n. plu., *hĕr'·năn·dī'·ĕ·ē* (after *Hernandez*, a Spanish botanist), a section or Sub-order of the Ord. Thymelæaceæ : **Hernandia**, n., *hĕr·năn'·dī·ă*, a genus of elegant and lofty-growing trees, whose bark, young leaves, and seeds are slightly purgative : **Hernandia sonora**, *sŏn·ōr'·ă* (L. *sonōrus*, sounding ; Sp. *sonōra*, a musical instrument), a species, the juice of whose leaves, it is said, is a powerful depilatory, destroying hair without pain.

hernia, n., *hĕrn'·ĭ·ă* (L. *hernia*, a rupture ; Gr. *hernos*, a branch, a sprout), the displacement of any viscus, or part of one, from its own cavity into an adjoining space: **hernia cerebri**, *sĕr'·ĕb·rī* (L. *cerebrum*, the brain), the hernia of the brain ; a protrusion of a portion of the brain and its membranes.

herpes, n., *hèrp'ēz* (L. *herpes*, a spreading eruption on the skin—from Gr. *herpo*, I creep along), a skin eruption consisting of clusters of vesicles upon an inflamed base: **herpes labialis,** *làb'ĭ·āl'ĭs* (L. *làbiālis*, pert. to a lip —from *labium*, a lip), herpes occurring on the upper lip : **h. zoster,** *zŏst'ĕr* (Gr. *zōstēr*, a girdle or belt), a variety of herpes also called 'shingles,' which is of neurotic origin.

hesperidium, n., *hĕs'pèr·ĭd'ĭ·ŭm* (L. *Hesperus*, Gr. *Hesperos*, Hesperus, western — in allusion to such fruit coming from the west of Europe), a fruit such as the orange, lemon, shaddock, in which the epicarp and mesocarp form a separate rind, the seeds being embedded amongst a mass of pulp.

heterocephalous, a., *hĕt'ĕr·ō·sĕf'·ăl·ŭs* (Gr. *heteros*, another; *kephalē*, the head), in *bot.*, having some flower - heads male, and others female, on the same plant.

heterocercal, a., *hĕt'ĕr·ā·sèrk'ăl* (Gr. *heteros*, another ; *kerkos*, a tail), applied to fishes having unequally lobed tails, as in the sharks and dogfish.

heterochromous, a., *hĕt'ĕr·ō·krōm'ŭs* (Gr. *heteros*, another ; *chroma*, colour), in *bot.*, having the central florets of a different colour from those of the circumference.

heterocysts, n. plu., *hĕt'ĕr·ō·sĭsts* (Gr. *heteros*, another ; *kustis*, a bag), in *bot.*, colourless large cells, incapable of division, occurring at intervals 'in the threads of Nostochineæ.

heterodromous, a., *hĕt'ĕr·ŏd'·rŏm·ŭs* (Gr. *heteros*, another, different; *drŏmos*, a course), in *bot.*, having spirals running in opposite directions ; running in different directions, applied to the arrangement of the leaves when these

follow a different direction in the branches from that pursued in the stem.

heterœcium, n., *hĕt'ĕr·ē'·shĭ·ŭm* (Gr. *heteros*, another, different; *oikos*, a house), applied to the potato fungus, so named on the supposition that it exists as a parasite on some other plant before it attacks the potato, and so the potato fungus has received various names accordingly : **heterœcism,** *hĕt'ĕr·ē·sĭzm*, the state or condition of a parasitic fungus, which is found in one stage of development on one body, and in another stage of development on quite a different body.

heterogamous, a., *hĕt'ĕr·ŏg'·ăm·ŭs* (Gr. *heteros*, another, different ; *gamos*, marriage), in *bot.*, having the essential parts of fructification on different spikelets arising from the same root ; having hermaphrodite and unisexual flowers on the same head, as in Compositæ ; **heterogamy,** n., *hĕt'ĕr·ŏg'·ăm·ĭ*, a change in the function of male and female flowers ; the state in which the sexual organs are arranged in some unusual manner.

heterogangliate, a., *hĕt'ĕr·ō·găng'·glĭ·āt* (Gr. *heteros*, another, different ; *gangglion*, a little tumour under the skin), in *zool.*, having a nervous system in which the ganglia are scattered and unsymmetrical, as in the Mollusca.

heterogeneous, a., *hĕt'ĕr·ō·jĕn'·ē·ŭs* (Gr. *heteros*, another, different; *gĕnos*, birth, race ; *gennăō*, I generate, I produce), of a different kind or nature ; confused and contradictory: **heterogenesis,** n., *hĕt'ĕr·ō·jĕn'·ĕs·ĭs* (Gr. *genesis*, origin, source), the doctrine that certain organisms are capable of giving origin to others totally different from themselves, and which show no tendency to revert to the parent form ; spontaneous generation, in which living cells are supposed to be produced by inorganic matter.

heterologous, a., *hĕt'·ĕr·ŏl'·ŏg·ŭs* (Gr. *heteros*, another, different ; *logos*, speech, appearance), in *anat.*, applied to growths which, originating in the development of indifferent formative cells, end in developing a tissue diverse from the matrix, as cartilage in the testicle, etc.; a synonym of ' heteroplastic.'

heteromerous, a., *hĕt'·ĕr·ŏm'·ĕr·ŭs* (Gr. *heteros*, another, different ; *mĕros*, a part, a portion), in *bot.*, applied to lichens where the thallus appears stratified by the crowding of the gonidia into one layer, and the hyphæ form two layers ; in *zool.*, applied to the coleopterous insects which have five joints in the tarsus of the first and second pairs of legs, and only four joints in the tarsus of the third pair : **Heteromerans**, n. plu., *hĕt'·ĕr·ŏm'·ĕr·ănz* (Gr. *mēros*, the upper part of the thigh), coleopterous insects whose legs have a different structure one from another.

heteromorphic, a., *hĕt'·ĕr·ō·mŏrf'·ĭk* (Gr. *heteros*, another, different; *morphē*, shape, form), differing in form or shape ; in *bot.*, having different forms of flowers as regards stamens and pistils, these being necessary for fertilization, as in Primula : **heteromorphism**, n., *hĕt'·ĕr·ō·mŏrf'·ĭzm*, a deviation from the natural form or structure : **heteromorphous**, a., *hĕt'·ĕr·ō·mŏrf'·ŭs*, having an irregular or unusual form : **heteromorphy**, n., *hĕt'·ĕr·ō·mŏrf'·ĭ*, deformity in plants ; heteromorphism.

Heterophagi, n. plu., *hĕt'·ĕr·ŏf'·ă·jī* (Gr. *heteros*, another; *phago*, I eat), those birds whose young are born in a helpless condition, and which require to be fed by the parents for a longer or shorter period ; birds that are foster-parents to young birds of a different kind, as to the young of the cuckoo.

heterophyllous, a., *hĕt'·ĕr·ō·fĭl'·lŭs* (Gr. *heteros*, another ; *phullon*, a leaf), in *bot.*, presenting two different forms of leaves : **heterophylly**, n., *hĕt'·ĕr·ō·fĭl'·lĭ*, the variation in the leaves of plants in external form.

heteroplastic, a., *hĕt'·ĕr·ō·plăst'·ĭk* (Gr. *heteros*, another ; *plastikos*, formed, fashioned—from *plasso* I form), in *anat.*, applied to those growths which are unlike the tissues from which they take their rise; syn. of ' heterologous,' which see.

heterorhizal, a., *hĕt'·ĕr·ō·rīz'·ăl* (Gr. *heteros*, another, different ; *rhiza*, a root), in *bot.*, having rootlets proceeding from various points of a spore during germination; rooting from no fixed point.

heterosporous, a., *hĕt'·ĕr·ō·spōr'·ŭs* (Gr. *heteros*, another; *spora*, spore, seed), in cryptogamic plants, having both microspores and macrospores on the same individual, as in Selaginella.

heterotaxy, n., *hĕt'·ĕr·ō·tăks'·ĭ* (Gr. *heteros*, another; *taxis*, arrangement), in *bot.*, the deviation of organs from their ordinary position.

heterotropal, a., *hĕt'·ĕr·ŏt'·rŏp·ăl* (Gr. *heteros*, another, different ; *tropos*, a turn, manner—from *trepō*, I turn), in *bot.*, lying across ; applied to the embryo of seeds when they lie in an oblique position; applied to the ovule when it is so attached to the placenta that the hilum is in the middle, and the foramen and chalaza at opposite ends, thus becoming transverse.

Hevea, n., *hĕv·ē'·ă* (not ascertained), a genus of plants, Ord. Euphorbiaceæ: **Hevea Brasiliensis**, *brăz·ĭl'ĭ·ĕns'·ĭs* (of or from

Brazil), the Para rubber tree, which yields caoutchouc.

hexagonenchyma, n., *hĕks'·ă·gŏn·ĕng'·kĭm·ă* (Gr. *hexagŏnios* or *hexagŏnos*, six-angled; *engchŭma*, an infusion—from *chuma*, tissue, juice), cellular tissue which when cut in any direction exhibits a hexagonal form.

hexagynous, a., *hĕks·ădj'·ĭn·ŭs* (Gr. *hex*, six; *gunē*, female), in *bot.*, having six styles or pistils.

hexandrous, a., *hĕks·ăn'·drŭs* (Gr. *hex*, six; *anēr*, a man, *andros*, of a man), having six stamens, as a flower.

hexapetalous, a., *hĕks'·ă·pĕt'·ăl·ŭs* (Gr. *hex*, six; *petalon*, a petal), in *bot.*, having six petals or flower leaves: **hexapetaloid**, a., *hĕks'·ă·pĕt'·ăl·ōyd* (Gr. *eidos*, resemblance), having six coloured parts like petals.

hexapod, n., *hĕks'·ă·pŏd* (Gr. *hex*, six; *pous*, a foot, *podos*, of a foot), a creature possessing six legs, as insects: **hexapodous**, a., *hĕks·ăp'·ŏd·ŭs*, having six legs.

hibernacula, n., *hĭb'·ĕr·năk'·ūl·ă* (L. *hibernācula*, winter quarters), a name applied to the leaf buds, as the winter quarters of the young branches; the winter quarters of a wild animal, or of a plant.

Hibisceæ, n. plu., *hĭb·ĭs'·sĕ·ē* (L. *hibiscum*, Gr. *hibiskos*, a species of wild mallow), a Tribe or Sub-order of the Ord. Malvaceæ: **Hibiscus**, n., *hĭb·ĭs'·kŭs*, a genus of plants producing showy flowers of a variety of colours in the species: **Hibisca rosa-sinensis**, *rōz'·ă·sīn·ĕns'·ĭs* (L. *rŏsa*, a rose; *Sinensis*, Chinese), a species possessing astringent properties, used by the Chinese to blacken their eyebrows and their shoes: **H. esculentus**, *ĕsk'·ūl·ĕnt'·ŭs* (L. *esc-ŭlentus*, good for food — from *esca*, food), whose fruit, from its abundant mucilage, a common ingredient in soups of hotter climates, under the name Ochro and Gombo: **H. cannabinus**, *kăn·năb'·ĭn·ŭs* (L. *cannăbĭnus*, of or belonging to hemp—from L. *cannabis*, Gr. *kannabis*, hemp), produces the Sunnee-hemp of India, yields a fibre like jute: **H. mutabilis**, *mūt·ăb'·ĭl·ĭs* (L. *mūtăbĭlis*, mutable—from *muto*, I change), a species which receives its name from the changing colour of its flowers, varying from a pale rose to a pink colour.

hiccough and **hiccup**, n., *hĭk'·ŭp* (Dut. *huckup*, F. *hoquet*, hiccough; Dut. *hikken*, to sob), a very troublesome affection, due to a short, abrupt contraction or convulsion of the diaphragm.

hickory, n., *hĭk'·ŏr·ĭ* (not ascertained), a nut-bearing American tree, whose wood possesses great strength and tenacity; the Carya alba, Ord. Inglandaceæ.

Hieracium, n., *hī'·ĕr·ā'·sĭ·ŭm* (Gr. *hiĕrax*, a hawk — said to be so called because eaten by the hawk, or its juice used by it for sharpening its sight), an extensive genus of pretty flowering plants, adapted for rockwork, Ord. Compositæ.

hiera picra, *hī'·ĕr·ă pĭk'·ră* (Gr. *hieros*, sacred ; *pikros*, bitter), a popular remedy for constipation, known by the name 'hickory pickory,' consisting of a mixture of equal parts of canella bark and aloes.

hilum, n., *hĭl'·ŭm* (L. *hĭlum*, a speck, a little thing), the eye of a seed ; the scar or spot in a seed indicating the point where the seed was attached to the pericarp, as the dark mark at the one end of a bean ; in *zool.*, **hilum** or **hilus**, a small fissure or aperture ; a small depression.

hip, n., *hĭp* (Dut. *heupe*, Norse *hupp*, the flank, the hip), the projection caused by the haunch bone and its covering flesh ; the upper fleshy part of the thigh.

Hippoboscidæ, n. plu., *hĭp'·pō·bŏs'·ĭd·ē* (Gr. *hippos*, a horse ; *bosko*, I feed), a family of dipterous insects, belonging to the Viviparous section of the Ord. Diptera, generally known by the name 'forest flies :' **Hippobosca**, n., *hĭp'·pō·bŏsk'·ă*, a genus of insects which live upon quadrupeds and birds : **Hippobosca equina**, *ĕ·kwin'·ă* (L. *equĭnus*, belonging to a horse—from *equus*, a horse), the horse fly.

hippocamp, n., *hĭp'·pō·kămp*, also **hippocampus**, n., *hĭp'·pō·kămp'·ŭs* (Gr. *hippos*, a horse ; *kampto*, I bend or curve ; *hippokampos*, L. *hippocampus*, the sea-horse), the sea-horse, a fabulous monster; a small fish of singular shape, with head and neck like a horse, called the Pipe-fish or Sea-horse; in *anat.*, one of the two convolutions of the brain resembling a ram's horn or the shape of a sea-horse, named respectively **hippocampus minor** and **hippocampus major**, that is, the lesser and greater hippocampus.

Hippocrateæ, n. plu., *hĭp'·pō·krāt'·ĕ·ē* (after *Hippocrates*, an ancient Greek physician, and one of the fathers of botany), a Tribe or Suborder of the Ord. Celastraceæ : **Hippocratea**, n., *hĭp'·pō·krāt'·ĕ·ă*, a genus of mostly climbing shrubs with very minute flowers : **Hippocratea comosa**, *kōm·ōz'·ă* (L. *comōsus*, hairy—from *cŏma*, the hair of the head), yields nuts which are oily and sweet.

Hippomane, n., *hĭp·pŏm'·ăn·ē* (Gr. *hippomănes*, furious with desire, a plant which is said by the ancients to have driven horses mad if eaten by them — from *hippos*, a horse ; *mănia*, madness), a genus of plants, Ord. Euphorbiaceæ: **Hippomane mancinella**, *măn'·sĭn·ĕl'·lă* (mod. L. *mancinella*, It. *mancinello*, the manchineel), the manchineel tree, growing 40 or 50 feet high in the W. Indian islands, yields a milky juice very acrid and poisonous, which applied to the skin excites violent inflammation and ulceration.

hippopathology, n., *hĭp'·pō·păth·ŏl'·ŏ·jĭ* (Gr. *hippos*, a horse; Eng. *pathology*), the doctrine or description of the diseases of horses ; the science of veterinary medicine.

Hippophae, n., *hĭp·pŏf'·ă·ē* (Gr. *hippos*, a horse ; *phăō*, I destroy, in allusion to the supposed poisonous qualities of the seed), a genus of ornamental trees, Ord. Elæagnaceæ : **Hippophaê rhamnoides**, *răm·nŏyd'·ēz* (Gr. *rhamnos*, the white thorn; *eidos*, resemblance), the sea buckthorn, furnished with sharp spines, fruit eaten, and has been used as a preserve.

hippophagy, n., *hĭp·pŏf'·ă·jĭ* (Gr. *hippos*, a horse ; *phago*, I eat), the practice of eating horse flesh: **hippophagi**, n. plu., *hĭp·pŏf'·ă·jī*, those who eat horse flesh.

hippuria, n., *hĭp·pūr'·ĭ·ă* (Gr. *hippos*, a horse ; *ouron*, urine), an excess of hippuric acid in the urine : **hippuric**, a., *hĭp·pūr'·ĭk*, denoting an acid ; a constituent of the urine, obtained in greatest abundance from the urine of horses or cows ; said to be also found in the blood of herbivora.

Hippuris, n., *hĭp·pūr'·ĭs* (Gr. *hippos*, a horse ; *oura*, a tail), a genus of curious aquatic plants, growing best in marshy places, so called from the stem resembling a mare's tail arising from the crowded whorls of very narrow, hair-like leaves ; Ord. Halorageaceæ.

hirsute, a., *hèr'·sūt* (L. *hirsutus*, rough, hairy), covered with long, stiffish hairs, thickly set ; hairy.

Hirudinea, n., *hèr'·ūd·ĭn'·ĕ·ă* (L. *hirūdo*, a leech, a blood-sucker, *hirūdĭnis*, of a leech), in *zool.*, the Order of Annelida including the Leeches.

hispid, a., *hĭsp'ĭd* (L. *hispĭdus*, bristly, rugged), rough ; covered with strong hairs or bristles.

histioid, a., *hĭs'tĭ-ŏyd* (Gr. *histos*, a web, a tissue ; *eidos*, resemblance), in *anat.*, tissue-like.

histogenesis, n., *hĭs'tō-jĕn'ĕs-ĭs*, also **histogeny**, n., *hĭs-tŏdj'ĕn-ĭ* (Gr. *histos*, a web, a tissue ; *gennăō*, I produce), the origin or formation of organic tissue : **histogenetic**, a., *hĭs'tō-jĕn-ĕt'ĭk*, promoting the formation of organic textures ; in *bot.*, applied to minute molecules supposed to be concerned in the formation of cells.

histology, n., *hĭs-tŏl'ō-jĭ* (Gr. *histos*, a web or tissue ; *logos*, discourse), the study of the tissues of the body, especially its minuter elements ; the study of microscopic tissues in animals or plants : **histological**, a., *hĭs'tō-lŏdj'ĭk-ăl*, relating to the description of minute tissues in animals or plants.

histolysis, n., *hĭs-tŏl'ĭs-ĭs* (Gr. *histos*, a web or tissue ; *lusis*, a solution—from *luō*, I dissolve), the disintegration of previously organized structures : **histolytic**, a., *hĭs'tō-lĭt'ĭk*, derived from the disintegration of previously organized structures ; of the nature of histolysis.

hives, n. plu., *hīvz* (as supposed to be shaped something like a *beehive;* may be a corruption of *heave*, to raise), variously applied to skin diseases among children, consisting of vesicles scattered over the body ; a popular name for chicken-pox.

holly, n., *hŏl'lĭ* (AS. *holegn*), an evergreen shrub having prickly leaves, and producing red berries; the leaves and bark said to possess tonic and febrifuge properties, while the berries are emetic and purgative ; the wood is esteemed in turnery, etc., and the bark furnishes bird - lime ; systematic name, Ilex aquifolium, Ord. Aquifoliaceæ.

hollyhock, n., *hŏl'lĭ-hŏk* (*holly*, a corruption of *holy*, as supposed to have been brought from the *Holy Land* ; AS. *hoc*, W. *hocys*, mallows), a tall, beautiful garden flowering plant, employed medicinally in Greece, yields fibres and a blue dye; the Althæa rosea, Ord. Malvaceæ.

Holocephali, n. plu., *hŏl'ō-sĕf'ăl-ĭ* (Gr. *holos*, whole ; *kephalē*, the head), in *zool.*, a Sub-order of the Elasmobranchii, comprising the Chimæræ.

holometabolic, a., *hŏl'ō-mĕt-ă-bŏl'ĭk* (Gr. *holos*, whole; *metăbŏle*, change), applied to insects which undergo a complete metamorphosis.

holosericeous, a., *hŏl'ō-sĕr-ĭsh'ŭs* (Gr. *holos*, whole ; *serikos*, L. *sericus*, silky), covered with minute silky hairs, best discovered by touch.

Holostomata, n. plu., *hŏl'ō-stŏm'ăt-ă* (Gr. *holos*, whole ; *stŏma*, a mouth, *stŏmăta*, mouths), a division of gasteropodous molluscs in which the aperture of the shell is rounded or entire.

Holothuroidea, n. plu., *hŏl'ō-thūr-ŏyd'ĕ-ă* (Gr. *holothourion*, a zoophyte resembling a sponge ; *eidos*, resemblance), an Order of Echinodermata, comprising the Trepangs.

Homaliaceæ, n. plu., *hŏm-ăl'ĭ-ā'sĕ-ē*, also **Homaliads**, n. plu., *hŏm-ăl'ĭ-ădz* (Gr. *homalos*, uniform, regular), the Homalia family, an Order of tropical trees and shrubs bearing flowers in spikes or racemes : **Homalium**, n., *hŏm-ăl'ĭ-ŭm*, a genus, so called because their stamens are regularly divided into three stamened fascicles.

homocarpous, a., *hŏm'ō-kărp'ŭs* (Gr. *homos*, alike ; *karpos*, fruit), having all the fruits of a flower-head alike.

homocercal, a., *hŏm'·ō·sérk'·ăl* (Gr. *hŏmos*, alike; *kerkos*, the tail), having equally-bilobate tails, as in the herring, the cod, etc.; composed of two equal lobes.

homochromous, a., *hŏm'·ō·krōm'·ŭs* (Gr. *hŏmos*, alike; *chrōma*, colour), having all the flowerets on the same flower-head of the same colour.

homodromous, a., *hŏm·ŏd'·rŏm·ŭs* (Gr. *hŏmos*, alike; *dromos*, a race-course), in *bot.*, running in the same direction, as spirals, or leaves on the stem and branches.

homœopathy, n., *hŏm'·ē·ŏp'·ăth·ĭ* (Gr. *homoios*, similar, like; *pathos*, suffering), a mode of treating diseases by the adminis-tration of medicines capable of exciting in healthy persons symptoms closely similar to those of the disease for which they are given; a theory of medical practice opposed to that commonly known as Allopathy.

homogamous, a., *hŏm·ŏg'·ăm·ŭs* (Gr. *homogamos*, married to-gether — from *homos*, alike, similar; *gamos*, marriage), in *bot.*, applied to composite plants having the flowers of the capitula all hermaphrodite.

homogangliate, a., *hŏm'·ō·găng'·glĭ·āt* (Gr. *homos*, like; *gangglion*, a knot), in *zool.*, having a nervous system in which the ganglia are symmetrically arranged.

homogeneous, a., *hŏm'·ō·jēn'·ĕ·ŭs* (Gr. *homos*, like; *genos*, kind), of the same kind or nature; having a uniform structure or substance; opposed to 'hetero-geneous.'

homologous, a., *hŏm·ŏl'·ŏg·ŭs* (Gr. *homŏlogos*, using the same words, of the same opinion—from *homos*, like, similar; *logos*, speech, ap-pearance), having the same ratio or proportion; constructed on the same plan, though differing

in form and function; in *anat.*, having a growth like normal tissues of the body, as opposed to 'heterologous'; in *chem.*, applied to analogous bodies whose com-positions differ by a constant difference : **homologue**, n., *hŏm'·ō·lŏg*, correspondence or equivalence of certain organs; a part in one animal which strictly represents a part in a different animal, as the arms in man, the wings in birds, and the pectoral fins in fishes : **homology**, n., *hŏm·ŏl'·ŏ·jĭ*, affinity dependent on structure or the essential corre-spondence of parts; the identity of parts which are apparently distinct; similarity of structure of different parts, as between the upper and lower limbs, exhibiting a community of plan.

homomorphy, n., *hŏm'·ō·mŏrf'·ĭ* (Gr. *hŏmos*, like, similar; *morphē*, shape, form), in *bot.*, the con-dition of the Compositæ when the disc florets assume the form of ray florets; the fertilization of the pistil by the pollen from its own flowers; self-fertilization : **homomorphic**, a., *hŏm'·ō·mŏrf'·ĭk*, having the pistil fertilized by the pollen from its own flowers : **homomorphous**, a., *hŏm'·ō·mŏrf'·ŭs*, in *zool.*, having a similar ex-ternal appearance or form.

homoomerous, a., *hŏm'·ō·ŏm'·ĕr·ŭs* (Gr. *homoios*, like, similar; *meros*, a part), in *bot.*, applied to lichens where the gonidia and hyphæ in the thallus appear about equally mingled.

homopetalous, a., *hŏm'·ō·pĕt'·ăl·ŭs* (Gr. *homos*, like; *petalon*, a leaf), in *bot.*, having all the petals formed alike; having all the florets alike in a composite flower.

homotropal, a., *hŏm·ŏt'·rŏp·ăl* (Gr. *homos*, like; *tropos*, a turning), in *bot.*, having the same general direction as the body of which it forms a part; applied to the slightly curved embryo when it has the same general direction as the seed.

homotype, n., *hŏm'·ō·tīp* (Gr. *homos*, like, similar; *tupos*, form, a type), that part of an animal which corresponds to another part; correspondence of parts which lie in series, as the bones of the foot with those of the hand: **homotypy,** n., *hŏm·ŏt'·ĭp·ĭ*, the state or condition of such correspondence: **homotypic,** a., *hŏm'·ō·tĭp'·ĭk*, pert. to; homologous.

honey-suckle, n., *hŭn'·ĭ-sŭk'·l* (Eng. *honey*, and *suckle*), a well-known climbing plant and flower; the common name of the plants of the genus Lonicera, Ord. Caprifoliaceæ; honey-suckle is sometimes applied to meadow clover, Trifolium pratense; the French honey-suckle is Hedysarum coronarium.

Honkeneja, n., *hŏng'·kĕn·ē'·jă* (an Iceland word), a genus of plants, Ord. Caryophyllaceæ: **Honkeneja peploides,** *pĕp·lŏyd'·ēz* (Gr. *peplos*, a covering, a robe; *eidos*, resemblance), a species which has been used as a pickle, and in Iceland as an article of food.

hops, n. plu., *hŏps* (Ger. *hopfen*, Dut. *hoppen*, hops), a climbing plant whose seeds or flowers are employed in imparting bitterness to beer and ale; the Humulus lupulus, Ord. Cannabinaceæ.

hordeolum, n., *hŏrd·ē'·ŏl·ŭm* (a dim. of L. *hordeum*, barley), inflammation of one of the meibomian glands in the margin of the eyelid, so called from its likeness in size and hardness to a small barley-corn; the stye.

Hordeum, n., *hŏrd'·ē·ŭm* (L. *hordeum*, barley), a genus of the cereal grains, the barleys and barley grasses, Ord. Gramineæ: **Hordeum vulgare,** *vŭlg·ār'·ē* (L. *vulgaris*, general, common), common barley: **H. hexastichum,**

hĕks·ăst'·ĭk·ŭm (Gr. *hex*, six; *stix*, order, rank, *stichos*, of order or rank), bere or bigg, a variety of barley.

horehound, n., *hōr'·hoŵnd* (AS. *hara-hune* — from *har*, hoary, grey; *hune*, consumption), a native wild plant, supposed to act as a tonic and expectorant, but not now used by physicians; the Marrubium vulgare, Ord. Labiata.

horn-beam, n., *hŏrn'·bēm* (Goth. *haurn*, horn; Ger. *baum*, Dut. *boom*, a tree), a tree whose wood is white, hard, and heavy, hence its name; the Carpinus betulus, Ord. Cupuliferæ or Corylaceæ.

horse-chestnut, n., the Æsculus hippocastanum, Ord. Sapindaceæ.

horse-radish, n., the Cochlearia Armoracia, Ord. Cruciferæ: **horse-radish tree,** the Moringa pterygosperma, Ord. Moringaceæ.

hortus siccus, *hŏrt'·ŭs sĭk'·kŭs* (L. *hortus*, a garden; *siccus*, dry), in *bot.*, a collection of dried plants preserved between paper or in books; a herbarium.

hospitalism, n., *hŏs'·pĭt·ăl·ĭzm* (L. *hospitālis*, hospitable — from *hospes*, a guest), the prejudicial influences of large hospital buildings upon sick residents, especially when the patients are numerous; the subject of hospital construction.

houseleek, n., *hoŵs'·lēk* (Eng. *house;* Icel. *laukr*, a leek), a well-known herb, the Sempervivum tectorum, Ord. Crassulaceæ.

Hoya, n., *hōy'·ă* (after *Thomas Hoy*, a botanist and gardener), a genus of plants, Ord. Asclepiadaceæ, which bear very handsome waxy flowers: **Hoya carnosa,** *kăr·nōz'·ă* (L. *carnōsus*, fleshy— from *căro*, flesh), the wax-flower, so named from the peculiar aspect of its blossoms.

humerus, n., *hŭm'·ĕr·ŭs* (L. *hŭmĕrus*, the shoulder), the arm from the shoulder to the elbow; the bone of that part, consisting of two parts, the scapula and the clavicle : **humeral**, a., *hūm'·ĕr·ăl*, pert. to the shoulder.

humifuse, a., *hūm'·ĭ·fūz* (L. *hŭmus*, the ground ; *fūsus*, spread), in *bot.*, spreading over the surface of the ground ; procumbent.

Humiriaceæ, n. plu., *hūm'·ĭr·ĭ·ā'·sĕ·ē* (formed probably from *Umiri*, where found), the Humiriads, an Order of plants of Brazil, which some place as a Sub-order under the Ord. Meliaceæ : **Humiria**, n., *hŭm·ĭr'·ĭ·ă*, a genus : **Humiria floribunda**, *flōr'·ĭ·bŭnd'·ă* (L. *flos*, a flower, *floris*, of a flower ; *abundans*, abounding), a species whose trunk, when wounded, yields a liquid yellow balsam, called balsam of Umiri : **H. balsamifera**, *băl'·săm·ĭf'·ĕr·ă* (L. *balsămum*, balsam ; *fero*, I bear), yields a balsam used for perfumery and in medicine.

humor or **humour**, n., *hūm'·ŏr* (L. *humor*, fluid of any kind, moisture ; F. *humeur*), any moisture or fluid of the body except the blood ; certain parts of the eye which abound in fluid: **humoral**, a., *hūm'·ŏr·ăl*, pert. to the fluids of the body or proceeding from them ; in *med.*, applied to that doctrine which ascribes all diseases to a degenerate or disordered state of the fluids of the body : **aqueous humor**, the watery matter which fills the space in the forepart of the eyeball between the cornea and iris.

Humulus, n., *hūm'·ŭl·ŭs* (L. *hŭmus*, the earth, the ground), a genus of creeping plants, Ord. Cannabinaceæ, constituting the well - known Hop, extensively cultivated in some parts of England, so named as it creeps along the ground if not supported : **Humulus lupulus**, *lōōp'·ŭl·ŭs* (dim. of L. *lŭpus*, a wolf), the common hops, the strobili

of the female plants of which constitute the hops ; employed as a tonic and narcotic in the form of extract, infusion, and tincture.

humus, n., *hūm'·ŭs* (L. *hŭmus*, earth, soil), vegetable mould, the product of decayed vegetation.

Hura, n., *hūr'·ă* (S. Amer. name), a genus of plants, Ord. Euphorbiaceæ : **Hura crepitans**, *krĕp'·ĭt·ănz* (L. *crepĭtans*, creaking, crackling), the sand-box tree or monkey's dinner-bell, the juice of which is very acrid ; the numerous parts of its fruit, when dry, separate from each other with great force.

husk, n., *hŭsk* (Dut. *hulsche*, the covering of seeds), the external covering of many fruits and seeds; the pericarp.

Hyacinthus, n., *hī'·ă·sĭnth'·ŭs* (L. *Hyăcinthus*, Gr. *Huakinthos*, a beautiful youth, beloved by Apollo, and accidentally killed by a blow of his quoits, and from whose blood the flowers sprang ; the blue iris, corn-flag, or gladiolus of the ancients), a beautiful and well-known genus of bulbous plants, Ord. Liliaceæ : **Hyacinthus orientalis**, *ōr'·ĭ·ĕnt·āl'·ĭs* (L. *orientālis*, oriental—from *oriens*, arising), the hyacinth, a popular spring flower having numerous garden varieties and various colours of flowers.

hyaline, a., *hī'·ăl·ĭn* (Gr. *hualos*, glass), consisting of or resembling glass ; in *med.*, clear and of a slight consistence like a jelly ; in *bot.*, transparent or colourless : n., a substance which originates the cell-nucleus, or the part where the cell-nucleus appears: **hyaloid**, a., *hī'·ăl·ŏyd* (Gr. *eidos*, resemblance), like glass ; transparent : n., an extremely thin and clear membrane.

hybrid, n., *hī'·brĭd* (L. *hybrida*, a hybrid, a mongrel—from Gr.

hubris, a wanton act, an outrage), an animal or plant the produce of different kinds or species ; a plant resulting from the fecundation of one species by another : **adj.**, having the origin or character of a hybrid : **hybridisation, n.**, *hī'·brĭd·ĭz·ā'·shŭn*, the act of rendering hybrid.

hydatids, n. plu., *hĭd'·ăt·ĭdz*, and **hydatides, n. plu.**, *hĭd·ăt'·ĭd·ēz* (Gr. *hudatis*, a vesicle, *hudatidos*, of a vesicle—from *hudōr*, water), little vesicles or bladders, with fluid or semi - fluid contents, found in the bodies of animals in a state of disease, and containing the larval forms of parasites : **hydatid mole**, the product of a morbid pregnancy consisting of bunches of mucoid vesicles, having a general resemblance to clusters of grapes.

Hydnocarpus, n., *hĭd'·nō·kȧrp'·ŭs* (Gr. *hudnon*, a tuber ; *karpos*, fruit), a genus of small trees, Ord. Bixaceæ : **Hydnocarpus venenatus**, *věn'·ěn·āt'·ŭs* (L. *venēnātus*, poisonous—from *venēnum*, poison), a species which produces a fruit of the size of an apple, which the Cingalese use to poison fish ; the seeds contain an oil used medicinally.

Hydnora, n., *hĭd·nōr'·ȧ* (see **Hydnum**), a genus of root parasites having a fungus-like aspect, Ord. Cytinaceæ: **Hydnora Africana**, *ȧf'·rĭk·ān'·ȧ* (*Africānus*, of or from Africa), a parasitic flowering plant of very singular construction, which attacks the roots of the Cistus, some succulent Euphorbiaceæ, and other plants.

Hydnum, n., *hĭd'·nŭm* (Gr. *hudnon*, a mushroom), a genus of mushrooms, Ord. Fungi : **Hydnum coralloides**, *kŏr'·ăl·ōyd'·ēz* (L. *corallum*, Gr. *korallion*, red coral), a species of mushroom which are eatable, found under the trunks of trees in moist situations.

hydra, n., *hĭd'·rȧ* (Gr. *hudra*, L. *hydra*, the hydra, a water snake; Gr. *hudor*, water), a water snake ; a fabulous monster serpent having many heads, slain by Hercules ; a fresh - water polype : **hydraform, a.**, *hĭd'·rȧ·fŏrm* (L. *forma*, shape), resembling the common fresh - water polype or hydra in form.

hydragogue, n., *hĭd'·rȧ·gŏg* (Gr. *hudor*, water ; *ago*, I lead), a medicine which produces copious watery stools.

Hydrangeæ, n. plu., *hĭd·rānj'·ě·ē* (Gr. *hudor*, water; *anggeion*, a vessel, a capsule), a Sub-order of the Ord. Saxifragaceæ : **Hydrangea, n.**, a genus of plants, pretty when in flower, so called from the capsules of some of the species appearing like a cap : **Hydrangea Thunbergii**, *tŭn·běrj'·ĭ·ī* (after *Thunberg*, a celebrated traveller and botanist), a species whose leaves furnish a tea of a very *recherché* character, bearing the name Ama-tsja in Japan.

hydranth, n., *hĭd'·rănth* (Gr. *hudra*, a water serpent ; *anthos*, a flower), the polypite or proper nutritive zöoid of the Hydrozoa.

hydrargyrum, n., *hĭd·rȧrj'·ĭr·ŭm* (Gr. *hudrarguros*, fluid silver—from *hudor*, water ; *arguros*, silver), quicksilver or mercury : **hydrargyria, n. plu.**, *hĭd'·rȧr·jĭr'·ĭ·ȧ*, one of the ill effects of mercury applied locally : **hydrargyriasis, n.**, *hĭd·rȧr'·jĭr·ĭ'·ȧs·ĭs*, a disease produced by the abuse of mercury.

Hydrastis, n., *hĭd·răs'·tĭs* (Gr. *hudor*, water), a genus of plants growing in moist situations, Ord. Ranunculaceæ : **Hydrastis Canadensis**, *kăn'·ăd·ěns'·ĭs* (of or from *Canada*), a species whose yellow roots are used as a tonic ; yellow root.

hydrate, n., *hĭd'·rāt* (Gr. *hudōr*,

water), a compound containing a definite proportion of water chemically combined : **hydrated,** a., *hĭd′·răt·ĕd,* combined with water in definite proportions : **hydration,** n., *hĭd·rā′·shŭn,* the act or state of becoming chemically combined with water.

hydraulic, a., *hĭd·rawl′·ĭk* (Gr. *hudōr,* water ; *aulos,* a pipe), relating to the conveyance of water through pipes ; worked by water : **hydraulics,** n. plu., *hĭd·rawl′·ĭks,* the science which treats of the application of the forces influencing the motions of fluids; the art of raising, conducting, and employing water for practical purposes.

hydrencephalocele, n., *hĭd′·rĕn·sĕf·ăl′·ō·sēl* (Gr. *hudōr,* water ; *engkephalon,* the brain ; *kēlē,* a tumour), a tumour occasioned by hernial protrusion of the membrane of the brain and the fluid contents of the cranium, through a deficiency in the latter.

hydro, *hĭd′·rō,* and **hydr,** *hĭd′·r* (Gr. *hudōr,* water), prefixes in scientific terms denoting the presence, action, or quality of water ; denoting the presence of hydrogen : **hydro-carbon,** *kârb′·ŏn* (Eng. *carbon*), a compound of hydrogen and carbon ; a term usually applied to bitumens, mineral resins, and mineral fats which are composed of hydrogen and carbon in varying proportions: **hydro-carburet,** n., *hĭd′·rō·kârb′·ŭr·ĕt,* a compound of hydrogen and carbon ; hydro-carbon.

hydrocaulus, n., *hĭd′·rō·kawl′·ŭs* (Gr. *hudra,* a water serpent ; *kaulos,* a stem), in *zool.,* the main stem of the cœnosarc of a hydrozoön.

hydrocele, n., *hĭd′·rō·sēl* (Gr. *hudōr,* water ; *kēlē,* a tumour), dropsy of the testicle ; a collection of serum in the external or serous covering of the testicle.

hydrocephalus, n., *hĭd′·rō·sĕf′·ăl·ŭs* (Gr. *hudor,* water ; *kephalē,* the head), a disease chiefly characterised by an accumulation of serous fluid in the central cavities of the brain, and frequently a result of tubercular disease ; dropsy or water in the head : **hydrocephalic,** a., *hĭd′·rō·sĕf·ăl′·ĭk,* relating to or connected with hydrocephalus.

Hydrocharidaceæ, n. plu., *hĭd′·rō·kăr′·ĭd·ā′·sĕ·ē* (Gr. *hudor,* water ; *charis,* grace, beauty), the Frogbit family, an Order of floating or aquatic plants found in various parts of the world : **Hydrocharis,** n., *hĭd·rŏk′·ăr·ĭs,* a genus of pretty aquatic plants, forming one of the prettiest ornaments of our still waters.

hydrochlorate, n., *hĭd′·rō·klōr′·āt* (Eng. *hydrogen* and *chlorine*), a compound of hydrochloric acid with a base : **hydrochloric,** a., *hĭd′·rō·klōr′·ĭk,* consisting of a combination of hydrogen and chlorine; denoting an acid known also as muriatic acid and spirit of salt.

Hydrocotyle, n., *hĭd′·rō·kŏt′·ĭl·ē* (Gr. *hudor,* water ; *kotulē,* a hollow, a cavity), a genus of plants, Ord. Umbelliferæ : **Hydrocotyle Asiatica,** *āzh′·ĭ·ăt′·ĭk·ă* (of or from *Asia*), a species used in medicine: **H. vulgaris,** *vŭl·gār′·ĭs* (L. *vulgaris,* general, common), a curious little native Umbellifer, called Pennywort, having round peltate leaves, growing in marshy situations, and reported injurious to sheep.

hydrocyanic, a., *hĭd′·rō·sĭ·ăn′·ĭk* (Gr. *hudōr,* water ; *kuanos,* dark-blue), denoting an acid consisting of hydrogen and cyanogen ; Prussic acid : **hydrocyanate,** n., *hĭd′·rō·sĭ′·ăn·āt,* a compound of hydrocyanic acid with a base.

hydrocysts, n. plu., *hĭd′·rō·sĭsts* (Gr. *hudra,* a water serpent ; *kustis,* a bladder, a cyst), in *zool.,* curious processes attached to the cœnosarc of the Physophoridæ, and termed feelers.

Hydrodictyon, n., *hĭd'·rō·dĭk'·tĭ·ŏn* (Gr. *hudōr*, water ; *diktuon*, a fishing-net), a genus of plants, Ord. Algæ or Hydrophyta, so named from the reticulated structure of the plants : **Hydrodictyon utriculatum,** *ŭt·rĭk'·ŭl·āt'·ŭm* (L. *utrĭcŭlus*, a small skin or leathern bottle), a species called 'water net,' which has the appearance of a green net composed of filaments enclosing pentagonal and hexagonal spaces.

hydrœcium, n., *hĭd·rē'·shĭ·ŭm* (Gr. *hudra*, a water serpent ; *oikos*, a house), the chamber into which the cœnosarc in many of the Calycophoridæ can be retracted.

hydrogen, n., *hĭd'·rō·jĕn* (Gr. *hudōr*, water ; *gennaō*, I produce), a metal which, in its gaseous form, is the lightest of all known bodies, producing water when combined with oxygen : **sulphuretted hydrogen,** a combination of hydrogen with sulphur, producing a gas having a smell like rotten eggs, found as a constituent of mineral waters.

Hydroida, n. plu., *hĭd·rŏyd'·ă* (Gr. *hudra*, a water snake ; *eidos*, resemblance), in *zool.*, the subclass of the Hydrozoa which comprises the animals most nearly allied to the hydra ; in *geol.*, an extensive genus of zoophytes.

hydrometra, n., *hĭd'·rō·mēt'·ră* (Gr. *hudōr*, water ; *mētra*, womb), an excessive secretion and accumulation of fluid within the cavity of the uterus.

hydronephrosis, n., *hĭd'·rō·nĕf·rŏz'·ĭs* (Gr. *hudōr*, water ; *nephros*, kidney), dropsy of the kidney, caused by any permanent obstruction of the ureter.

hydropathy, n., *hĭd·rŏp'·ăth·ĭ* (Gr. *hudor*, water ; *pathos*, feeling), the water cure : **hydropathic,** a., *hĭd'·rō·păth'·ĭk*, relating to the water cure.

hydropericardium, n., *hĭd'·rō·pĕr·ĭ·kârd'·ĭ·ŭm* (Gr. *hudōr*, water ; *peri*, round about ; *kardia*, the heart), an effusion of serum into the sac of the pericardium or membrane enclosing the heart ; dropsy of the pericardium.

hydrophobia, n., *hĭd'·rō·fōb'·ĭ·ă* (Gr. *hudor*, water ; *phobos*, fear, dread), a disease occurring in the human being after being bitten by any rabid animal, characterised by an aversion to water, and more or less general convulsions.

Hydrophyllaceæ, n. plu., *hĭd'·rō·fĭl·lā'·sĕ·ē* (Gr. *hudōr*, water ; *phullon*, a leaf), the Hydrophyllum family, an Order of trees and herbaceous plants, many of which have showy flowers, and some have glandular or stinging hairs : **Hydrophylleæ,** n. plu., *hĭd'·rō·fĭl'·lĕ·ē*, a Sub-order : **Hydrophyllum,** n., *hĭd'·rō·fĭl'·lŭm*, a genus.

Hydrophyllia, n. plu., *hĭd'·rō·fĭl'·lĭ·ă* (Gr. *hudra*, a water snake ; *phullon*, a leaf), in *zool.*, overlapping appendages or plates which protect the polypites in some of the oceanic Hydrozoa ; also termed 'bracts.'

Hydrophyta, n. plu., *hĭd·rŏf'·ĭt·ă* (Gr. *hudōr*, water ; *phuton*, a plant), the Sea-weed family ; the Algæ or cellular plants found both in salt and in fresh water : **hydrophyte,** n., *hĭd'·rō·fĭt*, a plant which lives and grows in water only.

hydrorhiza, n., *hĭd'·rō·rĭz'·ă* (Gr. *hudra*, a water snake ; *rhiza*, a root), in *zool.*, the adherent base or proximal extremity of any hydrozoön.

hydrosoma, n., *hĭd'·rō·sōm'·ă* (Gr. *hudra*, a water snake ; *soma*, body), in *zool.*, the entire organism of any hydrozoön.

hydrosulphuret, n., *hĭd'·rō·sŭlf'·ŭr·ĕt* (Eng. *hydrogen* and *sulphur*), a compound of hydrosulphuric acid with a base : **hyd-**

rosulphuric, a., *hĭd'·rō·sŭlf·ūr'·ĭk*, pert. to or derived from hydrogen and sulphur.

hydrotheca, n. plu., *hĭd'·rō·thēk'·ā* (Gr. *hudra*, a water snake; *thēkē*, a chest), in *zool.*, the little chitinous cups in which certain polypites are protected.

hydrothorax, n., *hĭd'·rō·thōr'·ăks* (Gr. *hudōr*, water; *thōrax*, the chest), a dropsical accumulation of fluid in the pleural sac; water in the chest.

Hydrozoa, n. plu., *hĭd'·rō·zō'·ă* (Gr. *hudra*, a water serpent; *zoon*, an animal), in *zool.*, gelatinous, oblong, or conical polypes organized like the hydra; the class of the Cœlenterata comprising animals constructed like the hydra.

hydruria, n., *hĭd·rôr'·ĭ·ă* (Gr. *hudōr*, water; *ouron*, urine), an excessive secretion of limpid, watery urine.

hygiene, n., *hĭ'·jĭ·ēn'* (L. *Hygēïa*, Gr. *Hugeia*, the goddess of health), that department of medical practice which treats of health, its preservation, restoration, and maintenance.

hygrometer, n., *hĭ·grŏm'·ĕt·ėr* (Gr. *hugros*, wet, moist; *logos*, discourse), an instrument for measuring the degree of moisture in the atmosphere : **hygrometric**, a., *hĭ'·grō·mĕt'·rĭk*, of or relating to the hygrometer ; in *bot.*, moving under the influence of moisture.

hygrophanous, a., *hĭ·grŏf'·ăn·ŭs* (Gr. *hugros*, wet; *phaino*, I show), in *bot.*, appearing watery when moist, but becoming opaque when dry.

Hygrophorus, n., *hĭ·grŏf'·ŏr·ŭs* (Gr. *hugros*, wet, moist; *phoreo*, I bear), a genus of plants, Ord. Fungi : **Hygrophorus pratensis**, *pră·tĕns'·ĭs* (L. *prātensis*, growing in meadows — from *prātum*, a meadow), a species of fungi, called the Herefordshire truffle.

hygroscope, n., *hĭ'·grō·skōp* (Gr.

hugros, water, moisture ; *skopeo*, I see or view), an instrument to show the moisture or dryness of the air : **hygroscopic**, a., *hĭ'·grō·skŏp'·ĭk*, pert. to ; applied to moisture not readily apparent.

hymen, n., *hĭm'·ĕn* (Gr. *humēn*, a thin membrane; Gr. *Humēn*, L. *Hymen*, the god of marriage, the son of Bacchus and Venus), the valvular fold of membrane which protects the virginal vagina : **hymeneal**, a., *hĭm'·ĕn·ē'·ăl*, pert. to marriage.

Hymenæa, n., *hĭm'·ĕn·ē'·ă* (Gr. *Humēn*, L. *Hymen*, the god of marriage), a genus of trees, Ord. Leguminosæ, Sub-ord. Cæsalpinieæ, whose species are highly ornamental, so named from its two leaflets: **Hymenæa Courbaril**, *kōr'·băr·ĭl* (unascertained), the West Indian locust tree ; the pods supply a nutritious matter, its inner bark is anthelmintic, and the plant yields a kind of resin called Animé.

hymenium, n., *hĭm·ēn'·ĭ·ŭm* (Gr. *humēn*, a membrane), in *bot.*, that portion of the fructification of a fungus in which the sporules are situated, usually more or less a membranous expansion ; the part which bears the fructification in Agarics : **hymenial**, a., *hĭm·ēn'·ĭ·ăl*, belonging to the hymenium: **hymenicolar**, a., *hĭm'·ĕn·ĭk'·ŏl·ăr* (L. *colo*, I inhabit), in *bot.*, inhabiting the hymenium.

Hymenomycetes, n., *hĭm·ĕn'·ō·mĭ·sēt'·ēz* (Gr. *humēn*, a membrane; *mukēs*, a fungus), a division of the Fungi in which the hymenium is naked; the spores appear in sets of four, borne on distinct sporophores, as seen in mushrooms.

hymenophorum, n., *hĭm'·ĕn·ŏf'·ŏr·ŭm* (Gr. *humēn*, a membrane ; *phoreo*, I bear), in *bot.*, the structure which bears the hymenium.

Hymenophylleæ, n. plu., *hĭm'·ĕn·ō·fŭl'·lĕ·ē* (Gr. *humēn*, a membrane;

phullon, a leaf), the Filmy Fern tribe, a Sub-order of the Ord. Filices or Ferns.

Hymenoptera, n. plu., *hīm'·ĕn·ŏp'·tĕr·ǎ* (Gr. *humēn*, a membrane ; *pteron*, a wing), an Order of insects characterised by the possession of four membranous wings, as in bees, ants, etc.

Hymenothalameæ, n. plu., *hīm'·ĕn·ō·thǎl·ăm'·ĕ·ē* (Gr. *humēn*, a membrane ; *thalămos*, a nest, a receptacle), a section or Sub-order of the Lichens, characterised by their open shields, and the nucleus bearing the sporangia on their surface.

hymenulum, n., *hīm·ĕn'·ŭl·ŭm* (a dim. from Gr. *humēn*, a membrane), in *bot.*, a shield containing asci.

hyo, *hī'·ō*, a prefix denoting connection with the hyoid bone : **hyoid**, a., *hī'·ōyd* (the Greek letter *v*, *ŭpsilon*, from the shape of the bone ; *eidos*, resemblance), the U - shaped bone situated between the tongue and the larynx : **hyoglossus**, n., *hī'·ō·glōs'·sŭs* (Gr. *glōssa*, tongue), a flat quadrate muscle, arising from the whole length of the great corner of the hyoid bone and the tongue.

Hyoscyamus, n., *hī'·ŏs·sī'·ăm·ŭs* (L. *hyoscyamus*, Gr. *huoskuamos*, henbane—from Gr. *hus*, a hog ; *kuamos*, a bean, in allusion to the fruit being eaten by swine), a genus of plants, Ord. Solanaceæ: **Hyoscyamus niger**, *nĭdj'·ĕr* (L. *nĭger*, black, dark), henbane, a biennial poisonous plant, with dingy yellow flowers, exhibiting beautiful purple reticulations, and having hairy viscous leaves ; a tincture of henbane is often used as a mild narcotic, and its oil is an energetic poison : **hyoscyamia**, n., *hī'·ŏs·sī·ăm'·ĭ·ǎ*, an alkaloid obtained from hyoscyamus, to which the plant owes its narcotic properties.

hypanthodium, n., *hīp'·ăn·thŏd'·ĭ·ŭm* (Gr. *hupo*, under ; *anthos*, a flower), a fleshy receptacle enclosing the flowers, as in the fig ; the receptacle of Dorstenia, bearing many flowers.

hyperæmia, n., *hīp'·ĕr·ēm'·ĭ·ǎ* (Gr. *huper*, over ; *haima*, blood), an excessive accumulation of blood in a part of the body ; a local or partial excess of blood.

hyperæsthesia, n., *hīp'·ĕr·ĕz·thē'·zhĭ·ā* (Gr. *huper*, above, over ; *aisthēsis*, perception, sensation), excessive or morbid sensibility, as intolerance of light, sound, etc.

hypercatharsis, n., *hīp'·ĕr·kăth·ărs'·ĭs* (Gr. *huper*, above, over ; *kathairo*, I purge), excessive purging of the bowels.

Hypericaceæ, n. plu., *hīp'·ĕr·ĭ·kā'·sĕ·ē* (Gr. *hupereikon*, the plant St. John's wort — from *ereikē*, heath, heather), the Tutsan or St. John's wort family, an Order of plants distributed very generally over all parts of the globe, which yield a resinous coloured juice, having purgative properties and resembling gamboge: **Hypericum**, n., *hīp·ĕr'·ĭk·ŭm*, an extensive genus, most of whose species produce showy plants: **Hypericum connatum**, *kŏn·nāt'·ŭm* (L. *connātus*, born with—from *nātus*, born), a species from which a gargle for sore throats is prepared in Brazil : **H. hircinum**, *hĕr·sĭn'·ŭm* (L. *hircĭnus*, of or from a goat—from *hircus*, a he-goat), a species having a fetid odour : **H. laxiusculum**, *lǎks'·ĭ·ŭsk'·ŭl·ŭm* (L. *laxus*, wide, loose ; *juscŭlum*, juice), a species, a decoction from whose leaves is esteemed a specific against the bite of serpents in Brazil : **H. perforatum**, *pĕrf'·ōr·āt'·ŭm* (L. *perforātus*, bored or pierced through), St. John's wort, much esteemed by the ancients as an anodyne.

hyperostosis, n., *hĭp'·ĕr·ŏs·tōz'·ĭs* (Gr. *huper*, over ; *osteon*, a bone), an unnatural growth or projection from a bone ; same as ' exostosis.'

hyperplasia, n., *hĭp'·ĕr·plās'·ĭ·ā* (Gr. *huper*, over ; *plasso*, I form), the excessive multiplication of the elements of a part.

hyperpyrexia, n., *hĭp'·ĕr·pĭr·ĕks'·ĭ·ā* (Gr. *huper*, over ; Eng. *pyrexia*), the temperature of any body when over 106° F.

hypertrophy, n., *hĭp·ĕr'·trŏf·ĭ* (Gr. *huper*, over ; *trophe*, food, nourishment), excessive growth of a part; an increase of size in the healthy structure of an organ, due to increased exercise or nutrition, as in the arms of a blacksmith, or in the limbs of an athlete ; in *bot.*, enlargement of organs.

hypha, n., *hīf'·ā*, **hyphæ**, n. plu., *hīf'·ē* (Gr. *huphē*, weaving), the filamentous tissue in the thallus of lichens : **hyphal**, a., *hīf'·ăl*, pert. to a filamentous tissue.

Hyphæne, n., *hīf·ēn'·ĕ* (Gr. *huphaino*, I weave), a genus of ornamental palm trees, Ord. Palmæ : **Hyphæne thebaica**, *thē·bā'·ĭk·ā* (L. *Thēbaĭcus*, of or from *Thebes*, in Egypt), the doom-palm of Egypt, whose pericarp has the taste of gingerbread, and is used as food.

hyphasma, n., *hīf·ăz'·mā* (Gr. *huphē*, weaving), in *bot.*, a weblike thallus of Agarics ; the mycelium of certain fungi ; same sense as 'hypha.'

Hypnum, n., *hĭp'·nŭm* (Gr. *hupnon*, moss or lichen), the most extensive genus among mosses, Ord. Musci or Bryaceæ, known by their prostrate, pinnated, bright green branches.

hypocarpogean, a., *hĭp'·ō·kărp'·ō·jē'·ăn* (Gr. *hupo*, under ; *karpos*, fruit ; *gē*, earth), in *bot.*, producing their fruit below ground, as in the ground nut.

hypochilium, n., *hĭp'·ō·kĭl'·ĭ·ŭm* (Gr. *hupo*, under ; *cheilos*, the lip), in *bot.*, the lower part of the labellum when it is divided, as in Orchids.

hypochondrium, n., *hĭp'·ō·kŏn'·drĭ·ŭm*, also **hypochondria**, n., *-drĭ·ā* (Gr. *hupochondria*, the viscera that lie under the cartilage of the ribs—from *hupo*, under ; *chondros*, cartilage), the part of the belly under the short ribs containing the liver and spleen ; a disease characterised by uneasiness about the region of the stomach and liver : **hypochondriasis**, n., *hĭp'·ō·kŏn·drĭ'·ăs·ĭs*, a form of insanity in which the patient converts an idea of purely mental origin into what appears to him to be a real material change ; a morbid self-consciousness similar in some respects to hysteria, but with the belief in the patient that he is suffering under numerous severe diseases: **hypochondriac**, a., *hĭp'·ō·kŏnd'·rĭ·ăk*, affected by severe depression of spirits: n., one who is suffering under severe depression of spirits ; a sufferer from hypochondriasis.

hypocotyledonary, a., *hĭp'·ō·kŏt'·ĭl·ēd'·ŏn·ăr·ĭ* (Gr. *hupo*, under ; Eng. *cotyledon*), in *bot.*, applied to peculiar thickened roots whose structure it is often difficult to determine, and which have the aspect of stems.

hypocrateriform, a., *hĭp'·ō·krăt·ĕr'·ĭ·fŏrm* (Gr. *hupo*, under ; *kratēr*, a cup ; L. *forma*, shape), in *bot.*, shaped like a saucer or salver, as the corolla of primula.

hypodermic, a., *hĭp'·ō·dĕrm'·ĭk* (Gr. *hupo*, under ; *derma*, the skin), applied or inserted under the skin : **hypoderma**, n., *hĭp'·ō·dĕrm'·ā*, in *bot.*, the layers of tissue lying beneath the epidermis, and serving to strengthen it : **hypodermis**, n., *hĭp'·ō·dĕrm'·ĭs*, in *bot.*, the inner layer of moss thecæ.

hypogastrium, n., *hĭp′·ō·găst′·rĭ·ŭm* (Gr. *hupo*, under ; *gastēr*, the belly), the lower anterior part of the abdomen, extending from the pubes to within about two inches of the umbilicus, and to each side as far as a line drawn upright from the anterior extremity of the crest of the haunch bone (ilium).: **hypogastric**, a., *hĭp′·ō·găst′·rĭk*, pert. to the middle part of the lower region of the belly.

hypogeous, a., *hĭp′·ō·jē′·ŭs*, also **hypogeal**, a., *hĭp′·ō·jē′·ăl* (Gr. *hupo*, under ; *gē*, the earth), in *bot.*, applied to the parts of plants growing beneath the surface of the soil : **hypogenous**, a., *hĭp·ŏdj′·ĕn·ŭs*, in *bot.*, growing beneath.

hypoglossal, a., *hĭp′·ō·glŏs′·săl* (Gr. *hupo*, under ; *glōssa*, the tongue), applied to the ninth pair of nerves, situated beneath the tongue.

hypogynous, a., *hĭp·ŏdj′·ĭn·ŭs* (Gr. *hupo*, under ; *gunē*, a female), in *bot.*, inserted below the ovary or pistil: **hypogyn**, n., *hĭp′·ō·jĭn*, a hypogynous plant.

hyponasty, n., *hĭp′·ō·năst·ĭ* (Gr. *hupo*, under ; *nastos*, pressed together, stuffed), in *bot.*, a form of nutation when the organs grow most rapidly on the dorsal side ; see 'epinasty.'

hypophloeodal, a., *hĭp′·ō·flē′·ŏd·ăl* (Gr. *hupo*, under ; *phloios*, bark), in *bot.*, existing beneath the epidermis of the bark.

hypophyllous, a., *hĭp′·ō·fĭl′·lŭs* (Gr. *hupo*, under ; *phullon*, a leaf), in *bot.*, situated under the leaf ; growing from the under side of a leaf.

hypophysis cerebri, *hĭp·ŏf′·ĭs·ĭs sĕr′·ĕb·rĭ* (Gr. *hupo*, under ; *phuō*, I grow ; L. *cĕrĕbrum*, the brain, *cerebri*, of the brain), the pituitary body ; a small reddish grey mass of a somewhat flattened oval shape, widest in the transverse direction, occupying the 'sella turcica' of the sphenoid bone.

hypospadias, n., *hĭp′·ō·spăd′·ĭ·ăs* (Gr. *hupo*, under ; *spădizo*, I pull or tear off), a malformation sometimes occurring in the under surface of the penis.

hyposporangium, n., *hĭp′·ō·spŏr·ănj′·ĭ·ŭm* (Gr. *hupo*, under; *spora*, seed ; *anggos*, a vessel), in *bot.*, the indusium of ferns growing from beneath the spore-case.

hypostome, n., *hĭp·ŏs′·tŏm·ē* (Gr. *hupo*, under ; *stoma*, a mouth), in *zool.*, the upper lip or labium of certain crustacea, as in the Trilobites.

hypothallus, n., *hĭp′·ō·thăl′·lŭs*, -thalli, n. plu., *-thăl′·lī* (Gr. *hupo*, under ; Gr. *thallos*, L. *thallus*, a young shoot or branch), delicate fungoid filaments, upon which a lichen thallus is first developed ; the mycelium of certain entophytic fungi, as Uredines.

hypothecium, n., *hĭp′·ō·thē′·shĭ·ŭm* (Gr. *hupo*, under ; *thekē*, a case), the cellular disc beneath the thalamium in lichens, which bears the thecæ.

hypothenar eminence, *hĭp·ŏth′·ĕn·ăr* (Gr. *hupo*, under ; *thĕnar*, the palm of the hand), the fleshy mass at the inner border of the hand, consisting of three muscles passing to the little finger.

hypoxanthin, n., *hĭp′·ŏks·ănth′·ĭn* (Gr. *hupo*, under ; *xanthos*, yellow), a peculiar organic compound found in the fluid of the spleen, and in very small quantity in muscle.

Hypoxidaceæ, n. plu., *hĭp·ŏks′·ĭd·ā′·sĕ·ē* (Gr. *hupo*, under ; *oxus*, sharp-pointed—referring to the base of the capsule), the Hypoxis family, an Order of herbaceous and usually stemless plants, some having bitter roots, and others edible tubers : **Hypoxis**, n., *hĭp·ŏks′·ĭs*, a genus of plants, natives of warm countries.

hypsometry, n., *hĭps·ŏm′·ĕt·rĭ* (Gr. *hupsos*, height ; *metron*, a measure), the method of ascertaining heights by the barometer, or by boiling water : **hypsometrical**, a., *hĭps'·ŏm·ĕt'·rĭk·ăl*, pert. to.

hypsophyllary, a., *hĭps'·ō·fĭl'·lăr·ĭ* (Gr. *hupsos*, top, summit ; *phullon*, a leaf), in *bot.*, applied to leaves which are bracts.

Hyptis, n., *hĭp'·tĭs* (Gr. *huptios*, lying on the back with the face upward—from *hupo*, under), a genus of shrubby plants, Ord. Labiatæ, so called because the limb of the corolla is turned on its back : **Hyptis membranacea**, *mĕm'·brān·ā'·sĕ·ă* (L. *membrāna*, skin or membrane), a species which attains the height of 20 or 30 feet in Brazil.

Hyracoidea, n. plu., *hĭr'·ăk·oȳd'·ĕ·ă* (Gr. *hurax*, a shrew ; *eidos*, resemblance), an Order of the Mammalia with the single genus Hyrax : **Hyrax**, n., *hĭr'·ăks*, the rock badger of the Cape : **hyraceum**, n., *hĭr·ās'·ĕ·ŭm*, a substance resembling castor in smell and properties, obtained from its urine.

hyssop, n., *hĭs'·sŏp* (Gr. *hussōpos*, L. *hyssopus*, hyssop), a garden plant having an aromatic smell and pungent taste, formerly used as a stomachic : **Hyssopus**, n., *hĭs·sŏp'·ŭs*, a genus of plants, Ord. Labiatæ : **Hyssopus officinalis**, *ŏf·fĭs'·ĭn·āl'·ĭs* (L. *officinālis*, official), the common hyssop; the hyssop in Scripture is supposed to be a species of caper, Capparis Ægyptiaca.

hysteranthous, a., *hĭst'·ĕr·ănth'·ŭs* (Gr. *husteros*, coming after ; *anthos*, a flower), in *bot.*, expanding after the flowers have opened, as leaves.

hysteria, n., *hĭs·tēr'·ĭ·ă*, also **hysterics**, n., *hĭs·tēr'·ĭks* (Gr. *husterikos*, caused by the womb—from *hustĕra*, the womb), a nervous disease or affection, not altogether peculiar to women, and not necessarily connected with the womb or ovaries, but due to an

imperfectly balanced mental and moral system : **hysteric**, a., *hĭs·tēr'·ĭk*, and **hysterical**, a., *hĭs·tēr'·ĭk·ăl*, affected with or liable to hysterics.

Hysterophyta, n. plu., *hĭs'·tĕr·ŏf'·ĭt·ă*, also **hysterophytes**, n. plu., *hĭs·tĕr'·ō·fītz* (Gr. *hustĕra*, the womb ; *phuton*, a plant), another name for the order Fungi ; plants living upon dead or living organic matter, as the Fungi.

Iceland moss, *īs'·lănd mŏs* (moss from *Iceland*), the Cetraria Islandica, Ord. Lichenes, a lichen used as a demulcent and tonic in the form of a decoction or jelly, found chiefly in northern regions, and used in Iceland and Lapland as food.

ice plant, *īs plănt*, the Mesembryanthemum crystallinum, Ord. Ficoideæ or Mesembryaceæ, a plant remarkable for the watery vesicles which cover its surface, having the appearance of particles of ice.

ichor, n., *īk'·ŏr* (Gr. *ichor*, matter, gore), a thin, watery, humor-like whey flowing from an ulcer : **ichorous**, a., *īk'·ŏr·ŭs*, like ichor; serous.

ichthyic, a., *īk'·thĭ·ĭk* (Gr. *ichthus*, a fish), relating to fishes : **ichthyoid**, a., *īk'·thĭ·oȳd* (Gr. *eidos*, resemblance), resembling a fish : **ichthyology**, n., *īk'·thĭ·ŏl'·ŏ·jĭ* (Gr. *logos*, discourse), that branch of zoology which treats of the structure, the classification, the habits, and the history of fishes : **Ichthyomorpha**, n. plu., *īk'·thĭ·ō·mŏrf'·ă* (Gr. *morphē*, shape), an Order of Amphibians, called also Urodela, comprising the fish-like newts : **Ichthyophthira**, n. plu., *īk'·thĭ·ŏf·thĭr'·ă* (Gr. *phtheir*, a louse), an Order of Crustacea comprising animals which are parasitic upon fishes : **Ichthyopsida**, n. plu., *īk'·thĭ·ŏps'·ĭd·ă* (Gr. *opsis*, appearance), the primary division of Vertebrata, comprising fishes and amphibia.

ichthyosis, n., *ĭk'·thĭ·ōz'·ĭs* (Gr. *ichthua*, the dried rough skin of the dog‑fish — from *ichthus*, a fish), a cutaneous disease in which the skin is dry, harsh, and rough, and apparently too tight for the body ; a form of the disease in which dry, hard, greyish or slate‑coloured scales appear on different parts of the body.

icosandria, n. plu., *ĭk'·ŏs·ănd'·rĭ·ă* (Gr. *eikosi*, twenty ; *hedra*, a seat, a basis), plants which have twenty or more stamens inserted on the calyx : **icosandrous,** a., *ĭk'·ŏs·ănd'·rŭs*, having twenty stamens.

icterus, n., *ĭk'·tĕr·ŭs* (Gr. *ikteros*, L. *ictĕrus*, jaundice), jaundice : **icterus neonatorum,** *nē'·ō·năt·ōr'·ŭm* (Gr. *neos*, new, fresh ; L. *natorum*, of the new‑born—from *nātus*, born), the jaundice of the new‑born ; yellow gum in new‑born infants.

idiocy, n., *ĭd'·ĭ·ŏs·ĭ* (Gr. *idiotes*, a private individual — from *idios*, proper, peculiar to oneself), a form of insanity where the mind from the first is imperfectly developed, and remains permanently in this undeveloped state : **idiot,** n., *ĭd'·ĭ·ŏt*, a human being more or less defective in regard to his mental or moral powers.

idiopathy, n., *ĭd'·ĭ·ŏp'·ăth·ĭ* (Gr. *idios*, peculiar; *pathos*, suffering), a morbid state or condition not dependent on or caused by any other: **idiopathic,** a., *ĭd'·ĭ·ō·păth'·ĭk*, not depending on any other disease ; arising without any apparent exciting cause ; the opposite of 'sympathetic.'

idiosyncrasy, n., *ĭd'·ĭ·ō·sĭng'·krăs·ĭ* (Gr. *idios*, peculiar ; *sungkrasis*, a mixing together), an unusual peculiarity of an individual in consequence of which he is affected in a different manner from the majority by one or several influences; that condition of mind or body commonly known as 'antipathy.'

idiot, see 'idiocy.'

Idiothalameæ, n. plu., *ĭd'·ĭ·ō·thăl·ăm'·ē·ē* (Gr. *idios*, peculiar ; *thalamos*, a receptacle), a section of the Lichens, having their shields closed at first and open afterwards, containing free spores in a nucleus composed of the gelatinous remains of the paraphyses and sporangia : **idiothalamous,** a., *ĭd'·ĭ·ō·thăl'·ăm·ŭs*, possessed of a colour or texture differing from the thallus in lichens.

Ignatia amara, *ĭg·nā'·shĭ·ă ăm·ār'·ă* (*St. Ignatius; amārus*, bitter), St. Ignatius's bean, producing strychnia ; also called Strychnos Ignatia.

ileo, *ĭl'·ĕ·ō*, denoting connection with the ileum, or some relation to it: **ileo‑cæcal,** *sēk'·ăl* (L. *cæcus*, blind), applied to two semi‑lunar folds of mucous membrane found at the termination of the ileum in the large intestine, forming the division between the cæcum and colon.

ileum, n., *ĭl'·ĕ·ŭm* (L. and Gr. *ileos*, a severe kind of colic—from Gr. *eileo*, I turn or twist), the lower portion of the small intestine, so called from its numerous convolutions : **ileus,** n., *ĭl'·ĕ·ŭs*, an obstruction in the bowels accompanied by vomiting, pain, and fever ; intussusception of the bowels ; iliac passion.

Ilex, n., *ĭl'·ĕks* (L. *ilex*, a kind of oak), a genus of elegant trees and shrubs, having evergreen prickly foliage, Ord. Aquifoliaceæ : **Ilex aquifolium,** *ăk'·wĭ·fōl'·ĭ·ŭm* (L. *ăcus*, a needle ; *fŏlium*, a leaf), the common holly, indigenous in Britain ; the leaves and bark are said to possess tonic and febrifuge properties, and its berries emetic and purgative : **I. Paraguensis,** *păr'·ă·gwĕns'·ĭs* (of or from *Paraguay*), a species which furnishes Yerba maté, or Paraguay tea : **I. vomitoria,** *vŏm'.ĭt·ōr'.ĭ·ă* (L. *vomitōrius*, that provokes vomiting), a species from whose leaves the black drink of the Creek Indians is prepared.

iliac, a., *ĭl'.ĭ·ăk* (L. and Gr. *ĭlĕos*, a severe kind of colic—from Gr. *eilĕō*, I turn or twist: L. *ilia*, the flanks, the entrails), pert. to the ileum, or to the bone called ilium : **iliac passion,** a vomiting of bilious and fæcal matter in consequence of obstruction in the intestinal canal ; colic : **iliac regions,** the sides of the abdomen between the ribs and the hips : **iliac crest,** an eminence on the ilium resembling lines, but broader and more prominent: **iliacus,** n., *ĭl·ĭ'·ăk·ŭs,* a flat radiated muscle which fills up the whole of the internal iliac fossa : **iliacus internus,** *ĭn·tĕrn'· ŭs* (L. *internus*, that which is within), a muscle situated in the cavity of the ilium : **iliacum os,** another name for the 'os innominatum,' which see : **ilium os,** *ĭl'.ĭ· ŭm ŏs* (L. *ilia*, the flanks ; *os*, a bone), the large, partly-flattened bone, forming the principal part of the pelvis, and entering into the composition of the hip-joint : **ilia,** n. plu., *ĭl'.ĭ·ă,* the flanks, the loins ; the part extending from the lowest ribs to the groin: **ilio,** *ĭl'.ĭ·ō,* a word denoting connection with the 'iliacum os.'

Ilicineæ, n. plu., *ĭl'.ĭ·sĭn'.ĕ·ē* (L. *ilex,* a kind of oak, *ilĭcis,* of an oak), the Holly family, an Order of plants, now generally called Aquifoliaceæ.

Illecebreæ, n. plu., *ĭl'.lĕ·sĕb'.rĕ·ē* (L. *illĕcĕbra,* an attraction, an allurement ; plants so named by Pliny), a section or sub-order of plants, Ord. Paronychiaceæ: **Illecebrum,** n., *ĭl·lĕs'.ĕb·rŭm,* a genus of pretty and interesting dwarf plants.

Illicium, n., *ĭl·lĭsh'.ĭ·ŭm* (L. *illĭcĭo,* I allure or attract), a genus of useful plants, Ord. Magnoliaceæ, so named from the agreeable perfume of the species : Illicium **anisatum,** *ăn'.ĭs·āt'.ŭm* (L. *anĭsum,* Gr. *anĭson,* the anise plant), the star anise, so called from its carpels being arranged in a star-like manner, and having the taste and odour of anise.

imago, n., *ĭm·āg'.ō* (L. *imāgo,* an image, an apparition), the third or perfect state of an insect, the first being the 'larva,' and the second the 'pupa.'

imbecile, n., *ĭm'.bĕs·ēl* (L. *imbēcillus,* feeble, weak), an idiot of a higher grade ; a weak-minded or facile person : **imbecility,** n., *ĭm'. bĕs·ĭl'.ĭ·tĭ,* a deficiency of mental and moral powers ; a state short of idiocy.

imbibition, n., *ĭm'.bĭb·ĭsh'.ŭn* (L. *imbĭbo,* I drink in—from *im,* into ; *bĭbo,* I drink), the action by which the passage of a fluid, or of gaseous matters, is affected through dead and living tissues ; endosmosis.

imbricate, a., *ĭm'.brĭk·āt,* also imbricated, a.,-āt·ĕd (L. *imbricatum,* to form like a gutter tile—from *imbrex,* a tile), in *bot.,* having parts overlying each other like tiles on a house; in *zool.,* applied to scales or plates which overlap one another like tiles : **imbricative,** a., *ĭm·brĭk'.āt·ĭv,* overlapping at the edge : **imbricated æstivation,** in *bot.,* the parts of the flower-bud alternatively overlapping each other, and arranged in a spiral manner.

immarginate, a., *ĭm·mărj'.ĭn·āt* (L. *im,* not ; *margo,* a border, *margĭnis,* of a border), in *bot.,* not having a border or margin.

impaction, n., *ĭm·păk'.shŭn* (L. *impactus,* driven into—from *im,* into ; *pango,* I drive), a disease in cattle, sheep, horse, fowls, etc., a fatal case of indigestion in which the food becomes closely impacted in the stomach ; becoming hard and dry, it is incapable of digestion, and the animal shortly dies; the stomach staggers.

impari-pinnate, a., *ĭm'·păr·ĭ·pĭn'·nāt* (L. *impar*, unequal; *pinnatus*, winged), unequally pinnate; a pinnate leaf ending in an odd leaflet.

Impatiens, n., *ĭm·pā'·shĭ·ĕnz* (L. *impătĭens*, that will not endure, impatient), a genus of very beautiful and singular plants, Ord. Balsaminaceæ, so named from the elastic valves of the capsules bursting when touched, and throwing out the seeds with great force.

imperforate, a., *ĭm·pĕr'·fŏr·āt* (L. *in*, into; *per*, through; *forātus*, bored), not bored or pierced through; without a terminal opening.

impetigo, n., *ĭm'·pĕt·ĭg'·ō* (L. *impetigo*, a scabby eruption—from *impĕto*, I attack), a skin disease, characterised by clusters of pustules which run into a crust; pustular eruptions: impetiginous, a., *ĭm'·pĕt·ĭdj'·ĭn·ŭs*, having the nature of or pert. to impetigo.

impregnation, n., *ĭm'·prĕg·nā'·shŭn* (L. *im*, in; *prægnātus*, pregnancy), the act of impregnating or rendering fruitful; fertilisation.

impressio colica, *ĭm·prĕs'·sĭ·ō kŏl'·ĭk·ă* (L. *impressio*, an impression; *colĭcus*, of or pert. to the colic), the colic impression; a shallow impression in front on the under surface of the right lobe of the liver.

inanition, n., *ĭn'·ăn·ĭsh'·ŭn* (L. *ĭnānĭs*, empty), starvation; a condition brought about by bad food, or food deficient in quantity.

inarching, n., *ĭn·ärtsh'·ĭng* (L. *in*, into; *arcus*, a bow; *arcuo*, I bend like a bow), a mode of grafting by bending two growing plants towards each other, and

causing a branch of the one to unite to a branch of the other.

inarticulate, a., *ĭn'·ärt·ĭk'·ŭl·āt* (L. *in*, not; *articulatus*, furnished with joints), in *bot.*, without joints or interruption to continuity.

incanescent, a., *ĭn'·kăn·ĕs'·sĕnt* (L. *incanescens*, becoming grey or hoary), in *bot.*, having a grey or hoary appearance.

incised, a., *ĭn·sīzd'* (L. *incīsus*, cut into—from *in*, into; *cædo*, I cut), in *bot.*, cut down deeply: incision, n., *ĭn·sĭzh'·ŭn*, a division of several tissues of the body, generally by a sharp-cutting instrument: incisive, a., *ĭn·sīz'·ĭv* having the quality of cutting; situated near the incisor teeth, or relating to them: incisors, n. plu., *ĭn·sīz'·ŏrs*, the four front teeth both in the upper and lower jaws, for cutting, dividing, or tearing the food before chewing or masticating it: incisura, n., *ĭn'·sīz·ūr'·ă*, a cut, gash, or notch.

included, a., *ĭn·klŏd'·ĕd* (L. *includo*, I shut up or in), in *bot.*, having the stamens enclosed within the corolla, and not pushed out beyond its tube.

incompatibles, n., *ĭn'·kŏm·păt'·ĭ·bls* (L. *in*, not; Eng. *compatible*), in *med.*, remedies which when mixed together destroy each other's effects, or materially alter them.

inconspicuous, a., *ĭn'·kŏn·spĭk'·ū·ŭs* (L. *in*, not; Eng. *conspicuous*), in *bot.*, small in size; not easily observed.

incontinence, n., *ĭn·kŏn'·tĭn·ĕns* (L. *in*, not; *continens*, keeping within bounds), want of restraint in the sexual appetite; inability to restrain natural discharges.

incrassate, a., *ĭn·krăs'·sāt* (L. *in*, into; *crassus*, thick, dense), thickened: incrassation, n., *ĭn'·krăs·sā'·shŭn*, the act of thickening.

incubation, n., *ĭn'·kŭb·ā'·shŭn* (L. *incubātus*, lain or rested upon— from *in*, on ; *cubo*, I lie down), in *med.*, the period during which a contagious disease lies latent before showing itself : **incubus**, n., *ĭn'·kŭb·ŭs* (L. *incŭbus*, the nightmare — from *incŭbo*, I lie upon), the nightmare ; any oppressive ·or stupefying influence.

incumbent, a., *ĭn·kŭm'·bĕnt* (L. *incumbens*, ·leaning or lying upon —from *in*, on ; *cubo* or *cumbo*, I lie down), in *bot.*, applied to cotyledons with the radicle on their back.

incurvate, a., *ĭn·kėrv'·āt* (L. *incurvātus*, bent or curved—from *in*, into ; *curvus*, bent, crooked), in *bot.*, curved inwards or upwards.

incus, ·n., *ĭnk'·ŭs* (L. *incus*, a smith's anvil), a small bone of the ear, so called from its supposed resemblance to an anvil.

indefinite, a., *ĭn·dĕf'·ĭn·ĭt* (L. *in*, not ; Eng. *definite*), in *bot.*, having an inflorescence with a centripetal expansion ; having more than twenty stamens ; having numerous ovules and seeds ; generally denoting uncertainty, or without limit.

indehiscent, a., *ĭn·dĕ·hĭs'·sĕnt* (L. *in*, not ; *dehisco*, I open, I gape, ·*dehiscens*, opening, gaping), in ¡*bot.*, not opening ; having no regular line of suture ; applied to fruits such as the apple, which do not split open.

independence, n., *ĭn'·dĕ·pĕnd'·ĕns* (L. *in*, not ; Eng. *dependence*), in *bot.*, ·the separation of organs usually entire.

indeterminate, a., *ĭn'·dĕ·tėrm'·ĭn·āt* (L. *in*, not ; Eng. *determinate*), in *bot.*, unlimited ; indefinite.

index finger, *ĭn'·dĕks fĭng'·gėr* (L. *indico*, I point out ; *index*, an informer), the forefinger, being that employed in pointing at an object.

indicator, n., *ĭn'·dĭk·āt'·ŏr* (L. *indicatus*, pointed out), in *anat.*, the muscle which extends the forefinger ; the extensor indicis.

indigenous, a., *ĭn·dĭdj'·ĕn·ŭs* (L. *indigĕna*, a native, born and bred in the same country or town), not exotic or introduced, applied to plants ; an aboriginal native in a country.

indigestion, n., *ĭn'·dĭ·jĕst'·yŭn* (L. *indigestus*, confused, disordered), a derangement of the powers of digestion ; a painful or imperfect change of food in the stomach ; dyspepsia.

indigo, n., *ĭn'·dĭg·ō* (F. *indigo*— from L. *indicus*, Indian), a beautiful blue dye, procured by fermentation from various species of Indigofera : **Indigofera**, n., *ĭn'·dĭg·ŏf'·ĕr·ă* (Eng. *indigo ;* L. *fero*, I bear), an extensive genus of elegant plants, Ord. Leguminosæ, Sub-ord. Papilionaceæ, most of whose species produce indigo, chiefly **Indigofera tinctoria**, *tĭngk·tōr'·i·ă* (L. *tinctōrius*, belonging to dyeing—from *tingo*, I dye), also from **I. anil**, *ăn'·ĭl* (Arab. *annil*, the indigo plant); **I. cærulea**, *sĕr·ŏl'·ĕ·ă* (L. *cærŭlĕus*, dark - blue); **I. argentea**, *ăr·jĕnt'·ĕ·ă* (L. *argentĕus*, made of silver—from *argentum*, silver), and many others ; the powdered leaf of **I. anil** has been used in hepatitis.

indumentum, n., *ĭn'·dū·mĕnt'·ŭm* (L. *indumentum*, a garment — from *indŭo*, I put on), the plumage of birds; in *bot.*, a hairy covering.

induplicate, a., *ĭn·dŭp'·lĭk·āt* (L. *in*, in ; *duplĭcātus*, doubled), in *bot.*, having the edges of the sepals or petals turned slightly inwards, in æstivation ; having the margins doubled inwards.

induration, n., *ĭn'·dūr·ā'·shŭn* (L. ·*indurātus*, hardened — from *in*, into ; *duro*, I harden), the hardening, or process of hardening of a part ; the hardening of tissues around a part formerly diseased.

indusia, n., *ĭn·dūz'·ĭ·ă*, **indusiæ,** n. plu., *-ĭ·ē* (L. *indusĭum*, a shirt, a woman's under garment —from *indŭo*, I put on), the cases or coverings of certain insects : **indusium,** n., *ĭn·dūz'·ĭ·ŭm*, in *bot.*, the epidermal covering of the fructification in some ferns ; a collection of hairs so united as to form a sort of cup, and which encloses the stigma of a flower.

indutive, a., *ĭn·dūt'·ĭv* (L. *indūtus*, a putting on—from *indŭo*, I put on), in *bot.*, applied to seeds which have the usual integumentary covering.

inequilateral, a., *ĭn·ĕk'·wĭ·lăt'·ĕr·ăl* (*in*, not ; Eng. *equilateral*), having the two sides unequal, as in the case of the shells of the ordinary bivalves ; not having the convolutions of the shells lying in the same plane, but obliquely wound round an axis, as in the Foraminifera.

inembryonate, a., *ĭn·ĕm'·brĭ·ŏn·āt* (L. *in*, not; Eng. *embryo*), in *bot.*, having neither embryo nor germ.

inenchyma, n., *ĭn·ĕng'·kĭm·ă* (Gr. *ines*, a fibre ; *engchuma*, what is poured in, juice, tissue), in *bot.*, cells in which there is a spiral elastic fibre coiled up in the inside, the cells generally consisting of membrane and fibre combined.

inequivalve, n., *ĭn·ēk'·wĭ·vălv* (L. *in*, not ; Eng. *equivalve*), a valve consisting of two unequal pieces or valves.

inermis, a., *ĭn·ĕrm'·ĭs* (L. *inermis*, unarmed), in *bot.*, unarmed ; without prickles or thorns.

infection, n., *ĭn·fĕk'·shŭn* (L. *infectus*, tainted, dyed—from *in*, into ; *facio*, I make), the act by which poisonous matter or exhalations produce disease in a healthy body ; see 'contagious.'

inferior, a., *ĭn·fēr'·ĭ·ŏr* (L. *inferior*, lower—from *infĕrus*, beneath, below), in *bot.*, growing below, as when one organ is below another ; applied to the ovary when it seems to be situated below the calyx, and to the part of a flower farthest from the axis; below, lower, inner, as opposed to 'superior,' which signifies above, upper, outer : **inferior extremities,** the legs as the lower parts of the body.

infiltration, n., *ĭn'·fĭl·trā'·shŭn* (L. *in*, into ; Eng. *filtration*), the act or process of passing into the textures of a body ; the liquid or substance which has so entered.

inflammation, n., *ĭn'·flăm·mā'·shŭn* (L. *inflammo*, I set on fire—from *in*, in or on ; *flamma*, a flame), redness and heat in some part of the body, accompanied with pain and swelling ; the succession of changes which occurs in a living tissue when injured, provided its structure and vitality are not destroyed.

inflated, a., *ĭn·flāt'·ĕd* (L. *inflātus*, blown into, swollen), in *bot.*, puffed out; distended.

inflexed, a., *ĭn·flĕkst'* (L. *inflexus*, bent, curved), in *bot.*, curved or bent upwards and inwards.

inflorescence, n., *ĭn'·flŏr·ĕs'·sĕns* (L. *inflorescens*, beginning to blossom —from *in*, in or on ; *floresco*, I blossom), a flowering or putting forth blossoms ; the mode in which the flowers are arranged on the axis.

influenza, n., *ĭn'·flŏŏ·ĕnz'·ă* (It. *influenza*, influence ; L. *influens*, flowing into), a specific epidemic fever, chiefly attacking the lining membrane of the nose, larynx, and bronchial tubes, and lasting from four to eight days.

infra-costales, n. plu., *ĭn'·fră·kŏst·āl'·ēz*, also **infra-costals,** n. plu., *-kŏst'·ălz* (L. *infra*, underneath, below; *costa*, a rib), in *anat.*, small bundles of fleshy and tendinous fibres, which vary in number and length, arising from the inner surface of one rib, and inserted into the inner surface of the first, second, or third rib below: **infra-maxillary**, a., *-măks'-ĭl·lăr·ĭ* (L. *maxilla*, the jaw), situated under the jaw, as certain nerves: **infra-orbital**, a., *-ŏrb'·ĭt·ăl* (L. *orbitum*, the orbit), situated underneath the orbit, as an artery: **infra-scapularis**, a., *-skăp'·ŭl·ār'·ĭs* (L. *scapŭla*, the shoulder-blade), situated underneath the shoulder-blade : **infra-spinatus**, a., *-spĭn·āt'·ŭs* (L. *spinātus*, the spine — from *spina*, a thorn), situated underneath a spinous process ; designating a muscle situated beneath the spine of the scapula, and inserted into the humerus.

infundibulum, n., *ĭn'·fŭn·dĭb'·ŭl·ŭm*, **infundibula**, n. plu., *-dĭb'·ŭl·ă* (L. *infundibulum*, a tunnel or funnel), in *anat.*, a name given to various parts of the body which more or less resemble a funnel ; in *zool.*, a tube formed by the coalescence or apposition of the epipodia in the Cephalopoda ; known also as the 'siphon' or 'funnel' : **infundibuliform**, a., *ĭn·fŭn'·dĭb·ŭl'·ĭ·fŏrm* (L. *forma*, shape), funnel-shaped.

infusion, n., *ĭn·fūzh'·ŭn* (L. *in*, into ; *fūsus*, poured, *infusĭŏ*, a pouring into), the operation of steeping a substance in hot or boiling water in order to extract its medicinal or other qualities.

infusoria, n. plu., *ĭn'·fŭz·ōr'·ĭ·ă* (L. *infusus*, poured into, crowded in—from *in*, into ; *fusus*, poured), very minute animal organisms, or animalcules, inhabiting water containing decaying vegetable or animal matter, so named from their being obtained in 'infusions' of vegetable matter that have been exposed to the air ; a class of Protozoa: **infusorial**, a., *ĭn'·fūz·ōr'·ĭ·ăl*, pert. to the infusoria ; obtained by infusion : **infusory**,

a., *ĭn·fūz'·ŏr·ĭ*, applied to a class of animalcules obtained in infusions ; containing infusoria.

ingesta, n. plu., *ĭn·jĕst'·a* (L. *ingestus*, poured or thrown into), things taken in, as food into the stomach ; substances introduced into the digestive organs.

inguinal, a., *ĭng'·gwĭn·ăl* (L. *inguen*, the groin, *inguinis*, of the groin), pert. to the groin ; connected with the groin or situated upon it.

inhumation, n., *ĭn'·hŭm·ā'·shŭn* (L. *in*, in or into ; *hŭmus*, the ground), the act of burying or placing in the ground ; a method of digesting a substance by burying the vessel containing it among dung or warm earth.

inject, v., *ĭn·jĕkt'* (L. *injectus*, thrown or cast into—from *in*, into ; *jactus*, thrown), to throw into : **injected**, a., *ĭn·jĕkt'·ĕd*, applied to a dead body, or a part, whose vessels have been filled by a composition forced into them : **injection**, n., *ĭn·jĕk'·shŭn*, the act of throwing or forcing a liquid into the vessels of a dead body ; the coloured liquid so thrown or forced into such vessels; a clyster, or method of administering remedies of various kinds, and of even feeding the patient by injecting medicinal or nutrient fluids into the lower bowel : **hypodermic injection**, a method of injecting various medicinal solutions beneath the skin by means of a syringe to which a hollow needle is attached.

innate, a., *ĭn'·nāt* (L. *innātus*, inborn, natural—from *in*, into ; *nātus*, born), in *bot.*, adhering to the apex; attached to the top of the filament, as anthers : **innato-fibrillose**, *ĭn·nāt'·ō·fĭb'·rĭl·lōz'* (L. *fibra*, a filament), clad with adherent fibrils.

inner aspect, in *anat.*, the inner appearance of a bone or a part.

innervation, n., *ĭn'·nĕrv·ā'·shŭn* (L. *in*, into ; *nervus*, a nerve), that vital process by which nervous energy is given to any part.

innoma, n., *ĭn·nōm'·ă*, properly **inoma**, *ĭn·ōm'·ă* (Gr. *ĭs*, a fibre, *ĭnos*, of a fibre), in *med.*, a new growth of connective tissue forming a distinct isolated mass, or fibrous tumour.

innominata arteria, *ĭn·nŏm'·ĭn·āt'·ă ăr·tēr'·ĭ·ă* (L. *in*, not; *nomen*, a name; *artēria*, an artery), the unnamed artery; the largest branch artery given off from the arch of the aorta : **innominate**, a., *ĭn·nŏm'·ĭn·āt*, also **innominata os**, *ŏs* (L. *ŏs*, a bone), a bone forming the pelvis, composed of three portions—the 'ilium,' or haunch-bone; the 'ischium,' or hip-bone; and 'os pubis,' or share-bone.

innovations, n. plu., *ĭn'·nŏv·ā'·shuns* (L. *innovātus*, renewed—from *in*, into; *novus*, new), in *bot.*, new growths or extensions of the stems of mosses; buds in mosses.

Inocarpus, n., *ĭn'·ō·kărp'·ŭs* (Gr. *ĭs*, a fibre, *ĭnos*, of a fibre; *karpos*, fruit), a genus of trees, Ord. Thymelæaceæ : **Inocarpus edulis**, *ĕd·ūl'·ĭs* (L. *edūlis*, eatable), a species whose seeds or nuts are eaten when roasted in the S. Sea Islands, and have the taste of chestnuts; the Otaheite chestnut : **inocarpous**, a., *ĭn'·ō·kărp'·ŭs*, having fibrous fruit.

inoculation, n., *ĭn·ŏk'·ŭl·ā'·shŭn* (L. *inoculatus*, ingrafted from one tree to another, as an eye or bud — from *in*, into; *oculus*, an eye), the introduction of the small-pox virus into a healthy system by puncturing or scratching the skin with a sharp-pointed instrument dipped in the matter in order to induce a mild type of the disease : **vaccination** is with the cow-pox virus, while **inoculation** is performed with the small-pox virus.

inoma, n., see 'innoma.'

Inoperculata, n. plu., *ĭn'·ō·pĕrk·* *ŭl·āt'·ă* (L. *in*, not, without; *operculum*, a lid), in *zool.*, the division of pulmonate 'Gasteropoda' in which there is no shelly or horny plate to close the shell when the animal is withdrawn within it : **inopercular**, a., *ĭn'·ō·pĕrk'·ŭl·ăr*, without an operculum or lid, as certain univalve shells.

inoscinic, a., *ĭn'·ŏs·sĭn'·ĭk* (Gr. *ĭs*, fibre, *ĭnos*, of fibre; *kineō*, I disturb, I change), applied to an acid obtained from muscular fibre : **inoscinate**, n., *ĭn·ŏs'·sĭn·āt*, the combination of inoscinic acid with a salifiable base.

inosculation, n., *ĭn·ŏs'·kŭl·ā'·shŭn* (L. *in*, into; *osculatus*, kissed— from *osculum*, a little mouth), the union, as two vessels in a living body; in *bot.*, grafting or budding.

inosite, n., *ĭn'·ŏs·īt* (Gr. *ĭs*, fibre, *ĭnos*, of fibre), a saccharine principle obtained from the juice of flesh, which is not susceptible of alcoholic fermentation : **inosuria**, n., *ĭn'·ŏs·ūr'·ĭ·ă* (Gr. *oureō*, I make water), the same substance when found in morbid urine.

insalivation, n., *ĭn·săl'·ĭv·ā'·shŭn* (L. *in*, into; *salivātio*, a filling with saliva—from *salivo*, I spit out), the process of mixing the saliva intimately with the food during mastication.

insane, a., *ĭn·sān'* (L. *insānus*, unsound in mind—from *in*, not; *sānus*, sound), deranged or unsound in mind : **insanity**, n., *ĭn·săn'·ĭt·ĭ*, unsoundness of mind; the state of mind which incapacitates for the proper management of property, or which renders the patient more or less an object of public danger; lunacy.

Insecta, n. plu., *ĭn·sĕkt'·ă* (L. *insectus*, cut into, *insecta*, things cut into—from *in*, into; *seco*, I cut), the class of articulate animals commonly known as insects, which commonly undergo transformations; a small creeping or flying animal, as the fly, bee, etc., whose body appears cut or almost divided into parts : **Insectivora,** n. plu., *ĭn'-sĕkt-ĭv'-ōr-ă* (L. *voro,* I devour), an Order of Mammals, such as the hedgehog and the mole, which live chiefly on insects : **insectivorous,** a., *ĭn'-sĕkt-ĭv'-ŏr-ŭs,* living upon insects.

Insessores, n. plu., *ĭn'-sĕs-sōr'-ēz* (L. *insessus,* seated or perched upon—from *in,* on ; *sedeo,* I sit), the Order of the perching birds, who live habitually among trees : **insessorial,** a., *ĭn'-sĕs-sōr'-ĭ-ăl,* pert. to the perching birds.

insolation, n., *ĭn'-sōl-ā'-shŭn* (L. *insolātus,* placed in the sun— from *in,* into ; *sol,* the sun), exposure to the sun's rays for drying or maturing, as fruits, drugs, etc. ; sunstroke.

inspiration, n., *ĭn'-spĭr-ā'-shŭn* (L. *inspiro,* I blow or breathe into— from *in,* into ; *spiro,* I breathe), the act of drawing air into the lungs.

inspissate, v., *ĭn-spĭs'-sāt* (L. *in,* into ; *spissātus,* made thick), to thicken, as a fluid by evaporation: **inspissated,** v., *ĭn-spĭs'-sāt-ĕd,* thickened, as juice by evaporation : **inspissation,** n., *ĭn'-spĭs-sā'-shŭn,* the operation of rendering a fluid thicker by evaporation.

insufflation, n., *ĭn'-sŭf-flā'-shŭn* (L. *in,* in ; *sufflātus,* blown up, puffed out), the act of blowing gas or air into a cavity of the body.

integument, n., *ĭn-tĕg'-ū-mĕnt* (L. *integumentum,* a covering—from *in,* in ; *tego,* I cover), the covering skin, membrane, shell, etc., which invests a body ; in *bot.,* the external cellular covering of plants.

intention, first, n., *ĭn-tĕn'-shŭn* (L. *intentus,* stretched out, extended), applied to a wound which heals without suppuration.

interaccessorii, n. plu., *ĭn'-tĕr-ăk'-sĕs-sōr'-ĭ-ī* (L. *inter,* between ; *accessus,* a coming to, an approach), another name for the muscles 'inter-transversales.'

interambulacra, n. plu., *ĭn'-tĕr-ăm'-būl-āk'-ră* (L. *inter,* between ; *ambulacrum,* that which serves for walking), in *zool.,* the unperforate places which lie between the perforate places, or 'ambulacra' in the shells or crusts of the sea-urchin and cidaris.

interarticular, a., *ĭn'-tĕr-ărt-ĭk'-ūl-ăr* (L. *inter,* between ; *articulus,* a little joint), in *anat.,* a term applied to the cartilages which lie within joints ; applied to certain ligaments, as that within the acetabulum.

intercalate, v., *ĭn-tĕr'-kăl-āt* (L. *intercalatum,* to proclaim that something has been inserted among — from *inter,* between ; *calo,* I call), to insert or place between: **inter'calated,** a., -*āt-ĕd,* interposed ; placed between : **intercalary,** a., *ĭn-tĕr'-kăl-ăr'-ĭ,* in *bot.,* applied to the growth of cell-wall, when a new deposition takes place in such a manner that an interposed piece of cell - wall from time to time appears.

intercellular, a., *ĭn'-tĕr-sĕl'-ūl-ăr* (L. *inter,* between ; *cellula,* a little storehouse), in *bot.,* lying between the cells, or the cellular tissue.

intercostal, a., *ĭn'-tĕr-cŏst'-ăl* (L. *inter,* between ; *costa,* a rib), in *anat.,* lying between the ribs.

interdigital, a., *ĭn'-tĕr-dĭdj'-ĭt-ăl* (L. *inter,* between ; *digitus,* a finger), in *anat.,* situated between the fingers ; pert. to the spaces between the fingers.

interfoliar, a., *ĭn'-tĕr-fōl'-ĭ-ăr* (L. *inter,* between ; *folium,* a leaf), in *bot.,* situated between two opposite leaves.

interlobar, a., *ĭn'-tĕr-lōb'-ăr* (L. *inter,* between; Gr. *lobŏs,* a lobe), situated between the lobes of organs.

interlobular, a., *ĭn′·tẽr·lŏb′·ŭl·ǎr* (L. *inter*, between ; *lobulus*, a little lobe), situated between the lobules of organs.

intermaxillæ, n. plu., *ĭn′·tẽr·măks·ĭl′·lē* (L. *inter*, between ; *maxillæ*, the jaws), the two bones which are situated between the two superior maxillæ in vertebrata ; also called 'præmaxillæ' : **inter-maxillary**, a., *ǎl′·lǎr·ĭ*, situated between the maxillary or jaw-bone.

intermission, n., *ĭn′·tẽr·mĭsh′·ŭn* (L. *inter*, between ; *missus*, sent), the period that intervenes between the end of one paroxysm of ague, and the beginning of the next ; also called **apyrexia**, *āp′·ĭr·ĕks′·ĭ·ǎ* (Gr. *a*, without, not ; *puresso*, I have a fever—from *pur*, fire).

intermittent, a., *ĭn′·tẽr·mĭt′·tĕnt* (L. *inter*, between ; *mittens*, sending), ceasing at intervals : n., a specific fever occurring in paroxysms, and characterised by a cold, a hot, and a sweating stage, followed by a period of complete absence from fever ; ague.

interneural, a., *ĭn′·tẽr·nūr′·ǎl* (L. *inter*, between ; Gr. *neuron*, a nerve), situated between the neural processes in spines ; applied to the sharp dermal bones in certain fish which support the rays of their fins on the upper or neural part.

internode, n., *ĭn′·tẽr·nōd* (L. *internodum*, the space between two knots or joints — from *inter*, between ; *nodus*, a knot), in a plant, the part of the stem lying between two nodes or leaf buds : **internodia**, n. plu., *-nōd′·ĭ·ǎ*, in *anat.*, the digital phalanges, or fourteen joints of the fingers and thumb.

inter-osseous, a., *ĭn′·tẽr·ŏs′·sĕ·ŭs* (L. *inter*, between ; *os*, a bone, *osseus*, belonging to a bone), a name applied to muscles situated between bones, as those between the metacarpal of the hand : **inter-osseus membrane**, n., the inter-osseous ligament which passes obliquely downwards from the ridge on the *radius*, or small bone of the arm, to that on the *ulna*, or large bone of the arm.

inter-peduncular, a., *ĭn′·tẽr·pĕd·ŭngk′·ŭl·ǎr* (L. *inter*, between ; mid. L. *pedunculus*, a little foot), in *anat.*, applied to a lozenge-shaped interval of the brain, situated immediately behind the diverging optic tracts, and between them and the peduncles of the cerebrum.

interpetiolar, a., *ĭn′·tẽr·pĕt′·ĭ·ŏl·ǎr* (L. *inter*, between ; *petiolus*, a little foot—from *pes*, a foot), in *bot.*, situated between the petioles or basis of opposite leaves.

interrupted, a., *ĭn′·tẽr·ŭpt′·ĕd* (L. *interruptus*, separated by breaking or rending), in *bot.*, having the usual continuity of a part destroyed : **interruptedly pinnate**, a., having a pinnate leaf in which pairs of small pinnæ occur between larger pairs.

interspinal, a., *ĭn′·tẽr·spīn′·ǎl*, also **interspinous**, a., *-spīn′·ŭs* (L. *inter*, between ; *spina*, a spine), in *anat.*, inserted between the spinous processes of the vertebræ : **interspinales**, n. plu., *-spīn·āl′·ēz*, short vertical fasciculi of fleshy fibres, placed in pairs between the spinous processes of the contiguous vertebræ.

interstaminal, a., *ĭn′·tẽr·stăm′·ĭn·ǎl* (L. *inter*, between ; Eng. *staminal*), in *bot.*, an organ placed between two stamens.

interstitial, a., *ĭn′·tẽr·stĭsh′·ǎl* (L. *interstitium*, distance or space between—from *inter*, between ; *sisto*, I stand), pert. to or containing interstices ; occupying the interstices of an organ.

inter-transversales, n. plu., *ĭn′·*

tĕr-trăns'-vĕrs-āl'-ēz (L. *inter*, between ; *transversus*, lying across, transverse), small muscles situated between the transverse processes of the vertebræ, developed most in the cervical region : **inter-transverse,** a., *trăns'-vĕrs*, applied to a few, thin-scattered fibres, interposed between the transverse processes.

intertrigo, n., *ĭn'-tĕr-trĭg'-ō* (L. *intertrīgo*, a fretting or galling of the skin — from *inter*, between ; *tĕro*, I rub), a local condition of the skin, called 'chafe' or 'fret,' consisting in redness and excoriation of a part of the skin, caused by friction.

interval, n., *ĭn'-tĕr-văl* (L. *inter*, between ; *vallum*, a wall), the period of time comprised between the beginning of one paroxysm of ague and the next, that is, the intermission and the preceding fit.

intervertebral, a., *ĭn'-tĕr-vĕrt'-ĕb-răl* (L. *inter*, between ; Eng. *vertebral*), in *anat.*, situated between the joints of the vertebræ or spine.

intestines, n., *ĭn-tĕst'-ĭnz* (L. *intestīnus*, inward, hidden—from *intus*, within), the long canal or tube which extends from the stomach to the anus, different portions of it having different names — (1) part nearest the stomach, the 'duodenum,' about twelve inches long ; (2) the 'jejunum,' about two feet long ; (3) the 'ileum,' several feet in length — which three portions make up the small intestines ; the large bowel or large intestine, as the continuation of the small intestines, commences in the right iliac region of the abdomen, as the 'cæcum,' and after a large curve it ends at the anus.

intextine, n., *ĭn-tĕks'-tĭn* (L. *intus*, within ; Eng. *extine*), in *bot.*, one of the inner coverings or membranes of the pollen grain,

situated between the extine and the exintine.

intine, n., *ĭn'-tĭn* (L. *intus*, within), in *bot.*, the inner covering of the pollen grain.

intrafoliaceous, a., *ĭn'-tră-fōl'-ĭ-ā'-shŭs* (L. *intra*, within ; *folium*, a leaf), in *bot.*, situated within the axil of a leaf so as to stand between the leaf and the stem.

intralobular, a., *ĭn'-tră-lŏb'-ŭl-ăr* (L. *intra*, within ; Eng. *lobular*), situated within lobules or little lobes.

intrarious, a., *ĭn-trăr'-ĭ-ŭs* (L. *intra*, within), in *bot.*, applied to the embryo when it is surrounded by the perisperm on all sides except its radicular extremity.

introrse, a., *ĭn-trŏrs'* (L. *introrsum*, within), in *bot.*, turned inwards or towards the axis of the part to which it is attached ; opening on the side next the pistil, as some anthers.

intussusception, n., *ĭn-tŭs'-sŭs-sĕp'-shŭn* (L. *intus*, within ; *susceptus*, taken or catched up), an invagination of a portion of the bowel, somewhat resembling the finger of a glove half turned inside out ; the act of taking foreign matter into a living body.

Inula, n., *ĭn'-ūl-ă* (L. *inula*, the plant elecampane), a genus of plants, Ord. Compositæ, Subord. Corymbiferæ, which are generally bitter, and some have an aromatic odour : Inula Helenium, *hĕl-ēn'-ĭ-ŭm* (after the celebrated *Helen* of ancient Troy), elecampane, whose root has stimulant and expectorant qualities : Inulin, n., *ĭn'-ūl-ĭn*, a white amylaceous matter, analogous to starch, found in the roots and tubers of I. Helenium.

inunction, n., *ĭn-ŭngk'-shŭn* (L. *in*, in ; *unctus*, smeared), the act of rubbing into a part of the surface of the body an ointment containing some remedial agent.

invaginate, v., *ĭn·vădj'·ĭn·āt* (L. *in*, into; *vāgĭna*, a scabbard, a sheath), to operate for hernia, in which after reduction, the skin is thrust into the canal by the finger of the operator, and there retained by sutures, etc. till adhesion ensue: **invagination**, n., *ĭn·vădj'·ĭn·ā'·shŭn*, the operation for hernia as above, also sometimes applied to intus-susception.

invermination, n., *ĭn·vĕrm'·ĭn·ā'·shŭn* (L. *in*, in; *vermĭno*, I have worms), the diseased condition of the bowels caused by worms.

inversion, n., *ĭn·vĕr'·shŭn* (L. *inversus*, turned bottom upwards —from *in*, in; *verto*, I turn), said of an organ which is completely or partially turned inside out, as the womb: **inverted**, a., *ĭn·vĕrt'·ĕd*, in *bot.*, having the radicle of the embryo pointing to the end of the seed opposite the hilum; having the ovules attached to the top of the ovary.

invertebral, a., *ĭn·vĕrt'·ĕb·răl* (L. *in*, not; *vertebra*, a joint in the backbone), without a vertebral column or spine bone: **invertebrate**, n., *ĭn·vĕrt'·ĕb·rāt*, an animal having no spinal bone: **adj.**, destitute of a backbone: **invertebrata**, n. plu., *ĭn·vĕrt'·ĕb·rāt'·ă*, the animals that are destitute of backbones and an internal skeleton.

involucels, n. plu., *ĭn·vŏl'·ūs·ĕls* (F. *involucelle*, an involucel; L. *involucrum*, a wrapper), in *bot.*, the collection of bractlets, or a sort of leaves, surrounding a secondary or partial umbel or flower head; secondary involucres.

involucre, n., *ĭn'·vŏl·ō'·kr* (L. *involucrum*, a wrapper—from *in*, into; *volvo*, I roll), in *bot.*, a collection of a sort of leaves round a cluster of flowers, or at some distance below them; the layer of epidermis covering the spore

cases in ferns: **involucral**, a., *ĭn'·vŏl·ō'·krăl*, belonging to the involucre.

involute, a., *ĭn'·vŏl·ōt*, also **involutive**, a., *ĭn'·vŏl·ōt'·ĭv* (L. *involutus*, inwrapped, enclosed — from *in*, into; *volvo*, I roll), in *bot.*, having the edges of leaves rolled inwards spirally on each side.

involution, n., *ĭn'·vŏl·ō'·shŭn* (L. *involutus*, inwrapped — from *in*, into; *volvo*, I roll), the return of an organ or tissue to its original state, as the womb after having expelled the child.

iodine, n., *ī'·ōd·ĭn* (Gr. *iōdēs*, resembling a violet in colour—from *ion*, violet; *eidos*, resemblance), a solid elementary substance of a greyish-black colour, obtained from marine plants, sea water, etc., whose vapour is of a beautiful violet colour; applied externally, it acts as an irritant: **iodide**, n., *ī'·ōd·ĭd*, a direct compound of iodine with a base: **iodism**, n., *ī'·ōd·ĭzm*, a morbid condition sometimes arising from the continued use of iodine, or some of its preparations.

Ionidium, n., *ī'·ōn·ĭd'·ĭ·ŭm* (Gr. *iōn*, a violet; *eidos*, resemblance), a genus of plants, Ord. Violaceæ, some of whose species are used in S. America as substitutes for ipecacuan.

ipecacuanha, n., *ĭp'·ĕ·kăk·ū·ăn'·ă*, also **ipecacuan**, n., *ĭp'·ĕ·kăk'·ū·ăn* (Brazilian or Portuguese), the root of a S. American plant, the 'Cephaëlis ipecacuanha,' used in *med.* as an emetic, etc., belonging to the same Order, the Rubiaceæ (Linn. Ord. Cinchonaceæ), which yields the Peruvian or cinchona bark.

Ipomœa, n., *ĭp'·ōm·ē'·ă* (Gr. *ips*, a worm which infests the vine; *homoios*, like, so named from its habit of creeping round other plants like a worm), a most beautiful genus of climbing plants, Ord. Convolvulaceæ: **Ipomœa purga**, pẽrg′-ă (L. *purgo*, I cleanse, I purify), the Jalap plant, a native of the Mexican Andes, whose root tubers in powder, or as a tincture, is an active irritant cathartic ; also called **Exogonium purga**: I. **Jalapa**, *jăl-ăp′-ă* (*Xalapa*, in Mexico, where it grows abundantly), a species which yields Mechoacan root, having purgative properties : I. **Orizabensis**, *ŏr-ĭz′-ăb-ĕns′-ĭs* (in Brazil), supplies a kind of Jalap, the Purgo Macho of the Mexicans : I. **simularis**, *sĭm′-ŭl-ār′-ĭs* (L. *simulo*, I make like), a species which furnishes Tampico Jalap : I. **Horsfalliæ**, *hŏrs-făwl′-lĭ-ē* (unascertained), a species admirably suited for training to a trellis, having beautiful bright scarlet flowers.

Iridaceæ, n. plu., *ĭr′-ĭd-ā′-sĕ-ē* (Gr. *iris*, the rainbow, the flag, *irĭdos*, of the rainbow), the iris or flower-de-luce family, an Order of herbaceous plants, so called in allusion to the variety and beauty of the flowers : **Iris**, n., *ĭr′-ĭs*, a genus of plants, a great favourite in the flower garden : **Iris Germanica**, *jẽr-măn′-ĭk-ă* (of or from Germany); **I. pallida**, *păl′-lĭd-ă* (L. *pallĭdus*, pale, pallid); I. **florentina**, *flōr′-ĕnt-ĭn′-ă* (L. *Flōrentĭnus*, Florentine — from *Flōrentĭa*, Florence), are species the root stock of which yields orris root which has a pleasant odour like violets, and an acrid taste, arising from the presence of a volatile oil : I. **pseudacorus**, *sŭd-ăk′-ōr-ŭs* (Gr. *pseudēs*, false ; *akŏros*, the sweet flag), the yellow water flag found in marshes, etc., whose seeds have been used as a substitute for coffee.

Iridæa, n., *ĭr′-ĭd-ē′-ă* (L. *iris*, the rainbow, the flag), a genus of the Algæ : **Iridæa edulis**, *ĕd-ū′-lĭs* (L. *edūlis*, eatable), an edible species of Algæ.

iris, n., *ĭr′-ĭs* (L. *iris*, the rainbow, the flag), the coloured circle which surrounds the pupil of the eye ; a structure partly vascular, partly muscular, loaded with pigments, stretched before the lens of the eye, separating the anterior from the posterior chambers ; in *bot.*, see under 'Iridaceæ :' **iritis**, n., *ĭr-īt′-ĭs*, inflammation of the iris of the eye.

Irish Moss, or **Carrageen**, the Sphærococcus crispus, also called **Chondrus crispus**, one of the Algæ which supplies a nutritious article of diet.

irrigation, n., *ĭr′-rĭ-gā′-shŭn* (L. *irrigatus*, watered, irrigated), a medical treatment of an injured or inflamed part in which cold water or a cooling lotion is made to drop continuously on its surface.

irritant, n., *ĭr′-ĭt-ănt* (L. *irritus*, not ratified or settled), a substance which, applied externally or internally, gives rise to a greater or less degree of inflammation.

Isatis, n., *ĭs-āt′-ĭs* (Gr. *isazō*, I make equal), a genus of plants, Ord. Cruciferæ, so called because it is believed by its simple application to destroy all roughness of the skin : **Isatis tinctoria**, *tĭngk-tōr′-ĭ-ă* (L. *tinctorius*, belonging to dyeing), woad which, when treated like indigo, yields a blue dye : I. **indigotica**, *ĭn′-dĭg-ŏt′-ĭk-ă* (L. *indĭgo*, a blue colouring matter), the Tein-Ching, or Chinese indigo.

ischium, n., *ĭsk′-ĭ-ŭm* (Gr. *ischion*, the hip), the hip-bone — a spinous process of the os innominatum : **ischial**, a., *ĭsk′-ĭ-ăl*, pert. to the hip-bone : **ischial tuberosity**, n., the round knob of bone forming that part of the ischium on which we sit ; also called **tuber-ischii**, n., *tŭb′-ẽr-ĭsk′-ĭ-ĭ* (L. *tuber*, a hump): **ischialgia**, n., *ĭsk′-ĭ-ălj′-ĭ-ă* (Gr. *algos*, pain), pain in or near the hip : **ischiatic**, a., *ĭsk′-ĭ-ăt′-ĭk*, of or pert. to the hip : **ischio**, *ĭsk'ĭ·ō*, attachment or connection with the ischium.

ischuria, n., *ĭsk·ūr'ĭ·ă*, also **ischury,** n., *ĭsk'ūr·ĭ* (Gr. *ischo*, I stop or retain; *ouron*, urine), the suppression or stoppage of urine: **ischuretic,** n., *ĭsk'ūr·ĕt'ĭk*, a medicine adapted to relieve ischuria: adj., having the power or quality of relieving ischuria.

isidoid, a., *ĭs'ĭd·ŏyd* (*isidos*, resembling coral—from Gr. *isos*, equal, similar; *eidos*, resemblance), in *bot.*, covered with a dense mass of conical soredia, as the surface of lichens: **isidiose,** a., *ĭs·ĭd'ĭ·ōz*, having powdery, coralline excrescences: **isidiiferous,** a., *ĭs·ĭd'ĭ·ĭf'ĕr·ŭs* (L. *fero*, I bear), having isidiose excrescences: **isidium,** n., *ĭs·ĭd'ĭ·ŭm*, coral-like soredia on the surface of some lichens.

isocheimal, a., *ĭs'ō·kīm'ăl*, also **isocheiminal,** a., *-kīm'ĭn·ăl* (Gr. *isos*, equal, similar; *cheima*, winter), of the same winter temperature; applied to imaginary lines drawn through places on the earth's surface which have the same mean winter temperature.

isochomous, a., *ĭs·ŏk'ŏm·ŭs* (Gr. *isos*, equal, similar; *chōma*, a heap, a mound), in *bot.*, applied to branches springing from the same plant, and at the same angle.

Isoetaceæ, n. plu., *ĭs'ō·ĕt·ā'sĕ·ē* (Gr. *isos*, equal; *ĕtos*, a year—the plants being the same throughout the year), the Quillwort family, an Order of plants, generally included under the Ord. Lycopodiaceæ: **Isoetes,** n., *ĭs'ō·ĕt'ēz*, a genus of curious little aquatic plants, found in some lakes in this country; moss-like plants, intermediate between ferns and mosses.

isomeric, a., *ĭs'ō·mĕr'ĭk* (Gr. *isos*, equal; *meros*, a part), formed of the same elements in the same proportions, but having different physical and chemical properties: **isomerism,** *ĭs·ŏm'ĕr·ĭzm*, identity in elements, but with difference of properties: **isomerous,** a., *ĭs·ŏm'ĕr·ŭs*, in *bot.*, having each of the organs of a flower composed of an equal number of parts.

Isonandra, n., *ĭs'ŏn·ănd'ră* (Gr. *isos*, equal; *anēr*, a male, *andros*, of a male), a genus of trees, Ord. Sapotaceæ: **Isonandra gutta,** *gŭt'tă* (L. *gutta*, a drop), the source of the Gutta Percha, a kind of caoutchouc, used largely in the manufacture of articles of daily use.

Isopod, n., *ĭs'ō·pŏd;* **Isopoda,** n. plu., *ĭs·ŏp'ŏd·ă* (Gr. *isos*, equal; *podes*, feet), an Order of Crustaceæ in which the feet are like one another, and equal: **isopodous,** a., *ĭs·ŏp'ŏd·ŭs*, having legs alike, and equal.

isosporous, a., *ĭs·ŏs'pōr·ŭs* (Gr. *isos*, equal; *poros*, a pore), in *bot.*, applied to cryptogamic plants which produce a single kind of spore, as ferns: **isosporeæ,** n. plu., *ĭs'ŏs·pōr'ĕ·ē*, those ferns, 'Ophioglossaceæ,' and 'Equisetaceæ,' which produce a single kind of spore, which in its turn gives origin to a prothallus furnished with chlorophyll and roots, and capable of independent existence.

isostemonous, a., *ĭs'ŏs·tĕm'ŏn·ŭs* (Gr. *isos*, equal; *stēmōn*, a thread or stem), in *bot.*, having the stamens and petals equal in number; having the stamens and floral envelopes the same in the number of their parts, or in the multiples of the parts.

isotheral, a., *ĭs·ŏth'ĕr·ăl* (Gr. *isos*, equal, similar; *thĕros*, summer), passing through places on the earth which have the same summer temperature.

isothermal, a., *ĭs'ō·thĕrm'ăl* (Gr. *isos*, equal; *thĕrmē*, heat), having the same mean annual temperature; applied to imaginary lines connecting all places on the earth which have the same mean temperature: **isotherm**, n., *ĭs'·ō·thẽrm*, one of those lines.

isotropic, a., *ĭs'·ō·trŏp'·ĭk* (Gr. *isos*, equal; *tropos*, a turning), applied to the condition of 'fibrils' which singly refract light; the condition of 'fibrils' which doubly refract light is called **anisotropic**, *ăn'·ĭs·ō·trŏp'·ĭk* (Gr. *anisos*, unequal; *tropos*, a turning).

issue, n., *ĭsh'·ū* (F. *issu*, born, sprung; Norm. F. *issir*, to go out), an artificially - produced wound, kept raw and open that there may be a constant flow of pus from the surface.

isthmus, n., *ĭst'·mŭs* (L. *isthmus*, Gr. *isthmos*, an isthmus), in *anat.*, the narrow intervening or uniting portion of organs: **isthmic**, a., *ĭst'·mĭk*, of or pert. to an isthmus: **isthmus faucium**, *făw'·shĭ·ŭm* (L. *fauces*, the upper part of the throat, *faucium*, of the upper part of the throat), the space between the soft palate and the root of the tongue.

itch, n., *ĭtsh* (AS. *gictha*, an itching, a scab), a very troublesome skin disease produced by the presence of the Acarus Scabiei, or itch parasite.

iter ad infundibulum, *ĭt'·ẽr ăd ĭn'·fŭnd·ĭb'·ūl·ŭm* (L. *iter*, a path, a way; *ad*, to; *infundĭbŭlum*, a funnel), the passage between the third ventricle of the brain and infundibulum: **iter a palato ad aurem**, *ă păl·āt'·ō ăd ăwr'·ĕm* (L. *a*, from; *palātum*, the palate; *auris*, the ear), the passage from the palate to the ear; the Eustachian tube: **iter a tertio ad quartum ventriculum**, *tẽr'·shĭ·ō ăd kwăwrt'·ŭm vĕnt·rĭk'·ŭl·ŭm* (L. *tertĭus*, a third; *quartus*, a fourth; *ventrĭcŭlus*, a ventricle of the heart), the passage between the third and fourth ventricles of the brain; the aqueduct of Silvius.

Ivory Palm, or **vegetable ivory**, the hard albumen of the 'Phytelophas Macrocarpa,' used in the same way as ivory.

ivy, n., *iv'·ĭ* (AS. *ĭfig*; Ger. *epheu*, ivy), a well - known evergreen climbing plant; the common ivy is the Hedra Helix, Ord. Araliaceæ.

Ixia, n., *ĭks'·ĭ·ă* (Gr. *ixia* and *ixos*, the mistletoe, bird lime), a genus of very handsome plants when in flower, Ord. Iridaceæ, so named from the viscous nature of some of the species: **ixous**, a., *ĭks'·ŭs*, having bird lime; viscous: **Ixodea**, n., *ĭks·ōd'·ĕ·ă*, the ticks, usually parasitic, on domestic animals, occasionally on man, Ord. Arachnida.

jactitation, n., *jăk'·tĭt·ă'·shŭn* (L. *jactitio*, I cast or toss to and fro), a tossing about the body; unconscious movements of a patient in the delirium of a fever.

jaggery, n., *jăg'·gẽr·ĭ* (an Indian name), a coarse dark sugar obtained from the cocoa-nut, and other palms, which when fermented produces arrack.

jalap, n., *jăl'·ăp* (*Xalapa* in Mexico, where found; F. *jalap*), the dried root of the plant Exogonium purga, also called the 'Ipomœa purga,' Ord. Convolvulaceæ, which in the form of powder is much used in medicine as a brisk purgative.

Janipha, n., *jăn·ĭf'·ă* (from *Janipaba*, the Brazilian name), a genus of interesting plants, Ord. Euphorbiaceæ: **Janipha Manihot**, *măn'·ĭ·ŏt* (a Brazilian name), a shrub much cultivated in tropical countries for its produce of starchy matter, made into Cassava bread: **J. læflingii**, *lĕf·lĭn'·jĭ·ĭ* (unascertained), a variety whose amylaceous matter is used as food under the name 'Sweet Cassava'; 'tapioca' is obtained from the starch of the Bitter Cassava.

Jasminaceæ, n. plu., *jăs'·mĭn·ă'·*

sĕ·ē (Arabic name *gasmin*), the jasmine or jessamine family, an Order of plants, much esteemed from the delicious fragrance emitted by several of the species, from which an essential oil is obtained, natives of the tropics : **Jasminum**, n., *jăs·mĭn'·ŭm*, an elegant and familiar genus of plants : **Jasminum officinale**, *ŏf· f'ĭs'·ĭn·āl'·ĕ* (L. *officinālis*, officinal); **J. grandiflorum**, *grănd'·ĭ·flōr'·ŭm*, (L. *grandis*, great, large; *flos*, a flower, *flōris*, of a flower) ; **J. odoratissimum**, *ōd'·ŏr·āt·ĭs'·ĭm·ŭm* (L. *odoratissimum*, very fragrant—from *odorātus*, sweet smelling, fragrant); and **J. sambac**, *săm'·băk* (a native name), are species from which the essential oil of jasmine is procured : **J. angustifolium**, *ăng·gŭst'·ĭ·fōl'·ĭ·ŭm* (*angustus*, small, narrow ; *folium*, a leaf), a species whose bitter root, ground small, and mixed with powdered Acarus calamus root, is considered good in India as an external application for ringworm : **Jasmine**, n., *jăs'·mĭn*, the English name for the genus ; also spelt **Jessamine**.

Jateorhiza, n., *jăt'·ĕ·ō·rīz'·ă* (Gr. *iatēr*, a physician ; *rhiza*, a root), a genus of plants, closely allied to Cocculus, Ord. Menispermaceæ: **Jateorhiza palmata**, *păl·māt'·ă* (L. *palmātus*, marked with the palm of the hand—from *palmus*, the palm of the hand), a plant of East Africa whose root, known as Calumba root, is used in the form of infusion or tincture, as a pure bitter tonic.

Jatropha, n., *jăt'·rŏf·ă* (Gr. *iatēr* or *iatros*, a physician ; *trophē*, food), a genus of valuable medicinal plants, Ord. Euphorbiaceæ, so named in allusion to their medicinal properties, and their use as food : **Jatropha curcas**, *kêrk'·ăs* (S. American name), physic or purging nut, a plant from whose seeds an oil is pro-

cured which has cathartic properties : **J. multifida**, *mŭlt·ĭf'·ĭd·ă* (L. *multifidus*, cleft or split into many parts—from *multus*, many; *findo*, I cleave or divide), a species from whose seeds a purgative oil is obtained, said also to be good as an external application for itch : **J. manihot**, produces tapioca, now called **Janipha manihot**, which see.

jaundice, n., *jăwnd'·ĭs* (F. *jaunisse*, the yellow disease—from *jaune*, yellow), a disease, or rather a symptom of disease, characterised by yellowness of the eyes, skin, etc., and general languor.

jejunum, n., *jĕ·jūn'·ŭm* (L. *jejūnus*, fasting, empty), the second portion of the small intestines, following the 'Duodenum,' so named as supposed to be empty after death.

jigger, n., *jĭg'·gêr*, another name for 'chigoe,' which see.

juga, n. plu., *jōg'·ă* (L. *jugum*, a yoke), in *bot.*, the ribs or ridges on the fruit of the umbelliferæ : **jugate**, a., *jōg'·āt*, having pairs of leaflets, as in compound leaves : **jugum**, n., *jōg'·ŭm*, a pair of opposite leaflets.

Juglandaceæ, n. plu., *jōg'·lănd·ā'·sĕ·ē* (L. *juglans*, a walnut said to be a corruption of *jovis glans*—from *jovis*, of Jupiter ; *glans*, a nut), the Walnut family, an Order of trees, yielding edible, oily nuts, and a valuable timber : **Juglans**, n., *jōg'·lănz*, an ornamental genus of tall, stately trees: **Juglans regia**, *rēdj'·ĭ·ă* (L. *regius*, royal—from *rex*, a king), the common walnut tree whose seeds yield a bland oil, used for olive oil : **J. nigra**, *nĭg'·ră* (L. *nĭgra*, black), the black walnut whose wood when polished is of a fine dark-brown colour.

jugular, a., *jōg'·ūl·ăr* (F. *jugulaire*, jugular ; L. *jugulum*, the collarbone, the neck), pert. to the neck cr throat ; applied to the large

P

vein of the neck ; applied to the ventral fins of fishes, placed beneath or in advance of the pectorals.

julep, n., *jŏl'·ĕp* (Pers. *jullab*—from *gulab*, rose water and julep; Sp. *julépe*, julep ; Mod. Gr. *zoulo*, I squeeze out juice), a mixture as of water and sugar, to serve as a vehicle for a medicine ; an alcoholic beverage compounded and flavoured.

Juncaceæ, n. plu., *jŭng·kās'·ĕ·ē* (L. *juncus*, a rush ; *jungo*, I join or weave), the Rush family, an Order of herbaceous plants : **Juncas,** n., *jŭng'·kăs*, a genus of plants found in moist situations, or growing among water used for domestic purposes : **Juncas glaucus,** *glāwk'·ŭs* (L. *glaucus*, Gr. *glaukos*, bluish-grey), the hard rush, used in the manufacture of rush fabrics : **J. effusus,** *ĕf·fūz'·ŭs* (L. *effusus*, poured out, shed), the soft rush : **J. conglomeratus,** *kŏn·glŏm'·ĕr·āt'·ŭs* (L. *conglomerātus*, rolled together — from *con*, together ; *glomero*, I wind into a ball), the hollow rush : **junciform,** a., *jŭn'·sĭ·fŏrm* (L. *forma*, shape), like a rush ; long and slender : **juncous,** a., *jŭng'·kŭs*, full of rushes.

Juncaginaceæ, n. plu., *jŭng·kădj'·ĭn·ā'·sĕ·ē* (L. *juncus*, a rush, and probably *ago*, I move, I drive), a Sub-order of the Ord. Alismaceæ or Water-Plantain family, found growing in ponds and marshes with minute green flowers ; some resemble rushes, others are floating plants.

Jungermannieæ, n. plu., *jŭng'·gĕr·măn·nĭ'·ĕ·ē* (after *Jungermann*, a German botanist), the Scale mosses, a Sub-order of plants, Ord. Hepaticæ : **Jungermannia,** n., *jŭng'·gĕr·măn'·nĭ·ă*, a genus of plants, usually found in little patches upon trees or rocks, or in damp places on the earth.

Juniperus, n., *jŏn·ĭp'·ĕr·ŭs* (L.

juniperus, the juniper tree), a well-known genus of shrubs, Ord. Coniferæ : **Juniperus communis,** *kŏm·mūn'·ĭs* (L. *commūnis*, common), the common juniper whose berries are used in the manufacture of Hollands or gin, and medicinally as a diuretic, as well as an oil procured from them: **J. Sabina,** *săb·īn'·ă* (L. *Sabīnus*, Sabine, because employed by the Sabine priests in their ceremonies), the plant Savin, the young branches and leaves of which contain an active, volatile oil, used as an anthelmintic and emenagogue : **J. Bermudiana,** *bĕr·mŭd'·ĭ·ān'·ă* (of or from *Bermuda*), a species whose wood furnishes Pencil Cedar : **juniper,** n., *jŏn'·ĭp·ĕr*, the English name of the 'J. communis.'

Justicia, n., *jŭs·tĭsh'·ĭ·a* (after *Justice*, a Scotch botanist), an extensive ornamental genus of flowering plants, Ord. Acanthaceæ ; a deep-blue dye is obtained from a species in China.

jute, n., *jōt* (an Indian name), the fibres of the 'Corchorus capsularis' and 'C. olitorius,' extensively used in the manufacture of coarse cloths and cordage, and in mixing with other fibres in finer cloths.

Kalmia, n., *kăl'·mĭ·ă* (after *Kalm*, a Swedish naturalist), a genus of very handsome hardy shrubs, Ord. Ericaceæ, some of whose species are poisonous and narcotic.

kamela, n., *kăm'·ĕl·ă* (Bengalee *kamala*), bright - red, semi-translucent, resinous glandules covering the surface of the tricoccous fruit of Rottlera tinctoria, Ord. Euphorbiaceæ, an Indian tree, used as a remedy against the tapeworm.

kelis, n., *kĕl'·ĭs* (Gr. *kēlis*, a stain, or *chēlē*, a claw or talon), another name for keloid; a disease of the skin, presenting a cicatrix-like appearance : **keloid**, n., *kĕl'·ŏyd* (Gr. *eidos*, resemblance), a disease, consisting of an indurated mass putting forth processes at its edges resembling crab's claws.

keratin, n., *kĕr'·ăt·ĭn* (Gr. *keras*, a horn), the substance of the horny tissues : **keratode**, n., *kĕr'·ăt·ōd* (Gr. *eidos*, resemblance), the horny substance making up the skeletons of many sponges : **keratosa**, n., *kĕr'·ăt·ōz'·ă*, the division of sponges having the skeleton composed of keratode : **kerato-cricoid**, *kĕr'·ăt·ō-krĭk'·ŏyd* (see 'cricoid'), a short slender bundle of muscle arising from the cricoid-cartilage.

kidneys, n., *kĭd'·nĭz* (Old Eng. *kidnere*, the kidneys—from Old Eng. *quid*, Icel. *koidr*, Scot. *kyte*, the belly ; Old Eng. *nere*, Ger. *niere*, the testicles, kidneys), literally the testicles of the belly; two oblong flattened bodies lying behind the intestines of an animal, which secrete the urine.

Kigelia, n., *kĭg'·ĕl·ĭ'·ă* (*kigelikeia*, the negro name), a genus of African trees, Ord. Bignoniaceæ : **Kigelia pinnata**, *pĭn·năt'·ă* (L. *pinnatus*, feathered, winged—from *pinna*, a wing), a tree yielding excellent timber in Africa, its long pendent fruit when roasted is there used as an external application for rheumatic complaints : **K. Africana**, *ăf'·rĭk·ān'·ă* (of or from *Africa*), a species whose bark is used on the Gold Coast for dysentery.

kind, n., *kīnd*, another name for 'genus,' which see.

kingdom, n., *kĭng'·dŭm*, one of the three great divisions of nature, animal, vegetable, and mineral.

kinic acid, *kĭn'·ĭk* (from *kina-kina*, a name for cinchona), an organic acid found in the bark of various species of cinchona, principally yellow and pale Peruvian bark.

kino, n., *kĭn'·ō* (an Indian name), the concrete exudation from Pterocarpus marsupium, a tree of the Indian forests, Ord. Leguminosæ, which forms a very powerful astringent ; 'P. erinaceus' furnishes African kino.

kirschwasser, n., *kĕrsh'·văs·sĕr* (Ger. cherry water—from *kirsche*, cherry ; *wasser*, water), cherry brandy, an alcoholic liquor distilled from a variety of Cerasus avium, Ord. Rosaceæ, the sweet black cherry.

kleistogamous, a., *klīst·ŏg'·ăm·ŭs* (Gr. *kleistos*, closed ; *gamos*, marriage), in *bot.*, having the fertilisation effected in closed flowers, as certain grasses.

kleptomania, n., *klĕp'·tō·mān'·ĭ·ă* (Gr. *kleptō*, I steal ; *mania*, madness), a morbid impulse or desire to steal or appropriate.

knaurs, n. plu., *nawrs*, or **gnaurs**, n. plu., *nawrs* (Dut. *knarren*, to growl ; Swed. *knorla*, to twist, to curl), a hard woody lump projecting from the trunk of a tree, as in the oak, thornbeam, etc.

kombe, n., *kŏm'·bē* (native name), the famous arrow poison of S. Africa, furnished by the Strophanthus kombe, Ord. Apocynaceæ.

koochla, n., *kŏtsh'·lă* (native name), the poison-nut tree of the Malabar and Coromandel coasts ; the 'Strychnos nux - vomica,' Ord. Loganiaceæ.

koumiss or **kumiss**, n., *kŏm'·ĭs* (Russ. *kumys*), a sparkling drink obtained by the Kalmucks, by fermenting the whey of mare's milk ; may also be made from cow milk.

kousso, n., *kows'·sō*, also **kusso**, n., *kŭs'·sō* (native name), the flowers of an Abyssinian tree, used in that country as a remedy for tapeworm ; the produce of Brayera anthelmintica, Ord. Rosaceæ.

Krameria, n., *krăm·ēr'·ĭ·ă* (after *Kramer*, a German botanist), a genus of ornamental shrubs, Ord. Polygalaceæ : **Krameria triandra**, *trĭ·ănd'·ră* (Gr. *treis*, three ; *anēr*, a male, a stamen, *andros*, of a male), a Peruvian plant which furnishes Rhatany-root, employed as an astringent in hæmorrhage and mucous discharges, and its infusion of a blood-red colour has been employed to adulterate port wine : **K. cistoidea**, *sĭst·ōyd'·ĕ·ă* (Gr. *kistē*, a box or chest ; *eidos*, resemblance), a Chilian plant which yields a kind of rhatany.

kreatin, n., see 'creatin.'

labellum, n., *lăb·ĕl'·lŭm* (L. *lab-ellum*, a little lip—from *labrum*, a lip), in *bot.*, one of the divisions of the inner whorl of the flower of Orchids ; the lip or lower petal of an Orchid, etc. : **label**, n., *lāb'·ĕl*, the terminal division of the lip of the flower in Orchids.

labia, labial, see 'labium.'

labia cerebri, *lăb'·ĭ·ă sĕr'·ĕb·rī* (L. *labia*, lips ; *cerebri*, of the cerebrum), the margins of the hemispheres of the brain which overlap the Corpus callosum.

labiate, a., *lăb'·ĭ·āt* (L. *labium*, a lip), lipped ; in *bot.*, applied to irregular gamopetalous flowers, with an upper and under portion separated more or less by a gap ; having two unequal divisions : **Labiatæ**, n. plu., *lăb'·ĭ·āt'·ē*, the Labiate family, an extensive Order of plants, in general fragrant and aromatic.

Labiatifloræ, n. plu., *lăb'·ĭ·ăt·ĭ·flōr'·ē* (L. *labium*, a lip ; *flōreo*, I blossom ; *Flora*, the goddess of flowers), one of the great sections into which De Candolle divides the extensive Ord. Compositæ, characterised by hermaphrodite flowers divided into two lips.

labium, n., *lāb'·ĭ·ŭm* (L. *labium*, a lip), in *bot.*, the lower lip of a labiate flower ; in *zool.*, the

lower lip of articulate ani labia, n. plu., *lāb'·ĭ·ă*, th divisions of irregular gamo ous flowers separated by a or gap.

laboratory, n., *lăb'·ŏr·ăt·ŏı laboratoire*, a laboratory *labor*, labour), a druggist's the workroom of a chemist

labrum, n., *lāb'·rŭm* (L. *labı* lip), the mouth cover, or l shield of an insect's mouth outer lip of a shell ; the up of articulate animals, a 'labium' is the lower lip.

Laburnum, n., *lăb·ĕrn'·ŭm* a beautiful ornamental shru 'Cytisus laburnum,' Ord. L inosæ, whose seeds are acri poisonous.

labyrinth, n., *lăb'·ĭr·ĭnth* (l *yrinthus*, any structure with winding passages), the ir ear, consisting of three part vestibule, the semicircular and the cochlea—so named the complexity of its shape

laccate, a., *lăk'·kāt* (Ger. *la lacca*, a varnish ; F. *laque*, or ruby colour), in *bot.*, a ing as if varnished, or sealing-wax.

Lacertilia, n. plu., *lăs'·ĕr* (L. *lacerta*, a lizard), an O Reptilia, comprising the and slow-worms.

lacertus, n., *lăs·ĕrt'·ŭs*, lace plu., *lăs·ĕrt'·ī* (L. *lacertu* muscular upper part of the a packet or bundle of mu fibres enclosed in a memb sheath ; another name fo ciculus.'

lachrymal, a., *lăk'·rĭm·ă lachryma*, a tear), pert. to generating or conveying tea

lacinia, n., *lăs·ĭn'·ĭ·ă*, la n. plu., *lăs·ĭn'·ĭ·ē* (L. *lacin* lappet or flap of a gar calycine segments, as i violet : **laciniate**, a., *lăs·* also **laciniated**, a., *lăs·ĭn'·ı* in *bot.*, irregularly cut into ı segments; fringed; also **laciniose**, a., *lăs·ĭn'·ĭ·ōz*, fringed: **laciniolate**, a., *lăs·ĭn'·ĭ·ŏl·āt* (dim. of *lacinia*), having very minute laciniæ: **lacinula**, n., *lăs·ĭn'·ŭl·ă* (dim.), the small inflexed point of the petals of Umbellifers.

Lacistemaceæ, n. plu., *lăs·ĭs'·tĕm·ā'·sĕ·ē* (probably Gr. *lakistos*, torn, rent, from the appearance of the shrubs), the Lacistema family, an Order of small trees or shrubs, natives of warm parts of America: **Lacistema**, n., *lăs'·ĭs·tĕm'·ă*, a genus.

lacquer, n., *lăk'·ér* (F. *laque*, rose or ruby colour; Pers. *lac*, lac; Sp. *lacre*, sealing - wax), a varnish from shell-lac ; the hard black varnish of Japan is procured from Stigmaria verniciflua, Ord. Anacardiaceæ.

lactation, n., *lăk·tā'·shŭn* (L. *lactatum*, to contain milk, to suck milk—from *lac*, milk, *lactis*, of milk), the period of suckling a child ; the act of giving milk : **lacteals**, n. plu., *lăk'·tē·ăls*, minute vessels or absorbents which arise in small conical projections of the mucous or lining membrane of the intestines, whose function is to absorb the various soluble portions of the digested food or chyme as it passes along the intestinal canal: **lactescence**, n., *lăk·tĕs'·ĕns*, a milky colour : **lactescent**, a., *lăk·tĕs'·ĕnt*, producing milk ; in *bot.*, yielding a milky juice : **lactic**, a., *lăk'·tĭk*, pert. to milk ; of or from milk or whey, as 'lactic acid' : **lactiferous**, a., *lăk·tĭf'·ér·ŭs* (L. *fero*, I bear), bearing or producing milk or milky juice : **lactin**, n., *lăk'·tĭn*, sugar of milk : **lactometer**, n., *lăk·tŏm'·ĕt·ér* (Gr. *metron*, a measure), an instrument for ascertaining the quality of milk.

Lactuca, n., *lăk·tūk'·ă* (L. *lactūca*, a lettuce—from *lac*, milk ; from their milky juice), a genus of plants, Ord. Compositæ, Sub-

order Chichoraceæ : **Lactuca sativa**, *săt·īv'·ă* (L. *satīvus*, that is sown or planted), the common lettuce, from which a milk - like juice exudes when broken : L. **virosa**, *vĭr·ōz'·ă* (L. *virōsus*, slimy, fetid—from *vīrus*, slime, stench), the wild or strong-scented lettuce; the 'Lectuarium' or lettuce opium is the inspissated juice of this and preceding, used for allaying pain and inducing sleep : **lactucin**, n., *lăk·tūs'·ĭn*, the active principle of the wild lettuce.

lacuna, n., *lăk·ūn'·ă*, **lacunæ**, n. plu., *lăk·ūn'·ē* (L. *lacuna*, a hole, a cavity), in *bot.*, a large space in the midst of a group of cells ; a depression ; a blank space ; in *anat.*, minute recesses or cavities in bone: **lacunar**, a., *lăk·ūn'·ăr*, pert. to or arising from lacunæ : **lacuna magna**, *măg'·nă* (L. *magnus*, great), in *anat.*, a large and conspicuous recess situated on the upper surface of the Fossa navicularis : **lacunose**, a., *lăk'·ūn·ōz'*, furrowed or pitted ; having cavities.

lacus lachrymalis, *lăk'·ŭs lăk'·rĭm·āl'·ĭs* (L. *lăcus*, a basin, a tank ; *lachrymālis*, pert. to tears—from *lachryma*, a tear), the tear-lake ; a triangular space situated between the eyelids towards the nose, into which the tears flow.

ladanum, n., *lăd'·ăn·ŭm*, or **labdanum**, n., *lăb'·dăn·ŭm* (L. *lādănum*, a resinous juice), a resinous matter obtained from the genus 'Cistus,' chiefly from the species 'Cistus creticus,' Ord. Cistaceæ.

Læmodipoda, n. plu., *lĕm'·ō·dĭp'·ŏd·ă* (Gr. *laimos*, the throat; *dis*, twice ; *podes*, feet), an Order of Crustacea, so named from having two feet placed so far forward as to be, as it were, under the throat.

lævigatus, a., see 'levigatus.'

Lagenaria, n., *lădj'·ĕn·ār'·ĭ·ă* (L. *lagēna*, a bottle), a genus of plants, Ord. Cucurbitaceæ, so named from the bottle-shaped fruit of some of the species : **Lagenaria vulgaris**, *vŭlg·ār'·ĭs* (L. *vulgāris*, common), the Bottle Gourd, the hard covering of whose fruit is used as a vessel or flask for containing fluid.

lageniform, a., *lădj·ēn'·ĭ·fŏrm* (L. *lagēna*, a bottle, a flask ; *forma*, shape), in *bot.*, having a shape like a Florence flask.

Lagerstrœmia, n., *lăg'·ēr·strĕm'·ĭ·ă* (after *Lagerstrœm* of Gottenburg), a very splendid genus of plants, Ord. Lythraceæ : **Lagerstrœmia reginæ**, *rĕ·jĭn'·ē* (L. *regīna*, a queen, *regīnæ*, of a queen) ; and **L. indica**, *ĭn'·dĭk·ă* (L. *indicus*, of or from India), produce flowers in panicles of a pale rose colour, gradually deepening to a beautiful purple.

Lagetta, n., *lădj·ĕt'·tă* (name in Jamaica), a genus of plants, Ord. Thymelæaceæ : **Lagetta lintearia**, *lĭnt'·ĕ·ār'·ĭ·ă* (L. *lintēārius*, of or pert. to linen—from *lintĕum*, linen cloth), a species whose inner bark, cut into thin pieces and macerated, assumes a beautiful net-like appearance, and is called lace-bark.

lambdoidal, a., *lăm·dōyd'·ăl* (Gr. letter Λ, called *lambda ; eidos*, resemblance), having the form of the Greek letter Λ.

lamella, n., *lăm·ĕl'·lă*, **lamellæ**, n. plu., *lăm·ĕl'·lē* (L. *lāmella*, a small plate or loaf—from *lāmĭna*, a plate), thin plates or scales, as those composing shells or bones ; in *bot.*, the gills of an Agaric ; the flat divisions of the stigma.

Lamellibranchiata, n., *lăm'·ĕl·lĭ·brăng'·kĭ·āt'·ă* (L. *lamella*, a small plate or scale ; Gr. *brangchia*, gills), the class of Mollusca, comprising the ordinary bivalves, which have lamellar gills : **lamellibranchiate**, a., *-brăng'·kĭ·āt*, having gills in symmetrical semicircular layers : **Lamellirosters**, n. plu., *lăm'·ĕl·lĭ·rŏst'·ĕrs* (L. *rostrum*, a beak), the flat-billed

swimming birds, such as ducks, geese, and swans : **lamellirostral**, a., *-rŏst'·răl*, having the margins of the back furnished with plates, as ducks and geese.

Lamiaceæ, n. plu., *lăm'·ĭ·ā'·sĕ·ē* (Gr. *laimos*, the neck, the throat, in allusion to the shape of their flowers), an extensive Order of plants, now named 'Labiatæ,' which see : **Lamium**, n., *lăm'·ĭ·ŭm*, a genus of plants.

lamina, n., *lăm'·ĭn·ă*, **laminæ**, n. plu., *lăm'·ĭn·ē* (L. *lamina*, a plate or leaf), a thin plate or scale ; a thin layer or coat lying over another ; the horny and sensitive folds by which the hoof wall is attached to the deeper-seated parts ; in *bot.*, the blade of the leaf ; the broad part of a petal or sepal : **laminated**, a., *lăm'·ĭn·āt·ĕd*, consisting of plates or layers disposed one over another : **lamination**, n., *lăm'·ĭn·ā'·shŭn*, arrangement in layers : **lamina cinerea**, *sĭn·ēr'·ĕ·ă* (L. *cinĕrĕus*, ash-coloured—from *cinis*, ashes), in *anat.*, a thin layer of grey substance extending backwards above the optic commissure, from the termination of the corpus callosum to the tuber cinereum : **lamina cribrosa**, *krĭb·rōz'·ă* (L. *cribrum*, a sieve), a sieve-like layer formed by the sclerotica at the entrance of the optic nerve, pierced by numerous minute openings for the passage of the nervous filaments : **lamina spiralis ossea**, *spĭr·āl'·ĭs ŏs'·sĕ·ă* (L. *spirālis*, spiral ; *ossĕus*, like bone, bony), a thin bony process projecting from the modiolus, consisting of two thin lamellæ of bone.

Laminaria, n. plu., *lăm'·ĭn·ār'·ĭ·ă* (L. *lamina*, a plate, a leaf), a genus of Ord. Algæ, so named from the flat blade-like form of the fronds, which have stalks of considerable size : **Laminaria digitata**, *dĭdg'·ĭt·āt'·ă* (L. *digitātus*, having fingers or toes—from *digitus*, a finger or toe), tangle, an esculent sea-weed, dried portions of which, from its property of absorbing moisture and thus increasing in bulk, are employed for the dilatation of narrow canals and apertures in dissections : **L. saccharina**, *săk'·kăr·in'·ă*, an esculent sea-weed, from which a sweet extract is obtained, in Iceland.

laminitis, n., *lăm'·in·it'·is* (L. *lamina*, a thin plate of metal), inflammation of the layers of the stomach, as in the horse.

lampas, n., *lăm'·păs* (Gr. *lampas*, a torch, a fiery meteor), among horses, the swelling of the gums and palate incidental to dentition, a term in use among horsemen.

lanceolate, a., *lăns'·ē·ōl·āt* (L. *lancĕolātus*, armed with a little lance or spear—from *lancĕa*, a lance or spear), having the form of a lance-head ; narrowly elliptical, and tapering to both ends.

lancet, n., *lăns'·ĕt* (F. *lancette*, dim. from *lance*, a lance), a small, sharp, two-edged knife, used by surgeons.

lancewood, n., *lăns'·wŏŏd*, a wood furnished by the Duguetia quiterensis, Ord. Anonaceæ.

lancinating, a., *lăns'·in·āt·ing* (L. *lancea*, a lance), piercing or seeming to pierce with a sudden shooting pain.

lansium, n., *lăns'·i·ŭm* (from native name), a genus of plants, Ord. Meliaceæ, which yields the Lansa, Langsat, or Ayer-ayer, a yellow fruit highly esteemed in the East.

Lantana, n., *lăn·tān'·ă* (unascertained), a genus of shrubs, Ord. Verbenaceæ, having an agreeable aromatic perfume, some of whose species are used as tea.

lanuginous, a., *lăn·ūdj'·in·ŭs*, also **lanuginose**, a., *lăn·ūdj'·in·ōz* (L. *lănūgĭnōsus*, woolly, downy—from *lănūgo*, a wool-like production, down), in *bot.*, woolly ;

covered with long curled, interlaced hairs: **lanugo**, n., *lăn·ūg'·ō*, the fine down or hair which covers the human fœtus during the sixth month.

Larch, n., *lârtsh* (L. and Gr. *larix*, It. *larice*, the larch), a forest tree, the Larix Europæa, Ord. Coniferæ, also called Abies larix ; the American larch Abies pendula.

lardaceous, a., *lârd·ā'·shŭs* (L. *lardum*, F. *lard*, lard), resembling lard or bacon.

Lardizabala, n., *lârd'·iz·ăb'·ăl·ă* (after *Lardizabala*, of S. America), a genus of hardy creepers, Ord. Berberidaceæ, which yield good edible fruit in Chili.

Larix, n., *lăr'·iks* (L. *larix*, the larch), a genus of forest trees, Ord. Coniferæ, Sub-ord. Abietineæ : **Larix Europæa**, *ŭr'·ŏp·ē'·ă* (pert. to Europe), the larch.

Larkspur, n., *lârk'·spėr* (*lark* and *spur*), a plant with showy flowers, usually of a vivid blue, genus Delphinium, Ord. Ranunculaceæ.

larva, n., *lârv'·ă*, **larvæ**, n. plu., *lârv'·ē* (L. *larva*, a ghost, a mask), an insect in the caterpillar or grub state after it has emerged from the egg.

laryngismus, n., *lăr'·ing·jiz'·mŭs* (Gr. *larunggismos*, shouting, vociferation—from *larunggizō*, I bawl out with open throat), a false or spasmodic croup, called 'child crowing,' from the crowing inspiration by which it is characterised ; a spasm of the glottis ; also **laryngismus stridulus**, *strid'·ūl·ŭs* (L. *strīdŭlus*, a creaking or hissing), same meaning as preceding ; due to destructive disease of vocal apparatus.

laryngo, n., *lăr·ing'·gō* (Gr. *larungx*, the upper part of the windpipe), a word indicating connection with the larynx.

laryngoscope, n., *lăr·ing'·gō·skōp* (Gr. *larungx*, the larynx ; *skopeo*,

I view or see), an instrument for exploring the larynx and upper part of the windpipe, consisting of a small reflecting mirror on a slender stem, upon which rays of artificial light may be thrown from another mirror : **laryngotomy**, n., *lăr′·ĭng·gŏt′·ŏm·ĭ* (Gr. *tomē*, a cutting), the operation of cutting into the larynx to permit breathing in cases of obstruction.

larynx, n., *lăr′·ĭnks* (Gr. *larungx*, the upper part of the windpipe, *larunggos*, of the upper part, etc.), the upper part of the trachea or windpipe, and concerned in mammals in the production of vocal sounds : **laryngeal**, a., *lăr′·ĭng·jē′·ăl* or *lăr·ĭng′·gĕ·ăl*, pert. to the larynx : **laryngitis**, n., *lăr′·ĭng·jīt′·ĭs*, inflammation of the larynx: **laryngo-tracheotomy**, *lăr·ĭng′·gō·trăk′·ĕ·ŏt′·ŏm·ĭ*(see 'tracheotomy'), the operation of opening the air-passage through the cricoid cartilage and upper ring of the trachea.

Lasiandra, n., *lăs′·ĭ·ănd′·ră* (Gr. *lasios*, hairy ; *anēr*, a male, *andros*, of a male), an elegant genus of shrubs, Ord. Melastomaceæ, having hairy stamens, and producing large purple blossoms in panicles.

latent, a., *lāt′·ĕnt* (L. *lătens*, concealing, hiding, *lătentis*, of concealing), not visible or apparent ; in *bot.*, applied to buds that remain in a dormant state.

lateral, a., *lăt′·ĕr·ăl* (L. *laterālis*, belonging to the side—from *lătus*, a side, *lătĕris*, of a side), arising from the side of the axis ; not terminal : **lateralis nasi**, *lăt′·ĕr·āl′·ĭs năz′·ĭ* (L. *năsus*, the nose, *năsĭ*, of the nose), the lateral of the nose; an artery derived from the facial as that vessel is ascending along the side of the nose.

lateritious, a., *lăt′·ĕr·ĭsh′·ŭs* (L. *lăter*, a brick, a tile, *lătĕris*, of a brick or tile), resembling brick dust in colour.

latex, n., *lāt′·ĕks* (L. *lătex*, a liquid or juice, *lătĭcis*, of a liquid), in *bot.*, a granular or viscid fluid contained in laticiferous vessels.

Lathræa, n., *lăth·rē′·ă* (Gr. *lathraios*, secret, private), a genus of curious little root-parasites, furnished with white fleshy scales in the place of leaves, Ord. Orobanchaceæ, so named as being found in concealed places : **Lathræa squamaria**, *skwŏm·ār′·ĭ·ă* (L. *squāma*, a scale), the tooth-wort, parasitical upon the roots of hazels, cherry laurels, and other trees.

Lathyrus, n., *lăth′·ĭr·ŭs* (Gr. *lathuros*, a kind of small vetch or pulse), a considerable genus of handsome plants when in flower, Ord. Leguminosæ, Sub-ord. Papilionaceæ : **Lathyrus cicera**, *sĭs′·ĕr·ă* (L. *cicĕra*, pulse — from *cĭcer*, the chick-pea); also **L. aphaca**, *ăf′·ăk·ă* (L. *aphaca*, Gr. *aphaka*, a kind of pulse), possess narcotic qualities in their seeds, etc. ; the seeds of the latter produce intense headaches if eaten in quantity : **L. tuberosus**, *tŭb′·ĕr·ōz′·ŭs* (L. *tuberōsus*, having fleshy knobs—from *tŭber*, a protuberance), a species whose roots produce wholesome food : **L. odoratus**, *ōd′·ŏr·āt′·ŭs* (L. *odorātus*, scent, smell), the sweet-pea of our gardens : **L. sativus**, *săt·ĭv′·ŭs* (L. *sativus*, sown or planted), the Gesse or Jarosse of the S. of Europe whose seeds are eaten.

laticiferous, a., *lăt′·ĭ·sĭf′·ĕr·ŭs* (L. *lătex*, a liquid, juice, *lătĭcis*, of a liquid ; *fero*, I bear), conveying latex, or elaborated sap ; having anastomising tubes containing latex.

latiseptæ, n. plu., *lăt′·ĭ·sĕp′·tē* (L. *lătus*, a side, *lătĕris*, of a side ; *septum*, a partition), in *bot.*, cruciferous plants which have a broad septum in their silicula.

latissimus dorsi, n., *lăt·ĭs′·sĭm·ŭs dŏrs′·ĭ* (L. *latissimus*, very broad

—from *latus*, broad ; *dorsum*, the back, *dorsi*, of the back), a flat muscle, situated on the back and side of the lower part of the trunk, which moves the arm backwards and downwards, or which brings forward the body when the hand is fixed.

latrines, n. plu., *lăt'-rĭnz* (F. *latrines*, a privy), a privy ; necessary conveniences or privies on a large scale.

laudanum, n., *lăwd'-ăn-ŭm* (L. *ladanum*, the resinous substance from the plant Cistus creticus ; said to be formed from L. *laude dignum*, worthy of praise, from its soothing qualities), a preparation of opium in spirits ; tincture of opium.

Lauraceæ, n. plu., *lăwr-ā'-sĕ.ē* (L. *laurus*, a laurel tree), the Laurel family, an Order of noble trees and shrubs, natives of the tropics in cool places, generally aromatic and fragrant, the species producing cinnamon, cassia, and camphor : **Laureæ**, n. plu., *lăwr'-ĕ-ē*, a Sub-order of the true laurel trees : **Laurus**, n., *lăwr'-ŭs*, a handsome and interesting genus of plants : **Laurus nobilis**, *nŏb'-ĭl-ĭs* (L. *nobilis*, famous, renowned), the common sweet bay ; the Victor's laurel whose leaves were used to crown the conquerors in the Olympic games; the common bay or cherry laurel is the ' Prunus, or Cerasus lauro-cerasus,' whose fresh leaves are employed in medicine, also called ' cherry laurel.'

Lavandula, n., *lăv-ănd'-ūl-ă* (It. *lavanda*, the act of washing, lavender — from *lăvo*, I wash, alluding to the uses made of its distilled water), a genus of plants, Ord. Labiatæ, much esteemed for the fragrance of their flowers : **Lavandula vera**, *vēr'-ă* (L. *vērŭs*, real, genuine), yields the best oil of Lavender : L. **latifolia**, *lăt'-ĭ-fŏl'-ĭ-ă* (L. *latus*, a side ; *fŏlĭum*, a leaf), furnishes

spike-oil : L. **stœchas**, *stĕk'-ăs* (Gr. *stoichas*, a species of lavender), a species of the S. of Europe, which also supplies an oil : **Lavender**, n., *lăv'-ĕnd-ĕr*, an odoriferous plant, the Lavandula vera, an under shrub having linear grey leaves, and close spikes of bluish flowers, from which the essential oil of lavender is distilled ; ' lavender' is tonic, stimulant, and carminative.

Lawsonia, n., *lăw-sōn'-ĭ-ă* (after *Dr. Isaac Lawson*), a genus of ornamental trees, Ord. Lythraceæ, producing flowers in panicles or racemes : **Lawsonia inermis**, *ĭn-ĕrm'-ĭs* (L. *inermis*, without weapons, unarmed), produces the ' Henna ' or ' Alhenna ' of the Arabs, used in Egypt for dyeing orange.

laxative, a., *lăks'-ăt-ĭv* (L. *laxus*, loose, open), a medicine which gently opens the bowels ; an aperient.

laxator tympani, *lăks-ăt'-ŏr tĭm'-păn-ĭ* (L. *laxātus*, stretched out, extended ; *tympănum*, a drum, *tympani*, of a drum), the *major*, a muscle that arises from the spinous process of the sphenoid bone, etc., and is inserted into the head of the malleus of the ear ; the *minor* arises from the upper and back part of the external meatus of the ear, both of these muscles are by some anatomists regarded as ligaments.

leader, n., *lēd'-ĕr* (Icel. *leida*, to lead), a popular name for a tendon ; in *bot.*, the terminal or primary shoot of a tree.

Lecanora, n., *lĕk'-ăn-ōr'-ă* (Gr. *lekanē*, a dish, a basin, in allusion to the form of the shields), a genus of Lichens comprising some valuable plants : **Lecanora tartarea**, *tăr-tăr'-ĕ-ă* (L. *Tartărĕus*, belonging to the infernal regions—from *Tartărus*, Tartarus), a species which supplies the dye Cubear.

Lecidea, n., *lē·sĭd′·ĕ·ă* (Gr. *lěkis*, a basin, a saucer; *eidos*, resemblance), an extensive genus of Lichens found at all seasons of the year.

lecotropal, a., *lĕk·ŏt′·rōp·ăl* (Gr. *lekos*, a dish; *tropē*, a turning), in *bot.*, shaped like a horse-shoe, as some ovules.

Lecythideæ, n. plu., *lĕs′·ĭ·thĭd′·ĕ·ē* (Gr. *lēkuthos*, an oil jar), a tribe or Sub-order of the Mystaceæ, so named from the form of the seed vessels: **Lecythis**, n., *lĕs′·ĭ·thĭs*, a genus of large trees of S. America, which furnish some of the nuts of commerce: **Lecythis ollaria**, *ŏl·lār′·ĭ·ă* (L. *ollārĭus*, of or belonging to pots—from *olla*, a pot), a species producing large fruits, commonly known as Monkey Pots: **L. usitata**, *ūz′·ĭt·āt′·ă* (L. *ŭsitātus*, used often), a species which produces the Sapucaia nuts, closely allied to Brazil nuts; also called **L. zabucajo**, *zăb′·ū·kă′·yō* (native name).

Ledum, n., *lēd′·ŭm* (Gr. *lēdon*, a species of Cistus), an ornamental genus of plants, Ord. Ericaceæ: **Ledum palustre**, *păl·ŭs′·trē* (L. *păluster*, swampy), a low shrub called the Labrador tea.

leeches, n. plu., *lētsh′·ĕs* (Icel. *lœknir*; Goth. *leikeis*, a leech—from Goth. *leikinon*, to heal; Bav. *lek*, medicine), worm-like animals found in ditches and swamps, used to abstract blood from inflamed parts; the 'Hirudo officinalis' and 'medicinalis,' Ord. Hirudinea.

leek, see 'house-leek.'

legume, n., *lĕg·ūm′*, also **legumen**, n., *lĕg·ūm′·ĕn* (L. *lĕgūmen*, pulse), a pod composed of one carpel opening usually by ventral and dorsal suture, as the pea; a dehiscent two-valved carpel: **legumin**, n., *lĕg·ūm′·ĭn*, an essential principle of the seeds of leguminous plants, and of oily seeds; casein: **Leguminosæ**, n. plu., *lĕg·ūm′·ĭn·*

ōz′·ē, the pea and bean tribe, an Ord. of herbaceous plants, shrubs, or trees: **leguminous**, a., *lĕg·ūm′·ĭn·ŭs*, pert. to the pea or bean tribe.

Lemneæ, n. plu., *lĕm′·nĕ·ē* (said to be corrupted from Gr. *lepis*, a scale), the duckweeds, a Sub-order of plants, Ord. Araceæ: **Lemna**, n., *lĕm′·nă*, the duckweeds, a curious genus of plants, floating as scales or small shield-like bodies on water, forming a green mantle.

lemniscus, n., *lĕm·nĭsk′·ŭs* (Gr. *lēmniskos*, a coloured band or fillet), in *anat.*, the fillet or bundle of fibres on each side of the peduncular system of the cerebrum.

lemon, n., *lĕm′·ŏn* (Sp. *limon*, Ar. *laymon*, a lemon), a fruit of an oblong form, the produce of the Citrus limonum, Ord. Aurantiaceæ, whose juice is anti-scorbutic, and is used in the manufacture of cooling drinks.

lens, n., *lĕnz*, **lenses**, n. plu., *lĕnz′·ĕs* (L. *lens*, a lentil), in the eye, a doubly convex transparent solid body with a rounded circumference; in an optical instrument a piece of glass of a convex, concave, or other shape for changing the direction of rays of light.

Lentibulariaceæ, n. plu., *lĕnt·ĭb′·ūl·ār·ĭ·ā′·sĕ·ē* (L. *lenticula*, the shape of a lentil—from *lens*, a lentil), the Butterwort family, an Order of plants, so named from the lenticular shape of the air bladders on the branches of Utricularia, one of the genera.

lenticel, n., *lĕnt′·ĭs·ĕl* (L. dim. of *lens*, a lentil, *lentis*, of a lentil), in *bot.*, a small process on the bark of the Willow and other plants, from which adventitious roots spring.

lenticular, a., *lĕnt·ĭk′·ŭl·ăr* (L. *lenticula*, the shape of a lens—from *lens*, a lentil, *lentis*, of a lentil), resembling a double convex lens : **lentiform**, n., *lĕnt'ĭ·fŏrm*, same sense.

lentignose, a., *lĕnt'ĭg·nōz'* (L. *lentignōsus*, full of freckles — from *lentigo*, a lentil-shaped or freckly spot), in *bot.*, covered with dots as if dusted : **lentigo**, n., *lĕnt·ĭg'·ō*, a freckly eruption on the skin : **lentiginous**, a., *lĕnt·ĭdj'·ĭn·ŭs*, freckly, scurfy.

Lentisk, n., *lĕnt·ĭsk'* (L. *lentiscus*, the mastich tree), the Pistacia lentiscus, Ord. Anacardiaceæ, a native of the Mediterranean coasts and islands, furnishes the concrete resinous exudation, called Mastich or Mastic.

Leopard's bane, *lĕp'·ărdz bān*, the Arnica Montana, Ord. Compositæ, Sub-order Corymbiferæ, called also mountain tobacco, a plant which is an acrid stimulant, frequently applied much diluted to bruises, etc.

Leopoldinia, n., *lē'·ō·pōld·ĭn'·ĭ·ă* (fem. of *Leopold*, after an empress of Brazil), a genus of fine palms of Brazil, Ord. Palmæ : **Leopoldinia piassaba**, *pĭ'·ăs·săb'·ă* (native name), a fibre used in manufactures under the name Piassaba.

Lepidium, n., *lĕp·ĭd'·ĭ·ŭm* (Gr. *lepidion*, a small scale, a plant — from *lepis*, a scale), a genus of plants, Ord. Cruciferæ, so called in allusion to the scale-like shape of the pods : **Lepidium sativum**, *săt·ĭv'·ŭm* (L. *sativum*, that is sown or planted), the well-known garden cress.

Lepidocaryinæ, n. plu., *lĕp'·ĭd·ō·kăr'·ĭ·ĭn'·ē* (Gr. *lepis*, a scale, *lepidos*, of a scale ; *karuon*, a nut), a Sub-order of trees of the Ord. Palmæ : **Lepidocaryum**, n., *lĕp'·ĭd·ō·kăr'·ĭ·ŭm*, a genus of the Palmæ.

Lepidoptera, n. plu., *lĕp'·ĭd·ŏp'·tĕr·ă* (Gr. *lepis*, a scale ; *pteron*, a wing), an Order of insects, comprising butterflies and moths, possessing four wings, which are usually covered with minute scales : **Lepidopteral**, a., *lĕp'·ĭd·ŏp'·tĕr·ăl*, of or pert. to the Lepidoptera.

lepidote, a., *lĕp'·ĭd·ōt* (Gr. *lepidōtos*, covered with scales — from *lepis*, a scale), in *bot.*, scurfy from minute scales ; covered with scales or scurf : **lepidota**, n. plu., *lĕp'·ĭd·ōt'·ă*, in *zool.*, an old name for the Ord. Dipnoi, which contains mud fishes.

lepiota, n. plu., *lĕp'·ĭ·ōt'·ă* (Gr. *lepis*, a scale), in *bot.*, the annules of some Fungi.

lepis, n., *lĕp'·ĭs* (Gr. *lepis*, a hair), a name applied to expansions of the epidermis in plants producing a scale or scurf whose surface is then said to be 'epidote.'

lepra, n., *lĕp'·ră* (Gr. *lepra*, leprosy ; *lepros*, rough, scaly), a term formerly applied to leprosy, now confined to a form of psoriasis, which see : **leprosy**, n., *lĕp'·rŏs·ĭ*, a disease of the skin of two kinds — the tuberculated one, in which the surface is marked with tubercles, and the anæsthetic, in which there is a number of spots having no feeling : **leprous**, a., *lĕp'·rŭs*, affected with leprosy ; covered with white scales, or with a white mealy substance : **leprose**, a., *lĕp'·rōz*, in *bot.*, scurf-like.

Leptosiphon, n., *lĕp'·tō·sĭf'·ŏn* (Gr. *leptos*, thin, slender ; *siphon*, a tube), a genus of very elegant annuals, Ord. Polemoniaceæ, so named from the slenderness of the tube of the corolla.

Leptospermeæ, n. plu., *lĕp'·tō·spĕrm'·ĕ·ē* (Gr. *leptos*, thin, slender ; *sperma*, seed), a Sub-order of the Ord. Myrtaceæ, having opposite or alternate leaves, usually dotted : **Leptospermum**, n., *lĕp'·tō·spĕrm'·ŭm*, a genus of the Myrtaceæ, having a neat foliage, and beautiful blossoms ; leaves of several species are used in Tasmania and Australia as tea : **Leptospermum lanigerum**, *lăn·ĭdj'·ĕr·ŭm* (L.

lanager, wool-bearing—from *lāna*, wool ; *gero*, I bear), a species whose leaves are used as tea.

lesion, n., *lĕzh'·ŭn* (L. *læsus*, hurt), a hurt ; an injury ; a morbid alteration in a function or structure.

lethal, a., *lēth'·ăl* (L. *lethālis*, mortal—from *lēthum*, death), deadly; mortal.

lethargy, n., *lĕth'·ăr·ji* (Gr. and L. *lethargia*, drowsiness — from *lēthē*, forgetfulness ; *argos*, idle), heavy, unnatural slumber; morbid drowsiness.

lettuce, n., *lĕt'·tis* (L. *lactuca*, a lettuce—from *lac*, milk), a garden salad plant of various kinds ; the common lettuce is the Lactuca sativa, Ord. Compositæ, Subord. Cichoraceæ.

leuchæmia, n., *lŏk·ēm'·i·ă* (Gr. *leukos*, white ; *haima*, blood), a morbid condition of the blood ; the same as leucocythæmia, which see.

leucin, n., *lŏs'·in* (Gr. *leukos*, white), a peculiar white substance derived from nitrogenous bodies.

leucocytosis, n., *lŏk'·ō·sit·ōz'·is* (Gr. *leukos*, white ; *kutos*, a cell), an increase in the number of white corpuscles in the blood in several morbid as well as physiological conditions—in the latter, after partaking of food for instance : leucocythæmia, n., *lŏk'·ō·sith·ēm'·i·ă* (Gr. *haima*, blood), the condition of the blood in which the white corpuscles are enormously increased in number.

Leucodendron, n., *lŏk'·ō·dĕn'·drŏn* (Gr. *leukos*, white ; *dendron*, a tree), a genus of splendid evergreen shrubs, having heads of yellow flowers, so called from their white leaves: Leucodendron argenteum, *ār·jĕn'·tĕ·ŭm* (L. *argentĕus*, made of silver –from *argentum*, silver), the silver tree or Witteboom of the Cape, having beautiful silky leaves.

leucoderma, n., *lŏk'·ō·dĕrm'·ă* (Gr. *leukos*, white ; *derma*, skin), a disease characterised by a mere discoloration of the skin, giving rise to no other symptoms.

Leucojum, n., *lŏk·ō'jŭm* (Gr. *leukos*, white ; *ĭŏn*, a violet), a genus of hardy bulbs producing spikes of pretty white flowers, like the snowdrop, Ord. Amaryllidaceæ : Leucojum vernum, *vĕrn'·ŭm* (L. *vernus*, belonging to spring — from *ver*, spring), the snow-flake.

Leucopogon, n., *lŏk'·ō·pōg'·ŏn* (Gr. *leukos*, white ; *pōgōn*, the beard, the limb of the corolla being bearded with white hairs), a genus of plants, Ord. Epacridaceæ : Leucopogon Richei, *ritsh'·ĕ·i* (a proper name), a fruit-bearing plant of Australia called Native Currant.

leucorrhœa, n., *lŏk'·ŏr·rē'·ă* (Gr. *leukos*, white ; *rheo*, I flow), the 'whites,' a disorder frequently met with in women, and the result either of debility, or of inflammatory changes in the genital organs.

leukæmia, n., *lŏk·ēm'·i·ă*, see 'leuchæmia.'

levator, n., *lĕv·āt'·ŏr* (L. *levātor*, a lifter—from *lĕvo*, I lift up), in *anat.*, a muscle which raises up a part ; the muscle which lowers a part being called 'depressor' : levator anguli oris, *ăng'·gŭl·i ŏr'·is* (L. *anguli*, of the angle, *os*, the mouth, *oris*, of the mouth), a muscle arising in the canine fossa, and inserted into the angle of the mouth : levatores costarum, *lĕv'·ăt·ŏr'·ēz kŏst·ār'·ŭm* (L. *costa*, a rib, *costārum*, of ribs), muscles which raise the ribs.

levis, a., *lĕv'·is* (L. *lēvis*, smooth), in *bot.*, even : levigatus, a., *lĕv'·ig·āt'·ŭs* (L.), made smooth ; having a smooth, polished appearance : levigation, n., *lĕv'·ig·ā'·shŭn*, the reduction of a hard substance by grinding or rubbing, with the aid of a little water, to an impalpable powder.

lianas, n. plu., *lī·ăn′·ăs*, or **lianes,** n. plu., *lī·ānz′* (probably a native name ; Sp. *liar*, to fasten), in *bot.*, luxuriant woody climbers, like those met with in tropical forests.

liber, n., *lĭb′·ér* (L. *liber*, the inner bark of a tree, a book), the fibrous inner bark of trees or plants ; the endophlœum : **libriform,** a., *lĭb′·rĭ·fŏrm* (L. *forma*, shape), having the form of fibrous bark.

lichen, n., *lĭk′·ĕn* or *lĭtsh′·ĕn* (L. *lichen*, a lichen), a skin eruption consisting of small pimples or papules, sometimes appearing in clusters, so named from its supposed resemblance to lichens : **lichenous,** a., *lĭk′·ĕn·ŭs*, of or belonging to the skin eruption called lichen.

Lichenes, n. plu., *lĭk·ēn′·ēz*, also **Lichens,** n. plu., *lĭk′·ĕnz* or *lĭtsh′·ĕnz* (L. *lichen*, a lichen, *lĭchēnis*, of a lichen ; Gr. *leichēn*, the liverwort), the Lichen family, an Order of plants forming a thallus which is either foliaceous, crustaceous, or pulverulent : **lichenic,** a., *lĭk·ĕn′·ĭk*, pert. to lichens : **lichenin,** n., *lĭk′·ĕn·ĭn*, the peculiar starch extracted from ' Cetraria islandica,' or Iceland moss : **lichenoid,** a., *lĭk′·ĕn·ŏyd* (Gr. *eidos*, resemblance), irregularly lobed, as the leafy lichens.

lienal, a., *lī·ēn′·ăl* (L. *lien*, the milt or spleen, *lĭēnis*, of the spleen), of or pert. to the spleen : **lienculus,** n., *lī·ĕng′·kŭl·ŭs* (L. dim.), a small or supplementary spleen.

ligament, n., *lĭg′·ă·mĕnt* (L. *ligămentum*, a band, a tie—from *ligo*, I bind), the strong fibrous substance which connects the ends of the moveable bones, and which sometimes protects the joints by a capsular envelope: **ligamentum nuchæ,** *nū′·kē* (mod. L. *nucha*, the nape of the neck, *nuchæ*, of

the nape— said to be from Arabic), the band of elastic fibres by which the weight of the head in Mammalia is supported : **ligamenta lata,** plurals, *lĭg′·ă·mĕnt′·ă lāt′·ă* (L. *lătus*, wide, broad), broad ligaments.

ligature, n., *lĭg′·ăt·ūr* (L. *ligātus*, bound), a cord or thread of silk, hemp, catgut, etc., employed to tie a blood-vessel or tumour.

ligneous, a., *lĭg′·nĕ·ŭs* (L. *lignum*, wood), woody ; resembling wood: **lignin,** n., *lĭg′·nĭn*, woody matter which thickens the cell-walls : **lignum vitæ,** *vīt′·ē* (L. *vīta*, life, *vitæ*, of life), ` the Guaiacum officinale, a beautiful W. Indian tree whose wood is prized for its hardness.

ligula, n., *lĭg′·ŭl·ă* (L. *lĭgŭla*, a little tongue, a shoe-strap), in *bot.*, the strap-shaped florets of Compositæ ; in *anat.*, a thin lamina : **ligulate,** a., *lĭg′·ŭl·āt*, in *bot.*, having strap-shaped florets as in the dandelion: **ligule,** n., *lĭg′·ŭl*, a tie ; a process arising from the petiole of grasses where it joins the blade : **ligulifloræ,** n. plu., *lĭg′·ŭl·ĭ·flŏr′·ē* (L. *flos*, a flower, *floris*, of a flower), composite plants having ligulate florets : **liguliflorate,** a., *-flŏr′·āt*, having ligulate florets.

Ligustrum, n., *lĭg·ŭst′·rŭm* (L. *lĭgustrum*, the plant privet), a genus of privets, Ord. Oleaceæ : **Ligustrum vulgare,** *vŭlg·ār′·ĕ* (L. *vulgāris*, common), the common privet, well suited for hedges, whose leaves are astringent : L. **lucidum,** *lōs′·ĭd·ŭm* (L. *lucĭdus*, clear, bright), yields a kind of waxy excretion, usefully employed in China : L. **ibota,** *ĭb′·ōt·ă* (unascertained), a Japan privet on which the wax insect feeds.

lilac, n., *lĭl′·ăk* (Sp. *lilac*, F. *lilas*, the lilac), the Syringa vulgaris, a shrub producing abundance of purple-coloured or white flowers : **lilacine,** n., *lĭl′·ăs·ĭn*, a principle in the bark of the lilac to which is due its febrifuge qualities.

Liliaceæ, n. plu., *lĭl'·ĭ·ā'·sĕ·ē* (L. *lilium*, a lily; Basque, *lili*, a flower), the Lily family, an Order of plants containing many showy garden flowers, as tulips, lilies, tube-roses, hyacinths, etc.: **Lilium**, *lĭl'·ĭ·ŭm*, the lilies, an ornamental and well-known genus: **Lilium Chalcedonicum**, *kăl'·sĕd·ŏn'·ĭk·ŭm* (from *Chalcēdon*, a town of Bithynia), said to be the lilies of the field of Scripture: **L. auratum**, *awr·āt'·ŭm* (L. *aurātus*, overlaid with gold—from *aurum*, gold), a lily of Japan having enormous white flowers, spotted with rich brown, and with numerous golden bands: **L. pomponium**, *pŏm·pōn'·ĭ·ŭm* (after *Pompōnĭus*, an ancient geographer), a species cultivated for its tubers in Kamtschatka as the potato in this country.

limb, n., *lĭm* (AS. *lim*, a limb—from *lime*, to join), the leg or arm of an animal; in *bot.*, the blade of the leaf; the broad part of a petal or sepal.

limbus luteus, *lĭm'·bŭs lŏt'·ĕ·ŭs* (L. *limbus*, a border that surrounds anything; *lūtĕus*, golden-yellow), a yellow spot in the axis of the ball of the eye.

lime, n., *lĭm* (Ger. *leim*, a viscous substance; L. *limus*, slime, mud), a white caustic earth used with sand as mortar or cement, obtained by burning limestone, marble, chalk, etc.; **lime water**, is used medicinally; the Linden tree, Tilia Europæa, so called from the glutinous juice of the young shoots—also said to be properly spelt 'line tree,' from the fact of its bark being used for making cordage; the fruit of the Citrus limetta.

Limonia, n., *lĭm·ōn'·ĭ·ă* (Gr. *leimōn*, a green field from its colour; said to be Arab *limoun*, the citron), a genus of plants nearly allied to

Citrus, Ord. Aurantiaceæ: **Limonia laureola**, *lăwr·ē'·ōl·ă* (L. *laurĕola*, a laurel-branch—from *laurĕă*, a laurel), a species found near the summit of lofty mountains.

Linaceæ, n. plu., *lĭn·ā'·sĕ·ē* (L. *lĭnum*, Gr. *lĭnon*, flax, lint), the Flax family, an Order of plants which yield mucilage and fibre: **Linum**, n., *lĭn'·ŭm*, an ornamental genus of plants, many having showy flowers: **Linaceæ grandiflorum**, *grănd'·ĭ·flōr'·ŭm* (L. *grandis*, great, large; *flos*, a flower, *flōris*, of a flower), a linum of North Africa having beautiful crimson flowers: **L. usitatissimum**, *ūz'·ĭt·ăt·ĭs'·ĭm·ŭm* (L. *ŭsĭtatissimum*, very common or familiar — from *ūsĭtātus*, common, familiar), the plant from the inner bark of whose stalk, after a process of steeping and breaking off the bark, the flax of commerce is procured; from the mucilaginous seeds, a demulcent and diuretic infusion is obtained: **linseed oil**, an oil obtained from the cotyledons of the seeds of L. usitatissimum used extensively in *med.*, the arts, etc.; the residue is made up into cakes, called oil-cake, for feeding cattle, and powdered receives the name linseed meal: **L. catharticum**, *kăth·ărt'·ĭk·ŭm* (Gr. *kathartĭkos*, purifying or cleansing), a species which has purgative properties, the active principle being called **linin**, n., *lĭn'·ĭn*: **L. selaginoides**, *sĕl'·ă·jĭn·ōyd'·ēz* (*sĕlāgo*, the upright club moss; Gr. *eidos*, resemblance), esteemed a bitter and aperient in Peru.

Linaria, n., *lĭn·ār'·ĭ·ă* (L. *linum*, flax), a genus of annuals well adapted for rock-work, Ord. Scrophulariaceæ: **Linaria vulgaris**, *vŭlg·ār'·ĭs* (L. *vulgāris*, common), a species having five-spurred flowers instead of one-spurred: **L. triornithophora**, *trĭ'·ŏr·nĭth·ŏf'·ŏr·ă* (Gr. *tris*, thrice;

ornis, a bird, *ornīthos*, of a bird ; *phoreo*, I bear), the form of whose flowers bears some resemblance to three little birds seated in the spur.

Linden, *lĭn'dĕn*, or **Lime tree,** the Tilia Europæa, Ord. Tiliaceæ, from whose tough fibrous inner bark are manufactured Russian mats ; the bark is also called 'bast' or 'bass.'

linea alba, *lĭn'ĕ·ă ălb'ă* (L. *linea*, a line ; *albus*, white), a white line formed by the meeting of the tendons of the abdominal muscles, which extend from the ensiform cartilage to the pubes : **linea aspera,** *ăs'pĕr·ă* (L. *asper*, rough, uneven), on the femur, a prominent ridge extending along the central third of the shaft posteriorly, and bifurcating above and below : l. **quadrata,** *kwŏd·rāt'ă* (L. *quadrātus*, square, four-cornered), a well-marked eminence passing vertically downwards for about two inches along the back part of the shaft of the femur : l. **splendens,** *splĕn'dĕnz* (L. *splendens*, bright, shining), a conspicuous, fibrous band running down in front over the interior medium fissure of the pia mater of the spinal cord.

lineæ semilunares, *lĭn'ĕ·ē sĕm'ĭ·lŏn·ār'ēz* (L. *lineæ*, lines ; *semi*, a half ; *lūna*, the moon), two curved tendinous lines on each side of the linea alba : **lineæ transversæ,** *trăns·vĕrs'ē* (L. *transversus*, turned or directed across—from *trans*, across ; *versus*, turned), three or four narrow transverse lines which intersect the rectus muscle.

linear, a., *lĭn'ĕ·ar* (L. *linea*, a line), narrow ; line-like ; in *bot.*, having very narrow leaves much longer than broad.

ling, n., *lĭng* (Icel. *ling*, any small shrub), common heather, the Culluna vulgaris, Ord. Ericaceæ.

lingual, a., *lĭng'gwăl* (L. *lingua*,

the tongue), connected with the tongue : **lingualis,** n., *lĭng·gwăl'ĭs*, a longitudinal band of muscular fibres situated on the under surface of the tongue, which contracts the tongue and compresses its point : **linguiform,** a., *lĭng'gwĭ·fŏrm* (L. *forma*, shape), in *bot.*, shaped like a tongue—also **lingulate,** a., *lĭng'gŭl·āt*, in same sense.

liniment, n., *lĭn'ĭ·mĕnt* (L. *linimentum*, an ointment — from *linĕre*, to besmear), a thick, oily, or other liquid substance, to be rubbed into the skin.

Linnæa, n., *lĭn·nē'ă* (after *Linnæus* the illustrious naturalist), a genus of elegant little plants, natives of northern regions, Ord. Caprifoliaceæ : **Linnæa borealis,** *bŏr'ĕ·āl'ĭs* (L. *borēālis*, northern), the two-flowered Linnæa.

linum, see 'linaceæ.'

lipoma, n., *lĭp·ōm'ă*, **lipomata,** n. plu., *lĭp·ŏm'·ăt·ă* (Gr. *lipos*, fat), fatty tumours or growths.

Liquidamber, n., *lĭk'·wĭd·ăm'bĕr* (L. *liquidus*, liquid ; *ambar*, amber, alluding to the gum which exudes from the trees), a genus of ornamental hard trees, Ord. Hamamelidaceæ : **Liquidambar orientalis,** *ŏr'ĭ·ĕnt·āl'ĭs* (L. *orientālis*, eastern—from *orior*, I arise), a species which yields liquid storax, used as a remedy for scabies : **L. altingiana,** *ăl·tĭnj'ĭ·ān'ă* (after *Alting*, a German botanist) ; **L. Formosana,** *fŏr'·mōz·ān'ă* (of or from *Formosa*) ; and **L. styraciflua,** *stĭr'ă·sĭf'lŏ·ă* (L. *styrax*, a resinous gum, storax, *styrăcis*, of storax ; *fluo*, I flow), the sweet gum tree, the three preceding yield resins which are used as fragrant balsams.

liquor, n., *lĭk'·ĕr* (L. *liquor*, a fluid —from *liqueo*, I melt), a fluid or liquid ; a natural fluid or secretion of the body ; a watery solution ; an extract ; a spirituous fluid : **liquor ammoniæ,** *ăm·mōn'ĭ·ē* (L.

ammoniæ, of ammonia), a solution of ammoniacal gas in water: l. **calcis**, *kăl'sĭs* (L. *calx*, lime ; *calcis*, of lime), lime-water : l. **Cotunnii**, *kō·tŭn'nĭ·ī* (first described by *Cotunnius*), a thin, slightly albuminous or serous fluid which separates the membranous from the osseous labyrinth in the vestibule and semicircular canals of the internal ear; the perilymph: l. **pericardii**, *pĕr'ĭ·kărd'ĭ·ī* (new L. of the pericardium), a serous fluid contained in the pericardium: l. **sanguinis**, *săng'gwĭn·ĭs* (L. *sanguis*, blood, *sanguĭnis*, of blood), the liquor of the blood ; the transparent colourless fluid part of the blood, in which the red corpuscles float during life : l. **seminis**, *sēm'ĭn·ĭs* (L. *sēmen*, seed, *semĭnis*, of seed), the transparent albuminous fluid containing the seed.

lirella, n., *lĭr·ĕl'lă* (dim. of L. *lira*, a ridge of land), in *bot.*, the apothecia of lichens when they are linear, as in Graphideæ: **lirellate**, a., *lĭr·rĕl'lāt*, like a furrow —also **lirelliform**, a., *lĭr·ĕl'lĭ·fŏrm* (L. *forma*, shape), formed like a furrow.

Liriodendron, n., *lĭr'ĭ·ō·dĕn'drŏn* (Gr. *leirion*, a lily ; *dendron*, a tree), a genus of trees, Ord. Magnoliaceæ, whose flowers bear some resemblance to the lily and tulip : **Liriodendron tulipifera**, *tūl'ĭp·ĭf'ĕr·ă* (F. *tulipe*, a tulip ; Pers. *tuliban*, a turban ; L. *fero*, I bear), the tulip tree, marked by its truncate leaves, used as a febrifuge, the wood used in ornamental and carved work.

Listera, n., *lĭst'ĕr·ă* (after *Dr. Lister*, an English naturalist), a genus of curious little plants, Ord. Orchidiaceæ, in which the viscid mass of the rostellum bursts with force, allowing the pollinia to escape.

lithate, n., *lĭth'āt* (Gr. *lithos*, a stone), a salt formed by lithic acid with a base ; the red or pink, sandy deposit which settles from the urine on cooling, often found in dyspepsia—also named 'urate': **lithia**, n., *lĭth'ĭ·ă*, an alkali, allied in its properties to potass, used as a remedy in gout : **lithic**, a., *lĭth'ĭk*, a term equivalent to uric, which see : **lithiasis**, n., *lĭth·ī'ăs·ĭs*, gravel or urinary calculi, deposits of solid elements in the parts of the urinary apparatus: **lithocysts**, n. plu., *lĭth'ō·sĭsts* (Gr. *kustis*, a cyst), in *zool.*, the sense organs or marginal bodies of such as the ' Lucernarida ': **lithology**, n., *lĭth·ŏl'ŏ·jĭ* (Gr. *logos*, discourse), a treatise on the stones or calculi found in the body : **litholysis**, n., *lĭth·ŏl'ĭs·ĭs* (Gr. *lŭsis*, a loosening or release), the treatment for the solution of the stone in the bladder.

lithontriptic, n., *lĭth'ŏn·trĭp'tĭk* (Gr. *lithos*, a stone ; *tribo*, I grind or wear by rubbing), a medicament supposed to act as a solvent of urinary calculi in the natural passages.

lithotomy, n., *lĭth·ŏt'ŏm·ĭ* (Gr. *lithos*, a stone ; *tomē*, a cutting), the operation of cutting into the bladder for the removal of a calculus or stone : **lithotrity**, n., *lĭth·ŏt'rĭ·tĭ* (L. *tritus*, bruised or ground), an operation in which the stone is crushed or broken, and removed without cutting.

litmus, n., *lĭt'mŭs* (Dut. *lakmoes*, an infusion of a lake or purple colour), a peculiar blue colouring matter extracted from lichens, the ' Rocella tinctoria,' ' R. fuciformis,' and 'R. hypomecha,' used as delicate tests for acids and alkalies, turned red by the former, and blue by the latter.

liver, n., *lĭv'ĕr* (AS. *lifere;* Ger. *leber*, the liver), the largest gland in the body, seated on the right side of the abdomen, below the diaphragm, one of whose functions is to secrete bile : **liver**·

fluke, an animal parasite found in the liver, the 'Fasciola hepatica,' common in the sheep, sometimes met with in the human body.

lixiviation, n., *lĭks·ĭv'·ĭ·ā'·shŭn* (L. *lixivius*, made into lye — from *lix*, lye), the operation or process of extracting alkaline salts from ashes by mixing or washing them with water, and then running off and evaporating the water.

Loasaceæ, n. plu., *lō'·ăs·ā'·sĕ·ē* (origin unascertained), the Chili nettle family, an Order of plants of America, distinguished by their stinging qualities : **Loasa**, n., *lō·ās'·ă*, a genus, highly interesting from the beauty of their curiously formed flowers : **Loasa placei**, *plās'·ĕ·ī* (unascertained), a species possessing powerful stinging properties.

lobe, n., *lōb* (Gr. *lobos*, the tip of the ear), in *bot.*, a large division of a leaf or seed ; a division of the anther ; a part or division of the lungs, liver, etc.: **lobate**, a., *lōb'·āt*, divided into small lobes ; having lobes or divisions; **lobulæ**, n. plu., *lŏb'·ūl·ē*, or **lobules**, n. plu., *lŏb'·ūlz*, subdivisions of a lobe ; very minute lobes.

Lobeliaceæ, n. plu., *lō·bēl'·ĭ·ā'·sĕ·ē*, also **Lobeliads**, n. plu., *lō·bēl'·ĭ·ădz* (after *Lobel*, the French physician and botanist to James I.), the Lobelia family, an Order of plants in which acridity prevails more or less : **Lobelia**, n., *lō·bēl'·ĭ·ă*, a genus of plants producing beautiful blossoms : **Lobelia inflata**, *ĭn·flāt'·ă* (L. *inflātus*, puffed up, inflated—from *in*, into; *flo*, I blow), Indian tobacco, a native of North America, used medicinally, chiefly as a sedative : **L. longiflora**, *lŏn'·jĭ·flōr'·ă* (L. *longus*, long ; *flos*, a flower, *flōris*, of a flower), one of the most venomous of plants : **L. syphilitica**, *sĭf'·ĭl·ĭt'·ĭk·ă* (Gr. *sun*,

with ; *phileo*, I love : or *sus*, a swine ; *philos*, dear), a plant whose root is acrid and emetic : **L. urens**, *ūr'·ĕnz* (L. *ūrens*, burning), a plant whose milky juice is said to be vesicant : **lobelina**, n., *lōb'·ĕl·īn'·ă*, a volatile alkaloid obtained from the Lobelia inflata.

lobule, n., *lōb'·ūl* (a dim. of Gr. *lobos*, a lobe, which see), a small lobe : **lobulate**, a., *lŏb'·ūl·āt*, divided into small lobes.

lobulus, n., *lŏb'·ūl·ŭs* (a mod. L. dim. of Gr. *lobos*, a lobe), a small lobe or division : **lobulus auris**, *awr'·ĭs* (L. *auris*, the ear, *auris*, of the ear), the lower dependent portion of the pinna of the ear : **l. spigelii**, *spĭ·jēl'·ĭ·ī* (after *Spigel*, a Belgian physician), a small lobe of the liver, on the left of the great lobe : **l. caudatus**, *kāwd·āt'·ŭs* (mod. L. *caudātus*, tailed— from L. *cauda*, a tail), the tailed appendage ; a small tail - like appendage to the lobulus spigelii: **lobuli testis**, *lŏb'·ūl·ī tĕs'·tĭs* (L. *testis*, a testicle, *testis*, of a testicle), the lobules of the testis ; the numerous lobules of which the glandular structure of the testis consists.

lochia, n., *lōk·ĭ'·ă* (Gr. *locheia*, child-birth), the discharges from the uterus and vagina after childbirth while the mucous membrane is returning to its primary condition.

lock - hospital, n., *lŏk - hŏs'·pĭt·ăl* (Dut. *locken*, Icel. *loka*, to shut, to fasten ; AS. *loc*, a place shut in : F. *loquet*, the latch of a door), a charitable institution for the treatment of venereal diseases.

NOTE.—The origin of this name is much disputed. There seems to be but little doubt that the name was first applied to a leper or lazar-house in the sense, as given in the root-words, 'of being shut off or isolated from all others.' The hospitals for venereal diseases, after the disappearance of leprosy from this

Q

country, appear to have replaced lazar-houses, or at least to have had the same name continued to them. We have also *loke*, a house for lepers; and in Ger. *lockern*, to play the rake or libertine.

lock-jaw, *lŏk-jăw,* the disease called 'trismus,' in which the jaws become locked or closely joined together by the persistent contraction of the voluntary muscles of the jaws.

locomotor ataxia, n., *lōk'·ō·mōt'·ŏr ăt·ăks'·ĭ·ă* (L. *locus*, a place; *mōtum*, to move; and *ataxia*, which see), the want of co-ordination in the movements of the arms, legs, or both, depending upon fascicular sclerosis of the posterior column of the spinal cord.

loculament, n., *lŏk'·ūl·ă·mĕnt,* also **loculus,** n., *lŏk'·ūl·ŭs* (L. *loculamentum*, a case, receptacle —from *loculus*, a little place), in *bot.*, a cavity in the pericarp containing the seed — called 'unilocular' with one cavity, and 'bilocular with two cavities, and so on; one of the cells of the anther: **loculicidal,** a., *lŏk'·ūl·ĭ·sīd'·ăl* (L. *cædo*, I cut), in *bot.*, having the fruit dehiscing through the back of the carpels.

locus cœruleus, *lōk'·ŭs sēr·ūl'·ĕ·ŭs* (L. *locus*, a place, a spot ; *cærŭl-ĕus*, dark blue), in the fourth ventricle of the cerebellum, a small eminence of dark-grey substance which presents a bluish tint through the thin stratum covering it : **locus niger,** *nĭdj'·ĕ·* (L. *niger*, black, dark), a mass of dark-grey matter in the cerebrum, situated in the interior of the crura: l. **perforatus,** *pĕr'·fōr·ăt'·ŭs* (L. *perforātus*, bored through), a whitish-grey substance situated between the crura cerebri, and perforated by apertures for vessels.

locusta, n., *lōk·ŭst'·ă* (L. *locusta*, a cray-fish, a locust), in *bot.*, a spikelet of grasses formed of one or more flowers : locust tree, the

Ceretonia siliqua, Ord. Leguminosæ, Sub-ord. Cæsalpinieæ.

lodicule, n., *lŏd'·ĭk·ūl* (L. *lōdĭcula,* a small coverlet), a scale at the base of the ovary of grasses, and of the grape vine.

Loganiaceæ, n. plu., *lŏg'·ăn·ĭ·ā'·sĕ·ē,* also **Loganiads,** n. plu., *lŏg·ăn'·ĭ·ădz* (after *Logan,* a botanist), the Nux Vomica family, an Order of plants possessing intensely poisonous properties, of which strychnos is an example : **Loganieæ,** n. plu., *lŏg'·ăn·ĭ'·ĕ·ē,* a Suborder : **Logania,** n., *lŏg·ăn'·ĭ·ă,* an interesting genus of shrubs producing their flowers in axillary or terminal branches.

Lolium, n., *lŏl'·ĭ·ŭm* (L. *lōlĭum,* darnel, tares), a genus of grasses, rye-grass, Ord. Gramineæ : **Lolium perenne,** *pĕr·ĕn'·nĕ* (L. *perennis,* that continues the whole year, perennial), the perennial rye-grass, an esteemed fodder-grass : L. **temulentum,** *tĕm'·ŭl·ĕnt'·ŭm* (L. *tēmŭlentus,* intoxicated), darnel grass, the supposed tares of Scripture, said to be narcotic.

Lomentaceæ, n. plu., *lŏm'·ĕnt·ā'·sĕ·ē* (L. *lomentum,* bean meal), a Sub-order or division of the Cruciferæ, founded on the seed vessels: **lomentum,** n., *lŏm·ĕnt'·ŭm,* a legume or pod with transverse partitions, each division containing one seed: **lomentaceous,** a., *lŏm'·ĕnt·ā'·shŭs,* furnished with a loment.

longipennatæ, n. plu., *lŏnj'·ĭ·pĕn·nāt'·ē* (L. *longus,* long ; *penna,* a wing), in *zool.*, a group of the natatorial birds : **longipennate,** a., *-pĕn'·nāt,* long-winged — applied to birds.

longirostres, n. plu., *lŏnj'·ĭ·rŏs'·trēz* (L. *longus,* long ; *rostrum,* a beak), in *zool.*, a group of the Wading birds, having long bills : **longirostral,** a., *lŏnj'·ĭ·rŏs'·trăl,* long-beaked—applied to birds.

longissimus dorsi, *lŏn·jĭs'·sĭm·ŭs*

dŏrs'ĭ (L. *longissimus*, very long —from *longus*, long; and *dorsum*, the back, *dorsi*, of the back), the very long muscle of the back ; a muscle which extends the vertebræ, and raises and keeps the trunk of the body erect.

longus colli, *lŏng'gŭs kŏl'lī* (L. *longus*, long ; *collum*, the neck, *colli*, of the neck), the long muscle of the neck ; a long flat muscle on the anterior surface of the spine, which supports and bends the neck.

Loniceræ, n. plu., *lŏn'ĭs-ĕr'ĕ-ē* (after *Lonicer*, a German botanist). a Sub-order of the Ord. Caprifoliaceæ, which embraces the true honeysuckles : **Lonicera,** n., *lŏn-ĭs'ĕr-ă*, a genus of very ornamental shrubs.

lophiostomate, a., *lŏf'ĭ-ŏs'tŏm-āt* (Gr. *lopheion*, a ridge, a crest ; *stoma*, a mouth), in *bot.*, having crested apertures or openings : **lophophore,** n., *lŏf'ō-fōr* (Gr. *phoreo*, I carry), in *zool.*, the disc or stage upon which the tentacles of the Polyzoa are placed.

Lophyropoda, n. plu., *lŏf'ĭ-rŏp'ŏd-ă* (Gr. *lopheion*, a crest, tuft of hair; *oura*, a tail; *podes*, feet), a section of the Crustacea, embracing those which have cylindrical, or conical ciliated or tufted feet.

Loranthaceæ, n. plu., *lŏr'ănth-ā'sĕ-ē* (Gr. *lōron*, a thong; *anthos*, a flower, alluding to the long linear form of the petals), the Mistletoe family, an Order of shrubs, usually parasitical, and growing into the tissues of other vegetables, many in Asia and America having showy flowers : **Loranthus,** n., *lŏr-ănth'ŭs*, a genus of parasitical plants including the well-known mistletoe: **Loranthus tetrandrus,** *tĕt-răn'drŭs* (L. *tetrans*, a fourth part), a species used in Chili to dye black.

lorica, n., *lŏr'ĭk-ă* (L. *lorīca*, a coat of mail ; *lōricātum*, to cover with a breastplate of metal), the protective case with which certain Infusoria are provided : **Loricata,** n. plu., *lŏr'ĭk·āt'ă*, the division of reptiles, comprising the Chelonia and Crocodilia, which are protected by an armour of bony plates : **loricate,** a., *lŏr'ĭk-āt*, covered with horny plates or scales ; covered by a shell or husk : **lorication,** n., *lŏr'ĭk-ā'shŭn*, the act of covering with a plate or crust for defence.

Lotus, n., *lōt'ŭs* (Gr. *lotos*, L. *lotus*, the lotus), an ornamental genus of leguminous creeping plants, Ord. Leguminosæ, Subord. Papilionaceæ : **Loteæ,** n. plu., *lōt'ĕ-ē*, a tribe of the Suborder: **Lotus** or **Lote-bush,** n., *lōt*, an Egyptian water-plant, sacred with the anc. Egyptians; the plant of the anc. classics, the 'Zizyphus lotus,' being the true lotus of the ancients : **Lotus corniculatus,** *kŏr·nĭk'ŭl·āt'ŭs* (L. *corniculātus*, horned — from *cornu*, a horn) ; also **L. major,** *mādj'ŏr* (L. *major*, greater), are species sometimes sown with white clover, etc., in laying down permanent pastures.

lubricous, a., *lŏb'rĭk-ŭs* (L. *lubrĭcus*, slippery), in *bot.*, smooth ; slippery.

Lucernarida, n. plu., *lŏs'ĕr-năr'ĭd-ă* (L. *lŭcerna*, a lamp), in *zool.*, an Order of the Hydrozoa.

Lucerne or **Lucern,** n., *lŏs'ĕrn* (F. *luzerne*), the Medicago sativa, Ord. Leguminosæ, Sub-ord. Papilionaceæ, a well-known artificial grass, much cultivated as food for cattle.

Lucuma, n., *lŏk·ūm'ă* (the native name in Peru), a genus of fruitbearing trees, Ord. Sapotaceæ : **Lucuma mammosa,** *măm·mōz'ă* (L. *mammosus*, having large breasts—from *mamma*, a breast), a species whose kernels contain prussic acid.

Luffa, n., *lŭf'fă* (Arabic *louff*), a

genus of plants producing a remarkable kind of gourd of a very disagreeable odour, Ord. Cucurbitaceæ : **Luffa Egyptiaca,** *ĕ'·jĭp·tĭ'·ăk·ă* (of or from *Egypt*), the towel gourd, its split fruit being used as a flesh brush.

Luhea, n., *lŏ·hŭ'·ă* (after *Luhe*, a German botanist), a genus of pretty plants, Ord. Tiliaceæ : **Luhea grandiflora,** *grănd'·ĭ·flŏr'·ă* (L. *grandis*, great, large ; *flos*, a flower, *flōris*, of a flower), a species whose bark is used in Brazil for tanning leather, and an infusion of whose flowers is used as an antispasmodic and expectorant.

lumbago, n., *lŭm·bāg'·ō* (mid. L. *lumbago*—from L. *lumbus*, the loin or haunch), a form of chronic rheumatism, chiefly affecting the loins : **lumbar,** a., *lŭm'·băr*, of or pert. to the loins : **lumbar region,** the lower part of the trunk : **lumbar vertebræ,** the bones of the spine of the lower part of the trunk.

lumbricales, n. plu., *lŭm'·brĭk·āl'·ēz* (L. *lumbricus*, an earthworm), four muscles of the hand and foot which assist in bending the fingers or toes, so named from their resemblance to earth-worms : **Lumbricus,** n., *lŭm'·brĭk·ŭs*, a genus of worms : **Lumbricus terrestris,** *tĕr·rĕst'·rĭs* (L. *terrestris*, of or belonging to the earth —from *terra*, the earth), the common earth-worm.

lunate, a., *lŏn'·āt* (L. *luna*, the moon), like a half moon ; crescent - shaped : **lunar caustic** (L. *luna*, old term for silver), nitrate of silver, used surgically as a caustic ; the Argenti nitras, or fused nitrate of silver, moulded into small sticks.

lunula, n., *lŏn'·ŭl·ă*, **lunulæ,** n. plu., *lŏn'·ŭl·ē* (L. *lūnŭla*, a little moon—from *luna*, the moon), a small portion of the nail near the root which is whiter than the rest, so named from its shape ;

the thinner portions of the arterial valves of the heart.

Lupinus, n., *lŏp·īn'·ŭs* (L. *lupīnus*, a kind of pulse), the lupine, a genus of very beautiful annual and herbaceous border flowers, Ord. Leguminosæ, Sub-ord. Papilionaceæ : **Lupinus albus,** *ălb'·ŭs* (L. *albus*, white), a species extensively cultivated in S. Europe for forage, the seeds or peas being used as food after their bitterness has been removed by boiling.

lupulin, n., *lŏp'·ŭl·ĭn* (L. *lupulus*, the hops—from *lupus*, the hop-plant), the bitter aromatic principle of hops : **lupulinic glands,** a., *lŏp'·ŭl·ĭn'·ĭk*, the name applied to the resinous glandular scales surrounding the fruit of the hopplants, also called ' lupulin.'

lupus, n., *lŏp'·ŭs* (L. *lupus*, a wolf), a general name applied to certain forms of obstinate inflammatory and ulcerative affections of the nose, cheeks, and lips, which often destroy soft parts, and cause much disfigurement, usually associated with scrofula, and occasionally with syphilis : **erythematous lupus** (see 'erythema'), a variety least troublesome in which slightly elevated deep-red or livid patches appear on the skin of the nose or face, which increase and run together, forming large purple patches, sometimes becoming covered with thick crusts of scarf skin : **lupus nonexedens,** *nŏn·ĕks'·ĕd·ĕnz* (L. *non*, not ; *exĕdens*, eating up, devouring), a variety of the disease in which there is no destruction of parts : **l. exedens,** the most severe form of the disease in which the parts affected, usually the nose or lips, are destroyed.

lusus naturæ, n., *lŏz'·ŭs năt·ūr'·e* (L. *lusus*, sport or freak ; *natura*, nature, *naturæ*, of nature), a term applied to anything unnatural in an animal at birth, or in a vegetable production ; a monstrosity.

Lychnis, n., *lĭk'·nĭs* (Gr. *luchnis*, a plant with red flowers, used in making garlands; said to be from Gr. *luchnos*, a torch, from the brilliancy of their flowers), an extremely beautiful genus of plants, Ord. Caryophyllaceæ, some of which are said to be poisonous; among the showiest and most beautiful are **Lychnis chalcedonica,** *kăl'·sēd·ŏn'·ĭk·ă* (of or from *Chalcedon*); **L. fulgens,** *fŭl'·gĕnz* (L. *fulgens*, flashing, shining); **L. grandiflora,** *grănd'·ĭ·flŏr'·ă* (L. *grandis*, great, large; *flos*, a flower, *flōris*, of a flower), and **L. cæli rosa,** *sēl'·ĭ rōz'·ă* (L. *cæli*, of the sky; *rosa*, a rose).

Lycoperdon, n., *lĭk'·ō·pěrd'·ŏn* (Gr. *lukos*, a wolf; *pěrdō*, I break wind, I explode backwards), the puff-balls, a genus of roundish tuber-like plants, which, when ripe, explode and emit their sporules like smoke, Ord. Fungi: **Lycoperdon giganteum,** *jĭg'·ăn·tē'·ŭm* (L. *gigantēus*, of or belonging to the giants — from *gigantes*, the giants), the common puff-balls, found in pastures, and on the stumps of trees.

Lycopersicum, n., *lĭk'·ō·pěrs'·ĭk·ŭm* (Gr. *lukos*, a wolf; *persikon*, the orange or lemon tree; *persikoi*, peaches), the tomatoes or love-apples, a genus of plants, cultivated for the sake of their fruit, Ord. Solanaceæ: **Lycopersicum esculentum,** *ĕsk'·ŭl·ĕnt'·ŭm* (L. *esculentus*, fit for eating), a species whose fruit is the edible tomato or love-apple.

Lycopodiaceæ, n. plu., *lĭk'·ō·pŏd'·ĭ·ā'·sĕ·ē*, also **Lycopods,** n. plu., *lĭk'·ō·pŏdz* (Gr. *lukos*, a wolf; *pous*, a foot, *podes*, feet, so named from the appearance of the roots), the Club-moss family, an Order of moss-like plants, intermediate between ferns and mosses, having chiefly creeping stems: **Lycopodium,** n., *lĭk'·ō·pŏd'·ĭ·ŭm*, a genus of moss-like plants, some of which

are emetic and cathartic; the powdery matter in the thecæ is inflammable, and has been used under the name Lycopod or vegetable brimstone: **Lycopodium clavatum,** *klăv·āt'·ŭm* (L. *clāvātus*, fastened or furnished with nails — from *clāvus*, a nail), a species whose spores, in the form of a yellow powder, are used for dusting excoriated surfaces, and putting in pill-boxes to preserve the pills from adhesion to one another: **L. giganteum,** *jĭg'·ăn·tē'·ŭm* (L. *gigantēus*, of or belonging to the giants), a species in whose ash potassium phosphate is found in large quantity.

Lycopus, n., *lĭk'·ŏp·ŭs* (Gr. *lukos*, a wolf; *pous*, a foot, *podes*, feet), a genus of plants, Ord. Labiatæ, so named from the appearance of their leaves: **Lycopus virginicus,** *věr·jĭn'·ĭk·ŭs* (L. *virginicum*, belonging to a virgin—from *virgo*, a virgin), the plant bugle-weed; and **L. Europæus,** *ŭr'·ō·pē'·ŭs* (of or belonging to *Europe*), the plant gipsy-wort, are used as astringents and sedatives.

lycotropal, a., *lĭk·ŏt'·rŏp·ăl* (Gr. *lukos*, the knocker of a door; *tropos*, a turning), in *bot.*, an orthotropal ovule curved like a horse-shoe.

lyencephala, n. plu., *lī'·ĕn·sĕf'·ăl·ă* (Gr. *luō*, I loose; *kephalē*, the head), Owen's primary division of mammals; the loose-brained implacentals.

Lygeum, n., *lĭdj·ē'·ŭm* (Gr. *lŭgīzō*, I bend or twist; *lugŏs*, a twig, a wand), a genus of plants, Ord. Gramineæ, so named from their flexibility; grasses with rushy leaves, much used in Spain, etc., for baskets, ropes, nets, mats, etc.: **Lygeum spartum,** *spăr'·tŭm* (L. *spartum*, Gr. *sparton*, a plant of Spain—from Gr. *spartos*, a rope or cord), a species yielding fibres, which are used in making paper and mats.

Lygodium, n., *lĭg·ōd'·ĭ·ŭm* (Gr. *lugōdēs*, flexible, pliant), snake's tongues, a climbing genus of ferns, Ord. Tilices, so named in allusion to the twining habit of the plants.

lymph, n., *lĭmf* (L. *lympha*, water), in animal bodies an alkaline, colourless fluid which fills the absorbents or lymphatics : **lymphatics**, n. plu., *lĭmf·ăt'·ĭks*, the minute absorbent vessels which carry lymph from all parts of the body, generally closely accompanying blood-vessels : **lymphadenoma**, n., *lĭmf·ăd'·ĕ·nōm'·ă* (L. *lympha*, spring-water ; *adēn*, an acorn, a gland, *adĕnos*, of a gland), a disease in which there is a gradual and it may be great enlargement of the lymphatic glands of the body, also **lymphoma**, n., *lĭmf·ōm'·ă*.

lymphosarcoma, n., *lĭmf'·ō·sărk·ōm'·ă* (L. *lympha*, spring-water ; *sarcoma*, which see), medullary sarcoma of the lymphatic glands, which may be either round-celled or spindle-celled.

lyra, n., *lī'·ă* (Gr. *lura*, a lyre), a triangular portion of the corpus callosum of the cerebrum, marked with transverse, longitudinal, and oblique lines : **lyrate**, a., *lī'·āt*, in *bot.*, applied to a leaf having a large terminal lobe, and several pairs of smaller lobes decreasing in size towards the base ; pinnatifid, having the upper lobe largest.

Lythraceæ, n. plu., *lĭth·rā'·sĕ·ē* (Gr. *luthron*, gore, black blood), the Loosestrife family, an Order of plants, so named in allusion to the purple colour of most of the flowers, many of the plants have astringent properties, some are used for dyeing : **Lythrum**, n., *lĭth'·rŭm*, a genus of perennial and annual plants, the former of which are very handsome : **Lythrum salicaria**, *săl'·ĭ·kār'·ĭ·ă* (L.

sălix, a willow, *salicis*, of a willow), the Loosestrife, or willow-strife, contains tannin and has been used in diarrhœa.

mace, n., *măs* (L. *macis*, a spice ; F. *macis*), the arillus or arillode, being an additional covering of the seed of the tree Myristica officinalis, Ord. Myristicaceæ, commencing at the exostome, natives of Moluccas ; nutmeg and mace are used as aromatic stimulants and condiments, and in large doses are narcotic.

maceration, n., *măs'·ĕr·ā'·shŭn* (L. *maceratum*, to soften by steeping —from *macer*, lean, thin), the process or operation of softening, or extracting the soluble portion of anything by steeping in a liquid, as cold water : **infusion** is performed by pouring a hot or boiling liquid over a substance, as in making tea ; **decoction**, by boiling a substance in a liquid.

Maclura, n., *măk·lōr'·ă* (after *Maclure*, a N. American botanist), a genus of very ornamental trees, Ord. Moraceæ : **Maclura tinctoria**, *tĭnk·tōr'·ĭ·ă* (L. *tinctōrius*, of or belonging to dyeing—from *tingo*, I dye), a species producing the yellow dye-wood called Fustic.

Macrochloa, n., *măk'·rŏk·lō'·ă* (Gr. *makros*, long ; *chloă*, green grass), a genus of plants having a long herbage, Ord. Gramineæ : **macrochloa tenacissima**, *tĕn'·ă·sĭs'·sĭm·ă* (L. *tenacissimus*, very tenacious or sticky—from *tenax*, tenacious), Esparto or Spanish grass which yields fibres extensively employed in making paper, and mats, etc.

Macrocystis, n., *măk'·rō·sĭst'·ĭs* (Gr. *makros*, long ; *kustis*, a bladder), a genus of sea-plants, Ord. Algæ : **Macrocystis pyrifera**, *pĭr·ĭf'·ĕr·ă* (L. *pyrum*, a pear ; *fĕro*, I bear), one of the Algæ or sea-weeds which in the Pacific and Northern Oceans reaches the enormous length of from 500 to 1500 feet, so named in reference to the extremity of its frond, swelling out like a pear.

Macrodactyli, n. plu., *măk'·rō· dăk'·tŭl·ĭ* (Gr. *makros*, long; *daktulos*, a finger), a group of the wading birds.

macroglossia, n., *măk'·rō·glŏs'·sĭ·ă* (Gr. *makros*, long; *glossa*, the tongue), an extraordinary hypertrophic enlargement of the tongue, in consequence of which it protrudes from the mouth.

macrophyllin, a., *măk'·rō·fĭl'·lĭn* (Gr. *makros*, long; *phullon*, a leaf), in *bot.*, having elongated leaflets.

Macropiper, n., *măk'·rō·pīp'·ĕr* (Gr. *makros*, long; *pippul*, a Bengali name), a genus of plants, Ord. Piperaceæ : **Macropiper methysticum,** *mĕth·ĭst'·ĭk·ŭm* (Gr. *methusis*, intoxication), the Kava of the S. Sea Islanders, used for preparing a stimulating beverage.

macropodous, a., *măk'·rŏp'·ŏd·ŭs* (Gr. *makros*, long; *pous*, a foot, *podes*, feet), in *bot.*, having the radicle large in proportion to the rest of the body, as in the thickened radicle of a monocotyledonous embryo.

macrospores, n. plu., *măk'·rō· spōrz* (Gr. *makros*, long; *spora*, seed), the large spores of Lycopods, etc.: **macrosporangia,** n., *măk'·rō·spōr·ănj'·ĭ·ă* (Gr. *anggos*, a vessel), the cells or thecæ which contain macrospores.

macrotherm, n., *măk'·rō·thĕrm* (Gr. *makros*, long; *therme*, heat), same as 'megatherm,' which see.

Macrura or **macroura,** n., *măk· rōr'·ă* (Gr. *makros*, long; *oura*, a tail), a family of decapod crustaceans having long tails, as the lobster, the shrimp, etc. : **macrourous,** a., *măk·rōr'·ŭs*, having long tails.

macula, n., *măk'·ŭl·ă*, **maculæ,** n.

plu., *-ŭl·ē* (L. *macula*, a spot), a spot, as on the skin ; skin diseases characterised by too much or too little pigment in the parts affected, as in the case of moles and freckles, which arise from too much colouring matter in the skin : **maculate,** a., *măk'·ŭl·āt*, stained ; spotted : **macula cribrosa,** *krĭb·rōz'·ă* (L. *cribrum*, a sieve), in the internal ear, the minute holes for the passage of the filaments of the auditory nerve : **m. germinativa,** *jĕrm'·ĭn· ăt·īv'·ă* (L. *germino*, I bud or germinate), the germinal spot, found in the vesicle of the ovum : **m. lutea,** *lōt'·ē·ă* (L. *lutum*, a yellow colour), the yellow spot, a part of the retina lying directly in the axis of vision.

madder, n., *măd'·dĕr* (AS. *maddere*), the root of the Rubia tinctoria containing the colouring matters, madder purple, orange, and red, Ord. Rubiaceæ.

Madia, n., *măd'·ĭ·ă* (Gr. *mados*, bald, smooth), a genus of rather handsome plants, natives of Chili, Ord. Compositæ, Sub-ord. Corymbiferæ : **Madia sativa,** *săt· īv'·ă* (L. *sativum*, sown or planted), a species cultivated for the bland oil extracted from its fruit.

Madrepore, n., *măd'·rĕ·pōr* (F. *madrépore*—from *madré*, spotted; *pore*, a pore), a genus of corals having numerous star-shaped cavities dotting the surfaces of its spreading branching form : **madreporiform,** a., *măd'·rĕ·pōr'·ĭ·fŏrm* (L. *forma*, shape), perforated with small holes, like a coral ; applied to the tubercle of the Echinoderms by which their ambulacral system generally communicates with the exterior.

magma, n., *măg'·mă* (L. and Gr. *magma*, the dregs of an unguent), a crude mixture of an organic or inorganic substance in a pasty state ; a thick ointment or confection.

magnesia, n., *măg·nēsh'i·ă* (*Magnesia*, a country of Thessaly; F. *magnésie*, magnesia), in the form of an oxide, is obtained by burning the carbonate, which then appears as a white powder having hardly any taste : **sulphate of magnesia**, Epsom salts, obtained from dolomite limestone, formerly from sea-water, or mineral-waters.

Magnoliaceæ, n., *măg·nōl'i·ā'sĕ·ē* (after *Pierre Magnol*, professor of medicine at Montpellier), the Magnolia family, an Order of splendid trees and shrubs, bearing large showy flowers with fine glossy leaves, and possessing bitter, tonic, and often aromatic properties : **Magnolieæ**, n. plu., *măg'·nōl·i'·ĕ·ē*, a Sub-order of shrubs and trees : **Magnolia**, n., *măg·nōl'i·ă*, a genus of plants, remarkable for their large odoriferous flowers, and tonic, aromatic qualities : **Magnolia glauca**, *glāwk'ă* (L. *glaucus*, bluish-grey), the Swamp Sassafras or beaver tree whose bark is used as a substitute for Peruvian bark : **M. Yulan**, *yôl'ăn* (probably of Chinese origin), a species with deciduous leaves, whose seeds are used in China as a febrifuge.

Maize, n., *māz* (from *mahiz*, a native word), Indian corn, the 'Zea mays,' Ord. Gramineæ.

Malachadenia, n., *măl'·ăk·ă·dēn'i·ă* (Gr. *malachē*, a mallow; *dēnaios*, lasting long), a genus of very singular plants, Ord. Orchidaceæ : **Malachadenia clavatus**, *klăv·āt'ŭs* (L. *clāvātus*, furnished with clubs—from *clāva*, a club), a plant having a very fetid odour resembling carrion.

malacosteon, n., *măl'·ăk·ŏs'·tĕ·ŏn* (Gr. *malakos*, soft; *osteon*, bone), an abnormally soft condition of bone.

Malacostraca, n. plu., *măl'·ăk·ŏs'·trăk·ă* (Gr. *malakos*, soft; *ostrakon*, a shell), a division of Crustacea, originally applied to the entire class because their shells were softer than those of the Mollusca : **malacostracous**, a., *-trăk·ŭs*, belonging to such Crustacea as the shrimp, lobster, etc.

malaise, n., *măl·āz'* (F. *malaise*, uneasiness), in *med.*, an indefinite feeling of uneasiness; being ill at ease.

malanders, n. plu., *măl'·ănd·ẽrz* (F. *malandres;* It. *malandra*, malanders — from *male*, badly; *andare*, to walk), chaps or scabs on the lowest parts of a horse's legs; sores on the inside of the forelegs.

malar, a., *măl'·ẽr* (L. *mala*, the cheek), pert. to the cheek.

malaria, n., *măl·ār'i·ă* (It. *male*, ill, bad; *aria*, air), exhalations from marshy districts which produce fevers and ague: **malar'ioid**, a., *-i·ŏyd* (Gr. *eidos*, resemblance), resembling malaria.

malic, a., *măl'·ĭk* (L. *mālum*, an apple), of or from apples, as malic acid.

malicorium, n., *măl'·ĭk·ōr'i·ŭm* (L. *mālicorĭum*—from *mālum*, an apple ; *corĭum*, skin), the tough rind of a pomegranate.

malignant, a., *măl·ĭg'·nănt* (L. *malignus*, envious, spiteful—from *malus*, bad), dangerous to life; dangerous in symptoms; **malignant tumour**, a tumour which though extirpated is sure to return, and probably diffuse itself more widely than before.

mallenders, n. plu., see 'malanders.'

malleolus, n., *măl·lē'·ŏl·ŭs*, **malleoli**, n. plu., *măl·lē'·ŏl·ī* (L. *malleolus*, a small hammer—from *malleus*, a hammer), the anklebone, so called from its resemblance to a mallet : **malleolus externus**, *ĕks·tẽrn'·ŭs* (L. *malleolus*, a small hammer ; *externus*, outward), the outward projection of the lower part of the leg-bone forming the ankle : **m. internus**, *ĭn·tẽrn'·ŭs* (L. *internus*, inward), the inward hump at the ankle: **malle′olar**, a., *-ōl·ăr*, of or belonging to the ankle.

malleus, n., *măl′·lĕ·ŭs* (L. *malleus*, a hammer), one of the three movable bones, viz. the *malleus*, the *incus*, and the *stapes*, which traverse the tympanum of the ear; the *malleus* consists of a head, neck, and three processes, so named from its fancied resemblance to a hammer.

Mallophaga, n. plu., *măl·lŏf′·ăg·ă* (Gr. *mallos*, a fleece; *phago*, I eat), an Order of insects, mostly parasitic upon birds.

Malpighian, a., *măl·pĭg′·ĭ·ăn* (after *Professor Malpighi* of Pisa, the discoverer or first describer), designating certain globular structures in the spleen and kidneys: **Malpighiaceæ**, n. plu., *măl·pĭg′·ĭ·ā′·sĕ·ē*, the Malpighia family, an Order of trees and shrubs, many species being astringent: **Malpighia**, n., *măl·pĭg′·ĭ·ă*, a genus of interesting plants: **Malpighia glabra**, *glāb′·ră* (L. *glăber*, without hairs or bristles, bald); and **M. punici-folia**, *pūn′·ĭs·ĭ·fōl′·ĭ·ă* (L. *pŭnĭcus*, belonging to Africa; *folium*, a leaf), are species whose fruit is called the Barbadoes cherry, used as a table-fruit: **malpighiaceous**, a., *măl·pĭg′·ĭ·ā′·shŭs*, applied to peltate hairs such as are seen on the Malpighiaceæ.

Malvaceæ, n. plu., *mălv·ā′·sĕ·ē* (L. *malva*, Gr. *malachē*, mallows —from Gr. *malassō*, I soften), the Mallow family, an extensive Order of plants which yield mucilage in large quantity, some furnish materials for cordage, and others yield cotton, so called from the emollient qualities of the species: **Malveæ**, n. plu., *mălv′·ĕ·ē*, a tribe or Sub-order: **Malva**, n., *mălv′·ă*, an extensive genus of plants: **Malva sylvestris**, *sĭl·vĕst′·rĭs* (L. *sylvestris*, woody—from *silva*, a wood), the

common mallow, used medicinally: **M. alcea**, *ăl′·sĕ·ă* (L. *alcĕa*, a species of mallows), possesses astringent properties, used by Chinese to blacken their eyebrows, and the leather of their shoes: **malvaceous**, a., *mălv·ā′·shŭs*, of or belonging to mallows.

mamillæ, n. plu., *măm·ĭl′·lē* (L. *mamilla*, a nipple or teat), in *bot.*, granular prominences on pollen-grains, and such like: **mamillated**, a., *măm′·ĭl·lāt·ĕd*, in the form of a hemisphere with a wart at the top.

mamma, n., *măm′·mă*, **mammæ**, n. plu., *măm′·mē* (L. *mamma*, the breast, a mother), the breasts in the male or female: **mam′miform**, a., *-fŏrm* (L. *forma*, shape), formed as breasts: **mam′-mifer**, n., *-mĭ·fĕr* (L. *fero*, I bear), one of the mammals: **mammal**, n., *măm′·măl*, an animal which suckles its young: **mammalia**, n. plu., *măm·māl′·ĭ·ă*, or **mammals**, n. plu., *măm′·măls*, the great class of vertebrate animals which suckle their young by teats or nipples: **mammary**, a., *măm′·măr·ĭ*, pert. to the breasts.

manakin, n., *măn′·ă·kĭn*, also **manikin**, n., *măn′·ĭ·kĭn* (F. *mannequin*, a layman, a manikin; Eng. *man* and *kin*, literally, a little man), a figure on which a student may practise the application of bandages, and also the operations of midwifery.

Manchineel, n., *măn′·tshĭn·ēl′* (It. *mancinello*, Sp. *manzanillo*), a large W. Indian tree, the Hippomane mancinella, Ord. Euphorbiaceæ, whose wood is hard, durable, and beautifully clouded, but whose sap is very acrid and poisonous, its application to the skin exciting violent inflammation followed by ulceration.

mandibles, n., *măn′·dĭ·blz* (L. *mandibulum*, a jaw), the upper pair of jaws in insects; applied to one of the pairs of the jaws in

Crustacea and spiders ; the beak of cephalopods ; the lower jaw of vertebrates.

mandioc, n., *mănd'.i̇·ŏk*, and **manioc,** n., *măn'.i̇·ŏk*, spellings of 'manihot,' which see.

Mandragora, n., *măn·drăg'·ŏr·ă* (L. and Gr. *mandragoras*, a mandrake—from Gr. *mandra*, a stable or cattle house ; *agŏreuō*, I speak, as indicating the sort of place where they grow best), a genus of plants, Ord. Solanaceæ : **Mandragora officinalis,** *ŏf·fis'·in·āl'·is* (L. *officinalis*, officinal), the mandrake which acts as a stimulant on the nervous system, and its forked root was long celebrated for such virtues ; a powerfully narcotic plant.

mandrake, n., *măn'·drāk*, a corruption of 'mandragora,' which see.

manganese, n., *măn'·găn·ēz* (new L. *manganesia*—from L. *magnes*, the magnet from its presumed resemblance), a metal of a greyish-white colour, very hard and difficult to fuse ; applied to the black-oxide of the metal : **manganesium,** n., *măn'·găn·ēz'·i̇·ŭm*, the chemical term for the metal manganese.

mange, n., *mānj* (F. *démanger*, to itch—from *manger*, to gnaw), the scab or itch in dogs, cattle, etc.

Mangifera, n., *măn·jif'·ĕr·ă* (*mango*, the native name of the tree ; L. *fero*, I bear), a genus of much esteemed tropical fruit-trees, Ord. Anacardiaceæ : **Mangifera indica,** *ind'·ik·ă* (L. *Indicus*, Indian), the common mango tree, whose fruit of a sweet perfumed flavour and grateful acidity is eaten within the tropics during the hot months with great avidity.

Mangold-wurzel, n., properly **mangel,** *măng'·gl·wĕr'·zl* (Ger. *mangel*, want, scarcity ; *wurzel*, root), the Beta campestris, Ord.

Chenopodiaceæ, a plant of the beet kind, having a large root.

mania, n., *mān'·i̇·ă* (Gr. *mania*, madness), a form of madness characterised by undue excitement of the mental powers, with or without delirium ; a general term to denote all kinds of insanity : **maniac,** n., *mān'·i̇·ăk*, a madman : **delirium** is a symptom of some bodily disease, as in a fever : **mania** is a mental alienation or derangement.

Manicaria, n., *măn'·ik·ār'·i̇·ă* (L. *manica*, a glove), a genus of fine palms, Ord. Palmæ : **Manicaria saccifera,** *săk·sif'·ĕr·ă* (L. *saccus*, a bag ; *fero*, I bear), a species whose spathes come off in the form of conical caps, and are used as coverings for the head in the W. Indies.

manicate, a., *man'·ik·āt* (L. *manicatus*, furnished with long sleeves —from *manica*, a glove), in *bot.*, covered with entangled hairs ; surrounded with matted scales which are easily removed from the surface in a mass.

manna, n., *măn'·nă* (Heb. *man hu*, what is this ? L. and Gr. *manna*), the hardened exudations from the barks of various trees, which form mild laxatives ; 'Tamarix gallica,' Ord. Tamaricaceæ, produces Mount Sinai manna ; 'Alhaji maurorum,' Ord. Leguminosæ, produces a kind of manna ; 'Ornus rotundifolia,' 'O. Europæa,' and other species of 'Ornus,' Ord. Oleaceæ, natives of S. Europe, produce the common manna of the shops ; a kind of manna is produced from the larch and the cedar of Lebanon : **mannite,** n., *măn'·nit*, the peculiar sweet principle of manna, called manna-sugar, which does not ferment ; is found also in mushrooms and sea-weeds.

mantle, n., *măn'·tl* (L. *mantellum*, a cloak), the outer soft integument of the Mollusca, largely developed and forming a cloak which protects the viscera; also called the 'pallium.'

manubrium, n., *măn·ūb′·rĭ·ŭm* (L. *manūbrium*, a handle—from *mănus*, a hand), in *anat.*, the upper piece of the sternum representing the handle, having a somewhat triangular form ; the polypite suspended from the roof of the swimming-bell of a medusa, also from the gonocalyx of a medusiform gonophore in the Hydrozoa ; in *bot.*, cells projected inwards from the centre of shields of the globule in Characeæ.

manus, n., *măn′·ŭs* (L. *mănus*, the hand), the hand of the higher Vertebrates.

manyplies, n. plu., *mĕn′·ĭ·plīz*, in Scot. **moniplies**, n. plu., *mŏn′·ĭ·plīz* (Eng. *many ; plies*, folds), the popular name for the Omasum, or third part of the stomach of ruminants, so named from its numerous flaps or folds.

Marantaceæ, n. plu., *măr′·ăn·tā′·sĕ·ē*, also called **Cannaceæ** (after *Maranti*, a Venetian physician), the Arrowroot family, an Order of plants containing much starch in the rhizomes and roots : **Maranta**, n., *măr·ăn′·tă*, a genus of interesting plants : **Maranta arundinacea**, *ăr·ŭnd′·ĭn·ā′·sĕ·ă* (L. *arundināceus*, pert. to or like a reed—from *arundo*, the reed-cane); also **M. Indica**, *ĭnd′·ĭk·ă* (L. *Indicus*, Indian), are species which produce the best arrowroot from their tuberous rhizomata.

marasmus, n., *măr·ăz′·mŭs* (Gr. *marasmos*, decay, weakness), a wasting disorder of children ; emaciation or wasting ; atrophy.

Marattieæ, n. plu., *măr′·ăt·tĭ′·ĕ·ē*, also **Marrattia** tribe, *măr·ăt′·tĭ·ă* (after *Maratti*, of Tuscany), a Sub-order of ornamental ferns, having their sporangia united in mass.

marcescent, a., *măr·sĕs′·ĕnt* (L.

marcescens, pining away or decaying), in *bot.*, gradually withering, but not falling off until the part bearing it is perfected.

Marcgravia, n., *mărk·grāv′·ĭ·ă* (after *Marcgrave*, a German botanist), a genus of curious shrubby and creeping plants, Ord. Ternstrœmiaceæ or Tea family, which are occasionally furnished with bracts folded and united so as to form ascidia, containing a sweet liquid which attracts insectivorous birds, as in the **Marcgravia nepenthoides**, *nĕp′·ĕnth·ōyd′·ēz* (Gr. *nepenthes*, a magic potion or drug — from *ne*, not ; *penthos*, grief, sadness ; *eidos*, resemblance); **M. umbellata**, *ŭm′·bĕl·lāt′·ă* (L. *umbella*, a little shadow —from *umbra*, a shade), a species whose stem, root, and leaves are regarded in the W. Indies as diuretic.

Marchantieæ, n. plu., *măr′·shăn·tĭ′·ĕ·ē* (after *Marchant*, a French botanist), the Liverworts, a section or Sub-order of the Ord. Hepaticæ, which have thecæ collected in heads, found growing on the ground or on trees in damp, shady places, and have a leathery structure : **Marchantia**, n., *măr·shăn′·tĭ·ă*, a genus of creeping plants, having green, cellular, fleshy fronds : **Marchantia hemispherica**, *hĕm′·ĭ·sfĕr′·ĭk·ă* (Gr. *hemi*, half; *sphaira*, a globe), a species recommended in dropsy.

margaric, a., *măr·găr′·ĭk* (Gr. *margaron*, a pearl), applied to an acid obtained by the action of potash upon cyanide of cetyl, and also obtained by saponification from vegetable wax, so named from presenting the appearance of pearly crystalline scales : **margarine**, n., *măr′·găr·ĭn*, one of the solid proximate principles of human fat.

marginate, a., *mărj′·ĭn·āt* (L.

margo, an edge, *marginis*, of an edge), in *bot.*, having a distinct edge or border of a different texture to the body.

Marrubium, n., *măr·rŏb'·ĭ·ŭm* (L. *marrŭbium*, the plant horehound), a genus of plants, Ord. Labiatæ : **Marrubium vulgare**, *vŭl·gār'·ĕ* (L. *vulgāris*, common, vulgar), white horehound, a plant having bitter tonic properties, a popular remedy for coughs and asthma.

Marsdenia, n., *mârz·dĕn'·ĭ·ă* (after *Marsden*, author of a history of Sumatra), a genus of interesting plants, Ord. Asclepiadaceæ : **Marsdenia flavescens**, *flăv·ĕs'·ĕnz* (L. *flavescens*, becoming golden yellow), a plant suited for covering rafters, pillars, and trellis-work : **M. tinctoria**, *tĭngk·tōr'·ĭ·ă* (L. *tinctorius*, of or belonging to dyeing — from *tingo*, I dye), a species yielding a dye similar to indigo.

Marsileaceæ, n. plu., *mâr'·sĭl·ĕ·ā'·sĕ·ē* (after *Count Marsigli*, of Bologna), also called **Rhizocarpeæ**, n. plu., *rīz'·ō·kârp'·ĕ·ē* (Gr. *rhiza*, a root ; *karpos*, fruit), the Pepperwort family, creeping or floating plants found in ditches and pools, and are apparently a link between ferns and club-mosses: **Marsilea**, n., *mâr·sĭl'·ĕ·ă*, a genus of curious aquatic plants: **Marsilea macropus**, *măk'·rō·pŭs* (Gr. *makros*, long ; *pous*, a foot), also called **M. salvatrix**, *sălv·āt'·rĭks* (L. a saviour), the Nardoo plant of Australia, the sporocarps of which have been used as food by travellers in that country.

marsupium, n., *măr·sŭp'·ĭ·ŭm* (L. *marsupium*, a pouch), the pouch of marsupial animals ; a dark-coloured membrane in the vitreous body of the eyes of birds : **Marsupialia**, n. plu., *mâr·sŭp'·ĭ·āl'·ĭ·ă*, an Order of Mammals in which the females generally have an abdominal pouch in which they carry their young, as the kangaroo.

masked, a., *măskt*, in *bot.*, same as ' personate.'

masseter, n., *măs·sēt'·ĕr* (Gr. *massētēr*, one that chews—from *massāomai*, I chew), a short, thick muscle at the posterior part of the cheek, which raises the lower jaw : **masseteric**, a., *măs'·sĕt·ĕr'·ĭk*, applied to an artery, a vein, or a nerve connected with the masseter muscle.

mastax, n., *măs'·tăks* (Gr. *mastax*, the jaw, the mouth), the muscular pharynx or buccal funnel, into which the mouth opens in most of the Rotifera.

mastication, n., *măst'·ĭk·ā'·shŭn* (L. *masticātum*, to masticate ; Gr. *mastax*, the jaw, *mastakos*, of the jaw), the process by which the food, when taken into the mouth, is chewed into small pieces by the teeth, and thoroughly mixed with the saliva : **masticatory**, a., *măst'·ĭk·āt'·ŏr·ĭ*, adapted for chewing : n., a substance to be chewed to increase the saliva.

mastich or **mastic**, n., *măst'·ĭk* (L. and Gr. *mastĭchē*, an odoriferous gum from the mastich tree ; *mastĭchāō*, I chew ; F. *mastic*), a concrete resinous exudation furnished by the Pistacia lentiscus, Ord. Anacardiaceæ, the Lentisk, a native of the islands and coasts of the Mediterranean, used as a masticatory for consolidating the gums and cleansing the teeth ; it enters into the composition of varnishes.

mastoid, a., *măst'·ōyd* (Gr. *mastos*, a breast ; *eidos*, resemblance), nipple-like : **mastitis**, n., *măst·īt'·ĭs*, inflammation of the breast : **mastodynia**, n., *măst'·ō·dĭn'·ĭ·ă* (Gr. *odunē*, pain), pain in the breast : **mastosis**, n., *măst·ōz'·ĭs*, an osseous tumour of the breast : **mastoid process**, the pointed portion of the temporal bone behind the organ of hearing, which consists of a thin external crust of bone containing large cellular spaces communicating with the middle ear.

mate, n., *măt'-ā*, Yerba maté or Paraguay tea, the leaves of the Ilex Paraguensis, Ord. Aquifoliaceæ, used extensively in S. America.

materia medica, *măt·ēr'·ĭ·ă mĕd'·ĭk·ă* (L. *mātĕrĭa*, stuff, matter ; *mĕdĭca*, healing, curative), the various substances, natural and artificial, which are employed in medicine ; the science which describes these substances and their properties.

matico, n., *măt'·ĭk·ō*, the leaves and unripe fruit of a kind of pepper plant of Peru, etc., which are aromatic, warm, and astringent, the Piper angustifolium, Ord. Piperaceæ ; the leaf is applied to small bleeding surfaces, or given for internal hæmorrhages.

matrix, n., *māt'·rĭks* (L. *matrix*, the womb), the womb ; the hollow or cavity in which anything is formed or cast ; in *bot.*, the body upon which a lichen or fungus grows.

mattulla, n., *măt·tŭl'·lă* (unascertained), the fibrous matter covering the petioles of palms.

maturation, n., *măt'·ūr·ā'·shŭn* (L. *matūrus*, ripe), the process of coming to maturity ; the act of ripening.

Mauritia, n., *măw·rĭsh'·ĭ·ă* (after *Prince Maurice*, of Nassau), a genus of splendid plants, Ord. Palmæ : **Mauritia vinifera**, *vĭn·ĭf'·ĕr·ă* (L. *vinum*, wine ; *fero*, I produce), the Buriti-palm, whose stem, when pierced, yields a reddish juice, having the taste of a sweet wine.

maw, n., *măw* (Dut. *maag*, Ger. *magen*, the stomach), the stomach of animals ; the craw or crop of fowls : **maw-worm**, the Asaris vermicularis, a parasite occasionally found in the maw or stomach.

maxilla, n., *măks·ĭl'·lă*, **maxillæ**, n. plu., -*ĭl'·lē* (L. *maxilla*, a jaw), in *anat.*, a jaw-bone ; the inferior pair or pairs of jaws in the Arthropoda ; the upper jaw-bones of Vertebrates : **maxillary**, a., *măks'·ĭl·lér·ĭ* or *măks·ĭl'·*, pert. to the jaw or jaw-bone.

maxilliped, n., *măks·ĭl'·lĭ·pĕd*, **maxillipedes**, n. plu., *măks'·ĭl·lĭp'·ĕd·ēz* (L. *maxilla*, a jaw ; *pedes*, feet), a jaw-foot ; the foot-like appendages of the mouth of a crab, a lobster, etc., which are converted into masticatory organs.

maximum, n., *măks'·ĭm·ŭm* (L. *maximum*, the greatest — from *magnus*, great), a term denoting the greatest quantity of effect ; opposed to **minimum**, the least quantity of effect; and contrasted with **medium**, a middle or mean between those extremes.

measles, n. plu., *mēz'·lz* (Dut. *mæselen*, measles — from *mæse*, a spot, a stain), a disease manifested by a peculiar crimson rash on the skin, chiefly affecting children ; a contagious febrile disorder, forming one of the group of the exanthemata : **measly**, a., *mēz'·lĭ*, having the character or appearance of measles—applied to the eruptions of typhus ; infected with measles—applied to pork which contains the parasite Cysticercus cellulosus.

meatus, n., *mē·āt'·ŭs* (L. *mĕātus*, a going, a passing), in *anat.*, a natural passage or canal wider than a duct ; an opening leading to a duct or cavity : **meatus auditorius externus**, *āwd'·ĭt·ōr'·ĭ·ŭs ĕks·tĕrn'·ŭs* (L. *audĭtōrĭus*, relating to hearing—from *audio*, I hear ; *externus*, external), the external auditory meatus ; the aperture of the ear forming a short canal which leads into the cavity of the tympanum : **m.**

urinarius, *ūr'·ĭn·ār'·ĭ·ŭs* (L. *urin-ārius*, urinary — from *ūrīna*, urine), the orifice of the urethra in both sexes: **meatuses**, n. plu., *mē·āt'·ŭs·ĕz*, those of the *nares* or nostrils, named respectively the *superior*, *middle*, and *inferior*.

meconic, a., *mēk·ŏn'·ĭk* (Gr. *mēkōn*, a poppy, the ink-bag of the cuttle-fish), applied to a peculiar acid contained in the juice of Papaver somniferum, Ord. Papaveraceæ, and in its concrete milky juice called opium : **meconate**, n., *mĕk'·ŏn·āt*, a salt consisting of meconic acid and a base.

meconium, n., *mĕk·ŏn'·ĭ·ŭm* (Gr. *mēkōnion*, L. *mēcōnium*, poppy juice — from Gr. *mēkōn*, the poppy, the ink-bag of the cuttle-fish), the inspissated juice of the poppy ; opium ; the first dark slimy discharge from the bowels of a newly-born infant.

medial, a., *mēd'·ĭ·ăl* (L. *medius*, the middle), in *bot.*, along the middle : **median**, a., *mēd'·ĭ·ăn*, relating to or connected with the middle of anything : **median line**, n., an ideal line dividing a body longitudinally into two equal parts.

mediastinum, n., *mēd'·ĭ·ăs·tīn'·ŭm* (L. *medĭastīnus*, one standing in the middle, a servant — from *medius*, the middle; *sto*, I stand), the space left in the median line of the chest by the non-approximation of the two pluræ, extending from the spine to the upper surface of the breast-bone : **mediastinal**, a., *mēd'·ĭ·ăs·tīn'·ăl*, of or connected with the mediastinum: **mediastinum testis**, *tĕs'·tĭs* (L. *testis*, of a testicle), a vertical process, from which is given off numerous septa, situated in the centre of a fibrous covering enclosing the body of the testis, called the 'tunica albuginea.'

mediate, a., *mēd'·ĭ·āt* (L. *medius*, middle), middle; situated between two extremes : **mediate auscult-**ation, auscultation through a stethoscope—opposed to 'immediate auscultation,' an auscultation made directly by the ear.

Medicago, n., *mĕd'·ĭk·āg'·ō* (Gr. *mēdikē*, name given by Dioscorides to a Median grass), a genus of plants, Ord. Leguminosæ, affording some fodder plants: **Medicago sativa**, *săt·īv'·ă* (L. *satīvum*, sown or planted), the Lucerne, cultivated as green food for horses and cattle: **M. lupulina**, *lŏp'·ŭl·īn'·ă* (new L. *lupŭlus*, the hop), the plant Nonsuch, introduced into pastures along with grasses and clovers, so named from having the appearance of the hop in its floral capitules.

medicament, n., *mĕd'·ĭk·ă·mĕnt* (L. *medĭcāmentum*, medicine, a drug ; *medico*, I heal or cure; It. and Sp. *medico*, a physician), a substance given for curing a disease or a wound : **medication**, n., *mĕd'·ĭk·ā'·shŭn*, the act or process of impregnating with a medicinal substance ; treatment by means of medicine : **medicinal**, a., *mĕd·ĭs'·ĭn·ăl*, having the properties of medicine ; used in medicine : **medicine**, n., *mĕd'·ĭs·ĭn* or *mĕd'·sĭn*, a substance administered for the cure or mitigation of disease : **medical jurisprudence**, the applications of medical science to the determination of certain questions in courts of law : **medico-legal**, pert. to law as affected by medical facts.

medick, n., *mĕd'·ĭk*, same as ' medicago,' which see.

medulla, n., *mĕd·ŭl'·lă* (L. *mĕdulla*, the marrow, the pith—from *medius*, the middle), the fat substance or marrow in the long bones ; the spinal cord ; the pith of plants ; the fibrous matter covering the petioles of palms : **medullary**, a., *mĕd·ŭl'·lăr·ĭ*, pert. to marrow or pith : **medulla oblongata**, *ŏb'·lŏng·gāt'·ă* (L. *oblongus*, long, oblong), the part continuous with the spinal cord within the skull resting on the basilar process of the occipital bone : **m. ossium**, *ŏs'·sĭ·ŭm* (L. *os*, a bone, *ossĭum*, of bones), the marrow lodged in the interior of the bones : **m. spinalis**, *spĭn·ā̆l'·ĭs* (L. *spīna*, the backbone), the spinal marrow or cord : **medullary rays**, in *bot.*, the rays of cellular tissue seen in a transverse section of exogenous wood, and which connects the pith with the bark: **m. sheath**, in *bot.*, a thin layer of vascular tissue which surrounds the pith in exogenous stems : **m. substance**, the interior white portion of the brain or kidney : **m. system**, the marrow bones, and the membranes which enclose the marrow : **medullated**, a., *mĕd·ŭl'·lāt·ĕd*, applied to the nerve fibres which form the white part of the brain, spinal cord, and nerves.

Iedusa, n., *mĕd·ūz'·ă*, **Medusæ**, n. plu., *mĕd·ūz'·ē* (L. *Medusa*, in anc. mythology a beautiful woman whose hair was turned into snakes), an Order of Hydrozoa ; sea animals, usually called sea-blubber, sea-nettles, or jelly-fish, whose usual form is that of a hemisphere with a marginal membrane and many trailing feelers, so named from the supposed resemblance of their tentacles or feelers to the snaky hair of Medusa: **medusiform**, a., *mĕd·ūz'·ĭ·fŏrm* (L. *forma*, shape), resembling the medusæ in shape : **medusoid**, a., *mĕd·ūz'·ŏyd* (Gr. *eidos*, resemblance), like a medusa ; medusiform : **n.**, one of the medusiform gonophores of the Hydrozoa.

Iegaspores, n. plu., *mĕg'·ă·spōrz* (Gr. *megas*, great ; *spora*, seed), the larger kind of reproductive spores found in Lycopods : **megasporangia**, n., *mĕg'·ă·spōr·ănj'·ĭ·ă*, same as ' macrosporangia.'

Iegatherms, n. plu., *mĕg'·ă·*

thĕrmz (Gr. *megas*, great; *thermē*, heat), plants requiring a high temperature ; also called 'macrotherms.'

megistotherms, n. plu., *mĕlj·ĭs'·tō·thĕrmz* (Gr. *megistos*, very great ; *thermē*, heat), plants requiring extreme or a very high degree of heat.

megrim, n., *mēg'·rĭm* (F. *migraine*, megrim; L. *hemicranium*, half the skull), a neuralgic pain confined to one side of the head ; sick headache.

Meibomian glands, *mī·bōm'·ĭ·ăn* (first described by *Meibomius*), glands situated upon the inner surface of the eyelids, between the tarsal cartilages and conjunctiva, presenting the appearance of parallel strings of pearls, about thirty in the upper cartilage, and somewhat fewer in the lower.

meiophylly, n., *mī·ŏf'·ĭl'·lĭ* (Gr. *meiōn*, less ; *phullon*, a leaf), in *bot.*, the suppression of one or more leaves in a whorl.

meiostemonous or **miostemonous**, a., *mī'·ō·stĕm'·ŏn·ŭs* (Gr. *meiōn*, less ; *stemon*, a stamen), in *bot.*, a term applied to stamens less in number than the parts of the corolla.

meiotaxy, n., *mī'·ō·tăks'·ĭ* (Gr. *meiŏn*, less ; *taxis*, arrangement), in *bot.*, the complete suppression in a plant of a set of organs, as the corolla or the stamens.

melæna, n., *mĕl·ēn'·ă* (Gr. *melan*, black), the discharge of matter, black like tar, from the bowels.

Melaleuca, n., *mĕl'·ă·lōk'·ă* (Gr. *melan*, black ; *leukos*, white), a genus of greenhouse plants, Ord. Myrtaceæ, producing splendid flowers—so named because the trunk is black, and the branches white : **Melaleuca leucadendron**, *lōk'·ă·dĕnd'·rŏn* (Gr. *leukos*, white ; *dendron*, a tree) ; **M. cajuputi**, *kădj'·ū·pūt'·ĭ* (Malay *caju-puti*, white tree), species, particularly the latter, whose leaves yield a volatile oil, called cajeput oil, of a green colour, a very powerful medicine ; the latter also named **M. minor**, *min'ŏr* (L. *minor*, less).

Melampyrum, n., *mĕl'ăm·pīr'ŭm* (Gr. *melampuron*, cow-wheat— from *melan*, black; *puros*, wheat), a genus of plants, Ord. Scrophulariaceæ, one of whose species, **Melampyrum arvense**, *ăr·vĕns'ĕ* (L. *arvum*, an arable field), is called cow-wheat, as being relished by cows.

melancholia, n., *mĕl'ăn·kōl'ĭ·ă* (Gr. *melangcholia*, black bile— from *melan*, black; *cholē*, bile), a variety of insanity characterised by dejection or depression of spirits.

melanosis, n., *mĕl'ăn·ōz'ĭs* (Gr. *melan*, black), the deposition of black or dark-brown colouring matter in various textures and organs of the body ; the disease, melanotic cancer, in which tumours containing black pigment are developed : **melanotic**, a., *mĕl'ăn·ŏt'ĭk*, of or pert. to melanosis.

Melanosporeæ, n. plu., *mĕl'ăn·ō·spōr'ĕ·ē* (Gr. *melan*, black; *spora*, seed), a Sub-ord. of Algæ of an olive-green or olive-brown colour, and cellulæ of filamentous structure, found in the sea.

Melanthaceæ, n. plu., *mĕl'ănth·ā'sĕ·ē* (Gr. *melan*, black ; *anthos*, a flower—in allusion to the dusky colour of the flowers), the Colchicum family, an Order of bulbous, tuberous, or fibrous rooted plants having medicinal properties, and sometimes bearing pretty flowers : **Melanthium**, n., *mĕl'ănth'ĭ·ŭm*, a genus of plants.

melasma, n., *mĕl·ăz'mă* (Gr. *melasma*, a black spot), a black spot on the lower extremities, especially of old people ; a disease in which is associated a peculiar degeneration of the supra-renal capsules with extreme cachexia, and a peculiar pigmentation or bronzing of the skin ; also called ' Addison's disease.'

Melastomaceæ, n. plu., *mĕl'ă·stōm·ā'sĕ·ē* (Gr. *melan*, black ; *stoma*, the mouth), an Order of very beautiful trees, shrubs, and herbs, many bearing sweet eatable berries, so named from the fruit of some dyeing the lips black : **Melastomeæ**, n., *mĕl'ă·stōm'ĕ·ē*, a Sub-order : **Melastoma**, n., *mĕl·ăs'tōm·ă*, a genus of plants having very showy flowers : **Melastoma elongata**, *ē·lŏng·gāt'ă* (L. *ēlongātus*, made long—from *e*, out ; *longus*, long), a species bearing large, beautiful flowers, varying from blue to purple and white.

Meliaceæ, n. plu., *mĕl'ĭ·ā'sĕ·ē* (Gr. *mĕlia*, an ash tree), the Melia family, an Order of plants which are bitter, tonic, and astringent : **Melia**, n., *mĕl'ĭ'ă*, a genus of trees : **Melia Indica** or **Azadirachta**, *ĭnd'ĭk·ă* or *ăz·ăd'ĭr·ăkt'ă* (L. *Indica*, of or from India ; Indian name), used in India as a febrifuge ; its fruit yields an oil much in household use, and as an antispasmodic ; its bark is tonic : **M. azedarach**, *ăz·ĕd'ăr·ăk* (an Indian name), native of China, an anthelmintic.

Melilotus, n., *mĕl'ĭl·ōt'ŭs* (L. *mel*, honey ; *lotus*, the lotus), the honey lotus, a genus of plants, Ord. Leguminosæ, the favourite haunts of bees ; the Melilot is cultivated as fodder for cattle.

Melissa, n., *mĕl·ĭs'să* (Gr. *melissa*, a bee), a genus of plants, Ord. Labiatæ : **Melissa officinalis**, *ŏf·fĭs'ĭn·ăl'ĭs* (L. *officinalis*, official—from *officina*, a workshop), common balm.

Melloca, n., *mĕl·lōk'ă* (unascertained), a genus of plants, Ord. Portulacaceæ : **Melloca tuberosa**, *tŭb'ĕr·ōz'ă* (L. *tūberōsus*, having fleshy knots—from *tŭber*, a protuberance), a plant much cultivated in the elevated districts of Peru, Bolivia, etc., for its esculent tubers, which have been recommended as a substitute for the potato.

Melocanna, n., *mĕl'·ō·kăn'·nă* (Gr. *melon*, an apple ; *kanna*, a reed), a genus of plants, Ord. Gramineæ: **Melocanna bambusoides**, *băm'· bŭz·ōyd'·ēz* (Malay *bambu*, a bamboo ; Gr. *eidos*, resemblance), one of the bomboo kind in whose joints is a substance called tab-asheer, which is composed of silica.

membrane, n., *mĕm'·brān* (L. *membrāna*, skin, a film or membrane), a thin layer or skin, serving to cover some part of an animal, or of a plant : **membranaceous**, a., *mĕm'·brăn·ā'·shŭs*, also **membranous**, a., *-brăn·ŭs*, having the consistence and structure of a membrane : **membrana basilaris**, *mĕm'·brăn·ă băs'· il·ār'·ĭs* (L. *băsilāris*, of or pert. to the base of a thing, as the skull—from *băsis*, the base), a thin membrane which extends over the bony wall of the cochlea, completing the scala tympani : **membrana fusca**, *fŭsk'·ă* (L. *fuscus*, dark, of a greyish-brown colour), a fine cellular web connected with the inner surface of the sclerotic : **m. germinativa**, *jĕr'·mĭn·ăt·īv'·ă* (L. *germíno*, I sprout, I germinate), the germinal membrane, the earliest development of the germ in fishes, and the amphibia : **m. limitans**, *lĭm'·ĭt·ănz* (L. *limĭtans*, limiting or bounding), an extremely thin and delicate structureless membrane which lines the inner surface of the retina, and separates it from the vitreous body : **m. nictitans**, *nĭkt'·ĭt·ănz* (L. *nĭctĭtans*, winking often— from *nicto*, I wink), the third eyelid found in all the mammalia except man, the quadrumana,

and the cetaceæ, attaining its greatest development in birds; a thin plate of cartilage between the two layers of the 'plica semilunaris,' which is the only trace of the third eyelid found in man: **m. pituitaria**, *pĭt·ū'·ĭt·ār'·ĭ·ă* (L. *pitŭīta*, slime, phlegm), the membrane which lines the cavity of the nose : **m. sacciformis**, *săk'·sĭ·fŏrm'·ĭs* (L. *saccus*, a sack ; *forma*, shape), the first of the synovial membranes; a membrane which covers the margin of the articular surface of the ulna, so named from its extreme looseness, forming a loose 'cul-de-sac' : **m. tympani**, *tĭm'·păn·ī* (L. *tympănum*, a drum), the membrane of the tympanum, separating the cavity of the tympanum from the external meatus.

meninges, n. plu., *mĕn·ĭn'·jēz* (Gr. *meningx*, a membrane), the membranes which envelope the brain — called the 'pia-mater' and 'dura-mater': **meningeal**, a., *mĕn'·ĭn·jē'·ăl*, of or pert. to the membranes of the brain : **meningitis**, n., *mĕn'·ĭn·jīt'·ĭs*, inflammation of the membranes covering the brain : **meningo**, *mĕn· ĭng'·gō*, denoting relation to, or connection with, membranes of the brain.

meniscus, n., *mĕn·ĭsk'·ŭs*, **menisci**, n. plu., *mĕn·ĭs'·sī*, or **meniscuses**, n. plu., *-ŭs·ēz* (Gr. *meniskos*, a little moon—from *mene*, the moon), a lens, concave on the one side, and convex on the other, having a sharp edge; in *anat.*, an appearance resembling the new moon, applied to inter-articular fibrous cartilages, as the 'glenoid': **meniscoid**, a., *mĕn·ĭsk'·ōyd* (Gr. *eidos*, resemblance), having the shape of a watch-glass.

Menispermaceæ, n. plu., *mĕn'·ĭ· spĕrm·ā'·sĕ·ē* (Gr. *meniskos*, a little moon ; *sperma*, seed), the Moon-seed family, an Order of

R·

plants whose seed or fruit is kidney or half-moon shaped, hence the name : **Menispermum**, n., a genus of hardy plants, possessing strong narcotic properties: **menispermin**, n., *-spèrm'·ĭn*, a non-poisonous substance obtained from the pericarp of the Anamirta cocculus.

menorrhagia, n., *mĕn'·ŏr·rādj'·ĭ·ă* (Gr. *mēn*, a month ; *mēnes*, a woman's menses ; *rhegnumi*, I break or burst forth), an immoderate flowing of the catamenia or menses.

menses, n. plu., *mĕn'·sēz* (L. *mensis*, a month), the monthly discharges of women.

menstrual, a., *mĕn'·strŏo·ăl* (L. *menstruālis*, every month—from *mensis*, a month), happening once a month ; lasting a month ; catamenial : **menstruant**, a., *mĕn'·strŏo·ănt*, subject to monthly flowings : **menstruation**, n., *mĕn'·strŏo·ā'·shŭn*, the periodical flow of the menses : **menstruous**, a., *mĕn'·strŏo·ŭs*, pert. to the monthly discharges of women.

menstruum, n., *mĕn'·strŏo·ŭm*, **menstruums**, n. plu., *-strŏo·ŭmz*, or **menstrua**, n. plu., *-strŏo·ă* (L. *menstruum*, that which lasts or continues a month—from *mensis*, a month), a solvent ; any liquor used in dissolving—so named by the old chemists, because they supposed that the moon had a mysterious influence on the making of their preparations for dissolving metals, etc.

mentagra, n., *mĕnt'·ăg·ră* (L. *mentum*, the chin ; Gr. *agra*, a seizure), a disease affecting the beard, moustache, whiskers, and inner part of the nostrils—caused by minute fungi, or vegetable parasites at the roots of the hair ; Tinea sycosis.

mentagraphyte, n., *mĕn·tăg'·ră·fĭt* (L. *mentum*, the chin; Gr. *agra*, a seizure ; Gr. *phuton*, a plant), cryptogamous plants, or mould,

supposed to be the cause of the cutaneous disease mentagra ; synonym of 'mentagra.'

Mentha, n., *mĕnth'·ă* (from *Menthe*, a nymph, fabled to have been changed into *mint* by Proserpine in a fit of jealousy), the mints, a well-known genus of useful herbs, Ord. Labiatæ, the species of which yield volatile oils : **Mentha piperita**, *pĭp'·ĕr·ĭt'·ă* (L. *pĭperĭtus*, of or pert. to pepper — from *pĭper*, pepper), peppermint, used as a carminative and diffusive stimulant in flatulent disorders: **M. pulegium**, *pŭl·ēdj'·ĭ·ŭm* (L. *pulēgĭum*, fleabane, pennyroyal—from *pūlex*, a flea, as the smell when burnt destroys fleas), pennyroyal ; **M. viridis**, *vĭr'·ĭd·ĭs* (L. *virĭdis*, green), spearmint, both plants act like peppermint, but are less powerful.

mentum, n., *mĕnt'·ŭm* (L. *mentum*, the chin), the basal portion of the labium or lower lip in insects.

Mentzelia, n., *mĕnt·zĕl'·ĭ·ă* (after *Mentzel*, a botanist of Brandenburg), a genus of curious plants, Ord. Loasaceæ : **Mentzelia hispida**, *hĭsp'·ĭd·ă* (L. *hispĭdus*, shaggy, hairy), a Mexican herb said to possess purgative qualities.

Menyantheæ, n. plu., *mĕn'·ĭ·ănth'·ĕ·ē* (Gr. *mēn*, a month, or *mēnŭō*, I show; *anthos*, a flower—referring to the duration of the flowers), a Sub-order of the Ord. Gentianaceæ: **Menyanthes**, n., *mĕn'·ĭ·ănth'·ēz*, a genus of plants : **Menyanthes trifoliata**, *trĭ'·fōl·ĭ·āt'·ă* (L. *tris*, three; *folĭum*, a leaf), buck-bean, marsh-trefoil, or bog-bean, whose leaves are used as a substitute for Gentian.

mephitic, a., *mĕf·ĭt'·ĭk* (L. *mephitis*, a noxious, pestilential exhalation), offensive to the smell; noxious; deadly : **mephitis**, n., *mĕf·ĭt'·ĭs*, or **mephitism**, n., *mĕf'·*

ĭt·ĭzm, any foul or noxious exhalation — generally applied to that caused by carbonic acid gas.

merenchyma, n., *mĕr·ĕng′·kĭm·ȧ* (Gr. *mĕris*, a part, a particle; *enychuma*, what is poured in—from *en*, in; *cheuma*, tissue: perhaps Gr. *mĕruō*, I wind or twine round), in *bot.*, cellular tissue composed of more or less rounded cells.

mericarp, n., *mĕr′·ĭ·kȧrp* (Gr. *meris*, a part; *karpos*, fruit), in *bot.*, the half of the fruit of an umbelliferous plant, like the hemlock.

merismatic, a., *mĕr′·ĭs·mȧt′·ĭk* (Gr. *merismos*, division—from *meris*, a part), in *bot.*, taking place by division or separation, as into cells or segments.

merispore, n., *mĕr′·ĭ·spōr* (Gr. *meris*, a part; *spora*, seed), in *bot.*, a cell capable of germination, formed by the division of an ascospore or a basidiospore.

meristem, n., *mĕr′·ĭ·stĕm* (Gr. *meristos*, separated, divisible—from *merizo*, I divide into parts), in *bot.*, tissue formed of cells which are all capable of dividing, and producing new cells; also called 'generative tissue.'

merithal, n., *mĕr′·ĭ·thȧl* (Gr. *meris*, a portion; *thallos*, a young shoot, a bough), in *bot.*, a term used for 'internode'; a term applied to the different parts of the leaf: **merithalli**, n. plu., *mĕr′·ĭ·thȧl′·ī*, the three principal parts of a plant — the *radicular merithal* corresponding to the root, the *cauline* to the stem, and the *foliar* to the leaf.

Merostomata, n. plu., *mĕr′·ō·stŏm′·ȧt·ȧ* (Gr. *meros*, the upper part of the thigh; *stoma*, a mouth), an Ord. of Crustacea in which the appendages placed round the mouth, and performing the office of jaws, have their free extremities developed into walking or prehensile organs.

Mertensia, n., *mĕr·tĕns′·ĭ·ȧ* (after *Mertens*, a professor of medicine at Bremen), a highly esteemed genus of plants producing brilliant flowers, Ord. Boragineæ: **Mertensia maritima**, *mȧr·ĭt′·ĭm·ȧ* (L. *maritimus*, belonging to the sea —from *mȧr′ĕ*, the sea), a species having the taste of oysters, hence called in Scotland the oyster plant.

Merulius, n., *mĕr·ōōl′·ĭ·ŭs* (origin unknown: said to be a corruption of *metul′·us*, the original name— from L. *mĕta*, anything of a conical or pyramidal form, so named from its shape), a genus of fungi, one of whose species causes the dry rot in wood: **merulius lacrymans**, *lȧk′·rĭ·mȧnz* (L. *lacrymans*, weeping, lamenting), the most destructive of the parasitical fungi, producing what is called the dry rot, and a pest to wood and wooden structures, sometimes penetrating thick walls and destroying the mortar and lath.

Mesembryaceæ, n., *mĕs·ĕm′·brĭ·ā′·sĕ·ē* (Gr. *mesĕmbrĭa*, mid-day), the Ficoideæ or Fig-marigolds, and Ice-plant family, an Ord. of plants, natives of hot sandy plains: **Mesembryeæ**, n. plu., *mĕs′·ĕm·brĭ′·ĕ·ē*, a Sub-order having numerous conspicuous petals: **Mesembryanthemum**, n., *mĕs·ĕm′·brĭ·ȧnth′·ĕm·ŭm* (Gr. *anthĕmon*, a flower), a genus of beautiful and well-known succulents: **Mesembryanthemum edule**, *ĕd·ū′·lĕ* (L. *ĕdūlis*, eatable), the Hottentot-fig, whose leaves are used as an article of diet: **M. crystallinum**, *krĭst·ȧl′·lĭn·ŭm* (L. *crystăllinus*, crystalline — from *crystallum*, a crystal), the Ice-plant, remarkable for the watery vesicles which cover its surface, and which have the appearance of particles of ice: **M. tripolium**, *trĭ·pōl′·ĭ·ŭm* (of or from *Tripoli* in Africa), has the property of expanding in a star-like manner when in water, and closing when dry.

mesencephalon, n., *mĕz'·ĕn·sĕf'·ăl· ŏn* (Gr. *mesos*, middle ; *engkĕph· ălon*, the brain), the middle primary vesicle of the brain, forming one of its principal structures, and comprising the Crura cerebri, the Corpora quadrigemina, and the Aqueduct of Sylvius.

mesentery, n., *mĕz'·ĕn·tĕr·ĭ*, **mesenteries**, n. plu., *-tĕr·ĭz* (Gr. *mesos*, middle ; *enteron*, an intestine), a membrane in the cavity of the abdomen which serves to retain the intestines and their appendages in their position ; a double fold of the peritoneum ; the vertical plates which divide the somatic cavity of a sea-anemone into chambers : **mesenteric**, a., *mĕz'·ĕn·tĕr'·ĭk*, of or belonging to the mesentery : **mesenteritis**, n., *mĕz·ĕn'·tĕr·īt'·ĭs*, inflammation of the mesentery.

mesial, a., *mēz'·ĭ·ăl* (Gr. *mesos*, middle), in *anat.*, middle ; dividing into two equal parts : **mesial line**, an imaginary plane dividing, from top to bottom, the head, neck, and trunk into right and left similar halves.

mesocæcum, n., *mĕz'·ō·sēk'·ŭm* (Gr. *mesos*, middle ; L. *cæcum*, a portion of the large intestines), in *anat.*, a duplicature of the peritoneum at the posterior part of the cæcum, which serves to connect the back part of the cæcum with the right iliac fossa.

mesocarp, n., *mĕz'·ō·kărp* (Gr. *mesos*, middle ; *kărpos*, fruit), in *bot.*, the middle layer of the pericarp or coat of the fruit.

mesocephalon, n., *mĕz'·ō·sĕf'·ăl·ŏn*, or *-kĕf'·ăl·ŏn* (Gr. *mesos*, middle ; *kephalē*, the head), in *anat.*, an eminence of transverse fibres above and in front of the medulla oblongata, below and behind the Crura cerebri, and between the lobes of the cerebellum ; also called the Pons Varolii.

mesochilium, n., *mĕz'·ō·kīl'·ĭ·ŭm* (Gr. *mesos*, middle ; *cheilos*, a lip), in *bot.*, the middle portion of the labellum of orchids.

mesocolon, n., *mĕz'·ō·kōl'·ŏn* (Gr. *mesos*, middle ; Eng. *colon*), in *anat.*, that part of the mesentery to which the colon is attached.

mesogastrium, n., *mĕz'·ō·găs'·trĭ· ŭm* (Gr. *mesos*, middle ; *gastĕr*, the belly), in *anat.*, the umbilical region of the abdomen ; a suspensory band of the stomach in early embryonic life, subsequently converted into a sac, called the 'great omentum' : **mesogastric**, a., *mĕz'·ō·găs'·trĭk*, that which attaches the stomach to the walls of the abdomen.

mesophlœum, n., *mĕz'·ō·flē'·ŭm* (Gr. *mesos*, middle ; *phloios*, bark), in *bot.*, the middle layer of the bark, situated between the liber and epiphlæum.

mesophyllum, n., *mĕz'·ō·fīl'·lŭm* (Gr. *mesos*, middle ; *phullon*, a leaf), in *bot.*, all the inner portion or parenchyma of leaves, situated between the upper and under epidermis.

mesopodium, n., *mĕz'·ō·pōd'·ĭ·ŭm* (Gr. *mesos*, middle ; *podes*, feet), the middle portion of the foot of Molluscs.

mesorchium, n., *mĕz·ŏrk'·ĭ·ŭm* (Gr. *mesos*, middle ; *orchis*, a testicle), a fold of the peritoneum which, in the fifth month of fœtal life, supports the testicle in its position in the lumbar region before it passes into the scrotum.

meso-rectum, n., *mĕz'·o·rĕk'·tŭm* (Gr. *mesos*, middle ; Eng. *rectum*), a narrow fold of the peritoneum which connects the upper part of the rectum with the front of the sacrum.

mesosperm, n., *mĕz'·ō·spĕrm* (Gr. *mesos*, middle ; *sperma*, seed), in *bot.*, the second membrane, or middle coat of a seed.

mesosternum, n., *měz'·ŏ·stěrn'·ŭm*
(Gr. *mesos*, middle ; *sternon*, the
breast-bone), the lower half of
the middle segment of the thorax
in insects ; the middle portion of
the sternum intervening between
the attachment of the second pair
of ribs, and the xiphoid cartilage.

mesotherms, n. plu., *měz'·ŏ·thěrmz*
(Gr. *mesos*, middle ; *thermē*,
heat), plants requiring but a
moderate degree of heat for their
perfect development.

mesothorax, n., *měz'·ŏ·thōr'·ăks*
(Gr. *mēsos*, middle ; *thōrax*, the
trunk, the breast), the middle
ring of the thorax in insects.

Mesua, n., *měs'·ū·ă* (after *Mesŭĕ*,
an ancient Arabian physician and
botanist in the eighth century), a
genus of trees, Ord. Guttiferæ or
Clusiaceæ : **Mesua ferrea,** *fěr'·rĕ·ă*
(L. *ferrĕus*, made of iron—from
ferrum, iron), a tree producing
beautiful orange and sweet-scented
flowers which, dried, are esteemed
for their fragrance, and used in
India in medicine, yields a hard
and durable timber.

metabolic, a., *mět'·ă·bŏl'·ĭk* (Gr.
metăbolē, change — from *meta*,
beyond ; *ballō*, I throw), pert. to
change or affinity ; applied to
chemical changes occurring in
living bodies : **metabolic force,**
vital affinity.

metacarpus, n., *mět'·ă·kărp'·ŭs* (Gr.
meta, beyond ; *karpos*, the wrist),
in *anat.*, that part of the hand
situated between the wrist or
carpus and the fingers or phalanges : **metacarpal,** a., -*kărp'·ăl*,
pert. to the metacarpus : **metacarpal bones,** the five long bones
which form the back of the hand
externally, and the palm internally : **metacarpal phalangeal,**
făl'·ănj·ē'·ăl (Gr. *phalangx*, a body
of soldiers), pert. to the bones of
the hand situated between the
wrist and the fingers, so named
from their arrangement.

metamorphosis, n., *mět'·ă·mŏrf'·*

ŏs·ĭs (Gr. *metamŏrphōsis*, a transformation—from *meta*, beyond,
change ; *morphē*, form, shape), a
transformation ; in *zool.*, the
change of form which certain
animals undergo in passing from
their younger to their fully-
grown condition ; in *bot.*, the
change of one organ into another,
as petals into stamens, or stamens
into pistils — sometimes called
metamorphy, n., *mět'·ă·mŏrf'·ĭ*.

metaphery, n., *mět·ăf'·ěr·ĭ* (Gr.
meta, beyond ; *phoreo*, I bear),
in *bot.*, the displacement of
organs.

metaplasm, n., *mět'·ă·plăzm* (Gr.
meta, beyond, change ; *plasma*,
that which has been formed, a
model), the matter which gives
the granular character to protoplasm.

metapodium, n., *mět'·ă·pŏd'·ĭ·ŭm*
(Gr. *meta*, beyond, after ; *podes*,
feet), the posterior lobe of the
foot in Mollusca.

metapophysis, n., *mět'·ă·pŏf'·ĭs·ĭs*
(Gr. *meta*, beyond ; *apophusis*,
a sprout, a process), the mam-
millary processes, according to
Owen.

metasperms, n. plu., *mět'·ă·spěrmz*
(Gr. *meta*, beyond ; *sperma*, seed),
in *bot.*, another name for 'angiosperms' : see under 'angiocarp-
us.'

metastasis, n., *mět·ăs'·tăs·ĭs* (Gr.
meta, beyond, over ; *stasis*, a
placing or sitting, a posture), a
change in the seat of a disease ;
in *bot.*, the sum of the changes
undergone by the products of
assimilation of the cells in plants :
metastatic, a., *mět'·ă·stăt'·ĭk*, of
or belonging to metastasis.

metastoma, n., *mět·ăs'·tŏm·ă* (Gr.
meta, beyond ; *stŏma*, a mouth),
the plate which closes the mouth
posteriorly in the Crustacea.

metatarsus, n., *mět'·ă·tărs'·ŭs* (Gr.
mĕta, beyond ; *tarsos*, the sole of
the foot), the bones which lie
between the tarsus or ankle and

the toes, consisting of five long bones, which corresponds to the 'metacarpus' of the hand : **metatarsal**, a., *mĕt'·å·tårs'·ål*, of or belonging to the metatarsus.

metathorax, n., *mĕt'·å·thōr'·åks* (Gr. *mĕta*, beyond ; *thorax*, the chest), the posterior or hinder ring of the thorax in insects.

metencephalon, n., *mĕt'·ĕn·sĕf'·ål·ŏn* (Gr. *mĕta*, beyond ; *engkephalon*, the brain), the posterior primary vesicle of the brain, comprising the medulla oblongata, the fourth ventricle, and the auditory nerve.

metria, n., *mēt'·rĭ·å* (Gr. *mētra*, the womb), child-bed or puerperal fever : **metric**, a., *mĕt'·rĭk*, of or belonging to the womb : **metritis**, n., *mĕt·rīt'·ĭs*, inflammation of the womb : **metralgia**, n., *mĕt·rǎl'·jĭ·å* (Gr. *algos*, pain, grief), also **metrodynia**, n., *mĕt'·rō·dĭn'·ĭ·å* (Gr. *odŭnē*, pain), pain in the womb : **metrophlebitis**, n., *mĕt'·rŏ·flĕb·ĭt'·ĭs* (Gr. *phlebs*, a vein ; *phlebos*, of a vein), inflammation of the veins of the womb.

Metroxylon, n., *mĕt·rŏks'·ĭl·ŏn* (Gr. *mētēr*, a mother ; *xulon*, wood), a genus of trees, Ord. Palmæ : **Metroxylon læve**, *lēv'·ē* (L. *lævis*, light), a species producing fine sago.

Mezereon, n., *mĕz'·ĕr·ē'·ŏn*, see 'Daphnæ.'

miasm, n., *mĭ·dzm'*, also **miasma**, n., *mĭ·dz'·må*, more usually in the plu. : **miasmata**, *mĭ·dz'·måt·å* (Gr. *miasma*, defilement), infection or pollution in the air arising from diseased, putrifying, or poisonous floating substances : **miasmal**, a., *mĭ·dz'·mål*, containing miasma : **miasmatic**, a., *mĭ'·dz·måt'·ĭk*, pert. to miasma, or containing it.

micrococcus, n., *mĭk'·rō·kŏk'·ŭs* (Gr. *mikros*, small ; *kokkos*, a kernel), any minute form or organism supposed to have life ; a genus of the Bacteria, the basis

of all yeast formations, and the source of fermentations.

microgonidium, n., *mĭk'·rō·gŏn·ĭd'·ĭ·ŭm* (Gr. *mikros*, small ; *gonos*, offspring, seed ; *eidos*, resemblance), in Algæ, a single small zoospore found in a germinating cell, formed at the expense of the contained plastic materials.

micrometer, n., *mĭk·rŏm'·ĕt·ėr* (Gr. *mikros*, small ; *metron*, a measure), an instrument for measuring microscopic objects.

microphylline, n., *mĭk'·rō·fĭl'·lĭn* (Gr. *mikros*, small ; *phullon*, a leaf), a material composed of minute leaflets or scales.

microphytes, n. plu., *mĭk'·rō·fĭtz* (Gr. *mikros*, small ; *phuton*, a plant), microscopic plants.

micropyle, n., *mĭk'·rō·pil* (Gr. *mikros*, small ; *pule*, a gate), in *bot.*, the opening or foramen of the ripened seed for the escape of the root of the embryo ; a minute pore.

microscope, n., *mĭk'·rō·skōp* (Gr. *mikros*, small ; *skopeo*, I view), an instrument which enables minute objects, and those invisible to the naked eye, to be seen distinctly : **microscopy**, n., *mĭk·rŏs'·kōp·ĭ*, investigations by means of the microscope.

microsporangia, n., *mĭk'·rō·spŏr·ănj'·ĭ·å* (Gr. *mikros*, small ; *spora*, seed ; *anggos*, a vessel), in *bot.*, cells or thecæ containing microspores.

microspores, n. plu., *mĭk'·rō·spōrz* (Gr. *mikros*, small ; *spora*, seed), in *bot.*, small reproductive spores in the capsules of Lycopods ; applied to certain vegetable parasites present in various cutaneous affections—also in same sense **microsporons**, n. plu., *mĭk·rŏs'·pŏr·ŏnz*.

microsporon furfur, *fėr'·fėr* (see *microspore* ; L. *furfur*, bran, scurf), a fungus, consisting of small globular sporules with short branching thalli, just large enough to contain them, which, growing in the epidermis, produces the disease known as ' chloasma,' or ' pityriasis versicolor.'

microtherms, n. plu., *mĭk'·rō·thĕrmz* (Gr. *mikros*, small; *thermē*, heat), in *bot.*, plants which require only a small degree of heat to bring them to perfection.

microzymes, n. plu., *mĭk'·rō·zīmz* (Gr. *mikros*, small ; *zumē*, fermenting matter), a general term for very minute organised particles, which present themselves in liquids fermenting or undergoing decomposition ; the minute organised particles which are supposed to be the contagious matter in zymotic diseases.

midrib, n., *mĭd'·rĭb* (*mid* and *rib*), in *bot.*, a large or central vein ; a continuation of the petiole.

midriff, n., *mĭd'·rĭf* (AS. *mid*, middle ; *hrif*, entrails ; Dut. *middelrift*, the diaphragm), in *anat.*, the muscular partition which separates the cavity of the chest from the belly ; the diaphragm.

migraine, n., *mĭg·rān'* (a French corruption of *hemicrania*), the brow-ague ; a painful disorder generally on one side of the forehead ; a megrim.

Mikania, n., *mĭk·ān'·ĭ·ă* (after *Professor Mikan*, of Prague), a genus of plants, Ord. Compositæ, Sub-ord. Corymbiferæ : **Mikania Guaco**, *gwăk'·ō* (from *Guaco*, S. America), a species which has been used to cure the bites of snakes.

miliaria, n., *mĭl'·ĭ·ār'·ĭ·ă* (L. *milium*, a grain called millet), little vesicles or blisters on the skin, containing a sero - albuminous fluid, which is simply retained perspiration, so named as resembling millet seeds ; miliary fever, associated with excessive heat of skin : miliary, a., *mĭl'·ĭ·*ăr·ĭ, accompanied with an eruption resembling millet seeds.

mimetic, a., *mĭm·ĕt'·ĭk* (Gr. *mimetikos*, imitative—from *mimos*, a farcical entertainment, a mime), applied to organs or animals which resemble each other in external appearance, but not in characteristic structure ; appearing like imitations of others.

Mimoseæ, n. plu., *mĭm·ōz'·ĕ·ē* (Gr. *mimos*, a mimic, an imitator), a Sub-order of Ord. Leguminosæ, which yield gum in quantity, and whose bark is frequently astringent—so named from many species mimicking animal sensibility in their leaves: **Mimosa**, n., *mĭm·ōz'·ă*, a genus of leguminous plants: **Mimosa sensitiva**, *sĕns'·ĭt·ĭv'·ă* (L. *sensus*, the faculty or power of perceiving or feeling), and **M. pudica**, *pŭd·ĭk'·ă* (L. *pŭdicus*, bashful, modest), are two species which are commonly called sensitive plants.

Mimulus, n., *mĭm'·ūl·ŭs* (Gr. *mimos*, a mimic), a genus of plants, Ord. Scrophulariaceæ, many of which are showy — so named from the ringent corollas of the species ; the two lamellæ are irritable, and close when irritated : **Mimulus guttatus**, *gŭt·tāt'·ŭs* (L. *gŭttātus*, spotted, speckled—from *gutta*, a drop), a species whose leaves are eatable as a salad : **M. luteus**, *lŏt'·ĕ·ŭs* (L. *lūtĕus*, yellowish—from *lutum*, a plant used in dyeing yellow), a species naturalised in many parts of Britain : **M. moschatus**, *mŏs·kāt'·ŭs* (mid L. *moschātus*, having a smell like musk—from Ar. *mosch*, musk), a plant cultivated on account of its musk-like odour.

Mimusops, n., *mĭm'·ŭs·ŏps* (Gr. *mĭmō*, an ape ; *ops*, the eye, the countenance), a genus of ornamental trees, Ord. Sapotaceæ—so named from the fancied resemblance of the flowers to the face of a monkey : **Mimusops elengi,** *ĕl·ĕnj'ĭ* (E. Indies), a species yielding a durable timber in Ceylon, whose fruit, the Surinam medlar, is eaten ; its seeds yield an oil, and its flowers a perfume : **M. hexandra,** *hĕks·ănd'·ră* (Gr. *hex,* six ; *anēr,* a man, *andros,* of a man, having six stamens), yields a durable wood : **M. kaki,** *kăk'·ĭ* (Indian name), produces an eatable fruit.

miostemonous, a., *mī'·ō·stĕm'·ŏn·ŭs* (Gr. *meiŏn,* less ; *stĕmŏn,* a stamen), in *bot.,* applied to a flower in which the stamens are neither equal to, nor a multiple of, the floral envelopes.

Mirabilis, n., *mĭr·ăb'·ĭl·ĭs* (L. *mĭr·ābĭlis,* wonderful, marvellous— from *mĭror,* I wonder or marvel at), a genus of plants, Ord. Nyctaginaceæ — so named from the appearance of their flowers : **Mirabilis jalapa,** *jăl·āp'·ă* (*Xalapa,* in Mexico), so named as having been considered at one time as the Jalap-plant, but this is now ascertained to be Exogonium purga, one of the Convolvulaceæ : **M. dichotoma,** *dĭk·ŏt'·ŏm·ă* (Gr. *dichotŏma,* cut in two halves —from *dicha,* doubly ; *temno,* I cut), the marvel of Peru, a common garden plant, called in W. Indies 'four o'clock flower,' from opening its blossoms at that hour P. M.

mitral, a., *mĭt'·răl* (L. and Gr. *mitra,* a headdress, a mitre), in *anat.,* applied to a valve attached to the circumference of the left auriculo-ventricular orifice, whose flaps are supposed to resemble the segment of a bishop's mitre : **mitriform,** a., *mĭt'·rĭ·fŏrm* (L. *forma,* shape), shaped like a mitre or bishop's official hat; conical ; hollow and open at the base.

modiolus, n., *mŏd·ī'·ŏl·ŭs* (L. *mŏdĭŏlus,* the nave of a wheel— from *modĭus,* a measure), in *anat.,* the central axis or pillar of the internal ear, conical in form, and extending from the base to the apex of the cochlea.

molar, a., *mōl'·ăr* (L. *mola,* a mill, *molāris,* a mill-stone), grinding ; having power to grind, as a tooth: **molars,** n. plu., *mōl'·ărz,* the grinders in man ; the teeth in diphyodont mammals which are not preceded by milk-teeth.

mole, n., *mōl* (L. *mola,* a false conception), in *anat.,* a mass of fleshy matter generated by some morbid process in the uterus ; a morbid development of the placenta : **hydatid mole,** one resembling a hydatid.

molecule, n., *mŏl'·ĕk·ūl* (F. *molécule,* a small particle of matter or air—from L. *moles,* a mass), a very minute particle of matter ; one of the elementary particles into which all bodies are supposed to be resolvable ; in *bot.,* a very minute body in which there is no obvious determinate external circle, or internal centre : **molecular,** a., *mōl·ĕk'·ūl·ăr,* pert. to molecules ; designating that force or attraction by which the minute particles or molecules of a body are held together in one mass.

NOTE. — **Molecule** means strictly the smallest quantity of an element, or of a compound, that can exist in the free state—probably in most cases consisting of two atoms ; **an atom** is an ultimate particle of matter.

moles, n. plu., *mōlz* (Ger. *mahl,* a stain, a spot ; Scot. *mail,* a spot in cloth ; Sw. *mal,* a mark), congenital marks of a light or dark brown, or of a black colour, on the human skin — also called 'liver stains,' 'mother's marks,' and 'pilous and pigmentary nævi.'

moles carnea, *mōl'·ēz kăr'·nĕ·ă* (L. *mōles,* a heavy mass ; *carnĕŭs,* fleshy—from *caro,* flesh), another name for the 'flexor accessorius,' a muscle connected with the tendon of the 'flexor longus digitorum,' which see.

nollities ossium, *mŏl·lĭsh'·ĭ·ĕz ŏs'·sĭ·ŭm* (L. *mollĭtĭes*, flexibility, softness; *os*, a bone, *ossium*, of bones), a fatty degeneration which takes place in bone, and which renders it more brittle and liable to bend, brought about by an absorption of the earthy matter.

Mollusca, n. plu., *mŏl·lŭsk'·ă*, also **molluscs,** n. plu., *mŏl'·lŭsks* (L. *molluscus*, soft—from *mollis*, soft; *mollusca*, a nut with a soft shell), the Sub-kingdom, forming one of the grand divisions of the animal kingdom, comprising the shellfish proper, the Polyzoa, the Tunicata, and the Lamp-shells—so named from the general soft nature of their bodies, and absence of internal skeleton: **Molluscoida,** n. plu., *mŏl'·lŭsk·öyd'·ă* (Gr. *eidos*, resemblance), the lower division of the Mollusca; certain mollusc - like animals, often compound, lower in structure than the true molluscs, and may have shelly or horny coverings: **molluscum,** n., *mŏl·lŭsk'·ŭm*, in *med.*, small, soft tumours, produced by distensions of the sebaceous glands by secretions—formerly applied to *fibroma* (L. *fibra*, a filament), a disease in which tubercles are formed by hypertrophy of the fibrous tissue of the skin.

Momordica, n., *mŏm·ŏrd'·ĭk·ă* (L. *mordĕo*, I bite, *mŏmŏrdi*, I have bitten), a genus of plants, Ord. Cucurbitaceæ: **Momordica elaterium,** *ĕl'·ăt·ēr'·ĭ·ŭm* (Gr. *elatērĭon*, that drives out or expels), the wild or squirting cucumber, so named on account of the force with which its seeds are expelled when ripe; the feculence subsiding from the juice constitutes the medicinal Elaterium, used in small doses as a violent cathartic in dropsical cases.

monadelphous, a., *mŏn'·ă·dĕlf'·ŭs* (Gr. *monos*, one, alone; *adelphos*, a brother), in *bot.*, having all the stamens united into one bundle by union of their filaments.

monads, n. plu., *mŏn'·ădz* (Gr. *mŏnas*, unity, a monad, *monădos*, of a monad—from *monos*, alone), microscopic organisms of the simplest structure; an indivisible thing; an ultimate particle; a primary cell.

monandrous, a., *mŏn·ănd'·rŭs* (Gr. *monos*, one, alone; *anēr*, a male, *andros*, of a male), in *bot.*, having only one stamen.

monembryony, n., *mŏn'·ĕm·brī'·ŏn·ĭ* (Gr. *monos*, one; *embruon*, an embryo), in *bot.*, the production of one embryo only: **monembryonic,** a., *mŏn·ĕm'·brī·ŏn'·ĭk*, having a single embryo.

moniliform, a., *mŏn·ĭl'·ĭ·fŏrm* (L. *monīle*, a necklace; *forma*, shape), beaded like a necklace; jointed so as to resemble a string of beads.

Monimiaceæ, n. plu., *mŏn'·ĭm·ĭ·ā'·sĕ·ē* (Gr. *monĭmŏs*, lasting, enduring), an Order of plants of S. America and Australia, of some the bark and leaves are aromatic and fragrant, and the fruit of others is eaten: **Monimia,** n., *mŏn·ĭm'·ĭ·ă*, a genus of plants.

Monk's - hood, n., a poisonous plant—so named from the cowl-like shape of the flowers; the Aconītum napellus, Ord. Ranunculaceæ.

monobasic, a., *mŏn'·ō·bāz'·ĭk* (Gr. *monos*, one; *basis*, base), in *chem.*, containing one equivalent of base to one of acid.

monocarpic, a., *mŏn'·ō·kărp'·ĭk*, also **monocarpous,** a., *-kărp'·ŭs* (Gr. *monos*, one, alone; *karpos*, fruit), in *bot.*, producing flowers and fruit once during life, and then dying.

monochlamydeæ, n. plu., *mŏn'·ō·klăm·ĭd'·ĕ·ē* (Gr. *monos*, one, alone; *chlamus*, a large cloak, *chlamŭdos*, of a cloak), in *bot.*, a large division of plants which have only one envelope in the flower : **mon′ochlamyd′eous**, a., *-ĭd′·ĕ·ŭs*, applied to a flower having a single envelope, which is the calyx.

monoclinous, a., *mŏn′·ō·klīn′·ŭs* (Gr. *monos*, alone ; *klīno*, I bend), in *bot.*, having both stamens and pistils in every flower.

monocotyledons, n. plu., *mŏn′·ō·kŏt′·ĭl·ēd′·ŏnz* (Gr. *monos*, alone ; *kotulĕdōn*, the hollow of a cup, cup-shaped), in *bot.*, a great division of plants which have only one cotyledon or seed lobe, as in oats or wheat, the seeds of which are entire : **mon′ocotyled′onous**, a., *-ēd′·ŏn·ŭs*, having one cotyledon or seed lobe.

monoculous, a., *mŏn·ŏk′·ūl·ŭs* (Gr. *monos*, one ; L. *oculus*, the eye), possessed of only one eye : **mon-ocule**, n., *mŏn′·ŏk·ūl*, a one-eyed insect.

monocystic, a., *mŏn′·ō·sĭst′·ĭk* (Gr. *monos*, one ; *kustis*, a bladder), consisting of only one cell or cavity ; unilocular.

Monodelphia, n. plu., *mŏn′·ō·dĕl′f′·ĭ·ă* (Gr. *monos*, alone, single ; *delphus*, a womb), the division of Mammals which have the uterus single : **monodelphous**, a., *mŏn′·ō·dĕlf′·ŭs*, bringing forth the young fit to live ; of or pert. to the Monodelphia ; (Gr. *adelphos*, a brother), in *bot.*, having all the filaments united into a tube around the style.

monodichlamydeous, a., *mŏn·ŏd′·ĭ·klăm·ĭd′·ĕ·ŭs* (Gr. *monos*, one ; *dis*, twice ; *chlamus*, a cloak or tunic), in *bot.*, having either one or both floral envelopes.

monœcious, a., *mŏn·ē′·shĭ·ŭs* (Gr. *monos*, one, single ; *oikos*, a house), in *zool.*, applied to individuals in which the sexes are united ; having male and female flowers on the same plants: **monœcism**, n., *mŏn′·ē·sĭzm*, the con-

dition where unisexual flowers are produced on different branches.

monogamia, n. plu., *mŏn′·ō·găm′·ĭ·ă* (Gr. *monos*, one, single ; *gamos*, marriage), a general name for plants which have their anthers united but their flowers not compound : **monogam**, n., *mŏn′·ō·găm*, a plant having a simple flower though the anthers are united.

monogastric, a., *mŏn′·ō·găst′·rĭk* (Gr. *monos*, one ; *gastēr*, the belly), having only one stomach.

monogynian, a., *mŏn′·ō·jĭn′·ĭ·ăn*, also **monogynous**, a., *mŏn·ŏdj′·ĭn·ŭs* (Gr. *monos*, one ; *gunē*, a female), in *bot.*, having only one pistil or stigma in a flower ; applied to plants having one style : **monogynœcial**, a., *mŏn′·ō·jĭn·ē′·shĭ·ăl* (Gr. *oikos*, a house), in *bot.*, applied to simple fruits formed by the pistil of one flower.

monolocular, a., *mŏn′·ō·lŏk′·ūl·ăr* (Gr. *monos*, one; L. dim. of *locus*, a place), syn. of 'monocystic,' or 'unilocular.'

monomania, a., *mŏn′·ō·mān′·ĭ·ă* (Gr. *monos*, one ; *mānĭa*, madness), a mental disease in which madness exists on one particular subject, or a limited number of subjects, while the mind is tolerably lucid on others ; a lunatic who has passed through the acute stage of the malady.

monomyary, n., *mŏn′·ō·mĭ′·ăr·ĭ* (Gr. *monos*, one, single ; *muōn*, a muscle, *muōnos*, of a muscle), one of the bivalves, the Lamellibranchiata, which have their shell closed by a single adductor muscle.

monopetalous, a., *mŏn′·ō·pĕt′·ăl·ŭs* (Gr. *monos*, one ; *petalon*, a leaf), in *bot.*, having the petals united by their edges into one organ ; gamopetalous.

monophyllous, a., *mŏn′·ō·fĭl′·lŭs* (Gr. *monos*, one ; *phullon*, a leaf or blade), having one leaf or membrane ; formed of one leaf only ; gamophyllous.

monophyodont, n., *mŏn'·ō·fī'·ŏd·ŏnt* (Gr. *monos*, one ; *phuo*, I produce ; *odous*, a tooth, *odontos*, of a tooth), one of the Mammals in which only a single set of teeth is developed, that is, who never shed their teeth.

monoplast, n., *mŏn'·ō·plăst* (Gr. *monos*, one ; *plastos*, formed), a naked non-vasicular body ; an animal cell destitute of envelope : **monoplastic,** a., *mŏn'·ō·plăst'·ĭk*, having one primary form.

monopodia, n., *mŏn'·ō·pōd'·ĭ·ă* (Gr. *monos*, one ; *pous*, a foot, *podos*, of a foot), a monstrosity having one foot only : **mon'opod'ium,** n., *-pōd'·ĭ·ŭm*, in *bot.*, an elongated axis giving off lateral structures having a similar nature : **mon'opod'ial,** a., *-pōd'·ĭ·ăl*, applied to a kind of inflorescence ; racemose.

monosepalous, a., *mŏn'·ō·sĕp'·ăl·ŭs* (Gr. *monos*, one ; *sepalon*, a sepal), in *bot.*, having the sepals which compose a flower united at their edges or margins so as to form a tube ; gamosepalous.

monosis, n., *mŏn·ōz'·ĭs* (Gr. *monos*, one, single), in *bot.*, the isolation of an organ from the rest.

monospermous, a., *mŏn'·ō·spĕrm'·ŭs* (Gr. *monos*, one ; *sperma*, seed), in *bot.*, one-seeded ; applied to a fruit having only one seed : **monosperm,** n., *mŏn'·ō·spĕrm*, a plant of one seed only.

monostachous, a., *mŏn·ŏs'·tăk·ŭs* (Gr. *monos*, one ; *stachus*, an ear or spike), in *bot.*, disposed or arranged in one spike only.

monostomum, n., *mŏn·ŏs'·tŏm·ŭm*, **monos'toma,** n. plu., *-tŏm·ă* (Gr. *monos*, one ; *stoma*, a mouth), a species of Frematode worm having one sucker only.

monothalamous, a., *mŏn'·ō·thăl'·ăm·ŭs* (Gr. *monos*, one ; *thalamos*, a chamber), possessing a single chamber only, applied to the shells of ' Foraminifera ' and ' Mollusca.'

monothecal, a., *mŏn'·ō·thēk'·ăl* (Gr. *monos*, one ; *thēkē*, a sheath or case), in *bot.*, having a single loculament.

Monotremata, n. plu., *mŏn'·ō·trĕm'·ăt·ă* (Gr. *monos*, one ; *trēma*, an orifice, an opening, *trēmatos*, of an opening), an Order of Mammals which have the intestinal canál opening into a ' cloaca,' common to the ducts of the urinary and generative organs, as in the Duck-mole and the Echidna : **monotrematous,** a., *mŏn'·ō·trĕm'·ăt·ŭs*, having only one external opening or ' cloaca ' for urine and other excrements.

Monotropaceæ, n. plu., *mŏn'·ō·trŏp·ā'·sĕ·ē* (Gr. *monos*, one ; *tropeo*, I turn), the Fir-rapes, a small Order of parasitic plants growing on the roots of fir-trees, several species of which are delightfully fragrant : **Monotropeæ,** n. plu., *mŏn'·ō·trŏp'·ĕ·ē*, a Sub-order : **Monotropa,** n., *mŏn·ŏt'·rŏp·ă*, a genus of plants, so named because all their flowers are turned one way ; curious parasitical plants growing on the roots of beech and pine trees in shady moist places.

monstrosity, n., *mŏn·strŏs'·ĭ·tĭ* (L. *monstrum*, anything strange or wonderful), an unnatural production ; in *bot.*, an abnormal development, applied particularly to double flowers.

monticulus cerebelli, *mŏnt·ĭk'·ŭl·ŭs sĕr'·ĕb·ĕl'·lī* (L. *monticŭlus*, a small mountain—from *mons*, a mountain ; *cerĕbĕllum*, a small brain), in *anat.*, the little mountain of the cerebellum ; the central projecting part of the superior vermiform process.

Moraceæ, n. plu., *mŏr·ā'·sĕ·ē* (L. *mŏrus*, Gr. *morĕă*, a mulberry tree), the Mulberry, Fig, and Bread Fruit family, an important Order of plants : **Moreæ,** n. plu., *mŏr'·ĕ·ē*, a Sub-order of plants, comprising the mulberries and figs : **Morus,** n., *mŏr'·ŭs,* a genus of plants : **Morus nigra,** *nĭg'·ră* (L. *nĭger,* black), the common black mulberry : **M. alba,** *ălb'·ă* (L. *albus,* white), the white mulberry, less esteemed than the black ; the leaves of both, especially the latter, are the favourite food of the silkworm, and the root of 'the white mulberry is anthelmintic.

morbid, a., *mŏrb'·ĭd* (L. *morbus,* disease, *morbĭdus,* sickly), diseased ; not sound and healthful : **morbidness,** n., *mŏrb'·ĭd·nĕss,* state of being diseased or unsound: **morbific,** a., *mŏrb·ĭf'·ĭk* (L. *facio,* I make), causing disease ; generating a sickly state : **morbid anatomy,** the study of the alterations in the structure of the body, or a part, produced by disease.

morbillous, a., *mŏrb·ĭl'·lŭs* (mid. L. *morbilli,* measles—from L. *morbus,* disease), pert. to the measles ; measly : **morbilli,** n. plu., *mŏrb·ĭl'·lī,* the measles.

morbus cœruleus, *mŏrb'·ŭs sĕr·ōōl'·ĕ·ŭs* (L. *morbus,* a disease ; *cœrŭlĕus,* dark-blue, azure), blue disease, arising from a congenital malformation of the heart or its great vessels—so named from the purple or livid colour of the skin : **morbus coxarius,** *kŏks·ār'·ĭ·ŭs* (L. *coxărĭus,* pert. to the hip—from *coxa,* the hip), hip disease ; a scrofulous disease, nearly allied to white swelling.

Morchella, n., *mor·kĕl'·lă* (Ger. *morchel,* the morel), a genus of eatable fungi found on the ground : **Morchella esculenta,** *ĕsk'·ŭl·ĕnt'·ă* (L. *esculentus,* fit for eating), an edible fungus: **Morel,** n., *mŏr·ĕl',* a genus of eatable fungi abounding with little holes, generally of the size of a walnut.

mordant, n., *mŏrd'·ănt* (L. *mordax,* biting, *mordeo,* I bite), a substance employed to give permanency or brilliancy to a dye ; any adhesive matter by means of which gold leaf is made to adhere to a surface.

moribund, n. or a., *mŏr'·ĭ·bŭnd* (L. *moribundus,* dying—from *mori,* to die), in a state of dying.

Morinda, n., *mŏr·ĭnd'·ă* (corruption of *Morus indica,* Indian mulberry), a genus of plants, Ord. Rubiaceæ— so named from the shape of their fruit and native country : **Morinda citrifolia,** *sĭt'·rĭ·fōl'·ĭ·ă* (L. *citrus,* a citron ; *folium,* a leaf), a plant whose root is employed in the East as a substitute for madder under the name Sooranjee: **morindin,** n., *mŏr·ĭnd·ĭn,* a peculiar colouring matter yielded by ' M. citrifolia.'

Moringaceæ, n. plu., *mŏr'·ĭng·gā'·sĕ·ē* (from *Muringo,* the native name in Malabar of the species), the Moringa family, a small Order of plants, some of which are pungent and aromatic : **Moringa,** n., *mŏr·ĭng'·gă,* a genus of plants: **Moringa pterygosperma,** *tĕr'·ĭ·gō·spĕrm'·ă* (Gr. *pterux,* a wing ; *sperma,* seed), the Horse-radish tree whose seeds are winged, and are called Ben-nuts ; from the seeds an oil is obtained, used by watchmakers ; the root is pungent and stimulant, resembling Horseradish.

morphia, n., *mŏrf'·ĭ·ă,* also **morphine,** n., *mŏrf'·ĭn* (Gr. *Morpheus,* the god of dreams), one of the alkaloids contained in opium.

morphology, n., *mŏrf·ŏl'·ŏ·jĭ* (Gr. *morphē,* form, shape ; *logos,* description), that department of botany which treats of the forms that different organs of plants assume and the laws which regulate their metamorphoses, tracing their primary forms to the leaf as a type ; applied to animals in same sense : **morphological,** a., *-ŏdj'·ĭk·ăl,* connected with or relating to morphology.

morphosis, n., *mŏrf·ŏz´·ĭs* (Gr. *morphē*, form, shape), in *bot.*, the order or mode of development in any organ of a plant.

mortification, n., *mŏrt´·ĭf·ĭk·ā´·shŭn* (mid. L. *mortificare*, to mortify —from *mors*, death ; *facio*, I make), the death of a part of the living body : **gangrene,** the stage in which the part is hot, swollen, and livid, but not quite dead : **sphacelus,** that stage in which the part is cold and dead: **mummification,** dry gangrene in which an extremity is dry and shrunken, but not quite dead : **sloughing** is the state in which the dead, soft parts come away gradually from the living parts : **necrosis,** the death of bone : **ramollissement,** the destruction and breaking down of brain tissue.

Morus, n., see 'Moraceæ.'

motor, n., *mōt´·ŏr* (L. *motum,* to move), that which gives motion : **adj.,** in *anat.,* producing a regulating motion, applied to certain nerves and muscles : **motorial,** a., *mōt·ōr´·ĭ·ăl,* giving motion : motor oculi, *ŏk´·ŭl·ĭ* (L. *oculus,* the eye, *oculi,* of the eye), the motor nerve of the eye, the third cerebral nerve, which supplies nearly all the muscles of the orbit: **motores oculorum,** *mōt·ōr´·ēz ŏk´·ŭl·ōr´·ŭm* (L.), the motors of the eyes: **motory,** a., *mōt´·ŏr·ĭ,* exciting or controlling motion.

Moxa, n., *mŏks´·ă* (F. *moxa,* but probably of Eastern origin), the woolly leaves of the Artemisia Moxa, Ord. Compositæ, Sub-ord. Corymbiferæ, used in China to form the inflammable cones or cylinders called 'Moxas,' which are employed as counter-irritants; a small cone of inflammable matter, chiefly used in Eastern countries as a counter-irritant by burning it above on the skin— supposed to be good in the cure of gout and other deep-seated pain.

mucedinous, a., *mū·sēd´·ĭn·ŭs* (Gr. *mukēs,* a mushroom, a mould), in *bot.,* like a mould.

mucilage, n., *mūs´·ĭl·ādj* (F. *mucilage*—from L. *mucus,* the discharge from the nose), a solution of gummy matter, as gum-arabic, in water; a slimy substance found in many vegetables : **mucic,** a., *mūs´·ĭk,* of or from gum: **muciparous,** a., *mūs·ĭp´·ăr·ŭs* (L. *pario,* I produce), secreting or producing mucus : **mucous,** a., *mūk´·ŭs,* of or pert. to mucus; slimy: **mucus,** n., *mūk´·ŭs,* the slimy, glairy substance secreted from the mucous membrane : **mucous membrane,** an extremely delicate membrane which lines the interior cavities of the human body ; the internal skin.

mucro, n., *mūk´·rō* (L. *mucro,* a sharp point or edge, *mucrōnis,* of a sharp point), a stiff or sharp point abruptly terminating an organ : **mucronate,** a., *mūk´·rŏn·āt,* having a mucro ; abruptly pointed by a sharp spine.

Mucuna, n., *mūk·ūn´·ă* (from the Brazilian name *Mucuna-guaca,* the cow-itch), a genus of plants, Ord. Leguminosæ, Sub-ord. Papilionaceæ: **Mucuna pruriens,** *prōr´·ĭ·ĕnz* (L. *pruriens,* itching), and **M. prurita,** *prōr·īt´·ă* (L. *prūrītus,* itched — from *prūrĭo,* I itch), the former species in the West, and the latter in the East Indies, have the name 'cowhage' or cow-itch applied to the hairs of their legumes ; they have irritating properties, and mixed with syrup, are used in treatment for intestinal worms.

mucus, see under 'mucilage.'

Mudar, n., *mūd´·ăr* (Indian name), a substance procured from the bark of the root of 'Calotropis procera' and ' C. gigantea,' used as a diaphoretic in India: **Mudarine,** n., *mūd´·ăr·ĭn,* a principle contained in Mudar which gelatinises on being heated, and becomes fluid on cooling.

mulberry, n., *mŭl'bĕr·rĭ* (Ger. *maulbeere*, L. *mōrus*, Gr. *morĕa*, a mulberry), the fruit of the 'Morus nigra' and the 'Morus alba,' Ord. Moraceæ : **mulberry calculus,** a stone in the bladder having a rugged surface.

mullein, n., *mŭl'lĭn* (F. *mouleine* or *molène*, Dan. *mol*, a moth), a name applied to species of Verbascum, Ord. Scrophulariaceæ ; the woolly leaves of the Great Mullein are emollient and slightly narcotic ; a wild hedge plant whose seed has been used to preserve clothes against moths.

muller, n., *mŭl'lĕr* (L. *mola*, a mill-stone), a stone held in the hand, used for grinding powders upon a stone slab ; also **mullet,** n., *mŭl'lĕt* (F. *molette*, Sp. *moleta*), used in same sense.

multicostate, a., *mŭlt'·ĭ·kŏst'·āt* (L. *multus*, many ; *costa*, a rib), in *bot.*, many-ribbed.

multicuspid, a., *mŭlt'·ĭ·kŭsp'·ĭd* (L. *multus*, many; *cuspis*, a spear-head, *cuspĭdis*, of a spear-head), having several tubercles or points; applied to the rough, grinding surfaces of the twelve molar teeth : **multicuspidati,** n. plu., *mŭlt'·ĭ·kŭsp·ĭd·āt'·ĭ*, the molar teeth, twelve in number, six in each jaw ; the 'bicuspids' are the small or false molars, and are eight in number.

multifid, a., *mŭlt'·ĭ·fĭd*, also **multifidous,** a., *mŭlt'·ĭf'ĭd·ŭs* (L. *multifidus*, cleft or split into many parts—from *multus*, many; *findo*, I cleave or split), having many clefts or divisions ; in *bot.*, applied to a simple leaf divided laterally, to about the middle, into numerous portions—when the divisions extend deeper it is called 'multipartite.'

multifidus spinæ, *mŭlt'·ĭf'ĭd·ŭs spĭn'·ē* (L. *multifĭdus*, many-cleft ; *spĭna*, a spine, *spĭnæ*, of a

spine), the many-cleft part of the spine ; in *anat.*, a number of fleshy and tendinous fasciculi, which fill up the groove on either side of the spinous processes of the vertebræ, from the sacrum to the axis.

multijugate, a., *mŭlt'·ĭdj'·ōōg·āt* (L. *multus*, many ; *jugum*, a yoke), in *bot.*, having many pairs of leaflets.

multilocular, a., *mŭlt'·ĭ·lŏk'·ūl·ăr* (L. *multus*, many ; *loculus*, a small compartment, a cell), having many cells or chambers.

multipartite, a., *mŭlt'·ĭp'·ărt·ĭt* (L. *multus*, many ; *partĭtus*, divided), in *bot.*, divided into several strips or portions; divided into many parts.

multiple, a., *mŭlt'·ĭ·pl* (L. *multus*, many ; *plĭco*, I fold), in *bot.*, numerous ; manifold ; applied to anthocarpous or polygynœcial fruits formed by the union of several flowers : n., a quantity or number which contains another an exact number of times without a remainder, thus 12 is a multiple of 6, 4, 3, or 2.

multipolar, a., *mŭlt·ĭp'·ŏl·ăr* (L. *multus*, many ; *polus*, a pole, a point), applied to nerve cells with many tail-like processes or prolongations.

multiseptate, a., *mŭlt'·ĭ·sĕpt'·āt* (L. *multus*, many ; *septum*, a hedge), in *bot.*, having numerous septa or partitions.

multivalve, n., *mŭlt'·ĭ·vălv* (L. *multus*, many ; *valvæ*, folding doors or valves), a shell composed of more valves or pieces than two.

multungula, n., *mŭlt·ŭng'·gŭl·ă* (L. *multus*, many ; *ungula*, a hoof), the division of the Perissodactyle ungulates, which have more than a single hoof on each foot : **multungulate,** a., *mŭlt·ŭng'·gŭl·āt*, having the hoof divided into more than two parts.

mumps, n. plu., *mŭmps* (low Ger. *mumms*, a swelling of the glands of the neck ; Icel. *mumpa*, to eat voraciously), infectious disorders accompanied by a painful swelling of the salivary glands ; known also by the name 'Cynanche Parotidea.'

muricate, a., *mūr'ĭk·āt*, also **muriculate**, a., *mŭr·ĭk'ŭl·āt* (L. *muricatus*, full of sharp points—from *murex*, a shell-fish armed with sharp prickles), in *bot.*, formed with sharp points ; covered with firm short points or excrescences.

muriform, a., *mŭr'ĭ·fŏrm* (L. *murus*, a wall ; *forma*, shape), in *bot.*, wall - like, applied to tissues presenting the appearance of bricks in a wall.

murrain, n., *mŭr'rān* (Sp. *morriña*, a disease among cattle ; old F. *morine*, the carcass of a dead beast ; Gr. *marainō*, I destroy), a term formerly applied to many forms of cattle plague, now restricted to the **aphtha epizootica**, *ĕp'ĭ·zō·ŏt'ĭk·ă* (*epizootic aphthæ*), the foot-and-mouth disease.

Musaceæ, n. plu., *mūz·ā'sĕ·ē* (after *Antonius Mūsa*, physician in ordinary to the anc. Roman king Augustus ; altered from Egyptian name *Mauz*), the Banana family, an Order of plants which furnish a very large supply of nutritious food to the inhabitants of warm countries, the tree also yielding other valuable products : **Musa**, n., *mūz'ă*, a genus of plants whose species produce, such as the 'Banana' and 'Plantain' : **Musa sapientum**, *săp'ĭ·ĕnt'ŭm* (L. *săpĭĕns*, tasting, *săpĭĕntum*, of good tastes, of the wise); the Musa-trees of the wise ; also **M. cavendishii**, *kăv'ĕn·dĭsh'ĭ·ī* (proper name of *Cavendish*), are species which furnish different kinds of Banana : **M. paradisiaca**, *păr'ă·dĭs·ĭ'·ăk·ă* (L. *părădĭsĭăcus*, of or belonging to

Paradise—from L. *Părădīsus*, Gr. *Paradeisos*, a park, Paradise), a species which produces the Plantain : **M. textilis**, *tĕks'·tĭl·ĭs* (L. *textĭlĭs*, woven, wrought), yields a kind of fibre, used in India in the manufacture of fine muslins, and produces Manilla hemp ; the juice of the fruit, and the lymph of the stem of the Musa, are slightly astringent and diaphoretic : **M. ensete**, *ĕns'·ĕt·ĕ* (unascertained), an Abyssinian species whose succulent interior is eaten, but the fruit is dry and full of seeds.

Musca, n., *mŭsk'·ă* (L. *musca*, a fly), a Linnæan genus of Dipterous insects : **Musca domestica**, *dŏm·ĕst'·ĭk·ă* (L. *domestĭcus*, belonging to the house—from *domus*, a house), the common house-fly : **M. vomitoria**, *vŏm'·ĭt·ōr'·ĭ·ă* (L. *vomĭtōrĭus*, that provokes vomiting—from *vomo*, I vomit), the large blow-fly : **M. volitans**, singular, *vŏl'·ĭt·ănz*, **Muscæ volitantes**, plural, *mŭs'·sē vŏl'·ĭt·ănt'·ēz*, a diseased condition, variously occasioned, in which there is an appearance of spots floating before the eyes with varying rapidity and in various directions, as if they were flies.

muscardine, n., *mŭsk'·ărd·ĭn* (F.), a disease affecting silkworms and very destructive to them, caused by the fungus 'Botrytis Bassiana' —so named from the fancied resemblance of the dead caterpillar to a little cake, or a kind of pastille.

Musci, n. plu., *mŭs'·sī* (L. *muscus*, moss), the Moss family, also called 'Bryaceæ,' an Order of plants, found in all regions, and are either terrestrial or aquatic : **muscicolous**, a., *mŭs·ĭk'·ŏl·ŭs* (L. *colo*, I inhabit), growing on mosses : **muscoid**, a., *mŭsk'·ŏyd* (Gr. *eidos*, resemblance), resembling or belonging to moss : **muscology**, n., *mŭsk·ŏl'·ŏ·jĭ* (Gr.

logos, a discourse), the study of mosses, or a treatise on them.

muscles, n., *mŭs'-lz* (L. *musculus,* a little mouse, a muscle of the body—from Gr. *mus*, used in same sense), the organs of motion in the body forming what is termed the flesh, made up of bundles of fibres, by the contraction of which, under the influence of the will, the individual is able to perform various movements ; the middle part of a muscle is termed its belly, and its extremities its origin and insertion : **muscular,** a., *mŭsk'-ŭl-ăr*, full of muscles ; performed by or dependent on muscles : **muscular action,** the motion of muscle either by contraction, or cessation of contraction, by which a part is moved, as a limb : **involuntary muscles,** those which contract and cease to contract independently of the will, as in the heart : **voluntary muscles,** those which move only in obedience to the will, as in any movement of a limb : **muscular tissue,** the fibrous or thready substance that makes up a muscle : **musculi papillares,** plurals, *mŭsk'-ŭl-ī păp'-ĭl-lār'-ēz* (L. *musculi,* muscles ; *păpillāris,* belonging to the nipple—from *păpilla,* the nipple), a few bundles of muscular bands directed from the apex towards the base of the ventricle ; **musculi pectinati,** plurals,*pĕct'-ĭn-āt'-ī* (L. *pectinatus,* combed or carded—from *pecten,* a comb), the muscular fasciculi, forming closely set reticulated bands in the appendix auriculæ of the heart, presenting an appearance like the parallel arrangement of the teeth of a comb.

musk, n., *mŭsk* (Gr. *moschos*, Ar. *mesk*, musk), a strong-scented substance obtained from the musk - deer inhabiting Central Asia, contained in a bag situated

on the belly of the male, stimulant and antispasmodic.

mustard, n., *mŭst'-ėrd* (Venetian *mostarda,* a sauce ; F. *moutarde,* mustard), a common condiment, the black is obtained from the seeds of 'Sinapis nigra,' while 'Sinapis alba' furnish white mustard, Ord. Cruciferæ, both yield an oil, used as a rubefaciant or counter-irritant.

muticous, a., *mŭt'-ĭk-ŭs* (L. *mutĭcus,* curtailed, docked, for *mutĭlus*), in *bot.*, without any pointed process or awn.

mycelium, n., *mī-sēl'-ĭ-ŭm* (Gr. *mukēs,* a mushroom), the cellular spawn of Fungi; the rudimentary filaments from which fungi are developed.

Mycina, n., *mīs-ĭn'-ă* (Gr. *mukēs,* a mushroom, a mould), in *bot.*, a variety of Lichen shield.

Mycoderma, n., *mīk'-ō-dĕrm'-ă* (Gr. *mukēs,* a mushroom ; *derma,* skin), a genus of Fungi, peculiar species of which are developed in vinegar, yeast, and flour ; vegetable parasites which constitute the crust of Favus.

mycology, n., *mīk-ŏl'-ŏ-jĭ* (Gr. *mukēs,* a mushroom ; *logos,* speech), the study of Fungi, or a treatise on them : **mycol'ogist,** n., *-ō-jĭst,* one devoted to the study of the Fungi.

mycrocysts, or **microcysts,** n. plu., *mīk'-rō-sĭsts* (Gr. *mikros,* small ; *kustis,* a bag), in *bot.,* swarm spores transformed from a plasmodium into cells with a cell-wall.

myelitis, n., *mī'-ĕl-ĭt'-ĭs* (Gr. *muĕl-os,* marrow), inflammation of the substance of the spinal cord, or of its membrane : **myeloid,** a., *mī'-ĕl-ŏyd* (Gr. *eidos,* resemblance), resembling marrow : **myeloid tumour,** a tumour consisting chiefly of peculiar many-nucleated cells, like the marrow of bones.

myelon, n., *mī'-ĕl-ŏn* (Gr. *muelos,* marrow), the spinal cord of vertebrates : **myelonal,** a., *mī-ĕl'-*

ŏn·ăl, of or pert. to the spinal marrow.

myentericus, n., mī'·ĕn·tĕr'·ĭk·ŭs (Gr. mus, a muscle; ĕntĕron, an intestine), in anat., a name applied to a peculiar nervous plexus, rich in ganglionic cells, situated between the circular and longitudinal muscular fibres of the intestine.

myitis, n., mī·ī·tʹ·ĭs (Gr. mus, a muscle), inflammation of a muscle: myocarditis, n., mī'·ō·kârd·ĭtʹ·ĭs (Gr. kardĭa, the heart), inflammation of the muscular substance of the heart: myodynia, n., mī'·ō·dĭn'·ĭ·ă (Gr. odŭnē, pain), pain in the muscles; cramp; also termed 'myosalgia.'

mylitta, n., mĭl·ĭtʹ·tă (Gr. mulē, a mill; mulĭtai, the grinders of the teeth), a genus of Fungi: mylitta Australis, āws·trălʹ·ĭs (of or from Australia), a species of Fungi, known in Australia as native bread.

mylo, prefix, mĭlʹ·ō (Gr. mulē, a mill; mulai, grinders), denoting connection with the muscles near the grinders: mylo-hyoid, a. (see 'hyoid'), a triangular muscle arising from the inside of the lower jaw between the molar teeth and the chin, which raises the hyoid-bone or depresses the jaw; denoting a branch of the dental artery which ramifies on the under surface of the mylo-hyoid muscle.

myography, n., mī·ŏgʹ·răf·ĭ (Gr. mus, a muscle of the body; grapho, I write), an anatomical description of the muscles.

myoid, a., mī'·ōўd (Gr. mus, a muscle; eidos, resemblance), composed of fibre cells or muscular fibres: myoides, n. plu., mī·ōўdʹ·ēz, a thin sheet of muscular fibres on the neck — see 'platysma myoides.'

myolemma, n., mī'·ō·lĕmʹ·mă (Gr. mus or muŏn, a muscle; lemma, a husk or rind), in anat., a tubular sheath enclosing muscular fibre, consisting of transparent and apparently homogeneous membrane; sarcolemma.

myolin, n., mī'·ōl·ĭn (Gr. muŏn, muscular part), the fluid contents of the cells of which an ultimate muscular fibre is composed.

myology, n., mī·ŏlʹ·ō·jĭ (Gr. mus, a muscle; logos, discourse), the doctrine of the muscles of the body; myography.

myoma, n., mī·ōmʹ·ă (Gr. muŏn, a muscle of the body, muŏnos, of a muscle), a fibrous tumour consisting of smooth muscular fibre.

myopia, n., mī·ōpʹ·ĭ·ă (Gr. muŏ, I shut or close; ops, the eye), short or near - sightedness: myope, n., mī'·ōp, or myops, n., mī'·ōps, a near-sighted person.

Myoporaceæ, n. plu., mī'·ō·pŏr·ā'·sĕ·ē (Gr. muo, I shut; poros, a pore), a Sub-order of plants, Ord. Verbenaceæ: Myoporum, n., mī·ŏpʹ·ŏr·ŭm, a genus of pretty shrubs: Myoporum platycarpum, plăt'·ĭ·kârp'·ŭm (Gr. plătus, broad, karpos, fruit), a species of Australia, which exudes a saccharine matter from its stem.

myosalgia, n., mī'·ŏs·ălj'·ĭ·ă (Gr. muŏn, a muscle of the body; algos, grief, pain), muscular pain; cramp.

myosin, n., mī'·ōs·ĭn (Gr. mus, a muscle), an albuminoid body extracted from muscular fibre.

myositis, n., mī'·ŏs·ītʹ·ĭs (Gr. muŏn, a muscle of the body), inflammation of the muscles, same as 'myitis'; 'myosalgia,' which see.

Myosotis, n., mī'·ŏs·ōtʹ·ĭs (Gr. mus, a mouse, muŏs, of a mouse; ous, an ear, ōtos, of an ear), a very beautiful genus of flowering plants—so named from a fancied resemblance in the leaves, Ord. Boraginaceæ: Myosotis palustris, păl·ŭs'·trĭs (L. pălŭstris, marshy — from pălus, a marsh), the forget-me-not.

S

myotome, n., *mī'·ŏ·tōm* (Gr. *mus*, a muscle ; *tomē*, a cutting), in *anat.*, the muscular section or segment of the skeleton : myotomy, n., *mī·ŏt'·ŏm·ĭ*, the anatomy of the muscles; the operation of dividing the muscles.

myriapoda, n. plu., *mĭr'·ĭ·ăp'·ŏd·ă*, also myriopoda, n. plu., *mĭr'·ĭ·ŏp'·ŏd·ă* (Gr. *murios*, ten thousand; *podes*, feet), a class of Arthropoda, comprising the centipedes, which have numerous feet.

Myricaceæ, n. plu., *mĭr'·ĭ·kā'·sĕ·ē* (Gr. *mūrīkē*, the shrub tamarisk), the Gale family, an Order of plants : Myrica, n., *mĭr·ĭk'·ă*, a genus of plants, which are aromatic, and yield resinous and oily matter : Myrica cerifera, *sĕr·ĭf'·ĕr·ă* (L. *cera*, wax ; *fero*, I bear), a species whose fruit called waxmyrtle, bay-myrtle, or candleberry, yields a greenish-coloured wax, used for candles : M. gale, *gāl'·ĕ* or *gāl* (old Eng. *gale*, Soot. *gaul*, Dutch or wild myrtle),Scotch or bog-myrtle, common in marshy grounds and damp heaths in Britain : M. sapida, *săp'·ĭd·ă* (L. *sapĭdus*, tasting, savouring—from *saplo*, I taste), a native of Nepaul and China, whose drupacious fruit is eaten.

Myristicaceæ, n. plu., *mĭr·ĭst'·ĭ·kā'·sĕ·ē* (Gr. *murismos*, a besprinkling with perfumes—from *murĭzō*, I perfume), the Nutmeg family, an Order of plants characterised by their acridity and aromatic fragrance : Myristica, n., *mĭr·ĭst'·ĭk·ă*, a very interesting genus of plants : Myristica officinalis, *ŏf·fĭs'·ĭn·āl'·ĭs* (L. *officinālis*, officinal), also named M. moschata, *mŏsk·āt'·ă* (mid. L. *moschātus*, having a smell like musk—from Ar. *mosch*, musk), M. fragrans, *frāg'·rănz* (L. *frāgrans*, emitting a smell), or M. aromatica, *ăr'·ŏm·ăt'·ĭk·ă* (L. *arŏmătĭcus*, fragrant), the most important species, attaining 30

feet in height, producing a drupacious fruit, the hard kernel of which is the well-known nutmeg; *nutmegs* yield a concrete oil: the mace, an arollode or additional covering of the seed, yields a fatty matter and a volatile oil—both are used as aromatic stimulants and condiments : M. tomentosa, *tŏm'·ĕnt·ōz'·ă* (L. *tomentum*, a stuffing for cushions), the kernels of whose fruit are used as aromatics under the name of wild-nutmegs.

Myrobalans, n., *mĭr·ŏb'·ăl·ănz* (L. *myrobălănum*, the fruit of a species of palm—from Gr. *muron*, perfume ; *balănos*, an acorn), the fruit of Terminalia Belerica, used as an astringent, and in dyeing, and the manufacture of ink.

myronic, a., *mĭr·ŏn'·ĭk* (Gr. *murŏn*, any odorous juice flowing from a plant—from *muro*, I drop or flow), denoting an acid, one of the components of black mustard seed, existing in the seed as a potassium salt : myrosin, n., *mĭr'·ŏs·ĭn*, an albuminous ferment, likewise contained in the seeds.

Myrospermum, n., *mĭr'·ŏ·spĕrm'·ŭm* (Gr. *muron*, any odorous juice from a plant—from *muro*, I drop or flow ; *sperma*, seed), a genus of plants, Ord. Leguminosæ, Sub-ord. Papilionaceæ, whose seeds and cells yield a strong-smelled balsam : Myrospermum Pereiræ, *pĕr·ĕr'·ē* (of or from *Pereira*), the balsam of Peru : M. toluiferum, *tŏl'·ū·ĭf'·ĕr·ŭm* (*Tolu*, and L. *fero*, I bear), yields the balsam of Peru—both preceding are used as stimulant expectorants.

Myroxylon, n., *mĭr·ŏks'·ĭl·ŏn* (Gr. *muron*, any odorous juice of a plant ; *xulon*, wood), another name for the genus Monospermum.

Myrsinaceæ, n. plu., *mĕrs'·ĭn·ā'·sĕ·ē* (Gr. *mursĭnē*, the myrtle tree), the Myrsine family, an Order of plants : **Myrsine**, n., *mèrs'·in·ē*, a genus of plants : **Myrsine bifaria**, *bĭf·ār'·ĭ·ă* (L. *bĭfārĭus*, divided into two parts—from *bis*, twice ; *fāri*, to speak), a species whose berries are said to possess cathartic properties.

Myrtaceæ, n. plu., *mèr·tā'·sĕ·ē* (Gr. *murtos*, L. *myrtus*, a myrtle tree), the Myrtle family, an Order of plants, which comprise the myrtle, the pomegranate, the rose-apple, the clove, and many plants producing beautiful flowers : **Myrteæ**, n. plu., *mèr'·tĕ·ē*, a Sub-order of plants : **Myrtus**, n., *mèr'·tŭs*, a genus : **Myrtus pimenta**, *pĭm·ĕnt'·ă* (Sp. *pimienta*, pepper), also called 'Eugenia pimenta,' Pimento, Allspice, or Jamaica pepper, the berried fruit of a tree which is a native of the W. Indies and Mexico ; it contains an acrid volatile oil, and is sometimes employed as a stimulant and carminative : **M. communis**, *kŏm·mūn'·ĭs* (L. *cŏmmūnis*, common), the common myrtle whose berries are used as food.

myrtiform, a., *mèr'·tĭ·fŏrm* (L. *myrtus*, myrtle ; *forma*, shape), having the shape of myrtle leaves or berries : **myrtiform fossa**, *fŏs'·să* (L. *fossa*, a ditch), in *anat.*, a depression on the facial surface just above the incisor teeth, also called 'incisive fossa.'

myxoma, n., *mĭks·ōm'·ă*, **myxomata**, n. plu., *mĭks·ōm'·ăt·ă* (Gr. *muxa*, mucus, slimy substance), a tumour composed of mucous tissue : **myxoamœbæ**, n. plu., *mĭks'·ŏ·ăm·ē'·bē* (Gr. *amoibos*, changing), swarm spores of myxomycetes.

myxomycetes, n., *mĭks'·ŏm·ĭ·sēt'·ēz* (Gr. *muxa*, a slimy substance ; *mukēs*, a fungus), a group of Thallophytes without chlorophyll, so named as the members of the group form creeping masses of naked protoplasm, which send up spore-bearing sporangia, whose spores are formed along with, and in the interstices of, thread-like filaments of varying character (the capillitium).

myxospores, n., *mĭks'·ō·spōrz* (Gr. *muxa*, mucus ; *spora*, a seed), the spores formed in the sporangia of the myxomycetes : **myxosporous**, a., *mĭks·ŏs'·pŏr·ŭs*, having myxospores, or pert. to them.

nacre, n., *nāk'·r* (F. *nacre*, mother-of-pearl), the beautiful, iridescent substance forming the inner covering of the shell of the pearl mussel or oyster, also called 'mother-of-pearl' : **nacreous**, a., *nāk'·rĕ·ŭs*, having a pearly lustre ; of the texture of mother-of-pearl.

nævus, n., *nēv'·ŭs*, **nævi**, n. plu., *nēv'·ī* (L. *nævus*, a mole on the body), congenital flat, or slightly elevated structures, occurring principally on the skin of the face, head, or neck, and composed of a plexus of the mere superficial vessels of the skin, which ceases to grow from the moment of birth —more serious vascular tumours are sometimes included under this head ; a congenital spot or mark varying in its appearance : **nævose**, a., *nēv·ōz'*, freckled ; having congenital marks : **nævoid**, a., *nēv'·ōyd* (Gr. *eidos*, resemblance), resembling a nævus.

Naiadaceæ, n. plu., *nā'·yăd·ā'·sĕ·ē*, also called **Potameæ** (Gr. *Naĭădĕs*, the Naiads or nymphs of the rivers and streams), the Naias or Pondweed family, an Order of plants living in fresh and salt water, one of the species of which is the lace-plant or lattice-plant of Madagascar, whose rhizome is used for food, and called the 'water yam' : **Naias**, n., *nā'·yăs*, a genus of the Order.

napiform, a., *năp'·ĭ·fŏrm* (L.

napus, a turnip ; *forma*, shape), turnip-shaped.

Narcissus, n., *năr·sĭs'·sŭs* (Gr. *Narkissos*, L. *Narcissus*, a man's name, a flower—from Gr. *narkē*, torpor, so called from the effect of its perfume on the nerves), a genus of favourite garden plants, Ord. Amaryllidaceæ, whose flowers grow upon a scape and have a cup at their mouth, including such species as Daffodils, Jonquils, and Tazettas, of soft and varied colours, and sweet scent : **Narcissus pseudo-narcissus**, *sŭd'·ō* (Gr. *pseudo*, false), the Daffodil whose flowers are said to be poisonous: **N. poeticus**, *pō·ĕt'·ĭk·ŭs* (L. *poēticus*, poetical —from *poēta*, a poet) ; **N. jonquilla**, *jŏng·kwĭl'·lă* (F. *jonquille*, one of the Daffodil species) ; **N. odorus**, *ŏd·ōr'·ŭs* (L. *odōrus*, sweet - smelling — from *odor*, smell) ; **N. pseudo-narcissus**.; **N. Tazetta**, *tăz·ĕt'·tă* (*Tazetta*, Spain), the bulbs of these and some other species are emetic.

narcotic, a., *năr·kŏt'·ĭk* (Gr. *narkotikos*, having the power to benumb — from *narkē*, torpor), having the power of producing drowsiness, sleep, or stupor : **narcotics**, n. plu., *năr·kŏt'·ĭks*, substances which procure sleep ; substances which may procure sleep by relieving pain : **narcotism**, n., *năr'·kŏt·ĭzm*, such effects as giddiness, headache, dimness of sight, partial stupor, produced by narcotic substances.

Nardostachys, n., *nărd·ŏs'·tăk·ĭs* (Heb. *nard*, Gr. *nardos*, spikenard of the ancients; Gr. *stachus*, an ear, a spike), a genus of plants, Ord. Valerianaceæ : **Nardostachys jatamansi**, *jăt'·ă·măns'·ī* (unascertained), the spikenard of the ancients, highly prized on account of its perfume.

nares, n. plu., *năr'·ēz* (L. *nāris*, a nostril, *nārēs*, nostrils), the openings of the nose, anterior and posterior ; the nostrils : **septum narium**, *sĕpt'·ŭm năr'·ĭ·ŭm* (L. *septum*, a fence, a wall ; *nārĭum*, of the nostrils), the internal walls of the nostrils, chiefly formed by the central plate of the ethmoid bone and the vomer.

Narthex, n., *nărth'·ĕks* (Gr. *narthĕx*, a plant resembling fennel), a genus of plants, Ord. Umbelliferæ : **Narthex asafœtida**, *ăs'·ă·fĕt'·ĭd·ă* (L. *asa*, a gum; *fœtĭdus*, fetid), a species which yields the asafœtida, a gum resin of highly offensive odour, much used in medicine.

nasal, a., *năz'·ăl* (L. *nāsus*, the nose), of or pert. to the nose ; formed or affected by the nose : **naso-**, prefix, *năz'·ō*, denoting connection with the nose.

nascent, a., *năs'·ĕnt* (L. *nascens*, being in its birth, gen. *nascentis* —from *nascor*, I am born), beginning to exist or grow ; in the moment of formation : **nascency**, n., *năs'·ĕns·ĭ*, the beginning of production.

nassology, n., *năs·sŏl'·ŏ·jĭ* (Gr. *nassō*, I stuff ; *logos*, discourse), the art of preparing specimens of animal bodies, or the art of stuffing them.

Nasturtium, n., *năs·tĕr'·shĭ·ŭm* (L. *nasus*, the nose ; *tortum*, to twist), Tropœolum majus, Ord. Tropœolaceæ ; the common Indian cress whose unripe fruit has been preserved and used as pickles.

natant, a., *năt'·ănt* (L. *natans*, swimming, gen. *natantis*—from *nato*, I swim), swimming; floating on the surface : **natatores**, n. plu., *năt'·ăt·ōr'·ēz*, the Order of the swimming birds : **natatory**, a., *năt'·ăt·ōr'·ĭ*, formed or adapted for swimming.

nates, n. plu., *năt'·ēz* (L. *nătēs*, the rump, the buttocks), the prominent parts formed by the glutei muscles; the buttocks: **nates cerebri**, *sĕr'·ĕb·rī* (L. *cerebrum*, the brain, *cerebri*, of the brain), the anterior tubercles of the quadrigeminal bodies of the brain ; the posterior are called the ' testes.'

natural selection, that process in nature by which the strongest, swiftest, etc., outlive, and take the place of the weaker, etc. ; the preservation of favoured races in the struggle for life ; survival of the fittest.

nausea, n., *năw′-zhĕ-ă* (L. *nausea*, Gr. *nausia*, sea-sickness — from Gr. *naus*, a ship), a sensation of sickness, inclining to vomit.

nautiloid, a., *nawt′-il-oyd* (Gr. *nautilos*, a nautilus ; *eidos*, resemblance), resembling the shell of the nautilus in shape.

navel, n., *năv′-ĕl* (Ger. *navel*, Dut. *navel*, Icel. *nabli*, the navel), the round depression in the centre of the lower part of the abdomen, indicating the place of detachment of the umbilical cord after birth: **navel string**, the umbilical cord.

navicular, a., *năv-ĭk′-ūl-ăr* (L. *navicula*, a little ship — from *navis*, a ship), hollowed or shaped like a boat.

necrogenous, a., *nĕk-rŏdj′-ĕn-ŭs* (Gr. *nekros*, dead ; *gennăō*, I produce), applied to cryptogamous parasitic plants which grow upon sickly and dead plants, and accelerate the death of the former.

necrosis, n., *nĕk-rōz′-ĭs* (Gr. *nek-rōsis*, a killing, deadness—from *nekros*, dead), the mortification and death of bone, but also applied to the death of other structures, the dead portions of bone are called ' sequestra'; dry gangrene ; a disease of plants marked by small black spots, followed by decay.

Nectandra, n., *nĕk-tănd′-ră* (perhaps Gr. *nĕktŏs*, swimming ; *andros*, of a male), a genus of trees, Ord. Lauraceæ : **Nec-**

tandra Rodiæi, *rŏd′-ĭ-ē-ī* (after Dr. Rodie, a botanist), a tree of British Guiana 60 feet high, yields Bibiru or Bebeeru-bark, the wood used in shipbuilding, etc., under the name ' Green - heart' : **Nectandrine**, n., *nĕk-tănd′-rĭn*, an alkaloid obtained by Dr. Douglas Maclagan along with Bebeerine from its bark: **N. puchury**, *pŭtsh′-ĕr-ĭ* (native name), a species whose cotyledons are imported from Brazil under the name Puchrim beans or Sassafras nuts.

nectar, n., *nĕk′-tăr* (L. *nectar*, Gr. *nektar*, the drink of the gods, honey), the sweet secretions of flowers ; any abnormal part of a flower : **nectary**, n., *nĕk′-tăr-ĭ*, **nectaries**, n. plu., *nĕk′-tăr-ĭz*, those parts of a flower which secrete a honey-like matter : **nectariferous**, a., *-ĭf′-ĕr-ŭs* (L. *fero*, I bear), having or bearing honey-like secretions.

necto-calyx, n., *nĕk′-tō-kăl′-ĭks* (Gr. *nēktŏs*, swimming ; *kalux*, cup), the swimming bell or disc of a Medusa or Jelly-fish.

Nelumboneæ, n., *nĕl′-ŭm-bōn′-ĕ-ē* (*nelumbo*, the Cingalese name), the Water - beans, a Sub - order of plants, Ord. Nymphæaceæ : **Nelumbium**, n., *nĕl-ŭm′-bĭ-ŭm*, a genus of water plants : **Nelumbium speciosum**, *spē′-shĭ-ōz′-ŭm* (L. *speciōsum*, full of beauty or display—from *species*, look, view, a sort), a species whose flower is supposed to be the Lotus figured on Egyptian and Indian monuments ; the fruit is said to be the Pythagorean bean ; the sacred bean of India : **N. Leichardti**, *lĭk-ărd′-tī* (after Leichardt, the traveller), the sacred bean of N. E. Australia.

Nematelmia, n., *nĕm′-ăt-ĕl′-mĭ-ă* (Gr. *nēma*, thread ; *hĕlmins*, a worm), the Division of Scolecida, comprising the round - worms, thread-worms, etc.

nemathecium, n., *něm'·ă·thē'·shǐ·ŭm* (Gr. *nēma*, a thread ; *thēkē*, a chest, a sac), in *bot.*, a case containing threads, as in some species of Sphærococcus.

nematocysts, n. plu., *něm·ǎt'·ō·sǐsts* (Gr. *nēma*, thread ; *kustis*, a bag), in *zool.*, the thread cells of the Cœlenterata.

Nematoidea, n. plu., *něm'·ǎt·oȳd'·ě·ǎ* (Gr. *nēma*, thread ; *eidos*, resemblance), an Order of Scolecida comprising the threadworms, vinegar eels, etc. : nematoid, a., *něm'·ǎt·oȳd*, long and slender ; thread - like : nematophores, n. plu., *něm·ǎt'·ō·fōrz* (Gr. *phoreo*, I carry), in *zool.*, cæcal processes on the cœnosarc of certain of the Sertularida, containing numerous thread-cells at their extremities.

Nemeæ, n. plu., *něm'·ě·ē* (Gr. *nēma*, thread), in *bot.*, a name applied to cryptogams in allusion to their germination by a protruded thread, without cotyledons : nemean, a., *něm·ē'·ǎn*, lengthened like a thread.

Nemertida, n. plu., *něm·ěrt'·ǐd·ǎ* (Gr. *nēměrtēs*, unerring, true), a division of the Turbellarian worms, usually called ' Ribbon-Worms.'

Nepenthaceæ, n. plu., *ně'·pěnth·ā'·sě·ē* (Gr. *nēpěnthěs*, a flower, supposed name for opium—from *ne*, not ; *penthos*, grief, sadness), the Pitcher plant family, an Order of plants, having alternate leaves slightly sheathing at the base, and having a foliaceous petiole which forms an ascidium at its extremity, with the lamina in the form of a lid : Nepenthes, n., *ně-pěnth'·ēz*, a genus of the plants : Nepenthes Wardii, *wârd'·ǐ·ǐ* (after Ward, a botanist), a species found in the Seychelles at a height of 2500 feet : N. Kennedyana, *kěn·něd'·ǐ·ān'·ǎ* (after *Kennedy*, a botanist), a tropical Australian species.

Nephelium, n., *něf·ēl'·ǐ·ŭm* (Gr.

nephelion, L. *něphělǐum*, a plant, the burdock), a genus of plants, Ord. Sapindaceæ, so named from the fruit having a resemblance to the heads of a burdock : Nephelium longan, *lǒng'·gǎn* (native name), and N. litchi, *lǐtsh'·ǐ* (native name), species in China producing excellent fruit, named respectively Longan and Litchi ; the kernel of the Longan powdered is sometimes made into paper.

nephria, n., *něf'·rǐ·ǎ* (Gr. *nephros*, a kidney), Bright's disease of the kidney : nephritic, a., *něf·rǐt'·ǐk* pert. to the kidneys; affected with disease of the kidneys, or relieving the disease : n., a medicine for the cure of disease of the kidneys : nephritis, n., *něf·rǐt'·ǐs*, inflammation of the kidney : nephralgia, n., *něf·rǎlj'·ǐ·ǎ* (Gr. *algos*, pain), pain in the kidney : nephroid, a., *něf'·roȳd* (Gr. *eidos*, resemblance), resembling a kidney in form or structure.

Nerium, n., *nēr'·ǐ·ŭm* (Gr. *nēros*, humid, moist), a genus of plants, Ord. Apocynaceæ : Nerium oleander, *ōl'·ě·ǎnd'·ěr* (It. *oleandro*, a corruption of *rhododendron*), the common oleander, a poisonous plant in all its parts ; the rose laurel, or S. Sea rose : N. odorum, *ǒd·ōr'·um* (L. *odorus*, sweet smelling—from *odor*, smell), a species whose leaves, and bark of the root, are applied externally in India as powerful repellents.

nervation, n., *něrv·ā'·shǔn* (L. *nervus*, a nerve), in *bot.*, the character or disposition of the nerves of a leaf or other foliaceous appendage : nerve, n., *něrv*, in *anat.*, one of the network of grey fibrous cords which are carried from the brain as their centre to all parts of the body, forming the organs of sensation and impressions ; in *bot.*, one of the fibrous bundles of the combined vascular and cellular tissue ramifying through leaves, etc., like veins and nerves in animals : **nervine, a.,** *nẽrv'ĭn,* good for the nerves : **n.,** anything that affects the nerves : **nervures, n.** plu., *nẽrv'ūrz,* the ribs which support the membranous wings of insects.

nervus superficialis cardiacus, *nẽrv'ŭs sŭp'ẽr·fĭsh·ĭ·āl'ĭs kȧrd·ĭ'·ăk·ŭs* (L. *nervus,* a nerve ; *super-ficialis,* superficial ; *cardĭăcus,* pert. to the stomach—from Gr. *kardia,* the heart or upper orifice of the stomach), the superficial cardiac nerve, runs down the neck behind the common carotid artery : **nervus cardiacus magnus,** *măg'nŭs* (L. *magnus,* great), the great cardiac nerve, descends on the right side, behind the common carotid artery, passing either in front of or behind the subclavian artery : **nervus cardiacus minor,** *mĭn'ŏr* (L. *minor,* less), the less cardiac nerve, passes down behind the subclavian artery.

neural, a., *nūr'ăl* (Gr. *neuron,* a nerve), connected with the nervous system : **neural arch,** the arch of a vertebra which protects a part of the nervous system : **neuralgia, n.,** *nūr·ălj'ĭ·ă* (Gr. *algos,* pain, grief), pains following the tracks of nerves.

neurapophysis, n., *nūr'ă·pŏf'ĭs·ĭs* (Gr. *neuron,* a nerve ; *apophusis,* a projecting part, a sprout), the spinous process of a vertebra ; the process formed at the point of junction of the neural arches, which aids in forming the canal that protects the spinal cord.

neurectomy, n., *nūr·ĕk'tŏm·ĭ* (Gr. *neuron,* a nerve ; *ek,* out ; *tom?,* a cutting), the excision of part of a nerve.

neurilemma, n., *nūr'ĭ·lĕm'mă* (Gr. *neuron,* a nerve ; *lemma,* skin, bark), in *anat.,* the delicate fibrous sheath of a nerve, which may be easily separated in the form of a tube, from the fibres it encloses : **neurilemmatous, a.,**

nūr'ĭ·lĕm'măt·ŭs, connected with the neurilemma.

neurin, n., *nūr'ĭn* (Gr. *neuron,* a nerve), the matter which composes the nervous system : **neuritis, n.,** *nūr·ĭt'ĭs,* inflammation of a nerve : **neuro-,** *nūr'ŏ,* indicating connection with a nerve or nerves.

neuroglia, n., *nūr·ŏg'lĭ·ă* (Gr. *neuron,* a nerve ; *glia,* glue), a delicate form of connective tissue found in the eye, and in the interior of the nervous centres.

neurology, n., *nūr·ŏl'ŏ·jĭ* (Gr. *neuron,* a nerve ; *logos,* discourse), a treatise on the nerves ; the doctrine of the nerves.

neuroma, n., *nūr·ŏm'ă* (Gr. *neuron,* a nerve), a tumour developed in the sheath of a nerve ; the true *neuroma* is composed of nerve-fibres, generally resembling those of the nerve trunk.

neuropodium, n., *nūr'ŏ·pŏd'ĭ·ŭm* (Gr. *neuron,* a nerve ; *podes,* feet), the neutral or inferior division of the foot tubercle of an Annelid ; the ventral oar.

neuroptera, n. plu., *nūr·ŏp'tẽr·ă* (Gr. *neuron,* a nerve ; *pteron,* a wing), an order of insects characterised by four membranous wings with finely reticulated nervures, as in Dragon-flies.

neurosis, n., *nūr·ōz'ĭs* (Gr. *neuron,* a nerve), a disease which depends on some perverted nervous influence ; nervous affections or diseases in which sense or motion or both are impaired without any apparent local disease : **neurotic, a.,** *nūr·ŏt'ĭk,* seated in the nerves or pert. to them : **n.,** a disease of the nerves, or a medicine used for strengthening them.

neurotome, n., *nūr'ŏ·tŏm* (Gr. *neuron,* a nerve ; *tŏmē,* a cutting), the nervous section or segment of the skeleton ; a long, narrow, two-edged scalpel employed in dissecting the nerves : **neurotomy,**

n., *nūr·ŏt'·ŏm·ĭ*, dissection of the nerves.

Nicotiana, n., *nĭk·ŏsh'·ĭ·ān'·ă* (after *Nicot*, a Frenchman who first introduced the seeds into France), a genus of plants in very extensive use as a popular narcotic, Order Solanaceæ : Nicotiana tobacum,. *tō·bāk'·ŭm* (of or from *Tobago*, in the W. Indies), the species chiefly used in Europe as tobacco, which is an energetic narcotic poison : **N. repanda,** *rĕ·pănd'·ă* (L. *repandus*, bent backward', turned up), a species from whose leaves the small Havannah cigars are prepared : **N. rustica,** *rŭst'·ĭk·ă* (L. *rustĭcus*, rustic, country),. species producing E. Indian,. Latakia, and Turkish tobacco : **N. Persica,** *pĕrs'·ĭk·ă* (of or from *Persia*), produces the finest Shiraz tobacco : Nicotina, *nĭk'·ŏt·ĭn·ă*, or nicotin, n., *nĭk'·ŏt·ĭn*, the peculiar oily-like alkaloid on which the active properties of the tobacco plant depends ; 'tobacco' is used in medicine as a sedative in the form of infusion, tincture, or wine.

nictation, n., *nĭk·tā'·shŭn*, also nictitation, n., *nĭk'·tĭ·tā'·shŭn* (L. *nictātum*, to make a sign with the eyes), the act of winking.

nidulant, a., *nĭd'·ŭl·ănt* (L. *nidulus*, a little nest—from *nidus*, a nest), nestling, as a bird in its nest; in *bot.*, embedded in pulp, as in a nest : nidularia, n. plu., *nĭd'·ŭl·ār'·ĭ·ă*, a genus of Fungi, growing on rotten leaves, shavings of wood, bark, etc., having the appearance of cups, which contain egg-like seeds ; the myceliums of certain Fungi.

nidus, n., *nĭd'·ŭs* (L. *nidus*, a nest), a place where parasites, worms, or insects lodge and lay their eggs.; a hatching place for infectious diseases : nidus hirundinis, *hĭr·ŭnd'·ĭn·ĭs* (L. *hĭrundo*, a swallow, *hĭrundĭnis*, of a

swallow), the nest of the swallow; a deep fossa of the cerebellum, lying between the posterior medullary velum, and the nodulus and uvula.

Nigella, n., *nĭg·ĕl'·lă* (L. *nĭgellus*, slightly black,. dark—from *nĭger*, black); a genus of plants, Ord. Ranunculaceæ, so named from their black seed : Nigella sativa, *săt·ĭv'·ă* (L. *sativus*, sown or planted), supposed to be the fitches of Scripture ; black cummin and fennel flower, the black aromatic seeds of which are used in the East as a substitute for pepper: N. sativa and N. arvensis, *ăr·vĕns'·ĭs* (L. *arvĕnsis*, field-inhabiting), are species whose seeds are employed in adulterating pepper : N. Damascena, *dăm'·ăs·sēn'·ă* (L. *dămascēnus*, of or from *Damascus*); N. Romana, *rōm·ān'·ă* (L. *romānus*, of or from *Rome*); and N. Hispanica, *hĭs·păn'·ĭk·ă* (L. *hĭspănĭcus*, of or from *Spain*), are species familiarly known by the name of Devil-in-a-bush.

nigrescent, a., *nĭg·rĕs'·sĕnt* (L. *nigrescens*,. growing black—from *nĭger*, black), growing dark or black ;. approaching black : nigricant, a., *nĭg'·rĭk·ănt*, becoming black.

Nipa, n., *nĭp'·ă* (native name in the Molluccas), a genus of trees, Ord. Pandanaceæ, species of which yield a kind of wine from their spadices : Nipa fruticans, *frŏt'·ĭk·ănz* (L. *frŭtĭcans*, putting forth shoots — from *frŭtex*, a shrub or bush), yields a fruit called the Atap in India.

nitidous, a., *nĭt'·ĭd·ŭs* (L. *nĭtĭdus*, shining), in *bot.*, having a smooth and polished surface ; glossy.

Nitraria, n., *nĭt·rār'·ĭ·ă* (L. *nitrum*, Gr. *nitron*, a mineral alkali), a genus of plants, Ord. Malpighiaceæ, curious dwarf-growing shrubs, with fleshy leaves, natives of Central Asia and Northern Africa, so named as first found near the nitre works of Siberia : **Nitraria tridentata**, *trĭd'·ĕnt·āt'·ă* (L. *trĭd-ens*, having three teeth or tines), a species found in the desert of Soussa near Tunis, is said to be the true Lotus-tree of the anc. Lotophagi.

nitre, n., *nĭt'·ér* (L. *nītrum*, Gr. *nitron*, F. *nitre*, a mineral alkali), saltpetre or nitrate of potash, a crystalline substance having the appearance of salt, used extensively in the manufacture of gunpowder : **nitrate**, n., *nĭt'·rāt*, a salt formed by the union of nitric acid with a base, as nitrate of soda : **nitrite**, n., *nĭt'·rīt*, a salt of nitrous acid with a base : **nitric acid**, a powerful acid, in its anhydrous state consisting of five parts of oxygen and one of nitrogen : **nitrous ether**, a spirit, called sweet spirit of nitre, whose basis is alcohol : **nitrous oxide**, a transparent, colourless gas, with a faint, sweetish smell and taste ; called also 'protoxide of nitrogen,' and popularly 'laughing gas,' used as an anæsthetic for minor operations.

nitrogen, n., *nĭt'·rō·jĕn* (Gr. *nitron*, nitre ; *gennăō*, I produce), an elementary gas which forms the base of nitric acid, and composes four-fifths by bulk of our atmosphere : **nitrogenous**, a., *nĭt·rŏdj'·ĕn·ŭs*, pert. to or containing nitrogen.

node, n., *nōd* (L. *nōdus*, a knot or nob ; *nōdōsus*, full of knots), a knot ; one of the two points where the orbit of a planet intersects the ecliptic ; in *bot.*, the part of the stem of a plant out of which the leaves grow ; an articulation or joining in a stem —the intervals between 'nodes' are called 'internodes'; in *surg.*, a tumour in connection with bone and its periosteum : **nodose**, a., *nōd·ōz'*, knotty ;

having knots or swollen joints : **nodosity**, n., *nŏd·ŏs'·ĭ·tĭ*, a knot of wood in the bark of certain trees formed of concentric layers ; in *surg.*, a calcareous concretion found in joints in gout, etc.

nodule, n., *nŏd'·ūl* (L. *nōdŭlus*, a little knot), in *anat.*, the anterior pointed termination of the inferior vermiform process which projects into the fourth ventricle of the cerebellum ; in *bot.*, any knot-like body ; in *geol.*, any irregular concretion of rockmatter collected around some central nucleus : **nodulose**, a., *nŏd'·ūl·ōz'*, in *bot.*, applied to roots having thickened knots at intervals.

noli-me-tangere, *nōl'·ĭ-mē-tănj'·ĕr·ĕ* (L. *noli*, do not wish ; *me*, me ; *tangere*, to touch), do not touch me ; the deeply-ulcerating lupus.

noma, n., *nōm'·ă* (Gr. *nŏmē*, corrosion — from *nemō*, I corrode), a gangrenous form of stomatitis ; may also affect the labia pudenda, resembling gangrene of the mouth : **nomæ**, n. plu., *nōm'·ē*, eating, corroding, or cancerous sores.

non compos mentis, *nŏn kŏmp'·ŏs mĕnt'·ĭs* (L. *non*, not; *compos*, able, possessed of ; *mens*, mind, *mentis*, of mind), not sound of mind ; not in his right senses ; incapable of conducting their own affairs owing to a morbid condition of intellect.

norma verticalis, *nŏrm'·ă vĕrt'·ĭk·āl'·ĭs* (L. *norma*, a rule ; *verticalis*, vertical — from *vertex*, the top or crown of the head), the examination of skulls by looking perpendicularly down upon them as a method of comparing skulls of different races.

normal, a., *nŏrm'·ăl* (L. *normālis*, according to rule—from *norma*, a rule), conforming to the usual standard ; adhering to the ordinary structure.

nosography, n., *nŏz·ŏg'·răf·ĭ* (Gr. *nosos*, disease ; *grapho*, I write), the scientific description of diseases.

nosology, n., *nŏz·ŏl'·ō·jĭ* (Gr. *nosos*, disease ; *logos*, discourse), the systematic arrangement of diseases ; in *bot.*, the study of diseases peculiar to plants ; the branch of medicine which treats of the systematic arrangement of diseases : **nosological**, a., *nŏz·ō·lŏdj'·ĭk·ăl*, pert. to : **nosologist**, n., *nŏz·ŏl'·ō·jĭst*, one skilled in the systematic arrangement of diseases.

nosophyta, n., *nŏz·ŏf'·ĭt·ă* (Gr. *nosos*, disease ; *phuton*, a plant), a disease caused by the growth or development of such parasitic plants as Fungi, in an animal tissue.

nostalgia, n., *nŏst·ălj'·ĭ·ă* (Gr. *nostos*, return, especially home ; *algos*, pain), a violent desire to return to one's native country ; home-sickness.

Nostochineæ, n. plu., *nŏs'·tō·kĭn'·ĕ·ē* (perhaps Gr. *nostos*, return, especially home; *ochĕō*, I bear or carry), a tribe or Sub-order of Algæ, composed of moving filaments immersed in a gelatinous matter: **Nostoc**, n., *nŏs'·tŏk*, a genus of Algæ forming a kind of mould ; one of the gelatinous, puckered, olive-coloured masses often found strewed on gravel and short grass after a few hours' rain.

Notobranchiata, n. plu., *nŏt'·ō·brăngk·ĭ·āt'·ă* (Gr. *nōtos*, the back; *brangchia*, gills), a division of the Annelida, so named from carrying their gills upon the back.

notochord, n., *nŏt'·ō·kŏrd* (Gr. *nōtos*, the back; *chordē*, a chord), in *zool.*, a delicate fibrous band or rod developed in the embryo of vertebrates immediately beneath the spinal cord, usually replaced in the adult by the vertebral column ; the 'chorda dorsalis.'

notoglossus, n., *nŏt'·ō·g[...]* (Gr. *nōtos*, the back ; *g[...]* a tongue), a muscle of tongue consisting mainl[...] longitudinal fibres, lying o[...] upper surface of the to[...] immediately beneath the m[...] membrane ; the 'lingualis s[...] ficialis.'

Notopodium, n., *nŏt'·ō·pŏd[...]* (Gr. *nōtos*, the back ; *podes,*[...] in *zool.*, the dorsal divisi[...] one of the foot-tubercles [...] Annelid ; the dorsal oar.

Notorhizeæ, n. plu., *nŏt'·ō·[...]* (Gr. *nōtos*, back ; *rhiza*, a [...] a Sub-order of the Crucife[...] named because in the plant[...] cotyledons are incumbent,[...] the radical dorsal, that is, a[...] to their back as in Shep[...] purses ; radicles on the ba[...] the cotyledons : **notorhiza[...]** *nŏt'·ō·rīz'·ăl*, having the ra[...] in the embryonic plant o[...] back of the cotyledons.

nucamentaceæ, n. plu., *n[...] mĕnt·ā'·sĕ·ē* (L. *nŭcāmenta*, t[...] which hang down from tr[...] the shape of nuts, fir-cones,[...] in *bot.*, one of the divisio[...] the Cruciferæ founded o[...] seed-vessels : **nucamentac**[...] a., *nŭk'·ă·mĕnt·ā'·shŭs*, havi[...] dry monospermal fruit, as c[...] Cruciferæ ; producing nuts.

nuciferous, a., *nū·sĭf'·ĕr·ŭ*[...] *nux*, a nut, *nūcis*, of a nut ;[...] I bear), bearing or prod[...] nuts.

nucleus, n., *nŭk'·lĕ·ŭs*, nucl[...] plu., *nŭk'·lĕ·ī* (L. *nucleus*, a [...] nut, a kernel—from *nux*, a[...] anything round which matt[...] accumulated ; the solid cen[...] any nodule or rounded mass[...] kernel of a nut ; the sol[...] vesicular body found in [...] cells ; the body which[...] origin to new cells ; in *zool.*[...] solid rod or band-shaped[...] found in the interior of ma[...] the Protozoa, having some[...]

the functions of an ovary: **nucleate**, a., *nŭk'·lĕ·āt*, also **nucleated**, a., *nŭk'·lĕ·āt·ĕd*, having a nucleus or central particle: **nuclear**, a., *nŭk'·lĕ·ăr*, pert. to or connected with a nucleus: **nucleolus**, n., *nŭk·lĕ'·ŏl·ŭs*, **nucleoli**, n. plu., *nŭk·lĕ'·ŏl·ī* (L. dim. little kernels), the minute solid particles in the interior of the nuclei of some cells; in *zool.*, the minute spherical particle attached to the exterior of the nucleus or ovary of certain Infusoria which performs the functions of a testicle: **nucleolated**, a., *nŭk·lĕ'·ŏl·āt'·ĕd*, of or pert. to a nucleolus.

uculanium, n., *nŭk'·ŭl·ān'·ĭ·ŭm* (L. *nŭcŭla*, a small nut—from *nux*, a nut), a superior pulpy fruit, the flesh of which contains several seeds; a term applied to the fruit of the Medlar which has nucules, and also to the 'Grape': **nucule**, n., *nŭk'·ŭl*, one of the numerous seeds of a nuculanium, as one of the hard carpels in the Medlar; one of the parts in the fructification in the Characeæ.

Iudibranchiata, n. plu., *nūd'·ĭ·brăngk·ĭ·āt'·ă* (L. *nudus*, naked; Gr. *brangchia*, gills), in *zool.*, an Order of the Gasteropoda which have no shells, and in which the gills are naked: **nudibranchiate**, a., *nūd'·ĭ·brăngk'·ĭ·āt*, pert. to the Gasteropoda or molluscous animals which have no shells, and have naked gills.

ummular, a., *nŭm'·ŭl·ăr* (L. *nummus*, a piece of money), flattened out like a piece of money; in heaps like rolls of money.

Iuphar, n., *nūf'·ăr* (Ar. *nauphar*), a genus of very beautiful water plants, Ord. Nymphæaceæ: **Nuphar luteum**, *lŏt'·ē·ŭm* (L. *lūtĕus*, yellowish—from *lūtum*, a plant used in dyeing yellow), the yellow pond lily whose stems

are said to be astringent, and the seeds contain a good deal of starch.

nutation, n., *nūt·ā'·shŭn* (L. *nutātĭo*, a nodding; *nūtans*, a nodding or wagging the head), a constant and involuntary movement of the head in one or more directions; in *bot.*, the curvature in an organ of a plant, produced by the unequal growth of different sides.

nutrition, n., *nūt·rĭsh'·ŭn* (L. *nūtrĭo*, I nurse or nourish), that function or process in a living body by which matter or food, already elaborated by organic actions, is converted into their different tissues, thus repairing waste and promoting growth.

nux vomica, *nŭks vŏm'·ĭk·ă* (L. *nux*, all fruits that have a hard shell; *vomĭcus*, pert. to vomiting —from *vomo*, I vomit), the nuts or fruit of the Strychnos nux-vomica, Ord. Loganiaceæ, which contains the alkaloids strychnia and brucia, and is a violent poison; a medicinal preparation made from it and highly poisonous.

Nyctaginceæ, n. plu., *nĭk'·tădj·ĭn·ā'·sĕ·ē* (Gr. *nux*, night, *nuktos*, of night; *agō*, I do, I act), a small Order of plants containing the 'Mirabilis' or Marvel of Peru, a very showy plant whose flowers are very fragrant in the evening; nearly all the plants of the Order have purgative qualities.

Nyctanthes, n., *nĭk·tănth'·ēz* (Gr. *nux*, night, *nuktos*, of night; *anthos*, a flower), a genus of plants, Ord. Jasminaceæ: **Nyctanthes arbor-tristis**, *ărb'·ŏr·trĭst'·ĭs* (L. *arbor*, a tree; *tristis*, sad), a tree valued on account of its fragrant flowers which expand at night, and fall off at the break of day.

nymphæ, n. plu., *nĭmf'·ē* (Gr. *numphē*, L. *nympha*, a bride, a nymph), two small folds of mucous membrane, one on each side of the vagina ; the labia minora.

Nymphæaceæ, n. plu., *nǐmf'·ē·ā'·sĕ·ē* (L. *nympha*, Gr. *numphē*, a water nymph), an Order of floating plants having handsome flowers, and generally peltate leaves, some being bitter and astringent, others sedative : **Nymphæa,** n. plu., *nǐmf'·ē·d*, a genus of water-lilies : **Nymphæa alba,** *ălb'·ă* (L. *albus*, white), the common white water-lily, a species whose rhizomes are better than oak-galls for dyeing grey, and are employed in tanning leather : **N. lotus,** *lōt'·ŭs* (Gr. *lotos*, L. *lotus*, the water-lily of the Nile), the Lotus Water-Lily, supposed to be the lily of the O. T.

nympha, n. plu., *nǐmfs* (L. *nympha*, a maid), in *zool.*, the active pupæ of certain insects.

nystagmos, n., *nǐs·tăg'·mŏs* (Gr. *nustagmos*, slumbering with nodding), an involuntary oscillatory movement of the eyeballs ; a condition of indistinct vision.

obcompressed, a., *ŏb'·kŏm·prĕst'* (L. *ob*, reversed ; Eng. *compressed*), in *bot.*, flattened in front and behind, not laterally.

obcordate, a., *ŏb·kŏrd'·āt* (L. *ob*, reversed ; *cor*, the heart, *cordis*, of the heart), heart-shaped, but inverted ; inversely cordate.

obesity, n., *ōb·ĕs'·ǐ·tǐ* (L. *obēsus*, fat, plump), excessive and unhealthy fatness ; corpulence.

oblique, a., *ŏb·lēk'* (L. *obliquus*, sidewise, slanting), deviating from straight line ; not parallel or perpendicular ; in *bot.*, unequal-sided.

obliquus descendens externus abdominis, *ŏb·lik'·wǔs dē·sĕnd'·ĕns ĕks·tĕrn'·ŭs ăb·dŏm'·ĭn·ĭs* (L. *obliquus*, sideways, slanting ; *descendens*, descending or falling ; *externus*, outward ; *abdōmen*, the

belly, *abdominis*, of the belly), the oblique descending outward muscle of the abdomen, which supports and compresses the abdomen, bends the body obliquely when the ribs are fixed, and raises the pelvis obliquely : **obliquus auris,** *āwr'·ĭs* (L. *auris*, the ear, *auris*, of the ear), the oblique fibres of the ear, a small band of fibres extending from the upper and back part of the concha of the ear to the convexity immediately above it.

obovate, a., *ŏb·ōv'·āt* (L. *ob*, reversed ; *ovātus*, egg-shaped), in *bot.*, ovate, but having the narrow end downwards.

obscured, a., *ŏb·skūrd'*, also **obscurate,** a., *ŏb'·skūr·āt* (L. *obscūrus*, dark, with but little light), in *bot.*, darkened; hidden.

obsolete, a., *ŏb'·sŏl·ēt* (L. *obsolētus*, grown out of use), in *bot.*, imperfectly developed or abortive ; having any part suppressed.

obstetrics, n. plu., *ŏb·stĕt'·rĭks* (L. *obstetrix*, a midwife—from *obsto*, I stand before or in the way), the art or science of midwifery ; the art of assisting women in childbirth, and treating their diseases during pregnancy : **obstetric,** a., *ŏb·stĕt'·rĭk*, pert. to midwifery : **obstetrician,** n., *ŏb'·stĕt·rĭsh'·ăn*, an accoucheur ; a midwife.

obturator, n., *ŏb'·tūr·āt'·ŏr* (L. *obturo*, I stop or close up), in *surg.*, a plug for closing an aperture ; in *anat.*, one of two muscles named respectively ' externus' and 'internus,' which move the thigh backwards, and roll it upon its axis.

obtuse, a., *ŏb·tūs'* (L. *obtusus*, blunt), in *bot.*, having a rounded or blunt termination.

obvallate, a., *ŏb·văl'·lāt* (L. *ob*, about ; *vallātus*, surrounded with a rampart), in *bot.*, walled up, applied to certain Fungi.

bvolute, a., *ŏb'·vŏl·ŏt* (L. *obvol-ūtus*, wrapped round—from *ob*, around ; *volvo*, I roll), in *bot.*, having the margins of one leaf alternately overlapping those of the leaf opposite to it.

occipitalis major, *ŏk·sĭp'·ĭt·āl'·ĭs mādj'·ŏr* (L. *occipitālis*, pert. to the *occiput* or back part of the head ; *mājor*, greater), the internal branch of the posterior cord of the second cervical nerve : occipitalis minor, *mĭn'·ŏr* (L. *minor*, less), a superficial ascending branch of the cervical plexus : occipito-frontalis, *frŏnt·āl'·ĭs* (L. *frontālis*, pert. to the forehead—from *frons*, the forehead, *frontis*, of the forehead), in *anat.*, applied to a pair of occipital, and a pair of frontal muscles, together with a thin aponeurosis, extending over the cranium, by the contraction of which the scalp is drawn backwards and the eyebrows elevated.

occiput, n., *ŏk'·sĭp·ŭt* (L. *occiput*, the back part of the head—from *ob*, against ; *caput*, the head), the bone forming the back part of the skull ; the hinder part of the head or skull—the opposite part being named the sinciput : occipital, a., *ŏk·sĭp'·ĭt·ăl*, pert. to or connected with the back part of the skull.

ocellated, a., *ŏs·ĕl'·lāt·ĕd* (L. *ocellus*, a little eye—from *oculus*, an eye), in *bot.*, having a broad round spot of one colour, with the spot of a different colour in the centre : ocelli, n. plu., *ŏs·ĕl'·lī*, in *zool.*, the simple eyes of many echinoderms, such as Spiders, Crustaceans, and Molluscs.

Ochnaceæ, n. plu., *ŏk·nā'·sĕ·ē* (Gr. *ochnē*, a wild pear), the Ochna family, an Order of plants, generally bitter, some used as tonics: Ochna, n., *ŏk'·nă*, a genus of very ornamental plants, remarkable for the large succulent

prolongation of the receptacle to which the carpels are attached.

ochrea, n., also ocrea, n., *ŏk'·rĕ·ă* (L. *ocrea*, a covering to protect the legs, a boot), in *bot.*, a tubular membranous stipule through which the stem seems to pass, as in Polygonaceæ : ochreate, a., *ŏk'·rē·āt*, sheathed after the manner of a boot.

ochroleucous, a., *ŏk'·rō·lŏk'·ŭs* (L. *ochra*, ochre ; L. *leucos*, Gr. *leukos*, white), in *bot.*, of a pale ochre colour.

octandrous, a., *ŏk·tănd'·rŭs* (Gr. *oktō*, L. *octo*, eight ; Gr. *anēr*, a male, *andros*, of a male), in *bot.*, having eight stamens.

octogynous, a., *ŏk·tŏdj'·ĭn·ŭs* (L. *octo*, Gr. *oktō*, eight ; Gr. *gunē*, a female), having eight styles.

octopod, n., *ŏk'·tŏ·pŏd*, octopoda, n. plu., *ŏk·tŏp'·ŏd·ă* (L. *octo*, Gr. *oktō*, eight ; *pous*, a foot, *podes*, feet), the tribe of cuttle-fishes with eight arms attached to the head ; a Crustacean or insect having eight feet or legs.

octosporous, a., *ŏk'·tŏ·spŏr'·ŭs* or *ŏk·tŏs'·* (L. *octo*, Gr. *oktō*, eight ; *spora*, seed), in *bot.*, eight-spored.

oculus, n., *ŏk'·ūl·ŭs* (L. *oculus*, an eye), in *bot.*, an eye ; a leaf bud : oculist, n., *ŏk'·ūl·ĭst*, a surgeon who practises only in diseases of the eye.

Ocymum, n., *ŏs'·ĭm·ŭm* (L. *ocymum*, Gr. *ōkimon*, the plant Basil), a genus of plants, Ord. Labiatæ, the Lamiaceæ of Lindley ; the plant Basil, used as a culinary vegetable, and to flavour dishes.

odontalgia, n., *ŏd'·ŏnt·ălj'·ĭ·ă* (Gr. *odous*, a tooth, *odontos*, of a tooth ; *algos*, pain), toothache : odontalgic, a., *ŏd'·ŏnt·ălj'·ĭk*, pert. to toothache : n., a remedy for toothache : odontology, n., *ŏd'·ŏnt·ŏl'·ŏ·jĭ* (Gr. *logos*, discourse), that branch of anatomy which treats of teeth.

Odontoceti, n. plu., *ŏd·ŏnt'·ō·sēt'·ī*

(Gr. *odous,* a tooth, *odontos,* of a tooth; *kētos,* a whale), the toothed whales, as distinguished from the whalebone whales.

odontoid, a., *ŏd·ŏnt'·ōўd* (Gr. *odous,* a tooth, *odontos,* of a tooth; *eidos,* resemblance), having the appearance of teeth; tooth-like; in *anat.,* applied to a 'process,' tooth-like in shape, which forms the centrum or body of the first cervical vertebra (atlas), and springs from the second (axis).

odontophore, n., *ŏd·ŏnt'·ō·fōr* (Gr. *odous,* a tooth, *odontos,* of a tooth; *phoreo,* I bear), the tongue or masticatory apparatus of Gasteropoda and Pteropoda, etc.

œdema, n., *ēd·ēm'·ă* (Gr. *oideo,* I swell; *oidēma,* a swelling), the swelling caused by effusion of serous or inflammatory fluid into the loose areolar tissue lying under the skin or mucous membrane: **œdematoid,** a., *ēd·ēm'·ăt·ōўd* (Gr. *eidos,* resemblance), resembling œdema.

Œdogonium, n., *ēd'·ō·gōn'·ĭ·ŭm,* **Œdogonia,** n. plu., *ēd'·ō·gōn'·ĭ·ă* (Gr. *oideō,* I swell; *gonos,* offspring—alluding to the mode of reproduction), a genus of minute Algæ, in which the greater part of the cells contain each a zoospore, provided anteriorly with a complete crown of cilia, and produced without sexual intercourse, the zoospore germinating and giving rise to a new plant in the same way as a bud does: **Œdogonium ciliatum,** *sĭl'·ĭ·āt'·ŭm* (L. *cilĭātum,* having an eyelash — from *cilĭum,* an eyelash), a species found attached to the leaves of aquatic mosses.

Œnanthe, n., *ēn·ănth'·ĕ* (Gr. *oinos,* wine; *anthos,* a flower—so named from its odour), a genus of plants, Ord. Umbelliferæ: **Œnanthe crocata,** *krŏk·āt'·ă* (L. *crŏcātus,* saffron yellow — from *crŏcus,* saffron), a poisonous species

called Hemlock-dropwort or Dead-tongue: **Œ. phellandrium,** *fĕl·lănd'·rĭ·ŭm* (Gr. and L. *phelland-rĭon,* a plant with leaves like ivy), a species called Water-drop having poisonous properties: **œnanthic,** a., *ēn·ănth'·ĭk,* designating a peculiar principle which gives wine its distinguishing aroma.

Œnothera, n., *ēn'·ŏth·ēr'·ă* (Gr. *oinos,* wine; *thēra,* a hunting, a search after), a genus of plants, Ord. Onagraceæ; the Evening Primrose, a genus of truly beautiful plants: **Œnothera biennis,** *bī·ĕn'·nĭs* (L. *bĭĕnnis,* biennial—from *bis,* twice; *annus,* a year), a species having edible roots, formerly eaten after meals as an incentive to wine-drinking.

œsophagus, n., *ēs·ŏf'·ăg·ŭs* (Gr. *oisophagos,* the gullet—from *oiō,* I bear or carry for another; *phago,* I eat), the canal through which food and drink pass to the stomach; the gullet—also spelt *esophagus:* **œsophagalgia,** n., *ēs·ŏf'·ăg·ălj'·ĭ·ă* (Gr. *algos,* pain, grief), pain of the œsophagus: **œsophagitis,** n., *ēs·ŏf'·ădj·īt'·ĭs* (*itis,* inflammation), inflammation of the œsophagus: **œsophagotomy,** n., *ēs·ŏf'·ăg·ŏt'·ŏm·ĭ* (Gr. *tomē,* a cutting), the operation of making an opening into the œsophagus: **œsophageal,** a., *ēs·ŏf'·ădj·ē'·ăl,* connected with the œsophagus.

officinal, a., *ŏf·fĭs'·ĭn·ăl* (L. *offĭc-ĭnālis,* officinal, by authority—from *offĭcīna,* a workshop), sold in the shops; applied to medicines or medicinal preparations, always kept ready for use by druggists, prepared, and of the strength as directed by the College of Physicians.

Oidium, n., *ō·ĭd'·ĭ·ŭm* or *ōўd'·ĭ·ŭm* (Gr. *oideō,* I swell up), an extensive genus of Fungi forming numerous moulds, consisting of minute interlacing filaments which cover a surface as if with a white down, as on bread, cheese, preserves, and fruits: **Oidium Tuckeri**, *tŭk'·ĕr·ī* (after *Tucker*), the oidium of Tucker ; the vine mildew, a fungus which has caused much destruction among grapes both in vineyards and hothouses : **O. albicans**, *ălb'·ĭk·ănz* (L. *albĭcans*, making white), the Fungi found in the patches known as aphthæ, muguet, or thrush, in the mouths of children, and similar growths on the lips and tongues of adults in certain fevers, the Fungi consisting of cylindrical, ramified, highly refracting threads, composed of long concatenated cells.

Olacaceæ, n. plu., *ŏl'·ăk·ā'·sĕ·ē* (L. *ŏlax*, having a smell, gen. *olācis*), the Olax family, an Order of plants : **Olax**, n., *ŏl'·ăks*, a genus of ornamental climbing plants : **Olax zeylanica**, *zī·lăn'·ĭk·ă* (new L. *zeylănĭca*, of or from *Ceylon*), a species whose wood is fetid with a saline taste ; employed in putrid fevers, and the leaves are used as a salad.

Oldenlandia, n., *ŏld'·ĕn·lănd'·ĭ·ă* (after *Oldenland*, a Danish botanist), a genus of interesting under shrubs, Ord. Rubiaceæ : **Oldenlandia umbellata**, *ŭm'·bĕl·lāt'·ă* (L. *umbella*, a little shadow — from *umbra*, a shadow), a species employed in the E. Indies as a substitute for madder.

Oldfieldia, n., *ŏld·fēld'·ĭ·ă* (after *R. A. Oldfield*), a genus of trees, Ord. Euphorbiaceæ : **Oldfieldia Africana**, *ăf'·rĭk·ăn'·ă* (L. *Afrĭcānus*, of or from *Africa*), the African oak, or African teak, a hard and ponderous wood, very durable where there is a free current of air.

Oleaceæ, n. plu., *ŏl'·ē·ā'·sĕ·ē* (L. *oleācĕŭs*, like the olive tree — from L. *ŏlĕă*, Gr. *elaia*, an olive, an olive tree), the Olive family, an Order of plants which are bitter, tonic, and astringent, and some yield a fixed oil : **Olea**, n.,

ŏl'·ĕ·ă, a very important genus of plants, chiefly on account of the oil, etc., obtained from some of them : **Olea Europæa**, *ūr'·ōp·ē'·ă* (of or from *Europe*), the olive tree of the O. Testament, growing naturally on the shores of the Mediterranean; its pericarp yields by expression olive oil, consisting of two oleaginous principles, Margarin and Elain ; olive oil has nutrient, emollient, and laxative properties, and is used in forming ointments, liniments, and plasters ; the bark has been used as a tonic, as also a resinous exudation from it, called ' olivile,' or olive gum ; Spanish or Castile soap is a combination of the oil with soda, and soft soap of oil with potash : **O. fragrans**, *frăg'·rănz* (L. *frăgrans*, emitting a smell), a species whose flowers are used by the Chinese, under the name Kwei-hwa, to perfume teas : **O. rotundifolia**, *rō·tŭnd'·ĭ·fōl'·ĭ·ă* (L. *rŏtundus*, round ; *fŏl·ĭum*, a leaf), and **O. Europæa**, yield a sweet exudation called manna—not, however, the manna of the Bible—nor the manna of commerce, which is the produce of the Manna or flowering Ash.

oleaginous, a., *ŏl'·ĕ·ădj'·ĭn·ŭs* (L. *oleagĭnus*, of or pert. to an olive tree—from *olĕa*, an olive tree), oily ; unctuous ; in *bot.*, fleshy and oily.

oleaster, n., *ŏl'·ĕ·ăst'·ĕr* (L. *oleaster*, the oleaster — from *olĕa*, the olive), the wild olive—so named as much resembling the olive ; the Ord. Elæagnaceæ, called the Oleaster family, which see.

olecranon, n., *ŏl·ĕk'·răn·ŏn* (Gr. *ōlekrānon*, the point of the elbow —from *ōlenĕ*, the elbow ; *kranon*, a helmet, the head), in *anat.*, the projecting part of the upper end of the ulna, forming the back of the elbow : **olecranoid**, a., *ŏl·ĕk'·răn·ŏyd* (Gr. *eidos*, resemblance), resembling an elbow.

olefiant, a., *ŏl'.ĕ.fī'.ănt* (L. *oleum*, oil; *facio*, I make), applied to a gas obtained by heating a mixture of two measures of sulphuric acid, and one of alcohol — so named from forming an oily liquid, called Dutch liquid, when mixed with chlorine.

oleic, a., *ŏl'.ĕ.ĭk* (L. *ŏlĕum*, oil), applied to the oily acid resulting from the action of linseed, or other oil, upon potash, or during the manufacture of soap: olein, n., *ŏl'.ĕ.ĭn*, the pure liquid portion of oil and fat: oleo-resin, *ŏl'.ĕ.ō-rĕz'.ĭn*, the natural mixture of a resin and an essential oil, forming the vegetable balsams and turpentines.

oleraceous, a., *ŏl'.ĕr.ā'.shŭs* (L. *ŏlĕrācĕŭs*, resembling herbs — from *olus*, any garden herbs for food), having the nature and qualities of pot-herbs; used as an esculent pot-herb.

oleum jecoris aselli, *ŏl'.ĕ.ŭm jĕk'. ŏr.ĭs ăs.ĕl'.lĭ* (L. *ŏlĕum*, oil; *jĕcur*, the liver, *jĕcŏris*, of the liver; *ăsĕllus*, a cod, *ăsĕlli*, of a cod), the oil of the liver of the cod; cod-liver oil.

olfactory, a., *ŏl.făk'.tŏr.ĭ* (L. *olfactum*, to smell, to scent—from *ŏlĕo*, I smell; *faciŏ*, I make), having the sense of smelling; olfactory nerves, nerves emerging from the brain, one on each side of the septum of the nose, which are distributed to the mucous membrane of the nares, and convey to the brain odorous sensations.

olibanum, n., *ŏl.ĭb'.ăn.ŭm* (Ar. *ol* or *al*, the; *lubin*, frankincense; Gr. *libanos*, the frankincense tree), a gum resin of a bitterish taste, and aromatic, forming a frankincense procured from the stems of several species of the genus Boswellia, Ord. Burseraceæ, inhabiting the hot and arid regions of eastern Africa, and south of Arabia.

oligandrous, a., *ŏl'.ĭg.ănd'.rŭs* (Gr. *ŏlĭgŏs*, few; *anēr*, a male, *andros*, of a male), in *bot.*, having less than twenty stamens.

oligochæta, n. plu., *ŏl'.ĭg.ō.kēt'.ă* (Gr. *ŏlĭgŏs*, few; *chaitē*, hair), in *zool.*, an order of Annelida, having few bristles or hairs, as in the earth-worms.

oligospermous, a., *ŏl'.ĭg.ō.spĕrm'.ŭs* (Gr. *ŏlĭgŏs*, few; *sperma*, seed), in *bot.*, having few seeds.

olivaceous, a., *ŏl'.ĭv.ā'.shŭs* (L. *olīva*, an olive), of a dusky-green or olive colour; having the qualities of olives: olivary, a., *ŏl'.ĭv.ăr.ĭ*, in the encephalon, a name applied to two prominent oval masses placed to the outer side of the pyramids, and sunk to a considerable depth in the substance of the 'medulla oblongata,' —so called from their shape: olivile, n., *ŏl'.ĭv.ĭl*, a resinous exudation procured from the bark of 'Olea Europæa,' used as a tonic.

omasum, n., *ŏm.ās'.ŭm* (L. *omāsum*, bullock's tripe), the third stomach, or manyplies, of ruminant animals; the 'psalterium.'

omentum, n., *ŏm.ĕnt'.ŭm* (L. *ŏmentum*, the membrane enclosing the bowels), a membranaceous covering of the bowels immediately above the intestines, and enclosing more or less fat; the caul: omenta, n. plu., *ŏm.ĕnt'.ă*, folds of the peritoneum, which proceed from one viscus to another, are three in number; the *great omentum*, consisting of four layers of peritoneum, two of which descend from the stomach, one from its anterior, and the other from its posterior surface; the *lesser omentum*, the duplicature of the peritoneum which extends between the transverse fissure of the liver and the lesser curvature of the stomach; the *gastro-splenic omentum*, the fold of the peritoneum which connects the concave surface of the spleen to the *cul-de-sac* of the stomach.

omo-hyoid, a., *ŏm'·ō·hī'·ŏyd* (Gr. *ŏmos*, a shoulder ; and *hyoid*), a muscle which passes across the side of the neck from the scapula to the hyoid bone, consisting of two fleshy bellies, united by a central tendon.

omphalic, a., *ŏmf·ăl'·ĭk* (Gr. *omphalos*, the navel, or navel-string), pert. to the navel : **omphalocele**, n., *ŏmf·ăl'·ō·sēl* (Gr. *kēlē*, a swelling or tumour), rupture or hernia of the navel : **omphalotomy**, n., *ŏmf'·ăl·ŏt'·ŏm·ĭ* (Gr. *tomē*, a cutting), the operation of cutting the umbilical cord or navel-string.

Omphalobium, n., *ŏmf'·ă·lōb'·ĭ·ŭm* (Gr. *omphalos*, a navel ; *lobos*, a pod), a genus of pretty plants, Ord. Connaraceæ : **Omphalobium Lamberti**, *lăm·bèrt'·ī* (of *Lambert*, after Lambert, a botanist), a species said to furnish zebra-wood.

omphalode, n., *ŏmf'·ăl·ōd*, also **omphalodium**, n., *ŏmf'·ăl·ōd'·ĭ·ŭm* (Gr. *omphalos*, the navel ; *hodos*, the way ; some say *eidos*, resemblance), in *bot.*, the central part of the hilum of a seed through which nourishing vessels pass : **omphaloid**, a., *ŏmf'·ăl·ŏyd* (Gr. *eidos*, resemblance), resembling the navel.

omphalo-mesenteric, a., *ŏmf'·ăl·ō·mĕs'·ĕn·tĕr'·ĭk* (Gr. *omphalos*, the umbilicus or navel ; and *mesenteric*), applied to vessels passing from the umbilicus to the mesentery at an early stage of the fœtus, and forming the first developed vessels in the germ.

Onagraceæ, n. plu., *ŏn'·ă·grā'·sĕ·ē* (L. *ŏnăgrus*, a wild ass), the Evening Primrose family, an Order of plants, whose species are common in gardens.

Oncidium, n., *ŏn·sĭd'·ĭ·ŭm* (Gr. *ongkos*, a swelling, a tumour), an extensive genus of very hand-

some plants, Ord. Orchidaceæ— so named because the species have warts, tumours, or other excrescences at the base of the labellum : **Oncidium altissimum**, *ălt·ĭs'·ĭm·ŭm* (L. *altissimus*, very high—from *altus*, high), a species which throws up from fifteen to twenty flower spikes, producing as many as 2000 flowers of a yellow colour spotted with brown : **O. lanceanum**, *lăns'·ĕ·ān'·ŭm* (L. *lancĕānum*, having a lance—from *lancĕā*, a lance, a spear), a much prized and singular flower : **O. nubigenum**, *nūb·ĭdj'·ĕn·um* (L. *nūbĭgĕna*, cloud-born — from *nūbes*, a cloud ; *genĭtus*, born, produced), a species growing on the Andes at an elevation of 14,000 feet : **O. papilio**, *păp·ĭl'·ĭ·ō* (L. *păpĭlĭŏ*, a butterfly), bears a striking resemblance to a butterfly on the wing.

Onobrychis, n., *ŏn·ŏb'·rĭk·ĭs* (Gr. *onos*, an ass ; *brucho*, I gnaw), a genus of plants, Ord. Leguminosæ, Sub-ord. Papilionaceæ, also called 'Sainfoin,' cultivated for the feeding of cattle.

onychia, n., *ŏn·ĭk'·ĭ·ă* (Gr. *onux*, a claw, a nail ; *onŭchos*, of a nail), a disease of the nail ; a whitlow.

Onygena, n., *ŏn·ĭdj'·ĕn·ă* (Gr. *onux*, a claw, a hoof; *genos* or *gŏnos*, off-spring), a genus of Fungi, many of whose species are found in autumn on the dung, feathers, and hoofs of particular animals.

oogonium, n., *ō'·ŏg·ōn'·ĭ·ŭm*, or *ŏg·ōn'·ĭ·ŭm*, **oogonia**, n. plu., *ō'·ŏg·ōn'·ĭ·ă* (Gr. *ōŏn*, an egg ; *gonos*, offspring), in *bot.*, bodies which are reckoned as equivalent to archegonia or sporangia in Fungi, in which, after the action of the antheridia, a fertilized spore is formed, called an 'oospore'; a kind of ovarian sac containing spores, which become oospores or zoospores when set free : **oogones**, n. plu., *ō·ŏg'·ŏn·ēz*, same as 'oogonia.'

T

oophoridium, n., ō·ŏf′·ŏr·ĭd′·ĭ·ŭm, oophoridia, n. plu., ō·ŏf′·ŏr·ĭd′· ĭ·ă (Gr. ōŏn, an egg ; phorĕō, I bear), spore-cases of club mosses containing large spores (macrospores) in their interior, which macrospores or megaspores have a cellular prothallium or endothallium in their interior bearing archegonia.

oophoritis, n. plu., ō·ŏf′·ŏr·īt′·ĭs (Gr. ōŏn, an egg ; phorĕō, I bear), non-puerperal inflammation of the ovaries, which may be either follicular or parenchymatous.

oosphere, n., ō′·ŏ·sfēr (Gr. ōŏn, an egg ; Gr. sphaira, L. sphœra, a globe, a ball), in bot., a germinal cell produced in the archegonia in Fungi.

oosporangia, n. plu., ō·ŏs′·pōr·ănj′· ĭ·ă (Gr. ōŏn, an egg ; spora, seed ; anggos, a vessel), in bot., sacs or spore-cases in some Algæ.

oospore, n., ō′·ŏs·pōr, oospores, n. plu., ō′·ŏs·pōrz (Gr. ōŏn, an egg ; spora, seed), in bot., a fertilized spore in Fungi ; spores produced in an oogonium.

ootheca, n., ō′·ō·thēk′·ă, oothecæ, n. plu., ō′·ō·thēs′·ē (Gr. ōŏn, an egg ; thēkē, a case), in bot., sometimes applied to thecæ of Ferns.

operculum, n., ō·pėrk′·ūl·ŭm, opercula, n. plu., ō·pėrk′·ūl·ă (L. ŏpercŭlum, a lid or cover—from opertŏ, I cover over), in bot., a cap, lid, or cover, as in certain seed vessels ; applied to the separable parts of the thecæ of Mosses ; in zool., the horny or shelly plate, developed in certain Molluscs upon the hinder part of the foot, which close the aperture of the shell when the animal is retracted within it, as in the periwinkle ; the gill-cover, or bony flap, covering and protecting the gills in many fishes : opercular, a., ō·pėrk′·ūl·ăr, having a lid or cover ; of the nature of a lid or cover : Operculata, n. plu., ō·pėrk′·ūl·āt′·ă, a division of pulmonate Gasteropoda, in which the shell is closed by an operculum : operculate, a., ō·pėrk′·ūl· āt, in bot., opening by a lid ; having a lid or cover, as a capsule.

Ophidia, n. plu., ŏf·ĭd′·ĭ·ă, also ophidians, n. plu., ŏf·ĭd′·ĭ·ănz (Gr. ophis, a serpent ; ŏphĭdĭŏn, a small serpent), the Order of reptiles comprising the snakes : ophidian, a., ŏf·ĭd′·ĭ·ăn, of or belonging to the Serpent Order.

Ophiobatrachia, n. plu., ŏf′·ĭ·ō· băt·răk′·ĭ·ă (Gr. ophis, a serpent, ophĭdion, a small serpent ; batrăchos, a frog), applied, sometimes, to the Order of snake-like Amphidians, as the Cæciliæ : Ophiomorpha, n. plu., ŏf′·ĭ·ō· mŏrf′·ă (Gr. morphē, shape), the Order of Amphibia which includes the Cæciliæ : ophiomorphous, a., ·mŏrf′·ŭs, having the form of a serpent.

Ophiocaryon, n., ŏf′·ĭ·ō·kăr′·ĭ·ŏn (Gr. ophis, a serpent ; karuon, a nut), a genus of plants, Order Sapindaceæ : Ophiocaryon paradoxum, păr′·ă·dŏks′·ŭm (Gr. părădoxos, marvellous, strange), the snake-nut tree of Demerara—so named from the embryo resembling a coiled-up snake.

Ophioglossaceæ, n. plu., ŏf′·ĭ·ō· glŏs·sā′·sĕ·ē, also Ophioglosseæ, n. plu., ŏf′·ĭ·ō·glŏs′·sĕ·ē (Gr. ophis, a serpent ; glossa, a tongue), a Sub-order of the Filices or Ferns, distinguished by the absence of a ring to the spore-cases, and by the straight, not circinate, vernation of their fronds : Ophioglossum, n., ŏf′·ĭ·ō·glŏs′·sŭm, a genus of Ferns called Adders' tongues— so named from the resemblance of their leaves to an adder's tongue : Ophioglossum vulgatum, vŭlg·āt′·ŭm (L. vulgātŭm, made common), the fern called Adders' tongue.

Ophiuroidea, n. plu., ŏf′·ĭ·ūr·ōyd′· ĕ·ă (Gr. ophis, a snake ; oura, a tail; *eidos*, form), an Order of Echinodermata, which includes the Brittle-stars, and the Sand-stars.

ophthalmia, n., *ŏf·thăl'·mĭ·ă* (Gr. *ophthalmos*, the eye), inflammation of any part of the eye, but generally restricted to the conjunction or thin mucous membrane which covers the front of the eyeball, and lines the inner surfaces of the lids: **ophthalmic, a.,** *ŏf·thăl'·mĭk*, pert. to the eye: **ophthalmoscope, n.,** *ŏf·thăl'·mō·skōp* (Gr. *skopĕō*, I view), an instrument for examining the interior of the living eye.

opiate, n., *ōp'·ĭ·āt* (L. *opĭum*, Gr. *ŏpĭŏn*, opium, the juice of the poppy), any preparation or medicine for inducing sleep or quiet, which contains opium: **opium, n.,** *ōp'·ĭ·ŭm*, the concrete milky juice, which speedily hardens and becomes brown, procured from the nearly ripe capsules of Papaver somniferum, and its varieties, Ord. Papaveraceæ; a substance much used in medicine as a narcotic or anodyne.

opisthocœlus, a., *ŏp·ĭs'·thŏ·sēl'·ŭs* (Gr. *ŏpisthĕn*, behind; *koilos*, hollow), having the anterior trunk vertebræ concave behind, as in certain Crocodilia.

opisthotonos, n., *ŏp'·ĭs·thŏt'·ŏn·ŏs* (Gr. *ŏpisthĕ*, backwards; *teinō*, I draw or stretch), tetanic spasms by which the whole body is bent backwards.

opium, see 'opiate.'

opodeldoc, n., *ŏp'·ŏd·ĕl'·dŏk* (a word coined by Paracelsus), the soap liniment, consisting of hard soap, camphor, rosemary, spirit, and water.

Opopanax, n., *ŏp·ŏp'·ăn·ăks* (Gr. *opopanax*—from *opos*, juice; *panax*, the plant All-heal), a genus of plants, Ord. Umbelliferæ: **Opopanax chironum,**

kīr·ōn'·ŭm (after *Chīron*, the son of Saturn, one of the fathers of medicine and botany), a plant, so named from the supposed virtue of the juice to cure all diseases: **opopanax,** a gum-resin procured from it, having a peculiar and disagreeable odour.

opponens pollicis, *ŏp·pōn'·ĕnz pŏl'·lĭs·ĭs* (L. *oppōnens*, setting or placing opposite; *pollex*, the thumb, *pŏllĭcĭs*, of the thumb), the opposing muscle of the thumb; a muscle arising partly from the annular ligament of the wrist, and inserted into the thumb, which brings the thumb inwards as if to oppose the fingers: **opponens minimi digiti,** *mĭn'·ĭm·ī dĭdj'·ĭt·ī* (L. *mĭnĭmi*, of the least; *dĭgĭtus*, the finger, *dĭgĭti*, of the finger), the opposing muscle of the little finger; a muscle which moves the fifth metacarpal bone forwards and outwards, increasing thus the cavity of the palm of the hand.

optics, n. plu., *ŏpt'·ĭks* (Gr. *optĭkos*, relating to sight—from *optomai*, I see), the science which treats of everything that pertains to light or vision, and the construction of such instruments as telescopes, microscopes, etc., in which light is the chief agent: **optic, a.,** *ŏpt'·ĭk*, relating to the sight, or the laws of vision; visual: **optic nerves,** the second pair of nerves which proceed directly from the brain, one to each eye, and are the nerves of vision.

Opuntia, n., *ŏp·ŭn'·shĭ·ă* (from the *Opuntii*, the inhabitants of anc. *Opus*, a town, Locris, Greece, where found), an interesting genus of plants, Ord. Cactaceæ, commonly called 'Indian figs' or 'prickly pears': **Opuntia cochinellifera,** *kŏtsh'·ĭn·ĕl·lĭf'·ĕr·ă* (Sp. *cochinilla*, a wood-louse; L. *fero*, I bear), one of the species on which the Coccus Cacti, cochineal insect, feeds.

ora serrata, *ŏr'·ă sĕr·răt'·ă* (L. *ōră*, extremity, border; *serrātus*, saw-shaped, serrated), in *anat.*, the finely indented border which terminates the outer edge of the ciliary processes of the choroid.

orbicular, a., *ŏrb·ĭk'·ūl·ăr* (L. *orbĭcŭlus*, a small disk — from *orbis*, a circle, a ring), in *bot.*, having a rounded leaf with a petiole attached to the centre of it; made in the form of an orb; completely circular.

orbicularis oris, *ŏrb·ĭk'·ūl·ār'·ĭs ōr'·ĭs* (L. *orbĭcŭlārĭs*, circular—from *orbis*, a circle; *ōs*, a mouth, *ōrĭs*, of a mouth), the circular muscle of the mouth; in *anat.*, a sphincter muscle, elliptic in form, composed of concentric fibres, which surround the orifice of the mouth: **orbicularis palpebrarum,** *pălp'·ĕb·rār'·ŭm* (L. *palpĕbra,* an eyelid, *palpĕbrārum,* of eyelids), the sphincter of the eyelids: o. **urethræ,** *ūr·ēth'·rē* (mod. L. *urēthră,* the urethra, *urethræ,* of the urethra—from Gr. *ouron,* urine), the sphincter of the urethra.

Orbiculus, n., *ŏrb·ĭk'·ūl·ŭs* (L. *orbĭcŭlus,* a small disk—from *orbis,* a circle), in *bot.*, one of the circular bodies found in the cups of Nidularia, a genus of Fungi.

orbit, n., *ŏrb'·ĭt* (L. *orbĭta,* a track, a path—from *orbis,* a circle), one of the two cavities in the skull containing the eyeballs: **orbital,** a., *ŏrb'·ĭt·ăl,* pert. to the orbits.

Orchidaceæ, n. plu., *ŏrk'·ĭd·ā'·sĕ·ē,* or **Orchids,** n. plu., *ŏrk'·ĭdz* (L. and Gr. *orchis,* a plant with roots in the form of testicles; Gr. *orchis,* a testicle), the Orchis family, an extensive Order of plants, distinguished by the peculiar forms of their flowers, etc., the flowers often resembling various insects, birds, and reptiles; many of the species possess mucilaginous prop-

erties, some are aromatic, and others antispasmodic and tonic: **Orchis,** n., *ŏrk'·ĭs,* a genus of these plants: **Orchis mascula,** *măsk'·ūl·ă* (L. *măscŭlus,* male); O. **papilionacea,** *păp·ĭl'·ĭ·ōn·ā'·sĕ·ă* (L. *păpĭlĭŏ,* a butterfly, *păpĭlĭŏnĭs,* of a butterfly); O. **mcrio,** *mŏr'·ĭ·ō* (L. *morĭo,* a dark-brown gem); O. **militaris,** *mĭl'·ĭt·ār'·ĭs* (L. *mĭlĭtārĭs,* soldier-like—from *mĭles,* a soldier); O. **coriophora,** *kŏr'·ĭ·ŏf'·ŏr·ă* (Gr. *kŏrĭŏn,* a coriander; *phorĕō,* I bear); O. **longicruris,** *lŏnj'·ĭ·krŏr'·ĭs* (L. *longicrūris,* of the long leg—from *longus,* long, *longi,* of long; *crus,* a leg, *crūris,* of a leg), are species which yield Salep, an article of diet for convalescents; the 'O. mascula' is supposed to be the 'long purples' of Shakespeare.

orchitis, n., *ŏrk·īt'·ĭs* (Gr. *orchis,* a testicle), inflammation of the testicle.

Order, n., *ŏrd'·ĕr,* a division of animals or plants above a Genus and below a Class; a collection or assemblage of Genera; see ' Genus.'

Oreodaphne, n., *ŏr'·ĕ·ō·dăf'·nē* (Gr. *ŏrŏs,* a mountain; *daphnē,* the laurel), a genus of plants, Ord. Lauraceæ: **Oreodaphne opifera,** *ŏp·ĭf'·ĕr·ă* (L. *ŏpĭfer,* aid-bringing —from *ops,* aid, power; *fero,* I bear), a species whose inner bark yields a large quantity of volatile oil.

organ, n., *ŏrg'·ăn* (L. *organum,* Gr. *organon,* an instrument), in *anat.*, a part of the living body by which some action, operation, or function is carried on; in *bot.*, any defined subordinate part of a vegetable structure, as a cell, a fibre, a leaf, a root, etc.: **organic,** a., *ŏrg·ăn'·ĭk,* also **organical,** a., *ŏrg·ăn'·ĭk·ăl,* pert. to or produced by living organs: **inorganic,** a., pert. to bodies without specific organs; dead matter: **organism,**

n., *ŏrg'.ăn·ĭzm*, a body possessing an organic structure : **organic bodies**, such bodies as possess life and sensation : **organic disease**, a disease marked by the altered structure of an organ : **organic remains**, the remains of organised bodies found in a fossil state in the crust of the earth.

organogeny, n., *ŏrg'·ăn·ŏdj'·ĕn·ĭ*, also **organogenesis**, *ŏrg'·ăn·ō·jĕn'·ĕs·ĭs* (Gr. *organon*, an instrument; *gennāō*, I produce ; *gĕnĕsis*, origin), in *bot.*, the development of organs from their primitive condition : **organography**, n., *ŏrg'·ăn·ŏg'·răf·ĭ* (Gr. *graphō*, I write), in *bot.*, the description or study of the structure of plants.

organology, n., *ŏrg'·ăn·ŏl'·ō·jĭ* (Gr. *organon*, an instrument ; *logos*, discourse), that branch of physiology which treats of the organs of animals ; organography.

Origanum, n., *ŏr·ĭg'·ăn·ŭm* (Gr. *ŏrŏs*, a mountain ; *ganŏs*, joy, delight), the Marjorams, a genus of well-known plants, Ord. Labiatæ—so named from the habitat of the plants : **Origanum vulgare**, *vŭlg·ār'·ĕ* (L. *vŭlgārĭs*, common, vulgar), wild marjoram, which yields a stimulant, acrid oil, sold as oil of thyme : O. **majorana**, *mădj'·ŏr·ān'·ă* (perhaps a corruption of L. *amārăcum*, marjoram), the sweet or knotted marjoram : O. **onites**, *ŏn·īt'·ēz* (Gr. *ŏnītĭs*, the sweet marjoram) ; also called O. **dictamnus**, *dĭk·tăm'·nŭs* (L. *dictamnus*, the plant Dittany—so named from growing abundantly on Mount *Dicte* in Crete), the Dittany of Crete, the pot-marjoram : O. **heracleoticum**, *hĕr'·ă·klē·ŏt'·ĭk·ŭm* (L. *hĕraclēŏtĭcus*, of or pert. to *Hēraclēa*, a city of Pontus named after *Hercules*, where best produced), the winter sweet marjoram, all of which are similar in properties.

ornithodelphia, n. plu., *ŏrn'·ĭth·ō·*

dĕlf'·ĭ·ă (Gr. *ornis*, a bird, *ornĭthos*, of a bird ; *delphus*, a womb), in *zool.*, the primary division of mammals, including the Monotremata.

ornithology, n., *ŏrn'·ĭth·ŏl'·ō·jĭ* (Gr. *ornis*, a bird, *ornĭthos*, of a bird ; *logos*, discourse), that branch of natural history which treats of the form, structure, habits, and uses of birds.

Ornus, n., *ŏrn'·ŭs* (L. *ornus*, the wild mountain ash), a genus of the Flowering Ash, called also Manna Ash from yielding the substance called 'manna,' Ord. Oleaceæ : **Ornus rotundifolia**, *rŏt·ŭnd'·ĭ·fōl'·ĭ·ă* (L. *rŏtundus*, round ; *fŏlĭum*, a leaf) ; and O. **Europæa**, *ūr'·ŏp·ē'·ă* (of or from *Europe*), yield the sweet exudation called 'manna,' but not the manna of the Israelites in the wilderness.

Orobanchaceæ, n. plu., *ŏr'·ō·băngk·ā'·sĕ·ē* (Gr. *ŏrŏbŏs*, a vetch ; *angchō*, I strangle, as supposed to kill plants on which they grow), the Broom-rape family, an Order of plants, generally astringent and bitter : **Orobanche**, n., *ŏr'·ō·băngk'·ē*, the Broom-rapes, a genus of curious parasitical plants, so named from the ravages they are supposed to commit on the broom tribe : **Orobanche rapum**, *rāp'·ŭm* (L. *rāpum*, a turnip), a species parasitical upon broom and furze : O. **ramosa**, *răm·ōz'·ă* (L. *rāmōsus*, branchy, ramose), a species parasitical upon hemp : O. **rubra**, *rŏb'·ră* (L. *rŭber*, red, ruddy), parasitical upon common thyme : O. **minor**, *mĭn'·ŏr* (L. *minor*, less), parasitical upon red clover : O. **hederæ**, *hĕd'·ĕr·ē* (L. *hĕdĕra*, the plant ivy, *hederæ*, of the plant ivy), parasitical upon the ivy : O. **elatior**, *ĕl·āt'·ĭ·ŏr* (L. *ĕlātus*, productive, *ĕlātĭor*, more productive) ; and O. **arenaria**, *ăr'·ĕn·ār'·ĭ·ă* (L. *arēnŏrĭus*, belonging to sand—from *arēna*, sand), are parasitic upon different species of Compositæ, as Centaury, and Milfoil : O. **major**, *mādj'-ŏr* (L. *mājŏr*, greater), a species powerfully astringent.

orthognathous, a., *ŏr-thŏg'-năth-ŭs* (Gr. *orthos*, straight, upright ; *gnathos*, a jaw), applied to the type of skull in which the jaw is overhung by the forehead ; having a vertical jaw.

orthoploceæ, n. plu., *ŏr'-thŏp-lŏ'-sĕ-ē* (Gr. *orthos*, straight ; *plŏkē*, a plait), in *bot.*, applied to the Cruciferæ which have conduplicate cotyledons.

orthopnœa, n., *ŏr'-thŏp-nē'-ă* (Gr. *orthos*, straight; *pnĕō*, I breathe), that condition of the respiration in which the difficulty of breathing is increased by stooping, or on lying down, under which, therefore, the patient has to sit more or less erect.

Orthoptera, n. plu., *ŏr-thŏp'-tĕr-ă*, also **Orthopters**, n. plu., *ŏr-thŏp'-tĕrz*, and **Orthopterans**, n. plu., *ŏr-thŏp'-tĕr-ănz* (Gr. *orthos*, straight ; *ptĕron*, a wing, *ptĕra*, wings), an Order of insects which have their two outer wings disposed in straight folds when at rest, as the grasshopper and house-cricket : **orthopterous**, a., *ŏr-thŏp'-tĕr-ŭs*, pert. to ; folding the wings straight.

orthospermæ, n. plu., *ŏr'-thŏ-spĕrm'-ē* (Gr. *orthos*, straight ; *sperma*, seed), in *bot.*, seeds which have the albumen flat on the inner face, neither involute nor convolute.

orthostichies, n. plu., *ŏr-thŏs'-tĭk-ĭz* (Gr. *orthos*, straight ; *stĭchos*, a series, a row), in *bot.*, applied to the several vertical rows formed by the leaves in a spiral phyllotaxis.

orthotropal, a., *ŏr-thŏt'-rī'p-ăl*, also **orthotropous**, a., *ŏr-thŏt'-rŏp-ŭs* (Gr. *orthos*, straight ; *tropos*, direction—from *trepŏ*, I turn), in *bot.*, having the embryo

in a seed lying straight towards the hilum or eye, as in the bean ; having the ovule with foramen opposite to the hilum.

Oryza, n., *ŏr-īz'-ă* (Gr. *ŏruza*, rice), a genus of the cereal grains known by the common name 'rice,' Ord. Graminaceæ ; the name by which rice was known to the anc. Greeks and Romans : **Oryza sativa**, *săt-īv'-ă* (L. *sătīvus*, that is sown or planted), the common rice of commerce.

os, n., *ŏs* (L. *ŏs*, a bone, *ŏssĭs*, of a bone, *ŏssă*, bones), a common prefix in anatomical terms denoting 'a bone' : **osseous**, a., *ŏs'-sĕ-ŭs*, composed of or resembling bone : **osseous tissue**, n., the substance of which bone is composed : **os calcis**, *ŏs kăls'-ĭs* (L. *călcis*, of the heel—from *calx*, the heel), the bone of the heel : **os femoris**, *ŏs fĕm'-ŏr-ĭs* (L. *fĕmŏris*, of the thigh—from *fĕmur*, the thigh), the thigh-bone : **os humeri**, *ŏs hūm'-ĕr-ī* (L. *hŭmĕrī*, of the shoulder—from *hŭmĕrus*, the shoulder), the shoulder-bone ; the large bone of the arm extending from the shoulder to the elbow : **os ilium**, n., *ŏs ĭl'-ĭ-ŭm* (see 'ilium'), the haunch-bone, forming part of the pelvis : **os innominatum**, *ŏs ĭn-nŏm'-ĭn-āt'-ŭm* (L. *in*, not ; *nomen*, name), the unnamed bone; a bone consisting of three parts—(1) the *ilium* or haunch-bone, (2) the *ischium* or hip-bone, and (3) the *pubis* or share-bone : **os ischium**, *ŏs ĭsk'-ĭ-ŭm* (see 'ischium'), the hip-bone forming part of the pelvis : **os magnum**, *ŏs măg'-nŭm* (L. *magnus*, great), the largest bone of the carpus, occupying the centre of the wrist : **os pedis**, *ŏs pĕd'-ĭs* (L. *pĕdĭs*, of a foot—from *pēs*, a foot), the large bone of the foot : **os pubis**, n., *ŏs pŭb'-ĭs* (L. *pubis*, of the pubes—from *pubes*, the pubes), the bone of the pubes ; the share-bone forming part of the pelvis : **os sacrum**, *ŏs săk'·rŭm* (L. *sacrum*, sacred), the bone which forms the basis of the vertebral column; see ' sacrum,' and ' pubis.'

oscula, n. plu., *ŏsk'·ūl·ă* (L. *osculum*, a little mouth—from *ŏs*, the mouth, *ōrĭs*, of the mouth), in *zool.*, the large apertures by which a sponge is perforated ; the suckers with which Tæniada are provided, as in Tape-worms, and Cystic-worms.

osmose, n., *ŏs·mōz'*, also **osmosis**, n., *ŏs·mōz'·ĭs* (Gr. *ōsmos*, a pushing influence), the tendency of fluids and gases of different kinds and densities to become diffused through a separating membrane when placed in contact with it ; the action produced by this tendency : **osmotic**, a., *ŏs·mŏt'·ĭk*, pert. to or having the nature of osmose ; see ' exosmose ' and ' endosmose.'

Osmundeæ, n. plu., *ŏs·mŭnd'·ĕ·ē* (after *Osmund*, who first found or used it), the Royal or Flowering Fern tribe, a Sub-order of plants, Ord. Filices : **Osmunda**, n., *ŏs·mŭnd'·ă*, a genus of ornamental ferns : **Osmunda royalis**, *rŏy·ăl'·ĭs* (mod. L. *regālis*, royal —from L. *rex*, a king), the Osmund royal, whose roots are said to have emmenagogue virtues.

osseous, see under ' os.'

ossicula, n. plu., *ŏs·sĭk'·ūl·ă* (L. *ossicŭlŭm*, a small bone—from *ŏs*, a bone, *ŏssĭs*, of a bone), small bones ; used to designate hard structures of small size, as the calcareous plates in the integument of the Star-fishes ; ossicles : **ossicula auditus**, *ăwd·ĭt'·ŭs* (L. *audītŭs*, the sense of hearing, *audītŭs*, of the sense of hearing), the three small bones of the ear —*malleus*, the outermost, is attached to the membrana tympani ; *stapes*, the innermost, is fixed in the fenestra ovalis ;

the *incus*, the third or middle, is connected with both by articular surfaces.

ossification, n., *ŏs'·sĭ·fĭk·ā'·shŭn* (L. *ŏs*, a bone, *ŏssĭs*, of a bone ; *faciŏ*, I make), the changing of any soft part of an animal body into bone or bony-like matter : **ossified**, a., *ŏs'·sĭ·fīd*, converted into bone, or a substance like it : **ossific**, a., *ŏs·sĭf'·ĭk*, bone-forming : **ossific dyscrasia**, a morbid condition accompanied by bony outgrowths in various parts of the body : see ' dyscrasia.'

osteal, a., *ŏst'·ĕ·ăl* (Gr. *ŏstĕŏn*, a bone), belonging to bone : **osteine**, n., *ŏst'·ĕ·ĭn*, osseous substance or bony-tissue : **ostitis**, n., *ŏst·ĭt'·ĭs*, the inflammation of bone.

osteoblasts, n. plu., *ŏst·ĕ'·ō·blăsts* (Gr. *ŏstĕŏn*, a bone ; *blastos*, a bud, a sprout), the granular corpuscles or cells which cover in a dense layer the osteogenic substance, and lie in its meshes, constituting the formative element of that class of bone not developed from cartilage.

osteo - chondroma, *ŏst'·ĕ·ō·kŏn·drōm'·ă* (Gr. *ŏstĕŏn*, a bone ; *chondros*, cartilage), a name applied to cartilaginous tumours.

osteoclasts, n. plu., *ŏst·ĕ'·ō·klăsts* (Gr. *ostĕŏn*, bone ; *klastos*, broken, fractured), large compound or giant cells, formed in the absorption of bone, and believed to be essential agents in the process of such absorption.

osteo-dentine, n., *ŏst'·ĕ·ō·dĕnt'·ĭn* (Gr. *ŏstĕŏn*, a bone ; *dens*, a tooth, *dentĭs*, of a tooth), the substance forming the teeth of vertebrate animals, and covered by the enamel ; a substance intermediate in structure between dentine and bone.

osteogen, n., *ŏst·ĕ'·ō·jĕn* (Gr. *ŏstĕŏn*, a bone ; *gennăō*, I produce), the soft, transparent matter in bone which becomes changed into bony tissue : **osteogenetic,** a., *ŏst'-ĕ-ō-jĕn-ĕt'-ĭk*, denoting the soft, transparent substance in bone which becomes ossified : **osteogeny,** n., *ŏst'-ĕ-ŏdj'-ĕn-ĭ*, also **osteogenesis,** n., *ŏst'-ĕ-ō-jĕn'-ĕs-ĭs*, the formation or growth of bone.

osteoid, a., *ŏst'-ĕ-ōyd* (Gr. *ŏstĕŏn*, a bone ; *eidos*, resemblance), like or resembling bone ; denoting a class of tumours growing from bone, which themselves contain bone.

osteology, n., *ŏst'-ĕ-ŏl'-ō-jĭ* (Gr. *ŏstĕŏn*, a bone ; *logos*, discourse), that part of anatomy which treats of the skeleton or bony fabric of different animals ; comparative anatomy.

osteoma, n., *ŏst'-ĕ-ōm'-ă* (Gr. *ŏstĕŏn*, a bone), an adventitious growth, consisting of a purely bony mass, set upon a bone, forming with it an organic whole ; an exostosis.

osteomalacia, n., *ŏst'-ĕ-ō-măl-āk'-ĭ-ă* (Gr. *ostĕŏn*, bone ; *malakos*, soft), a diseased softening of the bone.

osteomyelitis, n., *ŏst'-ĕ-ō-mī'-ĕl-ĭt'-ĭs* (Gr. *ostĕŏn*, bone ; *mŭĕlos*, marrow), inflammation of the red osseous medulla, and of the pulp contained in the cancelli of spongy bone.

osteophyte, n., *ŏst-ĕ'-ō-fīt* (Gr. *ostĕŏn*, bone ; · *phŭtŏs*, planted, grown—from *phŭō*, I produce), a term denoting a great variety of bony growths which are formed, for the most part, in an inflammatory exudation ; ' exostoses ' may be regarded as outgrowths from bone, while ' osteophytes ' seem only to be produced under the influence of a bone, often resulting, *e.g.*, from ossification of the exudation derived from the adjacent hyperæmic vessels.

osteoporosis, n., *ŏst'-ĕ-ō-pŏr-ōz'-ĭs* (Gr. *ostĕŏn*, bone; *pŏrŏs*, a callosity), a diseased state of bone which presents an increase of size of the bone-cells, and a consequent diminution of density, the surface of the bone being at the same time irregular and porous : **osteoporotic,** a., *ŏst'-ĕ-ō-pŏr-ŏt'-ĭk*, of or pert. to.

ostiolum, n., *ŏst-ĭ'-ŏl-ŭm* (L. *ostĭŏlum*, a little door), in *bot.*, the orifice through which spores are discharged ; the mouth of a perithecium.

ostitis, n., *ŏst-ĭt'-ĭs* (Gr. *ostĕŏn*, bone), a form of bone inflammation, which in its second stage passes on to hardening or sclerosis, or else to suppuration : **osteoid,** a., *ŏst'-ĕ-ōyd* (Gr. *eidos*, resemblance), having the appearance of bone.

ostium abdominale, *ŏst'-ĭ-ŭm ăb-dŏm'-ĭn-āl'-ĕ* (L. *ostĭum*, a door, an opening — from *ōs*, a mouth ; *abdōmĭnālis*, pert. to the abdomen), the orifice at the fimbriated extremity of the Fallopian tube : **ostium uterinum,** *ūt'-ĕr-ĭn'-ŭm* (L. *ŭtĕrĭnus*, uterine), the orifice at the uterine extremity of the Fallopian tube.

Ostracea, n. plu., *ŏst-rā'-shĭ-ă*, also **Ostraceans,** n. plu., *ŏst-rā'-shĭ-ănz* (L. *ostrĕă*, an oyster), the family of Bivalves of which the oyster or ' ostrea ' is the type.

Ostracoda, n. plu., *ŏst-răk'-ōd-ă* (Gr. *ŏstrăkŏn*, a shell), an Order of small Crustaceans enclosed in bivalve shells : **ostracoid,** a., *ŏst'-răk-ōyd* (Gr. *eidos*, resemblance), having the nature of shell.

otic, a., *ŏt'-ĭk* (Gr. *ous*, the ear, *ōtos*, of the ear), pert. to the ear ; employed in diseases of the ear : **otitis,** n., *ŏt-ĭt'-ĭs*, inflammation of the ear, the position of which is indicated by the adjectives *externa, media,* and *interna* : **otoconia,** n., *ŏt'-ŏ-kōn'-ĭ-ă* (Gr. *kŏnĭă*, dust), a small mass of calcareous particles or crystals of carbonate of lime, found in the membranous labyrinth of the ear : **otoliths**, n., *ŏt′·ō·lĭths* (Gr. *lithos*, a stone), the calcareous bodies connected with the sense of hearing, even in its most rudimentary form ; otoconia : otorrhœa, n., *ŏt′·ŏr·rē′·ă* (Gr. *rhēō*, I flow), a purulent discharge from the ears : **otalgia**, n., *ŏt·ălj′·ĭ·ă* (Gr. *algos*, pain), pain of the ear; ear-ache.

Ouvirandra, n. plu., *ŏv′·ĭr·ănd′·ră* (Polynesian *ouvi*, a yam, and *rano*, water), a most singular genus of aquatic plants, Ord. Naiadaceæ, whose leaves are without parenchyma, and consist of merely open network : **Ouvirandra fenestralis**, *fĕn′·ĕst·rāl′·ĭs* (L. *fĕnestrālis*, pert. to a window or opening — from *fĕnĕstra*, a window), has peculiar skeleton-like leaves, and is the lace plant or lattice plant of Madagascar, whose rhizome is used as food under the name of water-yam.

ovary, n., *ōv′·ăr·ĭ* (Sp. *ovario*, F. *ovaire*, an ovary ; L. *ŏvārĭum*, an ovary—from *ovum*, an egg), the part in the body of a female animal in which the eggs or first germs of future animals are lodged ; a hollow case in plants which encloses the young seeds : **ovarian**, a., *ōv·ār′·ĭ·ăn*, of or relating to the ovary : **ovarian vesicle** or **capsules**, the generative buds of the Sertularida : **ovariotomy**, n., *ōv·ār′·ĭ·ŏt′·ŏm·ĭ* (Gr. *tomē*, a cutting), the operation for removing the ovary : **ovaritis**, n., *ōv′·ăr·ĭt′·ĭs*, inflammation of the ovaries: **ovaralgia**, n., *ōv′·ăr·ălj′·ĭ·ă* (Gr. *algos*, pain), pain in the ovaries.

ovate, a., *ōv′·āt* (L. *ŏvātus*, shaped like an egg—from *ovum*, an egg), in *bot.*, having the shape of an egg, as in an egg-shaped leaf whose broader end is next the petiole or axis ; elliptical, being broadest at the base : **ovate-lanceolate**, a., a lanceolate leaf, somewhat ovate.

ovenchyma, n., *ōv·ĕng′·kĭm·ă* (L. *ovum*, Gr. *ōŏn*, an egg ; Gr. *engchuma*, an infusion, an injection—from Gr. *en*, in ; *cheuma*, anything poured out, tissue), in *bot.*, the tissue of plants composed of oval cells.

ovicapsule, n., *ōv′·ĭ·kăps′·ūl* (L. *ovum*, an egg ; *capsula*, a chest), the internal tunic of a developed Graafian vesicle of the ovary.

oviduct, n., *ōv′·ĭ·dŭkt* (L. *ovum*, an egg ; *ductus*, led), the duct or tube by which the semen is led to the ova ; the passage for the eggs in animals : **oviferous**, a., *ōv·ĭf′·ĕr·ŭs* (L. *fero*, I bear), or **ovigerous**, a., *ōv·ĭdj′·ĕr·ŭs* (L. *gero*, I bear), egg-bearing, applied to such animals as spiders, which carry about with them their eggs after exclusion from the body of the parent : **oviform**, a., *ōv′·ĭ·fŏrm* (L. *forma*, shape), egg-shaped : **oviparous**, a., *ōv·ĭp′·ăr·ŭs* (L. *parĭŏ*, I produce), producing by eggs, which are hatched after exclusion from the body of the parent.

oviposit, v., *ōv′·ĭ·pŏz′·ĭt* (L. *ŏvum*, an egg ; *posĭtum*, to place), to lay eggs : **oviposition**, n., *ōv′·ĭ·pŏz·ĭsh′·ŭn*, the laying or depositing of eggs : **ovipositor**, n., *ōv′·ĭ·pŏz′·ĭt·ŏr*, the organ possessed by some insects by whose means the eggs are placed in a position favourable for their development.

ovisac, n., *ōv′·ĭ·săk* (L. *ŏvum*, an egg ; *saccus*, a bag), the egg-bag or membrane which connects in one mass the eggs, spawn, or roe of crustaceans and many insects; the cavity in the ovary containing the ovum.

ovoid, a., *ōv′·ōyd* (L. *ŏvum*, an egg ; Gr. *eidos*, resemblance), having an egg-shape : n., a solid having an ovate figure.

ovoviviparous, a., *ōv′·ō·vĭv·ĭp′·ăr·ŭs* (L. *ŏvum*, an egg ; *vivus*, alive ; *parĭŏ*, I produce), applied to certain animals which retain their eggs within their bodies until they are hatched.

ovule, n., *ōv′·ŭl*, also **ovulum**, n., *ōv′·ŭl·ŭm*, **ovula**, n. plu., *ōv′·ŭl·ă* (dim. of L. *ōvum*, an egg), in *bot.*, the young seed contained in the ovary; the body borne by the placenta of a plant, which gradually changes into a seed; **ovula**, in *anat.*, certain serous vesicles found in the structure of the ovarium.

ovum, n., *ōv′·ŭm* (L. *ovum*, an egg), the germ produced within the ovary, capable, under certain conditions, of being developed into a new individual; one of the small cellular bodies in the ovary which, after impregnation, is developed into the future embryo.

oxalate, n., *ŏks′·ăl·āt* (L. *oxălis*, a kind of sorrel), in *chem.*, a salt of oxalic acid: **oxalic**, a., *ŏks·ăl′·ĭk*, pert. to sorrel, or procured from it: **oxalic acid**, a dry, poisonous acid, chiefly manufactured from wood sawdust.

Oxalidaceæ, n. plu., *ŏks′·ăl·ĭd·ā′·sē·ē* (L. *oxălis*, a kind of sorrel—from Gr. *oxus*, sour, acid), the Wood-sorrel family, an Order of plants often acid in their properties; some have large, tuberous, edible roots; some bear grateful fruit; while the leaves of others are highly sensitive, which last include sensitive plants: **Oxalis**, n., *ŏks′·ăl·ĭs*, a genus of plants of numerous species, well worthy of cultivation: **Oxalis acetosella**, *ăs′·ĕt·ō·sĕl′·lă* (dim. of mod. L. *acētōsa*, the sorrel or sour-dock—from L. *acētum*, vinegar), common wood-sorrel, so named from its acid taste, contains binoxalate of potash, often called salt of sorrel; the plant has been used as a refrigerant and antiscorbutic: **O. sensitiva**, *sĕns′·ĭt·īv′·ă* (L. *sensitīvus*, discerned by the senses—from *sentĭŏ*, I discern by the

senses), a species which has sensitive leaves: **O. crenata**, *krĕn·āt′·ă* (L. *crēnătus*, notched—from *crēna*, a notch); **O. esculenta**, *ĕsk′·ŭl·ĕnt′·ă* (L. *esculentus*, fit for food—from *ĕsca*, food); and **O. Deppei**, *dĕp′·pĕ·ī* (mod. L. *Deppēī* of *Deppĕŭs*), are species which yield tubers, used as a substitute for potatoes.

Oxycoccus, n., *ŏks′·ĭ·kŏk′·kŭs* (Gr. *oxus*, sour, acid; *kokkos*, a berry), a genus of plants, Ord. Vacciniaceæ, which include the cranberry: **Oxycoccus palustris**, *păl·ŭst′·rĭs* (L. *palustris*, marshy—from *pălus*, a marsh), the common cranberry, a native plant, producing crimson acid berries: **O. macrocarpus**, *măk′·rō·kărp′·ŭs* (Gr. *makros*, great; *karpos*, fruit), the American cranberry, which bears larger berries.

oxyde, or **oxide**, n., *ŏks′·ĭd* (Gr. *oxus*, sour, acid; *oxos*, vinegar), a compound of oxygen without the properties of an acid, as the rust of iron: **oxidise**, v., *ŏks′·ĭd·īz*, to convert into an oxide—also in same sense **oxygenise**, v., *ŏks′·ĭ·jĕn·īz′*: **oxydation**, n., *ŏks′·ĭd·ā′·shŭn*, the operation or process of converting a body into an oxide: **oxygen**, n., *ŏks′·ĭ·jĕn* (Gr. *gennăō*, I generate or produce), that elementary gaseous body which gives to air its power of supporting respiration and combustion, and which by its union with hydrogen forms water; a colourless, tasteless, and inodorous gas, which exists in the atmosphere in the proportion of twenty-one parts, to seventy-nine of nitrogen, by measure.

oxymel, n., *ŏks′·ĭ·mĕl* (Gr. *oxus*, sour, acid; *meli*, honey), a mixture of vinegar and honey.

oxytocic, a., *ŏks′·ĭt·ŏs′·ĭk* (Gr. *oxus*, sharp; *tŏkŏs*, childbirth), promoting delivery: n., an agent which promotes delivery.

oxyuris vermicularis, *ŏks′·ĭ·ūr′·ĭs*

vérm·ĭk'·ŭl·ār'·ĭs (Gr. *oxus*, sharp, rapid ; *oura*, a tail : *vermĭcŭlārĭs*, pert. to a worm—from *vermis*, a worm), a minute, white, thread-like parasitic worm, of separate sexes, the male about one-and - half - line in length, the female five or six : **oxyurides,** n. plu., *ŏks'·ĭ·ūr'·ĭd·ēz,* the thread-worms.

ozæna, n., *ŏz·ēn'·ă* (L. *ozæna,* Gr. *ozaina,* an offensive ulcer in the nose—from Gr. *ŏzē,* a stench), an offensive discharge from the nose, arising from various causes.

ozone, n., *ōz'·ŏn* (Gr. *ŏzō,* I emit an odour), a supposed modification of oxygen, existing both in air and water, developed by electrical action in thunderstorms, etc., and which emits a peculiar odour : **ozonised,** a., *ōz'·ŏn·īzd,* charged with or containing ozone.

Pacchioni glandulæ, *păk'·kĭ·ōn'·ī glănd'·ŭl·ē* (mod. L. *Pacchioni,* of *Pacchiōnus,* an Italian ; *glandulæ,* glandules), the bodies or glands of Pacchionus, their first describer ; numerous small pulpy - looking elevations, generally in clusters, upon the external surface of the dura - mater, coinciding with corresponding depressions on the inner surface of the skull.

Pachydermata, n. plu., *păk'·ĭ·dérm'·ăt·ă* (Gr. *pachus,* thick ; *derma,* skin, *dermătos,* of skin), the thick-skinned animals, an old Mammalian Order, to include such animals as the rhinoceros, the hippopotamus, and elephant : **pachydermatous,** a., *păk'·ĭ·dérm'·ăt·ŭs,* thick - skinned : **pachydermia,** n., *păk'·ĭ·dérm'·ĭ·ă,* a thickened state of the skin : **pachymeningitis,** n., *păk'·ĭ·mĕn·ĭng·jīt'·ĭs* (Gr. *mēningx,* a membrane, *mēninggos,* of a membrane), inflammation of the dura-mater.

Pacinian bodies, *pă·sĭn'·ĭ·ăn* (after *Pacini* of Pisa), certain small oval bodies, like little seeds,

found, in dissecting the nerves of the hand or foot, attached to their branches as they pass through the sub-cutaneous fat on their way to the skin.

Pæonia, n., *pē·ōn'·ĭ·ă* (after the physician *Pæon*), a fine genus of plants, Order Ranunculaceæ, much valued for their large, varied, and richly-coloured flowers, some varieties having double blossoms resembling large double roses : **Pæonia albiflora,** *ălb'·ĭ·flōr'·ă* (L. *albus,* white ; *flos,* a flower, *flōris,* of a flower), a species whose fleshy roots, cooked, are sometimes eaten by natives of N. Asia.

pagina, n., *pădj'·ĭn·ă* (L. *pagĭna,* a page or leaf), in *bot.,* the surface of a leaf ; any flat surface.

palæontology, n., *păl'·ē·ŏnt·ŏl'·ŏ·jĭ* (Gr. *palaios,* ancient ; *onta,* beings ; *logos,* discourse), that science or sub-division of geology which treats of the plants and animals found fossil in the crust of the earth.

palæophytology, n., *păl'·ē·ŏf'·ĭt·ŏl'·ŏ·jĭ* (Gr. *palaios,* ancient ; *phuton,* a plant ; *logos,* discourse), that branch of palæontology which treats of fossil plants.

palæozoic, a., *păl'·ē·ō·zō'·ĭk* (Gr. *palaios,* ancient ; *zōē,* life), in *geol.,* applied to the lowest division of stratified groups in which the earliest known forms of life appear.

palate, n., *păl'·āt* (L. *pălātus,* the palate), the upper part or roof of the mouth, consisting of two parts, the hard in front, and the soft behind ; in *bot.,* the projecting portion of the under lip of personate flowers : **palatal,** a., *păl'·āt·ăl,* applied to numerous glands which lie between the mucous membrane and the surface of the bone : **palatine,** a., *păl'·āt·ĭn,* same sense as 'palatal' ; contained or situated within the palate, as nerves or glands : **palato,** *păl·ăt'·ō,* connected with the palate : pal-ato-glossus, *glŏs'-sŭs* (Gr. *glossa*, the tongue), a muscle which passes between the soft palate and the side of the base of the tongue.

palea, n., *păl'-ē-ă* (L. *pălĕa*, chaff), in *bot.*, the small scale-plates, like chaff, in the receptacles of some composite flowers ; the part of the flower of grasses within the glume : **paleaceous**, a., *păl'-ē-ā'-shŭs*, resembling chaff ; covered with membranous scales like chaff.

palisade tissue, *păl'-ĭs-ād' tĭsh'-ū* (F. *palissade*, a stake, a hedge-row of trees ; L. *pălus*, a pole), in *bot.*, chlorophyll cells elongated in a direction vertical to the surface of the leaf, lying beneath the hypodermic layer in the leaves of Cycadaceæ and Coniferæ.

Paliurus, n., *păl'-ĭ-ūr'-ŭs* (L. *pali-ūrus*, Gr. *paliouros*, Christ's thorn), a genus of very handsome plants, Ord. Rhamnaceæ: **Pali-urus aculeatus**, *ă-kūl'-ē-āt'-ŭs* (L. *acūlĕātus*, thorny, prickly), Christ's thorn, common in the hedges of Judea, supposed to have formed the crown of thorns put on our Saviour's head.

pallescent, a., *păl-lĕs'-sĕnt* (L. *pallesco*, I grow pale), in *bot.*, growing pale : **pallid**, a., *păl'-lĭd* (L. *pallĭdus*, pale), of a pale, un-decided colour.

pallial, see ' pallium.'

palliobranchiata, n., *păl'-lĭ-ō-brăng'-kĭ-āt'-ă* (L. *pallium*, a mantle ; Gr. *brangchia*, gills of a fish), the old name for the ' Brachiopoda,' founded on the assumption that the system of tubes in the mantle constituted the gills : **palliobranch'iate**, a., *-kĭ-āt*, having gills developed from the mantle.

pallium, n., *păl'-lĭ-ŭm* (L. *pallĭum*, a mantle, a cloak), the fleshy covering lining the interior of the shells of bivalves : **pallial**, a., *păl'-lĭ-ăl*, pert. to a mantle or cloak : **pallial impressions**, the impressions or lines left in the shells of bivalves by the muscular margin of the mantle : **pallial shell**, a shell contained within the mantle, such as the bone of the cuttle-fish.

palma, n., *păl'-mă* (L. *palma*, Gr. *palamē*, the flat of the hand), in *anat.*, the palm or flat of the hand : **palmar**, a., *păl'-măr*, of or relating to the palm of the hand ; denoting two muscles of the hand : **palmaris longus**, *păl-măr'-ĭs lŏng'-gŭs* (L. *palmāris*, relating to the hand ; *longus*, long), a muscle arising from the inner condyle of the os humeri, finally fixed to the roots of all the fingers, and forming a flexor of the wrist : **palmaris brevis**, *brĕv'-ĭs* (L. *brĕvis*, short), a thin quad-rilateral muscle, placed beneath the integument on the ulnar side of the hand, and inserted into the skin on the inner border of the palm of the hand, which con-tracts the skin of the palm.

Palmæ, n. plu., *păl'-mē* (L. *palma*, the palm of the hand), one of the most interesting and valuable Orders of plants of the vegetable kingdom : **palmate**, a., *păl'-māt* (L. *palmātus*, marked with the palm of the hand), in *bot.*, having the shape of the open hand with the fingers apart, as in some leaves ; having leaves divided into lobes to about the middle : **Palma Christi**, *krĭst'-ĭ* (L. *Christ-us*, Christ, *Christi*, of Christ), a palm from whose seeds Castor-oil is expressed : **palmatifid**, a., *păl-măt'-ĭ-fĭd* (L. *findo*, I cleave ; *fidi*, I have cleft), having a leaf divided so as to resemble a hand; same as ' palmate.'

palmatipartite, a., *păl'-măt-ĭ-pârt'-ĭt* (L. *palmātus*, marked with the palm of a hand; *partītus*, divided, shared), in *bot.*, applied to a simple leaf having the subdivisions extending considerably more than half-way to the base ; cut nearly to the base in a palmate manner : **palmi-nerved**, a., *păl'-mĭ-nêrvd*, having the veins of the leaves arranged in a palmate manner.

Palmellaceæ, n. plu., *păl'-mĕl-lā'-sĕ-ē* (Gr. *palma*, a shaking, a vibration), a Sub-order or tribe of Algæ, composed of more or less rounded cells in a gelatinous matrix, as seen in the plant Redsnow : **Palmella**, n., *păl-mĕl'-lă*, a genus of plants comprising minute species found in marshy places, so named from their jelly-like nature.

palpation, n., *pălp-ā'-shŭn* (L. *palpātĭō*, a stroking—from *palpo*, I stroke or touch gently), examination by the sense of touch ; the mode of examining the physical condition of any part by the touch.

palpebræ, n. plu., *pălp'-ĕb-rē* (L.), a Latin word signifying the eyelids : **palpebral**, a., *pălp'-ĕb-răl*, pert. to the eyelids.

palpi, n. plu., *pălp'-ī* (L. *palpo*, I stroke or touch gently), the feelers of insects, attached to the head : **palpiform**, a., *pălp'-i-fŏrm* (L. *forma*, shape), having the form of palpi or feelers.

palpitation, n., *pălp'-ĭt-ā'-shŭn* (L. *palpĭtātĭō*, a frequent and rapid motion — from *palpo*, I stroke gently), an unnaturally rapid beating of the heart, obvious to the feeling of the individual, caused by disease, fear, or bodily exertion.

palsy, n., *păwl'-zĭ*, the common name for 'paralysis,' which see.

paludal, a., *păl-ūd'-ăl* (L. *pălus*, a swamp, *palūdis*, of a swamp), of or pert. to marshes or swamps.

pampiniform, a., *păm-pĭn'-ĭ-fŏrm* (L. *pampĭnus*, a tendril ; *forma*, shape), resembling a vine tendril.

panacea, n., *păn'-ă-sē'-ă* (L. *pănăcēă*, Gr. *panakeia*, a herb sup-posed to have power to heal all diseases—from Gr. *pan*, all, and *akeŏmai*, I heal or cure), a professed remedy for all diseases ; a universal medicine.

Panax, n., *păn'-ăks* (Gr. *pan*, all ; *ăkos*, a cure, a remedy), a genus of plants, Ord. Araliaceæ, a species of which yields the famous Ginseng root of the Chinese, used as a stimulant : **Panax quinquefolium**, *kwĭn'-kwĕ-fōl'-ĭ-ŭm* (L. *quinque*, five ; *fŏlĭum*, a leaf), a plant possessing qualities resembling those of the Ginseng.

pancratic, a., *păn-krăt'-ĭk* (Gr. *pan*, all; *kratos*, bodily strength), excelling in bodily strength or gymnastics : **Pancratium**, n., *păn-krā'-shĭ-ŭm*, a genus of handsome bulbous plants, Ord. Amaryllidaceæ, so named from their supposed medicinal virtues.

pancreas, n., *păn'-krĕ-ăs* (Gr. *pan*, all ; *kreas*, flesh), a fleshy gland in the abdominal cavity in front of the spine, and behind and below the stomach ; the sweetbread of cattle : **pancreatic**, a., *păn'-krĕ-ăt'-ĭk*, denoting a fluid secreted by the pancreas or sweetbread : **pancreatin**, n., *păn-krĕ'-ăt-ĭn*, an albuminoid principle present in pancreatic juice which has the property of converting starch into sugar.

Pandanaceæ, n. plu., *păn'-dăn-ā'-sĕ-ē* (said to be from *pandang*, a Malay word meaning 'conspicuous'), the Screw-pine family, an Order of plants nearly resembling palms : **Pandanus**, n., *păn-dăn'-ŭs*, a genus of plants whose species are remarkable for their aerial roots, with large cup-like spongioles : **Pandanus candelabra**, *kănd'-ĕl-āb'-ră* (L. *candēlāb-rum*, a branched candlestick), the chandelier tree of Guinea, so called from its mode of branching.

pandemic, a., *păn-dĕm'-ĭk* (Gr. *pan*, all ; *dēmos*, the people), a 'term to designate a disease spread over a whole continent, or several contiguous countries, such as cholera or influenza.

panduriform, a., *păn·dūr'·ĭ·fŏrm* (Gr. *pandoura*, L. *pandūra*, a musical instrument with three strings; L. *forma*, shape), in *bot.*, applied to the leaves of plants shaped like a fiddle.

panicle, n., *păn'·ĭ·kl* (L. *panicula*, a tuft on plants), in plants, a tuft or bunch of flowers or seeds, dense and close as in Indian corn, spreading or scattered as in oats, and in other forms; the down on reeds: **paniculate**, a., *păn·ĭk'·ūl·āt*, having the flowers in panicles; having branches variously subdivided.

Panicum, n., *păn'·ĭk·ŭm* (L. *pānis*, bread), a useful genus of grasses, Ord. Gramineæ: **Panicum miliaceum**, *mĭl'·ĭ·ā'·sē·ŭm* (L. *milĭā·cĕus*, of or pert. to millet—from *mĭlĭum*, millet), millet, frequently sown for feeding poultry, and used as a substitute for rice: **P. arborescens**, *ăr'·bŏr·ĕs'·ĕnz* (L. *arbŏrĕscens*, growing into a tree —from *arbor*, a tree), a species whose culm is little thicker than a goose's quill, and which yet attains the height of the loftiest forest tree.

panification, n., *păn'·ĭf·ĭk·ā'·shŭn* (L. *pānis*, bread; *facio*, I make), the changes by which the dough is converted into bread.

panniculus adiposus, *păn·ĭk'·ūl·ŭs ăd'·ĭp·ōz'·ŭs* (L. *pannĭcŭlus*, a flap or piece of cloth, a rag; *adĭpōsus*, fatty — from *adeps*, fat), the adipose tissue, forming a considerable layer underneath the skin, together with the sub-cutaneous areolar tissue: **panniculus carnosus**, *kăr·nōz'·ŭs* (L. *carnōsus*, fleshy—from *caro*, flesh), a fleshy covering; superficial muscle, or muscular bands, investing the greater part of the bodies of quadrupeds.

panspermism, n., *păn·spĕrm'·ĭzm* (Gr. *pan*, all; *sperma*, seed), in *bot.*, the universal diffusion of germs throughout the atmosphere.

Papaveraceæ, n. plu., *păp·ăv'·ĕr·ā'·sĕ·ē* (L. *păpāver*, the poppy, *păpāvĕris*, of the poppy), the Poppy family, an Order of plants possessing well-marked narcotic properties: **Papaver**, n., *păp·āv'·ĕr*, a genus of plants: **Papaver somniferum**, *sŏm·nĭf'·ĕr·ŭm* (L. *somnifer*, sleep-bringing — from *somnus*, sleep; *fero*, I bring), a species, and its varieties, which produce opium, a concrete milky juice procured from its nearly ripe capsules—the most important active principle in opium is the alkaloid called 'morphia'; other crystalline principles found in it are 'codeia,' 'narcotine,' 'thebaia,' and 'meconine,' etc.: **P. rhœas**, *rē'·ăs* (Gr. *rhĕō*, I flow, referring to its juice flowing from incisions), the red corn poppy or corn rose, whose petals are used in pharmacy, chiefly for their colouring matter: **papaveraceous**, a., *păp·ăv'·ĕr·ā'·shŭs*, resembling the poppy or pert. to it: **papaverous**, a., *păp·ăv'·ĕr·ŭs*, having the nature or qualities of the poppy.

Papayaceæ, n. plu., *păp'·ă·yā'·sĕ·ē* (said to be from Malay *papaya*), the Papaw family, an Order of plants: **Papaw tree**, or 'Carica Papaya,' yields an acrid milky juice, and an edible fruit.

Papilionaceæ, n. plu., *păp·ĭl'·ĭ·ŏn·ā'·sĕ·ē* (L. *păpĭlĭŏ*, a butterfly, *păpĭlĭōnis*, of a butterfly), a Suborder of the Order Leguminosæ, whose species have frequently beautiful showy flowers: **papilionaceous**, a., *păp·ĭl'·ĭ·ŏn·ā'·shŭs*, resembling a butterfly; applied to plants, as the pea, from the butterfly shape of their flowers.

papilla, n., *păp·ĭl'·lă*, **papillæ**, n. plu., *păp·ĭl'·lē* (L. *papilla*, a small pimple, a nipple), the minute elevations found on the tongue, the palm, or the surface of the fingers, etc., being the terminations of the nerves, producing the sense of taste and feeling; a minute soft prominence; in *bot.*, soft, superficial glands : **papillary**, a., *păp'·ĭl·lăr·ĭ*, pert. to a nipple, or to the papillæ : **papillate**, a., *păp·ĭl'·lāt*, also **papillose**, a., *păp'·ĭl·lōz'*, in *bot.*, covered with fleshy dots or points, as the stems of certain plants ; warty : **papillated**, a., *păp'·ĭl·lāt·ĕd*, same sense as preceding ; covered with small nipple-like prominences.

papilloma, n., *păp'·ĭl·lōm'·ă*, **papillomata**, n. plu., *păp'·ĭl·lōm'·ăt·ă* (a new L. formation from *papilla*, a teat or nipple), papillary growths, also called epidermic and epithelial tumours, from their seat in the body, which constitute a well-marked class of new formations, of which warts and callosities of the skin are minor instances.

pappus, n., *păp'·pŭs* (Gr. *pappos*, L. *pappus*, the woolly, hairy seeds of certain plants), in *bot.*, the hairs at the summit of the ovary or achene in Compositæ, consisting of the altered calyx ; the feathery crown on many singleseeded seed - vessels : **pappose**, a., *păp·pōz'*, downy, as the ripened seeds of the thistle, the dandelion, etc.

papula, n., *păp'·ūl·ă*, **papulæ**, n. plu., *păp'·ūl·ē* (L. *papula*, a pimple), a pimple ; a solid elevation of the true skin of minute size : **papule**, n., *păp'·ūl*, **papules**, n. plu., *păp'·ūlz*, same as preceding ; any small pimple : **papular**, a., *păp'·ūl·ăr*, also **papulous**, a., *păp'·ūl·ŭs*, covered with papulæ or pimples ; pimply.

papyraceous, a., *păp'·ĭr·ā'·shŭs* (L. *papyrus*, Gr. *papuros*, the paper plant), in *bot.*, paper-like in texture.

paracentesis, n., *păr'·ă·sĕnt·ēz'·ĭs* (Gr. *para*, side by side ; *kentĕŏ*, I pierce), the operation or art of perforating a part of the body to allow the escape of a fluid, usually called 'tapping' : **paracentesis abdominis**, *ăb·dŏm'·ĭn·ĭs* (L. *abdōmen*, the belly, *abdŏmĭnis*, of the belly), the paracentesis of the abdomen ; the operation of tapping the abdomen : **paracentesis thoracis**, *thōr·ăs'·ĭs* (L. *thōrax*, the breast, the chest, *thorācĭs*, of the chest), the operation of tapping the chest.

paraglobulin, n., *păr'·ă·glŏb'·ūl·ĭn* (Gr. *para*, beside, close to ; and Eng. *globulin*), a substance derived from the cellular structures of the body ; a form of *globulin*.

paralysis, n., *păr·ăl'·ĭs·ĭs* (Gr. *parălŭsis*, a loosening at the side, palsy—from *para*, beside ; *lusō*, I shall loose), a loss of motion, or sensation, or both, depending on central or local disease ; it is local or general, partial or complete, and includes *hæmiplegia*, a paralysis affecting one lateral half of the body, while *paraplegia* means paralysis affecting the body transversely, and may involve all four extremities : **locomotor ataxy**, a form of disease causing in-coordination of movement, and depending upon sclerosis of the posterior column of the spinal cord : see 'motor ataxy' : **paralysis agitans**, *ădj'·ĭt·ănz* (L. *agĭtans*, putting in motion), shaking palsy.

paralytic, a., *păr'·ăl·ĭt'·ĭk*, affected with, or inclined to, paralysis.

parametritis, n., *păr'·ă·mĕt·rīt'·ĭs* (Gr. *para*, beside ; *mētră*, the womb), inflammation by the side of the uterus, that is, inflammation of the sub-peritoneal connective tissue.

paranemata, n. plu., *păr'·ă·nĕm'·ăt·ă* (Gr. *para*, beside, close to ; *nēma*, a thread, *nēmăta*, threads), in *bot.*, the filaments found along with spores in the fructification of many Algæ.

paraphyses, n. plu., *păr·ăf'·ĭs·ĕz* (Gr. *para*, beside, about ; *phuō*, I grow), jointed or continuous filaments occurring in the fructification of Mosses and other Cryptogams ; abortive petals or stamens.

paraplegia, n., *păr'·ă·plēdj'·ĭ·ă* (Gr. *para*, beside, close to ; *plēgē*, a stroke), paralysis of the body transversely, affecting both sides ; see under ' paralysis.'

parapodia, n., *păr'·ă·pōd'·ĭ·ă* (Gr. *para*, beside ; *podes*, feet), the unarticulated, lateral, locomotive processes, or foot tubercles, of certain of the Annelida.

parapophyses, n. plu., *păr'·ă·pŏf'·ĭs·ĕz* (Gr. *para*, beyond ; *apophusis*, a process), in *anat.*, the processes which extend outwards, or outwards and downwards, from the body of the vertebræ in fishes ; a name given to the transverse processes of an ideal typical vertebra.

parasite, n., *păr'·ăs·īt* (Gr. *parasitos*, one who eats at another's expense at table), in *bot.*, a plant which grows upon another plant, and obtains nourishment from its juices ; an animal or vegetable which lives in or upon another animal, affecting the skin, hair, intestinal canal, or almost any internal organ : **parasitic**, a., *păr'·ăs·ĭt'·ĭk*, growing in or upon, and deriving support from another animal or plant : **parasitism**, n., *păr'·ăs·ĭt·ĭzm*, the condition of a parasite.

paraspermatia, n. plu., *păr'·ă·spérm·ā'·shĭ·ă* (Gr. *para*, beside ; *sperma*, seed), in *bot.*, bodies resembling spores, found in some Algæ.

parastichies, n. plu., *păr'·ă·stĭk'·ĭ·ĕz* (Gr. *para*, beside ; *stichos*, a row, a line), in *bot.*, the secondary spirals in a phyllotaxis.

paregoric, a., *păr'·ĕ·gŏr'·ĭk* (Gr. *parēgoria*, consolation, alleviation), a name applied to a compound tincture of opium, of which there are two forms, English and Scotch, the latter containing more than twice as much opium as the former ; assuaging pain.

pareira, n., *păr·īr'·ă* (Spanish), the wood of the stem and root of the ' Chondodendron tomentosum,' also called ' Cissampelos Pareira,' found in Peru and Brazil, Ord. Menispermaceæ, is tonic and diuretic, and is used in *chronic* or *atonic* inflammation of the bladder.

parenchyma, n., *păr·ĕng'·kĭm·ă* (Gr. *parengchuma*, a discharge of humour from the lungs, etc.—from *para*, beside ; *en*, in ; *chuma*, juice, tissue), in *anat.*, the secreting tissue of glands ; in *bot.*, the cellular tissue or pith of plants : **parenchymal**, a., *păr·ĕng'·kĭm·ăl*, also **parenchymatous**, a., *păr'·ĕng·kĭm'·ăt·ŭs*, pert. to or resembling parenchyma ; spongy ; full of pith.

Parideæ, n. plu., *păr·ĭd'·ē·ē* (L. *par*, equal—from the regularity of their parts), a tribe or Suborder of the Ord. Trilliaceæ : **Paris**, n., *păr'·ĭs*, a genus : **Paris quadrifolia**, *kwŏd'·rĭ·fōl'·ĭ·ă* (*Paris* of the Homeric mythology; L. *quadrus*, square ; *fŏlĭum*, a leaf), the herb paris or ' true-love,' is narcotic, and the juice of the berries has been used in inflammation of the eyes.

parietal, a., *păr·ĭ'·ĕt·ăl* (L. *parĭĕs*, a wall, *parĭĕtis*, of a wall), in *anat.*, constituting the sides or walls — applied to a large flat bone on each side of the head ; in *bot.*, growing from the side or wall of another organ—applied to the placentas on the wall of the ovary : **parietes**, n. plu., *păr·ĭ'·ĕt·ĕz*, in *anat.*, the enclosing walls of any cavity in the body ; in *bot.*, inside walls, as in an ovary, or fruit.

parietin, n., *păr·ĭ'ĕt·ĭn* (L. *pariēs*, a wall, *parĭĕtis*, of a wall), a yellow colouring matter found in 'Parmelia parietina,' Ord. Lichenes, also called 'Chrysophanic acid'; see 'Parmelia.'

parietosplanchnic, a., *păr·ĭ'ĕt·ōsplăngk'·nĭk* (L. *parĭēs*, a wall, *parĭĕtis*, of a wall; *splangchna*, bowels or entrails), denoting one of the nervous ganglia of the Mollusca, which supplies the walls of the body, and the viscera.

pari-pinnate, a., *păr'·ĭ-pĭn'·nāt* (L. *par*, equal; Eng. *pinnate*), in *bot.*, having a compound pinnate leaf, ending in two leaflets.

Paritium, n., *păr·ĭsh'·ĭ·ŭm* (*parĭti*, said to be its Malabar name), a genus of plants, Ord. Malvaceæ: **Paritium elatum**, *ĕl·āt'·ŭm* (L. *ēlātus*, productive), a species whose bark furnishes the Cuba bast: **P. tiliaceum**, *tĭl'·ĭ·ā'·sē·ŭm*, (L. *tĭlĭācĕus*, pert. to the Linden tree—from *tĭlĭa*, the Linden or Lime tree), the pariti of Malabar, yields a fibrous bark, which is made into fine matting, cordage, pack-thread, etc.

Parmelia, n., *păr·mēl'·ĭ·ă* (L. *parma*, Gr. *parmē*, a small round shield; Gr. *heilō*, I enclose), a genus of Lichens, found on rocks, trunks of trees, etc., several of which are used in dyeing: **Parmelia parietina**, *părĭ'·ĕt·ĭn'·ă* (L. *parĭēs*, a wall, *pariĕtis*, of a wall, so named from the places of their growth, as old walls, etc.), a species producing 'parietin,' which see.

Parmentiera, n., *păr'·mĕn·tĭ'·ĕr·ă* (Sp. *paramento*, ornament), a S. American genus of plants, Ord. Bignoniaceæ, bearing peculiar, fleshy, cylindrical fruit: **Parmentiera cereifera**, *sēr'·ĕ·ĭf'·ĕr·ă* (L. *cērĕus*, waxen — from *cēra*, wax; *fero*, I bear), a species found in Panama, called the

U

Candle-tree, whose fruit, often four feet long, somewhat resemble yellow wax candles: **P. edulis**, *ĕd·ŭl'·ĭs* (L. *edūlis*, eatable—from *edo*, I eat), a species whose fruit is eaten by the Mexicans.

Parnassia, n., *păr·năs'·sĭ·ă* (*Mount Parnassus*, the fabled abode of the gods, and therefore of grace and beauty), a genus of elegant plants, Ord. Hypericaceæ: **Parnassia palustris**, *păl·ŭs'·trĭs* (L. *păluster* or *pălustris*, marshy— from *pălus*, a marsh), Grass of Parnassus, has remarkable, glandlike bodies between the stamens.

paronychia, n., *păr'·ŏn·ĭk'·ĭ·ă* (Gr. *paronuchia*, a whitlow — from *para*, beside; *onux*, the nail), a whitlow or felon.

Paronychiaceæ, n. plu., *păr'·ŏn·ĭk'·ĭ·ā'·sĕ·ē* (Gr. *para*, beside; *onux*, a nail, a claw), the Knotwort family, an Order of plants having a slight degree of astringency: **Paronychia**, n., *păr'·ŏn·ĭk'·ĭ·ă*, a genus of plants, so named as supposed to cure whitlow.

parotid, a., *păr·ŏt'·ĭd* (Gr. *parōtis*, a tumour under the ears, *parōtĭdis*, of a tumour under the ears —from *para*, beside; *ous*, the ear), applied to two glands, one on each side, just below and in front of the ear, which secrete a great portion of the saliva, being most active during the process of mastication: **parotitis**, n., *păr'·ŏt·īt'·ĭs*, inflammation of the parotid glands; mumps.

parovarium, n., *păr'·ō·văr'·ĭ·ŭm* (Gr. *para*, beside; *ovārĭum*, an ovary), a group of scattered tubules lying transversely between the Fallopian tube and ovary.

paroxysm, n., *păr'·ŏks·ĭzm* (Gr. *paroxusmos*, excitement, exasperation — from *para*, beside; *oxunein*, to sharpen), a recurrence of the symptoms of a disease at equal or unequal intervals; a recurring increase and exacerbation of a disease.

parthenogenesis, n., *părth'ĕn·ō·jĕn'ĕs·ĭs* (Gr. *parthĕnos*, a virgin; *gennăō*, I produce), the reproduction of plants or animals by ovulation, yet without the immediate stimulus of the male principle ; in *bot.*, the production of perfect seed with embryo, without the application of pollen.

partite, a., *părt'ĭt*, also **parted**, a., *părt'ĕd* (L. *partītus*, divided or shared), cut down to near the base, the divisions being called 'partitions.'

parturition, n., *părt'ūr·ĭsh'ŭn* (L. *parturio*, I bring forth), the act of bringing forth, or of being delivered of young : **parturient**, a., *părt·ūr'ĭ·ĕnt*, bringing forth, or about to bring forth young.

Passifloraceæ, n. plu., *păs'sĭ·flōr·ā'sĕ·ē* (L. *passĭŏ*, a suffering, passion, *passus*, suffered ; *flos*, a flower, *flōris*, of a flower), the Passion-flower family, an Order of plants, so named on account of a fancied resemblance in the flowers to the appearance of the wounds of Christ, as the nails, blood, and pillar, presented at Calvary : **Passifloreæ**, n. plu., *păs'sĭ·flōr'ĕ·ē*, a tribe or Sub-order, forming climbing plants : **Passiflora**, n., *păs'sĭ·flōr'ă*, an interesting and elegant genus of plants: **Passiflora edulis**, *ĕd·ūl'ĭs* (L. *edūlis*, eatable) ; and P. **laurifolia**, *lāwr'ĭ·fōl'ĭ·ă* (L. *laurus*, the laurel ; *fŏlĭum*, a leaf), are species producing succulent and grateful fruit in hot climates : **P. quadrangularis**, *kwŏd·răng'gūl·ār'ĭs* (L. *quadrus*, square—from *quatuor*, four; *angulus*, a corner), a species whose roots are emetic and powerfully narcotic : **passiflorin**, n., *păs'sĭ·flōr'ĭn*, the peculiar and active principle of preceding.

passive, a., *păs'sĭv* (L. *passus*, suffered), applied to a morbid condition in which there is no special activity manifested ; denoting a failure of nutritive and formative powers of a part ; denoting the result of another morbid change in some organ or tissue on which its own tissue is dependent, *e.g.*, passive congestion or regurgitation of blood from a weak heart.

pastil, or **pastille**, n., *păs·tĭl'* (L. *pastillus*, an aromatic lozenge), an aromatic or medicated sugar-drop or lozenge ; a composition of aromatic resins, in the form of a small cone, burnt to clear and perfume the air of a room with its smoke.

Pastinaca, n., *păst'ĭn·āk'ă* (L. *pastĭnāca*, a carrot ; *pastĭnum*, a kind of dibble), a genus of plants, Ord. Umbellifera, so named from their shape : **Pastinaca sativa**, *săt·ĭv'ă* (L. *satīvus*, sown or planted), the parsnip, a well-known culinary vegetable.

patagium, n., *păt·ādj'ĭ·ŭm* (L. *patăgĭum*, an edging or border of a dress), the expansion of the integument by which bats, flying squirrels, etc., support themselves in the air.

patella, n., *păt·ĕl'lă* (L. *patella*, a small pan or plate), in *anat.*, the knee-cap or pan, the place where it moves upon the os femoris being called 'trochlea' ; a sesamoid bone developed in the tendon of insertion of the great extensor muscles of the thigh ; in *bot.*, a round or convex sessile apothecium in Lichens : **patellar**, a., *păt·ĕl'lăr*, of or pert. to a patella.

patent, a., *păt'ĕnt* (L. *patens* lying open), in *bot.*, spreading widely ; expanded.

pathetic, a., *păth·ĕt'ĭk* (Gr. *pathĕtĭkos*, liable to suffering—from *pathos*, suffering), the fourth nerve, being the motor of the superior oblique muscle of the eye, which turns up the eye, hence the name.

pathognomonic, a., *păth·ŏg'nō·*

nŏn'ĭk (Gr. *pathos*, feeling, suffering ; *gnōmōn*, one that knows), designating a characteristic symptom of a disease ; distinctive : **pathognomy**, n., *păth·ŏg'nŏm·ĭ*, the science of the signs by which the passions are indicated.

pathology, n., *păth·ŏl'·ō·jĭ* (Gr. (*pathos*, suffering ; *logos*, discourse), that part of medicine which treats of the nature of diseases, and their causes and symptoms : **vegetable pathology**, in *bot.*, the study of the functions of plants when vitiated by disease.

patulous, a., *păt'·ŭl·ŭs* (L. *patŭlus*, standing open—from *pateo*, I lie open), in *bot.*, slightly spreading open.

Paullinia, n., *pāwl·ĭn'·ĭ·ă* (after *S. Paulli* of Copenhagen), a genus of plants, Ord. Sapindaceæ, many of whose species are poisonous : **Paullinia sorbilis**, *sŏrb'·ĭl·ĭs* (L. *sorbĭlĭs*, that may be sucked or supped up — from *sorbĕŏ*, I suck up), a species from whose seeds Guarana bread, or Brazilian cocoa, is prepared : **P. pinnata**, *pĭn·nāt'ă* (L. *pinnatus*, winged, feathered—from *pinna*, a feather), a species which exhibits anomalous exogenous stems.

paunch, n., *pāwnsh* (F. *panse*, L. *pantex*, the paunch, the belly), the belly and its contents ; the largest stomach of a ruminant.

pectic, a., *pĕkt'·ĭk* (Gr. *pēktos*, coagulated, curdled), denoting an acid obtained by a small addition of potash to pectine, existing in many vegetable substances : **pectate**, n., *pĕkt'·āt*, a salt of pectic acid : **pectine**, n., *pĕkt'·ĭn*, the gelatinising principle of fruits and vegetables.

pectinate, a., *pĕkt'·ĭn·āt* (L. *pecten*, a comb, *pectĭnis*, of a comb), comb-like, applied to the gills of certain Gasteropods ; in *bot.*, divided into narrow segments like the teeth of a comb.

pectineus, n., *pĕkt'·ĭn·ē'·ŭs* (L. *pecten*, a comb or crest, *pectĭnis*, of a comb), a flat, quadrangular muscle, arising from the pectineal line of the os pubis : **pectineal**, a., *pĕkt'·ĭn·ē'·ăl*, or *pĕkt·ĭn'·ē·ăl*, of or pert. to the pectineus : **pectineal line**, a line forming a sharp ridge on the pubic bone of the pelvis.

pectoral, a., *pĕkt'·ŏr·ăl* (L. *pectus*, a breast, *pectŏris*, of a breast), connected with or placed upon the chest ; good for the chest or lungs : n., a medicine to relieve complaints of the chest : **pectoral fins**, the two fore fins near the gills of a fish : **pectoralis major**, *pĕkt'·ŏr·āl'·ĭs mādj'·ŏr* (L. *major*, greater), a broad, thick, triangular muscle, situated at the upper and fore part of the chest, in front of the axilla : **pectoralis minor**, *mīn'·ŏr* (L. *minor*, less), a thin, flat, triangular muscle, situated at the upper part of the thorax, beneath the pectoralis major.

pectoriloquy, n., *pĕkt'·ŏr·ĭl'·ō·kwĭ* (L. *pectus*, the breast, *pectŏris*, of the breast ; *loqui*, to speak), the apparent issuing of the voice from that part of the chest to which the ear or stethoscope is applied.

pectosic, a., *pĕkt·ŏz'·ĭk*, another name for 'pectic,' which see.

pectus, n., *pĕkt'·ŭs* (L. *pectus*, a breast), the breast ; the thorax or chest.

pedal, a., *pĕd'·ăl* (L. *pedālis*, of or belonging to a foot—from *pes*, a foot), connected with the foot.

pedate, a., *pĕd'·āt* (L. *pedātus*, footed — from *pedes*, feet), in *bot.*, having divisions like the feet ; having a palmate leaf of three lobes, the lateral lobes bearing other equally large lobes on the edges next the middle lobe : **pedatifid**, a., *pĕd·ăt'·ĭ·fĭd* (L. *findo*, I divide), in *bot.*, applied to a leaf whose parts are not entirely separate, but divided as a pedate one ; having the divisions of the lobes extending only half-way to the midrib : **pedatinerved,** a., *pĕd·ăt'·ĭ·nérvd* (L. *nervus,* a nerve), in *bot.,* having the veins of a leaf arranged in a pedate manner : **pedatipartite,** a., *pĕd·ăt'·ĭ·pârt'·ĭt* (L. *partītus,* divided), in *bot.,* having the venation of a leaf pedate, and the lobes almost free : **pedatisect,** a., *pĕd·ăt'·ĭ·sĕkt* (L. *sectus,* cut), having the veining of a leaf pedate, and the divisions of the lobes extending nearly to the midrib.

pedicel, n., *pĕd'·ĭs·ĕl* (L. *pediculus,* a small foot-stalk—from *pes,* a foot), a small, short foot-stalk of a leaf, flower, or fruit ; the foot-stalk or stem by which certain lower animals are attached : **pedicellate,** a., *pĕd'·ĭs·ĕl'·lāt,* supported by a pedicel.

pedicellariæ, n. plu., *pĕd'·ĭ·sĕl·lār'·ĭ·ē* (L. *pedicellus,* a louse), certain singular appendages found in many Echinoderms, attached to the surface of the body, and resembling a little beak or forceps supported on a stalk.

pedicle, n., *pĕd'·ĭ·kl* (see 'pedicel'), a little stem ; a pedicel.

Pedipalpi, n. plu., *pĕd'·ĭ·pălp'·ĭ* (L. *pedes,* feet ; *palpo,* I feel), an Order of Arachnida, comprising the scorpions, etc. : **pedipalpous,** a., *pĕd'·ĭ·pălp'·ŭs,* having feelers in the form of pincers, or armed with two claws, as the scorpions.

peduncle, n., *pĕd·ŭngk'·l* (L. *pĕd·ŭncŭlus,* a little foot, a foot-stalk—from *pes,* a foot), in *bot.,* a stem or stalk which supports one flower or fruit, or several ; in *zool.,* the muscular process by which certain Brachiopods are attached ; the stem which bears the body in barnacles ; in *anat.,* applied to different prolongations or appendices of the brain ; the constricted attachment or neck of a tumour : **pedunculate,** a., *pĕd·* *ŭngk'·ŭl·āt,* having a peduncle ; growing on a peduncle.

pelagic, a., *pĕl·ădj'·ĭk* (Gr. *pelagos,* the sea), growing in many distant parts of the ocean ; inhabiting the open ocean.

Pelargonium, n., *pĕl'·ăr·gōn'·ĭ·ŭm* (Gr. *pelargos,* a stork, a crane), a favourite and extensive genus of beautiful plants, Ord. Geraniaceæ, so named from the fancied resemblance of their capsules to the head and beak of a stork : **Pelargonium triste,** *trĭst'·ē* (L. *tristis,* sad, mournful), a species whose tuberous or moniliform roots are eatable.

pellicle, n., *pĕl'·lĭ·kl* (L. *pellicula,* a small skin—from *pellis,* skin), a thin skin or film ; in *bot.,* the outer, cuticular covering of plants.

pellitory, n., *pĕl'·lĭt·ŏr·ĭ* (Sp. *pelitre,* the pellitory of Spain), a plant from Spain, the 'Anacyclus pyrethrum' or 'Anthemis pyrethrum,' Ord. Compositæ, Sub-ord. Corymbiferæ, whose root is an irritant and sialogogue.

peloria, n., *pĕl·ōr'·ĭ·ă* (Gr. *pĕlōrĭos,* monstrous—from *pelōr,* a monster), in *bot.,* the five-spurred 'Linaria vulgaris,' instead of one-spurred, thus becoming symmetrical — so named as its first discovery was deemed marvellous ; the reversion of an irregular flower to the regular form : **pelorisation,** n., *pĕl'·ōr·ĭz·ā'·shŭn,* the act or process of the reversion of a flower, usually irregular, to the regular form.

pelta, n., *pĕlt'·ă* (L. *pelta,* a target), in *bot.,* the target-like apothecium of certain Lichens, the **peltidea,** *pĕlt·ĭd'·ē·ă,* without a distinct exciple : **peltate,** a., *pĕlt'·āt,* shield-like ; fastened to the stalk by a point within the margin : **peltate hairs,** hairs which are attached by their middle, or nearly so.

pelvis, n., *pĕlv'·ĭs* (L. *pelvis,* a

basin), in *anat.*, the bony cavity which contains the organs of generation, and made up of the 'two ossa innominata,' the 'sacrum,' and the 'coccyx'; the basal portion of the cup of crinoids ; the expanded part of the ureter which joins the kidney: **pelvic**, a., *pĕlv'·ĭk*, of or relating to the pelvis : **pelvic extremity,** the lowest part of the pelvis at the fork : **pelvic cellulitis,** *sĕl'·ŭl·ĭt'·ĭs* (L. *cella*, a store-room ; *itis*, inflammation), an inflammatory symptom of the cellular tissue which surrounds the bladder and womb.

pemphygus, n., *pĕmf'·ĭg·ŭs* (Gr. *pemphix*, a blister, *pemphĭgos*, of a blister), in *med.*, an eruption of bulla on the skin of various sizes.

Penæaceæ, n. plu., *pĕn'·ē·ā'·sĕ·ē* (after *P. Pena*, an early botanist and author), the Sarcocollads, or Sarcocol family, a small Order of evergreen shrubs : **Penæeæ,** n. plu., *pĕn·ē'·ĕ·ē*, a Sub-order : **Penæa,** n., *pĕn·ē'·ā*, a genus of plants : **Penæa sarcocolla,** *sărk'·ō·kŏl'·lă* (Gr. *sarx*, flesh ; *kolla*, glue), a species which with others is supposed to yield the gum-resin called Sarcocol, used by the Hottentots in dressing wounds.

pendulous, a., *pĕnd'·ŭl·ŭs* (L. *pendŭlus*, hanging—from *pendĕŏ*, I hang downwards), in *bot.*, inclined so that the apex is pointed vertically downwards ; applied to ovules which are hung from the upper part of the ovary.

penicillate, a., *pĕn'·ĭs·ĭl'·lāt* (L. *pĕnĭcillum*, a little tail, a painter's brush), in *bot.*, pencilled ; having a tufted stigma resembling a camel-hair brush, as in the nettle ; bordered or tipped with pencil-like hairs.

Penicillium, n., *pĕn'·ĭs·ĭl'·lĭ·ŭm* (L. *pĕnĭcillum*, a painter's brush or pencil), a genus of plants, Ord.

Fungi, so named from the form of their filaments : **Penicillium glaucum,** *glāwk'·ŭm* (L. *glaucus*, bluish-grey), one of the most common moulds occurring in organic infusions, on books, etc.

penis, n., *pēn'·ĭs* (L. *pēnis*, a tail), the male organ of generation.

penniform, a., *pĕn'·ni·form* (L. *penna*, a feather ; *forma*, shape), resembling the plume of a feather ; having the appearance of the feather of a pen.

penninerved, a., *pĕn'·ni·nĕrvd* (L. *penna*, a feather ; *nervus*, a nerve), having ribs running straight from the midrib to the margin ; having veins disposed like the parts of a feather.

pentacoccous, a., *pĕnt'·ă·kŏk'·ŭs* (Gr. *pente*, five ; *kokkos*, a kernel), splitting into fine cocci ; having five grains or seeds.

Pentadesma, n., *pĕnt'·ă·dĕz'·mă* (Gr. *pente*, five ; *dĕsmē*, a bundle, a bunch), a genus of handsome, lofty-growing trees, Ord. Guttiferæ, which have their stamens disposed in five bundles : **Pentadesma butyracea,** *bŭt'·ĭr·ās'·ĕ·ă* (L. *bŭtyrum*, butter), the butter and tallow tree of Sierra Leone, so named from the solid oil furnished by the fruit.

pentagonal, a., *pĕnt·ăg'·ŏn·ăl* (Gr. *pente*, five ; *gonĕă*, a corner or angle), in *bot.*, having five angles with convex spaces between them.

pentagynous, a., *pĕnt·ădj'·ĭn·ŭs* (Gr. *pente*, five ; *gunē*, a female), in *bot.*, having five styles.

pentamerous, a., *pĕnt·ăm'·ĕr·ŭs* (Gr. *pente*, five ; *mĕros*, a part), in *bot.*, composed of five parts ; having its different whorls in fives, or multiples of that number ; in *zool.*, having five joints on the tarsus of each leg : **pentamera,** n. plu., *pĕnt·ăm'·ĕr·ă*, a section of the beetle tribe having five joints on the tarsus of each leg.

pentandrous, a., *pĕnt·ănd'·rŭs* (Gr. *pente*, five ; *anēr*, a male, *andros*, of a male), in *bot.*, having five stamens, as a flower : **pentangular, a.,** *-ăng'·gŭl·ăr* (L. *angulus*, an angle or corner), having five angles : **pentapetalous, a.,** *-pĕt'·ăl'·ŭs* (Gr. *petalon*, a leaf), having five petals : **pentaphyllous, a.,** *-fĭl'·lŭs* (Gr. *phullon*, a leaf), having five leaves : **pentasepalous, a.,** *-sĕp'·ăl·ŭs* (Eng. *sepal*), having five sepals.

Pentastoma, n. plu., *pĕnt·ăs'·tŏm·ă* (Gr. *pente*, five ; *stŏma*, a mouth, *stŏmăta*, mouths), a genus of parasitic worms having five mouths or openings : **pentastomous, a.,** *pĕnt·ăs'·tŏm·ŭs*, having five mouths or openings.

pepo, n., *pĕp'·ō*, **peponida, n. plu.,** *pĕp·ŏn'·ĭd·ă* (L. *pĕpo*, Gr. *pĕpōn*, a large melon, *pĕpōnos*, of a large melon), a succulent, one-celled fruit, with seeds borne on three parietal placentas, which comprise the fruit of the Melon, Cucumber, and other Cucurbitaceæ.

pepsin, n., *pĕps'·ĭn* (Gr. *pepsis*, a digesting, a cooking—from *pepto*, I digest), the digestive principle secreted by the stomach, used as an aid to promote digestion, and obtained chiefly from pigs' and calves' stomachs : **peptone, n.,** *pĕp'·tōn*, a compound resulting from the action of pepsin, along with greatly diluted hydrochloric acid, on albuminous substances : **peptic, a.,** *pĕp'·tĭk*, relating to or promoting digestion.

percurrent, a., *pĕr·kŭr'·rĕnt* (L. *per*, through ; *currens*, running), in *bot.*, running through from top to bottom ; extending throughout the entire length.

percussion, n., *pĕr·kŭsh'·ŭn* (L. *percussus*, thrust or pierced through ; *percussĭŏ*, a beating or striking), the art of ascertaining the physical condition of internal organs by tapping the parts over them with a plessor, which may be the finger (immediate percussion); or by rapping with any plessor upon a pleximeter of ivory, etc., placed over the part (mediate percussion).

perenchyma, n., *pĕr·ĕng'·kĭm·ă* (Gr. *pēra*, a sac ; *engchuma*, what is poured in, an infusion ; *cheuma*, tissue), in *bot.*, cellular tissue containing starchy matter.

perennial, a., *pĕr·ĕn'·nĭ·ăl* (L. *perennis*, that lasts the year through — from *per*, through ; *annus*, a year), lasting through the year ; flowering for several years.

perennibranchiata, n. plu., *pĕr·ĕn'·nĭ·brăngk'·ĭ·āt'·ă* (L. *perennis*, that lasts the year through ; Gr. *brangchia*, gills), those Amphibia in which the gills are permanently retained throughout life : **perennibranchiate, a.,** *-brăngk'·ĭ·āt*, having the gills remaining throughout life, as certain Amphibians.

Pereskia, n., *pĕr·ĕsk'·ĭ·ă* (after *Pieresk*, a botanist of Aix in Provence), a genus of grotesque and ornamental plants, Ord. Cetaceæ : **Pereskia aculeata,** *ăk·ūl'·ĕ·āt'·ă* (L. *acūlĕātus*, thorny, prickly), a species which produces the Barbadoes gooseberry, used as an article of diet in W. Indies.

perfoliate, a., *pĕr·fōl'·ĭ·āt* (L. *per*, through ; *fŏlium*, a leaf), in *bot.*, applied to a leaf having the lobes of the base so united as to appear as if the stem ran through it.

perforans, a., *pĕrf'·ŏr·ănz* (L. *perfŏro*, I pierce through, *perfŏrans*, piercing, *perfŏrātus*, pierced), a muscle, so named from its perforating the tendon of the flexor sublimis : **perforatus, a.,** *pĕrf'·ŏr·āt'·ŭs*, a muscle, so named from its tendon being perforated by the tendon of the flexor profundus.

pergamentaceous, a., *pĕrg'·ȧ·mĕnt·ā'·shŭs* (L. *pergaména*, parchment), of the texture of parchment.

perianth, n., *pĕr'·ĭ·ănth* (Gr. *peri*, around; *anthos*, a flower), in *bot.*, a general name for the floral envelope; the external floral whorls which surround the stamens and pistil—in this sense including calyx and corolla.

periblem, n., *pĕr'·ĭ·blĕm* (Gr. *periblēma*, clothing, a cloak—from *peri*, around; *ballo*, I throw), in *bot.*, a collection of layers of cells beneath the dermatogen out of which the cortex arises.

pericambium, n., *pĕr'·ĭ·kăm'·bĭ·ŭm* (Gr. *peri*, around; new L. *cambium*, nutriment; L. *cambĭŏ*, I change), in *bot.*, the outer layer of the plerome.

pericardium, n., *pĕr'·ĭ·kȧrd'·ĭ·ŭm* (Gr. *peri*, round about; *kardĭȧ*, the heart), the membranous bag which surrounds and encloses the heart: **pericardiac**, a., *pĕr'·ĭ·kȧrd'·ĭ·ăk*, of or pert. to the pericardium: **pericarditis**, n., *pĕr'·ĭ·kȧrd·īt'·ĭs*, inflammation of the membrane which surrounds the heart.

pericarp, n., *pĕr'·ĭ·kȧrp* (Gr. *perikarpion*, the covering of seed—from *peri*, around; *karpos*, fruit), the part of the fruit immediately investing the seed; the covering, shell, or rind of fruits: **pericarpial**, a., *pĕr'·ĭ·kȧrp'·ĭ·ăl*, of or pert. to a pericarp.

perichætium, n., *pĕr'·ĭ·kē'·shĭ·ŭm* (Gr. *peri*, round about; *chaitē*, long loose flowing hair, as of the mane of a horse), the leaves that surround the base of the fruitstalk of some mosses: **perichætial**, a., *pĕr'·ĭ·kē'·shĭ·ăl*, pert. to the perichætium.

perichondrium, n., *pĕr'·ĭ·kŏnd'·rĭ·ŭm* (Gr. *peri*, round about; *chondros*, cartilage), in *anat.*, a fibrous membrane which covers cartilages.

pericladium, n., *pĕr'·ĭ·klȧd'·ĭ·ŭm* (Gr. *peri*, round about; *klados*, a branch), in *bot.*, the lowermost clasping portion of sheathing petioles; the large sheathing petiole of Umbelliferæ.

periclinium, n., *pĕr'·ĭ·klĭn'·ĭ·ŭm* (Gr. *peri*, round about; *klinē*, a bed), in *bot.*, the involucre of composite flowers.

pericranium, n., *pĕr'·ĭ·krān'·ĭ·ŭm* (Gr. *peri*, round about; *kranion*, the skull), the fibrous membrane which goes round or invests the skull, and corresponds to the periosteum of other bones.

periderm, n., *pĕr'·ĭ·dĕrm* (Gr. *peri*, round about; *derma*, skin), in *bot.*, the outer layer of bark; the stratified cork envelope which replaces the epidermis in parts of vigorous growth; in *zool.*, the hard cuticular layer which is developed by the Cœnosarc of certain of the Hydrozoa.

peridium, n., *pĕr'·ĭd'·ĭ·ŭm* (Gr. *perideo*, I wrap round), in *bot.*, the coat immediately enveloping the sporules of the lower tribes of acotyledons; a covering, as of a puff-ball: **peridiola**, n. plu., *pĕr'·ĭd·ĭ·ŏl·ȧ* (L. dim. of *peridium*), a number of small peridia inclosed in a general covering.

perigastric, a., *pĕr'·ĭ·găst'·rĭk* (Gr. *peri*, round about; *gaster*, the belly), in *zool.*, applied to the cavity which surrounds the stomach and other viscera, corresponding to the abdominal cavity in the higher animals.

perigone, n., *pĕr'·ĭ·gōn* (Gr. *peri*, round about; *goneus*, a parent; *gunē*, a female), a floral envelope; a synonym for *perianth*, especially when reduced to a single floral whorl: **perigonium**, n., *pĕr'·ĭ·gōn'·ĭ·ŭm*, a barren flower in mosses having involucral scales.

perigynium, n., *pĕr'·ĭ·jĭn'·ĭ·ŭm* (Gr. *peri*, round about; *gunē*, a female), in *bot.*, the covering of the pistil in the genus Carex ; the membranous perianth of sedges : **perigynous**, a., *pĕr·ĭdj'·ĭn·ŭs*, growing upon the calyx, or some part which surrounds the ovary in a flower ; applied to corolla and stamens when attached to the calyx.

perilymph, n., *pĕr'·ĭ·lĭmf* (Gr. *peri*, round about ; L. *lympha*, water), the clear fluid secreted by the serous membrane which lines the osseous labyrinth of the ear.

perimetritis, n., *pĕr'·ĭ·mĕt·rīt'·ĭs* (Gr. *peri*, round about ; *mētra*, the womb), the inflammation of the peritoneal covering of the uterus, usually involving neighbouring parts.

perimysium, n., *pĕr'·ĭ·mĭz'·ĭ·ŭm* (Gr. *peri*, round about ; *mus*, a muscle), an outward investment or sheath of areolar tissue, which surrounds an entire muscle, and sends partitions inwards between the fasciculi, furnishing to each of them a special sheath.

perinæum, n., *pĕr'·ĭn·ē'·ŭm* (Gr. *perinaion*, the part between the anus and the scrotum—from *peri*, round about ; *naiō*, I inhabit), the region of the lower part of the body, having the anus at its centre, bounded in front by the genitals, and at the sides by the inner surfaces of the thighs : **perinæal**, a., *pĕr'·ĭn·ē'·ăl*, of or pert. to the perinæum.

perineurium, n., *pĕr'·ĭ·nūr'·ĭ·ŭm* (Gr. *peri*, round about ; *neuron*, a nerve), the coarser sheathing of the nerves, and nervous cords— the general term for the sheathing being ' neurilemma.'

periodontal, a., *pĕr'·ĭ·ō·dŏnt'·ăl* (Gr. *peri*, round about ; *odous*, a tooth, *odontos*, of a tooth), applied to the lining membrane of a tooth-socket : **periodontitis**, n., *pĕr'·ĭ·ō·dŏnt·īt'·ĭs*, the inflammation of the lining membrane of a tooth-socket.

periosteum, n., *pĕr'·ĭ·ŏst'·ĕ·ŭm* (Gr. *peri*, round about ; *osteon*, a bone), the fibrous sensitive membrane immediately covering the bone, which performs an important part in its nourishment : **periosteal**, a., *pĕr'·ĭ·ŏst'·ĕ·ăl*, pert. to or connected with the periosteum : **periostitis**, n., *pĕr'·ĭ·ŏst·īt'·ĭs*, inflammation of the covering membrane of the bone.

periostracum, n., *pĕr'·ĭ·ŏst'·răk·ŭm* (Gr. *peri*, round about ; *ostrăkon*, a shell), in *zool.*, the layer of epidermis which covers the shell in most of the Mollusca.

periphery, n., *pĕr·ĭf'·ĕr·ĭ* (Gr. *periphereiă*, a circumference— from *peri*, round about ; *pherō*, I carry), in *bot.*, the outer stratum of cells in a cylindrical frond : **peripherical**, a., *pĕr'·ĭ·fĕr'·ĭk·ăl*, of or pert. to a periphery ; in *bot.*, having an embryo curved so as to surround the albumen, following the inner part of the covering of the seed.

periplast, n., *pĕr'·ĭ·plăst* (Gr. *peri*, round about ; *plasso*, I mould), in *zool.*, the intercellular substance or matrix in which the organised structures of a tissue are embedded.

Periploca, n., *pĕr·ĭp'·lŏk·ă* (Gr. *periplŏkē*, a plaiting, a coiling round—from *peri*, round about ; *plekō*, I twine), a handsome genus of plants, Ord. Asclepiadaceæ, so named from the habit of the species : **Periploca Mauritiana**, *măw·rĭsh'·ĭ·ān'·ă* (after Prince *Maurice*, of Nassau), a species, the source of the Bourbon Scammony, and a purgative : **P. Græca**, *grēk'·ă* (of or from *Greece*), a hardy climbing plant, valuable for covering naked walls.

peripneumonia, n., *pĕr·ĭp'·nŭ·mōn'·ĭ·ă* (Gr. *peripneumonia*, inflammation of the lungs—from *peri*, round about ; *pneumon*, a lung), inflammation of the lungs ; pneumonia.

perisarc, n., *pĕr'·ĭ·sȧrk* (Gr. *peri*, round about ; *sarx*, flesh), a general term for the chitinous envelope secreted by many of the Hydrozoa.

perisome, n., *pĕr'·ĭ·sōm* (Gr. *peri*, round about ; *sōma*, body), the coriaceous or calcareous integument of the Echinodermata.

perisperm, n., *pĕr'·ĭ·spĕrm* (Gr. *peri*, round about ; *sperma*, seed), in *bot.*, the innermost envelope of the seed ; the albumen or nourishing matter stored up with the embryo in the seed : **perispermic, a.**, *pĕr'·ĭ·spĕrm'·ĭk*, of or pert. to the perisperm.

perispheric, a., *pĕr'·ĭ·sfĕr'·ĭk*, also **perispherical, a.**, *-sfĕr'·ĭk·ăl* (Gr. *peri*, round about ; *sphaira*, a sphere), having the form of a ball ; globular.

perisporangium, n., *pĕr'·ĭ·spōr·ănj'·ĭ·ŭm* (Gr. *peri*, round about ; *spora*, seed ; *anggos*, a vessel), in *bot.*, the indusium of ferns when it surrounds the Sori.

perispore, n., *pĕr'·ĭ·spōr* (Gr. *peri*, round about ; *spora*, seed), the membrane or case surrounding a spore ; the mother-cell of spores in Algæ.

Perissodactyla, n. plu., *pĕr·ĭs'·sō·dăk'·tĭl·ȧ* (Gr. *perissos*, uneven ; *daktulos*, a finger), the hoofed quadrupeds, or Ungulata, in which the feet have an uneven number of toes.

peristaltic, a., *pĕr'·ĭ·stălt'·ĭk* (Gr. *peristaltikos*, drawing together all round—from *peri*, round about ; *stello*, I send), applied to the peculiar worm-like motion of the intestines by which their contents are gradually forced downwards ; circular contraction from above downwards.

peristome, n., *pĕr'·ĭs·tōm* (Gr. *peri*, round about ; *stoma*, a mouth), in *bot.*, the ring of bristles or toothed fringe situated around the orifice of the seed-vessels in mosses ; the opening of the sporangium of mosses after the removal of the calyptra and operculum ; in *zool.*, the space between the mouth and the margin of the calyx in Vorticella, or between the mouth and the tentacles in a sea-anemone ; the lip or margin of the mouth of a univalve shell : **peristomatic, a.**, *pĕr·ĭs'·tŏm·ăt'·ĭk*, of or pert. to a peristome ; in *bot.*, having cells surrounding a stoma.

perithecium, n., *pĕr'·ĭ·thē'·shĭ·ŭm*, **perithecia, n. plu.**, *pĕr'·ĭ·thē'·shĭ·ȧ* (Gr. *peri*, around ; *thēkē*, a box or case), in *bot.*, the envelope surrounding the masses of fructification in some Fungi and Lichens ; a hollow conceptacle in Lichens, containing spores, and having an opening at the end.

peritonæum, or peritoneum, n., *pĕr'·ĭ·tŏn·ē'·ŭm* (Gr. *peritonaion*, what is stretched round or over—from *peri*, round about ; *teino*, I stretch), a serous and smooth membrane which lines the whole internal surface of the abdomen, and envelopes more or less completely the several parts of the viscera, retaining them in their proper places, and at the same time allowing them to move freely when required: **peritoneal, a.**, *pĕr'·ĭ·tŏn·ē'·ăl*, of or pert. to the peritoneum : **peritonitis, n.**, *pĕr'·ĭ·tŏn·īt'·ĭs*, inflammation of the peritoneum.

peritropal, a., *pĕr·ĭt'·rŏp·ăl*, also **peritropous, a.**, *pĕr·ĭt'·rŏp·ŭs* (Gr. *peri*, round about ; *tropē*, a turning), in *bot.*, applied to the axis of a seed perpendicular to the axis of the pericarp to which it is attached.

perityphlitis, n., *pĕr'·ĭ·tĭf·līt'·ĭs* (Gr. *peri*, round about ; *tuphlos*, blind), inflammation around the cæcum.

perivascular, a., *pĕr'·ĭ·văsk'·ūl·ăr* (Gr. *peri*, round about ; L. *vasculum*, a small vessel), applied to canals which surround and enclose the blood-vessels of the brain and spinal cord; also called 'lymph' channels, from their containing lymph.

perivisceral, a., *pĕr'·ĭ·vĭs'·ĕr·ăl* (Gr. *peri*, round about; L. *viscĕra*, the inwards), in *zool.*, applied to the space surrounding the viscera.

peronate, a., *pĕr'·ŏn·āt* (L. *pērōn-ātus*, rough-booted), in *bot.*, thickly covered with woolly matter, becoming powdery or mealy externally.

peroneal, a., *pĕr'·ŏn·ē'·ăl* (Gr. *perŏnē*, the fibula or small bone of the leg), in *anat.*, belonging to or lying near the fibula, as certain muscles connecting it with the foot.

peroneus longus, *pĕr'·ŏn·ē'·us long'·gŭs* (Gr. *perŏnē*, the fibula or small bone of the leg ; L. *longus*, long), in *anat.*, a muscle situated at the upper part of the outer side of the leg : **peroneus brevis,** *brĕv'·ĭs* (L. *brĕvis*, short), a muscle lying beneath the 'peroneus longus,' and is shorter and smaller than it, both muscles extending the foot upon the leg, and serving to steady the leg upon the foot : p. **tertius,** *tĕr'·shĭ·ŭs* (L. *tertĭus*, the third), the smallest and lowest in attachment, passing from the leg to the fifth metatarsal bone.

Persea, n., *pĕrs'·ē·ă* (L. Gr. *persĕa*, a sacred Egyptian tree), a genus of ornamental trees, Ord. Lauraceæ : **Persea gratissima,** *grăt·ĭs'·sĭm·ă* (L. *grātus*, pleasing, agreeable, *gratissimus*, most pleasing), a species yielding a pear-shaped, succulent fruit, called Alligator pear, and containing a fixed oil.

persistent, a., *pĕr·sĭst'·ĕnt* (L. *per*, through ; *sisto*, I stand), in *bot.*, not falling off ; remaining attached to the axis until the part bearing it is matured.

personate, a., *pĕr'·sŏn·āt* (L. *persona*, a mask, a character), in *bot.*, a form of monopetalous corolla, where the orifice of the tube is closed by an inflated projection of the throat, the whole resembling a gaping mask ; having a fanciful resemblance to a face.

pertuse, a., *pĕr·tūs'* (L. *pertūsus*, perforated — from *per*, through ; *tūsus*, beat), pierced irregularly ; in *bot.*, having slits or holes : **pertusate,** a., *pĕr·tūs'·āt*, pierced at the apex.

pertussis, n., *pĕr·tūs'·ĭs* (L. *per*, intensive prefix ; *tussis*, a cough), hooping-cough.

perula, n., *pĕr'·ŭl·ă*, **perulæ,** n. plu., *pĕr'·ŭl·ē*, or **perules,** *pĕr'·ūlz* (L. *perula*, a little pocket), in *bot.*, the scales of the leaf-bud.

Peruvian or **Jesuits' bark,** n., *pĕr·ōv'·ĭ·ăn* (of or from *Peru*), the popular name for various species of Cinchona. Ord. Rubiaceæ, growing abundantly in Upper Peru : see 'Cinchoneæ.'

pes accessorius, *pĕs ăk'·sĕs·sōr'·ĭ·ŭs* (L. *pes*, a foot; *accessorius*, accessory or assistant—from *accessus*, an approach), a white eminence or cerebral convolution placed between the hippocampus major and minor : **pes anserinus,** *ăns'·ĕr·īn'·ŭs* (L. *anserīnus*, pert. to a goose), the goose's foot, forming the temporo-facial, and the cervico-facial division of the facial nerve, having numerous outspreading branches: p. **hippocampi,** *hĭp'·pō·kămp'·ī* (Gr. *hippos*, a horse ; *kampto*, I bend or curve), a number of rounded elevations with intervening depressions at the lower extremity of the lateral ventricle of the brain, so called as presenting a resemblance to the paw of an animal.

pessary, n., *pĕs'·săr·ĭ*, **pessaries,** n. plu., *pĕs'·săr·ĭz* (It. *pessario*, F. *pessaire*, L. *pessum*, Gr. *pesson*, a pessary), supports or medicaments for intravaginal use.

pestle, n., *pĕst'·l* or *pĕs'·l* (L.

pistillum, the pestle of a mortar), any heavy article for pounding and mixing substances in a mortar : **pestillation**, n., *pĕst'·ĭl·lā'·shŭn*, the act of pounding in a mortar.

petal, n., *pĕt'·ăl* (Gr. *petalon*, a leaf), in *bot.*, the leaf of an expanded flower; one of the separate parts of a corolla or flower : **petaline**, a., *pĕt'·ăl·ĭn*, of or pert. to a petal : **petalody**, n., *pĕt·ăl'·ŏd·ĭ* (Gr. *eidos*, resemblance), a state in which sepals become coloured like petals ; the conversion of parts of the flower into petals : **petaloid**, a., *pĕt'·ăl·ŏyd* (Gr. *eidos*, resemblance), having the appearance or colour of a petal ; in *zool.*, shaped like the petal of a flower : **petaloideæ**, n. plu., *pĕt'·ăl·ŏyd'·ĕ·ē*, a term applied to a sub-class of plants whose flowers have usually a perianth consisting either of verticillate leaves, or of a few whorled scales, in the former case sometimes separable into calyx and corolla, and often coloured.

petechiæ, n. plu., *pĕt·ĕk'·ĭ·ē* (It. *petecchie*), in *med.*, purple or dark-red spots which appear on the skin when there is much disorder in the blood, as in fevers of a malignant type : **petechial**, a., *pĕt·ĕk·ĭ·ăl*, of or connected with diseases having the characteristics of petechiæ, as 'petechial plague.'

petiole, n., *pĕt'·ĭ·ōl* (L. *petĭŏlus*, a little foot—from *pes*, a foot), in *bot.*, the footstalk of a leaf connecting the blade with the stem : **petiolar**, a., *pĕt·ĭ'·ōl·ăr*, also **petiolary**, a., *pĕt·ĭ'·ōl·ăr·ĭ*, pert. to or growing on a small stalk ; having a stalk or petiole: **petiolate**, a., *pĕt·ĭ'·ōl·āt*, growing on a petiole : **petiolule**, n., *pĕt·ĭ'·ōl·ūl*, the stalk of a leaflet in a compound leaf.

Petiverieæ, n. plu., *pĕt'·ĭ·vĕr·ĭ'·ĕ·ē* (after *Petiver*, a London apothecary), a Sub-order of plants having erect seeds, Ord. Phytolaccaceæ: **Petiveria**, n., *pĕt'·ĭv·ēr'·ĭ·ă*, a genus of ornamental plants: **Petiveria alliacea**, *ăl'·lĭ·ā'·sĕ·ă* (L. *allĭum*, garlic), the guinea-hen-weed, so named from these animals being partial to it.

Petroselinum, n., *pĕt'·rŏ·sĕl·ĭn'·ŭm* (Gr. *petros*, a rock ; *selinon*, parsley), a genus of plants, Ord. Umbelliferæ : **Petroselinum sativum**, *săt·ĭv'·ŭm* (L. *satīvum*, that is sown or planted), common parsley.

petrous, a., *pĕt'·rŭs* (L. *petrōsus*, full of rocks—from *petra*, a stone), hard ; stony ; in *anat.*, applied to a dense, solid mass of bone, forming a part of the temporal bone ; designating a ganglion situated in the lower border of the petrous portion of the temporal bone : **petrosal**, a., *pĕt·rŏz'·ăl*, in same sense as 'petrous' ; the ear-capsule bone in a fish : **petrosal nerve**, a branch of the Vidian nerve : **petro-occipital**, *pĕt'·rŏ·ŏk·sĭp'·ĭt·ăl*, connected with the petrous portion of the temporal bone, and with the occipital bone.

phænogamous, a., *fēn·ŏg'·ăm·ŭs* (Gr. *phaino*, I show, I manifest ; *gamos*, marriage), in *bot.*, having conspicuous flowers.

Phæosporeæ, n. plu., *fē'·ŏ·spŏr'·ĕ·ē* (Gr. *phaios*, dusky ; *spora*, seed), in *bot.*, a division of Melanosporeæ, or olive-coloured sea-weeds, which possess zoospores ; Algæ having an olive-green or olive-brown colour, and cellular or filamentous structure.

phagedæna, n., *fădj'·ĕd·ēn'·ă* (Gr. *phago*, I devour, I gnaw), a variety of ulceration which destroys the tissues more rapidly, and to a greater extent, than ordinary forms of ulcer ; gangrenous ulceration.

phaiophyll, n., *fī'·ŏ·fŭl* (Gr. *phaios*, brown ; *phullon*, a leaf), a group of colouring matters in the leaves of plants, comprising various browns, soluble in water.

phalanx, n., *făl'-ănks* (Gr. *phalanyx*, a line of battle, a battalion, *phalanggos*, of a line of battle), in *anat.*, a term applied to one of the small bones of a finger or toe, thus, a finger or toe has three phalanges : **phalanges**, n. plu., *făl-ănj'-ēz*, in *bot.*, bundles of stamens ; stamens divided into lobes like a partite or compound leaf ; in *anat.*, the bones of the fingers or toes, so named from their regularity, as soldiers in the ranks : **phalangeal**, a., *făl'-ăn-jē'-ăl*, pert. to the bones of the fingers or toes, which are arranged in rows.

Phalaris, n., *făl'-ăr-ĭs* (Gr. *phalăros*, white, brilliant), a genus of plants, Ord. Gramineæ, so named from their shining seeds : **Phalaris Canariensis**, *kăn-ăr'-ĭ-ĕns'-ĭs* (new L., of or from the *Canary Islands*), the source of the common canary-seed given to birds.

phanerogamous, a., *făn'-ĕr-ŏg'-ăm-ŭs* (Gr. *phaneros*, manifest ; *gamos*, marriage), applied to plants having conspicuous flowers, containing pistils and stamens—opposed to cryptogamic ; same as phænogamous : **phanerogam**, n., *făn-ĕr'-ŏg-ăm*, one of the plants which have conspicuous flowers.

pharmaceutic, a., *fărm'-ă-sūt'-ĭk*, also **pharmaceutical**, a., *-sūt'-ĭk-ăl* (Gr. *pharmăkeus*, a druggist ; *pharmăkeia*, the using of medicine), of or relating to pharmacy, or to the art of preparing medicines : **pharmaceutics**, n. plu., *-sūt'-ĭks*, the science of pharmacy, or of preparing medicines : **pharmaceutist**, n., *fărm'-ă-sūt'-ĭst*, one who practises pharmacy, or prepares medicines; an apothecary : **pharmacy**, n., *fărm'-ăs-ĭ*, the art of preparing and preserving substances to be used as

medicines; the occupation of a druggist : **pharmacopœia**, n., *fărm'-ăk-ō-pē'-yă* (Gr. *poiĕō*, I make), a book which contains authorised directions for the selection and preparation of substances to be used as medicines.

pharmacognosis, n., *fărm'-ă-kŏg-nōz'-ĭs*, also **pharmacognosy**, -*kŏg-nōz'-ĭ* (Gr. *pharmăkon*, medicine ; *gnōsis*, knowledge), the knowledge of drugs or medicines, their properties and operations ; the branch of Materia Medica which treats of simples, or unprepared medicines.

pharmacology, n., *fărm'-ăk-ŏl'-ō-jĭ* (Gr. *pharmăkon*, medicine; *logos*, discourse), a term used for Materia Medica ; more particularly the study of the action of drugs in the body.

pharynx, n., *făr'-ĭnks* (Gr. *pharungx*, the gullet or windpipe), the muscular pouch forming the back part of the mouth, and shaped like a funnel, terminating in the œsophagus or gullet : **pharyngeal**, a., *făr'-ĭn-jē'-ăl*, pert. to or connected with the pharynx : **pharyngitis**, n., *făr'-ĭn-jīt'-ĭs*, inflammation of the pharynx : **pharyngotomy**, n., *făr'-ĭng-gŏt'-ŏm-ĭ* (Gr. *tomē*, a cutting), the operation of making an incision into the pharynx : **Pharyngobranchii**, n. plu., *făr-ĭng'-gō-brăngk'-ĭ-ĭ* (Gr. *brangchia*, gills), an Order of Fishes comprising only the Lancelet.

Phaseoleæ, n. plu., *făz'-ĕ-ōl'-ĕ-ē* (L. *phasēlus*, an edible bean-pod, a light boat in the shape of a pod), a tribe of the Sub-ord. Papilionaceæ, so named from the fancied resemblance of the pods : **Phaseolus**, n., *făz-ĕ'-ŏl-ŭs*, a genus, mostly climbing plants, comprising the different varieties of kidney-bean, the flowers remarkable for the keel terminating in a twisted point : **Phaseolus multiflorus**, *mŭlt'-ĭ-flōr'-ŭs* (L.

multus, many; *flos*, a flower, *flōris*, of a flower), the scarlet-runner; and **P. radiatus**, *rād'·ĭ·āt'·ŭs* (L. *radĭātus*, rayed—from *radĭus*, a ray), are species whose roots are poisonous: **P. vulgaris**, *vŭlg·ār'·ĭs* (L. *vulgāris*, common), the common French or kidney-bean, or Haricot.

phelloderm, n., *fĕl'·lō·dĕrm* (Gr. *phĕllos*, the cork-tree; *derma*, skin), in *bot.*, the suberous cortical layer of epidermis formed on the inside of the cork Cambium: **phellogen**, n., *fĕl'·lō·jĕn* (Gr. *gĕnndō*, I produce), the cork Cambium.

phenic acid, *fĕn'·ĭk* (Gr. *phaino*, I show), carbolic acid; the hydrated oxide of phenyl; a product obtained chiefly from coal-tar: **phenyl**, n., *fĕn'·ĭl*, a radical hydrocarbon.

Philadelphaceæ, n. plu., *fĭl'·ă·dĕlf·ā'·sĕ·ē* (Gr. *philos*, dear, beloved; *adelphos*, a brother), the Syringa family, an Order of plants: **Philadelphus**, n., *fĭl'·ă·dĕlf'·ŭs*, a genus of handsome shrubs, producing elegant blossoms, having the appearance and smell of orange blossoms, but more powerful: **Philadelphus coronarius**, *kŏr'·ŏn·ār'·ĭ·ŭs* (L. *cŏrōnārĭus*, of or belonging to a wreath—from *cŏrōna*, a crown or wreath), the Syringa or Mock-orange, whose flowers have a strong orange odour, due to the presence of an oil.

phlebectasis, n., *flĕb·ĕk'·tăz·ĭs*, also **phlebectasia**, n., *flĕb'·ĕk·tāz'·ĭ·ă* (Gr. *phleps*, a vein, *phlĕbos*, of a vein; *ĕktăsis*, extension), dilatation or varicosity of a vein, or of part of a vein.

phlebitis, n., *flĕb·ĭt'·ĭs* (Gr. *phlĕps*, a vein, *phlĕbos*, of a vein), primary inflammation of a vein, which may be either acute or chronic: **endophlebitis**, n., *ĕn'·dō·flĕb·ĭt'·ĭs* (Gr. *endon*, within), inflammation of the inner coat of a vein: **mes-**

ophlebitis, n., *mĕs'·ō·flĕb·ĭt'·ĭs* (Gr. *mesos*, middle), inflammation in the middle coat of a vein: **periphlebitis**, n., *pĕr'·ĭ·flĕb·ĭt'·ĭs* (Gr. *peri*, round about), inflammation of the outer coat of a vein: **phlebotomy**, n., *flĕb·ŏt'·ŏm·ĭ* (Gr. *tomē*, a cutting), the operation of opening a vein to take blood from the body.

phleboidal, a., *flĕb·ōyd'·ăl* (Gr. *phleps*, a vein; *eidos*, resemblance), in *bot.*, applied to moniliform vessels; having the appearance of veins.

phlebolithes, n. plu., *flĕb'·ō·lĭthz* (Gr. *phleps*, a vein; *lithos*, a stone), concretions, termed vein-stones, which are found free in the cavity of the vessels, formed of concentric laminæ.

phlegm, n., *flĕm* (Gr. *phlegma*, inflammation, phlegm — from *phlego*, I burn), the bronchial mucus; the thick, viscid matter discharged by coughing: **phlegmatic**, a., *flĕg·măt'·ĭk*, abounding in phlegm; sluggish; heavy.

phlegmasia dolens, *flĕg·māzh'·ĭ·ă dōl'·ĕnz* (Gr. *phlegma*, inflammation; L. *dŏlĕns*, suffering, pain), white leg; inflammation of the veins and absorbents of the leg, often follows fevers, abortions, etc.

phlegmon, n., *flĕg'·mŏn* (Gr. *phlegma*, inflammation), a circumscribed inflammatory swelling, with increasing heat and pain, and tending to suppuration; an inflammatory tumour: **phlegmonous**, a., *flĕg'·mŏn·ŭs*, inflammatory; burning.

Phleum, n., *flē'·ŭm* (Gr. *phlĕōs*, an aquatic plant), a genus of agricultural grasses, Ord. Gramineæ, remarkable for the close, cylindrical form of their spike-like panicles: **Phleum pratense**, *prăt·ĕns'·ē* (L. *prātĕnsis*, growing in meadows), the Timothy or cat's-tail grass, early and productive, and freely introduced into pasture-lands.

phlœum, n., *flē'·ŭm*, also phlœm, n., *flĕm* (Gr. *phloĭos*, the bark of a tree), in *bot.*, the cellular portion of the bark, found immediately beneath the epidermis; the least portion of a fibro-vascular bundle, consisting at first of succulent thin-walled cells.

phlogiston, n., *flŏdj·ĭst'·ŏn* (Gr. *phlogistos*, burnt, set on fire—from *phlĕgo*, I burn), according to the theory of Stahl, a supposed principle or pure fire fixed in inflammable bodies, as distinguished from fire of combustion: phlogistic, a., *flŏdj·ĭst'·ĭk*, partaking of phlogiston; inflammatory.

phlorizin, n., *flŏr'·ĭz·ĭn* (Gr. *phloios*, bark; *rhiza*, a root), a white crystalline substance obtained from the bark of the roots of apple, pear, cherry, and plum trees, giving to the bark its bitter astringency: phloretin, n., *flŏr'·ĕt·ĭn*, a substance procured from phlorizin by dilute acids: phlorizein, n., *flŏr'·ĭz·ē'·ĭn*, a gum-like substance obtained from it by the action of oxygen and ammonia.

Phlox, n., *flŏks*, Phloxes, n. plu., *flŏks'·ĭz* (Gr. *phlox*, a flame), an extensive genus of elegant, favourite plants, Ord. Polemoniaceæ, so named from the appearance of the flowers, presenting lively red, purple, or white colours.

phlyctæna, n., *flĭk·tēn'·ă* (Gr. *phluktaina*, a vesicle), a small vesicle, containing a serous fluid: phlyctenoid, a., *flĭk·tēn'·ŏyd* (Gr. *eidos*, resemblance), bearing a resemblance to phlyctæna.

Phœnix, n., *fēn'·ĭks* (Gr. *phoinix*, a palm tree), a genus of noble palm trees, Ord. Palmæ, which includes the date: Phœnix dactylifera, *dăkt'·ĭl·ĭf'·ĕr·ă* (Gr. *daktulos*, a finger; L. *fero*, I bear), a lofty-growing palm of Arabia and Upper Egypt, having leaves

from six to eight feet long, from which many articles of domestic use are procured, as food, clothing, house-building, fibres and thread, ropes, and juice as wine: P. farinifera, *făr'·ĭn·ĭf'·ĕr·ă* (L. *farina*, meal; *fero*, I bear), a species which contains a farinaceous, nutritive substance in the heart of the stem: P. sylvestris, *sĭl·vĕst'·rĭs* (L. *silvĕstris*, woody—from *sĭlva*, a wood), produces the date sugar of Bengal.

phoranthium, n., *fŏr·ănth'·ĭ·ŭm* (Gr. *phorĕō*, I bear; *anthos*, a flower), in *bot.*, the receptacle of composite plants.

Phormium, n., *fŏrm'·ĭ·ŭm* (Gr. *phŏrmŏs*, a basket), a genus of very useful plants of New Zealand, etc., Ord. Liliaceæ, so named from the use made of it by the natives: Phormium tenax, *tĕn'·ăks* (L. *tĕnax*, holding fast, tenacious), the New Zealand flax, from which fibres are procured.

phosphate, n., *fŏs'·fāt* (Gr. *phos*, light; *phorĕō*, I bear), a combination of phosphoric acid with a base: phosphatic, a., *fŏs·făt'·ĭk*, pert. to phosphate: phosphide, n., *fŏs'·fīd*, a combination of phosphorus with a metal.

phosphorus, n., *fŏs'·fŏr·ŭs* (Gr. *phosphoros*, light-bringer—from *phos*, light; *phorĕō*, I bear), an elementary substance of a waxlike consistency, highly inflammable, always luminous in the dark in its ordinary state, obtained from bones: phosphorescence, n., *fŏs'·fŏr·ĕs'·ĕns*, the state of being luminous without sensible heat: phosphoric acid, an acid prepared from phosphorus by oxidation by means of nitric acid.

phosphuret, n., *fŏs'·fŭr·ĕt* (Eng. *phosphorus;* L. *uro*, I burn), a combination of phosphorus with a combustible body, or a metallic oxide: phosphuretted, a., *fŏs'·fŭr·ĕt'·ĕd*, combined with phosphorus.

photophobia, n., *fŏt'·ō·fŏb'·ĭ·ă* (Gr. *phos*, light, *photos*, of light; *phŏbĕō*, I dread), a dread or intolerance of light, a symptom common to many diseases of the eye.

phragma, n., *frăg'·mă*, phragmata, n. plu., *frăg'·măt·ă* (Gr. *phragma*, a fence or partition), in *bot.*, a transverse division or partition in fruits; a spurious dissepiment.

phragmacone, n., *frăg'·mă·kōn* (Gr. *phragma*, a fence or partition; *kōnŏs*, a cone), the chambered portion of the internal shell of a Belemnite.

phrenic, a., *frĕn'·ĭk* (Gr. *phrēn*, the heart or parts about it, *phrĕnos*, of the heart), of or pert. to the parts about the heart; the name of the nerve arising from the third, fourth, and fifth cervical nerves, which acts as motor of the diaphragm: phrenitis, n., *frĕn·ĭt'·ĭs*, inflammation of the brain or its membranes; delirium: phrenetic, a., *frĕn·ĕt'·ĭk*, liable to violent sallies of mental excitement: n., a person occasionally wild and erratic.

phthiriasis, n., *thĭr·ĭ'·ăs·ĭs* (Gr. *phtheiriăsis* — from *phtheir*, a louse), a diseased condition in which lice are bred on and infest the body; cutaneous invermination.

phthisis, n., *thĭs'·ĭs* (Gr. *phthisis*, a wasting—from *phthĭō*, I consume or waste away), pulmonary consumption, or wasting disease of the lungs, also called phthisis pulmonalis, *pŭl'·mŏn·āl'·ĭs* (L. *pulmo*, a lung, *pŭlmōnis*, of a lung), pulmonary phthisis: phthisic, n., *tĭz'·ĭk*, a wasting away; a person affected with phthisis: phthisical, a., *tĭz'·ĭk·ăl*, belonging to phthisis; consumptive: renal phthisis, scrofulous wasting of the kidney.

phycochrome, n., *fĭk'·ō·krōm* (Gr.

phukos, sea - weed; *chrōma*, colour), the colouring matter in Lichens, and in the lower Algæ.

phycocyanine, n., *fĭk'·ō·sĭ'·ăn·ĭn* (Gr. *phukos*, sea-weed; *kuănos*, blue), in *bot.*, the red colouring matter characteristic of Florideæ.

phycoerythrine, n., *fĭk'·ō·ĕr'·ĭth·rĭn* (Gr. *phukos*, sea - weed; *eruthros*, red), in *bot.*, the red colouring matter, soluble in water, found in Florideæ.

phycology, n., *fĭk·ŏl'·ō·jĭ* (Gr. *phukos*, sea-weed; *logos*, discourse), the study of Algæ or Sea-weeds.

phycophæine, n., *fĭk'·ō·fē'·ĭn* (Gr. *phukos*, sea-weed; *phaios*, brown), in *bot.*, a reddish-brown substance found in Algæ.

phycoxanthine, n., *fĭk'·ō·zănth'·ĭn* (Gr. *phukos*, sea-weed; *xanthos*, yellow), in *bot.*, the same as 'diatomine,' which see under 'Diatomaceæ.'

phylactolæmata, n. plu., *fĭl·ăk'·tō·lĕm'·ăt·ă* (Gr. *phulaktĭkos*, having the power to guard—from *phullasso*, I guard; *laimos*, the throat), the division of the Polyzoa in which the mouth is provided with the arched valvular process, called the 'epistome.'

phylla, n. plu., *fĭl'·lă* (Gr. *phullon*, a leaf), in *bot.*, the verticillate leaves which form the calyx or external envelope of the flower.

phyllaries, n. plu., *fĭl'·lăr·ĭz* (Gr. *phillurĕa*, a certain tree or shrub —from *phullon*, a leaf), in *bot.*, the leaflets forming the involucre of composite flowers.

phyllocysts, n. plu., *fĭl'·lō·sĭsts* (Gr. *phullon*, a leaf; *kustis*, a cyst), the cavities in the interior of the hydrophyllia of certain Oceanic Hydrozoa.

phyllodium, n., *fĭl·lōd'·ĭ·ŭm* (Gr. *phullon*, a leaf; *eidos*, appearance), in *bot.*, a leaf-stalk developed into a flattened expansion like a leaf: phyllody, n., *fĭl'-lŏd-ĭ*, the change of an organ into true leaves ; the substitution of true leaves for some other organ : **phylloid**, a., *fĭl'-loyd*, like a leaf : **phylloids**, n. plu., *fĭl'-loydz*, leaf - like appendages to the stems of Algæ.

phyllogen, n., *fĭl'-lō-jĕn* (Gr. *phullon*, a leaf ; *gennăō*, I produce), in *bot.*, the single terminal and central bud from which leaves are produced in Palms, and many herbaceous plants ; also called a 'phyllophor.'

phyllolobeæ, n. plu., *fĭl'-lō-lŏb'-ĕ-ē* (Gr. *phullon*, a leaf ; *lobos*, a lobe), in *bot.*, cotyledons green and leafy.

phyllomania, n., *fĭl'-lō-mān'-ĭ-ă* (Gr. *phullon*, a leaf ; *mania*, madness), in *bot.*, an abnormal or unusual development of leaf tissue.

phyllome, n., *fĭl'-lōm* (Gr. *phullon*, a leaf), in *bot.*, a leaf structure ; a structure morphologically equivalent to a leaf.

phyllomorphy, n., *fĭl'-lō-mŏrf'-ĭ* (Gr. *phullon*, a leaf ; *morphē*, form, shape), in *bot.*, the substitution of leaves for other organs ; same sense as 'phyllody.'

phyllophor, n., *fĭl'-lŏ-fōr* (Gr. *phullon*, a leaf ; *phorĕō*, I bear), the terminal bud or growing point in Palms ; same sense as 'phyllogen' : **phyllophorous**, a., *fĭl-lŏf'-ŏr-ŭs*, bearing or producing leaves.

phyllophytes, n. plu., *fĭl'-lō-fītz* (Gr. *phullon*, a leaf ; *phutōn*, a plant), plants of any kind in which leaves can be observed.

Phyllopoda, n. plu., *fĭl-lŏp'-ŏd-ă* (Gr. *phullon*, a leaf ; *podes*, feet), an Order of Crustaceæ having leaf-like feet : **phyllopodes**, n. plu., *fĭl-lŏp'-ōd-ēz*, in *bot.*, dead leaves in Isoetes.

phylloptosis, n., *fĭl'-lŏp-tōz'-ĭs* (Gr. *phullon*, a leaf ; *ptōsis*, a falling), in *bot.*, the fall of the leaf.

phyllotaxis, n., *fĭl'-lō-tăks'-ĭs* (Gr. *phullon*, a leaf ; *tasso*, I arrange ; *taxis*, order), in *bot.*, the arrangement of the leaves on the axis or stem.

phylloxera, n. plu., *fĭl'-lŏks-ēr'-ă* (Gr. *phullon*, a leaf ; *xēros*, dry, parched), an insect which infests the leaves and roots of the vine, forming leaf-galls, and causes much damage in wine-producing countries.

phyma, n., *fīm'-ă* (Gr. *phuma*, a tumour—from *phuo*, I produce), a boil or tumour : **phymosis**, n., *fīm-ōz'-ĭs*, a contracted prepuce which cannot be drawn back over the glands.

phyogemmaria, n. plu., *fī'-ō-jĕm-mār'-ĭ-ă* (Gr. *phuo*, I produce ; *gemma*, a bud), in *zool.*, the small gonoblastidia of Velella, Ord. Physophoridæ.

Physalis, n., *fīs'-ăl-ĭs* (Gr. *phusăllis*, a bubble), a genus of plants, Ord. Solanaceæ, so named in allusion to the calyx, which is also remarkable for being accrescent : **Physalis Peruviana**, *pĕr-ōōv'-ĭ-ān'-ă* (of or from *Peru*), a species whose fruit, called the Peruvian winter cherry, is eaten ; **P. edulis**, *ĕd-ūl'-ĭs* (L. *edūlis*, eatable), the Cape gooseberry.

physic, n., *fĭz'-ĭk* (Gr. *phusikos*, conformable to nature — from *phusis*, nature), the science or knowledge of medicine ; the art of healing diseases ; a remedy for disease ; a medicine : v., to treat with medicine : **physical**, a., *fĭz'-ĭk-ăl*, pert. to nature, or natural productions ; pert. to the body or material things : **physician**, n., *fĭz-ĭsh'-ăn*, one legally qualified to prescribe remedies for external or internal use in disease, as distinguished from a surgeon : **physicist**, n., *fĭz'-ĭs-ĭst*, one skilled in the natural sciences or physics : **physics**, n. plu., *fĭz'-ĭks*, the science which treats of the properties of matter, the laws of motion, and the phenomena of nature ; natural philosophy.

physiognomy, n., *fĭz′·ĭ·ŏg′·nŏm·ĭ* (Gr. *phusis*, nature ; *gnōmon*, one who knows; *gnōmē*, opinion), the art of determining the character and dispositions of a person by an examination of the features of the face ; in *bot.*, the general appearance of a plant, without any reference to its botanical characters ; **physiognomic**, a., *fĭz′·ĭ·ŏg·nŏm′·ĭk*, of or pert. to.

physiology, n., *fĭz′·ĭ·ŏl′·ŏ·jĭ* (Gr. *phusis*, nature ; *logos*, discourse), the science which treats of the vital actions or functions performed by the organs of plants and animals ; the science which treats of the history and functions of the human body, and its several parts and structures.

physometra, n., *fĭz′·ō·mēt′·ră* (Gr. *phūsāō*, I inflate or distend ; *mētra*, a womb), an accumulation of air in the uterus which causes an enlargement of the abdomen.

Physomycetes, n., *fĭz′·ō·mĭ·sēt′·ēz* (Gr. *phusa*, a bladder ; *mukēs*, a fungus), a division of the Fungi in which the thallus is floccose, and the spores are surrounded by a vesicular veil or sporangium, as in bread-mould.

Physophoridæ, n. plu., *fĭz′·ō·fŏr′·ĭd·ē* (Gr. *phusa*, an air-bladder ; *phoreō*, I bear), an Order of Oceanic Hydrozoa.

Physostigma, n., *fĭz′·ō·stĭg′·mă* (Gr. *phūsāō*, I inflate or distend ; *stigma*, a mark, a brand), a genus of plants, Ord. Leguminosæ, Sub-order Papilionaceæ : **Physostigma venenosum**, *vĕn′·ĕn·ōz′·ŭm* (L. *vĕnēnōsus*, very poisonous), a plant having a remarkable hooded stigma, yields the Calabar Ordeal Bean, which causes contraction of the pupil of the eye, is a violent poison.

Phytelephas, n., *fĭt·ĕl′·ĕf·ăs* (Gr. *phuton*, a plant ; *elĕphas*, ivory), a genus of plants, Ord. Palmæ : **Phytelephas macrocarpa**, *măk′·rō·kârp′·ă* (Gr. *mākros*, great ; *karpos*, fruit), the ivory palm, whose hard albumen is used in the same way as ivory.

phytochlor, n., *fĭt′·ō·klōr* (Gr. *phuton*, a plant ; *chlōrŏs*, green), the green colouring matter of plants ; chlorophyll.

phytoderma, n., *fĭt′·ō·dĕrm′·ă* (Gr. *phuton*, a plant ; *derma*, skin), any fungus or vegetable parasite growing on the skin : **phytodermata**, n. plu., *-dĕrm′·ăt·ă*, skin diseases caused by fungi.

phytogenesis, n., *fĭt′·ō·jĕn′·ĕs·ĭs* (Gr. *phuton*, a plant ; *gennăō*, I produce), the development of the plant.

phytography, n., *fĭt·ŏg′·răf·ĭ* (Gr. *phuton*, a plant ; *grapho*, I write), the description of plants.

phytoid, a., *fĭt′·ōȳd* (Gr. *phuton*, a plant ; *eidos*, resemblance), resembling a plant ; plant-like.

Phytolaccaceæ, n. plu., *fĭt′·ō·lăk·kā′·sĕ·ē* (Gr. *phuton*, a plant ; L. *lacca*, a plant, otherwise unknown ; Ger. *lack*, Pers. *lac*, a rose or ruby colour), the Phytolacca family, an Order of plants whose species have frequently much acridity, and some act as emetics or purgatives : **Phytolaccæ**, n. plu., *fĭt′·ō·lăk′·sĕ·ē*, a Sub-order or tribe : **Phytolacca**, n., *fĭt′·ō·lăk′·kă*, a genus said to be so named from their red juice : **Phytolacca decandra**, *dĕk·ănd′·ră* (Gr. *deka*, ten ; *anēr*, a male ; *andros*, of a male), the common poke, whose succulent fruit yields a red juice which has been used as a remedy in cases of chronic syphilitic pains, and the plant yields much potash.

phytology, n., *fĭt·ŏl′·ŏ·jĭ* (Gr. *phuton*, a plant ; *logos*, discourse), the science of the vegetable kingdom ; the study of plants ; botany.

phyton, n., *fĭt′·ŏn* (Gr. *phuton*, a

X

plant), in *bot.*, a rudimentary plant, as represented by a leaf.

phytophagous, a., *fīt·ŏf'·ăg·ŭs* (Gr. *phuton*, a plant ; *phago*, I eat), plant-eating ; herbivorous.

phytozoids, n. plu., *fīt'·ō·zō'·ĭdz* (Gr. *phuton*, a plant ; *zōŏn*, an animal ; *eidos*, resemblance), in *bot.*, peculiar bodies, rolled up in a circular or spiral manner, in the cellules of the antheridia in Hepaticæ and Mosses, which exhibit active movements at certain periods of their existence, and thus resemble animalcules ; Spermatozoids, or Antherozoids.

phytozoon, n., *fīt·ō·zō'·ŏn* (Gr. *phuton*, a plant ; *zōŏn*, an animal, *zōa*, animals), a plant-like animalcule, or one living in the tissues of plants : **phytozoa**, n. plu., *fīt'·ō·zō'·ă*, in *bot.*, moving filaments in the antheridia of Cryptogams.

pia-mater, n., *pī'·ă·māt'·ėr* (L., kind mother), a delicate, fibrous, and highly vascular membrane, which immediately invests the brain and spinal cord.

pica, n., *pīk'·ă* (L. and It. *pica*, a magpie, hunger), an appetite to eat or drink things unusual, such as coal, earth, etc.; a depraved appetite.

Picræna, n., *pīk·rēn'·ă* (Gr. *pikros*, bitter), a genus of plants, Ord. Simarubaceæ : **Picræna excelsa**, *ĕk·sĕls'·ă* (L. *excelsus*, elevated, lofty), a lofty forest tree of Jamaica, whose wood forms the Quassia of the shops, used in the form of an infusion and tincture as a slightly narcotic tonic, and anthelmintic.

picromel, n., *pīk'·rŏm·ĕl* (Gr. *pikros*, bitter ; *meli*, honey), a peculiar sweet bitter substance found in bile.

picrotoxin, n., *pīk'·rō·tŏks'·ĭn*, also **picrotoxia**, n., *pīk'·rō·tŏks'·ĭ·ă* (Gr. *pikros*, bitter ; Gr. *toxikon*, L. *toxicum*, the poison in which arrows were dipped), a crystalline, poison-

ous, narcotic principle, forming the active bitter ingredient in the berries of the Cocculus Indicus : **picrotoxic**, a., *pīk'·rō·tŏks'·ĭk*, of or pert. to.

pigment, n., *pĭg'·mĕnt* (L. *pigmentum*, a paint — from *pingo*, I paint), any colouring matter wherever found ; the term is mainly applied to colouring matter in certain positions of the body, as in the inner layer of the choroid.

pileate, a., *pīl'·ĕ·āt* (L. *pīlĕus*, a close-fitting felt cap), in *bot.*, having a cap like the head of a mushroom : **pileus**, n., *pīl'·ĕ·ŭs*, in *bot.*, the cap-like portion of the mushroom, bearing the hymenium on its under side : **pileoli**, n. plu., *pīl·ē'·ōl·ī* (dimin.), little pilei, several usually proceeding from the same common stem.

pileorhiza, n., *pīl'·ĕ·ō·rīz'·ă* (Gr. *pīlĕŏs*, a cap ; *rhiza*, a root), in *bot.*, a covering of the root, as in Lemna ; a cap found at the end of some roots.

piles, n. plu., *pīlz* (L. *pīla*, a ball of anything), a popular name for a disease of the veins at the extremity of the rectum, assuming a knotted or clustered form around the anus—called *bleeding piles* when there is a discharge of blood from them, and *blind piles* when there is none ; hæmorrhoids.

pili, n. plu., *pīl'·ī* (L. *pīlŭs*, hair), in *bot.*, fine slender hair-like bodies covering some plants : **piliform**, a., *pīl'·ĭ·fŏrm* (L. *forma*, shape), having the form of hairs : **Pilocarpus**, n., *pīl'·ō·kărp'·ŭs* (L. *pīlōsus*, hairy ; Gr. *karpos*, fruit), a genus of plants, Ord. Rutaceæ : **Pilocarpus pennatifolius**, *pĕn·năt'·ĭ·fōl'·ĭ·ŭs* (L. *penna*, a feather ; *fŏlium*, a leaf), used as a sudorific and sialagogue : **pilocarpine**, n., *pīl'·ō·kărp'·ĭn*, the active principle of preceding, and a powerful siala-gogue : **Pilocarpus jaborandi,** believed to be a different plant, but possessing similar physiological effects.

pilidium, n., *pĭl·ĭd'·ĭ·ŭm* (Gr. *pĭlĕŏs,* a cap ; *eidos,* resemblance), an orbicular lichen-shield.

pill, n., *pĭl* (L. *pĭla,* a ball), a medicine made up in the shape and size of a pea, to be swallowed whole : **pillule,** n., *pĭl'·ūl* (L. *pĭlŭlă,* a little ball), a small pill.

pilose, a., *pĭl·ōz'* (L. *pĭlōsus,* hairy, shaggy—from *pĭlus,* a hair), in *bot.,* covered with long, distinct hairs ; abounding in hairs.

Pimenta, n., *pĭm·ĕnt'·ă* (Sp. *pimiento*), a genus of plants, Ord. Myrtaceæ : **Pimenta officinalis,** *ŏf·fĭs'·ĭn·āl'·ĭs* (L. *offĭcĭnālis,* officinal), a tree, a native of the W. Indies and Mexico, producing Pimento, Allspice, or Jamaica pepper, so named from the aromatic odour of the fruit, contains an acrid volatile oil, and is sometimes employed as a stimulant and carminative.

Pimpinella, n., *pĭmp'·ĭn·ĕl'·lă* (corrupted from *bipinnate,* referring to the leaves being twice pinnate ; It. *pimpinella,* the herb pimpernel), a genus of plants, Ord. Umbelliferæ : **Pimpinella anisum,** *ăn·ĭz'·ŭm* (L. *ănĭsum,* anise), a species from whose fruit a carminative and aromatic oil is obtained.

pinakenchyma, n., *pĭn'·ăk·ĕng'·kĭm·ă* (Gr. *pĭnax,* a table ; *engchuma,* an infusion), in *bot.,* the muriform tissue of the medullary rays of woody stems, whose flattened, much shortened cells assume a tabular form.

Pinckneya, n., *pĭnk·nē'·ă* (after *Mr. Pinckney,* an American), a genus of plants, Ord. Rubiaceæ : **Pinckneya pubens,** *pūb'·ĕnz* (L. *pūbens,* pubescent), a tree which yields the fever-bark of Carolina, having long downy leaves.

pinenchyma, n., *pĭn·ĕng'·kĭm·ă,* same as 'pinakenchyma,' which see ; in *bot.,* tissue composed of tabular cells.

pin-eyed, a., *pĭn'·īd'* (pin, and *eye*), in *bot.,* having long styles with stigma visible at the top of the floral tube, as in the flower of Primula.

Pinguicula, n., *pĭn·gwĭk'·ŭl·ă* (L. *pinguĭcŭlus,* somewhat fat—from *pinguis,* fat), a genus of beautiful little plants, Ord. Lentibulariaceæ, whose species are called butterworts from giving consistence to milk, so named in reference to the greasy appearance of their foliage,—the leaves secrete a viscid fluid which detains insects : **Pinguicula grandiflora,** *grănd'·ĭ·flōr'·ă* (L. *grandis,* great, large ; *flos,* a flower, *flōris,* of a flower), a species peculiar to Ireland : **P. alpina,** *ălp·īn'·ă* (of or from the Alps), a species peculiar to Scotland.

pinna, n., *pĭn'·nă,* **pinnæ,** n. plu., *pĭn'·nē* (L. *pinna,* a feather or fin), a general name applied to the fin of a fish, or to the feather or wing of a bird ; in *bot.,* the leaflet of a pinnate leaf ; in *anat.,* the auricle or outer ear, projecting beyond the head : **pinnate,** a., *pĭn'·nāt,* in *bot.,* having the leaves divided in a feathery manner ; in a compound leaf, having several leaflets attached to each side of a central rib ; feather-shaped, or possessing lateral processes : **pinnatifid,** a., *pĭn·năt'·ĭ·fĭd* (L. *findo,* I cleave ; *fidi,* I cleft), in *bot.,* having leaves cut into lateral segments to about the middle, like those of the common groundsel : **pinnatipartite,** a., *pĭn·năt'·ĭ·pärt'·ĭt* (L. *partītus,* divided), having leaves cut into lateral segments, the divisions extending nearly to the central rib.

pinnatisect, a., *pĭn·năt'·ĭ·sĕkt* (L. *pinna,* a fin ; *sectus,* cut), in *bot.,* having leaves divided nearly to the midrib in a pinnate manner.

Pinnigrada, n. plu., *pĭn'·nĭ·grād'·ă* (L. *pinna*, a feather; *gradior*, I walk), the group of Carnivora, comprising the Seals and Walruses, adapted for aquatic life, and so named from the use of their fins or flaps for locomotion.

pinnules, n. plu., *pĭn'·nŭlz*, also **pinnulæ**, n. plu., *pĭn'·nŭl·ē* (L. *pinnula*, a little fin or feather), in *bot.*, the small pinnæ of a bipinnate or tripinnate leaf; the secondary divisions of a pinnate leaf; in *zool.*, the lateral processes of the arms of Crinoids.

Pinus, n., *pīn'·ŭs* (Gr. *pinos*, L. *pinus*, a pine tree), a genus of important trees, Ord. Coniferæ, which yield valuable products besides their timber, as turpentine, resin, tar, and pitch : **Pinus sylvestris**, *sĭl·vĕst'·rĭs* (L. *sĭlvĕstris*, woody—from *silva*, a wood), the Scotch fir, which yields common turpentine ; essence of spruce, used in making spruce-beer, is obtained by boiling the leaves in water ; the Norwegians prepare bark-bread from the inner bark: **P. pinaster**, *pĭn·ăst'·ĕr* (L. *pĭnăster*, a wild pine), the Cluster pine : **P. maritima**, *măr·ĭt'·ĭm·ă* (L. *mărĭtĭmus*, belonging to the sea —from *mărĕ*, the sea), the Bordeaux pine : **P. palustris**, *păl·ŭst'·rĭs* (L. *paluster*, marshy—from *pălus*, a marsh), the Swamp pine: **P. tæda**, *tēd'·ă* (L. *tæda*, the pitch - pine tree), the Loblolly or Frankincense pine ; the two preceding yield the Thus or Common Frankincense : **P. pumilio**, *pŭm·ĭl'·ĭ·ŏ* (L. *pŭmĭlio*, a dwarf, a pigmy), yields Hungarian balsam : **P. pinea**, *pĭn'·ĕ·ă* (L. *pĭnĕŭs*, of the pine—from *pinus*, a pine), the Stone pine, the source of Carpathian balsam.

Piperaceæ, n. plu., *pĭp'·ĕr·ā'·sĕ·ē* (L. *pĭper*, pepper, *pĭpĕris*, of pepper), the Pepper family, an Order of plants, natives of the hottest parts of the world, having pungent, acrid, and aromatic properties : **Piper**, n., *pĭp'·ĕr*, an interesting genus of plants : **Piper officinarum**, *ŏf·fĭs'·ĭn·ār'·ŭm* (L. *officina*, a workshop, *officinārum*, of workshops), the Piper of the laboratories ; an Indian creeper, whose dried fruiting spikes constitute long - pepper : **P. nigrum**, *nĭg'·rŭm* (L. *nĭger*, black), an Indian creeper, whose dried unripe fruit or drapes constitute black pepper ; the ripe fruit, when deprived of its outer fleshy covering, constitutes white pepper ; these peppers, hot aromatic condiments, are used medicinally as tonic, stimulant, febrifuge, and stomachic : **P. cubeba**, *kŭb·ēb'·ă* (Ar. *cubabah*), an Indian and Javan climbing plant, the cubeb pepper, used extensively in arresting discharges from mucous membranes : **P. clusii**, *klūz'·ĭ·ī* (after *C. Clusius*, a botanist), a species which yields the African cubebs, W. African black pepper : **P. angustifolium**, *ăng·gŭst'·ĭ·fōl'·ĭ·ŭm* (L. *angustus*, narrow ; *fŏlĭum*, a leaf), a shrub growing in the moist woods of Bolivia, Peru, etc., whose leaves and unripe fruit are called Matico, possesses aromatic, fragrant, and astringent qualities, and the property of checking hæmorrhage: **P. lanceæfolium**, *lăns'·ĕ·ē·fōl'·ĭ·ŭm* (L. *lancĕa*, a lance ; *fŏlĭum*, a leaf), also yields Matico: **P. methysticum**, *mĕth·ĭst'·ĭk·ŭm* (Gr. *methustĭkŏs*, intoxicating — from *methu*, wine), the plant from which the intoxicating liquor 'Ava,' 'Cava,' or 'Kava,' is prepared by fermentation, much used by S. Sea islanders, and peculiar to them ; also used as a remedy for syphilis : **piperin**, n., *pĭp'·ĕr·ĭn*, a white crystallisable substance extracted from black pepper.

Pisces, n. plu., *pĭs'·ēz* (L. *piscis*, a fish), the class of vertebrates comprising the Fishes: **pisciform**, a., *pĭs'·ĭ·fŏrm* (L. *forma*, shape), having the shape of a fish.

Piscidia, n., *pĭs·ĭd'·ĭ·ă* (L. *piscis*, a fish; *cædo*, I kill), a genus of plants, Ord. Leguminosæ, Sub-ord. Papilionaceæ, so named as used for stupefying fish: **Piscidia erythrina**, *ĕr'·ĭ·thrĭn'·ă* (Gr. *ĕrŭthrŏs*, red), the Jamaica Dogwood, used as a poison to catch fish.

pisiform, a., *pĭz'·ĭ·fŏrm* (L. *pisum*, a pea; *forma*, shape), pea-shaped; having the form or structure of a pea; applied to the smallest bone of the carpus.

Pistacia, n., *pĭs·tāsh'·ĭ·ă* (Gr. *pistăkiă*, the Pistachio nuts), a genus of plants, Ord. Anacardiaceæ: **Pistacia vera**, *vēr'·ă* (L. *vērus*, real, genuine), the Pistacia or Pistachio nut - tree, extensively cultivated in Syria and the East, and in the South of Europe; the green-coloured, oily kernels are used as food: **P. terebinthus**, *tĕr'·ĕ·bĭnth'·ŭs* (L. *tĕrĕbinthus*, the turpentine tree), a tree, a native of the S. of Europe, of N. of Africa, and of the East, which yields a resinous substance called Chian or Cyprian turpentine, and has diuretic and excitant properties: **P. lentiscus**, *lĕnt·ĭsk'·ŭs* (L. *lĕntiscus*, the mastich tree), furnishes the concrete resinous exudation called Mastich or Mastic, which see.

pistil, n., *pĭst'·ĭl* (L. *pistillum*, a pestle; F. *pistil*), in *bot.*, the seed-bearing organ, occupying the centre of a flower, consisting of an upper part or *stigma*, a central part or *style*, and a lower part or *ovarium*, containing the young seeds, called *ovules*: **pistillary**, a., *pĭst'·ĭl·lăr·ĭ*, connected with or pert. to a pistil: **pistillate**, a., *pĭst'·ĭl·lāt*, having a pistil — applied to a female flower or plant: **pistillidium**, n., *pĭst'·ĭl·lĭd'·ĭ·ŭm* (Gr. *eidos*, resemblance), an organ in the higher Cryptogams having female sexual functions: **pistilliferous**, a., *pĭst'·ĭl·lÿ''·ĕr·ŭs* (L. *fero*, I bear), having a pistil without stamens, as a female flower; same as 'pistillate': **pistillody**, n., *pĭst·ĭl·lŏd·ĭ* (Gr. *eidos*, resemblance), the change of any organ of a flower into carpels.

Pisum, n., *pĭz'·ŭm* (L. *pisum*, the pea), a genus of plants, Ord. Leguminosæ, Sub-ord. Papilionaceæ: **Pisum sativum**, *săt·ĭv'·ŭm* (L. *sativus*, that is eaten), a hardy annual producing the common pea.

pitch, n., *pĭtsh* (Ger. *pech*, Gr. *pitta*, pitch), the officinal variety, termed Burgundy Pitch, is the resinous exudation from the spruce-fir, used as a plaster.

pitcher, n., *pĭtsh'·ĕr* (F. *pichet*, It. *pitero*, an earthen pot), in a plant, a leaf which has the petiole or leaf - stalk expanded into a vase-like form, the blade being attached as a lid: **Pitcher-plants**, the genus Nepenthes.

Pittosporaceæ, n. plu., *pĭt'·tō·spŏr·ā'·sĕ·ē* (G. *pitta*, pitch; *spora*, seed), the Pittosporum family, an Order of plants, so named because the seeds are covered with a resinous pulp: **Pittosporum**, n., *pĭt·ŏs'·pŏr·ŭm*, a genus of very handsome shrubs, having glossy foliage and pretty flowers.

pituitary, a., *pĭt·ū'·ĭt·ăr·ĭ* (L. *pĭtūĭtă*, slime, phlegm), that secretes or conveys phlegm or mucus: **pituite**, n., *pĭt'·ū·ĭt*, phlegm or mucus: **pituitary gland**, a gland situated on the lower side of the brain, supposed by the ancients to secrete the mucus of the nostrils: **pituitary membrane**, the fine membrane lining the nostrils.

pityriasis, n., *pĭt'·ĭ·rī'·ăs·ĭs* (Gr. *pituron*, scurf or bran), a skin disease, exhibiting a slight redness of limited portions of the skin, on which small, thin scales are formed and fall off—called *dandriff* when it affects children: **pityroid**, a., *pĭt'·ĭr·ŏyd* (Gr. *eidos*, resemblance), bran or scurf-like.

placenta, n., *plăs·ĕnt'·ă* (L. *placenta*, a cake ; Gr. *plakous*, a flat cake), a flat, round body formed in the womb during pregnancy, and serving to connect the circulation of the mother and child — coming away a few minutes after delivery, it is commonly known as the afterbirth ; in *bot.*, the cellular part of the carpel bearing the ovule : **placentary**, n., *plăs·ĕnt'·ăr·ĭ*, in *bot.*, a placenta bearing numerous ovules : **placental**, a., *plăs·ĕnt'·ăl*, pert. to the placenta : n., a mammal having a placenta : **placentation**, n., *plăs'·ĕnt·ā'·shŭn*, in *bot.*, the manner in which the seeds are attached to the pericarp; the manner in which the placentæ are developed.

placoid, a., *plăk'·ŏyd* (Gr. *plax*, a plate, *plakos*, of a plate ; *eidos*, resemblance), applied to the irregular bony plates, grains, or spines found in the skin of various fishes, as skates, rays, dog-fish, and sharks : n., an Order of fishes.

plagiostomi, n. plu., *plădj'·ĭ·ŏs'·tŏm·ĭ* (Gr. *plagĭŏs*, transverse ; *stoma*, a mouth), the Sharks and Rays in which the mouth is transverse, being placed on the under surface of the head.

Planarida, n. plu., *plăn·ăr'·ĭd·ă* (Gr. *planos*, wandering), a Suborder of the Turbellaria.

plantæ tristes, n. plu., *plănt'·ē trĭst'·ēz* (L. *plantæ*, plants ; *trĭstĭs*, sad, melancholy, *tristes*, plu.), melancholy plants, so named because they expand their flowers in the evening only, as some species of Hesperis, Pelargonium, etc.

Plantaginaceæ, n. plu., *plănt·ădj'·ĭn·ā'·sĕ·ē* (L. *plantāgo*, a plantain—from *planta*, the sole of the foot, from the leaves lying flat on the ground), the Ribwort family, an Order of plants having radical ribbed leaves, and the species are frequently bitter and astringent : **Plantago**, n., *plănt·āg'·ō*, a genus of plants, usually called Plantains : **Plantago maritima**, *măr·ĭt'·ĭm·ă* (L. *mărĭtĭmus*, belonging to the sea —from *măr'ĕ*, the sea), a species found on the sea-shores, and on the highest mountains in Scotland : **P. major**, *mādj'·ŏr* (L. *major*, greater), the Way-bred, which follows the footsteps of man in his migrations, the spiklets are used for feeding birds : **P. decumbens**, *dĕ·kŭm'·bĕnz* (L. *dēcumbens*, lying down), a native of Egypt, India, etc., whose seeds are used to form a demulcent drink in diarrhœa.

plantar, a., *plănt'·ăr* (L. *plantāris*, belonging to the sole of the foot —from *planta*, the sole of the foot), pert. to the sole of the foot : **plantaris**, a., *plănt·ăr'·ĭs*, applied to a muscle arising from the external condyle of the femur, and which extends to the foot.

plantigrade, a., *plănt'·ĭ·grād* (L. *planta*, the sole of the foot ; *grădĭor*, I walk), placing the sole of the foot to the ground in walking : n., an animal that does so, as the bear.

planula, n., *plăn'·ūl·ă* (L. *plānŭla*, a little plane—from *plānus*, flat), the oval ciliated embryo of certain of the Hydrozoa.

planum temporale, *plăn'·ŭm tĕmp'·ŏr·āl'·ĕ* (L. *plānus*, plane ; *tempŏrālis*, lasting but for a time), the temporal plane ; the parietal part of the temporal ridge of the skull, which bounds a surface somewhat flatter than the rest, forming part of the temporal fossa.

plasma, n., *plăz'mă* (Gr. *plasma,* a thing moulded or formed, a model), the colourless fluid part of the blood in which the corpuscles float ; liquor sanguinis.

plasmodium, n., *plăs·mōd'·ĭ·ŭm* (Gr. *plasma,* a thing moulded or formed ; *eidos,* resemblance), in *bot.,* a membranous protoplasmic body, formed by the coalescence of swarm spores in myxosporous Fungi.

plastic, a., *plăst'ĭk* (Gr. *plastĭkos,* suitable for being fashioned or formed—from *plassō,* I form or mould), having power to give form to matter : **plastic element,** an element which contains the germ of a higher form : **plastic force,** the force which gives to matter a definite organic form.

plastron, n., *plăst'·rŏn* (Gr. *emplastron,* a plaster ; F. *plastron,* a breastplate), that part of the bony covering of turtles and tortoises, etc., which covers the lower or ventral portion.

Platanaceæ, n. plu., *plăt'·ăn·ā'·sĕ·ē* (Gr. *platanos,* L. *plătănus,* the plane tree — from Gr. *platus,* wide, broad), the Plane family, an Order of trees so named from their wide-spreading branches : **Platanus,** n., *plăt'·ăn·ŭs,* a genus of plane trees : **Platanus orientalis,** *ŏr'·ĭ·ĕnt·āl'·ĭs* (L. *orĭentālis,* eastern), the Oriental plane, having broad palmate leaves like the sycamore : **P. occidentalis,** *ŏk'·sĭ·dĕnt·āl'·ĭs* (L. *occĭdĕntālis,* western), and **P. acerifolia,** *ăs'·ĕr·ĭ·fōl'·ĭ·ă* (L. *acer,* the maple tree ; *fōlĭum,* a leaf), are cultivated as showy trees under the name of Planes.

Platyelmia, n. plu., *plăt'·ĭ·ĕl'·mĭ·ă* (Gr. *platus,* broad ; *helmins,* an intestinal worm), the division of the Scolecida, comprising the Tapeworms.

Platylobeæ, n. plu., *plăt'·ĭ·lŏb'·ĕ·ē* (Gr. *platus,* broad ; *lobos,* a lobe), in *bot.,* a general name for the tribes Pleurorhizeæ and Notorhizeæ, meaning that the cotyledons are plane or flat : **platyphyllous,** a., *plăt'·ĭ·fĭl'·ŭs* (Gr. *phullon,* a leaf), in *bot.,* broadleaved.

Platyrhina, n. plu., *plăt'·ĭ·rīn'·ă* (Gr. *platus,* broad ; *rhines,* nostrils), a group of the Quadrumana : **platyrhine,** a., *plăt'·ĭ·rīn,* broad-nosed ; applied to the new-world monkeys, which have their nostrils separated from each other by a broad septum.

platysma myoides, *plăt·ĭs'·mă mĭ·ōyd'·ēz* (Gr. *platusmos,* enlargement ; *mus,* a muscle ; *eidos,* resemblance), a pale-coloured, thin sheet of muscular fibres, extending over the front and sides of the neck, and lower portions of the side part of the face, which assists in drawing the angle of the mouth downwards and outwards ; it is the sole remains in man of the 'muscular panniculosus,' or skin muscle, of animals.

Plectranthus, n., *plĕk·trănth'·ŭs* (Gr. *plektron,* a cock's spur ; *anthos,* a flower), a genus of plants, Ord. Labiatæ, so named in reference to the shape of the flowers : **Plectranthus graveolens,** *grăv·ē'·ŏl·ĕnz* (L. *grăvĕŏlens,* strong-smelling), the Patchouli plant of the East Indies, used as a perfume, and yields a volatile oil of a yellowish-green colour.

pleiomazia, n., *plī'·ō·māzh'·ĭ·ă* (Gr. *pleion,* more ; *māzŏs,* the breast), an excess in the number of mammæ, rarely observed in men, more commonly in women.

pleiomorphy, n., *plī'·ō·mŏrf'·ĭ* (Gr. *pleion,* more ; *morphē,* shape), in *bot.,* the renewed growths in arrested parts of irregular flowers : **pleiotaxy,** n., *plī'·ō·tăks'·ĭ* (Gr. *taxis,* arrangement), the multiplication of whorls : **pleiotracheæ,** n. plu., *plī'·ō·trăk'·ĕ·ē* (Gr. *tracheia,* the windpipe ; *trachus,* rough), numerous fibres united together, as in the banana, and assuming the aspect of a broad riband; spiral vessels with several fibres united.

plenus, a., *plĕn'ŭs* (L. *plēnus,* full), in *bot.,* double, when applied to the flower.

pleospora, n., *plē·ŏs'·pōr·ă* (Gr. *plĕŏs,* full; *spora,* a spore, seed), another name for 'Cladosporium herbarum,' a disease in silkworms, caused by a Fungus.

plerome, n., *plĕr'·ōm* (Gr. *plēr-ōma,* fulness, completion), the state of being filled up or completed; in *bot.,* the central portion of the primary meristem immediately enclosed and overarched by the periblem.

plessor, n., *plĕs'·ŏr,* also **plexor,** n., *plĕks'·ŏr* (Gr. *plēssō,* I strike), any instrument used in percussion.

plethora, n., *plĕth'·ōr·ă* (Gr. *plēthōra,* fulness, abundance), redundant fulness of the blood-vessels; excess in the amount or quality of food and drink partaken of.

pleura, n., *plŏr'·ă* (Gr. *pleura,* the side, a rib), a serous membrane or sac covering each lung, and also lining the inside of the chest or thorax: **pleuræ,** n. plu., *plŏr'·ē,* two independent serous membranes forming two shut sacs, quite distinct from each other, which line the right and left sides of the thoracic cavity, each pleura consisting of a *visceral* and a *parietal* portion: **pleura pulmonalis,** *pŭl'·mŏn·āl'·ĭs* (L. *pulmo,* a lung, *pulmōnis,* of a lung), the visceral portion of the pleura which covers the lung: **pleura costalis,** *kŏst·āl'·ĭs* (L. *costa,* a rib), the parietal portion of the pleura which lines the ribs and intercostal spaces, and covers the upper convex surface of the diaphragm, etc.: **pleural,** a., *plŏr'·ăl,* connected with the pleura: **pleurisy, n.,** *plŏr'·ĭs·ĭ,*

also **pleuritis, n.,** *plŏr·ĭt'·ĭs,* the inflammation of the inner membrane of the thorax: **pleuritic,** a., *plŏr·ĭt'·ĭk,* pert. to or affected with pleurisy: **pleural cavity,** the sac of the pleura.

pleurapophysis, n., *plŏr'·ă·pŏf'·ĭs·ĭs,* -**pophyses,** n. plu., -*pŏf'·ĭs·ēz* (Gr. *pleura,* the side, a rib; *apophũsis,* a process), the true ribs.

pleurenchyma, n., *plŏr·ĕng'·kĭm·ă* (Gr. *pleura,* the side; *engchũma,* what is poured in, tissue), in *bot.,* woody tissue of plants, consisting of tough, slender tubes, out of which the wood is mainly formed.

pleurisy, see under 'pleura.'

pleurocarpi, n. plu., *plŏr'·ō·kărp'·ĭ* (Gr. *pleura,* a rib or side; *karpos,* fruit), in *bot.,* mosses with the fructification proceeding laterally from the axils of the leaves: **pleurocarpous,** a., *plŏr'·ō·kărp'·ŭs,* having the fructification springing from the axils of the leaves, or from the side of the stem.

pleurodynia, n., *plŏr'·ō·dĭn'·ĭ·ă* (Gr. *pleura,* a side; *ŏdũnē,* pain), pain in the side, arising from various causes; myalgia.

pleuron, n., *plŏr'·ŏn* (Gr. *pleuron,* a rib), the lateral extension of the shells of Crustacea.

pleuro-pneumonia, n., *plŏr'·ō·nŭm·ŏn'·ĭ·ă* (Gr. *pleura,* the side; *pneumŏnĭa,* disease of the lungs—from *pneumōn,* the lungs—from *pnĕō,* I breathe), an inflammatory disease of the pleura and lungs.

pleurorhizeæ, n. plu., *plŏr'·ō·rĭz'·ē·ē* (Gr. *pleura,* the side; *rhiza,* a root), in *bot.,* Cruciferous plants in which the cotyledons are applied by their faces, and the radicle folded on their edges, so as to be lateral, while the cotyledons are accumbent.

pleurothotonos, a., *plŏr'·ō·thŏt'·ŏn·ŏs* (Gr. *pleurŏthĕn,* from the side; *pleura,* the side, and *ŏthĕn,* whence; *teinō,* I bend, *tŏnŏs,* tension), applied to lateral convulsions, sometimes seen in cases of tetanus, the patient throwing himself from side to side : **adj.**, bent or stretched from one side.

plexiform, a., *plĕks'ĭ-fŏrm* (L. *plexum*, to interweave, to twist; *forma*, shape), in the form of network ; complicated ; entangled.

pleximeter, n., *plĕks-ĭm'ĕt-ĕr* (Gr. *plexis*, a striking ; *metron*, a measure), a plate of ivory, india-rubber, or the like, and very frequently the fingers of the left hand, used in mediate percussion of the chest or abdomen.

plexus, n., *plĕks'ŭs*, **plexuses**, n. plu., *plĕks'ŭs-ĕs* (L. *plexus*, twisted), in *anat.*, a network of interlacing blood-vessels, or nerves.

plica, n., *plīk'ă* (L. *plĭco*, I fold or plait), in *bot.*, a diseased state of plants in which the buds, instead of developing true branches, become short twigs, and these in their turn produce others, thus forming an entangled mass ; in *med.*, a disease in man in which the hair becomes matted and the scalp exceedingly sensitive—also called **plica polonica**, *pŏl-ŏn'ĭk-ă*, as being peculiar to Poland, Lithuania, and Tartary : **p. semilunaris**, *sĕm'ĭ-lŏn-ār'ĭs* (L. *semi*, half ; *luna*, the moon), the semilunar fold ; a vertical fold of the conjunctiva resting on the eyeball, forming the rudiment of the third eyelid, the membrana nictitans, found in some animals.

plicate, a., *plīk'āt* (L. *plĭcātus*, folded or doubled up), in *bot.*, folded like a fan ; regularly disposed in folds, as in the vernation of some leaves : **plicative**, a., *plĭk'āt-ĭv*, plaited or folded, applied to æstivation : **plicatulate**, a., *plĭk-ăt'ŭl-āt* (dim. of *plicate*), disposed in very small folds.

Plumbaginaceæ, n. plu., *plŭm-bădj'ĭn-ā'sĕ-ē* (L. *plumbago*, the plant lead-wort or flea-wort—from *plumbum*, lead), the Sea-pink family, an Order of plants, some of which are acrid, and others have tonic qualities : **Plumbago**, n., *plŭm-bāg'ō*, a genus of pretty free-flowering plants : **Plumbago Europæa**, *ūr'ō-pē'ă* (of or from *Europe*), has been employed for the relief of toothache, while the root is so acrid as to be employed for causing issues, and by beggars to raise ulcers to excite pity ; also used internally in small doses as an emetic : **P. rosea**, *rōz'ĕ-ă* (L. *rŏsĕŭs*, pert. to roses—from *rōsa*, a rose), and **P. scandens**, *skăn'dĕnz* (L. *scandens*, climbing), are two species whose roots when fresh are most energetic blistering agents, the latter in San Domingo being called ' Herbe-du-diable.'

plumbism, n., *plŭm'bĭzm* (L. *plumbum*, lead), the condition of an individual whose system has been brought under the influence of lead poison, as plumbers and painters.

plumose, a., *plŏm-ōz'* (F. *plume*, L. *pluma*, a feather, a small, soft feather), in *bot.*, applied to hairs or plants that have branches arranged like the beard on a feather ; feathery.

plumule, n., *plŏm'ūl* (L. *plumula*, a little feather), in *bot.*, the rudimentary or first bud of an embryo, usually enclosed by the cotyledons.

plurilocular, a., *plŏr'ĭ-lŏk'ŭl-ăr* (L. *plus*, more, *pluris*, of more ; *loculus*, a little place), in *bot.*, having several divisions containing seeds, as the orange and lemon ; many-celled : **pluripartite**, a., *plŏr'ĭ-pàrt'ĭt* (L. *partītus*, parted or divided), having an organ deeply-divided into several nearly distinct portions : **pluri-**

septate, -*sĕpt'.āt* (L. *septum*, a hedge), having many septa.

pluteus, n., *plŏt'.ĕ.ŭs* (L. *plŭtĕus*, a pent-house, a shed), the larval form of the Echinoidea.

pneumatic, a., *nū.măt'.ĭk* (Gr. *pneuma*, air), pert. to air, or air-breathing organs ; filled with air : **pneumatica**, n., *nū.măt'.ĭk.ă*, a diseased state of the respiratory functions ; an agent that acts on them : **pneumatics**, n. plu., *nū.măt'.ĭks*, the science which treats of the mechanical properties of air, and of similar elastic fluids : **pneumatocele**, n., *nū.măt'.ō.sēl* (Gr. *kēlē*, a tumour), a tumour or distension filled with air : **pneumatocyst**, n., *nū.măt'.ō.sĭst* (Gr. *kustis*, a cyst), the air-sac or float of certain of the Ocean Hydrozoa : **pneumatophore**, n., *nū.măt'.ō.fōr* (Gr. *phorĕō*, I bear), the proximal dilatation of the cœnosarc in the Physophoridæ which surrounds the pneumatocyst.

pneumatosis, n., *nū'.măt.ŏz'.ĭs* (Gr. *pneuma*, air, breath, *pneumătŏs*, of air), in *med.*, a windy swelling.

pneumogastric, a., *nū'.mō.găst'.rĭk* (Gr. *pneumōn*, the lungs; *gastēr*, the belly), of or pert. to the lungs and stomach ; denoting the principal nerve of the stomach, which extends over the viscera of the chest and abdomen, and regulates the functions of respiration and digestion ; also called 'vagus' from its wide distribution.

pneumonia, n., *nū.mōn'.ĭ.ă*, also **pneumonitis**, n., *nū'.mōn.it'.ĭs* (Gr. *pneumōn*, the lungs), inflammation of the lungs : **pneumonic**, a., *nū.mŏn'.ĭk*, pert. to the lungs ; pulmonic : **n.**, a medicine for affections of the lungs.

pneumo-skeleton, n., *nū'.mō.skĕl'.ĕt.ŏn* (Gr. *pneumōn*, the lungs ; Eng. *skeleton*), the hard structures connected with the breathing organs of certain animals, as the shell of Mollusca.

pneumothorax, *ăks* (Gr. *pne thōrax*, the tr condition in w the cavity of ing collapse which state tl to use the lu side.

poculiform, a. *pocŭlum*, a ct in *bot.*, cup-sh

podagra, n., *1* Gr. *podagra*, from *pous*, a *agra*, a seizi feet : **podagr** pert. to the g

podetium, n., etia, n. plu. *pous*, a foot, *1* a stalk-like e branched, risi in some lichen *pŏd.ēsh'.ĭ.ĭ.fŏr* resembling a *1*

podocarp, n., *pous*, a foot, *kărpos*, fruit) supporting the n., *pŏd'.ō.kăr* plants, Ord. carpus totar Zealand name supplies good Zealand.

podogynium, (Gr. *pous*, a foot ; *gunē*, a fleshy and soli to support the

podophyllum, (Gr. *pous*, a foot ; *phullon*, Foot, a genu Ranunculaceæ peltatum, *pĕlt* armed with a shaped shield shield), the M in America a cathartic : **P.** probable nativ species bearing a similar fruit : **podophyllum**, n., the dried underground stem of the P. peltatum, known also as the American May Apple, or Mandrake : **podophyllin**, n., *pŏd'·ō·fĭl'·lĭn*, a resin extracted from P. peltatum, also called 'vegetable mercury' from its influence on the liver.

podosperm, n., *pŏd'·ō·spérm* (Gr. *podes*, ropes at the corners of a sail ; *sperma*, seed), in *bot.*, the filament or thread by which the ovule adheres to the placenta ; the funiculus.

Podostemaceæ, n. plu., *pŏd'·ō·stĕm·ā'·sĕ·ē* (Gr. *podes*, ropes at the corners of a sail ; *stĕmma*, a garland or wreath), the Podostemon family, an Ord. of aquatic plants which flower and ripen their seed under water, and their ashes furnish salt : **Podostemon**, n., *pŏd'·ō·stĕm'·ŏn*, a genus of floating plants, with the habit of Liverworts or Scale Mosses.

poephaga, n., *pō·ĕf'·ăg·ă* (Gr. *poē*, grass, herbage ; *phago*, I eat), in *zool.*, a group of the Marsupials.

pogon, n., *pōg'·ŏn* (Gr. *pōgōn*), in *bot.*, a beard.

Pogostemon, n., *pōg'·ō·stĕm'·ŏn* (Gr. *pōgōn*, a beard ; *stĕmōn*, the thread or warp of a web, a stamen), a genus of plants, Ord. Labiatæ: **Pogostemon suavis**, *swāv'·ĭs* (L. *suāvis*, sweet, pleasant) ; also P. **Patchouly**, *păt'·tshŏl·ĭ* (native name), the Patchouly plant of the East Indies, used as a perfume, and yields a volatile oil of a yellowish green colour.

polarity, n., *pōl·ăr'·ĭ·tĭ* (L. *polus*, the end of an axis ; Gr. *polos*, a pivot on which anything turns), that property of bodies, or particles of all kinds of matter, which causes them, when at liberty to move freely, to arrange themselves in certain determinate directions—to point, as it were, to given poles.

Polemoniaceæ, n. plu., *pŏl'·ĕ·mŏn·ĭ·ā'·sĕ·ē* (L. *pŏlĕmōnĭa*, Gr. *polĕmōnĭon*, Greek Valerian—said to be from Gr. *polĕmŏs*, war, as kings quarrelled and made war for the honour of the discovery of its virtues), the Phlox family, an Order of plants, many of which have showy flowers, and are commonly cultivated : **Polemonium**, n., *pŏl'·ĕ·mōn'·ĭ·ŭm*, a genus of plants : **Polemonium cœruleum**, *sĕr·ūl'·ĕ·ŭm* (L. *cœrŭlĕŭs*, dark-blue, cerulean), Greek Valerian, or Jacob's Ladder, has bitter properties, esteemed by Russians as useful in hydrophobia when its leaves are applied as poultices.

pollen, n., *pŏl'·lĕn* (L. *pollen*, fine flour), in *bot.*, the fecundating or fertilising powder contained in the anthers of flowers, and afterwards dispersed on the stigma : **pollen cells**, the cavities of anthers : **pollen tubes**, the thread - like tubular processes developed from the pollen grains after they have become attached to the stigma : **pollen mass**, or **pollinia**, n., *pŏl·lĭn'·ĭ·ă*, an agglutinated mass of pollen, occurring in some orders of plants, as in Orchids : **pollination**, n., *pŏl'·lĭn·ā'·shŭn*, the conveyance of the pollen from the anthers to the stigma in Angiosperms, or to the nucleus in Gymnosperms : **pollinodium**, n., *pŏl'·lĭn·ōd'·ĭ·ŭm* (Gr. *eidos*, resemblance), another name for the 'Antheridium,' which see.

pollex, n., *pŏl'·lĕks* (L. *pollex*, the thumb), the thumb in man ; the innermost of the five normal digits of the anterior limb of the higher vertebrates.

polyadelphous, a., *pŏl'·ĭă·ă·dĕlf'·ŭs* (Gr. *polus*, many ; *adelphos*, a brother), in *bot.*, having stamens united by their filaments so as to form more than two bundles ; having stamens in many bundles.

polyandrous, a., *pŏl'·ĭ·ănd'·rŭs* (Gr. *polus*, many ; *anēr*, a male, *andros*, of a male), having more than twenty stamens, as a flower.

Polyanthes, n., *pŏl'·ĭ·ănth'·ēz* (Gr. *polus*, many ; *anthos*, a flower), a genus of showy garden flowers, Ord. Liliaceæ : Polyanthes tuberosa, *tūb'·ĕr·ōz'·ă* (L. *tūbĕrōsus*, having fleshy knots—from *tūber*, a protuberance), the tube-roses, prized for their fragrance, and beauty of their flowers.

Polyanthus, n., *pŏl'·ĭ·ănth'·ŭs* (Gr. *polus*, many ; *anthos*, a flower), a favourite garden flower, a variety of the Primrose, Primula vulgaris, Ord. Primulaceæ.

polycarpic, a., *pŏl'·ĭ·kărp'·ĭk*, also polycarpous, a., *pŏl·ĭ·kărp'·ŭs* (Gr. *polus*, many ; *karpos*, fruit), in *bot.*, having the carpels distinct and numerous, each flower bearing several fruit ; applied to plants which flower and fruit many times in the course of their life.

polychroite, n., *pŏl'·ĭ·krō'·ĭt* (Gr. *polus*, many ; *chrŏizō*, I colour), the yellow colouring matter of Saffron, which exhibits a variety of colours when acted upon by various re-agents, and is totally destroyed by the action of the solar rays : polychrome, n., *pŏl'·ĭ·krōm* (Gr. *chrōma*, colour), a substance obtained from the bark of the horse-chestnut, and from quassia-wood, which gives to water the quality of exhibiting a curious play of colours when acted upon by reflected light.

polycotyledon, *pŏl'·ĭ·kŏt'·ĭl·ēd'·ŏn* (Gr. *po'us*, many ; *kotŭlēdōn*, a hollow vessel), a plant the seeds of which have more than two lobes : polycotyledonous, a., *-ĭl·ēd'·ŏn·ŭs*, applied to an embryo having many cotyledons, as in Firs ; having more than two cotyledons or seed-lobes : polycotyledony, n., *pŏl'·ĭ·kŏt'·ĭl·ēd'·ŏn·ĭ*, an accidental increase in the number of cotyledons.

Polycystina, n., *pŏl'·ĭ·sĭst'·ĭn·ă* (Gr. *polus*, many ; *kustis*, a bladder), an Order of Protozoa with foraminated siliceous shells.

polydipsia, n., *pŏl'·ĭ·dĭps'·ĭ·ă* (Gr. *polus*, much ; *dipsa*, thirst), excessive thirst.

polyembryony, n., *pŏl'·ĭ·ĕm·brĭ'·ŏn·ĭ* (Gr. *polus*, many ; *embrŭŏn*, an embryo), in *bot.*, an increase in the number of embryos in a seed : polyembryonic, a., *-brĭ·ŏn'·ĭk*, having more than one embryo.

Polygalaceæ, n. plu., *pŏl'·ĭ·găl·ā'·sĕ·ē* (Gr. *polus*, much ; *găla*, milk), the Milkwort family, an Order of plants, generally bitter, their roots yielding a milky juice : Polygala, n., *pŏl·ĭg'·ăl·ă*, an extensive genus, all whose species are showy : Polygala senega, *sĕn'·ĕg·ă* (of or from Senegal), the Senega or Snakeroot, used in large doses as an emetic and cathartic, in smaller doses as a stimulant, sudorific, and expectorant, named snakeroot as a supposed antidote to the bite of the rattlesnake : polygalic acid, *pŏl·ĭg·ăl'·ĭk*, an acrid principle obtained from it.

polygamous, a., *pŏl·ĭg'·ăm·ŭs* (Gr. *polus*, many ; *gamŏs*, a marriage), in *bot.*, bearing hermaphrodite flowers, as well as male and female flowers, on the same plant.

Polygonaceæ, n. plu., *pŏl'·ĭ·gŏn·ā'·sĕ·ē* (Gr. *polus*, many ; *gonu*, the knee), the Buckwheat family, an Order of plants, so named from the numerous joints of the stems, have astringent and acid properties : Polygoneæ, n. plu., *pŏl'·ĭ·gŏn'·ē·ē*, a tribe or Suborder : Polygonum, n., *pŏl·ĭg'·ŏn·ŭm*, a genus of plants : Polygonum bistorta, *bĭs·tŏrt'·ă* (L. *bis*, twice ; *tortus*, twisted), a species whose root or rhizome, which contains much tannin, some galic acid, and some starch, is a powerful astringent, so called from its *double twist*: **P. hydropiper**, *hĭd'-rō-pīp'-ér* (Gr. *hudor*, water; L. *piper*, pepper), the water-pepper, whose leaves are acrid and vesicant: **P. tinctorium**, *tĭngk·tōr'-ĭ-ŭm* (L. *tinctōrĭus*, of or belonging to dyeing —from *tingo*, I dye), a species yielding a blue dye: **P. aviculare**, *ăv·ĭk'-ŭl·ār'-ĕ* (L. *avĭcŭlāris*, belonging to the *avĭcŭlă*, a small bird), a species whose fruit is emetic and purgative: **P. cymosum**, *sĭm·ōz'-ŭm* (L. *cymōsus*, full of shoots—from *cyma*, the young sprout of a cabbage), a species on the Himalaya, used as spinach.

polygynia, n. plu., *pŏl'-ĭ-jĭn'-ĭ-ă* (Gr. *polus*, many; *gunē*, a female), plants which have several distinct styles: **polygynous**, a., *pŏl·ĭdj'-ĭn-ŭs*, having many pistils or styles.

polygynœcial, a., *pŏl'-ĭ-jĭn-ē'-shĭ-ăl* (Gr. *polus*, many; *gunē*, a female; *oikos*, a house), in *bot.*, having multiple fruits formed by the united pistils of many flowers.

polymerous, a., *pŏl-ĭm'-ĕr-ŭs* (Gr. *polus*, many; *mĕrŏs*, a part), composed of many parts.

polymorphic, a., *pŏl'-ĭ-mŏrf'-ĭk*, also **polymorphous**, a., *-mŏrf'-ŭs* (Gr. *polus*, many; *morphē*, form), assuming various forms or shapes: **polymorphy**, n., *pŏl'-ĭ-mŏrf'-ĭ*, the existence of several forms of the same organ on a plant.

polynucleated, a., *pŏl'-ĭ-nŭk'-lĕ-āt-ĕd* (Gr. *polus*, many; L. *nuclĕus*, a kernel), containing many nuclei.

polypary, n., *pŏl-ĭp'-ăr-ĭ*, also **polyparia**, n., *pŏl'-ĭp-ār'-ĭ-ă* (Gr. *polus*, many; L. *parĭŏ*, I produce), coral, so called because produced by polypes; the hard chitinous covering secreted by many of the Hydrozoa.

polype or **polyp**, n., *pŏl'-ĭp*, **polypes**, n. plu., *pŏl'-ĭps* (L. *polypus*, Gr. *polupous*, a polypus —from Gr. *polus*, many; *pous*,

a foot), in *zool.*, *strictly*, the single individual of a simple Actinozoön, as a sea-anemone; also applied to the separate zoöids of a compound Actinozoön; *loosely*, one of those radiate or worm-like water-animals which are furnished with many tentacula or foot-like organs surrounding the mouth or free orifice.

polypetalous, a., *pŏl'-ĭ-pĕt'-ăl-ŭs* (Gr. *polus*, many; *pĕtălon*, a leaf), in *bot.*, having the corolla composed of separate petals; having the petals free or distinct; syn. of 'eleutheropetalous,' and 'apopetalous.'

polyphyllous, a., *pŏl'-ĭ-fĭl'-lŭs* (Gr. *polus*, many; *phullon*, a leaf), in *bot.*, having a calyx or involucre composed of separate leaflets; many-leaved: **polyphylly**, n., *pŏl'-ĭ-fĭl'-lĭ*, the increase in the number of organs or leaves in a whorl.

polypide, n., *pŏl'-ĭp-ĭd* (L. *polypus*, a polypus), in *zool.*, the separate zoöid of a Polyzoön: **polypite**, n., *pŏl'-ĭp-ĭt*, the separate zoöid of a Hydrozoön; a fossil coral: **polypidom**, n., *pŏl-ĭp'-ĭd-ŏm* (L. *domus*, a house), one of the stems or fabrics containing the polypes or animals which construct them; a coral; the dermal system of a colony of the Hydrozoa or Polyzoa.

Polypodieæ, n. plu., *pŏl'-ĭ-pŏd-ĭ'-ĕ-ē* (Gr. *polus*, many; *pous*, a foot, *podos*, of a foot), a tribe or Sub-order of Ferns, so named from their numerous root-like feet: **Polypodium**, *pŏl'-ĭ-pōd'-ĭ-ŭm*, a genus of ornamental Ferns: **Polypodium crassifolium**, *krăs'-sĭ-fōl'-ĭ-ŭm* (L. *crassus*, thick; *fŏlĭum*, a leaf), a species said to be sudorific and anti-rheumatic: **P. phymatodes**, *fĭm'-ăt-ōd'-ēz* (Gr. *phumatōdēs*, affected with tumours or hard swellings—from *phūma*, a shoot, an excrescence), a species whose bruised fronds are used to perfume cocoa-nut oil.

polyporous, a., *pŏl·ĭp'·ŏr·ŭs* (Gr. *polus*, many ; *poros*, a passage), in *bot.*, having many pores ; applied to the Fungi found in pastures, on old trees, etc., whose under surfaces are full of pores.

polypus, n., *pŏl'·ĭp·ŭs*, **polypi,** n. plu., *pŏl'·ĭp·ī* (Gr. *polus*, many ; *pous*, a foot), a pear - shaped tumour, attached by its thin end or stalk to some mucous membrane.

polysepalous, a., *pŏl'·ĭ·sĕp'·ăl·ŭs* (Gr. *polus*, many ; Eng. *sepal*), in *bot.*, applied to plants where the sepals of a calyx form no cohesion ; having a calyx composed of separate sepals ; syn. of ' eleutherosepalous,' and ' aposepalous.'

polysperm, n., *pŏl'·ĭ·spĕrm* (Gr. *polus*, many ; *sperma*, seed), in *bot.*, a pericarp containing numerous seeds : **polyspermal,** a., *pŏl'·ĭ·spĕrm'·ăl*, also **polyspermous,** a., *-spĕrm'·ŭs*, containing many seeds.

polysporous, a., *pŏl'·ĭ·spōr'·ŭs* (Gr. *polus*, many ; *spora*, seed), in *bot.*, having many seeds.

polystemonous, a., *pŏl'·ĭ·stĕm'·ŏn·ŭs* (Gr. *polus*, many ; *stēmōn*, a stamen), in *bot.*, having the stamens more than double the sepals or petals, in number.

polystome, n., *pŏl·ĭs'·tŏm·ĕ* (Gr. *polus*, many ; *stoma*, a mouth), in *zool.*, an animal having many mouths, as among certain of the Protozoa : **polystomous,** a., *pŏl·ĭs'·tŏm·ŭs*, in *zool.*, having more than two suckers or mouths ; in *bot.*, having many suckers in the same fibril or root.

polysymmetrical, a., *pŏl'·ĭ·sĭm·mĕt'·rĭk·ăl* (Gr. *polus*, many ; Eng. *symmetrical*), in *bot.*, having a member which can be divided by several planes into portions, each the reflected image of the other.

polythalamous, a., *pŏl'·ĭ·thăl'·ăm·ŭs* (Gr. *polus*, many ; *thalamos*, a chamber), in *zool.*, having many chambers, as in the shells of Foraminifera and Cephalopoda.

Polytrichum, n., *pŏl'·ĭ·trĭk'·ŭm* (Gr. *polus*, many ; *thrix*, hair, *trichos*, of hair), a very pretty genus of Mosses, Ord. Musci or Bryaceæ, with rigid leaves and a hairy calyptra : Polytrichum commune, *kŏm·mūn'·ĕ* (L. *commūnis*, common), a species made into dusting-brooms, called silk-brooms.

polyuria, n., *pŏl'·ĭ·ūr'·ĭ·ă* (Gr. *polus*, much ; *ouron*, urine), an excessive flow of urine, as in diabetes ; see ' hydruria.'

Polyzoon, n., *pŏl'·ĭ·zō'·ŏn*, **Polyzoa,** n. plu., *pŏl'·ĭ·zō'·ă* (Gr. *polus*, many ; *zōŏn*, an animal), a division of the Molluscoida, comprising compound animals, as the sea-mat ; a numerous class of plant-like animals, chiefly inhabitants of the sea, also called ' Bryozoa': **Polyzoarium,** n., *pŏl'·ĭz·ō·ār'·ĭ·ŭm*, the dermal system of a colony of the Polyzoa ; see ' Polypidom.'

pome, n., *pŏm* (L. *pomum*, an apple), a fleshy many-celled fruit, as the apple and pear : **Pomeæ,** n. plu., *pōm'·ĕ·ē*, a Sub-order of the Ord. Rosaceæ, forming the Pomaceæ of Lindley: **pomaceous,** a., *pōm·ā'·shŭs*, consisting of or pert. to apples : **pomum Adami,** *ăd·ām'·ī* (*Ādāmi*, of Adam), the apple of Adam ; the prominence in the neck formed by the thyroid cartilage, in the neck of the male especially.

pompholyx, n., *pŏmf'·ŏl·ĭks* (Gr. *pompholux*, a bubble, a blister), another name for ' pemphygus,' which see.

pons, n., *pŏnz* (L. *pons*, a bridge), in *anat.*, a form of communication between two parts: pons hepatis, *hĕp'·ăt·ĭs* (L. *hēpar*, the liver, *hēpātis*, of the liver), the prolongation of the 'hepatic substance' which often partially bridges over the 'umbilical fissure' of the liver : **p. Tarini**, *tăr·in'i* (after *Tarin*), the bridge of Tarin ; a greyish matter connecting together the diverging 'crura cerebri' : **p. Varolii**, *văr·ōl'i·i* (after *Varolius*), the bridge of Varolius ; an eminence of transverse fibres, above and in front of the 'medulla oblongata,' and between the lateral lobes of the cerebellum.

Pontederaceæ, n. plu., *pŏnt'ĕ·dĕr·ā'sĕ·ē* (after *Pontedera*, professor of botany at Padua), the Ponte-deria family, an Order of aquatic or marsh plants : **Pontederia**, n., *pŏnt'ĕ·dēr'i·ă*, a genus of aquatic plants.

popliteal, a., *pŏp'lĭt·ē'ăl*, also *pŏp·lĭt'ĕ·ăl* (L. *poples*, the ham of the knee, *poplĭtis*, of the ham of the knee), in *anat.*, pert. to the ham or back part of the knee-joint : **popliteus**, a., *pŏp'lĭt·ē'ŭs*, applied to an oblique muscle placed below the knee, arising by a thick tendon from the fore-part of the popliteal groove.

poppy, n., *pŏp'pi* (AS. *popig*, L. *papaver*, the poppy), a gay flowering plant of several species, some of which yield opium ; the 'Papaver somniferum,' Ord. Papaveraceæ, is the opium poppy.

Populus, n., *pŏp'ūl·ŭs* (L. *pŏp'ŭlus*, a poplar tree), a genus of trees called Poplars, Ord. Salic-aceæ : **Populus alba**, *ălb'ă* (L. *albus*, white), the Abele or White Poplar tree : **P. tremula**, *trĕm'ūl·ă* (L. *trĕmŭlus*, quivering, shaking), the Aspen : **P. fastigi-ata**, *făst·ĭdj'i·āt'ă* (L. *făstĭgĭātus*, pointed at the top—from *fastĭgĭum*, a projecting point), and **P. dilatata**, *dĭl'ăt·āt'ă* (L. *dĭlāt-ātŭs*, spread out, enlarged—from *dĭlāto*, I spread out), are species which are called the Lombardy

Poplars : **P. nigra**, *nĭg'ră* (L. *nĭger*, black), and **P. balsam-ifera**, *băl'săm·ĭf'ĕr·ă* (L. *balsămum*, balsam ; *fero*, I bear), are species whose buds, covered with a resinous exudation, called Tacamahac, are said to be diuretic and antiscorbutic ; Poplars secrete a saccharine substance called 'Populine.'

pore, n., *pōr* (Gr. *poros*, L. *porus*, a passage or channel), a very minute opening or interstice, as in the skin : **poriform**, a., *pōr'i·fŏrm* (L. *forma*, shape), resembling a pore : **porifera**, n. plu., *pōr·ĭf'ĕr·ă* (L. *fero*, I bear), the Foraminifera or Sponges, from their numerous openings or pores : **pore-capsules**, in *bot.*, dry dehiscent capsules, splitting by the detachment of small valves from the pericarp, as in Papaver : **porous vessels**, in *bot.*, pitted or dotted vessels.

porrect, a., *pŏr·rĕkt'* (L. *porrectum*, to reach out or extend), in *bot.*, extended forward, as to meet something.

porrigo, n., *pŏr·rĭg'ō* (L. *porrigo*, the scurf), formerly applied to any affection of the head where there were scabs.

porta, n., *pŏrt'ă* (L. *porta*, a gate), in the liver the transverse fissure where the 'vena portæ' divides into two principal branches : **portal**, a., *pōrt'ăl*, pert. to the system of the 'vena portæ' of the liver : **portal vein**, the vein which receives the venous blood on its way from the stomach, spleen, and intestines, carrying the blood on to the liver to be distributed through that organ.

portio dura, *pŏr·sht·ō dūr'ă* (L. *portĭŏ*, a portion ; *dūrŭs*, hard), the *facial nerve*, is the *hard portion* of the seventh pair of cranial nerves, and forms the 'motor nerve' of all the muscles of expression in the face : **portio mollis**, *mŏl'lis* (L. *mŏllĭs*, soft), the *auditory nerve*, is the *soft portion* of the seventh pair of cranial nerves, and forms the special nerve of the organ of hearing.

Portulacaceæ, n. plu., *pŏrt'·ŭl·ăk· ā'·sĕ·ē* (L. *porto*, I carry ; *lac*, milk), the Purslane family, an Order of plants, so named from their juicy nature : **Portulaca**, n., *pŏrt'·ŭl·āk'·ă*, (L. *portūlāca*, purslane), a genus of plants : **Portulaca oleracea**, *ŏl'·ĕr·ā'·sĕ·ă* (L. *olerācĕŭs*, herb - like — from *olus*, a kitchen herb), common Purslane, used as a potherb from its cooling and antiscorbutic qualities.

porus opticus, *pŏr'·ŭs ŏpt'·ĭk·ŭs* (Gr. *poros*, a pore ; *optĭkŏs*, relating to the sight), the round disc where the optic nerve expands, and having in its centre the point from which the vessels of the retina branch.

posology, n., *pōz·ŏl'·ō·jĭ* (Gr. *posos*, how much ; *logos*, discourse), the branch of medicine which treats of quantity or doses: **posological**, a., *pōz'·ŏl·ŏdj'·ĭk·ăl*, of or pert. to quantities or doses in medicine.

post-anal, a., *pŏst·ān'·ăl* (L. *post*, after or behind ; *ānŭs*, the fundament), situated behind the anus : **post-œsophageal**, a., *ĕs·ŏf'·ădj·ē'·ăl* (Gr. *oisophagos*, the gullet), situated behind the gullet : **post-oral**, *ōr'·ăl* (L. *ōs*, a mouth, *ōris*, of a mouth), situated behind the mouth : **post-pharyngeal**, a., situated behind the pharynx.

posterior, a., *pŏst·ēr'·ĭ·ŏr* (L. *posterior*, hinder—from *postĕrus*, coming after), coming after ; hinder ; in *bot.*, applied to the part of the flower next the axis— same as Superior : **posteriors**, n. plu., the hinder parts of an animal.

posticæ, n. plu., *pŏst'·ĭs·ē* (L. *postĭcus*, behind), in *bot.*, a name

applied to anthers when they open on the outer surface ; same as 'extrorse,' which see.

Potameæ, n. plu., *pŏt·ăm'·ĕ·ē* (Gr. *potămos*, a river), the Naias or Pondweed family, an Order of plants ; another name for 'Naiadaceæ,' which see : **Potamogeton**, n., *pŏt'·ăm·ŏdj·ēt'·ŏn* (Gr. *geitōn*, bordering, contiguous), a genus of plants whose species mostly grow wholly in water : **Potamogeton natans**, *nāt'·ănz* (L. *nătans*, swimming), a species whose roots are said to be eaten in Siberia.

potass, n., *pŏt·ăs'* (*potassa*, a Latinised form of *potash* ; F. *potasse*, potashes), the hydrated oxide of the metal potassium, or kalium, much used in medicine : **liquor potassæ**, *lĭk'·ŏr pŏt·ăs'·sē* (L. *liquor*, a fluid ; *potassæ*, of potassa), a solution of potass, a colourless and very acrid fluid, prepared from carbonate of potass by adding quicklime.

Potentilleæ, n. plu., *pŏt'·ĕnt·ĭl'·lĕ·ē* (L. *potens*, powerful, *potentis*, of powerful), a Sub-order of plants, Ord. Rosaceæ, so named from the supposed medicinal qualities of some of the species : **Potentilla**, n., *pŏt'·ĕnt·ĭl'·lă*, a genus of plants : **Potentilla tormentilla**, *tŏr'·mĕnt·ĭl'·lă* (L. *tormentum*, pain or torment), a species whose root was supposed to relieve pain in the teeth, also tonic and astringent.

præcipitate, n., *prē·sĭp'·ĭt·āt* (L. *præcĭpĭto*, I throw down headlong), any substance thrown down, from its state of solution in a liquid, to the bottom of a vessel, generally in a pulverised form.

præfloration, n., *prē'·flōr·a'·shŭn* (L. *præ*, before ; *flos*, a flower, *flōris*, of a flower), another term for 'æstivation' : **præfoliation**, n., *prē'·fōl·ĭ·ā'·shŭn* (L. *fŏlium*, a leaf), another name for 'vernation.'

præ-molars, n. plu., *prē-mŏl'ărz* (L. *pre*, before ; *molārēs*, the grinders), the molar teeth of Mammals, which come after the molars of the milk-set of teeth, and occupy the same places ; the bicuspid teeth in man : **præ-œsophageal**, a., situated in front of the gullet : **præ-sternum**, n., the anterior portion of the breast-bone, extending as far as the point of articulation of the second rib.

præmorse, a., *prē·mŏrs'* (L. *præmŏrsus*, bitten into—from *præ*, before ; *mŏrsus*, bitten), in *bot.*, applied to a rhizome or root terminating abruptly, as if bitten off.

Prangos, n., *prăng'gŏs* (a native name), a genus of plants, Ord. Umbelliferæ: **Prangos pabularia**, *păb'ŭl·ār'ĭ·ă* (L. *pābŭlārĭus*, belonging to *pābŭlum*, food for man or beast), a plant of S. Tartary, an excellent fodder for cattle.

pre, as in premolar, etc., see under 'præ.'

precordium, n., *prē·kŏrd'ĭ·ŭm*, **precordia**, n. plu., *-ĭ·ă* (L. *præ*, before ; *cor*, the heart, *cordis*, of the heart), the region of the chest which lies in front of the heart ; parts about the heart : **precordial**, a., *prē·kŏrd'ĭ·ăl*, of or relating to the precordia.

preformative, a., *prē·fŏrm'ăt·ĭv* (L. *præ*, before, in front ; Eng. *formative*), in *anat.*, applied to the fine, pellucid, homogeneous membrane which covers the entire pulp of the tooth.

prehensile, a., *prē·hĕns'ĭl* (L. *prehendo*, I lay hold of, *prehensus*, laid hold of), adapted for seizing or laying hold, as the hands in man, or the tails of some monkeys : **prehension**, n., *prē·hĕn'shŭn*, a seizing or grasping, as with the hand.

prepuce, n., *prēp'ūs* (F. *prépuce*, the prepuce ; L. *præpūtĭum*, the foreskin), the membranous or cutaneous fold covering the 'glans penis' ; the foreskin.

presbyopia, n., *prĕs'bĭ·ŏp'ĭ·ă* (Gr. *presbus*, old ; *ōps*, the eye), a defect of the eyesight, generally met with in advanced life, in which the lens is so flattened that the near point of vision has receded to beyond eight inches, instead of being at three, as in early life.

prescription, n., *prĕ·skrĭp'shŭn* (L. *præ*, before ; *scrīptus*, written), a written statement by a physician or surgeon, in which he directs what medicine or medicines are to be taken by a patient, the dose, and how often.

pressirostres, n. plu., *prĕs'ĭ·rŏst'rēz* (L. *pressus*, flattened ; *rostrum*, a beak), a group of the grallatorial birds, having a compressed or flattened beak : **pressirostral**, a., *-rŏst'răl*, pert. to.

prevertebral, a., *prē·vĕrt'ĕb·răl* (L. *præ*, before ; Eng. *vertebral*), situated immediately in front of the vertebræ.

prickles, n. plu., *prĭk'klz* (Dut. *prik*, a stab or prick ; low Ger. *prikken*, to pick or stick), in *bot.*, sharp conical elevations of the epidermis, of a nature similar to hairs.

primary, a., *prĭm'ăr·ĭ* (L. *prīmus*, the first), first in place, rank, or importance ; in *bot.*, applied to the principal division of any organ : **primaries**, n. plu., *prĭm'ăr·ĭz*, the stiff feathers or quills in the last joint of the wing of a bird : **primine**, n., *prĭm'ĭn*, the first or outermost covering of an ovule.

primordial, a., *prĭm·ŏrd'ĭ·ăl* (L. *prīmŏrdĭum*, first beginning—from *prīmus*, first ; *ordior*, I commence), elementary ; original, in *bot.*, earliest formed ; applied to the first true leaves given off

Y

by the young plant, also to the first fruit produced on a raceme or spike : **primordial utricle**, in *bot.*, the lining membrane of cells in their early state : **p. vesicle**, the elementary ovule of animals.

Primulaceæ, n. plu., *prĭm'·ŭl·ā'·sĕ·ē* (L. *prĭmŭlus*, the first, dim. from *prĭmus*, first), the Primrose family, an Order of plants, among which acridity more or less prevails : **Primula**, n., *prĭm'·ŭl·ă*, a genus of showy garden flowers, so named as being a very early flowering plant : **Primula auricula**, *awr·ĭk'·ŭl·ă* (L. dim. from *auris*, the ear), a yellow plant, native of Swiss Alps, from which all the fine forms of auriculas are derived : **P. veris**, *vēr'·ĭs* (L. *vēr*, spring, *vērĭs*, of spring), the Cowslip, the flowers of which are said to be narcotic : **P. elatior**, *ĕl·āt'·ĭ·ŏr* (L. *ēlātus*, productive, *ēlātĭor*, more productive), the Oxlip : **P. vulgaris**, *vŭlg·ār'·ĭs* (L. *vulgāris*, common, vulgar), the Primrose : **P. farinosa**, *făr'·ĭn·ōz'·ă* (L. *farĭnōsus*, mealy— from *farīna*, meal), the Bird's-eye Primrose : **P. Scotica**, *skŏt'·ĭk·ă* (of or from *Scotland*), the Scottish Primrose.

princeps cervicis, *prĭn'·sĕps sĕrv·ĭs'·ĭs* (L. *prĭnceps*, the first, chief ; *cervix*, the neck, *cervīcis*, of the neck), applied to the large branch artery (arteria princeps cervicis) which descends along the back part of the neck, and divides into a superficial and a deep branch.

Pringlea, n., *pring·glē'·ă* (after *Sir John Pringle*, who wrote on scurvy), a genus of plants, Ord. Cruciferæ : **Pringlea antiscorbutica**, *ăn'·tĭ·skŏr·bŭt'·ĭk·ă* (Gr. *anti*, against ; mid. L. *scorbūtus*, the disease scurvy), the Kerguelen's-land cabbage, so named from its properties.

Prionium, n., *prī·ōn'·ĭ·ŭm* (Gr. *priōn*, a saw), a genus of plants, Ord. Xyridaceæ : **Prionium**

palmita, *pălm'·ĭt·ă* (L. *palmes*, a young branch, *palmĭtĭs*, of a young branch), the Palmite, a remarkable, aquatic, juncaceous plant of S. Africa, having a very thick stem.

prismenchyma, n., *prĭz·mĕng'·kĭm·ă* (Gr. *prisma*, a prism ; *engchumos*, juicy ; *cheuma*, tissue), in *bot.*, tissues formed of prismatic cells : **prismenchymal**, a., *-kĭm·ăl*, of or pert. to.

probang, n., *prō·băng'* (from *probe*, and *bang*, in the sense of pushing), a slender piece of whalebone, with a piece of ivory or sponge at the extremity, used for pushing bodies down the gullet or œsophagus into the stomach, or for ascertaining the permeability of that passage.

probe, n., *prōb* (L. *probo*, I test or try), a small, slender rod for examining a wound, ulcer, or cavity.

proboscis, n., *prōb·ŏs'·sĭs* (L. *proboscis*, Gr. *proboskis*, a trunk), the snout or trunk of an elephant; the spiral trunk of Lepidopterous insects ; the projecting mouth of certain Crinoids ; the central polypite in the Medusæ : **Proboscidea**, n. plu., *prōb'·ŏs·sĭd'·ĕ·ă*, the Order of Mammals comprising the Elephants.

procambium, n., *prō·kăm'·bĭ·ŭm* (new L. *cambĭum*, nutriment ; L. *cambĭŏ*, I change), in *bot.*, the prosenchymatous, complete, cellular tissue of a future fibrovascular bundle ; see 'cambium.'

process, n., *prŏs'·ĕs* or *prŏs'·ĕs* (L. *processus*, an advance, process— from *pro*, forward ; *cessum*, to go or move along), in *bot.*, any prominence, projecting part, or small lobe ; the principal divisions of the inner peristome of Mosses ; in *anat.*, a projecting part of a bone ; any protuberance: **processes of bone**, in *anat.*, the eminences on the surfaces of bones, which are of various kinds, and named accordingly, such as :

heads, processes, round in form, which belong to the moveable articulations or joints ;

condyles, processes which are broader in one direction than in the others ;

impressions, irregular eminences, not much elevated ;

lines, unequal eminences, long, but not very prominent ;

crests, eminences resembling lines, but broader, and more prominent ;

prominences, prominent elevations, rounded, broad, and smooth ;

tuberosities, elevations, rounded and rough ;

spinous processes, have the form of a spine ;

styloid, resemble a style or pen ;

corocoid, resemble a crow's beak ;

odontoid, resemble a tooth ;

mastoid, resemble a nipple.

procidentia uteri, *prŏs'·ĭd·ĕn'·shĭ·ă ūt'·ĕr·ī* (L. *prōcĭdentĭa,* a falling down—from *pro,* forward ; *cado,* I fall ; *ūtĕrus,* the womb, *ūtĕri,* of the womb), the prolapse or falling down of the womb ; the protrusion of the uterus beyond the vulva.

procœlous, a., *prō·sēl'·ŭs* (Gr. *pro,* before, forward ; *koilos,* hollow), applied to vertebræ, the bodies of which are hollow or concave in front.

procumbent, a., *prō·kŭmb'·ĕnt* (L. *procumbens,* leaning forward), prostrate ; in *bot.,* lying upon or trailing along the ground.

proembryo, n., *prō·ĕm'·brĭ·ō* (Gr. *pro,* before ; Eng. *embryo),* in *bot.,* a free cell in the embryonal vesicle, which divides into eight cells by vertical and transverse septa, constituting together a short cylindrical cellular body ; the first part produced by the spore of an acrogen in germinating ; a prothallus.

profunda corvicis, *prō·fŭnd'·ă* *sĕrv·ĭs'·ĭs* (L. *profundus,* deep ; *cervix,* the neck, *cervĭcĭs,* of the neck), the deep artery of the neck ; the deep cervical branch of the subclavian artery : **profunda femoris,** *fĕm'·ŏr·ĭs* (L. *fĕmur,* the thigh, *femŏris,* of the thigh), the deep artery of the thigh ; the deep femoral artery.

proglottis, n., *prō·glŏt'·tĭs* (Gr. *pro,* for ; *glotta,* the tongue), the generative segment or jo'nt of a tapeworm : **proglottides,** n. plu., *prō·glŏt'·tĭd·ēz,* the sexually free and mature segments of Tænia solium.

prognathous, a., *prŏg·nāth'·ŭs,* also **prognathic,** a., *prŏg·nāth'·ĭk* (Gr. *pro,* before ; *gnathos,* jaw, the cheek), having prominent or projecting jaws, as in the Negro and Hottentot.

prognosis, n., *prŏg·nōz'·ĭs* (Gr. *prognosis,* foreknowledge — from *pro,* before ; *gignosko,* I know), the foretelling the result of any disease, based upon a consideration of its signs and symptoms.

progressive atrophy, see ' atrophy,' a gradual and systematic advancing atrophy of muscles ; fatty degeneration.

progressive locomotor ataxia, a disease characterised in walking by the peculiar gait, as that of a drunken man, arising from the loss of the faculty of co-ordination of the limbs, and harmonising the movements of independent parts ; see 'locomotor ataxia.'

prolapse, n., *prō·lăps',* also **prolapsus,** n., *prō·lăps'·ŭs* (L. *prolapsus,* slidden or fallen down), in *surg.,* a protrusion or falling down of a part, especially the gut or womb, so as to become partly external and uncovered : **prolapsus ani,** *ān'·ī* (L. *ānus,* the fundament), an affection analogous to invagination, in which a fold of the mucous membrane comes down, or in which both the mucous and muscular tunics descend, forming a tumour of a sausage or pyriform shape : **prolapsus uteri**, *ūt'·ĕr·ī* (L. *ŭtĕrus*, the womb), the protrusion of the womb at the vulva, or below its natural level in the pelvic cavity.

prolegs, n. plu., *prŏ'·lĕgz* (L. *pro*, for ; Eng. *legs*), the fleshy pediform organs, often retractile, which assist various larvæ in their movements ; the false abdominal feet of caterpillars.

proliferous, a., *prŏ·lĭf'·ĕr·ŭs* (L. *proles*, offspring ; *fero*, I carry), in *bot.*, bearing abnormal buds ; having an unusual development of supernumerary parts, as when flower-buds become viviparous, or when leaves produce buds : **prolification**, n., *prŏ·lĭf'·ĭk·ā'·shŭn*, the condition in which the axis is prolonged beyond the flower, and bears leaves, ending in an abortive flower-bud, as in the Rose and Geum.

pronation, n., *prōn·ā'·shŭn* (L. *pronus*, hanging downwards, stooping), in *anat.*, the act by which the palm of the hand is turned downwards, with the thumb towards the body ; the position of the hand so turned : **pronator**, n., *prōn·āt'·ŏr*, one of the two muscles used in the act of turning the palm downwards : **pronator teres**, *tĕr'·ēz* (L. *teres*, a taper), a muscle which rolls the radius inwards, together with the hand : **pronator quadratus**, *kwŏd·rāt'·ŭs* (L. *quadrātus*, square), a small flat quadrilateral muscle, extending transversely across the front of the radius and ulva : **prone**, a., *prōn*, prostrate; lying flat on the earth.

propagulum, n., *prŏp·ăg'·ŭl·ŭm* (L. *propāgo*, the slip or shoot of a plant, offspring), in *bot.*, an offshoot or germinating bud attached by a thickish stalk to the parent plant ; a runner ending in an expanded bud : **propagula**, n. plu., *prŏp·ăg'·ŭl·ă*,

powdery grains of the soredia of Lichens.

prophylactic, a., *prŏf'·ĭl·ăk'·tĭk* (Gr. *prophulaktikos*, preservative — from *pro*, before ; *phulasso*, I preserve), in *med.*, defending from disease ; preventive : n., a medicine which preserves against disease.

propodium, n., *prŏ·pōd'·ĭ·ŭm* (Gr. *pro*, before ; *podes*, feet), the anterior part of the foot in Molluscs.

proptosis, n., *prŏp·tōz'·ĭs* (Gr. *proptōsis*, a falling down or forwards — from *pro*, before ; *ptōsis*, a fall), a protrusion of the eyeball.

proscolex, n., *prŏ·skōl'·ĕks* (Gr. *pro*, before ; *skōlēx*, a worm), the first embryonic stage of a tapeworm.

prosencephalon, n., *prŏs'·ĕn·sĕf'·ăl·ŏn* (Gr. *pros*, before ; *engkephalon*, the brain), one of the five fundamental parts of the brain, comprising the cerebral hemispheres, corpus callosum, corpora striata, fornix, lateral ventricles, and olfactory nerve ; the cerebrum proper in fishes.

prosenchyma, n., *prŏs·ĕng'·kĭm·ă* (Gr. *pros*, before, addition ; *engchĕō*, I pour in ; *cheuma*, juice, tissue), in *bot.*, fusiform tissue forming wood ; tissue formed of elongated pointed cells: **prosenchymatous**, a., *prŏs'·ĕng·kĭm'·ăt·ŭs*, of or pert. to prosenchyma.

Prosobranchiata, n. plu., *prŏs'·ō·brăng·kĭ·āt'·ă* (Gr. *prŏsō*, in front, in advance of ; *brangchia*, gills of a fish), a division of Gasteropodous Molluscs, in which the gills are situated in advance of the heart.

prosoma, n., *prŏ·sōm'·ă* (Gr. *pro*, before ; *sōma*, a body ; *sōmăta*, bodies), in *zool.*, the anterior part of the body.

prostate, a., *prŏs'·tāt* (Gr. *prostătēs*, one who stands before—

from *pro*, before ; *stasis*, a setting or standing), applied to a pale, firm, glandular body, resembling a horse - chestnut in shape and size, which surrounds the neck of the bladder and commencement of the urethra : **prostatic**, a., *prŏ·stăt´·ĭk*, of or pert. to the prostate gland : **prostatitis**, n., *prŏs´·tăt·ĭt´·ĭs*, the inflammation of.

protandrous, a., *prŏt·ănd´·rŭs* (Gr. *prōtos*, first ; *anēr*, a male, *andros*, of a male), in *bot.*, having stamens reaching maturity before the pistil.

Proteaceæ, n. plu., *prŏt´·ĕ·ā´·sĕ·ē* (L. *Proteus*, a self-transforming sea-god), the Protea family, an Order of plants, so named from their great diversity of appearance : **Protea**, n., *prŏt´·ĕ·ă*, a genus of magnificent evergreen shrubs, producing peculiar flowers : **Protea mellifera**, *mĕl·lĭf´·ĕr·ă* (L. *mĕl*, honey, *mĕllis*, of honey ; *fero*, I produce), the sugar-bush, a species so named from the honey furnished by its flowers : **P. grandiflora**, *grănd´·ĭ·flōr´·ă* (L. *grandis*, great ; *flos*, a flower, *flōris*, of a flower), a species whose bark, called Wagenboom, is used in diarrhœa at the Cape.

protenchyma, n., *prŏt·ĕng´·kĭm·ă* (Gr. *prōtos*, first ; *engcheō*, I pour in ; *cheuma*, juice, tissue), in *bot.*, the fundamental tissue out of which by differentiation other tissues arise.

prothallus, n., *prŏ·thăl´·lŭs*, also **prothallium**, n., *prŏ·thăl´·lĭ·ŭm* (Gr. *pro*, before ; *thallos*, a sprout), in *bot.*, the first results of the germination of the spore in the higher Cryptogams, as ferns, horse-tails, etc.

prothorax, n., *prŏ·thŏr´·ăks* (Gr. *pro*, before ; *thorax*, the chest), in *zool.*, the anterior ring of the thorax of insects.

Protococcus, n., *prŏt´·ō·kŏk´·ŭs* (Gr. *prōtos*, first ; *kŏk·kŏs*, a berry), a genus of Algæ : **Protococcus nivalis**, *nĭv·ā´·lĭs* (L. *nĭvālis*, of or belonging to snow—from *nix*, snow), and **P. viridis**, *vĭr´·ĭd·ĭs* (L. *virĭdis*, green), are species which occur in red and green snow.

protogynous, a., *prŏt·ŏdj´·ĭn·ŭs*, also **proterogynous**, a., *prŏt´·ĕr·ŏdj´·ĭn·ŭs* (Gr. *prōtos*, first ; *gunē*, a woman), in *bot.*, having the pistil reaching maturity before the stamens.

protophyte, n., *prŏt´·ō·fĭt* (Gr. *prōtos*, first ; *phuton*, a plant), a production lowest in the scale of the vegetable kingdom : **protophyta**, n. plu., *prŏt·ŏf´·ĭt·ă*, the lowest division of plants.

protoplasm, n., *prŏt´·ō·plăzm*, also **protoplasma**, n., *prŏt´·ō·plăz´·mă* (Gr. *prōtos*, first ; *plasma*, what has been formed, a model), in *bot.*, deposits upon the inner walls of the cells of cellular tissue, from which the cell-nuclei are formed ; the elementary basis of organised tissues : **protoplast**, n., *prŏt´·ō·plăst* (Gr. *plastos*, formed), the thing first formed ; a first formed nucleated cell in an organised body.

protopodite, n., *prŏt·ŏp´·ŏd·ĭt* (Gr. *prōtos*, first ; *pous*, a foot, *podes*, feet), in *zool.*, the basal segment of the typical limb of a Crustacean.

protospores, n., *prŏt´·ō·spōrz* (Gr. *prōtos*, first ; *spora*, a seed), in *bot.*, the spores of the first generation.

protovertebræ, n. plu., *prŏt´·ō·vért´·ĕb·rē* (Gr. *prōtos*, first ; and *vertebræ*), the primitive vertebræ of early fœtal life, appearing early as dark spots, soon forming quadrangular laminæ on each side of the chorda dorsalis, which, however, do not coincide with the permanent vertebræ.

protoxide, n., *prŏt·ŏks´·ĭd* (Gr. *prōtos*, first ; and *oxide*), in *chem.*, a compound containing one equivalent of oxygen combined with one of a base, that is, the first oxide, and so of numerous other words similarly formed.

Protozoa, n. plu., *prŏt'·ō·zō'·ă*, also **Protozoans**, n. plu., *prŏt'·ō·zō'·ănz* (Gr. *prōtos*, first; *zoōn*, an animal), the lowest division of the animal kingdom, apparently occupying a sort of neutral ground between animals and vegetables: **protozoic**, a., *prŏt'·ō·zō'·ĭk*, belonging to the Protozoa; containing the first traces of life: **protozoon**, n., *prŏt'·ō·zō'·ŏn*, also **protozoan**, n., *-zō'·ăn*, one of the Protozoa.

proventriculus, n., *prŏv'·ĕn·trĭk'·ŭl·ŭs* (L. *pro*, in front of; *ventrĭcŭlus*, the stomach—from *venter*, the belly), the cardiac portion of the stomach of birds.

proximal, a., *prŏks'·ĭm·ăl* (L. *proxĭmus*, next, nearest), toward or nearest, as to a body or centre; in *zool.*, applied to the slowly-growing, comparatively - fixed extremity of a limb, or of an organism: **proximal part**, the part toward or nearest.

proximate, a., *prŏks'·ĭm·āt* (L. *proxĭmus*, nearest), nearest; immediate: **proximate cause**, that which immediately precedes and produces the effect, though not the only operating cause; opposed to 'remote or immediate': **proximate principles**, in *chem.*, distinct compounds which exist ready formed in animals, as albumen, fat, etc., and in vegetables, as sugar, starch, etc.

pruinose, a., *prŏ'·ĭn·ōz'* (L. *pruĭnōsus*, full of hoar-frost — from *prŭina*, hoar-frost), in *bot.*, covered with glittering particles, as if fine globules of dew had been congealed upon it.

Prunus, n., *prŏn'·ŭs* (L. *prūnum*, a plum), a genus of plants, Ord. Rosaceæ: **Prunus domestica**, *dŏm·ĕst'·ĭk·ă* (L. *domĕstĭcus*, of or belonging to the house—from

dŏmus, the house), the Plum tree and its varieties, which, when dried, constitute prunes: P. **laurocerasus**, *lăwr'·ō·sĕr'·ăs·ŭs* (L. *laurus*, the laurel; *cerăsus*, the cherry tree), the Cherry-laurel, or common Bay - laurel, have been used as anodyne and hypnotic remedies: P. **Lusitanica**, *lŏz'·ĭt·ăn'·ĭk·ă* (*Lusitania*, old name for Portugal), the Portugal laurel, cultivated as an evergreen: P. **spinosa**, *spĭn·ōz'·ă* (L. *spĭnōsus*, thorny, prickly—from *spīna*, a spine), the Sloe, whose leaves have been employed to adulterate tea.

prurigo, n., *prōr·īg'·ō* (L. *prūrīgo*, an itching, *prūrĭŏ*, I itch), a skin disease characterised by intolerable itching: **pruritus**, n., *prōr·īt'·ŭs* (L. *prūrītus*, an itching), itching, forming the main symptom of the disease prurigo.

prussic, a., *prŭs'·ĭk* (from *Prussia*), applied to a deadly poison originally obtained from Prussian-blue, existing in the laurel, and in kernels of various fruits —prepared commercially from prussiate of potass; hydrocyanic acid.

psalterium, n., *săwlt·ēr'·ĭ·ŭm* (L. *psaltērĭum*, a stringed instr. of the lute kind), the manyplies or third cavity of the stomach of a ruminant animal; in *anat.*, a part of the brain, consisting of lines impressed on the under surface of the posterior part of the body of the fornix.

psammoma, n., *săm·mōm'·ă* (Gr. *psammŏs*, loose earth, sand), a tumour usually found in the brain and its appendages, whose characteristic feature is the occurrence of calcareous matter, or 'brain sand,' in the centre of small concentric lobules.

pseudembryo, n., *sŭd·ĕm'·brĭ·ō* (Gr. *pseudēs*, lying, false; and *embryo*), the larval form of an Echinoderm.

pseudobranchia, n., *sŭd'·ō·brăngk'·ĭ·ă* (Gr. *pseudēs*, false; *brangchia*, gills of a fish), in certain fishes, a supplementary gill, which, receiving arterialised blood only, does not assist in respiration.

pseudo-bulb, n., *sŭd'·ō·bŭlb* (Gr. *pseudēs*, false; and *bulb*), in *bot.*, a swollen aerial of many Orchids, resembling a tuber; a bulb in appearance only.

pseudocarp, n., *sŭd'·ō·kărp* (Gr. *pseudēs*, false; *karpos*, fruit), in *bot.*, applied to such fruit as the strawberry, in which other parts are incorporated with the ovaries in forming the fruit.

pseudohæmal, a., *sŭd'·ō·hēm'·ăl* (Gr. *pseudēs*, false; *haima*, blood), in *zool.*, applied to the vascular system of the Annelida: **pseudo-hearts**, certain contractile cavities connected with the arterial system of Brachiopoda, formerly looked upon as hearts: **pseudo-navicellæ**, n. plu., *năv'·ĭ·sĕl'·lē* (L. *năvĭcula*, a little ship—from *nāvis*, a ship), the embryonic forms of the Gregorinidæ, so named from their resemblance to the Navicula.

pseudopodium, n., *sŭd'·ō·pŏd'·ĭ·ŭm*, **pseudopodia**, n. plu., *sŭd'·ō·pŏd'·ĭ·ă* (Gr. *pseudēs*, false; *pous*, a foot, *podes*, feet), in *bot.*, the leafless prolongation of the leafy stem in Mosses, bearing the sporangium; in *zool.*, the extensions of the body-substance which are put forth by the Rhizopoda at will, serving for locomotion and prehension: **pseudopodial**, a., *sŭd'·ō·pŏd'·ĭ·ăl*, of or pert. to.

pseudospermous, a., *sŭd'·ō·spĕrm'·ŭs* (Gr. *pseudēs*, false; *sperma*, seed), in *bot.*, bearing single seeded seed - vessels, resembling seeds, as in Achenes; having a false seed cr carpel.

pseudova, n. plu., *sŭd·ōv'·ă* (Gr. *pseudēs*, false; L. *ovum*, an egg), the egg-like bodies from which the young of the viviparous Aphis are produced.

Psidium, n., *sĭd'·ĭ·ŭm* (Gr. *psidias*, so named by the anc. Greeks), a genus of fruit-bearing plants, Ord. Myrtaceæ: **Psidium pyriferum**, *pĭr·ĭf'·ĕr·ŭm* (L. *pīrum*, a pear; *fero*, I produce); and P. **pomiferum**, *pŏm·ĭf'·ĕr·ŭm* (L. *pomum*, an apple; *fero*, I produce), species which produce the pulpy edible fruits called Guavas: P. **Cattleyanum**, *kăt'·l·yăn'·ŭm* (after *William Cattley*, an English patron of Botany), has a fruit of a fine claret colour, bearing some resemblance in consistence and flavour to the strawberry.

psoas, n., *sō'·ăs* (Gr. *psoa*, the loins), in *anat.*, applied to two muscles of each loin, lying along the sides of the lumbar vertebræ, viz. the **psoas magnus**, *măg'·nŭs* (L. *magnus*, great), and the p. **parvus**, *pârv'·ŭs* (L. *parvus*, little), the great psoas, and the little psoas.

psoriasis, n., *sōr·ĭ'·ăs·ĭs* (Gr. *psorĭasis*, the being itchy or mangy—from *psōra*, scab, itch), a dry, scaly disease of the skin, characterised by slightly raised red patches, covered by white, shining, opaque scales: **psora**, n., *sōr'·ă*, the itch; a rough scaliness of the skin: **psoric**, a., *sōr'·ĭk*, of the nature of itch: n., a remedy for itch.

psychology, n., *sī·kŏl'·ō·jĭ* (Gr. *psuchē*, the soul; *logos*, a word), the doctrine of man's spiritual nature; the science conversant about the phenomena of the mind, or of the conscious subject: **psychical**, a., *sīk'·ĭk·ăl*, relating to or connected with the soul, spirit, or mind.

Psychotria, n., *sīk·ŏt'·rĭ·ă* (Gr. *psuchē*, the soul, life; *iătreia*, healing), a genus of plants, Ord. Rubiaceæ, so named in reference to the powerful medical qualities of some of the species: **Psychotria**

emetica, *ĕm·ĕt'·ĭk·ă* (L. *ĕmĕtĭca*, an incitement to vomit), the large black striated Ipecacuanha, inferior to true Ipecacuanha : **P. cephaelis**, *sĕf'·ă·ĕl'·ĭs* (Gr. *kephalē*, the head, their flowers being disposed in heads) ; and **P. Randia**, *răndˈ·ĭ·ă* (after *J. Rand*, a London botanist), species which act so violently as to produce poisonous effects.

pteridographia, n., *tĕr'·ĭd·ō·grăf'·ĭ·ă*, also **pteridography**, n., *tĕr'·ĭd·ŏg'·răʒ·ĭ* (Gr. *ptĕrĭs*, the ferns ; *grapho*, I write), a treatise on Ferns.

Pteris, n., *tĕr'·ĭs* (Gr. *pterux*, a wing), an ornamental genus of Ferns, so named in allusion to the appearance of the leaves, Ord. Filices : **Pteris aquilina**, *ăk'·wĭl·ĭn'·ă* (L. *aquila*, an eagle), the well - known bracken of this country : **P. esculenta**, *ĕsk'·ŭl·ĕnt'·ă* (L. *escŭlĕntus*, fit for eating—from *esca*, food), a species occasionally used as food in different countries.

Pterocarpus, n., *tĕr'·ō·kărp'·ŭs* (Gr. *ptĕrŏn*, a wing ; *karpos*, fruit), a genus of plants, mostly fine ornamental trees, Ord. Leguminosæ, Sub-ord. Papilionaceæ, so named from their pods being girded with broad wings : **Pterocarpus erinaceus**, *ĕr'·ĭn·ā'·sĕ·ŭs* (L. *ērĭnācĕus*, a hedgehog), African Kino, used as a powerful astringent : **P. marsupium**, *măr·sūp'·ĭ·ŭm* (L. *marsūpĭum*, a pouch, a purse), a tree of the Indian forests yielding the concrete exudation called 'Kino' : **P. santalinus**, *săn'·tăl·ĭn'·ŭs* (Ar. *zandal*), yields the red sandalwood, used as a dye : **pterocarpous**, a., *tĕr'·ō·kărp'·ŭs*, having winged fruit.

Pteropoda, n., *tĕr·ŏp'·ŏd·ă* (Gr. *ptĕrŏn*, a wing ; *pous*, a foot, *podos*, of a foot), a class of the Mollusca which swim by means of fins attached near the head :

pteropodous, a., *tĕr·ŏp'·ŏd·ŭs*, wing-footed ; having a wing-shaped expansion attached near the head for swimming.

pterygo, *tĕr'·ĭg·ō* (Gr. *pterux*, a wing), a prefix denoting attachment to, or connection with, the pterygoid processes of the sphenoid bone : **pterygoid**, a., *tĕr'·ĭg·ōyd* (Gr. *eidos*, resemblance), in *anat.*, applied to the wing-like processes of the sphenoid bone.

ptosis, n., *tōz'·ĭs* (Gr. *ptosis*, a falling), paralysis of the upper eyelid, which falls and covers the eye, the patient being unable to open the eye except by means of his fingers.

ptyalin, n., *tī'·ăl·ĭn* (Gr. *ptuĕlŏn*, saliva), a ropy organic matter, being the active principle of saliva: **ptyalism**, n., *tī'·ăl·ĭzm*, an increased and involuntary flow of saliva.

puberty, n., *pūb'·ĕrt·ĭ* (L. *pubertas*, the age of maturity—from *pubes*, of ripe age, the privy parts), the age at which persons are capable of begetting, or bearing children ; maturity: **pubes**, n., *pūb'·ēz*, in *anat.*, the external part where the generative organs are situated, which at puberty begins to be covered with hair : **pubic**, a., *pūb'·ĭk*, pert. to or connected with the pubes : **pubic arch**, *ărtsh*, the bony arch formed over the concave border of the pelvis : **pubis os**, *pūb'·ĭs ŏs* (L. *pūbis*, the pubes, or of the pubes ; *ŏs*, a bone), the pubic or share-bone, forming part of the os innominatum : **os symphysis**, *sĭmf'·ĭs·ĭs* (Gr. *sumphusis*, a growing together), the connection of the bones forming the pubic arch : **pubescence**, n., *pūb·ĕs'·ĕns* (L. *pubescens*, reaching the age of puberty, ripening), the state of puberty ; in *bot.*, the downy substance on plants : **pubescent**, a., *pūb·ĕs'·ĕnt*, arriving at maturity; in *bot.*, covered with soft hair or down.

NOTE.—Although *pubes* is the correct Latin nominative, and *pubis* the genitive, late Latin authors have written *pubis* for *pubes*, and accordingly we now find both *pubes* and *pubis* used in the nominative, and *pubis* the genitive.

pubo-, *pūb'ō* (L. *pubes*, the signs of manhood, the privy parts), of or connected with the pubis os or share-bone : **pubo-femoral,** *fĕm'ŏr-ăl*, applied to a ligament which enters into the formation of the capsule of the hip joint : **pubo - prostatic,** *prŏ·stăt'ik*, applied to the anterior ligaments of the bladder.

pudendum, n., *pūd·ĕnd'ŭm*, **pudenda,** n. plu., *pūd·ĕnd'ă* (L. *pŭdendus*, of which one ought to feel ashamed—from *pŭdĕŏ*, I am ashamed), the external organs or parts of generation ; the *labia majora ;* the vulva : **pudendal,** a., *pūd·ĕnd'ăl*, applied to a branch of the small sciatic nerve ; pert. to the pudenda.

pudic, a., *pūd'ik* (L. *pudīcus*, shamefaced, modest), pert. to the pudenda or private parts ; applied to arteries, nerves, and veins connected with the generative organs.

puerperal, a., *pū·ĕr'pĕr·ăl* (L. *puerpera*, a woman in child-bed —from *puer*, a child ; *parĭo*, I bring forth), relating to or following child-birth ; applied to a fever coming after child-birth.

pulex irritans, *pūl'ĕks ĭr·ĭt'ănz* (L. *pūlex*, a flea, *pūlīcis*, of a flea ; *ĭrrĭtans*, exciting, inflaming), the common flea, a torment common to men and animals : **Pulicidæ,** n. plu., *pūl·ĭs'ĭd·ē*, the genus or group of insects which comprises the fleas.

Pulmo-gasteropoda, *pŭl'mō-găst'ĕr·ŏp'ŏd·ă* (L. *pŭlmo*, a lung, *pŭlmōnis*, of a lung ; Gr. *gastēr*, the stomach ; Gr. *pous*, a foot, *podos*, of a foot), in *zool.*, a

division of the Mollusca, comprising those creatures which breathe air directly by means of a pulmonary sac or chamber ; also called **Pulmonifera,** *pŭl'mŏn·ĭf'ĕr·ă* (L. *fero*, I bear) : **Pulmonaria,** n. plu., *pŭl'mŏn·ār'ĭ·ă*, a division of Arachnida which breathe by means of pulmonary sacs : **pulmonate,** a., *pŭl'mōn·āt*, possessing lungs : **pulmograde,** a., *pŭl'mō·grād* (L. *gradi*, to walk), having a lung-like movement ; moving by the alternate expansion and contraction of the body, especially of the disc, as in the case of the Medusæ: **pulmonary,** a., *pŭl'mŏn·ār·ĭ*, pert. to or affecting the lungs.

pulsation, n., *pŭls·ā'shŭn* (L. *pŭlsātŭm*, to beat, to strike ; *pŭlsus*, a beating), the beating or throbbing of the heart, or of an artery ; vibration : **pulse,** n., *pŭls*, the beating or throbbing of an artery, perceptible to the touch, and caused by the action of the heart.

pulverulent, a., *pŭl·vĕr'ŭl·ĕnt* (L. *pulvĕrŭlentus*, full of dust —from *pulvis*, dust), also **pulveracious,** a., *pŭl'vĕr·ā'shŭs* (L. *pulvĕrĕus*, full of dust), and **pulverous,** a., *pŭl'vĕr·ŭs*, in *bot.*, covered with dust or fine powdery matter ; powdery.

pulvinate, a., *pŭl'vĭn·āt* (L. *pulvīnātus*, cushion-shaped—from *pulvīnus*, a cushion), in *bot.*, shaped like a cushion or pillow : **pulvinuli,** n. plu., *pŭl·vĭn'ŭl·ĭ*, excrescences on the surface of the thallus of certain Lichens : **pulvinus,** n., *pŭl·vĭn'ŭs*, in *bot.*, a cellular swelling at the point where the leaf-stalk joins the axis ; a sort of cushion at the base of some leaves : **pulvinar,** n., *pŭl·vĭn'ăr*, in *anat.*, a cushion-like prominence on each 'thalamus opticus' of the brain.

puncta vasculosa, plu., *pŭngk'tă văsk'ŭl·ōz'ă*, also **puncta cruenta,**

krô·ĕnt'·ă (L. *punctus*, a sting, a point ; *vascŭlum*, a small vessel ; *crŭentus*, stained with blood), in *anat.*, the numerous minute red dots, produced by the escape of blood from divided blood-vessels, which stud the surface of the white central mass of the cerebrum : **punctum lachrymale**, *pŭngk'·tŭm lăk'·rĭm·ăl'·ĕ* (L. *punctus*, pierced, punctured ; *lachrĭma*, a tear), a small aperture which perforates each papilla of the papilla lachrymale: **punctum cæcum**, *sēk'·ŭm* (L. *cœcus*, blind), a point in the retina from which the optic nerve fibres radiate, so named because insensible to light : **p. vegetationis**, *vĕdj'·ĕt·ā'·shĭ·ōn'·ĭs* (L. *vegetātĭŏ*, a quickening, vegetation, *vegetātĭōnis*, of a quickening), in *bot.*, the point of vegetation or growth in a plant.

punctate, a., *pŭngk'·tāt*, also **punctated**, a., *pŭngk'·tāt·ĕd* (L. *punctum*, a point, a small hole), in *bot.*, having the surface covered with small holes or dots ; dotted.

Punica, n., *pŭn'·ĭk·ă* (L. *pūnĭcus*, of or from the *Pœni* or *Carthaginians*), a genus of plants, Ord. Myrtaceæ : **Punica granatum**, *grăn·āt'·ŭm* (L. *grānātus*, having many grains or seeds — from *grānum*, a seed ; *granātum*, a pomegranate), the pomegranate tree, which produces dark scarlet flowers, used as an astringent, and the rind of the fruit and the bark of the root used as anthelmintics, especially in tapeworm.

pupa, n., *pūp'·ă*, **pupæ**, n. plu., *pūp'·ē* (L. *pupa*, a doll or puppet), the third or last state but one of insect existence—the first being the *egg*, the second the *caterpillar*, the third the *pupa* or *chrysalis*, and the fourth or perfect insect state the *imago*.

pupil, n., *pūp'·ĭl* (L. *pupilla*, a little girl—from *pupa*, a girl, a

doll ; It. *pupilla*, the eye-ball), the opening in the iris of the eye through which the rays of light pass to the retina.

purgative, a., *pėrg'·ăt·ĭv* (L. *purgo*, I purify), having the power of evacuating the bowels : n., a medicine that causes frequent evacuations of the bowels.

purples, n. plu., *pėrp'·lz*, also called **ear-cockles**, or **peppercorn**, a disease affecting the grains of wheat, in which the grains become first of a dark-green, and ultimately of a black colour, caused by the animal parasite *vibrio trilici*, or eel of the wheat.

purpura, n., *pėrp'·ūr·ă* (L. *purpura*, the shell-fish which yields purple), a disease accompanied by an eruption of spots on the skin called petechiæ, or patches called ecchymoses, caused by hæmorrhage into the skin, and which vary in tint from bright red to violet : **purpuric**, a., *pėrp·ūr'·ĭk*, denoting an acid of a purple colour, obtained from excrement of the boa-constrictor, and also from urinary calculi ; of or pert. to purpura : **purpuric fever**, a fever occasionally accompanying purpura.

purulent, a., *pūr'·ūl·ĕnt* (*pūrŭlentus*, full of corrupt matter— from *pūs*, the viscous matter of a sore ; *pūris*, of the matter of a sore), consisting of pus or corrupt matter : **purulence**, n., *pūr'·ūl·ĕns*, the formation of pus or matter : **pus**, n., *pŭs*, the fluid matter contained in abscesses, and discharged from the surfaces of ulcers and granulating wounds, healthy pus being of a white or pale-yellow colour: **ichorous pus**, the fœtid and dirty fluid discharged from foul and unhealthy ulcers, or from abscesses in those of a vitiated constitution.

pustule, n., *pŭst'·ūl* (L. *pustŭla*, a blister or pimple — from *pūs*, matter from a sore), a small elevation of the skin or cuticle containing pus : **pustular**, a., *pŭst'·ŭl·ăr*, covered with or resembling pustules : **pustula maligna**, *pŭst'·ŭl·ă măl·ĭg'·nă* (L. *mălignus*, of an evil nature), anthrax or carbuncular fever.

putamen, n., *pūt·ām'·ĕn* (L. *pŭt·āmen*, a pod or shell), in *bot.*, the hard endocarp, or bony stone, of some fruits, as the peach.

putrid, a., *pŭt'·rĭd* (L. *pŭtrĭdus*, decayed — from *pŭtris*, rotten), rotten ; corrupt : **putrid fever**, formerly applied to *typhus fever*, but now used to designate any very bad form of scarlet or typhus fever, or small-pox.

pyæmia, n., *pī·ēm'·ĭ·ă* (Gr. *puon*, pus ; *haima*, blood), a disease supposed to be due to the introduction of pus into the blood, or of some morbid poison—is often accompanied with inflammation of one or more veins, and the formation of abscesses in other parts of the body than those originally affected ; blood poisoning.

pycnide, n., *pĭk'·nĭd·ē* (Gr. *puknos*, dense), a wart-like, minute, cellular, reproductive body in the thallus of Lichens : **pycnidia**, n. plu., *pĭk·nĭd'·ĭ·ă*, cysts containing stylospores, found in Lichens and Fungi.

pyelitis, n., *pī'·ĕl·īt'·ĭs* (Gr. *puĕlos*, a basin, a trough), a disease of the kidney, in which pus is formed in that organ, or in the ureter.

pylorus, n., *pĭl·ōr'·ŭs* (Gr. *pulŏrŏs*, a gate-keeper—from *pulē*, a gate), the lower and right hand orifice of the stomach leading to the intestines : **pyloric**, a., *pĭl·ōr'·ĭk*, pert. to the pylorus.

pyramid, n., *pĭr'·ăm·ĭd* (L. *pyrămis*, a pyramid, *pyrămĭdis*, of a pyramid—of Egyptian origin), a conical and laminated projection on the under surface of the cereb-

ellum ; a small conical eminence on the posterior wall of the tympanum : **pyramidalis abdominis**, *pĭr'·ăm·ĭd·āl'·ĭs ăb·dŏm'·ĭn·ĭs* (L. the pyramidal of the abdomen), a muscle arising from the pubes, which assists the lower part of the rectus : **pyramidalis nasi**, *nāz'·ī* (L. *nāsus*, the nose, *nāsi*, of the nose), a muscle of the nose ; also applied to the conoidal division of the kidney, seen in the section of that organ.

pyrena, n., *pĭr·ēn'·ă*, **pyrenæ**, n. plu., *pĭr·ēn'·ē* (Gr. *pūrēn*, the kernel or stone of fruit), in *bot.*, stony coverings of the seeds, as in the medlar ; the putamen : **pyrenous**, a., *pĭr·ēn'·ŭs*, full of fruit stones.

pyrenocarpous, a., *pĭr·ēn'·ō·kärp'·ŭs* (Gr. *pūrēn*, the stone of fruit ; *karpos*, fruit), having fructification, like certain Lichens.

Pyrethrum, n., *pĭr·ēth'·rŭm* (L. *pyrethrum*, Spanish chamomile ; Gr. *pur*, fire), a genus of plants, Ord. Compositæ, Sub-ord. Corymbiferæ ; the pellitory of Spain, whose roots, hot to the taste, are used in medicine : **Pyrethrum parthenium**, *pärth·ēn'·ĭ·ŭm* (Gr. *parthĕnos*, a virgin), common feverfew, is aromatic and stimulant.

pyrexia, n., *pĭr·ĕks'·ĭ·ă*, **pyrexiæ**, n. plu., *pĭr·ĕks'·ĭ·ē* (Gr. *purĕtŏs*, a fever—from *pur*, fire), fever, or the febrile condition ; febrile diseases.

pyriformis, a., *pĭr'·ĭ·fŏrm'·ĭs* (L. *pyrum*, a pear ; *forma*, shape), pear-shaped ; applied to a muscle which moves the thigh, taking its rise from the hollow of the sacrum, it is inserted into the cavity at the root of the trochanter major : **pyriform**, a., *pĭr'·ĭ·fŏrm*, having the shape of a pear : **pyridium**, n., *pĭr·ĭd'·ĭ·ŭm*, a synonym of 'pome.'

pyroligneous, a., *pĭr'·ō·lĭg'·nĕ·ŭs* (Gr. *pur*, fire, *puros*, of fire ; L.

lignum, wood), applied to wood vinegar, and to crude acetic acid: **pyrolignite**, n., *pĭr′·ō·lĭg′·nĭt*, a salt of pyroligneous acid.

pyrosis, n., *pĭr·ōz′·ĭs* (Gr. *purosis*, a burning — from *pur*, fire), a disease of the stomach, characterised by pain, with a copious eructation of a watery, and often acrid, fluid, known as 'water-brash'; *gastralgia*, pain in the stomach, is employed to designate 'heartburn,' and *pyrosis*, the 'acid eructations' which commonly accompany it.

pyroxylin, n., *pĭr·ŏks′·ĭl·ĭn*, also **pyroxyle**, n., *pĭr·ŏks′·ĭl* (Gr. *pur*, fire; *xulon*, wood), gun-cotton; any explosive substance obtained by steeping a vegetable fibre in nitric or nitro-sulphuric acid, and afterwards carefully washing it among pure water, and drying it: **pyroxylic**, a., *pĭr′·ŏks·ĭl′·ĭk*, applied to a product of the destructive distillation of wood, as wood-naphtha.

Pyrrhosa, n., *pĭr·rōz′·ă* (Gr. *purrhos*, red, fiery), a genus of plants, Ord. Myristicaceæ: **Pyrrhosa tingens**, *tĭnj′·ĕnz* (L. *tingens*, dyeing), a species which furnishes a red pigment.

Pyrularia, n., *pĭr′·ŭl·ār′·ĭ·ă* (unascertained), a genus of plants, Ord. Santalaceæ: **Pyrularia oleifera**, *ōl′·ĕ·ĭf′·ĕr·ă* (L. *olĕum*, oil; *fĕro*, I produce), Buffalo tree or oil nut, whose large seeds yield a fixed oil.

Pyrus, n., *pĭr′·ŭs* (L. *pyrum*, a pear; *pyrus*, a pear tree), a genus of plants, Ord. Rosaceæ, Sub-ord. Pomeæ: **Pyrus malus**, *māl′·ŭs* (L. *mālus*, an apple tree), the native species of the Apple, from which the cultivated species have been derived by grafting: **P. communis**, *kŏm·mūn′·ĭs* (L. *cŏmmūnis*, common), the native species of the Pear: **P. Cydonia**, *sĭd·ōn′·ĭ·ă* (from being a native of

Kydon, in the island of Crete), the Quince, also called 'Cydonia vulgaris': **P. sorbus**, *sŏrb′·ŭs* (L. *sorbus*, the sorb or service tree), the Service tree: **P. aria**, *ār·ĭ′·ă* (Gr. *āĕrĭos*, lofty), the White Bean tree: **P. aucuparia**, *āwk′·ŭ·pār′·ĭ·ă* (L. *aucŭpārĭus*, having the power to catch birds—from *aucŭpor*, I go a bird - catching—from *ăvis*, a bird; *căpto*, I take), the Mountain Ash or Rowan, from whose fruit a jelly is made.

pyxidium, n., *pĭks·ĭd′·ĭ·ŭm* (L. *pyxis*, Gr. *puxis*, a box), in *bot.*, a fruit dividing into an upper and lower half, the former acting as a kind of lid.

quadratus, n., *kwŏd·rāt′·ŭs* (L. *quadratus*, squared), the name of several muscles, so called from their square or oblong shape: **quadratus femoris**, *fĕm′·ŏr·ĭs* (L. *femur*, the thigh; *femōris*, of a thigh), a muscle at the upper part of the thigh, which moves the thigh backwards: **q. lumborum**, *lŭm·bōr′·ŭm* (L. *lumbōrum*, of the loins or haunch — from *lumbus*, the loins or haunch), a muscle connected with the haunch bone, and inserted into the last rib, which inclines the loins on one side: **q. menti**, *mĕnt′·ī* (L. *mentum*, the chin, *menti*, of the chin), a muscle which depresses the lower lip.

quadriceps, n., *kwŏd′·rĭ·sĕps* (L. *quadriceps*, having four heads or tops — from *quatuor*, four; and *caput*, the head), a collective designation for four muscles of the thigh, so named from their similarity of action,—they are, the *rectus femoris*, the *vastus externus*, the *vastus internus*, and the *cruræus*.

quadrifarious, a., *kwŏd′·rĭ·fār′·ĭ·ŭs* (L. *quadrifarius*, fourfold—from *quatuor*, four), in *bot.*, in four rows ; proceeding from all the sides of the branch.

quadrifid, a., *kwŏd'·rĭ·fĭd* (L. *quadrĭfĭdus*, four - cleft — from *quatuor*, four ; *findo*, I cleave), in *bot.*, four-cleft ; cut down into four parts to about the middle.

quadrifurcate, a., *kwŏd'·rĭ·fĕrk'·āt* (L. *quadrans*, a fourth part ; *furca*, a two-pronged fork), in *bot.*, doubly forked ; divided into two pairs : **quadrijugate**, a., *kwŏd·rĭdj'·ūg·āt* (L. *jugum*, a yoke), in *bot.*, a compound leaf with four pairs of leaflets.

quadrigeminous, a., *kwŏd'·rĭ·jĕm'·ĭn·ŭs*, also **quadrigeminal**, a., *-jĕm'·ĭn·ăl* (L. *quadrans*, a fourth part ; *gemini*, twins), four-fold ; having four similar parts : **quadrifoliate**, a., *kwŏd'·rĭ·fōl'·ĭ·āt* (L. *fŏlĭum*, a leaf), in *bot.*, having four leaflets diverging from the same point : **quadrigeminal bodies**, four neighbouring eminences on the upper surface of the ' pons varolii ' in the brain.

quadrilocular, a., *kwŏd'·rĭ·lŏk'·ūl·ăr* (L. *quadrans*, a fourth part ; *loculus*, a little space), in *bot.*, having four cells or chambers : **quadripartite**, a., *kwŏd·rĭp'·ărt·ĭt* or *kwŏd'·rĭ·pârt'·ĭt* (L. *partītus*, divided), deeply divided into four parts.

Quadrumana, n. plu., *kwŏd·rŏm'·ăn·ă* (L. *quadrans*, a fourth part ; *mănus*, the hand), the Order of Mammals which have four hand-like extremities, as in the monkey tribe : **quadrum'anous**, a., *-ăn·ŭs*, having four hands.

qualitative, a., *kwŏl'·ĭt·āt·ĭv* (L. *quālĭtas*, a quality or property— from *quālis*, of what sort or kind), in *chem. analysis*, intended merely to determine the nature or quality of component parts of any compound.

quantitive, a., *kwŏnt'·ĭt·ĭv* (L. *quantĭtas*, greatness—from *quant-*

us, how great), relating to quantity ; in *chem.*, having regard to the quantity of the ingredients in any given compound.

quaquaversal, a., *kwă'·kwă·vĕrs'·ăl* (L. *quaqua*, on every side ; *versus*, turned), dipping on all sides ; directed every way.

quarantine, n., *kwŏr'·ăn·tēn* (It. *quarantina*, quarantine ; L. *quadraginta*, forty), the time during which a ship arriving from an infected port, home or foreign, must refrain from communicating with the shore, except under medical control, and at a fixed place ; originally extending over forty days, but now much more restricted.

quartan, a., *kwăwrt'·ăn* (It. and L. *quartana*, the quartan ague ; L. *quartus*, fourth), occurring every fourth day, applied to a form of ague.

quartine, a., *kwăwrt'·ĭn* (L. *quartus*, the fourth), in *bot.*, the fourth coat of the ovule, which is often changed into albumen.

Quassia, n., *kwŏsh'·ĭ·ă* (after a negro, *Quassy*, who first discovered its qualities), a genus of plants, Ord. Simarubaceæ : **Quassia amara**, *ăm·ār'·ă* (L. *amārus*, bitter), a tall shrub of Guiana, etc., which originally yielded Quassia wood ; the Quassia of the shops is obtained from Picræna excelsa, a large forest tree of W. Indian islands ; used in medicine in infusion and tincture as a tonic and anthelmintic, frequently mixed in beer for hops, against the law : **quassin**, n., *kwŏs'·ĭn*, the bitter crystalline principle of Quassia.

quaternate, a., *kwŏt·ĕrn'·āt* (L. *quaterni*, four each—from *quatuor*, four), arranged in fours ; in *bot.*, having leaves growing in fours from one point.

queasy, a., *kwēz'·ĭ* (Icel. *quasa*, to pant ; *queisa*, colic), sickish at stomach ; squeamish : **queasi-**

ness, n., *kwēz'·ĭ·nĕs*, nausea; inclination to vomit.

Quercus, n., *kwĕrk'·ŭs* (L. *quercus*, an oak tree), a highly important genus of trees, Ord. Cupuliferæ or Corylaceæ: **Quercus pedunculata**, *pĕd·ŭngk'·ŭl·āt'·ă* (L. *pedunculus*, a little foot—from *pes*, a foot, *pedis*, of a foot), the Common Oak, containing much tannin, and used as an astringent: **Q. sessiliflora**, *sĕs'·sĭl·ĭ·flōr'·ă* (L. *sessĭlis*, fit for sitting upon, low, dwarf; *flos*, a flower, *flōris*, of a flower), a British species, having sessile fruit, and yields best timber: **Q. ægilops**, *ēdj'·ĭl·ŏps* (L. *ægĭlops*, an oak with edible fruit), a species whose acorn cups, called Valonia, are used by dyers: **Q. infectoria**, *ĭn'·fĕk·tōr'·ĭ·ă* (L. *infectōrius*, that serves for dyeing—from *infector*, a dyer), a native of Asia Minor, producing galls which are used as powerful astringents, and in dyeing, tanning, and making ink: **Q. suber**, *sūb'·ĕr* (L. *sūber*, the cork tree), a species whose bark constitutes cork: **Q. tinctoria**, *tĭng·tōr'·ĭ·ă* (L. *tĭnctōrius*, of or belonging to dyeing—from *tingo*, I die), the Quercitron, whose bark yields a yellow dye: **Q. ilex**, *ĭl'·ĕks* (L. *ĭlex*, the holm oak), the Evergreen Oak.

Quilaieæ, n. plu., *kwĭl·ĭ'·ĕ·ē* (Spanish), a Sub-order of the Ord. Rosaceæ: **Quilaia**, n., *kwĭl·ĭ'·ă*, a genus of plants: **Quilaia saponaria**, *săp'·ŏn·ār'·ĭ·ă* (L. *săpo*, soap, *săpōnis*, of soap), a species whose bark, as well as that of other species, is used as a substitute for soap.

quinary, a., *kwĭn'·ăr·ĭ* (L. *quini*, five each—from *quinque*, five), in *bot.*, composed of five parts, or of a multiple of five: **quinate**, a., *kwĭn'·āt*, in *bot.*, applied to five similar parts arranged together, as five leaflets coming off from one point.

quincunx, n., *kwĭng'·kŭngks* (L. *quincunx*, five - twelfths, an arrangement in five—from *quinque*, five; *uncia*, a twelfth part), in *bot.*, the arrangement of the leaves of a bud into five, of which two are exterior, two interior, and the fifth covers the interior with one margin, and has its other margin covered by the exterior: **quincuncial**, a., *kwĭng·kŭn'·shĭ·ăl*, arranged in quincunx.

quinia, n., *kwĭn'·ĭ·ă*, or **quinine**, n., *kwĭn'·ĭn* (Sp. *quina*, Peruvian bark), the most important constituent of Cinchona bark, largely employed in medicine, chiefly in the form of the sulphate, as an antiperiodic and antipyretic: **quinicine**, n., *kwĭn'·ĭs·ĭn*, an alkaloid resembling quinia and quinidine, from either of which it may be prepared: **quinidine**, n., *kwĭn'·ĭd·ĭn*, an alkaloid found in quinia: **quinism**, n., *kwĭn'·ĭzm*, the appearances produced by much overdosing with quinia, or its salts.

NOTE.—**antiperiodic**, n., *ăn'·tĭ·pĕr'·ĭ·ŏd'·ĭk* (Gr. *anti*, against; *periŏdos*, a circuit), a remedy which removes the periodicity of disease: **antipyretic**, n., *ăn'·tĭ·pĭr·ĕt'·ĭk* (Gr. *anti*, against; *purĕtos*, fever), a medicine for inflammation; adj., antiphlogistic.

quinquecostate, a., *kwĭng'·kwĕ·kŏst'·āt* (L. *quinque*, five; *costa*, a rib), in *bot.*, having five ribs on the leaf.

quinquefarious, a., *kwĭng'·kwĕ·fār'·ĭ·ŭs* (new L. *quinquefārĭus*—from L. *quinque*, five), in *bot.*, in five directions; opening into five parts.

quinquefid, a., *kwĭng'·kwĕ·fĭd* (L. *quinque*, five; *fidi*, I have split, *findo*, I split), in *bot.*, five-cleft; cut into five parts as far as the middle: **quinquelocular**, a., *kwĭng'·kwĕ·lŏk'·ŭl·ăr* (L. *loculus*, a little place or cell), in *bot.*, having five cells, as a pericarp:

quinquepartite, a., *kwĭng'·kwĕ·pärt'·ĭt*, or *kwĭng·kwĕp'·ärt·ĭt* (L. *partītus*, divided), in *bot.*, divided deeply into five parts.

quinsy, n., *kwĭn'·zĭ* (corrupted from F. *squinancie*, quinsy ; L. *cynanche*, a bad kind of sore throat), suppurative inflammation of the tonsils and adjacent parts of the fauces or back part of the mouth.

quintine, n., *kwĭnt'·ĭn* (L. *quintus*, the fifth), in *bot.*, the fifth coat of the ovule ; the embryo sac.

quotidian, a., *kwōt·ĭd'·ĭ·ăn* (L. *quŏtidiānus*, every day — from *quŏtus*, how many ; *dĭēs*, a day), occurring every day, or returning daily, though not at the same hour, as an ague: n., a particular form of ague.

rabies, n., *rāb'·ĭ·ēz* (L. *rābĭēz*, madness), canine madness ; an obscure disease, probably resulting from congestion of the central nervous system: **rabid**, a., *răb'·ĭd*, affected with canine madness, or pert. to it: *hydrophobia* is supposed to follow the bite of a mad dog, but it is doubtful if *rabies* is inoculable.

race, n., *rās* (F. *race*, It. *razza*, race, family ; old H. Ger. *reiza*, a line), in *bot.*, a permanent variety ; a particular breed.

raceme, n., *răs·ēm'* (L. *răcēmus*, the stalk of a cluster of grapes), in *bot.*, an inflorescence having a common axis or stem bearing stalked flowers, as in the hyacinth, the currant, etc.: **racemation**, n., *răs'·ĕm·ā'·shŭn*, a cluster, as of grapes ; the cultivation of clusters : **racemose**, a., *răs'·ĕm·ōz*, bearing flowers in racemes or clusters.

rachis, n., *rāk'·ĭs* (Gr. *rhachis*, the spine or back-bone), in *bot.*, the part of a culm which runs up through the ear of corn ; the stalk or axis bearing the flowers in plants ; in *zool.*, the vertebral column: **rachitis**, n., *răk·ĭt'·ĭs*,

the diseased state of the bones called rickets ; inflammation of the spine: **rachitic**, a., *răk·ĭt'·ĭk*, pert. to the muscles of the back ; rickety.

radial, radiant, see **radius**.

Radiata, n. plu., *rād'·ĭ·āt'·ă* (L. *rădĭātus*, furnished with spokes), one of the lowest divisions of the animal kingdom, whose parts are disposed around a central axis, like the star-fish,—the animals formerly so included are now placed under separate sub-kingdoms, as Cœlenterata, Echinodermata, Infusoria, etc.: **radiate**, a., *rād'·ĭ·āt*, in *bot.*, disposed like the spokes of a wheel ; belonging to the Radiata or rayed animals.

radical, a., *răd'·ĭk·ăl* (L. *rādix*, a root, *rădĭcis*, of a root), in *bot.*, pert. to or arising from the root ; applied to leaves close to the ground ; clustered at the base of a flower stalk : **radicle**, n., *răd'·ĭk·l*, the young root of the embryo ; small rooting fibres : **radicular**, a., *răd·ĭk'·ŭl·ăr*, of or pert. to the radicle : **radicular merithral**, *mĕr·ĭth'·răl* (Gr. *meros*, a part ; *thallos*, a young shoot), in *bot.*, the part corresponding to the root.

Radiolaria, n. plu., *rād'·ĭ·ōl·ār'·ĭ·ă* (L. *rădĭŭs*, a staff, a beam or ray), a division of the Protozoa.

radius, n., *rād'·ĭ·ŭs* (L. *rădius*, a spoke, a ray), the small bone of the forearm, which chiefly forms the wrist joint, and carries the thumb, so called from its fancied resemblance to the spoke of a wheel ; in *bot.*, the ray or outer part of the heads of Composite flowers : **radio-carpal**, *rād'·ĭ·ō·kärp'·ăl*, applied to the joint at the wrist uniting the radius with the carpus : **radio-ulnar**, *ŭl'·năr*, applied to the joint at the point where the radius and ulna unite, as at the wrist or the elbow : **radial**, a., *rād'·ĭ·ăl*, of or pert. to the radius : **radial aspect**, an aspect towards the side where the radius is placed: **radialis indicis**, *răd'·ĭ·ăl·ĭs ĭn'·dĭs·ĭs* (L. *rădiālis*, radial; *index*, the forefinger, *ĭn·dĭcĭs*, of the forefinger), the radial artery of the forefinger of the hand, runs along the radial side of the index finger : **radiant**, a., *răd'·ĭ·ănt*, in *bot.*, having flowers which form a ray-like appearance, as in Umbelliferæ, and Viburnum.

Rafflesiaceæ, n. plu., *răf'·flēzh'·ĭ·ă'·sĕ·ē* (after *Sir Stamford Raffles*), an Order of singular flowering Fungi: **Rafflesia**, n., *răf·flēzh'·ĭ·ă*, a genus of gigantic parasites, the perianth being sometimes three feet in diameter, and capable of holding twelve pints of fluid : **Rafflesia patma**, *păt'·mă* (unascertained), a species employed in Java as an astringent and styptic : **R. Arnoldi**, *ăr·nōld'·ĭ* (after *Arnold*, a botanist), a species weighing sometimes more than 14 lbs., parasitic on Cissus Augustifolia.

rale, n., *râl* (F. *râle*, a rattling in the throat), every kind of noise attending the breathing in the bronchia and vesicles of the lungs different from the sound of the breathing in health ; also called ' rhonchus.'

ramal, a., *răm'·ăl* (L. *rāmus*, a branch), in *bot.*, belonging to branches ; growing on a branch or originating on it.

ramenta, n., *răm·ĕnt'·ă* (L. *ramenta*, scrapings, shavings), in *bot.*, the thin, brown, leafy scales with which the stems of some plants, especially ferns, are covered : **ramentaceous**, a., *răm'·ĕnt·ā'·shŭs*, covered with ramenta or scales.

ramification, n., *răm'·ĭ·fĭk·ā'·shŭn* (L. *rāmus*, a branch ; *facio*, I make), in *bot.*, the subdivisions of roots or branches ; the manner in which a tree produces its branches ; in *anat.*, the issuing or spreading of small vessels from a large one.

ramollissement, n., *răm'·ŏl·lĭs'·mĕnt* (F. *ramollir*, to soften ; L. *mollio*, I soften), in *anat.*, a diseased condition of a part of the body in which it becomes softer than natural, usually limited in its application to the nervous system.

ramose, a., *răm·ōz'*, or **ramous**, a., *răm'·ŭs* (L. *rāmus*, a branch), in *bot.*, producing branches ; very much branched: **ramus**, n., *răm'·ŭs*, in *anat.*, each half or branch of the lower jaw or mandible of vertebrates, of a quadrilateral form ; the thin, flattened part of the ischium : **rami**, n. plu., *răm'·ĭ*, the two parts into which the pubes is divisible, namely, a horizontal and a perpendicular ramus : **ramulus**, n., *răm'·ŭl·ŭs*, a small branch : **ramulous**, a., *răm'·ŭl·ŭs*, having many small branches.

ranine, a., *răn'·īn* (L. *rāna*, a frog), a continuation of the lingual artery which runs along the under surface of the tongue: **ranula**, n., *răn'·ŭl·ă* (L. *rānŭlus*, a little tongue), a tumour situated below the tongue, of a bluish colour, and cystic.

Ranunculaceæ, n. plu., *răn·ŭng'·kūl·ā'·sĕ·ē* (L. *rānuncŭlus*, a little frog — from *rāna*, a frog), the Crowfoot family, an Order of plants having narcotico - acrid properties, and usually more or less poisonous : **Ranunculeæ**, n. plu., *răn'·ŭng·kūl'·ĕ·ē*, a tribe or Sub-order of plants : **Ranunculus**, n., *răn·ŭng'·kūl·ŭs*, an extensive genus of plants, so named as found in moist places frequented by frogs: **Ranunculus sceleratus**, *sĕl'·ĕr·āt'·ŭs* (L. *scelerātus*, polluted); **R. Alpestris**, *ălp·ĕst'·rĭs*, (new L. *Alpĕstris*, of or from the *Alps*); **R. bulbosus**, *bŭlb·ōz'·ŭs* (L. *bulbōsus*, bulbous — from *bŭlbus*, a bulb); **R. gramineus**, *grăm·ĭn′·ĕ·ŭs* (L. *grāmĭnĕus*, grassy—from *grāmĕn*, grass); **R. acris**, *ăk′·rĭs* (L. *ācĕr*, sharp, masc.; *ācrĭs*, sharp, fem.); and **R. flammula**, *flăm′·mŭl·ă* (L. *flammŭla*, a little flame—from *flamma*, a flame), are species which are all acrid, the acridity entirely disappearing by drying : **R. repens**, *rĕp′·ĕnz* (L. *rēpens*, creeping); **R. aquatilis**, *ăk′·wăt·ĭl′·ĭs* (L. *ăquătĭlis*, growing in or found in or near water—from *ăqua*, water); **R. lingua**, *lĭng′·gwă* (L. *lingua*, a tongue); **R. ficaria**, *fĭk·ār′·ĭ·ă* (L. *fĭcārĭus*, belonging to a fig—from *ficus*, a fig), are species which are bland.

raphe, n., *răf′·ē* (Gr. *rhaphē*, a seam), in *bot.*, applied to parts which appear as if they had been sewn together ; in seeds, the channel of vessels which connects the chalaza with the hilum ; in umbelliferous plants, the line of junction of the two halves of which their fruit is composed ; in *anat.*, the raised seam-like line which runs along the perinæum to the anus.

raphides, n. plu., *răf′·ĭd·ēz* (Gr. *rhaphis*, a needle, *rhaphĭdos*, of a needle), in *bot.*, minute crystals, like needles, found in the tissues of plants : **raphidian**, a., *răf·ĭd′·ĭ·ăn*, pert. to the raphides.

Raptores, n. plu., *răp·tōr′·ēz* (L. *raptores*, robbers—from *rapto*, I plunder), the Order of the birds of prey.

rash, n., *răsh* (It. *raschĭa*, itching), an eruption in the skin.

Rasores, n. plu., *răz·ōr′·ēz* (L. *răsōres*, scrapers), the Order of the scratching or scraping birds, as common fowls.

ratitæ, n. plu., *răt·ĭt′·ē* (L. *rătĭt·us*, marked like a raft — from *rătis*, a raft), cursorial birds which do not fly, and have therefore a raft-like sternum without a median keel.

rattles, n. plu., *răt′·lz* (Dut. *rat-*

elen, to make rattling sounds), the noise in the throat caused by the air passing through the mucus filling the air passages, which often precedes death.

re-agent, n., *rē·ādj′·ĕnt* (*re* and *agent*), in *chem.*, a substance employed to detect the presence of other bodies.

receptacle, n., *rĕ·sĕpt′·ă·kl* (L. *receptaculum*, a magazine or storehouse—from *re*, back; *capĭo*, I take), in *bot.*, that part of the fructification which bears or receives other parts, as the expanded top of the peduncle of a dandelion, the inner surface of a fig, etc.; a chamber in which secretions are stored.

receptaculum chyli, *rē′·sep·tăk′·ŭl·ŭm kīl′·ĭ* (L. *rĕceptăcŭlum*, a magazine or storehouse ; *chylum*, a Latinised form of Gr. *chulos*, juice or humour, L. *chyli*, of juice), a small chamber or cavity lying in the abdomen behind the aorta, and in front of the second lumbar vertebra, which receives the chyli from the lacteals of the intestine, and various lymphatics : **receptacula seminis**, *sĕm′·ĭn·ĭs* (L. *sēmĕn*, seed, *sĕmĭnis*, of seed), the receptacles of the semen ; organs in earthworms which receive the male reproductive fluid: **receptaculi arteriæ**, *ăr·tēr′·ĭ·ē* (L. *arteriæ*, arteries), the arteries of the receptaculum, see ' arteria.'

receptive spot, *rē·sĕpt′·ĭv*, in *bot.*, the point in the oosphere of ferns where the antherozoids effect an entrance.

reclinate, a., *rĕ·klīn′·āt* (L. *reclinātus*, bent back, reclined), in *bot.*, curved down from the horizontal; having the leaves folded longitudinally from apex to base in the bud: **reclination**, n., *rĕk′·lin·ā′·shŭn*, in *surg.*, an operation for the cure of cataract.

recrudescence, n., *rē′·krô·dĕs′·sĕns* (L. *recrudescens*, breaking out again—from *re*, again ; *crudesco*,

I become hard), in *bot.*, the reproduction of a young shoot from the tip of a ripened spike of a seed.

rectembryeæ, n. plu., *rĕk'tĕm·brĭ'ĕ·ē* (L. *rectus,* right ; *embruon,* the fœtus), in *bot.*, the embryo straight in the axis of the seed.

rectification, n., *rĕk'tĭ·fĭk·ā'shŭn* (L. *rectus,* right ; *factus,* made), in *chem.*, the repeated distillation of a spirit in order to make it finer and purer : **rectify,** v., *rĕk'tĭ·fĭ,* to refine or purify a substance by repeated distillations.

rectinervis, a., *rĕk'tĭn·ĕrv'ĭs* (L. *rectus,* straight ; *nervus,* a nerve), in *bot.*, straight and parallel veined ; also **rectinervate,** *rĕk'tĭn·ĕrv'āt.*

rectiserial, a., *rĕk'tĭ·sēr'ĭ·ăl* (L. *rectus,* straight ; *series,* a row), in *bot.*, disposed in a rectilinear or straight series — applied to leaves.

rectivenius, a., *rĕk'tĭ·vēn'ĭ·ŭs* (L. *rectus,* straight ; *vena,* a vein), straight and parallel veined ; same as 'rectinervis'; also **rectivenous,** a., *rĕk'tĭ·vēn'ŭs.*

recto, *rĕk'tō* (L. *rectus,* straight), of or connected with the rectum : **recto-uterine,** a., *-ūt'ĕr·ĭn,* the posterior ligaments of the uterus: **recto-vesical fascia,** *vĕs'ĭk·ăl făs'sĭ·ă,* a fascia which lies between, and connects the rectum and urinary bladder.

rectum, *rĕk'tŭm* (L. *rectus,* straight), the third and terminal portion of the large intestine ending at the anus, so named because formerly supposed to be straight, which it is not : **rectus,** n., *rĕk'tŭs,* in *bot.*, applied to the stem and other straight parts of plants ; in *anat.*, a name for several muscles of the body, which are so called from the rectilinear direction of their fibres : **rectus femoris,** *rĕk'tŭs fĕm'ŏr·ĭs* (L. *femur,* the thigh, *femŏris,* of the thigh), a muscle arising by two heads

from the ilium and acetabulum, and inserted into the patella, which extends the leg, etc. : **rectus abdominis,** *rĕk'tŭs ăb·dŏm'ĭn·ĭs* (L. *abdōmen,* the abdomen, *abdomĭnis,* of the abdomen), a muscle which begins at the pubes, and is inserted into the three lower true ribs, and the ensiform cartilage ; it pulls down the ribs in respiration, etc. : **rectus cruris,** *rĕk'tŭs krŏr'ĭs* (L. *crus,* a leg, *cruris,* of the leg), the straight muscle of the leg ; a muscle which extends the leg in a powerful manner by the intervention of the patella like a pulley.

recurrent, a., *rĕ·kŭr'rĕnt* (L. *recurro,* I return, I recur—from *re,* back ; *curro,* I run), returning from time to time ; seeming to return or reascend towards the origin ; applied to tumours which return after removal.

recurved, a., *rĕ·kĕrvd'* (L. *re,* back ; *curvus,* crooked), in *bot.*, bent backwards.

reduction, n., *rĕ·dŭk'shŭn* (L. *re,* back ; *ductus,* led), in *surg.*, the operation of restoring displaced parts to their natural position.

reduplicate, a., *rĕ·dūp'lĭk·āt* (L. *redŭplĭcātus,* redoubled — from *re,* again ; *dŭplĭco,* I double), in *bot.*, applied to a form of æstivation in the edges of the sepals or petals, which are turned outwards ; also **reduplicative,** a., *-āt·ĭv.*

reflexed, a., *rĕ·flĕkst'* (L. *re,* back ; *flexus,* bent), in *bot.*, curved backwards.

refrigerant, n., *rĕ·frĭdj'ĕr·ănt* (L. *refrigĕro,* I make cool or cold— from *re,* back ; *frigus,* cold, coolness), a medicine which cools or abates heat : **refrigeration,** n., *rĕ·frĭdj'ĕr·ā'shŭn,* the lowering the temperature of a body : **refrigerator,** n., *rĕ·frĭdj'ĕr·āt·ŏr,* a vessel for cooling liquids.

regeneration, n., *rĕ·jĕn'·ĕr·ā'·shŭn* (L. *re*, again ; *genero*, I beget), the renewal of a portion of lost or removed tissue.

regimen, n., *rĕdj'·ĭ·mĕn* (L. *regimen*, direction — from *rego*, I rule), in *med.*, the strict regulation of diet and habits, with the view of preserving or restoring health.

region, n., *rēdj'·ŭn* (L. *rĕgĭo*, a boundary line), a definite space on the surface of the body, or a division of the organs, as abdominal region, gluteal region, etc.

regma, n., *rĕg'·mă* (Gr. *rhegma*, a rupture), in *bot.*, a seed vessel, the two valves of which open by an elastic movement, as in Euphorbia.

regurgitation, n., *rĕ·gĕrj'·ĭt·ā'·shĕn* (L. *re*, again ; *gurges*, a raging abyss, a stream), the act of flowing or pouring back by the same orifice or place of entrance ; the natural and easy vomiting of food by infants.

rejuvenescence, n., *rĕ·jŏv'·ĕn·ĕs'·sĕns* (L. *re*, again ; *juvenesco*, I become young), a renewal of youth : **rejuvenescence of a cell,** in *bot.*, the formation of one new cell from the whole of the protoplasm of a cell already in existence.

relapse, n., *rĕ·lăps'* (L. *relapsus*, sunk or fallen back — from *re*, back ; *lapsus*, a slipping), a return of a disease after convalescence : **relapsing fever,** a contagious disease, characterised by one or more relapses after apparent convalescence, chiefly met with as an epidemic in periods of scarcity and famine ; also called 'famine fever.'

relaxation, n., *rē'·lăks·ā'·shŭn* (L. *re*, back ; *laxo*, I loose or slacken), in *med.*, a lessening of the normal and healthy tone of the body.

reliquiæ, n. plu., *rĕ·lĭk'·wĭ·ē* (L. *reliquiæ*, the remains), the remains of the dead ; in *bot.*, the remains of withered leaves attached to the plant ; in *pathology*, the permanent evidence of past morbid processes.

remittent, a., *rĕ·mĭt'·ĕnt* (L. *re*, back ; *mitto*, I send), in *med.*, applied to diseases whose symptoms alternately diminish and return : **remittent fever,** a malarious fever, having irregular repeated exacerbations, known also by various other names, as 'jungle fever,' 'bilious fever,' etc.

renal, a., *rēn'·ăl* (It. *renale*, renal ; L. *renes*, the kidneys), relating to or connected with the reins or kidneys : **reniform,** a., *rēn'·ĭ·fŏrm* (L. *forma*, shape), in shape like a kidney.

rennet, n., *rĕn'·nĕt* (Ger. *rennen*, to run ; Dut. *runnen*, to curdle), an infusion of the inner membrane of a calf's stomach, used for coagulating milk.

repand, a., *rĕp'·ănd* (L. *repandus*, bent backwards, bent up—from *re*, back ; *pandus*, bent), in *bot.*, applied to a leaf when its margin is undulated, and unequally dilated.

repent, a., *rĕp'·ĕnt* (L. *rĕpens*, creeping, *repentis*, of creeping), in *bot.*, lying flat upon the ground, and remitting roots along the under surface.

replicate, a., *rĕp'·lĭk·āt* (L. *rĕplĭcātus*, folded or rolled back — from *re*, back ; *plĭco*, I fold), in *bot.*, doubled down so that the upper part comes in contact with the lower.

replum, n., *rĕp'·lŭm* (L. *replum*, a door-cheek, the leaf of a door), in *bot.*, a longitudinal division in a pod formed by the placenta, as in Cruciferæ ; the persistent portion of some pericarps after the valves have fallen away.

Reptilia, n. plu., *rĕp·tĭl'·ĭ·ă* (L. *reptilis*, a reptile—from *repto*, I crawl), the class of the Vertebrata comprising the tortoises, snakes, lizards, crocodiles, etc.

Resedaceæ, n. plu., *rĕs·ĕd·ā'·sĕ·ē* (L. *rĕsēda*, a plant ; *rĕsēdo*, I calm, I heal), the Mignonette family, an Order of plants : **Reseda**, n., *rĕs·ēd'·ă*, a genus of plants, so named because considered by the anc. Latins as good for bruises : **Reseda luteola**, *lŏt'·ĕ·ōl'·ă* (L. *lūtĕŏlus*, yellowish —from *lūtum*, a plant which dyes yellow), the plant Weld, which yields a yellow dye : **R. odorata**, *ōd'·ōr·āt'·ă* (L. *odorātus*, having a smell or perfume—from *ōdŏr*, scent, smell), the fragrant mignonette of our gardens.

resolution, n., *rĕs'·ŏl·ū'·shŭn* (L. *re*, back ; *sŏlūtus*, loosed), in *med.*, the dispersion or disappearance of a tumour, or inflammatory process.

resonance, n., *rĕs'·ŏn·ăns* (L. *rĕs·ŏnans*, resounding or re-echoing —from *re*, back ; *sŏno*, I sound), the property of certain parts of the body to transmit sound, usually of the voice, which may either be normal, exaggerated, or impaired.

respiration, n., *rĕs'·pĭr·ā'·shŭn* (L. *rĕspīro*, I respire—from *re*, back; *spīro*, I breathe), the process by which the air enters and emerges from the lungs, thus effecting the aëration of venous blood : **respirator**, n., *rĕs'·pĭr·āt·ŏr*, an instrument worn over the mouth by those of weak lungs, or having a liability to colds, in order to warm and dry the inspired air : **respiratory**, a., *rĕs·pĭr'·ăt·ŏr·ĭ*, pert. to or serving for respiration : **respiratory murmur**, the continuous sounds heard in auscultation, produced by the air entering into, and being expelled from, the lungs in a healthy state.

Restiaceæ, n. plu., *rĕst'·ĭ·ā'·sĕ·ē* (L. *restis*, a rope, a cord), the Restio or Cord-rush family, an Order of sedge - like plants, so

named because used for cordage at the Cape : **Restio**, n., *rĕst'·ĭ·ŏ*, a genus of plants having tough, wiry stems, used for making baskets and brooms.

restiform, a., *rĕst'·ĭ·fŏrm* (L. *restis*, a rope ; *forma*, shape), having the form or appearance of a rope.

resupinate, a., *rĕ·sūp'·ĭn·āt* (L. *rĕsūpīnātus*, bent or turned back —from *re*, back ; *sūpīno*, I bend backwards), in *bot.*, so turned or twisted that the parts naturally the undermost become the uppermost, and *vice versâ ;* turned upside down.

rete, n., *rēt'·ē* (L. *rēte*, a net, a snare), a net ; network : **rete mirabile**, *mĭr·ăb'·ĭl·ĕ* (L. *mirābĭle*, wonderful), in *anat.*, an arrangement of blood-vessels at the base of the brain of quadrupeds : **r. mucosum**, *mŭk·ōz'·ŭm* (L. *mūcōsus*, slimy, mucous), the soft under-layer of the epidermis or scarf - skin, which gives the colour to the skin : **r. vasculosum testis**, *vǎsk'·ūl·ōz'·ŭm tĕst'·ĭs* (L. *vasculōsus*, pert. to a small vessel—from *vascŭlum*, a small vessel), the vascular net of the testicle ; the close network of tubes lying in the substance of corpus Highmorianum, along the back part of the testicle.

retention, n., *rĕ·tĕn'·shŭn* (L. *retĕntĭo*, a holding back—from *re*, back ; *tenĕo*, I hold), the undue holding back of any natural excretion, as the urine or sweat.

reticulate, a., *rĕ·tĭk'·ūl·āt*, also **reticulated**, a., *-āt·ĕd* (L. *rĕtĭcŭlātus*, net-like—from *rĕtĭc'·ŭlum*, a little net), in *bot.*, having distinct veins or lines crossing like network : **reticular**, a., *rĕ·tĭk'·ūl·ăr*, having interstices like network : **reticularia**, n. plu., *rĕ·tĭk'·ūl·ār'·ĭ·ă*, those Protozoa, such as the Foraminifera, in which the pseudopodia run into one another and form a network : **reticulum**, n., *rĕ·tĭk'·ŭl·ŭm*, the second stomach of the ruminant animal, often called the honey-comb bag, from the numerous polygonal cells which cover its surface ; in *bot.*, the debris of cross-fibres about the base of the petioles in palms.

retiform, a., *rēt'·ĭ·fŏrm* (L. *rete*, a net ; *forma*, shape), having the structure of a net.

retina, n., *rĕt'·ĭn·ă* (L. *rete*, a net), one of the coats of the eye, resembling fine network, which receives the impressions resulting in the sense of vision : **retinitis**, n., *rĕt'·ĭn·īt'·ĭs*, inflammation of the retina.

retinaculum, n., *rĕt'·ĭn·ăk'·ŭl·ŭm* (L. *retinaculum*, a holdfast, a band), in *bot.*, the viscid matter by which the pollen-masses in Orchids, etc., adhere to a prolongation of the anther : **retinacula**, n. plu., *rĕt'·ĭn·ăk'·ŭl·ă*, in *anat.*, bands which hold the tendons close to the bones of the wrist, ankle, etc. ; the fold of membrane continued from each commissure of the ilio-cæcal and ilio-colic valves round on the inner side of the cæcum.

retinervis, a., *rĕt'·ĭn·ĕrv'·ĭs* (L. *rete*, a net ; *nervus*, a nerve), in *bot.*, having reticulated veins or nerves ; also called **retivenius**, a., *rĕt'·ĭ·vēn'·ĭ·ŭs* (L. *vena*, a vein).

retractor, a., *rĕ·trăkt'·ŏr* (L. *retractus*, withdrawn — from *re*, back ; *tractus*, drawn), a name for those muscles which, by their contraction, withdraw the parts to which they are attached.

retrahens aurem, *rē'·tră·hĕnz äwr'·ĕm* (L. *rētrăhens*, drawing back ; *auris*, nom., the ear, *aurem*, obj., the ear), the smallest of the three muscles placed immediately beneath the skin around the external ear ; see 'attollens aurem.'

retro-peritoneal, a., *rĕt'·rō·pĕr'·ĭ·tŏn·ē'·ăl* (L. *retro*, back ; Eng. *peritoneal*), a name for the layer of tissue which forms the parietal portion of the serous membrane of the abdominal cavity, connected loosely with the fascia lining the abdomen and pelvis ; also called 'sub-peritoneal.'

retro-pharyngeal, a., *rĕt'·rō·făr'·ĭn·jē'·ăl* (L. *retro*, behind, backward ; Eng. *pharyngeal*), pert. to the parts behind the pharynx ; denoting an abscess formed at the back part of the pharynx.

retrorse, a., *rĕ·trŏrs'* (L. *retrorsum*, backwards — from *retro*, backwards ; *versus*, turned), turned or directed backwards.

retuse, a., *rĕ·tūs'* (L. *rĕtūsus*, blunted), in *bot.*, having the extremity broad, blunt, and slightly depressed ; appearing as if bitten off at the end.

revolute, a., *rĕv'·ŏl·ōt*, also **revolutive**, a., *rĕv'·ŏl·ōt·ĭv* (L. *revolutus*, revolved—from *re*, back ; *volvo*, I roll), in *bot.*, rolled backwards from the margins upon the under surface, usually applied to the edges of leaves ; having the edges rolled back spirally in vernation.

Rhabdophora, n. plu., *răb·dŏf'·ŏr·ă* (Gr. *rhabdos*, a rod ; *phorĕō*, I bear), a name for the Graptolites, because they commonly possess a chitinous rod or axis supporting the perisarc.

rhachitis, *răk·īt'·ĭs*, see 'rachitis,' but the former is the proper spelling.

Rhamnaceæ, n. plu., *răm·nā'·sĕ·ē* (Gr. *rhamnos*, the white-thorn), the Buckthorn family, an Order of plants, many of which have active cathartic properties : **Rhamnus**, n., *răm'·nŭs*, a genus of plants : **Rhamnus catharticus**, *kăth·ărt'·ĭk·ŭs* (Gr. *kathartikos*, purifying or cleansing), common or purging Buckthorn, whose black succulent berries are used as a hydragogue cathartic in dropsy, and whose greenish juice, mixed with lime, forms the colour called sap - green : **R. frangula,** *frăng'·gŭl·ă* (L. *frango*, I break, so named from the brittleness of its branches), Black Alder, is emetic and purgative, the wood supplying charcoal for gunpowder, and crayons for artists : **R. infectorius,** *ĭn'·fĕk·tŏr'·ĭ·ŭs* (L. *infectōrius*, that serves for dyeing—from *infector*, a dyer), a species whose berries are called French berries, and have been used for dyeing : **R. dahuricus,** *dă·ŭr'·ĭk·ŭs* (native name), produces a red - wood, called by the Russians 'Sandal-wood': **R. chlorophorus,** *klōr·ŏf'·ŏr·ŭs* (Gr. *chloros*, green ; *phoreō*, I bear) ; and **R. utilis,** *ŭt'·ĭl·ĭs* (L. *utĭlis*, useful), are species from which the Chinese prepare their beautiful green dye, Lo-kao, called in this country Chinese Green-Indigo.

rheum, n., *rŏm* (Gr. *rheuma*, that which flows—from *rhĕō*, I flow), the increased secretions of the mucous glands caused by a cold.

Rheum, n., *rē'·ŭm* (*Rha*, old name of river Volga, from whose banks originally brought), the Rhubarb plant, a genus of well-known plants, Ord. Polygonaceæ: **Rheum officinale,** *ŏf·fĭs'·ĭn·āl'·ĕ* (L. *offĭc-ĭnālis*, officinal, by authority—*officĭna*, a workshop), the officinal rhubarb plant, originally from Thibet : **R. palmatum,** *păl·māt'·ŭm* (L. *palmātus*, marked with the palm of the hand), at one time considered the true rhubarb plant · **R. undulatum,** *ŭn'·dŭl·āt'·ŭm* (L. *ĭndŭlātus*, undulated —from *unda*, a wave), a species which yields much of the French rhubarb : **R. compactum,** *kŏm·păkt'·ŭm* (L. *compactus*, pressed —from *con*, together ; *pactus*, driven in), a species also yielding

French rhubarb, but cultivated in Britain for its acid petioles : **R. rhaponticum,** *ră·pŏnt'·ĭk·ŭm* (*Rha*, old name of river Volga ; L. *ponticus*, of or relating to the *Pontus* or Black Sea), a species used in France and Britain as R. compactum : **R. hybridum,** *hĭ'·brĭd·ŭm* (L. *hybrĭdus*, of or pert. to a hybrid—from *hybrida*, a mongrel, a hybrid), common rhubarb, cultivated in Germany for its root, and in Britain for its stalks : **R. leucorhizum,** *lōk'·ŏr·ĭz'·ŭm* (Gr. *leukos*, white ; *rhiza*, a root), a Siberian and Altai species, said to yield imperial or white rhubarb ; rhubarb contains raphides of oxalate of lime, along with tannin, gallic acid, resin, and a peculiar yellow coloured principle called 'rhubarberin'; rhubarb is employed as a cathartic, astringent, and tonic.

rheumatism, n., *rŏm'·ăt·ĭzm* (L. *rheumatismus*, rheum, catarrh—from *rheuma*, a watery fluid), a painful disease affecting the muscles and joints, causing swelling and stiffness : **rheumatic,** a., *rŏm·ăt'·ĭk*, pert. to or affected with rheumatism : **rheumatoid,** a., *rŏm'·ăt·ōÿd* (Gr. *eidos*, resemblance), having the appearance of rheumatism : **rheumatoid arthritis,** a chronic disease of the bones and joints, having a great resemblance to rheumatism, but is really a different disease. NOTE. —This is that form of rheumatism which cripples and deforms. It may attack all the joints, but has no tendency to shorten life.

rhinal, a., *rīn'·ăl* (Gr. *rhīs*, the nose), of or pert. to the nose : **rhinalgia,** n., *rīn·ălj'·ĭ·ă* (Gr. *algos*, pain, grief), pain in the nose : **rhinoplastic,** a., *rīn'·ō·plăst'·ĭk* (Gr. *plastĭkos*, suitable for being fashioned or formed), an operation by which a piece of skin can be taken from a healthy part of the body and placed on a part injured or destroyed—a nose partly destroyed has thus been made presentable : **rhinoscope**, n., *rĭn′·ō·skōp* (Gr. *skopĕō*, I view), an instrument, consisting of an adapted oval or circular mirror, by which the back part of the nostrils may be examined : **rhinoscopy**, n., *-ŏs′·kōp·ĭ*, the examination of the back parts of the soft palate, the nose, etc., by means of the rhinoscope.

Rhizanths, n. plu., *rĭz′·ănths* (Gr. *rhiza*, a root ; *anthos*, a flower), in *bot.*, same as 'rhizogen,' which see ; a class of plants occupying a position between the flowering and non-flowering species.

rhizinæ, n. plu., *rĭz′·ĭn·ē* (Gr. *rhiza*, a root), in *bot.*, minute fibrils on the under surface of the thallus of some lichens, by which they adhere : **rhizinose**, a., *rĭz′·ĭn·ōz*, having root-like filaments or rhizinæ.

rhizocarp, n., *rĭz′·ō·kârp* (Gr. *rhiza*, a root ; *karpos*, fruit), in *bot.*, applied to Marsilea, as producing spore-cases on root-like processes : **Rhizocarpeæ**, *rĭz′·ō·kârp′·ĕ·ē*, the Pepperwort family, another name for the ' Marsileaceæ,' which see : **rhizocarpous**, a., *rĭz′·ō·kârp′·ŭs*, having perennial roots and annual stems.

Rhizogens, n. plu., *rĭz′·ō·jĕnz* (Gr. *rhiza*, a root ; *genŏ*, I produce), a class of plants growing on the roots of other plants ; such plants as Rafflesia, which consist of a flower and root only : **rhizogen**, a., producing a root and a flower only : **rhizoids**, n. plu., *rĭz′·oÿdz* (Gr. *eidos*, resemblance), the root-like outgrowths of many Algæ.

rhizomes, n. plu., *rĭz′·ōmz*, also **rhizomata**, n. plu., *rĭz·ŏm′·ăt·ă* (Gr. *rhizōmă*, a root, a race), in *bot.*, thick stems running along and partly underground, and sending forth shoots above and roots below ; filamentous bodies

attaching foliaceous lichens to their supporting substance.

Rhizomorpha, n., *rĭz′·ō·mŏrf′·ă* (Gr. *rhiza*, a root ; *morphē*, form), a genus of Fungi, so named from the appearance of the plants, whose species have the property of giving out a sort of phosphorescent light, found in cellars and coal mines : **rhizomorphoid**, a., *rĭz′·ō·mŏrf′·oÿd* (Gr. *eidos*, resemblance), root-like in form.

Rhizophaga, n. plu., *rĭz·ŏf′·ăg·ă* (Gr. *rhiza*, a root ; *phago*, I eat), a group of the Marsupials : **rhizophagous**, a., *rĭz·ŏf′·ăg·ŭs*, living or feeding on roots.

Rhizophoraceæ, n. plu., *rĭz′·ō·fŏr·ā′·sĕ·ē* (Gr. *rhiza*, a root ; *phorĕō*, I bear), the Mangrove family, an Order of plants whose bark is often astringent, and is sometimes used in dyeing : **Rhizophora**, n., *rĭz·ŏf′·ŏr·ă*, a genus of remarkable trees of tropical countries, whose branches throw out roots freely, the roots descending into the mud : **Rhizophora mangle**, *măng′·gl* (Malay, *mangle*, the mangrove), the Mangrove, forms thickets at the muddy mouths of rivers, the tree having the appearance of being supported on many stalks.

Rhizopoda, n. plu., *rĭz·ŏp′·ŏd·ă* (Gr. *rhiza*, a root ; *pous*, a foot ; *podos*, of a foot), the division of Protozoa comprising all those capable of emitting pseudopodia.

rhizotaxis, n., *rĭz′·ō·tăks′·ĭs* (Gr. *rhiza*, a root ; *taxis*, a putting in order), in *bot.*, the regularity in the arrangement of roots ; also **rhizotaxy**, n., *rĭz′·ō·tăks′·ĭ*.

Rhododendron, n., *rōd′·ō·dĕnd′·rŏn* (Gr. *rhodon*, a rose ; *dendron*, a tree), a genus of handsome, elegant, and showy shrubs, Ord. Ericaceæ : **Rhododendron chrysanthum**, *krĭs·ănth′·ŭm* (Gr. *chrusos*, gold ; *anthos*, a flower), a Siberian species, whose poisonous, narcotic qualities are well-marked.

Rhodosporeæ, n. plu., *rōd'·ō·spōr'·ĕ·ē* (Gr. *rhodon*, a rose; *spora*, seed), a Sub-order of Algæ, constituting rose or purple-coloured sea-weeds, with fronds formed of a single row of articulated cells, or of several rows of cells combined into a flat expansion.

Rhodymenia, n., *rōd'·ĭ·mēn'·ĭ·ä* (Gr. *rhodon*, a rose; *humēn*, a thin membrane), a genus of sea plants, Ord. Algæ: **Rhodymenia palmata,** *păl·māt'·ä* (L. *palmātus*, marked with the palm of the hand), the sea-weed called Dulse.

rhomboid, n., *rŏm'·bŏyd* (L. *rhombus*, Gr. *rhombos*, a magical wheel; *eidos*, resemblance), a four-sided figure having its opposite sides equal, but its angles not right angles: **rhomboideus,** n., *rŏm·bŏyd'·ĕ·ŭs*, the name of two muscles, 'rhomboideus minor' and 'r. major,' which are placed parallel to one another, and are separated only by a slight interval; the rhomboidei extend obliquely from the spinous processes of the lowest cervical, and some of the upper dorsal vertebræ, to the base of the scapula.

rhonchus, n., *rŏngk'·ŭs* (L. *rhonchus*, Gr. *rhongchos*, a snoring), a wheezing, snoring, sibilant, chirping, or whistling sound, usually low-toned, produced in the air passages by the narrowing of their calibre, heard on auscultation.

Rhus, n., *rŭs* (Gr. *rhousios*, reddish-brown; *rhous*, a tree whose bark and fruit are used in tanning), a genus of plants, Ord. Anacardiaceæ, the fruit and leaves of some species becoming a reddish-brown in autumn: **Rhus toxicodendron,** *tŏks'·ĭk·ō·dĕnd'·rŏn* (L. *toxicum*, Gr. *toxikon*, poison; Gr. *dendron*, a tree), Poison-oak, found in N. America, leaves used in medicine as a stimulant, and, like other species,

yields an acrid milky juice: **R. radicans,** *răd·ĭk'·ănz* (L. *rādicans*, striking or taking root), Poison-ivy or Poison-vine: **R. venenata,** *věn'·ĕn·āt'·ä* (L. *věnēnātus*, furnished with poison— from *venēnum*, poison), Poison-sumach, or Poison-elder, possesses acrid, poisonous properties: **R. coriaria,** *kŏr'·ĭ·ār'·ĭ·ä* (L. *corĭārius*, belonging to leather — from *corĭum*, leather); **R. typhina,** *tĭf·īn'·ä* (Gr. *tuphē* or *tiphē*, spelt or German wheat); **R. glabra,** *glăb'·rä* (L. *glāber*, smooth, without hair), are species which are extensively used in tanning, and their fruit is acid: **R. cotinus,** *kŏt'·ĭn·ŭs* (L. *cŏtĭnus*, a shrub yielding a purple dye), called wig-tree in France, from the hairy appearance of its abortive pedicles, yields the yellow dye-wood Toung Fustic: **R. succedanea,** *sŭk'·sĕd·ān'·ĕ·ä* (L. *succēdānĕus*, that supplies the place of something), the species whose fruit produce Japan-wax, imported from that country: **R. vernicifera,** *vĕrn'·ĭs·ĭf'·ĕr·ä* (L. *verno*, I spring, I bloom; *fero*, I bear), a small Japanese tree, yielding the famous lacquer so extensively employed by the inhabitants of that country.

rhythm, n., *rĭthm* (Gr. *rhuthmos*, measured motion), the harmony and due relation which exists between the different movements of an organ in health: **rhythmic,** a., *rĭth'·mĭk*, also **rhythmical,** a., *rĭth'·mĭk·äl*, denoting the regular healthy discharge of the functions of an organ, as the pulsations of the heart.

rib, n., *rĭb* (Dut. *ribbe*, a rib, a beam), in *anat.*, one of the curved bony hoops or bars of the thorax which protect the lungs, the heart, etc.; in *bot.*, the central longitudinal nerve or vein of a leaf: **true ribs,** the seven ribs which are attached to the sternum or breast bone, as distinguished from the five **false ribs**, which are not so attached ; the last two false ribs are called **floating ribs**, because they are not attached to anything in front.

Ribesiaceæ, n. plu., *rīb·ēz'·ĭ·ā'·sĕ·ē* (*ribes*, an Arabic name for an acid-leaved species of Rheum), the Gooseberry and Currant family, now more usually called 'Grossulariaceæ,' which see : **Ribes**, n., *rīb'·ēz*, a genus of plants : **Ribes grossularia**, *grŏs'·ŭl·ār'·ĭ·ă* (mid. L. *grossŭla*, a gooseberry ; L. *grossŭlus*, a small, unripe fig), the various kinds of Gooseberry : **R. rubrum**, *rŏb'·rŭm* (L. *rubrus*, red, ruddy) ; and **R. nigrum**, *nīg'·rŭm* (L. *nĭger*, black), the Red and Black Currants, the latter possessing tonic and stimulant properties.

Ricinus, n., *rĭs'·ĭn·ŭs* (L. *rīcīnus*, the tick of the sheep, a plant), a genus of plants, Ord. Euphorbiaceæ—so named from the shape of its seeds : **Ricinus communis**, *kŏm·mūn'·ĭs* (L. *commūnis*, common), a species from whose seeds castor oil is expressed, also called 'Palma Christi'—see under 'Palmæ.'

rickets, n. plu., *rĭk'·ĕts* (Gr. *rhachitis*, disease of the spine—from *rhachis*, the spine), a constitutional disease, characterised chiefly by a curvature of the shafts of the long bones of the arms and legs, and enlargement of their articular extremities—the result of deficient appropriation of earthy principles by their structures.

rictus, n., *rĭkt'·ŭs* (L. *rictus*, the mouth wide open), in *bot.*, among labiate or lipped corollas, the condition of the lower lip pressed against the upper, so as to leave only a chink between them, as in Frogsmouth.

rigescent, a., *rĭ·jĕs'·sĕnt* (L. *rigescens*, growing stiff or numb ; gen., *rigescentis*), in *bot.*, having a rigid or stiff consistence.

rigor, n., *rĭg'·ŏr* (L. *rigor*, stiffness, rigidity), a sudden coldness with involuntary shivering, symptomatic of the beginning of a disease, especially a fever : **rigor mortis**, *mŏrt'·ĭs* (L. *mors*, death, *mortis*, of death), the stiffening of the body after death.

rima, n., *rīm'·ă* (L. *rima*, a cleft, a crack), in *anat.*, a cleft ; an elliptic interval : **rima glottidis**, *glŏt'·tĭd·ĭs* (Gr. *glōttis*, the opening of the windpipe, *glōttĭdĭs*, of the opening of the windpipe—from *glōtta*, the tongue), the narrow aperture of the glottis.

rimose, a., *rĭm·ōz'* (L. *rīmōsus*, full of cracks—from *rīma*, a cleft, a crack), in *bot.*, marked by chinks or cracks, mostly parallel, as the bark of a tree : **rimulose**, a., *rĭm'·ŭl·ōz*, having a small marks or chinks.

ringent, a., *rĭnj'·ĕnt* (L. *ringens*, opening wide the mouth), in *bot.*, applied to a labiate flower in which the upper lip is much arched, and the lips are separated by a distinct gap.

ringworm, n., *rĭng'·wĕrm*, a skin eruption, caused by the action of a vegetable parasite, occurring chiefly on the scalp and arms, and upper part of the chest, is very contagious.

risorius, a., *rĭz·ōr'·ĭ·ŭs* (L. *rĭsor*, a laugher, *rĭsōris*, of a laugher ; *rĭsŭs*, laughter), the smiling muscle ; denoting a muscle of the cheek, consisting of a narrow bundle of fibres, which arises in the fascia over the Masseter muscle : **risus Sardonicus**, *săr·dŏn'·ĭk·ŭs* (*Sardinia*, where first seen, from the effects of eating a species of ranunculus there growing), a singularly convulsive grin or laugh, observed in cases of 'Tetanus.'

Robinia, n., *rŏb·ĭn'·ĭ·ă* (after *Robin*, an old botanist of France), a genus of plants remarkably handsome when in flower, Ord. Leguminosæ, Sub-ord. Papilionaceæ : **Robinia pseudo-acacia**, *sŭd'ō - ăk·kā'·shĭ·ă* (Gr. *pseudēs*, false ; Eng. *acacia*), a species often cultivated in Britain, as the locust tree, producing a durable wood.

Roccella, n., *rŏk·sĕl'·lă* (Port. *roccha*, a rock, so named from its habitat), a genus of plants, Ord. Lichenes : **Roccella tinctoria**, *tĭngkt·ōr'·ĭ·ă* (L. *tinctōrĭus*, of or belonging to dyeing—from *tingo*, I dye), from the Canaries ; **R. fuciformis**, *fūs'·ĭ·fŏrm'·ĭs* (Gr. *phukos*, L. *fucus*, sea-weed, rock-lichen ; L. *forma*, shape) ; and **R. hypomecha**, *hĭp·ŏm'·ĕk·ă* (unascertained), are species which furnish valuable dyes, under the name of Orchil or Archil, the general name of the dye being Litmus.

Rodentia, n. plu., *rōd·ĕn'·shĭ·ă* (L. *rōdens*, eating away, gnawing ; gen. *rodentis*), an Order of the Mammals, so named from the habit of gnawing or nibbling, as the rat, the rabbit, etc.: **rodent**, a., *rōd'·ĕnt*, gnawing : n., one of the gnawers : **rodent ulcer**, a malignant form of ulceration, allied to cancer, generally on the upper part of the face.

root, the subterranean or descending axis of a plant : **root-cap**, in *bot.*, a mass of tissue, covering as a helmet the *punctatum vegetationis* of a root : **root-sheath**, the coleorhiza : **root-stock**, the rhizome of a plant.

Rosaceæ, n. plu., *rōz·ā'·sĕ·ē* (L. *rosa*, Gr. *rhodon*, a rose), the Rose family, an extensive Order of plants, well known for their beauty, fragrance, and grateful products, such as the strawberry and blackberry : **Rosa**, n., *rōz'·ă*, a genus of plants unrivalled for the fragrance and beauty of its flowers : **Rosa spinosissima**, *spĭn'·ō·sĭs'·sĭm·ă* (L. *spinōsus*, thorny,

prickly—from *spĭna*, a thorn), the very thorny rose ; the species from which the varieties of the Scotch roses have been derived : **R. canina**, *kăn·ĭn'·ă* (L. *cănīnus*, of or pert. to a dog—from *cănis*, a dog), the Dog-rose, is beat into a pulp after the hairy achenes have been removed, and used, sweetened, as an acidulous refrigerant and astringent : **R. Gallica**, *găl'·lĭk·ă* (L. *gallĭcus*, of or pert. to the *Gaul*—from *Gălli*, the Gauls or French), the red, French, and Provence rose, whose petals are employed, in the form of infusion, as a tonic and slight astringent : **R. centifolia**, *sĕnt'·ĭ·fōl'·ĭ·ă* (L. *centum*, a hundred ; *fŏlĭum*, a leaf), the petals of the hundred-leaved or Cabbage-rose, and its varieties : **R. Damascena**, *dăm'·ăs·sēn'·ă* (of or from *Damascus*), the petals of the Damask-rose ; **R. moschata**, *mŏsk·āt'·ă* (mid. L. *moschātus*, having a smell like musk — from Arabic *mosch*, musk), the petals of the Musk-rose, and others, are employed in the production of rose-water, and the oil or attar of roses : **rosaceous**, a., *rōz·ā'·shŭs*, arranged in a circular form, as the petals of a single rose.

rose, n., *rōz* (L. *rosa*, a rose), the popular name in Scotland for Erysipelas, so named from its colour : **roseola**, n., *rōz·ē'·ŏl·ă* (mid. L. *rosēŏla*, a little rose), in *med.*, a rose-coloured rash of several varieties.

rosette, n., *rōz·ĕt'* (F. *rosette*—from L. *rosa*, a rose), in *bot.*, a cluster of leaves disposed in close circles.

Rosmarinus, n., *rŏs'·măr·ĭn'·ŭs* (L. *ros*, dew ; *mărīnus*, belonging to the sea—from *mărĕ*, the sea), a genus of pretty shrubs, so named from their maritime habitat : **Rosmarinus officinalis**, *ŏf·fĭs'·ĭn·āl'·ĭs* (L. *offĭcĭnālis*, officinal, by authority—from *offĭcīna*, a workshop), Rosemary, whose flowering tops furnish an oil which is tonic, stimulant, and carminative, much used in perfumery, as in the composition of Eau-de-Cologne and Hungary-water.

rostel, n., *rŏst'.ĕl*, also **rostellum**, n., *rŏst.ĕl'.lŭm* (L. *rostellum*, a little beak — from *rŏstrum*, a beak, a bill), in *bot.*, that part of the heart of a seed which descends and becomes the root; a peculiar body in Orchids, bearing the glands of the pollen mass, with its viscid balls attached; in *anat.*, a beak-shaped process: **rostellate**, a., *rŏst.ĕl'.lāt*, having a small beak, or little elongated neck.

rostrate, a., *rŏst'.rāt* (L. *rostrum*, the bill or snout of an animal), in *anat.*, having a process resembling the beak of a bird; in *bot.*, furnished with beaks; having a long, sharp point: **rostrum**, n., *rŏst'.rŭm*, the beak, or suctorial organ, formed by the appendages of the mouth in certain insects; the frontal spine of the Crustacea; in *anat.*, a triangular spine in the middle line of the anterior surface of the sphenoid bone of the skull; the reflected portion of the bend or genu which the 'corpus callosum' forms in its course.

rotate, a., *rōt.āt'* (L. *rŏtātum*, to turn a thing round like a wheel —from *rŏta*, a wheel), in *bot.*, having a regular gamopetalous corolla, with a short tube and spreading limb: **rotation**, n., *rŏt.ā'.shŭn*, in *anat.*, the revolving motion of a bone round its axis; in *bot.*, the internal circulation of the fluids in the cells of plants: **rotate-plane**, or **rotato-plane**, *rōt.āt'.ō-*, in *bot.*, wheel-shaped, and flat, without a tube: **rotation of gyration**, in *bot.*, a peculiar circulation of the cell sap, as seen in characeæ, and others.

rotator, n., *rōt.āt'.ŏr* (L. *rŏtātum*,

to turn a thing round like a wheel—from *rŏta*, a wheel), in *anat.*, a muscle which gives a circular or rolling motion to a part: **rotatory movement**, a movement which is circular: **rotatores spinæ**, *rŏt'.ăt.ōr'.ēz spīn'.ē* (L. *spina*, the spine or backbone), the rotators of the spine; eleven pairs of small muscles, eleven on each side of the spine, each pair passing from the transverse processes of one vertebra, and inserted into the vertebra next above: **rotatoria**, n. plu., *rŏt'.ăt.ōr'.ĭ.ă*, has the same sense as **Rotifera**, which see.

Rotifera, n. plu., *rŏt.ĭf'.ĕr.ă* (L. *rŏta*, a wheel; *fero*, I carry), a class of the Scolecida, microscopic animals, characterised by a ciliated trochal disc: **rotiferous**, a., *rŏt.ĭf'.ĕr.ŭs*, having or bearing organs like wheels.

Rottlera, n., *rŏt'.lĕr.ă* (after *Dr. Rottler*, a Dane), a genus of plants, Ord. Euphorbiaceæ: **Rottlera tinctoria**, *tĭngk.tōr'.ĭ.ă* (L. *tinctōrĭus*, of or belonging to dyeing —from *tingo*, I dye), a small tree of Abyssinia, India, etc., whose ruby-like glands on its fruit are brushed off, and the powder administered for tape-worm.

rotula, n., *rŏt'.ūl.ă* (L. *rŏtula*, a little wheel—from *rŏta*, a wheel), in *anat.*, the patella or knee-pan, situated at the front of the knee-joint.

rubefacient, n., *rŏb'.ē.fā'.shĭ.ĕnt* (L. *ruber*, red; *facĭo*, I make), in *med.*, an irritant substance which, applied to the skin, gives rise to heat, redness, etc., as if there existed a slight local inflammation, as a mustard poultice, useful in dissipated, slight local pains.

rubeola, n., *rŏb.ē'.ŏl.ă* (L. *ruber*, red), applied to measles, now restricted to an eruptive disease presenting the characters of both measles and scarlet fever : **rubel-oid**, a., *rŏb'·ĕl·ŏўd* (Gr. *eidos*, resemblance), resembling the eruptive disease rubeola.

Rubiaceæ, n. plu., *rŏb'·ĭ·ā'·sĕ·ē* (L. *ruber*, red—in allusion to the colour of the roots), the Madder and Peruvian Bark family, an Order of plants possessed of tonic, febrifuge, and astringent properties, and which furnish important substances to the materia medica : **Rubia**, n., *rŏb'·ĭ·ă*, an interesting genus of plants : **Rubia tinctoria**, *tĭngk·tōr'·ĭ·ă* (L. *tinctōrĭus*, of or belonging to dyeing—from *tingo*, I dye), a species whose root is the Madder of commerce, which contains three colouring matters, viz. madder purple, orange, and red : **R. munjista**, *mŭn·jĭst'·ă* (native name), also called **R. cordifolia**, *kŏrd'·ĭ·fŏl'·ĭ·ă* (L. *cor*, the heart, *cordis*, of the heart ; *fŏlĭum*, a leaf), another species from which similar dyes are obtained : **rubian**, n., *rŏb'·ĭ·ăn*, an intensely bitter, amorphous, yellow substance procured from Madder.

rubiginose, a., *rŏb·ĭdj'·ĭn·ōz* (L. *rubīginōsus*, abounding in rust—from *rūbīgo*, rust), in *bot.*, of a brownish-red tint ; having the colour of rust.

Rubus, n., *rŏb'·ŭs* (L. *rŭbus*, a bramble-bush), an extensive and interesting genus of plants, comprising the rasp, black, and dewberries, etc., Ord. Rosaceæ : **Rubus idæus**, *ĭd·ē'·ŭs* (unascertained), the Raspberry and its varieties : **R. chamæmorus**, *kăm·ē'·mŏr·ŭs* (Gr. *chamai*, on the ground ; *mŏrĕa*, the mulberry tree), a species whose fruit is acid and pleasant, known as the Cloudberry, so named from the high situations where found.

rudimentary, a., *rŏd'·ĭ·mĕnt'·ăr·ĭ* (L. *rudĭmĕntum*, a first attempt or trial), in *bot.*, an early stage of

development ; in an imperfectly developed condition.

Ruellia, n., *rŏ·ĕl'·lĭ·ă* (after *John Ruelle*, a French botanist), a genus of pretty flowering plants, Ord. Acanthaceæ : **Ruellia anisophylla**, *ăn·ĭs'·ō·fĭl'·lă* (Gr. *anĭsŏn*, anise ; *phullon*, a leaf), a species whose style exhibits a peculiar irritability ; a deep - blue dye, called 'Room,' is obtained from a species of Ruellia.

rufescent, a., *rŏf'·ĕs'·sĕnt* (L. *rūfĕsco*, I grow red—from *rūfus*, red), in *bot.*, becoming reddish-brown : **rufous**, a., *rŏf'·ŭs*, of a red-brown colour.

ruga, n., *rŏg'·ă*, **rugæ**, n. plu., *rŏdj'·ē* (L. *ruga*, a plait or wrinkle, *rugæ*, plaits or wrinkles), in *anat.*, the folds into which the mucous membrane of some organs are thrown by the contraction of the external coats, as the rugæ of the stomach, or of the vagina : **Rugosa,** n. plu., *rŏg·ōz'·ă*, an Order of Corals : **rugose**, a., *rŏg·ōz'*, rough with wrinkles ; covered with wrinkled lines : **rugulose**, a., *rŏg'·ŭl·ōz* (dim. of L. *ruga*), finely wrinkled.

rumen, n., *rŏm'·ĕn* (L. *rūmen*, the throat or gullet, *rūmĭnĭs*, of the throat or gullet), the first cavity of the complex stomach of Ruminants, often called the paunch : **Ruminants**, n. plu., *rŏm'·ĭn·ănts*, those animals which ruminate or chew the cud, as the ox, sheep, cow, camel, etc. ; also called **Ruminantia**, n. plu., *rŏm'·ĭn·ăn'·shĭ·ă* : **ruminate**, a., *rŏm'·ĭn·āt*, in *bot.*, applied to the hard albumen of some seeds presenting a mottled appearance ; having mottled albumen.

Rumex, n., *rŏm'·ĕks* (L. *rŭmex*, sorrel), a genus of plants, comprising the dock, Ord. Polygonaceæ : **Rumex acetosa**, *ăs'·ĕt·ōz'·ă* (mod. L. *acētōsus*, the sorrel or sour dock — from L. *ăcētum*, vinegar), common sorrel, which contains pure oxalic acid ; and R. acetosella, *ăs'·ĕt·ō·sĕl'·lă* (dim. of mod. L. *acētōsus*), sheep's sorrel, the leaves of both of which are acid and astringent : R. aquaticus, *ăk·wăt'·ĭk·ŭs* (L. *ăquātĭcus*, growing in water—from *ăqua*, water), the water-dock ; and R. hydrolapathum, *hĭd'·rō·lăp'·ăth·ŭm* (Gr. *hudor*, water ; *lapathŏn*, the dock), the great water - dock, as well as other species, have their roots employed as astringents and alteratives : R. Alpinus, *ălp·ĭn'·ŭs* (L. *alpīnus*, of or from the *Alps*), a species whose roots, under the name of Monk's rhubarb, were formerly employed as purgatives : R. scutatus, *skūt·āt'·ŭs* (L. *scūt-ātus*, armed with a shield—from *scūtŭm*, a shield), the French sorrel, have larger and more succulent leaves than the common sorrel, much used in French cookery.

runcinate, a., *răn'·sĭn·āt* (L. *runcīno*, I plane off—from *runcīna*, a plane, a large saw), in *bot.*, having a pinnatifid leaf with a triangular termination, and sharp divisions pointing down-wards, as in the dandelion ; toothed like a large pit-saw, as a leaf : runcinately-lyrate, in *bot.*, lyrate with the lobes hooked back.

runner, n., *răn'·nẽr* (from *run*), in *bot.*, a slender prostrate stem, which roots at the nodes, as in the strawberry.

rupia, n., *rōp'·ĭ·ă* (Gr. *rhupŏs*, dirt, filth), a syphilitic form of skin disease, characterised by superficial ulcerations and con-ical scabs.

rupture, n., *răp'·tūr* (L. *ruptum*, to burst, to rend), in *mẽd.*, a tumour caused by the protrusion of a part of the bowels ; hernia : rupturing, n., in *bot.*, an irreg-ular manner of bursting.

rust, n., *răst* (Ger. *rost*, Dut.

roest, rust), oxide of iron ; an orange powder, exuding from the inner chaff scales of growing grain, forming yellow or brown spots and blotches, caused by the parasite ' Uredo Rubigo.'

Rutaceæ, n. plu., *rōt·ā'·sĕ·ē* (Gr. *rhutē*, L. *rūta*, the plant rue), the Rue family, an Order of plants hav-ing a peculiar odour, very power-ful and penetrating : Ruteæ, n. plu., *rōt'·ē·ē*, a Sub-order, having albuminous seeds : Ruta, n., *rōt'·ă*, a genus of plants : Ruta grave-olens, *grăv·ē·ŏl·ĕnz*(L. *grăvĕŏlens*, strong-smelling), the common or garden rue, whose leaves and unripe fruit are used as a stimulant, an antispasmodic, an anthelmintic, and an emmenagogue.

Sabal, n., *săb'·ăl* (unascertained), a genus of Palmæ or Palms : Sabal umbraculifera, *ŭm·brăk'·ūl·ĭf'·ĕr·ă* (L. *umbrāculum*, a thing that furnishes shade—from *um-bra*, a shade ; *fero*, I bear), the Fan Palm or Bull Palm of the W. Indies.

sac, n., *săk* (L. *saccus*, a sack), in *anat.*, a sac, bag, or pouch : sac of the embryo, in *bot.*, the sac of the nucleus within which the embryo is formed : saccate, a., *săk'·kāt*, forming a bag or sac, seen in some petals ; in the form of a bag : sacciform, a., *săk'·sĭ-fŏrm* (L. *forma*, shape), like a bag.

Saccharum, n., *săk'·kăr·ŭm* (Gr. *sacchăron*, a sweet juice, sugar), a valuable genus of grasses, Ord. Gramineæ : Saccharum Sinense, *sĭn·ĕns'·ĕ* (L. *Sinenis*, Chinese—from *Sina*, China), the species in China which yields sugar : S. viola-ceum, *vĭ'·ōl·ā'·sĕ·ŭm* (L. *vĭŏlācĕus*, violet-coloured—from *vĭŏla*, the violet), the W. Indian sugar-cane : S. officinarum, *ŏf'·fĭs'·ĭn·ār'·ŭm* (L. *officīna*, the workshop, *officīnārum*, of workshops), the common sugar-cane.

saccule, n., *săk'·ŭl* (L. *sacculus*, a little bag—from *saccus*, a bag), a little sac ; a cyst or cell : **saccular**, a., *săk'·ŭl·ăr*, of or pert. to a little sac or cyst : **sacculus laryngis**, *lăr·ĭnj'·ĭs* (mod. L. *larynx*, the upper part of the windpipe, *laryngis*, of the larynx ; Gr. *larungx*), the little pouch of the larynx ; a membranous sac, conical in form, placed between the superior vocal cord, and the inner surface of the thyroid cartilage.

sacrum, n., *săk'·rŭm* (L. *sacer* or *sacrum*, sacred), in *anat.*, the bone which forms the termination or basis of the vertebral column, also called **os sacrum**, the sacred bone : **sacral**, a., *săk'·răl*, of or pert. to the sacrum : **sacral aspect**, the appearance towards the region where the sacrum is situated : **sacro**, *săk'·rō*, denoting parts connected with the sacrum: **sacro - coccygeal**, *kŏk'·sĭdj·ē'·ăl* (see ' coccyx '), a name for two ligaments, — the *inferior*, consisting of a few irregular fibres, the *posterior*, of a flat band of ligamentous fibres of a pearly tint : **sacro - iliac**, *ĭl'·ĭ·ăk* (L. *ilia*, the flank, the entrails), applied to the joints which connect the sacrum with the ilium : **sacro-lumbalis**, *lŭm·băl'·ĭs* (L. *lumbālis*, pert. to the *lumbus* or loin), a large muscle which passes from the ilium to the lower ribs : **sacro-sciatic**, *sĭ·ăt'·ĭk* (L. *ischia*, the hip-bones, of which *sciatica* is a mere corruption), applied to the ligaments connected with the ischium or hip-bone.

Sageretia, n., *sădj'·ĕr·ēsh'·ĭ·ă* (after *Sageret*, an eminent French agriculturist), a genus of plants, Ord. Rhamnaceæ : **Sageretia theezans**, *thēz'·ănz* (a native name), a plant whose leaves are used as a substitute for tea by the poorer classes in China.

sagittal, a., *sădj·ĭt'·tăl* (L. *sagitta*, an arrow), arrow-like ; resembling an arrow : **sagittate**, a., *sădj·ĭt'·tāt*, in *bot.*, shaped like the head of an arrow : **sagittal suture**, in *anat.*, the suture which unites the parietal bones of the skull.

Saguera, n., *săg'·ū·ĕr'·ă*, and **Sagus**, n., *săg'·ŭs* (*sagu*, Malay name for various palms), genera of the Ord. Palmæ : **Sagus farinifera**, *făr'·ĭn·ĭf'·ĕr·ă* (L. *farīna*, meal ; *fero*, I bear), a native of Malacca ; and **Saguera saccharifer**, *săk·kăr'·ĭf·ĕr* (L. *saccharum*, sugar ; *fero*, I bear), found in the eastern islands of the Indian Ocean, are Sago Palms, which produce fine sago.

sal, n., *săl* (L. *sal*, salt), a common prefix among the older chemists, denoting a compound having definite proportions of an acid with an alkali, an earth, or a metallic oxide : **sal-ammoniac**, muriate of ammonia; a compound of ammonia and hydrochloric acid : **sal-mirabile**, *mĭr·ăb'·ĭl·ĕ* (L. *mirăbĭlis*, wonderful), Glauber's salt ; sulphate of soda : **sal-prunella**, *prŏn·ĕl'·lă* (Ger. *prunelle*—from L. *prūna*, a burning or live coal), a name given to nitre when fused and cast into cakes or balls: **sal-volatile**, *vŏl·ăt'·ĭl·ĕ* (L. *volātĭlis*, winged, swift), the volatile salt ; the popular name for ammonia, and popularly pronounced *săl-vŏl'·ăt·ĭl*.

Salacia, n., *săl·ā'·sĭ·ă* (in Roman mythology, *Salacĭa*, the wife of Neptune), a genus of plants, Ord. Celastraceæ : **Salacia pyriformis**, *pĭr'·ĭ·fŏrm'·ĭs* (L. *pyrum*, a pear ; *forma*, shape), a species which produces an eatable fruit about the size of a Bergamot pear, a native of Sierra Leone.

salcus spiralis, *sălk'·ŭs spīr·āl'·ĭs* (L. *salcus*, a furrow ; *spirālis*, spiral — from *spīra*, a coil, a spire), a grooved border which terminates the 'limbus laminæ spiralis ' of the cochlea.

Salicaceæ, n. plu., *săl'·ĭk·ā'·sĕ·ē*

(L. *salix*, a willow tree, *salĭcĭs*, of a willow tree), the Willow family, an Order of well-known trees and shrubs, comprising the willow and poplar : **Salix**, n., *săl'ĭks*, an extensive genus of ornamental trees and shrubs : **Salix capræa**, *kăp·rē'ă* (L. *căprĕa*, a wild she-goat); **S. alba**, *ălb'ă* (L. *albus*, white); **S. Russelliana**, *rŭs'sĕl·i·ăn'ă* (*Russell*, proper name); **S. fragilis**, *frădj'ĭl·ĭs* (L. *frăgĭlis*, easily broken); **S. pentandra**, *pĕnt·ănd'ră* (Gr. *pente*, five ; *anēr*, a male, *andros*, of a male); **S. vitellina**, *vĭt'ĕl·lin'ă* (L. *vĭtĕllinus*, of a yellow colour—from *vĭtellus*, the yolk of an egg); **S. purpurea**, *pėr·pŭr'ĕ·ă* (L. *purpŭrĕus*, purple); and **S. helix**, *hĕl'ĭks* (Gr. *hĕlĭx*, a winding or spiral body), are species of willow whose bark yields a crystalline bitter substance called **salicin**, *săl'ĭs·ĭn*, used as an antipyretic, and tonic, has also wonderful anti-rheumatic properties : **S. fragilis**, yields a saccharine exudation : **S. Babylonica**, *băb'ĭl·ŏn'ĭk·ă* (of or from *Babylon*), the weeping willow : **S. arctica**, *ărk'tĭk·ă* (new L. *arctĭcus*, of or from the north—from Gr. *arktos*, a bear, the north); and **S. polaris**, *pŏl·ār'ĭs* (L. *polāris*, of or pert. to the pole — from *pŏlus*, the pole), are species which extend to the Arctic regions : **S. herbacea**, *hėrb·ā'sĕ·ă* (L. *herbācĕus*, grass-green—from *herba*, a herb, grass), a small creeping willow, abundant on the Scotch mountains : **salicylic acid**, *săl'ĭ·sĭl'ĭk*, an acid obtained by the action of fused potassa on salicin.

saliva, n., *săl·iv'ă* (L. *salīva*, spittle), the frothy fluid which gathers in the mouth, and which, when discharged from it, is called spittle : **salivine**, n., *săl·iv'ĭn*, a peculiar animal extractive substance obtained from saliva : **salivary**, a., *săl'ĭv·ăr·ĭ*, secreting or conveying saliva : **salivation**, n., *săl'ĭv·ā'shŭn*, an increased flow of saliva in the mouth, with swelling of the mucous membrane, generally caused by the action of mercury.

salpinx, n., *săl'pĭngks* (Gr. *salpingx*, a trumpet), in *anat.*, the Eustachian tube, or channel of communication between the mouth and the ear : **salpingitis**, *săl'ping·jit'ĭs*, inflammation of the Eustachian tube ; inflammation of the Fallopian tubes.

Salsola, n., *săl'sŏl·ă* (L. *salsus*, salted, salt), a genus of plants found chiefly on the sea-shore, Ord. Chenopodiaceæ, many yielding kelp and barilla.

Salvadoraceæ, n. plu., *sălv'ăd·ōr·ā'sĕ·ē* (Sp. *Salvador*, a saviour), the Salvadora family, an Order of plants acrid and stimulant, and some like mustard : **Salvadora**, n., *sălv'ăd·ōr'ă*, a genus of plants : **Salvadora Persica**, *pėrs'ĭk·ă* (of or from *Persia*), supposed by some to be the mustard tree of Scripture.

Salvia, n., *sălv'ĭ·ă* (L. *salvo*, I save, alluding to the healing qualities of sage), an extensive genus of extremely showy flowering plants, Ord. Labiatæ: **Salvia officinalis**, *ŏf·fis'ĭn·āl'ĭs* (L. *offĭcĭnālis*, officinal, by authority—from *offĭcīna*, a workshop), the common sage, is often used in the form of tea as a stomachic : **S. pomifera**, *pŏm·ĭf'ėr·ă* (L. *pomum*, an apple ; *fero*, I bear), a species producing sage apples, being only galls arising from the punctures of certain insects.

Samadera, n., *săm'ăd·ēr'ă* (unascertained), a genus of plants, Ord. Simarubaceæ : **Samadera Indica**, *ĭn'dĭk·ă* (L. *Indĭcus*, Indian), a species whose bark is bitter and tonic, containing a principle like Quassia.

samara, n., *săm'ăr·ă* (L. *samara*, the seed of the elm), in *bot.*, a winged, indehiscent fruit, as in the elm, ash, and maple : **samaroid**, a., *săm'·ăr·ŏyd* (Gr. *eidos*, resemblance), having a seed-vessel like a samara.

Sambucus, n., *săm·būk'·ŭs* (L. *sambūcus*, an elder-tree), a genus of plants, Ord. Caprifoliaceæ : **Sambucus nigra**, *nīg'·ră* (L. *nīger*, black), the common Elder, whose fruit is employed in the manufacture of a kind of wine, and whose juice, and the inner bark, possess purgative qualities.

Samydaceæ, n. plu., *săm'·ĭd·ā'·sĕ·ē* (unascertained), the Samyda family, an Order of plants : **Samyda**, n., *săm'·ĭd·ă*, a genus of tropical trees.

sanguify, v., *săng'·gwĭ·fī* (L. *sanguis*, blood ; *facĭo*, I make), to form or produce blood ; to convert chyle into blood : **sanguification**, n., *săng'·gwĭ·fĭk·ā'·shŭn*, the conversion of chyle into blood : **sanguine**, a., *săng'·gwĭn*, warm or ardent in temper ; containing or abounding with blood : **sanguivorous**, a., *săng·gwĭv'·ŏr·ŭs* (L. *voro*, I devour), drinking blood, and subsisting on it : **sanguineous**, a., *săng·gwĭn'·ē·ŭs*, resembling blood ; abounding in blood.

Sanguinaria, n., *săng'·gwĭn·ār'·ĭ·ă* (L. *sanguis*, blood, *săngŭĭnis*, of blood, the plants yielding a red juice when broken), a genus of plants, Ord. Papaveraceæ : **Sanguinaria Canadensis**, *kăn'·ăd·ĕns'·ĭs* (new L. *Canadensis*, of or pert. to *Canada*), the plant Blood-root or Pucoon, has emetic and purgative properties : **Sanguisorba**, n., *săng'·gwĭ·sŏrb'·ă* (L. *sorbĕŏ*, I suck up), a Sub-order of plants, Ord. Rosaceæ, some of whose species were supposed to be powerful vulneraries.

sanitary, a., *săn'·ĭt·ăr'·ĭ* (L. *sānĭtas*, health—from *sānus*, sound in body), a term applied to any measures taken for the preserv-

ation of health ; pert. to arrangements connected with the prevention of disease : **sanatory**, a., *săn'·ăt·ŏr'·ĭ* (L. *sānătum*, to heal, to restore to health—from *sānus*, sound in body), a term applied to any measures taken for the restoration of health after it has been lost ; pert. to arrangements connected with the cure of disease : **sanitary** thus applies to preventive measures : **sanatory** applies to curative measures : **sanatorium**, n., *săn'·ăt·ōr'·ĭ·ŭm*, places to which persons may retire for a time for the benefit of their health : **sanitarium**, n., *săn'·ĭt·ār'·ĭ·ŭm*, **sanitaria**, n. plu., *-ār'·ĭ·ă*, places where a high state of health may be maintained, as hill stations for troops in tropical climates.

sanitary and **sanitarium**, see under 'sanatory' : **sanity**, n., *săn'·ĭt·ĭ* (L. *sanĭtas*, health), a sound state of mind.

Santalaceæ, n. plu., *sănt'·ăl·ā'·sĕ·ē* (L. *săndălis*, a species of palm tree ; Ar. *zandal*), the Sandal-wood family, an Order of plants, some are astringent, others yield a perfume : **Santalum**, n., *sănt'·ăl·ŭm*, a genus of trees : **Santalum album**, *ălb'·ŭm* (L. *albus*, white), and other species, yield Sandal-wood, which is used medicinally, and as a perfume : **S. Persicari**, *pĕrs'·ĭk·ār'·ĭ* (new L. *Persicārum*, Persia), the Sandal-wood of Persia ; a dwarf kind of Australian Sandal-wood, whose bark furnishes an amylaceous food.

santonin, n., *sănt'·ōn·ĭn* (Gr. *santŏnĭŏn*, wormwood), a crystalline substance obtained from the unexpanded flower-heads of certain species of Artemisia, especially A. Santonina, a good remedy for round worms ; wormseed.

saphena, n., *săf·ēn'·ă* (Gr. *saphēnēs*, clear, manifest), in *anat.*, a name of two conspicuous veins of the lower extremities, extending from the knee to the ankle and foot ; also applied to a nerve in same region : **saphenous, a.,** *săf·ēn'·ŭs,* applied to superficial veins and nerves of the thigh and leg.

sapid, a., *săp'·ĭd* (L. *sapĭdus,* having taste — from *sapĭo,* I taste), tasteful ; that affects or stimulates the palate : **sapidity, n.,** *săp·ĭd'·ĭ·tĭ,* taste ; the quality of affecting the organs of taste.

Sapindaceæ, n. plu., *săp'·ĭn·dā'· sĕ·ē* (L. *sapo-ĭndĭcus,* Indian soap—from *săpo,* soap ; *indĭcus,* of or from India), the Soapwort family, an Order of plants, many yield edible fruits, others are poisonous : **Sapindus,** *săp·ĭn'·dŭs,* a genus of plants whose species contain a saponaceous principle : **Sapindus saponaria,** *săp'·ŏn·ār'· ĭ·ă* (L. *săpo,* soap, *sapōnis,* of soap), a species whose berries, called soap berries, are used as a substitute for soap in the West Indies : **sapindaceous, a.,** *săp'·ĭn· dā'·shŭs,* of or pert. to the Sapindaceæ.

sapodilla, n., *săp'·ŏd·ĭl'·lă* (Sp. *sapotĭlla*), a tree and its fruit, native of W. Indies and S. America, whose wood is a fancy wood.

Saponaria, n., *săp'·ŏn·ār'·ĭ·ă* (L. *săpo,* soap, *sapōnis,* of soap), a genus of plants, Ord. Caryophyllaceæ, which produce some very beautiful species : **Saponaria officinalis,** *ŏf·fĭs'·ĭn·āl'·ĭs* (L. *officīnālĭs,* by authority—from *officīna,* a workshop), a species whose leaves are said to produce a lather like soap when agitated in water, and equally efficacious in removing grease spots : **saponine,** n., *săp'· ŏn·ĭn,* a peculiar, poisonous substance existing in many of the species of the Order : **saponify, v.,** *săp·ŏn'·ĭ·fĭ* (L. *facĭo,* I make), to convert into soap : **saponification,** n., *săp·ŏn'·ĭ·fĭk·ā'·shŭn,* conversion into soap.

Sapotaceæ, n. plu., *săp'·ŏt·ā'·sĕ·ē,* or **Sapota plums,** *săp·ōt'·ă* (*sapōta,* a W. Indian name of several fruits), the Sapodilla family, an Order of plants, many yield edible fruits, some an oily matter, while others act as tonics, astringents, and febrifuges.

Saprolegnieæ, n. plu., *săp'·rō·lĕg· nī'·ĕ·ē* (Gr. *sapros,* putrid ; *legnon,* a fringe or border), a Sub-order of the Algæ, colourless, aquatic, filamentous plants growing on decaying organic matter : **saprophytes,** n. plu., *săp'·rō·fītz* (Gr. *phuton,* a plant), plants growing on decaying vegetable matter.

Sarcina, n., *săr'·sĭn·ă* (L. *sarcĭna,* a package), a genus of Fungi or minute Cryptogamic plants, sometimes found in vomited matter : **Sarcina ventriculi,** *vĕn·trĭk'·ŭl·ī* (L. *ventrĭcŭlus,* the belly), the Sarcina of the belly, nuclei or cells placed in close opposition, forming organisms found in the vomited contents of the stomach in many morbid conditions, and occasionally in other parts ; also named **Sarcinula ventriculi,** *săr· sĭn'·ŭl·ă* (L. *sarcĭnŭla,* a little package).

sarcocarp, n., *sărk'·ō·kărp* (Gr. *sarx,* flesh ; *karpos,* fruit), in *bot.,* the fleshy part of certain fruits, usually that eaten ; also called 'sarcoderm,' which see.

sarcocele, n., *sărk'·ō·sēl* (Gr. *sarx,* flesh ; *kēlē,* a tumour), a fleshy and firm tumour on a testicle.

sarcode, n., *sărk'·ōd* (Gr. *sarkodes,* fleshy—from *sarx,* flesh ; *eidos,* resemblance), the simple glutinous substance which constitutes the body or vital mass of the Protozoa, or lowest forms of animal life ; animal protoplasm : **sarcoids,** n. plu., *sărk'·ŏydz,* the separate amœbiform particles which in the aggregate make up the flesh of a sponge.

sarcoderm, n., *sărk'·ō·dĕrm* (Gr. *sarx,* flesh ; *derma,* skin), in *bot.*, the fleshy covering of a seed lying between the internal and external covering; also called 'sarcocarp.'

sarcolemma, n., *sȧrk'·ō·lĕm'·mȧ* (Gr. *sarx*, flesh; *lemma*, skin, rind), in *anat.*, the proper tubular sheath of muscular fibre.

sarcolobeæ, n. plu., *sȧrk'·ō·lōb'·ĕ·ē* (Gr. *sarx*, flesh; *lobos*, a lobe), in *bot.*, thick and fleshy cotyledons, as in the bean and pea.

sarcoma, n., *sȧrk·ōm'·ȧ* (Gr. *sark·ōma*, a fleshy excrescence—from *sarx*, flesh), any firm fleshy tumour or excrescence not inflammatory; a growth chiefly composed of tissue resembling the immature or embryonic form of connective tissue; now applied to a solid malignant tumour, distinguished from cancers by the arrangement of the cellular elements of which it consists—the latter may be 'round,' 'spindle - shaped,' or polynucleated: **sarcomata,** n. plu., *sȧrk·ōm'·ȧt·ȧ*, or **sarcomatous tumours,** are generally innocent growths, but many are in every respect as malignant as true cancer: *sarcomata* assume different forms, known by such names as 'spindle-celled sarcoma'; 'recurrent fibroid - tumour'; 'fibro - plastic tumour'; and 'fibro - nucleated tumour,' the last being the commonest and best known.

sarcoptes, n. plu., *sȧrk·ŏp'·tēz,* also **sarcocopta,** n. plu., *sȧrk'·ō·kŏp'·tȧ* (Gr. *sarx*, flesh; *kopto*, I wound, I injure), the itch insect.

sarcosis, n., *sȧrk·ōz'·ĭs* (Gr. *sark·osis*, the producing of flesh — from *sarx*, flesh), the generation of flesh: **sarcotic,** a., *sȧrk·ŏt'·ĭk,* that promotes the growth of flesh: **sarcous,** a., *sȧrk'·ŭs,* having elements that produce flesh; of or pert. to muscle or flesh.

sarcosperm, n., *sȧrk'·ō·spĕrm* (Gr. *sarx*, flesh; *sperma*, seed), same

sense as 'sarcoderm,' which see.

sardonic laughter, *sȧr·dŏn'·ĭk* (so named from the *sardonica* herb, which is said when eaten to produce convulsive motions of the cheeks and lips; Gr. *sardonios*, a scornful, bitter laugh), a convulsive horrible grin; the 'risus sardonicus.'

sarmentum, n., *sȧr·mĕnt'·ŭm,* **sarmenta,** n. plu., *sȧr·mĕnt'·ȧ* (L. *sarmentum*, a twig), in *bot.*, the slender woody stem of climbing plants; a flagellum or runner giving off leaves and roots at intervals, as the strawberry: **sarmentous,** a., *sȧr·mĕnt'·ŭs,* having a running naked stem.

Sarraceniaceæ, n. plu., *sȧr'·rȧ·sēn·ĭ·ā'·sē·ē* (after *Dr. Sarrasin*, a French physician), the Side-saddle flower, Water-pitcher, or Trumpet-leaf flower, an Order of plants having radical leaves, whose petioles are so folded as to form ascidia or hollow tubes: **Sarracenia,** n., *sȧr'·rȧ·sēn'·ĭ·ȧ,* a genus of plants called the Trumpet-leaf: **Sarracenia purpurea,** *pẽr·pūr'·ĕ·ȧ* (L. *pŭrpŭrĕus,* purple); **S. flava,** *flāv'·ȧ* (L. *flāvus,* golden, yellow); **S. rubra,** *rŏb'·rȧ* (L. *rubrus,* red); and **S. Drummondii,** *drŭm·mŏnd'·ĭ·ī* (after *Drummond*, a botanist), have pitchers with open mouths and erect lids, into which rain-water can easily enter; while **S. variolaria,** *vȧr·ĭ'·ŏl·ȧr'·ĭ·ȧ* (L. *vȧrĭo,* I diversify), and **S. psittacina,** *sĭt'·tȧ·sĭn'·ȧ* (L. *psittăcīnus,* pert. to a parrot—from *psittăcus,* a parrot), are species which have their mouths closed by a lid, through which rain can hardly enter.

sarsaparilla, n., *sȧrs'·ȧ·pȧr·ĭl'·lȧ* (Sp. *zarzaparilla* — from *zarza,* a bramble; *parrilla,* a vine), a medicinal substance obtained from various species of the genus Similax, Order Similaceæ,—the part employed is the rhizome or roots, and is used as a tonic and alterative.

artorius, a., *sărt·ŏr'·ĭ·ŭs* (L. *sartor*, a tailor, *sartōris*, of a tailor), in *anat.*, applied to the muscle of the thigh which enables the legs to be thrown across each other, or to be bent inwards obliquely.

assafras, n., *săs'·să·frăs* (F. *sassafras*, L. *saxum*, a rock; L. *frango*, I break), a genus of plants, Ord. Lauraceæ: **Sassafras officinale**, *ŏf·fĭs'·ĭn·āl'·ĕ* (L. *officinālis*, officinal, by authority), an American tree whose root, wood, and flowers are employed in medicine, the root being used as an aromatic, stimulant, and diaphoretic, and contains a volatile oil; the name of the substance so obtained.

atellite, a. n., *săt'·ĕl·īt* (L. *satelles*, an attendant, *sătellītis*, of an attendant), in *anat.*, applied to the veins which accompany the brachial artery as far as the head of the cubit.

auria, n. plu., *săwr'·ĭ·ă*, also **saurians**, n. plu., *săwr'·ĭ·ănz* (Gr. *sauros*, a lizard), any lizard-like reptiles; sometimes restricted to the crocodiles and lacertilians: **sauroid**, a., *săwr'·ōyd* (Gr. *eidos*, resemblance), having some of the characteristics of the saurians.

aurobatrachia, n. plu., *săwr'·ō·băt·rāk'·ĭ·ă* (Gr. *sauros*, a lizard; *bătrăchos*, a frog), the Order of the tailed Amphibians: **Sauropsida**, n. plu., *săwr·ŏps'·ĭd·ă* (Gr. *opsis*, appearance, sight), the two classes of the Birds and Reptiles taken together.

aururaceæ, n. plu., *săwr'·ŭ·rā'·sĕ·ē* (Gr. *sauros*, a lizard; *oura*, a tail), the Lizard-tail family, an Order of plants, so named in allusion to the appearance of the flower-spike: **Saururus**, n., *săwr·ōr'·ŭs*, a genus of plants, said to have acrid properties.

avin, n., *săv'·ĭn* (F. *savinier*, L.

sabina), the fresh and dried tops of Juniperus Sabina, Ord. Coniferæ, which contain an active volatile oil, used as an anthelmintic and emmenagogue; in large doses, is an irritant poison.

Saxifragaceæ, n. plu., *săks'·ĭ·frăg·ā'·sĕ·ē* (L. *saxifrăgus*, stonebreaking—from *saxum*, a rock; *frango*, I break), the Saxifrage family, an Order of plants, some are astringent, some for tanning, others bitter tonics: **Saxifrageæ**, n. plu., *săks'·ĭ·frādj'·ĕ·ē*, a Suborder of plants: **Saxifraga**, n., *săks·ĭf'·răg·ă*, an extensive genus of beautiful Alpine plants, having reputed medical qualities in diseases of the stone: **saxifragous**, a., *săks·ĭf'·răg·ŭs*, having power to dissolve vesical calculi.

scabies, n., *skāb'·ĭ·ēz* (L. *scābies*, scurf, scab), the itch; a contagious vesicular eruption, caused by, or accompanied with, the 'acarus scabiei,' or itch parasite: **scabies equi sarcoptica**, *ēk'·wĭ săr·kŏp'·tĭk·ă* (L. *equus*, a horse; *equi*, of a horse; Gr. *sarx*, flesh; *kŏptō*, I pierce), the sarcoptic itch of the horse: **s. equi dermatodectica**, *dèrm'·ăt·ō·dĕk'·tĭk·ă* (Gr. *derma*, skin; *dektĭkos*, receiving or containing, capacious), a skin disease of horses, characterised by irritation, scurf, and denudation of hair.

Scabiosa, n., *skāb'·ĭ·ōz'·ă* (L. *scăbies*, the itch), a genus of plants, Ord. Dipsacaceæ, which are said to cure the itch: **Scabiosa succisa**, *sŭk·sĭz'·ă* (L. *succisus*, cut off, cut down), yields a green dye, and has astringent qualities: **scabrous**, a., *skāb'·rŭs*, also **scabrid**, a., *skāb'·rĭd*, rough; covered with very short, stiff hairs: **scabriusculus**, a., *skāb'·rĭ·ŭsk·ŭl·ŭs*, somewhat rough.

Scævola, n., *sēv'·ŏl·ă* (L. *scæva*, the left hand; *scævus*, left), a genus of ornamental plants, Ord. Goodeniaceæ, so named from the form of the corolla : **Scævola taccada**, *tăk·kād'·ă* (Sp. *tacada*, marked or spotted ; a Malay name), a species whose leaves are eaten, as pot-herbs, the pith is soft and spongy : **S. bela-modogam**, *bĕl'·ă-mŏd'·ŏg·ăm* (unascertained), is emollient, and used in India to bring tumours to a head.

scala media, *skāl'·ă mēd'·ĭ·ă* (L. *scāla*, a ladder ; *mĕdĭus*, internal, middle), the middle ladder ; a ladder-like canal in the cochlea of the ear : **scala tympani**, *tĭm'·păn·ĭ* (L. *tympănum*, a drum), the ladder or staircase of the drum ; a canal in the cochlea of the ear : **s. vestibuli**, *vĕs·tĭb'·ūl·ĭ* (L. *vestĭb-ŭlum*, a fore-court, an entrance), the ladder of the vestibule, a canal in the cochlea of the ear communicating with the vestibule.

scalariform, a., *skăl·ăr'·ĭ·fŏrm* (L. *scāla*, a ladder ; *forma*, shape), ladder-shaped ; in *bot.*, applied to vessels or tissue having bars like a ladder, as seen in ferns.

scald-head, *skăld* (Dan. *skolde*), the common name for porrigo and eczema : **scalds**, n. plu., burns.

scalenus, n., *skăl·ēn'·ŭs* (Gr. *skăl-ēnos*, oblique, unequal), a group of muscles on each side of the neck, which bend the head and neck, named **scalenus posticus**, *pŏst·ĭk'·ŭs* (L. *postĭcus*, that is behind) ; **s. medius**, *mēd'·ĭ·ŭs* (L. *mĕdĭus*, middle) ; and **s. anticus**, *ănt·ĭk'·ŭs* (L. *antĭcus*, that is before).

scales, n. plu., *skălz*, in *bot.*, rudimentary or metamorphosed leaves.

scalp, n., *skălp* (L. *scalpo*, I cut, I carve ; Dut. *schelp* or *schulp*, a shell), the skin and subcutaneous tissues of the top of the head on which the hair grows.

scalpel, n., *skălp'·ĕl* (L. *scalpellum*, a scalpel—from *scalpo*, I carve, I scrape), in *anat.*, a knife used in dissecting, and in surgical operations : **scalpelliform**, a., *skălp·ĕl'·lĭ·fŏrm*, in *bot.*, shaped like the blade of a scalpel.

scalpriform, a., *skălp'·rĭ·fŏrm* (L. *scalprum*, a knife or chisel ; *forma*, shape), applied to certain teeth which have cutting edges, as in the incisors of the Rodents : **scalprum**, n., *skălp'·rŭm*, the cutting edge of the incisor teeth.

scammony, n., *skăm'·mŏn·ĭ* (L. *scammōnia*), a gummy resinous exudation, used as a drastic purgative, and obtained from the root of the Convolvulus Scammonia.

scandent, a., *skănd'·ĕnt* (L. *scand-ens*, climbing, *scandentis*, of climbing), in *bot.*, climbing by means of supports, as a plant upon a wall or rock.

Scansores, n. plu., *skăn·sōr'·ēz* (L. *scansōrĭus*, of or for climbing—from *scansum*, to climb), the Order of the climbing birds : **scansorial**, a., *skăn·sōr'·ĭ·ăl*, climbing ; formed for climbing.

scape, n., *skāp* (L. *scapus*, Gr. *skapos*, a stem, a stalk), in *bot.*, a naked flower-stalk bearing one or more flowers arising from a short axis, as in the cowslip and hyacinth : **scapiform**, a., *scăp'·ĭ·fŏrm* (L. *forma*, shape), resembling a scape.

scaphognathite, n., *skăf·ŏg'·năth·ĭt* (Gr. *skăphē*, a boat ; *gnathos*, a jaw), the boat-shaped appendage of the second pair of maxillæ in the Lobster, whose function is to spoon out the water from the branchial chamber.

scaphoid, a., *skăf'·ŏyd* (Gr. *skăphē*, a boat or skiff ; *eidos*, resemblance), resembling a boat : n., in *anat.*, one of the bones of the carpus, and also one of the tarsus.

scapula, n., *skăp'·ūl·ă* (L. *scapula*, the shoulder - blade—from Gr. *skăphē*, a skiff, from its hollowness), the shoulder - blade ; the shoulder - blade of the pectoral arch of Vertebrates; the row of plates in the cup of Crinoids : **scapular**, a., *skăp'·ŭl·ăr*, of or pert. to the scapula : **scapulary**, n., *skăp'·ŭl·ăr·ĭ*, a broad bandage with two flaps passed over the shoulders.

scar, n., *skăr* (Gr. *eschără*, the scab on a wound ; Dan. *skaur*, a notch ; F. *escarre*, a scar), a mark left by a wound that has healed ; a cicatrix.

scarf-skin, n., *skărf-skĭn* (Bav. *schurffen*, to scratch off the outside of a thing; see 'scurf'), the outer thin integument of the skin ; the cuticle ; the epidermis.

scarification, n., *skăr'·ĭ·fĭk·ā'·shŭn* (L. *scărĭfĭco*, I scratch open), in *surg.*, the act of cutting the cuticle or external skin only, with a lancet, as to draw blood from the minute vessels only, or to permit the fluid to escape in the case of dropsy.

scarious, a., *skăr'·ĭ·ŭs* (F. *scarieux*, membranous ; Eng. *scar*), in *bot.*, having the consistence of a dry scale ; having a thin, dry, shrivelled appearance.

scarlet fever, *skăr'·lĕt fĕv'·ĕr* (F. *ecarlate*, It. *scarlatto*, scarlet), an acute febrile disease, characterised by a scarlet rash upon the skin, and a sore throat, often with swellings of various glands : **scarletina**, n., *skăr'·lĕt·ēn'·ă* (It. *scarlattina*, scarlet fever), another name for scarlet fever, and not a different type of the disease.

schindylesis, n., *skĭn'·dĭl·ēz'·ĭs* (Gr. *schindŭlēsis*, a fissure — from *schizō*, I cleave), that form of articulation in which a thin plate of bone is received into a cleft or fissure formed by the separation of two laminæ of another, as in the articulation of the rostrum of the sphenoid, and perpendicular plate of the ethmoid with the vomer.

schizocarp, n., *skīz'·ō·kărp* (Gr. *schizō*, I cleave ; *karpos*, fruit), in *bot.*, a dry seed - vessel, splitting into two or more one-seeded mericarps.

Schneiderian membrane, *snĭ·dēr'·ĭ·ăn* (after the discoverer), the mucous membrane lining the nose.

Schonleinii achorion, *skŏn·līn'·ĭ·ī ăk·ōr'·ĭ·ŏn* (Gr. *achōr*, scald-head, *achōrŏs*, of a scald-head ; after the discoverer), the parasitic plant which forms the crusts in ' Porrigo-favosa.'

sciatica, n., *sĭ·ăt'·ĭk·ă* (mid. L. *sciătĭca*, sciatica — from Gr. *ischias*, a pain in the hips, from *ischion*, the hip - joint), a neuralgic affection of the hip ; hip-gout : **sciatic**, a., *sĭ·ăt'·ĭk*, of or pert. to rheumatic or neuralgic affections of the hip.

Scilleæ, n. plu., *sĭl'·lĕ·ē* (L. *scilla*, Gr. *skilla*, the sea - onion, the squill), a tribe or Sub-order of plants, Ord. Liliaceæ : **Scilla**, *sĭl'·lă*, an extensive genus of interesting bulbous plants, some being used as purgatives, stimulants, emetics, and diaphoretics : **Scilla maritima**, *măr·ĭt'·ĭm·ă* (L. *marĭtĭmus*, belonging to the sea— from *mărĕ*, the sea), a species whose bulb supplies the officinal squill, grows on Mediterranean coast, used in medicine as a powder, tincture, vinegar, or syrup : **scillitina**, n., *sĭl'·ĭt·ĭn'·ă*, a bitter, crystalline principle, obtained from the S. maritima or squill.

scion, n., *sī'·ŏn* (F. *scion*, or *sion*, a young and tender plant), in *bot.*, a graft or branch ; a branch of one tree inserted into the stem of another ; a shoot of the first year.

scirrhus, n., *skĭr'·rŭs* (Gr. *skirrhos*, L. *scirrus*, a hard swelling), a hard tumour on any part of the body; a term generally restricted to a hard form of cancer : **scir-**

rhoma, n., *sklr·rōm'·ă*, a tumour of a marble-like appearance and consistence.

Scitamineæ, n. plu., *sĭt'·ăm·ĭn'·ĕ·ē* (L. *scĭtāmentum*, a delicacy), another name for Zingiberaceæ, the Ginger family, an Order of plants.

sclerenchyma, n., *sklĕr·ĕng'·kĭm·ă* (Gr. *sklēros*, hard; *enchuma*, what is poured in, tissue), the calcareous tissue of which a coral is composed; in *bot.*, tissue of thickened and hard cells or vessels.

sclerites, n. plu., *sklĕr'·ītz* (Gr. *sklēros*, hard), the calcareous spicules scattered in the soft tissues of certain Actinozoa.

sclerobasic, a., *sklĕr'·ō·bāz'·ĭk* (Gr. *sklēros*, hard; *basis*, a foundation, a pedestal), having foot-secretion; applied to the coral produced by the outer surface of the integument in certain Actinozoa, forming a solid axis invested by the soft parts of the animal—called the sclerobase, *sklĕr'·ō·bāz*.

scleroderma, n., *sklĕr'·ō·dĕrm'·ă* (Gr. *sklēros*, hard; *dĕrma*, skin), a diseased condition in which the skin hardens and indurates: **sclerodermic**, a., *sklĕr'·ō·dĕrm'·ĭk*, applied to the corallum deposited within the tissues of certain Actinozoa; having tissue-secretion: **sclerodermite**, n., *sklĕr'·ō·dĕrm'·īt*, the hard skeleton in the Crustacea; the corallum deposited with the tissues of certain Actinozoa.

sclerogen, n., *sklĕr'·ō·jĕn* (Gr. *sklēros*, hard; *gennāō*, I produce), in *bot.*, the thickening or woody matter deposited in the cells of plants.

scleroma, n., *sklĕr·ōm'·ă* (Gr. *sklerōma*, an induration—from *sklēros*, hard), in *med.*, hardness of texture; the hardened part of a body.

sclerosis, n., *sklĕr·ōz'·ĭs* (Gr. *sklēros*, hard; *sklērŏtēs*, hardness), the hardening of a part by an increase of its connective tissue

resulting from inflammatory action: **sclerotal**, a., *sklĕr·ŏt'·ăl*, the eye-capsule bone of a fish: **sclerotic**, a., *sklĕr·ŏt'·ĭk*, hard or firm — applied to the external membrane of the eye: n., the outer dense coat of the eye, forming the white of the eyeball; in *med.*, a substance that hardens parts—also **sclerotica**, n., *sklĕr·ŏt'·ĭk·ă*, in same sense: **sclerotome**, *sklĕr'·ō·tōm* (Gr. *tomē*, a cutting), a section of the skeleton of the body: **sclerotomy**, n., *sklĕr·ŏt'·ŏm·ĭ*, an incision of the sclerotic.

scobiform, a., *skŏb'·ĭ·fŏrm* (L. *scobs* or *scŏbis*, sawdust; *forma*, shape), in *bot.*, in the form of filings; like fine sawdust.

scobina, n., *skŏb·ĭn'·ă* (L. *scŏbĭna*, a rasp or file), in *bot.*, the immediate support to the spikelets of grasses: **scobinate**, a., *skŏb·ĭn'·āt*, having the surface rough like a file.

Scolecida, n. plu., *skŏl'·ĕ·sĭd'·ă* (Gr. *skōlĕx*, a worm, *skōlĕkos*, of a worm), a division of the Annuloida: **scolecite**, n., *skŏl'·ĕs·īt*, in *bot.*, a vermiform body, consisting of a row of short cells branching from the mycelium in Discomycetes: **scolex**, n., *skŏl'·ĕks*, the embryonic stage of a tapeworm; formerly called a Cystic worm; the non-sexual Cysticercus.

Scolymus, n., *skŏl'·ĭm·ŭs* (Gr. *skolos*, a pine or thorn), a genus of plants, Ord. Compositæ, Sub-ord. Cynarocephalæ: **Scolymus Hispanicus**, *hĭs·păn'·ĭk·ŭs* (of or from *Hispanica* or Spain), the Spanish oyster plant, whose tubers are used like potatoes.

scorbutus, n., *skŏrb·ūt'·ŭs* (mid L. *scōrbūtus*, the scurvy), a disease characterised by extreme debility, swollen gums, and purple-like spots on the skin, induced by privation and mal-nutrition, often from the want of vegetables; scurvy: **scorbutic**, a., *skŏrb·ūt'·ĭk*, affected with scurvy ; resembling scurvy.

Scorodosma, n., *skŏr'·ō·dŏs'·mă* (Gr. *skŏrŏdon*, garlic), a genus of plants, Ord. Umbelliferæ : **Scorodosma fœtidum**, *fĕt'·ĭd·ŭm* (L. *fœt'ĭdus*, fetid, stinking), a species found on the Sea of Aral, which yields a substance similar to asafœtida.

scorpioid, a., *skŏrp'·ĭ·ŏyd*, also **scorpioidal**, a., *-ŏyd'·ăl* (Gr. *skorpion*, a scorpion ; *eidos*, resemblance), in *bot.*, rolled in a circinate manner, or resembling the tail of a scorpion ; having a peculiar twisted cymose inflorescence, as in Boraginaceæ : **scorpioid cyme**, flowers arranged alternately, or in a double row, along one side of a false axis, the bracts forming a double row on the other side.

Scorzonera, n., *skŏr'·zŏn·ēr'·ă* (Sp. *escorzonēra*, viper grass ; Prov. Sp. *scorza*, a viper ; *nera*, black), a genus of plants, Ord. Compositæ, Sub-ord. Chichoraceæ, said to be good against the bites of vipers : **Scorzonera Hispanica**, *hĭs·păn'·ĭk·ă* (of or from *Hispānĭa* or Spain), the viper's grass, cultivated for its root, of the shape of a carrot, which has medicinal qualities in indigestion.

scrobiculate, a., *skrŏb·ĭk'·ūl·āt* (L. *scrobĭcŭlus*, a little furrow—from *scrŏbis*, a ditch), in *bot.*, marked with little pits or depressions : **scrobiculus cordis**, *skrŏb·ĭk'·ūl·ŭs kŏrd'·ĭs* (L. *cor*, the heart, *cordis*, of the heart), the depression at the upper part of the belly immediately below the ensiform cartilage ; the pit of the stomach.

scrofula, n., *skrŏf'·ūl·ă* (L. *scrōf-ŭla*, scrofula — from *scrōfa*, a breeding sow, from the belief that swine were subject to a similar complaint), a constitutional disease, exhibiting itself by hard indolent tumours of the glands, particularly about the neck, and various other symptoms, and by a liability to many diseases: **scrofula-derma**, *dĕrm'·ă* (Gr. *derma*, skin), cutaneous scrofula.

Scrophulariaceæ, n. plu., *skrŏf'·ūl·ār'·ĭ·ā'·sĕ·ē* (L. *scrophula*, scrofula), the Figwort family, an Order of plants, so named from their supposed use in the cure of scrofula, and having many beautiful and useful species : **Scrophularia**, n., *skrŏf'·ūl·ār'·ĭ·ă*, a genus of plants : **Scrophularia nodosa**, *nŏd·ōz'·ă* (L. *nŏdōsus*, full of knots — from *nōdus*, a knot), the knotted figwort, whose leaves have irritant qualities, and have been used as emetic and cathartic remedies ; also to skin diseases and tumours as an ointment or a fomentation.

scrotum, n., *skrōt'·ŭm* (L. *scrotum*, the scrotum), the sac or bag which contains the testicles : **scrotal**, a., *skrōt'·ăl*, of or pert. to the scrotum.

scurf, n., *skĕrf* (Ger. *schorf*, Dut. *schorfte*, scurf, scab), branny scales on the scalp; the epithelial scales of the skin as shed: **scurfy**, a., *skĕrf'·ĭ*, having scurf, or covered with it.

scurvy, n., *skĕrv'·ĭ* (mid L. *scōrb-ūtus*, F. *scorbut*, scurvy), a disease characterised by livid spots of various sizes on the skin, and by a general debility, caused by confinement and improper food, chiefly affecting sailors on long voyages, formerly very fatal, but now prevented or cured by the free use of lime juice and similar substances.

scuta, n. plu., *skū'·ă* (L. *scutum*, a shield), any shield-like plates, especially those developed in the integument of many reptiles : **scutate**, a., *skūt'·āt*, protected by large scales ; buckler-shaped : **scute**, n., *skūt*, a scale as of a fish or reptile.

scutellum, n., *skūt·ĕl'·lŭm* (L.

scutellum, a small shield—from *scutum*, a shield), in *bot.*, the smaller cotyledon on the outside of the embryo of wheat, placed lower down than the other more perfect cotyledon; a round flattened lichen-shield, with a rim derived from the thallus : **scutelliform**, a., *skŭt·ĕl'·lĭ·fŏrm* (L. *forma*, shape); also **scutellate**, a., *skŭt·ĕl'·lāt*, shaped like a little shield.

scutiform, a., *skŭt'·ĭ·fŏrm* (L. *scutum*, a shield; *forma*, shape), in *bot.*, applied to the peculiar leaf in Rhizocarpeæ; having the form of a shield.

scybala, n., *sĭb'·ăl·ă* (Gr. *skubalon*, dung, ordure), the fæces or contents of the bowels, when passed in hard small masses, like marbles or the excretions of sheep, denoting an unhealthy and costive habit.

scyphus, n., *sīf'·ŭs*, **scyphi**, n. plu., *sīf'·ī* (L. *scyphus*, Gr. *skuphos*, a cup or goblet), in *bot.*, the cup of a Narcissus; a funnel-shaped corolla; the funnel-shaped expansion of the podetia in some Lichens : **scyphiferous**, a., *sīf·ĭf'·ĕr·ŭs* (L. *fero*, I bear), bearing scyphi, as some Lichens.

Scytosiphon, n., *sīt'·ō·sīf'·ŏn* (Gr. *skutos*, skin, leather; *siphōn*, a tube), a genus of plants, Ord. Algæ or Hydrophyta, so named because their fronds are coriaceous and tubular : **Scytosiphon filum**, *fĭl'·ŭm* (L. *filum*, a string, a cord), a species attaining in the British seas a length of 40 or 50 feet.

sebaceous, a., *sĕb·ā'·shŭs* (L. *sebum*, tallow or suet), containing or secreting fatty matter: **sebaceous glands**, glands at the roots of hairs, which secrete an oily matter for their lubrication.

secreting, a., *sĕ·krēt'·ĭng* (L. *secretus*, severed, separated), separating or producing from the blood, or its constituents, substances different from the blood itself;

in plants, separating substances from the sap : **secretion**, n., *sĕ·krēsh'·ŭn*, one of the substances separated from the blood, etc., such as saliva, bile, and urine ; a separated portion of a nutritive fluid different from it in qualities: **secretory**, a., *sĕ·krēt'·ŏr·ĭ*, performing the office of secretion.

sectile, a., *sĕk'·tĭl* (L. *sectilis*, that may be cut, cleft—from *seco*, I cut), that may be cut or sliced, as with a knife ; in *bot.*, cut into small pieces : **section**, n., *sĕk'·shŭn*, a part separated from the rest ; a division.

secular, a., *sĕk'·ūl·ăr* (L. *secularis*, of or belonging to a generation—from *seculum*, a generation), in *geol.*, applied to great natural processes, whose results become appreciable only after the lapse of ages.

secund, a., *sĕk'·ŭnd* (L. *secundus*, second, next), in *bot.*, all turned to one side, as flowers or leaves on a stalk arranged on one side only : **secundine**, n., *sĕk'·ŭnd·ĭn*, in *bot.*, the second coat of the ovule, lying within the primine ; in *anat.*, the fœtal membranes collectively.

sedative, a., *sĕd'·ăt·ĭv* (L. *sedatus*, settled, composed), diminishing or allaying irritability or pain : n., a medicine which diminishes or allays irritability or pain.

Sedum, n., *sĕd'·ŭm* (L. *sedes*, a seat), a genus of plants, Ord. Crassulaceæ, found growing upon stones, rocks, walls, and roofs of houses, admirable for ornamenting rock-work : **Sedum acre**, *āk'·rĕ* (L. *acris* or *acre*, biting, sharp), the biting Stone-crop, having acrid properties.

segment, n., *sĕg'·mĕnt* (L. *segmentum*, a piece cut off—from *seco*, I cut), in *bot.*, the division of a frond : **segmentation**, n., *sĕg'·mĕnt·ā'·shŭn*, the division or splitting into segments or portions : **segmenting**, n., *sĕg·mĕnt'* *ing*, splitting into segments or divisions.

segregate, a., *sĕg'-rĕg-āt* (L. *sēg-rĕgātum*, to set apart, to separate—from *se*, aside ; *gregāre*, to collect into a flock), in *bot.*, separated from each other ; having no organic connection though frequently associating together.

Selachia, n. plu., *sĕl-āk'-ĭ-ă*, also **Selachii**, n. plu., *sĕl-āk'-ĭ-ĭ* (Gr. *selāchos*, a cartilaginous fish), the Sub-order of Elasmobranchii, comprising the Sharks and Dog-fishes.

Selaginaceæ, n. plu., *sĕl-ădj'-ĭn-ā'-sĕ-ē* (L. *selāgo*, a plant resembling the savin tree, gen. *selāgĭnĭs*), a small Order or group of herbaceous or shrubby plants, nearly related to the verbenas—also called Globulariaceæ : **Selago**, n., *sĕl-āg'-ō*, a genus of pretty plants : **Selaginella**, n., *sĕl-ădj'-ĭn-ĕl'-lă*, a genus of plants, Ord. Lycopodiaceæ.

sella Turcica, *sĕl'-lă tėr'-sĭk-ă* (L. *sella*, a seat; *Turcicus*, of or from *Turkey*), the part of the sphenoid bone supposed to resemble a Turkish saddle ; also called **sella equina**, *ĕk-wīn'-ă* (L. *equinus*, pert. to a horse); and **s. sphenoides**, *sfĕn-ŏyd'-ēz* (Gr. *sphēn*, a wedge; *eidos*, resemblance), a deep depression of the sphenoid bone which lodges the pituitary body.

Semecarpus, n., *sĕm'-ĕ-kârp'-ŭs* (Gr. *semeion*, a mark or sign ; *karpos*, fruit), a genus of plants, Ord. Anacardiaceæ, the black acrid juice of whose nuts is used by the natives in marking cotton cloths: **Semecarpus anacardium**, *ăn'-ă-kârd'-ĭ-ŭm* (Gr. *ana*, like ; *kardia*, the heart), the marking nut tree which supplies the Sylhet varnish.

semen, n., *sēm'-ĕn* (L. *semen*, seed—from *sero*, I sow), the seed of animals; the fluid secreted in the testicles ; **seminal**, a., *sĕm'-ĭn-ăl*, radical ; in *bot.*, applied to the cotyledons or seed-leaves, or to portions of the generative apparatus.

semi-amplexicaul, a., *sĕm'-ĭ-ăm-plĕks'-ĭk-āwl* (L. *semi*, half ; Eng. *amplexicaul*), in *bot.*, partially clasping the stem.

semi-anatropal, a., *sĕm'-ĭ-ăn-ăt'-rŏp-ăl* (L. *semi*, half; Eng. *anatropal*), in *bot.*, half-anatropal—applied to ovules.

semi-flosculous, a., *sĕm'-ĭ-flŏsk'-ŭl-ŭs* (L. *semi*, half ; Eng. *flosculous*), having the florets ligulate, as in the Dandelion.

semi-lunar, a., *sĕm'-ĭ-lŏn'-ăr* (L. *semi*, half ; *luna*, the moon), having the form of a half moon : **semi-lunar cartilage**, two plates of cartilage situated around the margin of the head of the tibia.

semi-membranosus, a., *sĕm'-ĭ-mĕm'-brān-ōz'-ŭs* (L. *semi*, half ; *membrana*, skin or membrane), half-membranous ; one of the muscles of the thigh which bend the leg—so named from the flat membrane-like tendon at its upper part.

seminiferous, a., *sĕm'-ĭn-ĭf'-ĕr-ŭs* (L. *semen*, seed ; *fero*, I bear), secreting and conveying the seminal fluid ; in *bot.*, bearing seed.

semi-nude, a., *sĕm'-ĭ-nŭd'* (L. *semi*, half ; Eng. *nude*), in *bot.*, having the ovules or seeds exposed, as in Mignonette.

semi-penniform, a., *sĕm'-ĭ-pĕn'-nĭ-fŏrm* (L. *semi*, half ; *penna*, a feather ; *forma*, shape), in *anat.*, applied to certain muscles bearing some resemblance to the plume of a feather.

semi-spinalis, a., *sĕm'-ĭ-spīn-āl'-ĭs* (L. *semi*, half ; *spīna*, a spine), in *anat.*, applied to the muscles which connect the transverse and articular processes to the spinous processes of the vertebræ : **semi-spinalis dorsi**, *dŏrs'-ī* (L. *dorsum*, the back, *dorsi*, of the back), half-spinal muscle of the back :

consists of thin, narrow fasciculi, interposed between tendons of considerable length: **s. colli,** *kŏl'.lĭ* (L. *collum*, the neck, *colli*, of the neck), the half-spinal muscle of the neck, thicker than the preceding.

semi-tendinosus, a., *sĕm'.ĭ-tĕnd'.ĭn. ōz'.ŭs* (L. *semi*, half; *tendo*, I stretch), half-tendinous; one of the dorsal muscles of the thigh, which arises from the tuber ischii, and is inserted in the tibia.

Sempervivum, n., *sĕm'.pĕr-viv'.ŭm* (L. *semper*, always; *vivo*, I live), a genus of plants, Ord. Crassulaceæ—so named from the well-known tenacity of life of the house-leek, one of the species: **Sempervivum tectorum,** *tĕk·tōr'.ŭm* (L. *tectum*, a house, *tectorum*, of houses), the common house-leek, having thick fleshy leaves arranged in the form of a double rose, commonly met with on the tops of out-houses and cottages, said to possess cooling properties: **S. glutinosum,** *glŏt'.ĭn·ōz'.ŭm* (L. *glutinōsus*, gluey—from *gluten*, glue), a species whose fresh leaves are employed by the fishermen of Madeira to rub their nets with, after being steeped in an alkaline liquor, thus rendering them as durable as if tanned: **S. cæspitosus,** *sĕs'.pĭt·ōz'.ŭs* (of or pert. to a turf—from *cæspes*, a turf, a sod cut out), a species which exhibits a wonderful vitality, growing after being kept dry for eighteen months.

Senecio, n., *sĕn·ē'.shĭ·ō* (L. *senex*, an old man), a genus of plants, Ord. Compositæ, remarkable as being the most extensive in point of species in the vegetable kingdom, so named from their naked receptacle resembling a bald head: **Senecio vulgaris,** *vŭlg·ār'.ĭs* (L. *vulgāris*, common, vulgar), the plant groundsel: **S. Jacobæa,** *jăk'.ōb-ē'.ă* (from L. *Jacobus*, James), the ragwort or ragweed:

S. cineraria, *sĭn'.ĕr·ār'.ĭ·ă* (L. *cinĕrēs*, ashes, from the soft white down on its leaves), extensively used in planting flower-beds for the sake of contrast—also called **S. maritima,** *măr·ĭt'. ĭm·ă* (L. *marĭtĭmus*, of or belonging to the sea—from *mărĕ*, the sea).

Senega, or snake root; see ' Polygalaceæ.'

Senna, see under ' Cassia.'

sensorium, n., *sĕns·ōr'.ĭ·ŭm* (L. *sensus*, perception—from *sentĭo*, I discern by the senses), the central seat of sensation, or of consciousness, supposed to be in the brain; the organ which receives the impressions made on the senses: **sensorial, a.,** *sĕns· ōr'.ĭ·ăl*, of or pert. to the sensorium: **sensory, a.,** *sĕns'.ōr·ĭ*, having direct connection with the nerves of sensation: **n.,** in *anat.*, those parts of the neural axis with which the sensory nerves are connected.

sepal, n., *sĕp'.ăl* (a term invented by the change of the *pet* in Gr. *petalon* into *sep*, thus making *sepalon;* L. *sēpes*, a hedge or fence), in *bot.*, one of the leaf-like divisions of the calyx or cup which encloses the corolla or blossom of a flower: **sepaloid, a.,** *sĕp'.ăl·ōyd* (Gr. *eidos*, resemblance), having the appearance of a sepal: **sepalody, n.,** *sĕp·ăl'.ōd·ĭ* (Gr. *hŏdŏs*, a way), the conversion of petals, or parts of the flower, into sepals.

sepiostare, n., *sĕp'.ĭ·ŏs·tār'*, also **sepiostarium, n.,** *sĕp'.ĭ·ŏs·tār'.ĭ·ŭm* (Gr. *sēpĭa*, the cuttle-fish; *ŏstĕon*, a bone), the internal shell of the Sepia, usually called the cuttle-bone.

septa, and septate, see ' septum.'

septemfid, a., *sĕp'.tĕm·fĭd* (L. *septem*, seven; *fĭdi*, I cleft), in *bot.*, having seven divisions in a leaf, extending about half-way through it: **septempartite, a.,** *sĕp'.tĕm·*

părt'.ĭt (L. *partitus*, divided), having seven divisions in a leaf, with radiating venation, which may extend to near the base.

septenate, a., sĕp·tĕn'.āt (L. *septēni*, seven each), in *bot.*, having parts in sevens; applied to a compound leaf with seven leaflets coming off from one point.

septic, a., sĕp'.tĭk (Gr. *sēptĭkos*, that causes putrefaction — from *sēpō*, I make putrid or rotten), having the power to promote putrefaction: septicity, n., sĕp·tĭs'.ĭ·tĭ, the tendency to promote putrefaction: septicæmia, n., sĕp'.tĭ·sēm'.ĭ·ă (Gr. *haima*, blood), an acute disease, resembling pyæmia in its general characters, supposed to be caused by the absorption into the blood of putrid matter from the surface of a wound or ulcer; also called ichorrhæmia, ĭk'.ŏr·rēm'.ĭ·ă (Gr. *ichōr*, corrupted matter; *haima*, blood), and septic pyæmia.

septicidal, a., sĕp'.tĭ·sĭd'.ăl (L. *septum*, a partition; *cœdo*, I cut or divide), in *bot.*, applied to seed vessels which open by dividing through the septa or partitions of the ovary: septifragal, a., sĕp·tĭf'.răg·ăl (L. *frango*, I break), in *bot.*, applied to a dehiscence which takes place along the lines of suture, the valves at the same time separating from the dissepiments, which are not subdivided.

septum, n., sĕp'.tŭm, septa, n. plu., sĕp'.tă (L. *septum*, a partition), in *bot.*, any partition separating a body, as a fruit, into two or more cells in the direction of its length; separating partitions across, or in the direction of its breadth, are called *phragmata;* in *anat.*, the membrane or plate separating from each other two adjacent cavities or organs; one of the partitions or walls of a chambered shell: septate, a., sĕp'.tāt, separated or divided by partitions: septulum, n., sĕp'.tŭl·ŭm (dim. of *septum*), a division between small spaces or cavities: septulate, a., sĕp'.tŭl·āt, in *bot.*, having spurious transverse dissepiments: septula renum, rĕn'.ŭm (L. *septula*, partitions; *rēnes*, the kidneys, rēnum, of the kidneys), the prolongations sent inwards of the cortical substance of the kidneys: septum lucidum, lŏs'.ĭd·ŭm (L. *lucĭdus*, full of light, clear), one of the partitions which separate the lateral ventricles of the brain from each other: septum nasi, nāz'.ī, also septum narium, nār'.ĭ·ŭm (L. *nasus*, the nose, *nasi*, of the nose; *nāris*, the nostril, nārĭum, of the nostrils), the cartilaginous partition separating the nostrils: s. pectiniforme, pĕk'.tĭn·ĭ·fŏrm'.ĕ (L. *pectin*, a comb; *forma*, shape), a partition which divides incompletely the cavity of the 'corpus cavernosum' into two lateral portions: s. posticum, pŏst·ĭk'.ŭm (L. *postĭcus*, posterior), a partition separating the sub-arachnoid space on the dorsal surface of the cord: s. scroti, skrŏt'.ī (L. *scrōtum*, the scrotum, the cod), the partition which separates the two testes of the scrotum: s. transversum, trăns·vĕrs'.ŭm (L. *transversus*, transverse), the diaphragm, a membrane which separates the thorax from the abdomen; the partition separating the cerebrum from the cerebellum; a certain incomplete partition of the semicircular canals of the ear.

sequela, n., sĕk·wēl'.ă (L. *sequēla*, a result or consequence), a diseased state of the body following on an attack of some other disease.

sequestrum, n., sĕk·wĕst'.rŭm (L. *sequestrātum*, to remove, to separate from anything), a dead portion of bone which separates from the sound part.

sericeous, a., sĕr·ĭsh'.ŭs (L. *sericus*, silken—from *Sēres*, a people of Eastern Asia, the Chinese), in *bot.*, covered with fine, close-pressed hairs ; silky.

serolin, n., *sĕr'·ō·lĭn* (L. *sĕrum*, whey ; *ŏlĕum*, oil), a peculiar fatty matter found in the blood.

serous, a., *sēr'·ŭs* (L. *sĕrum*, whey), watery ; like whey : **serosity**, n., *sĕr·ŏs'·ĭ·tĭ*, the watery part of blood when coagulated : **serous membrane**, in *anat.*, a closed membranous bag having its internal surface moistened with serum, and lining some cavity of the body which has no outlet : **serum**, n., *sēr'·ŭm*, the thin watery substance like whey which separates from the blood when coagulation takes place.

serpentiform, a., *sĕrp·ĕnt'·ĭ·fŏrm* (L. *serpens*, a serpent, *serpentis*, of a serpent ; *forma*, shape), resembling a serpent in shape : **serpentary**, n., *sĕrp'·ĕnt·ăr·ĭ*, the Virginia snake-root, the 'Aristolochia serpentaria,' an infusion and a tincture of whose roots are used as stimulants.

serrate, a., *sĕr'·rāt*, or **serrated**, a., *sĕr'·rāt·ĕd* (L. *serratus*, saw-shaped—from *serra*, a saw), in *bot.*, notched on the edge like a saw, as a leaf : **biserrate**, a., *bĭ·sĕr'·rāt*, having alternately large and small teeth on the edge : **serratus magnus**, *sĕr·āt'·ŭs măg'·nŭs* (L. *magnus*, great), in *anat.*, the great saw-shaped muscle of the lateral thoracic region, arising by fleshy serrations from the upper ribs, and inserted into the whole length of the scapula : **serration**, n., *sĕr·rā'·shŭn*, a formation resembling a saw : **serrulate**, a., *sĕr'·ŭl·āt* (L. *serrula*, a little saw), having very fine notches like a saw : **serrature**, n., *sĕr'·răt·ūr*, a saw-like notching on the edge of anything.

Sertularida, n. plu., *sĕrt'·ŭl·ăr'·ĭd·ă* (dim. of L. *sertum*, a wreath of flowers), an Order of the Hydrozoa : **Sertularia**, n. plu., *sĕrt'·ŭl·ăr'·ĭ·ă*, a genus of compound

tubular Polypes, in which the cells are arranged on two sides of the stem, either opposite or alternate.

serum, see under 'serous.'

sesamoid, a., *sĕs'·ăm·ōyd* (Gr. *sēsamon*, the grain sesame; *eidos*, appearance), in *anat.*, applied to one of the small bones formed at the articulations of the great toes, and sometimes at the joints of the thumbs.

Sesamum, n., *sĕs'·ăm·ŭm* (Gr. *sēsamon*, L. *sēsămum*, the sesame, an oily plant), a genus of plants, Ord. Bignoniaceæ : **Sesamum orientale**, *ŏr'·ĭ·ĕnt·āl'·ĕ* (L. *oriēntālis*, eastern), a species producing Teel seeds, which yield a bland oil, used in adulterating oil of almonds.

sessile, a., *sĕs'·sĭl* (L. *sessilis*, of or pert. to sitting—from *sĕdĕo*, I sit), sitting directly upon the body to which it belongs without a support or foot-stalk, as a sessile leaf ; sitting close.

seta, n., *sĕt'·ă*, setæ, n. plu., *sĕt'·ē* (L. *seta*, a thick, stiff hair), in *bot.*, a bristle or sharp hair ; the bristle-like stalk that supports the theca, capsule, or sporangium of Mosses ; the awn or beard of grasses which proceeds from the extreme of a husk or glume ; the glandular points of the rose, etc. ; in *zool.*, bristles or long stiff hairs, as on caterpillars, or the Crustaceans : **setaceous**, a., *sĕt·ā'·shŭs*, resembling a bristle; bristle-shaped : **setiferous**, a., *sĕt·ĭf'·ĕr·ŭs* (L. *fero*, I bear), also **setigerous**, a., *sĕt·ĭdj'·ĕr·ŭs* (L. *gero*, I bear), producing bristles ; supporting bristles : **setiform**, a., *sĕt'·ĭ·fŏrm* (L. *forma*, shape), having the shape of a bristle : **setose**, a., *sĕt·ōz'*, set or covered with bristles ; bristly.

Setaria, n. plu., *sĕt·ăr'·ĭ·ă* (L. *seta*, a bristle), a genus of plants, Ord. Gramineæ, whose involucrum is bristly: **Setaria Germanica**, *gĕrm·ăn'·ĭk·ă* (L. *Germanĭcus*, of or from *Germany*), a species which yields German millet.

seton, n., *sĕt'n* (It. *setone*, a seton; L. *seta*, a bristle), in *surg.*, an artificial discharge of matter occasioned by the introduction of some foreign body, such as horse hairs, fine thread, or a pea, under the skin.

setuliform, a., *sĕt·ūl'·ĭ·fŏrm* (L. *sētŭla*, a little bristle—from *seta*, a bristle), in *bot.*, thread - like: **setulose,** a., *sĕt'·ūl·ōz'*, resembling a little bristle.

shaking palsy, 'paralysis agitans,' which see.

sheath, n., *shēth* (Ger. *scheide*, Icel. *skeidir*, a sheath), in *bot.*, a petiole when it embraces the branch from which it springs, as in grasses: **sheath - winged,** having cases for covering the wings, as in many insects.

shingles, n. plu., *shĭng'·glz* (L. *cingulum*, a girdle), the popular name for herpes-zoster, an eruptive disease, characterised by groups of vesicles on an inflamed base, these groups usually following the course of a nerve.

Shorea, n., *shōr·ē'·ă* (after *Sir I. Shore*), a genus of plants, Ord. Diptero - carpaceæ, consisting of large resinous trees which produce terminal panicles of sweet-smelling yellow flowers: **Shorea robusta,** *rōb·ŭst'·ă* (L. *robustus*, of oak-wood, hard), native of India, supplies the valuable timber called Sal, and yields the Dhoom or Dammar pitch, used for incense in India.

sialagogue, n., *sĭ·ăl'·ăg·ŏg* (F. *sialagogue* — from Gr. *siălon*, saliva; *agō*, I lead), a medicine which increases the flow of saliva.

sibilant, a., *sĭb'·ĭl·ănt* (L. *sibilans*, hissing), making a hissing or whistling sound: **sibilant rhonchus** (L. *rhonchus*, a snoring), low whistling sounds, produced in the smaller bronchial tubes

during inspiration or expiration when their calibre is diminished.

sigmoid, a., *sĭg'·mōyd* (the Gr. letter Σ or ς, called Sigma; *eidos*, resemblance), curved like the Greek letter Sigma; in *anat.*, applied to several structures of the body; in *bot.*, curved in two directions, like the letter S, or the Greek ς.

silica, n., *sĭl'·ĭk·ă* (L. *silex*, a pebble, *silĭcis*, of a pebble), the earth of flints; a substance constituting the characteristic ingredient of a great variety of minerals; an inorganic element of plants: **silicate,** a., *sĭl'·ĭk·āt*, a salt of silicic acid: **silicated,** a., *sĭl'·ĭk·āt·ĕd*, combined or impregnated with silica: **siliceous,** a., *sĭl·ĭsh'·ŭs*, partaking of the nature and qualities of silica; composed of flint.

silicle, n., *sĭl'·ĭ·kl*, also **silicula,** n., *sĭl·ĭk'·ūl·ă*, and **silicule,** n., *sĭl'·ĭk·ūl* (L. *silicŭla*, a little pod—from *silĭqua*, a pod or husk), a short pod with a double placenta and replum; a siliqua as broad as long: **siliculose,** a., *sĭl·ĭk'·ūl·ōz*, bearing silicles; bearing husks.

silique, n., *sĭl'·ĭk*, also **siliqua,** n., *sĭl'·ĭk·wă* (L. *silĭqua*, a pod or husk), in *bot.*, a pod-like fruit, consisting of two long cells, divided by a partition, having seeds attached to each side, as in the seed-pods of the cabbage, turnip, and wallflower: **siliquose,** a., *sĭl'·ĭk·wōz*, bearing siliques: **siliquiform,** a., *sĭl·ĭk'·wĭ·fŏrm* (L. *forma*, shape), shaped like a silique.

Simarubaceæ, n. plu., *sĭm·ăr'·ŭb·ā'·sĕ·ē* (*Simaruba*, the native name in Guiana), the Quassia and Simaruba family, an Order of plants, which are all intensely bitter: **Simaruba,** n., *sĭm·ăr'·ŭb·ă*, a genus of valuable plants from their medical properties: **Simaruba amara,** *ăm·ăr'·ă* (L. *amārus*, bitter), a species the bark of whose root is used as a bitter tonic and astringent, especially in advanced stages of diarrhœa and dysentery, found in Cayenne and W. Indies — also called S. officinalis, ŏf·yĭs'·ĭn·ăl'·ĭs (L. offĭc-ĭnālĭs, by authority).

Sinapis, n., sĭn·āp'·ĭs (L. sĭnāpĭs, mustard), a genus of plants, Ord. Cruciferæ: Sinapis nigra, nīg'·rā (L. nĭger, black; nĭgra, fem.), a species whose seeds furnish table mustard, and which contain a bland fixed oil, a peculiar bitter principle, and myronic acid: S. alba, ălb'·ā (L. ălbus, white), a species furnishing white mustard, and containing more fixed oil than black mustard, is cultivated as a salad: sinapin, n., sĭn·āp'·ĭn, or sinapisin, n., sĭn·āp'·ĭs·ĭn, a principle in 'S. alba' analogous to myronic acid, found in 'S. nigra': sinigrin, n., sĭn'·ĭg·rĭn, a crystallisable substance found in mustard: sinapism, n., sĭn'·āp·ĭzm, a poultice of which mustard is the basis.

sinciput, n., sĭn'·sĭ·pŭt (L. sinciput, the fore part of the head—from semi, half; caput, the head), the forepart of the head, the back part being called the occiput.

sinistral, a., sĭn'·ĭs·trăl (L. sĭnĭs-tra, fem., on the left hand; sĭnis-ter, masc.), left-handed; applied to the direction of the spiral in certain shells when they turn to the left: sinistrorse, a., sĭn'·ĭs·trŏrs', in bot., applied to a spiral directed towards the left.

sinuate, a., sĭn'·ū·āt, also sinuated, a., sĭn'·ū·āt'·ĕd (L. sĭnuātum, to swell out in curves—from sĭnus, a bent surface, a curve), in bot., cut so as to have a broken and wavy margin, as the margin of a leaf: sinuous, a., sĭn'·ū·ŭs, tortuous; having a wavy or flexuous margin, as a leaf.

sinus, n., sĭn'·ŭs, (L. sĭnus, a bent surface, a curve), in anat., a cavity in a bone wider at the bottom than at the entrance; in surg., an elongated cavity containing pus; a dilated vein or blood receptacle; in bot., a groove or cavity; the indentation or recess formed by the lobes of leaves: sinuses, n. plu., sĭn'·ūz·ĕs, hollows or cavities, as in the bones, or in the dura - mater: sinus pocularis, pŏk'·ŭl·ār'·ĭs (L. pōcŭlum, a cup or goblet), a cup-like cavity in the male urethra leading into the prostatic vesicle: s. rhomboidalis, rŏm'·bŏyd·āl'·ĭs (L. rhombŏĭdes, a rhomboid — from L. rhombus, a rhombus, and Gr. eidos, resemblance), a lozenge-shaped cavity at the hinder extremity of the medullary canal: s. terminalis, tĕrm'·ĭn·āl'·ĭs (L. terminālis, terminal—from term-ĭnus, a boundary), a venous canal encircling the vascular area in the embryo: s. urogenitalis, ūr'·ō·jĕn·ĭt·āl'·ĭs (Gr. ouron, urine; L. gĕnĭtālĭs, generative), a sinus situated in front of the termination of the intestine forming a separation, which produces a distinct passage for the genito-urinary organs, formerly opening into a cloaca: s. venosus, vĕn·ōz'·ŭs (L. vĕnōsus, full of veins—from vēna, a vein), the main portion of the auricles of the heart, as distinguished from the auricular appendages: osseous sinuses, cavities in bones containing air: venous sinuses, hollows in the membrane of the dura - mater of the brain, which contain blood serving the purpose of veins.

siphon, n., sĭf'·ŏn (Gr. siphōn, L. sīpho, a hollow reed or tube), a bent pipe or tube whose arms are of unequal length, chiefly employed to draw off liquids from casks, etc.; applied to the respiratory tubes in the Mollusca, and to other tubes of different functions: siphonium, n., sĭf·ōn'·ĭ·ŭm, a bony air-tube in some birds.

Siphonia, n., *sĭf·ōn'·ĭ·ă* (Gr. *siphōn*, a tube), a genus of plants, Ord. Euphorbiaceæ, so named from the use made of their exudation : Siphonia elastica, n., *ĕ·lăst'·ĭk·ă* (mid. L. *elastĭcus*, elastic), a species which contains much caoutchouc, and supplies the bottle india-rubber.

Siphonophora, n. plu., *sĭf'·ŏn·ŏf'·ŏr·ă* (Gr. *siphōn*, a tube ; *phorĕō*, I bear), a division of the Hydrozoa: Siphonostomata, n. plu., *sĭf·ŏn·ŏs·tŏm'·ăt·ă* (Gr. *stoma*, a mouth), a division of the Gasteropodous Molluscs, in which the aperture of the shell is not entire, but has a notch or tube for the emission of the respiratory siphon.

siphuncle, n., *sĭf·ŭng'·kl* (L. *siphunculus*, a little pipe—from *sipho*, a tube), any small tube or tubular passage ; the tube-like perforation which passes through the septa and chambers of such shells as the nautilus, the ammonite, etc.; the tube which connects together the various chambers of the shell of certain Cephalopoda : Siphunculoidea, n. plu., *sĭf·ŭng'·kŭl·ōyd'·ĕ·ă* (Gr. *eidos*, resemblance), a class of Anarthropoda.

Sirenia, n. plu., *sĭr·ēn'·ĭ·ă* (L. *siren*, Gr. *seiren*, a siren), an Order of Mammalia, comprising the Dugongs and Manatees.

sitiology, n., *sĭt'·ĭ·ŏl'·ō·jĭ* (Gr. *sĭtos* or *sĭtĭon*, bread ; *logos*, a discourse), the doctrine or consideration of aliments ; dietetics.

Sium, n., *sĭ'·ŭm* (Gr. *seiō*, I quiver, from its motion in the water), a genus of plants, Ord. Umbelliferæ, which thrive best in very moist soil : Sium sisarum, *sĭs·ăr·ŭm* (Gr. *sĭsărŏn*, L. *siser*, a plant with an esculent root, skirret), a species whose succulent roots were formerly esteemed in cookery, under the name of 'skirret.'

slashed, a., *slăsht* (an imitative word), in *bot.*, deeply gashed ; divided by very deep incisions.

slough, n., *slŭf* (AS. *slog* ; Icel. *slög*, anything cast off or thrown away), the dead structure of flesh that separates from a wound, or during mortification.

smegma, n., *smĕg'·mă* (L. *smegma*, Gr. *smēgma*, a detergent, soap), the white substance often seen upon the skin of new-born infants: smegma preputii, *prē·pŭsh'·i·ī* (L. *præpūtĭum*, the foreskin or prepuce, *præputĭi*, of the foreskin), the secretions of Tyson's glands which surround the base of the glans penis.

Smilaceæ, n. plu., *smĭl·ā'·sĕ·ē* (L. *smĭlax*, bindweed, *smĭlăcis*, of bindweed), the Sarsaparilla family, an Order of plants having mucilaginous and demulcent properties : Smilax, n., *smĭl'·ăks*, a genus of plants, the roots of various of the species constituting sarsaparilla or sarza, as the following—Smilax officinalis, *ŏf·fĭs'·ĭn·āl'·ĭs* (L. *officinālis*, by authority, officinal); S. medica, *mĕd'·ĭk·ă* (L. *mĕdĭcus*, medical); S. syphilitica, *sĭf'·ĭl·ĭt'·ĭk·ă* (new L. *syphĭlĭtĭcus*, of or pert. to syphilis); S. papyracea, *păp'·ĭr·ā'·sĕ·ă* (L. *papyrus*, the paper reed); and S. Brasiliensis, *brăz·ĭl'·ĭ·ĕns'·ĭs* (of or from *Brazil*), the roots of all of them are mucilaginous, bitterish, and slightly acrid ; sarsaparilla is used in decoction and infusion as a tonic and alterative, in cachexia, and syphilis : S. China, *tshĭn'·ă* (of or from *China*), a species which yields the china-root, used as a remedy in syphilis.

smut, n., *smŭt* (Ger. *schmutz*, dirt, mud), a powdery matter, having a peculiarly fœtid odour, which occupies the interior of diseased grains of wheat and other cereals, caused by a parasitic fungus called 'Uredo caries' or 'fœtida'—also called 'bunt,' 'pepper-brand,' or 'blight'; a sooty powder, having no odour, found in oats and barley, caused by the parasitic fungus 'Urego segetum'—also called 'dust-brand.'

snuffles, n. plu., *snŭf'·lz* (Dut. *snuffelen*, to breathe through the nose), obstruction of the nose through mucus.

soboles, n., *sŏb'·ŏl·ĕz* (L. *soboles*, a sprout, a shoot), in *bot.*, a creeping underground stem.

socia parotidis, *sō'·shĭ·ă păr·ŏt'·ĭd·ĭs* (L. *socia*, a companion ; *părōtis*, a tumour near the ears, *parŏtĭdis*, of a parotis), in *anat.*, a small detached portion of the parotid gland, which occasionally exists as a separate lobe, just beneath the zygomatic arch.

soda, n., *sōd'·ă* (Ger., Sp. *soda*), an alkali obtained from the ashes of certain sea-plants, or from common salt : liquor sodæ, *lĭk'·ŏr sōd'·ē* (L. *sodæ*, of soda), the liquor of soda, that is, a solution of caustic soda, made by heating carbonate of soda with slaked lime : carbonate of soda, the proper name of soda as above, used chiefly for cleanliness, and soap-making: bicarbonate of soda, is only slightly alkaline, and not caustic, used in the preparation of effervescing drinks, and in making 'medicinal soda-water': sulphate of soda, Glauber's salt, found in certain mineral waters, and in sea-water : sulphite of soda, important for its sulphurous acid : nitrate of soda, a very deliquescent salt, used in making the arseniate of soda or nitric acid, and as a manure : phosphate of soda, a tasteless purging salt, obtained by adding to a solution of bone earth in sulphuric acid, carbonate of soda to neutralisation : chlorinated soda, a combination of soda and chlorine, constituting a bleaching solution : citro-tartarate of soda, a substance which in the granulated form is com-

monly called 'citrate of magnesia': sodium, n., *sōd'·ĭ·ŭm*, the metallic base of soda, soft, of a silvery lustre, and lighter than water : chloride of sodium, common salt : soda-water, an effervescing beverage, containing a weak solution of bicarbonate of soda, and highly charged with carbonic acid gas.

Solanaceæ, n. plu., *sōl'·ăn·ā'·sĕ·ē* (L. *sōlānum*, the plant nightshade), the Nightshade family, an Order of plants, often possessing narcotic qualities, some species having these qualities so highly developed as to become poisonous, contains the potato and tobacco plants : Solaneæ, n. plu., *sōl·ān'·ĕ·ē*, a Sub-order of plants : Solanum, n., *sōl·ān'·ŭm*, an extensive genus of plants, many having a showy, ornamental appearance: Solanum dulcamara, *dŭlk'·ăm·ār'·ă* (L. *dulcis*, sweet ; *amārus*, bitter), Bitter-sweet or Woody Nightshade, has diaphoretic properties, a decoction of the twigs useful in certain cutaneous diseases, and the scarlet berries are not poisonous : S. nigrum, *nĭg'·rŭm* (L. *nĭgrum*, black), a species whose black berries have been used in tarts, but the plant is a virulent poison : S. tuberosum, *tŭb'·ĕr·ōz'·ŭm* (L. *tŭbĕrōsus*, having fleshy knobs—from *tūber*, a protuberance), the well-known Potato plant, producing nutritious, starchy tubers : S. melongena, *mĕl·ŏnj'·ĕn·ă* (Gr. *mēlon*, an apple; *gĕnos*, birth, production), yields the Aubergine, an edible fruit ; the mad apple : S. laciniatum, *lăs·ĭn'·ĭ·āt'·ŭm* (L. *lacĭnātus*, jagged, indented—from *lacĭnĭa*, a flap, a lappet), the Kangaroo apple, eaten in Tasmania : S. ovigerum, *ōv·ĭdj'·ĕr·ŭm* (L. *ŏvum*, an egg ; *gero*, I bear), produces the fruit Egg apple : S. vescum, *vĕsk'·ŭm* (L. *vescus*, small, feeble, fine), the Gunyang of Australia, used as a potato : **S. indigofera,** *ĭn'dĭg·ŏf'ĕr·ă* (*indigo*, and L. *fero*, I produce), cultivated in Brazil for the sake of its indigo-dye : **S. gnaphalioides,** *năf·ăl'ĭ·ŏyd'ēz* (Gr. *gnaphălĭŏn*, the plant cudweed ; *eidos*, resemblance : *gnaphălŏn*, soft down), the juice of the fruit used by Peruvian ladies to tint their cheeks: **S. saponaceum,** *săp'ŏn·ās'ĭ·ŭm* (L. *săponācĕus*, of or pert. to soap—from *săpō*, soap), a species whose fruits are used in Peru instead of soap to whiten linen: **S. marginatum,** *mărj'ĭn·āt'ŭm* (L. *margĭnātus*, furnish with a border—from *margo*, an edge, a border), employed in Abyssinia for tanning leather : **solania,** n., *sŏl·ān'ĭ·ă*, a white alkaloid substance, highly poisonous, obtained from S. dulcamara, greened potatoes, and other species of Solanum — also called **solanin,** n., *sŏl'ăn·ĭn*, and **solanina,** n., *sŏl'ăn·ĭn'ă*.

solar, a., *sōl'ăr* (L. *sōl*, the sun, *sōlis*, of the sun), in *anat.*, having branches or filaments like the rays of the sun: **solar plexus,** *plĕks'ŭs* (L. *plexus*, twisted), a great network of nerves and ganglia, situated behind the stomach, which supplies all the viscera in the abdominal cavity.

soleaform, a., *sŏl·ē'ă·fŏrm* (L. *solĕa*, a sandal ; *forma*, shape), in *bot.*, slipper-shaped.

Solenostemma, n., *sōl·ĕn'ō·stĕm'mă* (Gr. *sōlēn*, a tube ; *stĕmma*, a garland, a wreath), a genus of plants, Ord. Asclepiadaceæ : **Solenostemma argel,** *ăr'jĕl* (may be connected with Sp. *argel*, Algiers), a species whose leaves are used to adulterate Alexandrian senna.

soleus, n., *sŏl·ē'ŭs* or *sōl'ĕ·ŭs* (L. *sōlĕă*, a sandal, a sole-fish), in *anat.*, a muscle of the leg shaped like the sole-fish ; also called ' gastrocnemius internus.'

Solidungula, n. plu., *sŏl'ĭd·ŭng'ŭl·ă*, also **Solidungulates,** n. plu., **2 B**

-ŭl·ātz (L. *solĭdus*, solid; *ungŭla*, a hoof), the group of hoof quadrupeds,which comprises the horse, ass, and zebra, having each foot a single solid hoof only ; also called ' Solipedia.'

solution, n., *sŏl·ō'shŭn* (L. *solūtum*, to loose ; *solvo*, I loose, I melt), a liquid which contains one or more solid substances diffused throughout it ; in *bot.*, the separation of whorls which are usually adherent : **solution of continuity,** in *surg.*, the accidental separation of connected parts : **solvent,** n., *sŏlv'ĕnt*, a fluid in which a solid may be dissolved ; anything which can dissolve or render liquid another.

somatic, a., *sōm·ăt'ĭk* (Gr. *sōma*, a body, *sōmătos*, of a body), connected with the body: **somatocyst,** n., *sōm·ăt'ō·sĭst* (Gr. *kustis*, a cyst), a peculiar cavity in the cœnosarc of the Calycophoridæ : **somatomes,** n. plu., *sōm'ăt·ōmz* (Gr. *tŏmē*, a cutting), the vertebral segments of the body: **somatotomy,** n., *sōm'·ăt·ŏt'·ŏm·ĭ*, another name for anatomy : **somite,** n., *sōm'·ĭt*, a single segment in the body of an articulate animal.

sophisticate, v., *sōf·ĭst'ĭk·āt* (Gr. *sŏphĭstĭkos*, fallacious—from *sŏphŏs*, skilful, artful), to adulterate ; to debase by something spurious or foreign : **sophistication,** n., *sōf·ĭst'ĭk·ā'shŭn*, adulteration.

soporific, a., *sōp'·ŏr·ĭf'·ĭk* (L. *sopor*, a heavy sleep ; *facĭo*, I make), that has the quality of inducing sleep : n., a medicine which causes sleep.

soredia, n., *sōr·ēd'ĭ·ă* (Gr. *sōros*, a heap or pile), in *bot.*, powdery cells on the surface of the thallus of some Lichens: **sorediferous,** a., *sōr'·ĕd·ĭf'·ĕr·ŭs* (L. *fero*, I bear), bearing soredia.

Sorghum, n., *sŏrg'·ŭm* (from *Sorghi*, its Indian name), a genus of plants, Ord. Gramineæ : **Sorghum vulgare,** *vŭlg·ār'ĕ* (L.

vulgāris, common), Guinea Corn.

sori, n. plu., *sōr'ī* (Gr. *sōros*, a heap, a pile), in *bot.*, the patches of fructification on the back of the fronds of ferns : **sorus,** n. sing., *sōr'ŭs*, in *bot.*, a cluster of sporangia in ferns : **sorosis,** n., *sōr·ōz'īs*, a kind of fleshy fruit, resulting from the consolidation in one mass of many flowers, as in the pine-apple.

Soymida, n., *soȳm'īd·ă* (its native name), a genus of plants, Ord. Cedrelaceæ : **Soymida febrifuga,** *fĕb·rĭf'ūg·ă* (L. *fĕbris*, a fever ; *fŭgo*, I drive away), the Rohuna of Hindustan, a kind of mahogany whose bark is a useful tonic in intermittent fevers, and in typhus.

spadix, n., *spād'īks*, **spadices,** n. plu., *spăd·ĭs'ēz* (L. *spadix*, a palm branch broken off together with its fruits, a date or nut-brown colour ; *spādĭcis*, of a date or nut-brown colour, etc.), in *bot.*, a form of inflorescence in which the flowers are closely arranged around a thick fleshy axis, and the whole wrapped in a large leaf, called a *spathe*, as in the arum : **spadiceous,** a., *spăd·ĭsh'ŭs*, of a clear reddish-brown colour, resembling a spadix.

spanæmia, n., *spăn·ēm'ĭ·ă* (Gr. *spanos*, scarce ; *haima*, blood), a diseased condition of the blood, characterised by a deficiency in its red globules ; the opposite condition to *plethora* : **spanæmic,** a., *spăn·ĕm'ĭk*, having the property of impoverishing the blood ; having an impoverished or thin state of blood.

spasm, n., *spăzm* (Gr. *spasmos*, L. *spasmus*, a cramp, a spasm), the violent and uncontrollable action of a particular set of muscles : **spasms** are of two sorts, *tonic* and *clonic ;* in **tonic spasms** (see 'tonic') the muscles of a part contract violently, and remain rigid and immovable during a shorter or longer interval, independent of the will ; in **clonic spasms** (see 'clonic') there are regular alternations of sudden contractions and relaxations ; in common language, spasms are grips and violent internal pains, dependent on indigestion or constipation : **spasmodic,** a., *spăz·mŏd'ĭc*, of or pert. to spasms.

spathe, n., *spāth*, also **spatha,** n., *spāth'ă* (L. *spătha*, Gr. *spăthē*, a broad blade or flat piece of wood), in *bot.*, a large membranous bract, or kind of leaf, forming a sheath to cover a spadix ; a calyx-like sheath, found as a covering in numerous flowers : **spathed,** a., *spātht*, having a spathe or calyx like a sheath : **spathaceous,** a., *spāth·ā'shŭs*, having the appearance and membranous consistence of a spathe : **spathellæ,** n. plu., *spāth·ĕl'lē* (L. dim.), small spathes surrounding separate parts of the inflorescence : **spathose,** a., *spāth·ōz'*, resembling a spathe.

spathulate, a., *spăth'ŭl·āt* (L. *spăthŭla*, a broad piece, a spoon), in *bot.*, spoon-shaped ; applied to a leaf having a linear form, enlarging suddenly into a rounded extremity.

spawn, n., *spăwn* (Bav. *span*, Dut. *spenne*, milk drawn from the breast), the cellular axis of Fungi, on which ultimately the fructification is developed ; the mycelium of frogs, etc.

species, n., *spēsh'ēz* (L. *species*, a particular sort — from *specĭo*, I look at, I behold), a group of individuals alike or identical with each other — that is, the individuals having no permanent or marked difference—an accidental or minor difference in an individual being termed a *variety ;* an assemblage of individuals having characters in common, and coming from an original stock or protoplast, as in a field of wheat : **specific,** a., *spĕs·ĭf'ĭk* (L. *facĭo*,

esignates the
ites it ; in the
:ture of plants,
; second name,
h follows that
h double name
ie of the species
distinguished
ipecific centre,
express the
ion which each
igin, and from
.duals became
character, the
inguishing one
y other species
specific name,
ecific remedy,
y found usually
tive of a partic-

trŭm, spectra,
(L. spectrum,
n image), the
g seen after the
)sed ; the pris-
:d in a darkened
itting a ray of
nto it through
:trum may be
ray proceeding
; body, as from
iet : spectrum
)r art of ascer-
.cter and com-
us bodies, or of
lies when in a
on, by causing
rom the body
nalysed to pass
each substance
n having its
ic system of

pĕk'-ŭl-ŭm (L.
ir—from specĭō,
, an instrument
more perfectly
of the body, in
nteriors may be
ned.
pĭr'-ănth-ĭ (Gr.
r curl ; anthos,

a flower), in bot., the twisted
growth of the parts of a flower.

sperm, n., spėrm (Gr. sperma,
seed, spermătos, of seed—from
speïro, I sow), animal seed: **sperm
cell,** a cell which impregnates, as
opposed to a germ cell, which
has been impregnated : **spermoderm,** n., spėrm'-ŏ-dėrm (Gr.
derma, skin), in bot., the outer
covering of a seed : **spermaceti,**
n., spėrm'-ă-sēt'-ĭ (Gr. ketos, L.
cetus, any large fish, a whale), a
white, brittle, semi-transparent
substance obtained from the head
of the sperm whale, and from
sperm-oil : **spermatic,** a., spėrm·
ăt'-ĭk, pert. to or consisting of
seed or semen ; seminal : **spermatic cord,** a cord made up of the
vessels and nerves which pass to
and from the testis.

spermagones, n. plu., spėrm-ăg'-
ŏn-ēz (Gr. sperma, seed ; gonos,
offspring), reproductive bodies in
the form of very minute hollow
sacs, found on the thallus of
Lichens ; capsules or cysts in
Lichens, Fungi, etc., containing
spermatia.

spermarium, n., spėrm-ār'-ĭ-ŭm
(Gr. sperma, seed), the organ in
which spermatozoa are produced :
spermatia, n., spėrm-ā'-shĭ-ă, in
bot., motionless spermatozoids in
the conceptacles of Fungi, supposed to be possessed of fertilising
power.

spermatheca, n., spėrm'-ă-thēk'-ă
(Gr. sperma, seed ; thēkē, a receptacle), a receptacle or sac
in which ejected semen is stored
up, as among some insects.

spermatic, see under 'sperm.'

spermatophore, n., spėrm-ăt'-ō-fŏr,
spermatophores, n. plu., -fŏrz,
or **spermatophora,** n. plu., spėrm'-
ăt-ŏf'-ŏr-ă (Gr. sperma, seed ;
phorĕō, I bear), in anat., cases
of albuminous matter in which
the bundles of spermatozoa are
packed.

spermatozoon, n., spėrm'-ăt-ō-zō'-

ŏn, **spermatozoa**, n. plu., -zō'ă, (Gr. *sperma*, seed ; *zoön*, an animal), one of the filamentary bodies developed in the semen, consisting of an enlarged extremity called *body*, and a vibratile filamentary appendage called *tail*, which are essential to impregnation : **spermatozoids**, n. plu., *spĕrm·ăt'·ō·zōȳdz* (Gr. *eidos*, resemblance), in *anat.*, same sense as spermatozoa ; in *bot.*, moving filaments contained in the antheridia of Cryptogams, supposed to possess a fecundative power—also known as *phytozoa*, and *antherozoids*.

spermoderm, see under ' sperm.'

spermogone, n., *spĕr·mŏg'·ŏn·ē*, an inaccurate spelling for ' spermagone,' which see.

sphacelus, n., *sfăs'·ĕl·ŭs* (Gr. *sphakĕlŏs*, gangrene), that stage in mortification in which the part is dead and cold ; see ' mortification ' : **sphacelate**, v., *sfăs'·ĕl·āt*, to affect with gangrene ; to decay and become carious, as a bone : **sphacelation**, n., *sfăs'·ĕl·ā'·shŭn*, the process of becoming gangrenous.

sphæraphides, n. plu., *sfĕr·ăf'·ĭd·ēz* (Gr. *sphaira*, a globe ; *rhaphis*, a needle, *rhaphĭdŏs*, of a needle), in *bot.*, globular clusters of raphides, or globular aggregations of minute crystals, as found in phanerogamous plants.

sphærenchyma, n., *sfĕr·ĕng'·kĭm·ă* (Gr. *sphaira*, a globe ; *engchuma*, juice, tissue — from *engchĕŏ*, I pour in), in *bot.*, tissue composed of spherical cells.

Sphæria, n. plu., *sfĕr'·ĭ·ă* (Gr. *sphaira*, a globe), an extensive genus of very minute plants, Ord. Fungi, found at all seasons on many decaying bodies, such as leaves, fir cones, trunks of trees, etc.: **Sphæria Sinensis**, *sĭn·ĕns'·ĭs* (*Sĭnĕnsis*—from *Sina*, an old name of *China*), a species found on a caterpillar, which constitutes a

celebrated Ch
ertsii, *rōb·ĕrt*
a botanist),
on larvæ in
Taylori, *tāl·*
a botanist), a
Australian ca
era, *sŏb'·ŏl·ĭf*
sprout, a sh
S. **entomorh**
(Gr. *entŏmă*,
root) ; S. **m**
(L. *mĭlĭtāris*
mĭles, a soldie
grow on anim
Sphærococcus,
(Gr. *sphaira*,
seed or fruit
plants, Ord.
cus crispus, *l*
curled, wrinl
Irish Moss, w
tritious articl
oides, *lĭk'·ĕn·*
lichen ; Gr.
Ceylon Moss
article of die
kărt'·ĭl·ădj·ĭn
ĕus, cartilagi
āgo, cartilage
used as a sub
swallows' nes
Sphæroplea,
(Gr. *sphaira*
swim), a ger
Algæ, in one
Sphæroplea a
(L. *annŭlus*,
produce stella
spring first d
into four or ei
zoospores ; t
about, then
give rise to y
' baculiform.'
Sphagneæ, n.
and Gr. *sph*
fragrant mos
bog mosses, (
aceæ, aquatic
imbricated l
n., *sfăg'·nŭm*,
whose species

at all seasons, and have nerveless leaves of a singularly whitish colour: **sphagnous**, a., *sfăg'·nŭs*, pert. to bog moss.

sphalerocarpum, n., *sfăl'·ĕr·ō·kărp'·ŭm* (Gr. *sphălĕros*, unsteady, faithless ; *karpos*, fruit), in *bot.*, a small indehiscent, one-seeded fruit, enclosed within a fleshy complex pericarp.

sphenoid, a., *sfēn'·ōyd*, also **sphenoidal**, a., *sfēn·ōyd'·ăl* (Gr. *sphēn*, a wedge, *sphēnos*, of a wedge ; *eidos*, resemblance), wedge-like, as applied to a bone of the skull, which wedges in and locks together most of the other bones: **spheno**, *sfēn'·ō*, indicating connection with the sphenoid bone : **spheno-maxillary**, *măks'·ĭl·lăr·ĭ*, in *anat.*, applied to a fissure and also to a fossa.

spheroid, n., *sfēr'·ōyd* (Gr.*sphaira*, a globe ; *eidos*, resemblance), a round body or solid figure not perfectly spherical : **spheroidal**, a., *sfēr·ōyd'·ăl*, having the form of a spheroid.

sphincter, n., *sfĭngk'·tĕr* (Gr. *sphingkter*, that binds tightly or contracts—from *sphinggo*, I bind tight), in *anat.*, a muscle which contracts or shuts an orifice or opening which it surrounds : **sphincter ani**, *ān'·ĭ* (L. *ānus*, the anus, *ānī*, of the anus), the sphincter at the distal end of the rectum : **s. vesicæ**, *vĕs·ĭs'·ē* (L. *vēsīca*, the bladder, *vēsīcæ*, of the bladder), the sphincter muscle at the mouth of the bladder : **s. oris**, *ōr'·ĭs* (L. *ōs*, the mouth, *ōris*, of the mouth), the sphincter muscle of the mouth, etc.

sphygmograph, n., *sfĭg'·mō·grăf* (Gr. *sphugmos*, the pulse; *grapho*, I write), an instrument, consisting of a combination of a delicate spring and lever, which, when applied over an artery, traces the form of the pulsations on a slip of paper or a bit of smoked glass.

spicate, a., *spĭk'·āt* (L. *spica*, an ear of corn), in *bot.*, having a spike or ear, as of corn : **spicula**, n. plu., *spĭk'·ūl·ă* (L. *spīcŭlum*, a little sharp point), in *bot.*, little spikes ; pointed, needle - shaped bodies : **spicular**, a., *spĭk'·ūl·ăr*, having sharp points : **spiculum**, n., *spĭk'·ūl·ŭm*, in *anat.*, a small pointed piece of bone, or other hard matter: **spicule**, n., *spĭk'·ūl*, a minute, slender granule or point ; a spikelet.

Spigelia, n., *spĭ·jēl'·ĭ·ă* (after *Spigelius*, a botanical writer, 1625), a genus of plants, Ord. Loganiaceæ, having showy flowers when in blossom: **Spigelia Marilandica**, *măr'·ĭ·lănd'·ĭk·ă* (probably from Maryland), a species whose root, the Carolina Pink-root, is used as an anthelmintic in the United States: S. **anthelmia**, *ănth·ĕl'·mĭ·ă* (Gr. *anti*, against; *elmins*, a tapeworm), the Guiana Pink-root, used in Demerara as an anthelmintic, and which possesses narcotic qualities.

spike, n., *spĭk* (L. *spīca*, an ear of corn ; Swed. *spik*, a nail), in *bot.*, an inflorescence consisting of numerous flowers, sessile on an axis or single stem, as in the wheat and lavender : **spikelets**, n. plu., *spĭk'·lĕts*, in *bot.*, small clusters of flowers, forming secondary spikes or locustæ of grasses.

spina bifida, *spĭn'·ă bĭf'·ĭd·ă* (L. *spīna*, the spine ; *bĭfĭdus*, cleft into two parts—from *bis*, twice ; *findo*, I cleave or split), a congenital swelling situated over some part of the spine, generally in the region of the loins, due to the deficient or arrested growth of the posterior arches of one or more vertebral bones : **spina ventosa**, *vĕnt·ōz'·ă* (L. *ventōsus*, full of wind—from *ventus*, the wind), a morbid condition of bone in which the cellular structure between the external and internal walls of a bone are abnormally distended into a cavity, which may contain air.

Spinacia, n., *spĭn·ā'·sĭ·ă* (L. *spīna*, a thorn), a genus of plants, Ord. Chenopodiaceæ, so named from their prickly fruit: **Spinacia oleracea**, *ŏl'·ĕr·ā'·sĕ·ă* (L. *olerāceus*, herb-like—from *olus*, a kitchen herb), spinach, a well-known pot-herb: **spinaceous**, a., *spĭn·ā'·shŭs*, pert. to spinach, or to the species of the genus, Spinacia.

spinalis cervicis, *spĭn·āl'·ĭs sĕrvĭs'·ĭs* (L. *spīnālis*, spinal ; *cervix*, the neck, *cervīcis*, of the neck), the spinal muscle of the neck, consisting of a few irregular bundles of fibres, arising from the spines of the fifth and sixth cervical, and inserted into the spine of the axis : **spinalis dorsi**, *dŏrs'·ī* (L. *dorsum*, the back, *dorsi*, of the back), the spinal cord of the back, a long narrow muscle placed at the inner side of the longissimus dorsi, and closely connected with it.

spine, n., *spīn*, also **spinus**, n., *spīn'·ŭs* (L. *spīna*, a thorn, a spine), the vertebral column or backbone, so called from its series of thorn-like processes ; in *bot.*, an abortive branch with a hard sharp point : **spinal**, a., *spīn'·ăl*, of or relating to the backbone : **spinal column**, the connected vertebræ of the back : **spinal cord**, the greyish-white matter lodged in the interior of the spinal column or backbone : **spinal meningitis** (see under 'meninges'), inflammation of the membranes of the spinal cord : **spinitis**, n., *spĭn·īt'·ĭs*, inflammation of the spine.

spinescent, a., *spĭn·ĕs'·sĕnt* (L. *spīna*, a thorn), bearing spines : **spinose**, a., *spĭn·ōz'*, also **spinous**, a., *spīn'·ŭs*, full of spines; thorny; spinescent.

spinneret, n., *spĭn'·nĕr·ĕt* (Icel. *spinna*, Ger. *spinnen*, to spin), among certain insects, an organ

with which they form their silk or webs, as spiders and caterpillars.

spiracle, n., *spĭr'·ă·kl* (L. *spīrācŭlum*, an air-hole—from *spīro*, I breathe), the breathing pores, or apertures of the breathing tubes of insects ; the single nostril of the hag-fishes ; the blow-hole of cetaceans.

Spiræeæ, n. plu., *spĭr·ē'·ĕ·ē* (Gr. *speirāō*, I wind round or about), a Sub-order of the Ord. Rosaccæ: **Spiræa**, n., *spĭr·ē'·ă*, an extensive genus of handsome plants in flower, among which is the fragrant-blossomed Meadow-sweet.

spiral, a., *spĭr'·ăl* (L. *spīra*, Gr. *speira*, a coil, a fold), winding like a screw : **spiral vessels**, in *bot.*, vessels which have spiral fibres coiled up inside tubes.

spirillum, n., *spĭr·ĭl'·lŭm*, **spirilla**, n. plu., *spĭr·ĭl'·lă* (L. *spīra*, a coil, a fold), in *bot.*, moving filaments in the antheridia of Cryptogams ; spermatozoids ; in *phys.*, organisms in the blood of persons suffering from relapsing fever.

spiroid, n., *spĭr'·ŏyd* (Gr. *speira*, a coil, a fold ; *eidos*, resemblance), resembling a spiral : **spiroidea**, n. plu., *spĭr·ŏyd'·ĕ·ă*, spiral vessels—see 'spiral': **spirolobeæ**, n. plu., *spĭr'·ŏ·lōb'·ĕ·ē* (Gr. *lobos*, a lobe), in *bot.*, Cruciferæ which have the cotyledons folded transversely, and the radicle dorsal.

Spiroptera, n. plu., *spĭr·ŏp'·tĕr·ă* (L. *spīra*, a coil, a convolution ; Gr. *ptĕrŏn*, a wing), a genus of intestinal parasites whose species are found in various animals : **spiropterous**, a., *spĭr·ŏp'·tĕr·ŭs*, in *anat.*, having a spiral tail with membranous wing-like expansions.

splanchnic, a., *splăngk'·nĭk* (Gr. *splangchnon* an entrail), in *anat.*, belonging to the viscera or entrails ; applied to three sympathetic nerves which supply parts of the viscera, named respectively the greater, the lesser, and smallest : **splanchnica**, n. plu., *splăngk'·nĭk·ă*, medicines for bowels ; diseases affecting the bowels : **splanchnology**, n., *splăngk·nŏl'·ŏ·jĭ* (Gr. *logos*, discourse), in *anat.*, that branch of anatomy which treats of the organs of digestion, the organs of respiration, the urinary organs, and the organs of generation : **splanchno-skeleton**, *splăngk'·nŏ-skĕl'·ĕt·ŏn*, in *zool.*, the hard structure occasionally developed in connection with the internal organs or viscera.

spleen, n., *splēn* (L. and Gr. *splēn*, the milt or spleen), a spongy viscus near the large extremity of the stomach, on the left side of the abdominal cavity, is supposed to be connected with the lymphatic system; the spleen was formerly supposed to be the seat of melancholy, anger, and vexation.

splenculus, n., *splĕnk'·ŭl·ŭs*, **splenculi**, n. plu., *splĕnk'·ŭl·ī* (dim. of L. *splēn*, the milt or spleen), in *anat.*, small detached, roundish nodules, occasionally found in the neighbourhood of the spleen, and similar to it in substance ; supplementary spleens.

splenial, a., *splēn'·ĭ·ăl* (L. *splēnium*, a plaster, a patch, a splint), in *anat.*, applied to a bone of the skull in certain vertebrata ; denoting an osseous plate connected with the mandible of a Reptile.

splenic, a., *splĕn'·ĭk* (L. and Gr. *splēn*, the milt or spleen), of or belonging to the spleen: **splenitis**, n., *splĕn·īt'·ĭs*, inflammation of the spleen: **splenic apoplexy**, congestion and extravasation of the spleen, occurring suddenly in plethoric animals, but may occur from any cause : **splenic fever**, a malignant and highly contagious disease of cattle : **splenisation**, n., *splĕn'·ĭz·ā'·shŭn*,

a change produced in the lungs by inflammation, giving to them the appearance of the substance of the spleen.

splenius, a., *splēn'·ĭ·ŭs* (L. and Gr. *splēn*, the spleen ; L. *splēnium*, a patch or pad), a muscle of the back, so named from its having the form of a strap which binds down the parts lying under it : **splenius capitis**, *kăp'·ĭt·ĭs* (L. *căput*, the head, *căpĭtis*, of the head), one of the two dividing branches of the splenius muscle, which arises from the spines of the seventh cervical and two upper dorsal vertebræ : **splenius colli**, *kŏl'·lī* (L. *collum*, the neck, *colli*, of the neck), the other dividing branch of the splenius muscle, attached inferiorly to the spinous processes of the third, fourth, fifth, and sixth dorsal vertebræ.

splint, n., *splĭnt* (Ger. *splint*, a pin or peg), a thin piece of wood or metal, generally padded with a soft material, two or more pieces being employed in the case of fractures, or severe sprains, to bind the parts together, and keep them in absolute rest, the better to permit the healing powers of nature to effect a cure : **splint-bone**, the fibula, or small bone of the leg, so called from its resemblance to a surgical splint.

Spondias, n., *spŏn'·dĭ·ăs* (Gr. *spondĭas*, a kind of wild plum), a genus of plants, Ord. Anacardiaceæ, so named from the appearance of its fruit: **Spondias birrea**, *bĭr'·rē·ă* (from a native name), supplies an edible kernel in Abyssinia and in Senegal, the fruit is employed in the preparation of an alcoholic drink : **S. dulcis**, *dŭls'·ĭs* (L. *dulcis*, sweet), a native of the Society Islands, whose fruit, the Wi, is compared in flavour to the pine-apple : **S. lutea**, *lŏt'·ĕ·ă* (L. *lūtĕus*, golden-yellow—from *lūtum*, a plant used in dyeing yellow); **S. mombin,** *mŏm'bĭn* (unascertained); **S. tuberosa,** *tŭb'ĕr·ōz'ă* (L. *tŭberōsus,* having fleshy knots—from *tŭber,* a protuberance), are species producing fruits called Hog-plums, peculiar in taste, chiefly used to fatten swine ; the leaves of 'S. mombin' are astringent, and the fruit laxative ; and the fruit of 'S. tuberosa' is employed in fevers: **S. mangifera,** *măn·jĭf'·ĕr·ă* (L. *mango,* the mango fruit ; *fero,* I yield), yields a yellowish-green fruit, eaten in India, and used as a pickle in the unripe state : **S. venulosa,** *vĕn'·ŭl·ōz'ă* (L. *vĕnŭl-ōsus,* full of veins—from *vēna,* a vein), has aromatic astringent properties.

Spongida, n. plu., *spŭnj'ĭd·ă* (L. *spongia,* Gr. *sponggia,* a sponge ; Gr. *eidos,* resemblance), a division of the Protozoa, known as sponges: **spongioles,** n. plu., *spŭnj'·ĭ·ōlz* (dim. *ole*), also **spongelets,** n. plu., *spŭnj'·ĕ·lĕtz* (dim. *lets*), in *bot.*, the cellular extremities of young roots, constituting the absorbing parts of the roots : **spongiose,** a., *spŭnj'·ĭ·ōz',* having a spongy texture : **spongy,** a., *spŭnj'·ĭ,* full of small cavities or concelli.

spongiopiline, n., *spŭnj'·ĭ·ŏp'·ĭl·ĭn* (L. *spongia,* a sponge ; Gr. *pĭlŏs,* felt ; L. *pĭlus,* hair), a useful and efficient substitute for a poultice, consisting of a mass of felted shreds of wool and sponge with an india-rubber backing.

sporadic, a., *spŏr·ăd'·ĭk* (Gr. *sporadĭkos,* dispersed, scattered—from *speiro,* I sow seed), scattered ; applied to diseases which occur in single and scattered cases ; opposed to 'epidemic,' and 'endemic' ; in *bot.,* applied to plants confined to limited localities.

spores, n. plu., *spōrs,* also **sporules,** n. plu., *spōr'·ŭlz* (Gr. *spora,* a seed), in *bot.,* the minute grains in flowerless plants which perform

the functions of seeds, as in Ferns and Club mosses ; cellular germinating bodies in Cryptogamic plants ; in *zool.,* the reproductive gemmules of certain sponges : **sporaceous,** a., *spŏr·ā'·shŭs,* convertible into spores : **sporangium,** n., *spŏr·ănj'·ĭ·ŭm,* **sporangia,** n. plu., *spŏr·ănj'·ĭ·ă* (Gr. *anggos,* a vessel), hollow, flask-shaped organs, like ovaries, found in Cryptogamic plants, containing spores ; spore-cases : **sporangium,** n., a spore-case producing spores in the centre : **sporangiferous,** a., *spŏr'·ăn·jĭf'·ĕr·ŭs* (L. *fero,* I bear), bearing or producing spores : **sporangioles,** n. plu., *spŏr·ănj'·ĭ·ōlz* (dim. *ole*), very minute sporangia.

spore-sacs, n. plu., *spŏr-săks'* (*spore* and *sac*), in *zool.,* the simple generative buds of certain Hydrozoa, not having the medusoid structure developed.

sporidium, n., *spŏr·ĭd'·ĭ·ŭm,* **sporidia,** n. plu., *spŏr·ĭd'·ĭ·ă* (Gr. *spora,* seed ; *eidos,* resemblance), in *bot.,* a cellular germinating body in Cryptogamics, containing two or more cells ; reproductive cells produced within asci or sporangia.

sporocarp, n., *spŏr'·ō·kărp* (Gr. *spora,* seed ; *karpos,* fruit), in *bot.,* the ovoid sac containing the organs of reproduction in Marsileaceæ : **sporophore,** n., *spŏr'·ō·fōr* (Gr. *phorĕō,* I bear), in *bot.,* a stalk supporting a spore; in *plu.,* filamentous processes supporting spores in Fungi.

sporophyllum, n., *spŏr'·ō·fĭl'·lŭm,* **sporophylla,** n. plu., *spŏr'·ō·fĭl'·lă* (Gr. *spora,* a seed ; *phullon,* a leaf), in *bot.,* small leafy lobes, which contain tetraspores.

sporozoid, a., *spŏr'·ō·zōÿd* (Gr. *spora,* seed ; *eidos,* resemblance), in *bot.,* a moving spore furnished with cilia or vibratile processes.

sporules, see 'spores.'

sprain, n., *sprān* (old F. *espreindre,*

to press, to strain ; probably only a corruption of Eng. *strain*, to squeeze), a sudden and excessive strain of the muscular fascia, tendons, or ligaments.

spur, n., *spér* (AS. *spura*, Ger. *sporn*, Gael. *spor*, a spur), the same as 'calcar' : **spurred, a.,** *spérd*, same as 'calcarate ; ' see 'calcar.'

squama, n., *skwăm'·ă*, **squamæ,** n. plu., *skwăm'·ē* (L. *squăma*, the scale of a fish or serpent), in *bot.*, a scale ; a part arranged like a scale, as tracts on the receptacle of Compositæ : **squamæform, a.,** *skwăm'·ĕ·fŏrm* (L. *forma*, shape), scale-like : **Squamata,** n. plu., *skwăm·ăt'·ă* (L. *squămătus*, scaly), the division of Reptiles, among which the integument develops horny scales, while there are no dermal ossifications : **squamate,** a., *skwăm'·ăt*, scale-like ; scaly.

squama occipitis, *skwăm'·ă ŏk·sĭp'·ĭt·ĭs* (L. *squăma*, a scale ; *occĭput*, the back part of the head, *occĭp·ĭtis*, of the back part of the head), in *anat.*, a region of the occipital bone.

squamo-parietal, a., *skwăm'·ō·păr·ĭ'·ĕt·ăl*, one of the three sutures at the side of the skull which is arched : **squamo-sphenoidal, a.,** *sfĕn·ōyd'·ăl*, the outer portion of an irregular suture, occurring between the outer extremity of the basilar suture and the spheno-parietal : **squamo-zygomatic, a.,** *zĭg'·ŏm·ăt'·ĭk*, a suture which forms a centre of ossification in the fœtal skull.

squamose, a., *skwăm·ōz'*, and **squamous, a.,** *skwăm'·ŭs* (L. *squăma*, a scale), in *bot.*, covered with scales ; squamate ; in *anat.*, applied to a portion of the temporal bone: **squamosal, a.,** *skwăm·ōz'·ăl*, in *anat.*, applied, in the lower vertebrata, to one of the bones of the skull.

squamulæ, n. plu., *skwăm'·ūl·ē* (dim. of L. *squăma*, a scale), in

bot., minute membranous scales, occasionally occurring in the flowers of grasses : **squamulose, a.,** *skwăm'·ūl·ōz'*, having minute scales.

squarrose, a., *skwŏr·rōz'* (mid. L. *squarra*, roughness of the skin ; *squarrōsus*, covered with scurf), in *bot.*, covered with projecting parts or jags, as leaves ; having scales, small leaves, or projections, spreading widely from the axis on which they are crowded.

Stachytarpheta, n., *stăk'·ĭ·tărf·ĕt'·ă* (Gr. *stachus*, an ear or spike of corn ; *tarpheios*, thick, dense), a genus of plants, Ord. Verbenaceæ : **Stachytarpheta mutabilis,** *mŭt·ăb'·ĭl·ĭs* (L. *mŭtābĭlis*, changeable), a handsome, ever-flowering shrub, whose leaves have been imported from S. America to adulterate tea ; it is also used for tea.

Stackhousiaceæ, n., *stăk·hŏwz'·ĭ·ă'·sĕ·ē* (after *Mr.* Stackhouse, a British botanist), the Stackhousia family, an Order of plants of Australia : **Stackhousia,** n., *stăk·hŏwz'·ĭ·ă*, a genus of plants.

staggers, n. plu., *stăg'·gérz* (Dan. *staggre*, Prov. Ger. *staggeren*, to stagger), a disease in horses and cattle attended with reeling or giddiness.

Stagmaria, n., *stăg·mār'·ĭ·ă* (Gr. *stagma*, a fluid, a liquor), a genus of plants, Ord. Anacardiaceæ : **Stagmaria verniciflua,** *vérn'·ĭs·ĭ·flŏ'·ă* (F. *vernis*, mid. L. *vernix*, varnish ; L. *fluo*, I flow), a species which is the source of the hard black varnish called Japan Lacquer.

stamen, n., *stăm'·ĕn* (L. *stāmen*, the standing thing, as a thread from the distaff, or the warp in the upright looms of the ancients —from *sto*, I stand), in *bot.*, the male organ of the flower, situated within the petals, and consisting of stalks or filaments, and anthers containing pollen : **staminal, a.,** *stăm'·ĭn·ăl*, of or pert. to a stamen :

staminate, a., *stăm'·ĭn·āt*, also **staminiferous**, a., *stăm'·ĭn·ĭf'·ĕr·ŭs* (L. *fero*, I bear), bearing stamens; applied to a male flower, or to plants bearing male flowers.

staminidia, n. plu., *stăm'·ĭn·ĭd'·ĭ·ă* (L. *stāmen*, a stamen, *stāmĭnis*, of a stamen), in *bot.*, same as 'Antheridia,' which see.

staminodium, n., *stăm'·ĭn·ōd'·ĭ·ŭm*, **staminodia**, n. plu., *-ōd'·ĭ·ă* (L. *stāmen*, a stamen; *hŏdos*, a way), in *bot.*, rudimentary or abortive stamens; stamens which become sterile by the degeneration or non-development of the anthers: **staminody**, n., *stăm'·ĭn·ōd'·ĭ*, the conversion of parts of the flower into stamens, either perfect or imperfect.

stapedius, n., *stăp·ēd'·ĭ·ŭs* (mid. L. *stāpes*, a stirrup), in *anat.*, a small muscle inserted into the neck of the stapes posteriorly: **stapes**, n., *stăp'·ēz*, a stirrup-like bone of the middle ear, forming the third and innermost bone of the chain ossicles, stretching across the middle ear.

Stapelia, n., *stăp·ēl'·ĭ·ă* (after *Dr. Stapel*, of Amsterdam), an extensive genus of plants, Ord. Asclepiadaceæ, having a grotesque appearance, and singularly beautiful star-like flowers, often having a fetid odour, and hence called carrion-flowers, as they attract blow-flies, which deposit maggots on them, and these by their movements are alleged to cause fertilisation of the plants.

Staphyleaceæ, n. plu., *stăf·ĭl'·ē·ā'·sĕ·ē* (Gr. *stăphŭlē*, a grape, a bunch), the Bladder-nut family, so named from the flowers and fruit being disposed in clusters: **Staphylea**, n., *stăf'·ĭl·ē'·ă*, a genus whose species have inflated bladder-like pericarps.

staphyloma, n., *stăf·ĭl·ōm'·ă* (Gr. *stăphŭlē*, a grape), an unnatural protrusion of the tunics of the eye-ball; a protrusion of a portion of the sclerotic.

stasimorphy, n., *stăs'·ĭ·mŏrȳ'·ĭ* (Gr. *stasis*, a standing; *morphē*, form, shape), in *bot.*, a deviation in form, arising from an arrest of growth.

stasis, n., *stăs'·ĭs* (Gr. *stasis*, a stationary posture), in *med.*, a stagnation of the blood, or animal fluids.

Staticeæ, n. plu., *stăt·ĭs'·ĕ·ē* (Gr. *stătĭkē*, capable of stopping, astringent — from *statĭzō*, I stand at), a tribe or Sub-order of the Order Plumbaginaceæ, so named in allusion to the powerful astringency of some species: **Statice**, n., *stăt'·ĭs·ē*, a genus of plants: **Statice Caroliniana**, *kăr'·ō·lĭn·ĭ·ān'·ă* (of or from *Carolina*), a species whose root is one of the most powerful vegetable astringents.

statoblasts, n. plu., *stăt'·ō·blăsts* (Gr. *stătŏs*, stationary; *blastos*, a bud), in *zool.*, certain reproductive buds developed in the interior of Polyzoa, but not set at liberty till the death of the parent organism.

stearin, n., *stē'·ăr·ĭn* (Gr. *stĕar*, suet, tallow: F. *stearine*), the solid fatty principle of animal fat: **stearic**, a., *stē·ăr'·ĭk*, pert. to stearin, or obtained from it, as *stearic acid*.

stearoptene, n., *stē'·ăr·ŏp'·tēn* (Gr. *stĕar*, suet; *optănō*, I inspect or view), a solid crystalline matter deposited from many essential oils, allied to camphor.

steatoma, n., *stē'·ăt·ōm'·ă* (Gr. *stĕar*, suet; *stĕătōma*, fat), a tumour containing a fatty or granular material; an 'atheroma, which see.

Steganophthalmata, n. plu., *stĕg'·ăn·ŏf·thăl'·măt·ă* (Gr. *stĕgănos*, covered; *ŏphthălmos*, the eye), in *zool.*, certain Medusæ having the 'sense organs,' or 'marginal bodies,' protected by a sort of hood ; now separated from the Medusæ, and placed as a separate division under the name Lucernarida.

stellate, a., *stĕl'·lāt* (L. *stella*, a star), in *bot.*, arranged like a star ; radiating : **stelliform**, a., *stĕl'·lĭ·fŏrm* (L. *forma*, shape), radiating like a star ; stellate ; in *zool.*, star-shaped.

stellerida, n. plu., *stĕl·lĕr'·ĭd·ă* (L. *stella*, a star), a name sometimes given to the Order of Star-fishes.

stellulæ, n. plu., *stĕl'·ŭl·ē* (dim. of L. *stella*, a star), in *anat.*, a name given to any cluster of small veins or vessels which have a stellate arrangement.

stem, n., *stĕm* (AS. *stemn*, Ger. *stamm*, the stem of a tree), the body of a tree or plant ; the ascending axis of a plant ; a prostrate or underground shoot.

stemmata, n. plu., *stĕm'·măt·ă* (Gr. *stĕmma*, a garland), in *zool.*, the simple eyes or ocelli of certain animals, such as insects, spiders, and crustacea.

stenophyllous, a., *stĕn·ŏf'·ĭl·ŭs* (Gr. *stĕnŏs*, narrow ; *phullon*, a leaf), in *bot.*, narrow-leaved.

stercoraceous, a., *stĕrk'·ŏr·ā'·shŭs*, also **stercoral**, a., *stĕrk'·ŏr·ăl* (L. *stercorōsus*, full of dung — from *stercus*, dung), pert. to or resembling dung ; fæcal.

Sterculiaceæ, n. plu., *stĕrk'·ŭl·ĭ·ā'·sĕ·ē* (L. *Stercŭlĭus*, the god which presides over manure—from *stercus*, manure), the Sterculia and Silk - cotton family, an Order of plants, some are mucilaginous and demulcent, some used as food, and others supply a material like cotton : **Sterculieæ**, n. plu., *stĕrk'·ŭl·ĭ·ĕ·ē*, a tribe or Sub-order : **Sterculia**, n., *stĕrk·ŭl'·ĭ·ă*, a genus of plants, the leaves and flowers of some species being fetid : **sterculia tomentosa**, *tŏm'·ĕnt·ōz·ă* (L. *tomentum*, a stuffing for cushions ; Sp. *tomentōso*, pert

to horse hair) ; also **S. acuminata**, *ăk·ūm'·ĭn·āt'·ă* (L. *acuminātus*, made sharp - pointed — from *acūmen*, a point), are species whose seeds in Africa are called Kola, and are used there to sweeten water.

sterigmata, n. plu., *stĕr·ĭg'·măt·ă* (Gr. *stĕrĭgma*, a prop or support ; *stĕrĭgmăta*, props), in *bot.*, cells bearing naked spores ; cellular filaments to which spores or spermatia are attached, as in the Spermagones of Lichens.

sterile, a., *stĕr'·ĭl* (L. *sterĭlis*, barren), in *bot.*, incapable of producing seeds ; applied to male flowers not bearing fruit : **sterility**, n., *stĕr·ĭl'·ĭ·tĭ*, inability of male flowers to bear fruit ; in animals, the inability of either sex to propagate their species.

Sternbergia, n., *stĕrn·bĕrg'·ĭ·ă* (after *Count Sternberg*, a botanist), a genus of plants, Ord. Amaryllidaceæ : **Sternbergia lutea**, *lōt'·ĕ·ă* (L. *lūtĕus*, yellowish—from *lūtum*, a plant used in dyeing yellow), supposed to be the 'lily of the fields' referred to by Christ.

sternum, n., *stĕrn'·ŭm* (Gr. *stĕrnon*, the breast), the flat bone of the breast to which the ribs are attached in front ; the breast-bone : **sternal**, a., *stĕrn'·ăl*, of or pert. to the sternum : **sternalis brutorum**, *stĕrn·āl'·ĭs brŏt·ŏr'·ŭm* (L. *sternālis*, sternal ; *brūtum*, a brute), the sternal-bone of the brutes ; a muscle of the thorax constant in some of the brutes, occasionally present in man : **sterno**, *stĕrn'·ŏ*, denoting attachment to, or connection with, the sternum : **sterno - clavicular**, applied to a ligament extending from the sternum to the clavicle or collar - bone : **sterno - hyoid**, (see under 'hyo'), applied to the thin, narrow, riband - like muscle arising from the inner extremity of the clavicle : **sterno-**

mastoid, or **sterno-cleido-mastoid**, a large, thick muscle, which passes obliquely across the side of the neck, enclosed between the two layers of the deep cervical fascia : **sterno-thyroid**, a muscle arising from the posterior surface of the first bone of the sternum, and inserted into a part of the thyroid cartilage.

sternutation, n., *stḗrn'·ŭt·ā'·shŭn* (L. *sternuo*, I sneeze), the act of sneezing : **sternutatory**, a., *stḗrn·ŭt'·ăt·ŏr·ĭ*, having the quality of provoking sneezing.

stertor, n., *stḗrt'·ŏr* (L. *sterto*, I snore), the loud snoring which accompanies inspiration in certain diseases : **stertorous**, a., *stḗrt'·ŏr·ŭs*, applied to the loud snoring of apoplexy.

stethoscope, n., *stḗth'·ō·skŏp* (Gr. *stēthŏs*, the breast ; *skŏpĕō*, I view), a tube or solid instrument, of any material, and of various shapes, used by medical men in listening to the sounds produced by the action of the organs in the chest or other cavities of the body ; an instrument employed in mediate auscultation : a stethoscope may be single, binaural, or double.

sthenic, a., *sthĕn'·ĭk* (Gr. *sthĕnos*, strength), attended·with a morbid increase of vital action ; opposed to *asthenic*, or diseases of debility.

stichidium, n., *stĭk·ĭd'·ĭ·ŭm*, **stichidia**, n. plu., *stĭk·ĭd'·ĭ·ă* (Gr. *stichidion*, a little bladder ; also may be, Gr. *stichos*, a row or series), in *bot.*, case-like receptacles for the spores of some Algæ ; free spore-cases in Algæ having the spores arranged in rows.

stigma, n., *stĭg'·mă*, **stigmata**, n. plu., *stĭg'·măt·ă* (Gr. *stigma*, a mark made with a sharp-pointed instrument—from *stĭzō*, I mark with points), in *bot.*, *sing.*, the naked upper portion of the pistil on which the fertilising pollen

falls ; the breathing pore of an insect ; in *bot.*, *plu.*, the points of the basidia in some Fungi ; in *zool.*, the breathing pores or spiracles of insects, and Arachnida :

stigmaria, n. plu., *stĭg·mār'·ĭ·ă*, in *geol.*, fossil root stems having regular pitted or dotted surfaces.

Stilaginaceæ, n. plu., *stĭl'·ă·jĭn·ā'·sĕ·ē* (Gr. *stūlos*, a column or pillar), the Stilago family, an Order of plants, some yielding edible fruits, and others used as pot herbs : **Stilago**, n., *stĭl·āg'·ō*, a genus of ornamental trees.

Stillingia, n., *stĭl·ĭnj'·ĭ·ă* (after *Dr.Stillingfleet*, an Eng. botanist), a genus of plants, Ord. Euphorbiaceæ : **Stillingia sebifera**, *sĕb·ĭf'·ĕr·ă* (L. *sēbum*, tallow, fat ; *fero*, I bear), the tallow tree of China, used in making candles, and the plant also yields a bland oil.

stipate, a., *stĭp'·āt* (L. *stīpātus*, crowded or pressed together), in *bot.*, pressed together ; crowded : **stipation**, n., *stĭp·ā'·shŭn*, an accumulation in the cavities or tissues.

stipe, n., *stĭp* (L. *stīpes*, a stock, a stalk, *stīpĭtis*, of a stalk), in *bot.*, the stem of palms and tree-ferns ; the stalk of fern fronds ; the stalk or stem bearing the pileus in Agarics : **stipels**, n. plu., *stĭp'·ĕlz*, small leaflets at the base of the pinnules of compound leaves : **stipitate**, a., *stĭp'·ĭt·āt*, in *bot.*, supported on a stalk ; stalked : **stipitiform**, a., *stĭp·ĭt'·ĭ·fŏrm* (L. *forma*, shape), resembling a stalk or stem.

stipule, n., *stĭp'·ŭl* (L. *stĭpŭla*, a stem, a stalk), in *bot.*, a leaflet at the base of other leaves, having a lateral position, and more or less changed in form or texture ; a process developed at the base of a petiole : **stipulary**, a., *stĭp'·ŭl·ăr·ĭ*, in *bot.*, occupying the place of stipules, such as tendrils : **stip-**

ulate, a., *stĭp'·ŭl·āt*, furnished with stipules.

stole, n., *stōl*, also stolon, n., *stōl'·ŏn* (L. *stŏlo*, a twig or shoot springing from the stock of a tree), in *bot.*, a lax trailing and rooting branch, given off at the summit of the root, and then turning downwards and taking root at intervals; in *zool.*, one of the connecting processes of sarcode in Foraminiferæ; also the processes sent out by the cœnosarc of certain Actinozoa; the connecting tube among the social Ascidians.

stoloniferous, a., *stōl'·ŏn·ĭf'·ĕr·ŭs* (L. *stŏlo*, a shoot from the stock of a tree; *fero*, I bear), in *bot.*, having creeping runners, which root at the joints; see 'stolon.'

Stomapoda, n. plu., *stŏm·ăp'·ŏd·ă* (Gr. *stŏma*, the mouth; *pous*, the foot, *podes*, feet), an Order of Crustaceans, which have thoracic or true feet in connection with the mouth.

stomata, n. plu., *stŏm'·ăt·ă*, and stomates, n. plu., *stŏm'·ātz* (Gr. *stŏma*, the mouth, *stomăta*, mouths), in *bot.*, minute openings in the epidermis of plants, especially in the leaves: stomatitis, n., *stŏm'·ăt·īt'·ĭs*, in *med.*, inflammation of the mouth.

stomatode, n., *stŏm'·ăt·ōd* (Gr. *stŏma*, a mouth; *hodos*, a way), in *zool.*, possessing a mouth, as in the so-called stomatode Protozoa.

stool, n., *stŏl* (L. *stŏlo*, a shoot, a sucker; Ger. *stuhl*, a stock; Manx *sthol*, a sprout or branch), in *bot.*, a plant from which layers are propagated by bending down some of its branches to the ground in order to permit them to root in the earth; the root or stump of a timber tree which throws up shoots.

storax, see 'Styrax.'

stramonium, see 'Datura.'

strangulated, a., *străng'·gŭl·āt'·ĕd* (L. *strangŭlo*, I throttle), in *bot.*,

contracted and expanded irregularly: strangulation, n., *străng'·gŭl·ā'·shŭn*, a forcible obstruction of the air passages; the condition of any part or organ too closely constricted: strangury, n., *străng'·gŭr·ĭ*, difficult and painful urination.

stratum bacillorum, *străt'·ŭm băk'·sĭl·ōr'·ŭm* (L. *strātum*, a layer; *băcillum*, a small rod or wand, *băcillōrum*, of small rods), the external columnar layer of the retina, consisting of innumerable thin rods placed vertically side by side like palisades, and of larger bodies interspersed, named cones.

Strepsiptera, n. plu., *strĕp·sĭp'·tĕr·ă* (Gr. *strĕpho*, I twist; *ptĕron*, a wing), an Order of insects in which the anterior wings are represented by twisted rudiments: strepsipterous, a., *-tĕr·ŭs*, having the first pair of wings represented by twisted rudiments: Strepsirhina, n. plu., *strĕps'·ĭ·rīn'·ă* (Gr. *rhĭs*, the nose, *rhīnos*, of the nose, *rhīnĕs*, nostrils), a group of the quadrumana; also called Prosimiæ.

stria, n., *strī'·ă*, striæ, n. plu., *strī'·ē* (L. *stria*, a furrow, a channel), in *bot.*, a narrow line or mark: striæ, lines or streaks on the surface of a body: striated, a., *strī'·āt·ĕd*, marked or impressed with thread-like lines: stria terminalis, *tĕrm'·ĭn·āl'·ĭs* (L. *termĭnālis*, terminal), the terminal streak; in *anat.*, a narrow whitish band running along the inner border of each corpus striatum of the brain: striæ longitudinalis, *lŏnj'·ĭt·ŭd'·ĭn·āl'·ēz* (L. *longĭtudĭnālis*, longitudinal), longitudinal streaks; in *anat.*, two white tracts, placed close to each other, in the corpus callosum of the brain.

stricture, n., *strĭkt'·ŭr* (L. *strictus*, drawn together, bound or tied tight), in *med.*, a spasmodic or morbid contraction of any passage of the body, generally applied to the contraction of the urethra, or channel by which the urine passes from the body.

strigæ, n. plu., *strĭdj'·ē* (L. *strīga*, a row or ridge left in ploughing ; *strĭgæ*, ridges), in *bot.*, little, upright, unequal, stiff hairs, swelled at their bases : **strigose**, a., *strĭg·ōz'*, covered with sharp ridged hairs.

strobila, n., *strōb'·ĭl·ă* (Gr. *strŏbĭl·os*, a top, a fir cone), in *zool.*, the adult tapeworm with its generative segments ; also applied to one of the stages in the life of the Lucernarida.

strobile, n., *strŏb'·ĭl*, also **strobilus**, n., *strŏb'·ĭl·ŭs* (Gr. *strŏbĭlos*, L. *strŏbĭlus*, anything shaped like a top, a cone), in *bot.*, a multiple fruit in the form of a head or cone, as in the hop and pine.

stroma, n., *strōm'·ă* (Gr. *strōma*, anything spread out for resting, a stratum), in *anat.*, the substance or tissue which forms a foundation or basis, or affords mechanical support ; in *bot.*, the arborescent or cup-shaped receptacle containing the perithecia in large numbers, as in certain Fungi.

strombuliform, a., *strŏm·bŭl'·ĭ·fŏrm* (L. *strŏmbus*, a kind of spiral snail-shell ; *forma*, shape), in *bot.*, twisted in a long spire.

Strongyle, n., *strŏnj'·ĭl*, or **Strongylus**, n., *strŏnj'·ĭl·ŭs*, **Strongyli**, n. plu., *strŏnj'·ĭl·ī* (Gr. *strŏnggŭl·os*, round, globular), a genus of internal parasites found in the heart and kidney : **Strongylus gigas**, *gīg'·ăs* (L. *gīgas*, a giant), a formidable large round worm, of a blood-red colour, which infests the kidneys : S. **armatus**, *ărm·āt'·ŭs* (L. *armātus*, armed), the needle-worm.

Strophanthus, n., *strŏf·ănth'·ŭs* (Gr. *strophanthos*, a twisted thing, a cord ; *anthos*, a flower),

a genus of very beautiful shrubs, Ord. Apocynaceæ, the segments of the corolla being long, narrow, and twisted : **Strophanthus kombe**, *kŏmb'·ē* (native name), furnishes the kombe arrow poison of S. Africa : S. **hispidus**, *hĭsp'·ĭd·ŭs* (L. *hispĭdus*, shaggy, hairy), supplies an arrow poison in W. Africa.

strophioles, n., *strŏf'·ĭ·ōlz* (L. *strophĭŏlum*, a small wreath or garland), in *bot.*, cellular bodies not dependent on fertilisation, which are produced at various points on the testa of seeds ; swollen fungus-like excrescences on the surface of some seeds about the hilum : **strophiolate**, a., *strŏf'·ĭ·ōl·āt*, having little fungus-like excrescences around the hilum.

strophulus, n., *strŏf'·ŭl·ŭs* (Gr. *strophĕō*, I turn), red-gum, a simple form of skin eruption occurring in infants.

struma, n., *strōm'·ă* (L. *strūma*, a scrofulous tumour), a diseased state, having, with other characteristics, a tendency to a swelling of the glands in various parts of the body ; a scrofulous swelling or tumour ; in *bot.*, a cellular swelling at the point where a leaflet joins the midrib : **strumous**, a., *strōm'·ŭs*, scrofulous.

Strychneæ, n. plu., *strĭk'·nĕ·ē* (L. *strychnus*, Gr. *struchnos*, a kind of nightshade), a Sub-order of the Ord. Loganiaceæ : **Strychnos**, n., *strĭk'·nŏs*, a genus of valuable plants from their medicinal properties, which, however, are highly poisonous : **Strychnos Nux-vomica**, *nŭks-vŏm'·ĭk·ă* (L. *nux*, a nut; *vomĭcus*, of or pert. to vomiting), the poison-nut or koochla, which supplies Nux-vomica, obtained from the seeds ; all parts of the plant are intensely bitter, especially the seeds and bark : S. **Ignatia**, *ĭg·nā'·shĭ·ă* (after *St. Ignatia*), St. Ignatia's bean, or 'Ignatia amara ; **S. colubrina**, *kŏl'·ŭb·rīn'·ă* (L. *cŏlŭber*, a serpent); **S. lagustrina**, *lăg'·ŭs·trīn'·ă* (unascertained), snakewood, are other species from which strychnia is obtained : **S. Tieute**, *tī·ūt'·ĕ* (a native name), the source of a Java poison called 'Upas Tieute': **S. toxifera**, *tŏks·ĭf'·ĕr·ă* (L. *toxĭcum*, poison ; *fero*, I bear); and **S. Guianensis**, *gwī·ăn·ĕns'·ĭs* (of or from *Guiana*), species which are supposed to yield the Hoorali or Ourari poison of Guiana : **S. potatorum**, *pŏt'·ăt·ōr'·ŭm* (L. *pōtātus*, a drinking, a draught), called clearing-nut, and used in India for purifying water; and **S. pseudo-quina**, *sūd'·ō-kwīn'·ă* (Gr. *pseudēs*, false ; Sp. *quina*, Peruvian bark), are used as tonics and febrifuges, and do not possess the characteristic poisonous qualities in large quantities : **strychnia**, n., *strĭk'·nĭ·ă*, one of the alkaloids contained in the seeds of S. Nux-vomica: **strychnic**, a., *strĭk'·nĭk*, denoting an acid obtained also from the seeds: **strychnism**, n., *strĭk'·nĭzm*, the toxical symptoms induced by the use of strychnia.

stupe, n., *stūp* (L. *stūpa*, Gr. *stupē*, tow), in *med.*, flax, or a cloth, dipped in a warm medicament and applied to a sore, a wound, or part ; a fomentation : **stupa**, n., *stūp'·ă*, in *bot.*, a tuft or mass of hair, or fine filament, matted together : **stupose**, a., *stūp·ōz'*, in *bot.*, having a tuft of hair ; composed of matted filaments.

stupor, n., *stūp'·ŏr* (L. *stupĕō*, I am stupefied), that state of partial insensibility often preceding coma.

sturdy, n., *stĕrd'·ĭ* (Gael. *stuird*, a disease in sheep), a parasitic disease of the brain of sheep, characterised by dulness and stupor.

stye, n., *stī* (Icel. *stigje*, low Ger. *stieg*, a pustule at the corner of the eye), an inflamed pustule in one or other, or both, eyelids.

style, n., *stīl* (L. *stylus*, a stake, a pale), in *bot.*, the stalk interposed between the ovary and the stigma ; the prolongation of an ovary bearing the stigma : **styliform**, a., *stīl'·ĭ·fŏrm* (L. *forma*, shape), pointed in shape.

Stylidiaceæ, n. plu., *stīl·ĭd'·ĭ·ā'·sĕ·ē* (Gr. *stūlos*, a column, a pillar), the Stylidium or stylewort family, an Order of plants found at the southern point of S. America : **Stylidium**, n., *stīl·ĭd'·ĭ·ŭm*, a genus of plants ; in the species, the column formed by the union of the filaments and style possess a peculiar irritability.

stylo-glossus, *stīl'·ō-glŏs'·ŭs* (*stylo*, from Gr. *stulos*, a column, a style, denoting connection with the styloid process of the temporal bone ; Gr. *glōssa*, a tongue), in *anat.*, the shortest of three muscles which spring from the styloid process of the temporal bone, situated partly under the tongue : **stylo-hyoid**, *hī'·ōyd* (see 'hyoid'), a ligament, consisting of a thin fibrous cord, which extends from the point of the styloid process to the lesser corner of the hyoid bone ; a small branch of the facial nerve : **stylo-mastoid**, *măst'·ōyd* (see 'mastoid '), the small branch given off by the posterior articular artery, which enters the stylo-mastoid foramen in the temporal bone ; a foramen in the temporal bone : **stylo-maxillary**, *măks'·ĭl·lăr·ĭ* (see 'maxilla,' a jaw), a ligament consisting of a strong thickened band of fibres connected with the cervical fascia, and which separates the parotid from the sub-maxillary gland : **stylo-pharyngeus**, *făr'·ing·gē'·ŭs* (Gr. *pharungx*, the pharynx), a muscle arising from the styloid process of the temporal bone, and passing to the side of the pharynx.

styloid, a., *stil'·ōўd* (Gr. *stulos*, a style, a column ; *eidos*, resemblance), in *anat.*, shaped like a style or pen, applied to such processes as the ulna and temporal bone.

stylopod, n., *stil'·ō·pŏd*, also **stylopodium**, n., *stil'·ō·pŏd'·ĭ·ŭm* (Gr. *stulos*, a style, a column ; *pous*, a foot, *podos*, of a foot), in *bot.*, a fleshy disc bearing the styles in Umbelliferæ : **stylospores**, n. plu., *stĭl'·ō·spōrz* (Gr. *spora*, seed), the spores borne upon a stem ; the spore-like bodies borne on a cellular stalk in the Picnides of Lichens.

styptic, n., *stĭp'·tĭk* (L. *styptĭcus*, Gr. *stŭptĭkos*, astringent), a substance which arrests local bleeding, such as cold water and ice, and astringents.

Styracaceæ, n. plu., *stĭr'·ăk·ā'·sĕ·ē* (L. *styrax*, Gr. *sturax*, a resinous gum, storax), the Storax family, an Order of plants, which possess, in general, stimulant, aromatic, and fragrant properties: **Styraceæ**, n. plu., *stĭr·ā'·sĕ·ē*, a tribe or Sub-order : **Styrax**, n., *stĭr'·ăks*, a genus of handsome flowering and useful plants : **Styrax benzoin**, *bĕn'·zō·ĭn* (said to be from Ar. *benzoah*), a lofty tree which yields the concrete balsamic exudation called Benzoin, used as a stimulant expectorant, and for fumigation and incense : **S. officinale**, *ŏf·fĭs'·ĭn·āl'·ĕ* (L. *officinālis*, officinal, by authority—from *officīna*, a workshop), a tree of Syria and Arabia, the source of the balsamic resinous substance called Storax, employed as a pectoral remedy.

sub-acute, a., *sŭb·ăk·ūt'* (L. *sub*, under, and *acute*), acute in a moderate degree ; neither acute nor chronic.

sub-anconeus, a., *sŭb'·ăng·kōn'·ĕ·ŭs*, or *-ăng'·kŏn·ē'·ŭs* (L. *sub*, under ; L. *ancon*, Gr. *angkon*, the curvature of the arm, the elbow), a small

muscle consisting of one or two slender fasciculi, which arise from the humerus, and pass to the elbow-joint.

sub-arachnoid, a., *sŭb'·ăr·ăk'·nŏўd* (L. *sub*, under, somewhat, and *arachnoid*), in *anat.*, a space between the arachnoid and pia-mater ; the space which the visceral layer leaves as a loose sheath around the spinal cord.

sub-calcareous, a., somewhat calcareous.

sub-caudal, a., beneath the tail.

sub-central, a., nearly central, but not quite.

subclavian, a., *sŭb·klāv'·ĭ·ăn* (L. *sub*, under ; *clāvis*, a key ; *clavicula*, a small key, the collar-bone), in *anat.*, lying under the clavicle or collar-bone, as an artery or a vein : **subclavius**, *sŭb·klāv'·ĭ·ŭs*, a long thin spindle-shaped muscle, placed in the interval between the clavicle and the first rib.

subcrurius, n., *sŭb·krōr'·ĭ·ŭs* (L. *sub*, under ; *crūs*, the leg, *crūris*, of the leg), a small band of muscular fibres extending from the lower part of the femur to the knee-joint.

subcutaneous, a., *sŭb'·kūt·ăn'·ĕ·ŭs* (L. *sub*, under, and *cutaneous*), situated or placed immediately under the skin or cutis.

suberate, a., *sŭb'·ĕr·āt* (L. *sūber*, the cork tree, *sūbĕris*, of the cork tree), in *chem.*, a salt formed by suberic-acid with a base : **suberic**, a., *sŭb·ĕr'·ĭk*, pert. to cork ; applied to an acid produced by the action of nitric acid on cork and fatty bodies : **suberous**, a., *sŭb'·ĕr·ŭs*, in *bot.*, having a corky texture ; applied to the epiphlœum or external layer of bark.

sub-genus, somewhat less than a genus, formed by grouping certain species which happen to agree more nearly with each other in some important particulars than do the other species of the genus: **sub-order and tribes** consist of certain genera more nearly allied in particular characters than others : **sub-class** consists of certain orders having general characters more nearly allied than the others.

subiculum, n., *sŭb·ĭk'·ŭl·ŭm* (L. *sŭbĭcŭlum,* an under-layer), in *bot.,* the filamentous mycelium of certain Fungi ; the Hypothallus.

sub-involution, n., *sŭb·ĭn'·vōl·ŭ'·shŭn* (L. *sub,* somewhat, and *involution*), the state or condition of the womb when it does not return to its usual size after delivery, but is somewhat larger and heavier.

subjacent, a., *sŭb·jās'·ĕnt* (L. *sub,* somewhat, beneath ; *jacens,* lying), lying under or on a lower situation, though not exactly beneath.

subject, n., *sŭb'·jĕkt* (L. *subjectus,* laid or placed under), in *anat.,* a dead body for dissection : **subjective,** a., *sŭb·jĕkt'·ĭv,* derived from one's own consciousness, in distinction from external or objective observation : **subjective sensations,** sensations which originate in the brain.

sublimation, n., *sŭb'·lĭm·ā'·shŭn* (L. *sublĭmātus,* lifted up on high —from *sublīmis,* high), the operation of bringing a solid substance into the state of a vapour by heat and condensing it again.

sublingua, n., *sŭb·lĭng'·gwă* (L. *sub,* somewhat ; *lingua,* the tongue), a tongue-shaped organ : **sublingual,** a., *sŭb·lĭng'·gwăl,* situated under the tongue.

sublobular, a., *sŭb·lŏb'·ŭl·ăr* (L. *sub,* somewhat ; Gr. *lŏbos,* the tip of the ear ; mid L. *lŏbŭlus,* a small lobe), small veins of the liver on which the lobules rest, and into which the intra-lobular veins pour their blood.

submaxillary, a., *sŭb·măks'·ĭl·lăr·ĭ* (L. *sub,* under ; *maxilla,* a jaw), placed under the jaw ; applied to

a ganglion connected by filaments with the gustatory nerve ; a gland next in size to the parotid, situated immediately below the base and the inner surface of the inferior maxilla.

submental, a., *sŭb·mĕnt'·ăl* (L. *sub,* under ; *mentum,* the chin), situated under the chin ; applied to an artery and a vein running beneath the chin.

submucous, a., *sŭb·mūk'·ŭs* (L. *sub,* under, and *mucous*), applied to a coat of the small intestine, connected more firmly with the mucous than with the muscular coat, between which two it is placed.

sub-occipital, a., *sŭb'·ŏk·sĭp'·ĭt·ăl* (L. *sub,* under, and *occipital*), applied to a branch of the first spinal nerve which runs under the back of the head.

sub-pedunculate, a., *sŭb'·pĕd·ŭngk'·ŭl·āt* (L. *sub,* under ; *pedŭncŭlŭs,* a little foot, a foot stalk), supported upon a very short stem.

sub-peritoneal, a., *sŭb·pĕr'·ĭt·ŏn·ē'·ăl* (L. *sub,* under, and *peritoneal*), in *anat.,* a layer of areolar tissue, distinct from the abdominal fasciæ, by which the parietal portion is connected loosely with the fascia lining the abdomen and pelvis.

sub-scapularis, n., *-skăp'·ŭl·ār'·ĭs,* **sub-scapulares,** plu., *-skăp'·ŭl·ār'·ēz* (L. *sub,* under ; *scapula,* the shoulder-blade), a muscle arising from all the internal surface of the scapula, and inserted into the humerus, which muscle pulls the arm backwards and downwards : **sub-scapular,** a., *-skăp'·ŭl·ăr,* denoting the large branch of the axillary artery arising near the lowest margin of the scapula : **sub-scapularis fossa,** *fŏs'·să* (L. *fossa,* a ditch), a shallow depression on the anterior surface of the scapula.

subserous, a., *sŭb·sēr'·ŭs* (L. *sub,*

2 C

under, and *serous*), applied to the connective tissue beneath the serous membranes.

sub-sessile, -*sĕs'ĭl* (L. *sub*, somewhat, and *sessile*), in *bot.*, nearly sessile ; almost without a stalk.

substantia cinerea gelatinosa, *sŭb·stăn'shĭ·ă sĭn·ēr'ĕ·ă jĕl·ăt'ĭn·ŏz'ă* (L. *substantia*, a substance ; *cĭnĕrĕŭs*, ash-coloured ; mid. L. *gelatinōsus*, gelatinous), the grey matter, of a peculiar semi-transparent aspect, in the back part of the posterior horn of the spinal cord : **substantia spongiosa**, *spŭnj'ĭ·ŏz'ă* (L. *spongĭōsus*, spongy, porous), the remaining and greater part of the same grey matter.

subulate, a., *sŭb'ŭl·āt* (L. *sŭbŭla*, an awl), in *bot.*, shaped like a cobbler's awl.

succedaneum, n., *sŭk'sĕd·ān'ĕ·ŭm* (L. *succēdānĕus*, that which supplies the place of something— from *sub*, under ; *cedo*, I go), that which is used for something else ; a substitute ; an amalgam for filling teeth : **succedaneous**, a., *sŭk'sĕd·ān'ĕ·ŭs*, supplying the place of something else : **caput succedaneum**, *kăp'ŭt* (L. *căput*, the head), a puffy tumour of the infant scalp, produced during parturition.

succisus, a., *sŭk·sīz'ŭs* (L. *succīsus*, lopped off), in *bot.*, abrupt, appearing as if it were cut off ; premorse.

succulent, a., *sŭk'kŭl·ĕnt* (L. *succus*, juice or moisture ; F. *succulent*), in *bot.*, having juicy and soft stems and leaves ; soft and juicy.

sudamens, n., *sŭd·ăm'ĕnz* (L. *sūdo*, I sweat or perspire), minute vesicles, containing fluid, appearing abundantly on the chest in cases of rheumatic fever and other diseases, accompanied by profuse perspiration.

sudorific, n., *sūd'ŏr·ĭf'ĭk* (L. *sūdor*, sweat ; *făcĭo*, I make), a remedy which causes and pro-

motes perspiration ; a diaphoretic : **sudoriferous**, a., *sūd'ŏr·ĭf'ĕr·ŭs*, bearing or conveying sweat.

suffrutex, n., *sŭf'frŏt·ĕks* (L. *sub*, under ; *frŭtex*, a shrub, *frŭtĭcis*, of a shrub), in *bot.*, an undershrub, not exceeding the length of the arm : **suffruticose**, a., *sŭf·frŏt'ĭk·ōz*, shrubby underneath ; having the characters of a small shrub.

sulcate, a., *sŭlk'āt* (L. *sulcus*, a furrow), in *bot.*, furrowed or grooved ; having a deeply furrowed surface : **sulciform**, a., *sŭls'ĭ·fŏrm* (L. *forma*, shape), furrowed ; same as sulcate.

sulcus, n., *sŭlk'ŭs*, **sulci**, n. plu., *sŭls'ī* (L. *sulcus*, a furrow), in *anat.*, a groove on the surface of bones, and other parts ; the depressions which separate the convolutions of the brain : **sulcus frontalis**, *frŏnt·āl'ĭs* (L. *frons*, the forehead, the front, *frontis*, of the front), a groove which lodges the commencement of the longitudinal sinus.

sulphate, n., *sŭlf'āt* (L. *sulphur*, brimstone), in *chem.*, a salt formed by sulphuric acid with any base, as sulphate of lime : **sulphur**, n., *sŭlf'ĕr*, one of the elementary substances, occurring as a greenish-yellow, brittle, solid body, crystalline in structure, of a peculiar odour when rubbed, burns with a bluish flame, and emits most suffocating fumes ; employed in medicine in two forms — sublimed sulphur, and precipitated sulphur or milk of sulphur : **sulphuret**, n., *sŭlf'ăr·ĕt*, a compound of sulphur with hydrogen, or with a metal : **sulphuretted**, a., *sŭlf'ŭr·ĕt·ĭĕd*, combined with sulphur : **sulphuric**, a., *sŭlf·ūr'ĭk*, pert. to or obtained from sulphur : **sulphurous**, a., *sŭlf'ūr·ŭs*, containing or resembling sulphur : **sulphuric acid**, a powerful acid formed by one equivalent of sulphur combined with three of oxygen, much used in the arts and medicine; popularly named *oil of vitriol*: **sulphurous acid**, an acid forming the fumes evolved from sulphur when burned in air: **sulphuretted hydrogen**, a gas having the fetid odour of rotten eggs, composed of one equivalent of sulphur and one of hydrogen.

supercarbonate, n., *sŭp'ĕr·kârb'ŏn·āt* (L. *super*, above, in excess, and *carbonate*), a substance which holds the greatest quantity of the carbonate which can be held: **superphosphate**, n., *-fŏs'fāt* (see 'phosphate'), any substance containing the greatest quantity of phosphoric acid which can combine with the base.

supercilia, n. plu., *sŭp'ĕr·sĭl'ĭ·ā* (L. *super*, above; *cilium*, an eyelid, *cilia*, eyelids), the eyebrows, consisting of two arched eminences of integument, which surmount the upper circumference of the orbit on each side, and support numerous short, thick hairs: **superciliary**, a., *-sĭl'ĭ·ăr·ĭ*, situated above the eyebrow.

superficialis colli, *sŭp'ĕr·fĭsh·ĭ·āl'ĭs kŏl'lī* (L. *superficiālis*, superficial; *collum*, the neck, *colli*, of the neck), in *anat.*, the superficial plexus of the neck, being a branch of the cervical plexus, arises from the second and third cervical nerves.

superior, a., *sŭp·ēr'ĭ·ĕr* (L. *superior*, higher—from *super*, above), in *bot.*, placed above another organ—applied especially to indicate the position of the ovary with respect to the calyx; in *anat.*, higher; more elevated.

supertuberation, n., *sŭp'ĕr·tūb'ĕr·ā'shŭn* (L. *super*, over, above; *tūber*, a hump or excrescence), in *bot.*, the growth of young potatoes from old ones still attached to the shaw and growing.

supervolute, a., *sŭp'ĕr·vŏl·ūt'*, also **supervolutive**, a., *-vŏl·ūt'ĭv* (L.

super, above, over; *vŏlūtus*, rolled or twisted), in *bot.*, having a plaited and rolled arrangement in the bud; rolled upon itself in vernation.

supine, a., *sŭp·īn'* (L. *sŭpīnus*, bent or thrown backwards, lying on the back), lying on the back, or with face upwards: **supinate**, a., *sŭp'ĭn·āt*, in *bot.*, leaning or inclining with exposure to the sun: **supination**, n., *sŭp'ĭn·ā'shŭn*, in *anat.*, the movement of the arm by which the palm of the hand is turned upwards; the opposite is called *pronation*: **supinator**, n., *sŭp'ĭn·āt'ŏr*, a muscle which turns the palm of the hand upwards: **supinator brevis**, *brĕv'ĭs* (L. *brevis*, short), a broad muscle, of a hollow cylindrical form, and curved round the upper third of the radius: **supinator longus**, *lŏng'gŭs* (L. *longus*, long), the most superficial muscle on the radial side of the forearm, passing from the humerus to the radius.

suppository, n., *sŭp·pŏz'ĭt·ŏr·ĭ* (L. *suppositus*, placed under—from *sub*, under; *pōno*, I place), a solid medicinal agent for introduction into the rectum.

suppression, n., *sŭp·prĕsh'ŭn* (L. *suppressus*, held or kept back—from *sub*, under; *pressus*, pressed), in *bot.*, the complete non-development of organs; in *med.*, arrest of a normal secretion.

suppuration, n., *sŭp'pūr·ā'shŭn* (L. *suppurātus*, having matter gathered underneath—from *sub*, under; *pus*, the white and viscous matter of a sore), the matter formed in a sore; matter or pus: **suppurative**, a., *sŭp'pūr·āt'ĭv*, tending to suppurate: n., a medicine or application which promotes the formation in a sore of pus.

supra-acromial, *sŭp'rā·ăk·rōm'ĭ·ăl* (L. *sŭprā*, on the upper side or top; *sŭpĕrus*, upper, on high;

and *acromial*), the name of an artery, and also of a nerve lying above the acromium of the artery: **supra-clavicular**, a., *klăv·ĭk'·ŭl·ăr* (see 'clavicular'), applied to two branches of nerves arising from the third and fourth cervical nerves: **supra-decompound**, *dē'·kŏm·poŭnd* (see 'decompound'), in *bot.*, very much divided and sub-divided: **supra - maxillary**, *măks'·ĭl·lăr·ĭ*, or *măks·ĭl'·lăr·ĭ* (see 'maxillary'), applied to a branch of the facial nerve passing over the side of the maxilla to the angle of the mouth: **supra-renal**, *-rēn'·ăl* (see 'renal'), situated above the kidneys: **supra-spin-atus**, *spīn·āt'·ŭs* (L. *spina*, a spine; *spīnātus*, the back-bone), a muscle situated above the spine of the scapula, and inserted into the humerus, which raises the arm, etc.: **supra-spinous**, *spīn'·ŭs* (see 'spinous'), applied to the superior and smaller division of the posterior surface of the scapula; designating the ligaments which connect the spines of the vertebræ: **supra-sternal**, *stẽrn'·ăl* (L. *sternum*, the breast-bone), designating a branch of the cervical nervous plexus: **supra-trochlear**, *trŏk'·lĕ·ăr* (L. *trochlĕa*, a pulley), applied to a branch of the ophthalmic nerves, which is prolonged to the inner angle of the orbit, close to the point at which the pulley of the upper oblique muscle is fixed to the orbit.

sural, a., *sūr'·ăl* (L. *sūra*, the calf of the leg), in *anat.*, pert. to the calf of the leg.

surculus, n., *sẽrk'·ŭl·ŭs* (L. *surc-ŭlus*, a young twig, a shoot), a shoot thrown off underground, and only rooting at its base; a sucker from the neck of a plant beneath the surface.

suspended, a., *sŭs·pĕnd'·ĕd* (L. *sub*, under; *pendĕo*, I hang), in *bot.*, applied to an ovule hanging from a point a little below the apex of the ovary: **suspensor**, n., *sŭs·pĕns'·ŏr*, in *surg.*, a band to suspend the scrotum; in *bot.*, the cord which suspends the embryo, and is attached to the radicle in the young state: **suspensory**, n., *sŭs·pĕns'·ŏr·ĭ*, anything which suspends or holds up: **suspensor-ium**, n., *sŭs'·pĕns·ŏr'·ĭ·ŭm*, the apparatus by which the lower jaw is suspended to the upper jaw.

sustentacular, a., *sŭs'·tĕn·tăk'·ŭl·ăr* (L. *sustentācŭlum*, a prop, a support), applied to a kind of connective tissue which serves as a supporting framework to the peculiar elements and nourishing blood - vessels of certain organs and textures: **sustentaculum li-enis**, *lĭ·ēn'·ĭs* (L. *lĭēn*, the spleen, *lĭēnis*, of the spleen), the support of the spleen; a fold of peritoneum extending from the diaphragm to the colon: **sustentaculum tali**, *tāl'·ī* (L. *tālus*, the ankle-bone, *tālī*, of the ankle-bone), the support of the ankle-bone; a flattened process which projects inwards near the anterior extremity of the os calcis.

suture, n., *sūt'·ūr*, also **sutura**, n., *sūt·ūr'·ă* (L. *sūtūra*, a seam — from *sūtum*, to sew or stitch), the method of keeping the parts of a wound together by sewing, etc.; the line of junction of two parts which are immovably connected together; the line where the whorls of a univalve shell join one another; an immovable articulation of bone, as in the bones of the skull; in *bot.*, the line of junction of two parts: **sutural**, a., *sūt'·ūr·ăl*, in *bot.*, applied to that form of dehiscence or separation of fruits which takes place at the sutures.

Swietenia, n., *swēt·ēn'·ĭ·ă* (after *Swieten*, a Dutch botanist), a genus of interesting and valuable plants: **Swietenia mahagoni**, *mă·hăg'·ŏn·ĭ* (a native name), a species which supplies the well-known mahogany wood.

swimmerets, n. plu., *swĭm'·mĕr·ĕts* (from Eng. *swim*), the limbs of the Crustaceæ which are adapted for swimming.

syconus, n., *sĭk'·ŏn·ŭs* (Gr. *sūkŏn*, a fig), in *bot.*, a multiple succulent hollow fruit, as in the fig; a fruit, such as the fig, which encloses the fruits: **sycosis**, n., *sĭk·ōz'·ĭs*, a parasitic disease of the hair follicles of the chin or upper lip.

sympathy, n., *sĭm'·păth·ĭ* (Gr. *sumpatheia*, conformity of feeling — from *sun*, together ; *pathos*, suffering), that influence or correspondence which arises in one part of the body from the existence of disease or irritation in another part, as the headache of indigestion, the pain in the right shoulder in disease of the liver, or the affection of one eye from disease of the other: **sympathetic**, a., *sĭm'·păth·ĕt'·ĭk*, dependent on sympathy or irritation; in *anat.*, applied to a system of nerves consisting of one or more ganglia, or a series of them; the sympathetic nerves are, in man, chiefly disposed in plexuses, as the cardiac, the solar, and the hypogastric, etc.

symphysis, n., *sĭmf'·ĭs·ĭs* (Gr. *sumphusis*, a growing together — from *sun*, together; *phuo*, I grow), in *anat.*, the union of bones by an intervening cartilage, so as to form an immovable joint, or only slightly movable.

Symplocarpus, n., *sĭm'·plŏ·kărp'·ŭs* (Gr. *sumplokē*, a connection, an intertwining ; *karpos*, fruit), a curious genus of plants, Ord. Araceæ, bearing large handsome leaves: **Symplocarpus fœtidus**, *fĕt'·ĭd·ŭs* (L. *fœtĭdus*, stinking), the skunk-cabbage, has a very disagreeable odour, the rhizomes and seeds have been employed as antispasmodics.

Symplocos, n., *sĭm'·plŏk·ŏs* (Gr. *sumplokē*, a connection), a genus of plants, Ord. Styracaceæ, some of whose species are used as dyes, others as tea: **Symploceæ**, n. plu., *sĭm·plŏs'·ĕ·ē*, a Sub-order or tribe.

sympodium, n., *sĭm·pŏd'·ĭ·ŭm* (Gr. *sun*, together; *pous*, a foot, *podes*, feet), in *anat.*, a monster fœtus having its feet grown together ; in *bot.*, in forked branching, when the primary axis consists of the bases of consecutive bifurcations or branchings : also called **pseudaxis**, n., *sūd·ăks'·ĭs* (Gr. *pseudēs*, false, and *axis*), a false axis.

symptom, n., *sĭm'·tŏm* (Gr. *sumptōma*, what happens with another thing — from *sun*, together; *ptōma*, a fall), a token or mark which indicates disease, and specifically the kind of disease. NOTE.— Strictly speaking, a symptom is evidence of disease appreciable only by the patient (*subjective*), in contradistinction to signs or *objective* evidence.

synacme, n., *sĭn·ăk'·mē* (Gr. *sun-akmāzō*, I flourish at the same time with another), in *bot.*, the condition of stamens and pistils when they reach maturity at the same time.

Synandræ, n. plu., *sĭn·ănd'·rē* (Gr. *sun*, together ; *anēr*, a male, *andros*, of a male), in *bot.*, a division of gamopetalous Dicotyledons, having the carpels unequal in number to the parts of the other whorls, while the stamens are synantherous.

Synantheræ, n. plu., *sĭn·ănth'·ĕr·ē* (Gr. *sun*, together, with ; *anthēros*, flowery — from *anthos*, a flower), in *bot.*, a name sometimes given to the Order of plants Compositæ : **synantherous**, a., *sĭn·ănth'·ĕr·ŭs*, having anthers united so as to form a tube round the style.

synanthous, a., *sĭn·ănth'·ŭs* (Gr.

sun, together ; *anthos*, a flower), in *bot.*, having flowers united together : **synanthy**, n., *sĭn·ănth'ĭ*, the adhesion of several flowers.

synaptase, n., *sĭn'ăp·tāz* (Gr. *sun-aptos*, joined, united—from *sun*, together ; *aptō*, I connect or tie to), a substance called emulsin, a nitrogenous compound, found in certain oily seeds, as in almonds : **synapticulæ**, n. plu., *sĭn'ăp·tĭk'ŭl·ē* (dim.), transverse props, sometimes found in corals, extending across the loculi like the bars of a grate.

synarthrosis, n., *sĭn'ărth·rōz'ĭs* (Gr. *sunărthrŏs*, connected by a joint—from *sun*, together ; *arthron*, a joint), in *anat.*, a union of bones without motion, or but little motion.

syncarpium, n., *sĭn·kărp'ĭ·ŭm* (Gr. *sun*, together ; *karpos*, fruit), in *bot.*, an aggregate fruit having the carpels of a multiple ovary formed into a solid mass, with a slender receptacle : **syncarpous**, a., *sĭn·kărp'ŭs*, having the carpels united so as to form one ovary or pistil : **syncarpy**, n., *sĭn'kărp·ĭ*, the accidental adhesion of several fruits.

synchondrosis, n., *sĭn'kŏn·drōz'ĭs* (Gr. *sun*, together ; *chŏndrŏs*, cartilage or gristle), in *anat.*, the connection of bones by means of a plate of cartilage, as the ʻsacro-iliac-synchondrosis.'

synchronous, a., *sĭn'krŏn·ŭs* (Gr. *sun*, together, with ; *chrŏnŏs*, time), occurring at the same time; simultaneous.

syncope, n., *sĭn'kōp·ĕ* (Gr. *sŭng-kŏpē*, a cutting, a shortening—from *sun*, together ; *kŏptō*, I cut off), partial or total loss of consciousness from temporary failure of the normal action of the heart.

syngenesious, a., *sĭn'jĕn·ĕz'ĭ·ŭs* (Gr. *sun*, with ; *genesis*, generation, birth), in *bot.*, having the stamens united in a cylindrical form by the anthers.

synochreate, a., *sĭn·ŏk'rē·āt* (Gr. *sun*, with, together ; L. *ocrĕātus*, furnished with greaves or leggings), in *bot.*, having stipules uniting on the opposite side of the stem, enclosing it in a sheath.

synoicous, a., *sĭn·ōyk'ŭs* (Gr. *sun*, together ; *oikos*, a house), in *bot.*, having antheridia and archegonia on the same receptacle.

synostosis, n., *sĭn'ŏs·tōz'ĭs*, or *sĭn·ŏst'ŏs·ĭs* (Gr. *sun*, together ; *ŏstĕŏn*, a bone), in *anat.*, the premature obliteration of certain of the sutures of the skull.

synovia, n., *sĭn·ōv'ĭ·ă* (Gr. *sun*, together ; Gr. *ōŏn*, L. *ovum*, an egg), a viscid, transparent fluid, having a yellowish or faintly reddish tint, and a slightly saline taste, secreted in the cavity of joints for keeping them moist : **synovial membrane**, *sĭn·ōv'ĭ·ăl mĕm'brān* (L. *membrāna*, skin or membrane), a thin membrane which covers the extremities of bone joints, and the surface of the ligaments connected with the joints, and secretes the peculiar fluid called synovia ; **synovitis**, n., *sĭn'ŏv·ĭt'ĭs*, inflammation of the synovial membrane.

synspermous, a., *sĭn·spĕrm'ŭs* (Gr. *sun*, with ; *sperma*, seed), in *bot.*, having several seeds united : **synspermy**, n., *sĭn'spĕrm·ĭ*, the union of several seeds.

synsporous, a., *sĭn·spōr'ŭs* (Gr. *sun*, together ; *spora*, a seed), propagating by conjugation of cells, as in Algæ.

syntonin, n., *sĭn'tŏn·ĭn* (Gr. *sun*, together ; *tŏnŏs*, a tension, a bracing—from *teinō*, I stretch), a peculiar fibrin obtained from muscular fibre ; musculin.

syphilis, n., *sĭf'ĭl·ĭs* (Gr. *sus*, a sow, a swine ; *philos*, dear : Gr. *supheios*, a hog-stye), a form of venereal disease, a virulent and specific affection, the result of contagion.

Syringa, n., *sĭr·ĭng'·gă* (G. *suringx*, a pipe), a genus of trees having long straight branches filled with medulla, Ord. Oleaceæ : **Syringa vulgaris,** *vŭlg·ār'·ĭs* (L. *vulgāris*, common, vulgar), common lilac, whose bark is used as a febrifuge.

systole, n., *sĭst'·ōl·ē* (Gr. *sustŏlē*, a drawing together — from *sun*, together ; *stello*, I send), the contraction of the heart expelling the blood, and carrying on the circulation ; the contraction of any contractile cavity.

Tabernæmontana, n., *tăb'·ĕr·nē·mŏnt·ān'·ă* (after *Dr. Tabernæmontānus*, a great physician and botanist), a genus of interesting plants, Ord. Apocynaceæ, bearing sweet-scented flowers : **Tabernæmontanus utilis,** *ūt'·ĭl·ĭs* (L. *ūtĭlis*, profitable), Cow-tree or milk-tree of Demerara, juice used as milk.

tabes, n., *tāb'·ēz* (L. *tabes*, a wasting away), a wasting away gradually of the body, accompanied by languor and depressed spirits, with no apparent disease of the viscera : **tabes mesenterica,** *mĕs'·ĕn·tĕr'·ĭk·ă* (Gr. *mesĕntĕrĭŏn*, the mesentery), a tubercular disease of the mesenteric glands, generally a disease of childhood, characterised by emaciation and loss of appetite, and tenderness and distention of the abdomen.

tabulæ, n. plu., *tăb'·ūl·ē* (L. *tabula*, a tablet), horizontal plates or floors found in some corals.

tænia, n., *tēn'·ĭ·ă* (Gr. *tainia*, L. *tænia*, a ribbon), the intestinal flattened worm, usually called the tape - worm : **Tæniada,** n. plu., *tēn·ĭ'·ăd·ă*, the division of Scolecida, comprising the tape-worm : **tænioid,** a., *tēn'·ĭ·ōyd* (Gr. *eidos*, resemblance), shaped like a ribbon, as in the tape - worm : **tænia hippocampi,** *hĭp'·pō·kămp'·ĭ*

(L. *hĭppocămpus*, a sea-horse— from Gr. *hippos*, a horse ; *kampto*, I bend or curve), the tænia of the hippocampus ; a narrow white band running along the inner edge of the eminence, hippocampus major, of the lateral ventricles of the brain : **t. semicircularis,** *sĕm'·ĭ·sĕrk·ŭl·ār'·ĭs* (L. *semi*, half ; *circulāris*, circular), a narrow flat band lying between the optic thalamus, and the corpus striatum of the brain : **t. solii,** *sōl'·ĭ·ī* (L. *sŏlĭum*, a seat), the tape-worm at the seat ; a tape-worm of a flat, ribbon-like shape, from 6 to 10 or 20 feet in length, having at a part of the head a double row of hooks : **t. mediocanellata,** *mĕd'·ĭ·ō·kăn'·ĕl·lāt'·ă* (L. *medĭus*, the middle ; *canellatus*, reed-like—from *canna*, a reed), the reed - like tænia ; the most common tape-worm, exceeds in breadth, etc., the tænia solii, and has no hooklets, but has sucking discs instead : **t. echinococca,** *ĕ·kĭn'·ō·kŏk'·ă* (Gr. *echinos*, the hedgehog ; *kŏkkŏs*, a grain or berry), the grain and hedgehog tænia ; a small tape-worm, about a quarter of an inch long, so named from its fancied resemblance to these objects, found in the liver.

Talauma, n., *tăl·āwm'·ă* (a native name), a genus of plants, Ord. Magnoliaceæ, producing beautiful and fragrant flowers : **Talauma fragrantissima,** *frăg'·rănt·ĭs'·sĭm·ă* (L. *fragrans*, sweet-scented), a species which supplies the organ-nut of Brazil.

Taliacotian operation, *tăl'·ĭ·ă·kō'·shĭ·ăn*, the operation of forming a new nose, first performed by *Taliacotius*, a celebrated Chinese surgeon, who flourished about the beginning of the Christian era.

talo-scaphoid, *tăl'·ō·skăf'·ōyd* (L. *tālus*, the ankle - bone, and *scaphoid*), a membranous band of fibres, situated on the dorsum of the foot, extending from the extremity of the astragalus to the scaphoid bone.

talus, n., *tāl'ŭs* (L. *talus*, the ankle-bone, a die), in *anat.*, the 'astragalus,' which see.

Tamaricaceæ, n. plu., *tăm·ăr'·ĭ·kā'·sĕ·ē* (said to be after the river *Tamaris*, now *Tambro*, near the Pyrenees, on whose banks they grow : L. *tămărix*, the tamarisk), the Tamarisk family, an Order of plants, which have a bitter astringent bark, some yielding a quantity of sulphate of soda when burnt : **Tamarix**, n., *tăm'·ăr·iks*, a genus of very elegant shrubs : **Tamarix Gallica**, *găl'·ĭk·ă* (L. *gallĭcus*, of or from *Gallia* or *Gaul*) ; and **T. mannifera**, *măn·nĭf'·ĕr·ă* (L. *manna*, manna ; *fero*, I bear), are species which yield the saccharine substance Tamarisk, or Mount Sinai manna, caused by the puncture of an insect, the coccus manniparus : **T. Orientalis**, *ōr'·ĭ·ĕnt·āl'·ĭs* (L. *ŏrientālis*, eastern—from *orĭor*, I arise), a species of N. W. India, which furnish galls, used for oak-galls.

Tamarindus, n., *tăm'·ăr·ĭnd'·ŭs* (Ar. *tamar-hindi*, the Indian date), a genus of plants, Ord. Leguminosæ, Sub-order Caesalpinieæ : **Tamarindus Indica**, *ĭnd'·ĭk·ă* (L. *indĭcus*, of or from *India*), the Tamarind tree, from whose pericarp a laxative pulp is procured, forming a delicious confection.

Tanacetum, n., *tăn'·ăs·ēt'·ŭm* (F. *tanaisie*, the tansy, said to be a corruption of Gr. *athanasia*, immortality), a genus of plants, Ord. Compositæ, Sub-Ord. Corymbiferæ : **Tanacetum vulgare**, *vŭlg·ār'·ē* (L. *vulgāris*, common), the Tansy, whose leaves have stimulant, antispasmodic properties, containing also a bitter resin, and an aromatic volatile oil.

Tanghinia, n., *tăng·hĭn'·ĭ·ă* (tang-

hin, a Madagascar name), a genus of remarkable plants, Ord. Apocynaceæ : **Tanghinia venenata**, *vĕn'·ĕn·āt'·ă* (L. *venēnātus*, furnished with poison — from *vĕnēnum*, poison, a potion), a plant, the seeds of which, called Tangena nuts, supply the famous Tanghin poison, formerly used in Madagascar as an ordeal for criminals, and for witchcraft : **Tanghin**, n., *tăng'·hĭn*, the poison so called.

tannic, a., *tăn'·ĭk* (F. *tan*, Bret. *tann*, oak, bark of oak ; Ger. *tanne*, a fir tree), denoting a peculiar acid found in oak bark, and more abundantly in gall nuts, very astringent, converting the skins of animals into leather : **tannin**, n., *tăn'·ĭn*, another name for tannic acid, a powerful antiseptic or preservative from putrefaction.

tapetum, n., *tăp·ēt'·ŭm* (L. *tapētĕ*, or *tăpētum*, a carpet), in *anat.*, a silvery layer forming the lining on a greater or less extent of the back part of the choroid membrane of the eye, instead of the usual dark pigment in fishes and many mammals.

tape-worm, see ' tænia.'

taphrenchyma, n., *tăf·rĕng'·kĭm·ă* (Gr. *tăphrŏs*, a trench, a pit ; *en*, in ; *cheuma*, juice, tissue), in *bot.*, pitted vessels ; ' bothrenchyma,' which see.

tapping, n., *tăp'·ĭng* (from Eng. *tap*), the surgical operation of removing fluid from the body, as in dropsy ; paracentesis.

tap-root, in *bot.*, a conical root with branches striking off from it.

Taraxacum, n., *tăr·ăks'·ăk·ŭm* (Gr. *tărăxis*, a disorder of the bowels —from *tarassō*, I disturb), a genus of plants, Ord. Compositæ, Subord. Cichoraceæ ; **taraxacum dens-leonis**, *dĕns'·lĕ·ōn'·ĭs* (L. *dens*, the tooth ; *lĕō*, the lion, *lĕōnis*, of the lion), dandelion, whose roots yield a milky juice, and has been used as a diuretic and alterative; the root is prepared and mixed with coffee as chicory is, or is often used alone medicinally: **taraxacine**, n., *tăr·ăks'·ăs·ĭn*, a bitter crystalline principle obtained from the dandelion.

tarsus, n., *tărs'·ŭs* (Gr. *tarsos*, the sole of the foot, or its upper surface, also the edge of the eyelid), that part of the foot to which the leg is articulated, the front of which is called the instep; a thin layer of cartilage in the substance of each eyelid: **tarsi**, n. plu., *tărs'·ī*, the articulated feet of insects: **tarsalia**, n. plu., *tărs·āl'·ĭ·ă*, the bones of the tarsus: **tarsal**, a., *tărs'·ăl*, pert. to the instep: **meta-tarsus**, n., *mĕt'·ă·tărs'·ŭs* (Gr. *meta*, beyond, and *tarsos*), the front of the foot between the tarsus and the toes: **tarso-meta-tarsus**, the single bone in the leg of a bird, produced by the anchylosis of the lower and distal portion of the tarsus with the meta-tarsus: **tarso-meta-tarsal**, pert. to an articulation of the tarsus with the meta-tarsus.

tartar, n., *tărt'·ăr* (F. *tartre*, Sp. *tartaro*, tartar; mid. L. *tartărum*), a whitish saline substance, tartrate of potass, which, in the form of a crust, gathers on the sides of casks and vats containing wine; a white crust which gathers on the teeth of man: **tartareous**, a., *tărt·ār'·ĕ·ŭs*, in *bot.*, having a rough and crumbling surface: **tartaric**, a., *tărt·ăr'·ĭk*, of or from tartar; denoting an acid found in tartar, and in the juice of grapes and other fruit: **tartar emetic**, or **tartrate of antimony**, a preparation of antimony, which is a powerful emetic and depressant.

taurocholic, a., *tăwr'·ō·kŏl'·ĭk* (Gr. *tauros*, a bull; *chŏlē*, bile), denoting an acid procured from the bile of the ox, and found in quantity in the bile of man.

Taxineæ, n. plu., *tăks·ĭn'·ĕ·ē* (L. *taxus*, the yew tree; Gr. *taxis*, an arrangement, the leaves being arranged on the branches like the teeth of a comb), the Yew family, a Sub-ord. of plants, Ord. Coniferæ: **Taxus**, n., *tăks'·ŭs*, a genus of ornamental trees: **Taxus baccata**, *băk·kāt'·ă* (L. *bacca*, a berry, *baccātus*, furnished with berries), the Yew, forming a valuable timber tree; it yields resin, and its leaves and berries are narcotico-acrid.

taxis, n., *tăks'·ĭs* (Gr. *taxis*, order, arrangement), the process by which parts which have left their natural position in the body are reduced or replaced by the hand without the aid of instruments.

taxonomy, n., *tăks·ŏn'·ŏm·ĭ* (Gr. *taxis*, an arranging; *nŏmŏs*, law), the department of natural history which treats of the laws and principles of classification: **taxonomist**, n., *tăks·ŏn'·ŏm·ĭst*, one skilled in these laws and principles of classification.

Tectibranchiata, n. plu., *tĕk'·tĭ·brăng'·kĭ·āt'·ă* (L. *tectus*, covered; Gr. *brangchia*, gills), an Order of Molluscs having the branchiæ or gills covered, or partly covered, by the mantle.

Tectona, n., *tĕk·tŏn'·ă* (from its native name *tekka*), a genus of valuable timber trees, Ord. Verbenaceæ: **Tectona grandis**, *grănd'·ĭs* (L. *grandis*, great), the Teak tree of India, whose wood, very hard and durable, is used for shipbuilding.

tegmen, n., *tĕg'·mĕn* (L. *tegmen*, a covering), in *bot.*, the second covering of the seed.

tegmentum, n., *tĕg·mĕnt'·ŭm* (L. *tegmentum*, a covering), in *anat.*, the upper part of the main body of the peduncular fibres of the cerebrum: **tegmenta**, n. plu., *tĕg·mĕnt'·ă*, in *bot.*, the scaly coats which cover leaf-buds.

tegument, n., *tĕg'·ŭm·ĕnt* (L. *tegūmĕntum*, a covering), any natural covering or envelope : **tegumentary**, a., *tĕg'·ŭm·ĕnt'·ăr·ĭ*, connected with the tegument or skin.

tela, n., *tēl'·ă* (L. *tēla*, a web), in *anat.*, applied to any web-like tissue : **tela choroidea**, *kōr·ōyd'·ĕ·ă* (Gr. *chŏrĭŏn*, skin or leather ; *eidos*, resemblance), the choroid web ; in *anat.*, the membrane which connects the two choroid plexuses of the two sides of the cerebrum together.

teleangiectasis, n., *tĕl'·ĕ·ănj'·ĭ·ĕk'·tăs·ĭs* (Gr. *tēlē*, distant, remote ; *anggeion*, a vessel ; *ekteinō*, I distend), the expansion of the remote vessels ; a disease of the capillaries, called 'aneurism by anastomosis,' or 'erectile tumour' ; a congenital affection, presenting a cutaneous swelling of a circumscribed form.

Teleostei, n. plu., *tĕl'·ĕ·ŏst'·ĕ·ī* (Gr. *teleŏs*, perfect ; *ŏstĕon*, bone), the Order of the Bony-fishes.

teleutospores, n. plu., *tĕl·ūt'·ō·spōrz* (Gr. *teleutē*, an end, a conclusion ; *spora*, seed), the spores of the preceding generation ; long two-celled spores ending the vegetation of Puccinia, and beginning a new generation in spring.

telson, n., *tĕls'·ŏn* (Gr. *tĕlsŏn*, the end, extremity), the last joint in the abdomen of Crustaceæ, forming a supposed segment without appendages.

temporal, a., *tĕmp'·ŏr·ăl* (L. *temporālĭs*, lasting but for a time — from *tempus*, time, *tĕmpŏrĭs*, of time), pert. to or relating to the temples, as the temporal bone, the temporal arteries, etc.: **temporo-facial**, *tĕmp'·ŏr·ō·făsh'·ĭ·ăl* (L. *făcĭĕs*, the face), the larger of the two temporal branches of the facial nerve : **t. maxillary**, *măks'·ĭl·lăr·ĭ*, or *măks·ĭl'·lăr·ĭ*, applied to the articulation of the lower

jaw by its condyle on each side with the smooth surface of the temporal bone ; one of the veins of the head : **t. parietal**, *păr·ī'·ĕt·ăl*, a suture which joins the temporal and parietal bones of the skull.

tendo Achillis, *tĕnd'·ō ăk·ĭl'·lĭs* (F. *tendon*, L. *tendo*, a tendon, the end of a muscle—from *tendo*, I stretch ; *Achillis*, of *Achilles*), the tendon of Achilles, a strong tendon which is inserted into the heel, so called from the heel having been the only vulnerable part in Achilles, from a wound in which in battle he died.

tendon, n., *tĕnd'·ŏn* (F. *tendon*, L. *tendo*, the end of a muscle, a tendon—from *tendo*, I stretch ; Gr. *tenon*, a tendon—from *teino*, I stretch), a fibrous cord at the extremity of a muscle, by which the muscle is attached to a bone : **tendon of insertion**, the part of the tendon by which a muscle is attached to a bone : **tendon of the biceps** (see ' biceps '), the tendon at the end of the biceps muscle.

tendril, n., *tĕnd'·rĭl* (F. *tendron*, the tender shoot of a plant ; old F. *tendrillon*, a tendril : L. *tĕner*, tender), the twisting claws of a climbing plant by which it attaches itself to an object for support.

tension, n., *tĕn'·shŭn* (L. *tensus*, drawn tight), the act of stretching or straining ; the state of being stretched to its full length : **tensor**, n., *tĕns'·ŏr*, a muscle which stretches any part : **tensor palati**, *păl·āt'·ī* (L. *pălātum*, the palate, *pălātĭ*, of the palate), the tensor of the palate : **t. tarsi**, *tărs'·ī* (Gr. *tarsos*, the edge of the eyelid, the sole of the foot), one of the small muscles of the eyelids : **t. vaginæ femoris**, *vădj·ĭn'·ē fĕm'·ŏr·ĭs* (L. *tensor*, a stretcher ; *vāgīna*, a sheath, *văgīnæ*, of a sheath ; and *fĕmur*, the thigh, *fĕmŏris*, of a thigh), a muscle which assists in the adduction of the thigh, also assisting to some extent in its rotation inwards.

tentacle, n., *těnt'·ă·kl*, tentacles, n. plu., *těnt'·ă·klz*, also **tentaculum,** n., *těnt·ăk'·ūl·ŭm,* tentacula, n. plu., *-ūl·ă* (new L. *tenĭăcŭlum,* a feeler—from L. *tento,* I handle or touch), slender flexible organs proceeding from the heads of many smaller animals, used for the purpose of feeling, exploring, prehension, or attachment, etc., as in snails, insects, crabs, etc. : **tentacular,** a., *těnt·ăk'·ūl·ăr,* resembling the feelers of a snail.

tentorium, n., *těnt·ōr'·ĭ·ŭm* (L. *těntōrium,* a tent—from *tendo,* I stretch), an elevated part in the middle of the dura-mater, declining downwards, and corresponding in form with the upper surface of the cerebellum ; also called **tentorium cerebelli,** *sěr'·ěb·ěl'·ī,* a roof of dura - mater thrown across the cerebellum.

tenuirostres, n. plu., *těn'·ū·ĭ·rŏst'· rēz* (L. *těnŭĭs,* slender ; *rostrum,* a beak), a group of the perching birds, characterised by their slender beaks : **tenuirostral,** a., *-rŏst'·răl,* slender-beaked.

Tephrosia, n., *těf'·rŏz'ĭ·ă* (Gr. *těphros,* ash-coloured), a genus of plants, Ord. Leguminosæ, Sub-order Papilionaceæ, so named from the colour of the foliage of some species : **Tephrosia Apollinea,** *ăp'·ŏl·lĭn'ĕ·ă* (L. *Apŏllĭnĕus,* of or pert. to *Apollo*), a species whose leaves are purgative, and occasionally mixed with senna : **T. toxicaria,** *tŏks'·ĭk·ār'·ĭ·ă* (Gr. *toxĭkŏn,* L. *toxĭcum,* poison in which arrows were dipped), a species whose leaves and branches, well-pounded, and thrown into a river, powerfully affect the water, and intoxicate the fish ; cultivated in Jamaica for its intoxicating qualities.

teratology, n., *těr'·ăt·ŏl'·ō·jĭ* (Gr.

těras, a sign or wonder, *těrătos,* of a sign or wonder ; *logos,* discourse), that branch of physiology which treats of malformations and monstrosities in animals or plants.

tercine, n., *těrs'·n* (F. *tercine,* L. *tertius,* the third), in *bot.,* the third coat of the ovule, forming the covering of the central nucleus.

teres, n., *těr'·ēz* (L. *těres,* long and round, tapering as a tree), the name of two muscles, the **teres minor** (L. *minor,* less), and the **teres major** (L. *major,* greater), arising from the scapula and inserted into the humerus.

terete, a., *těr·ēt'* (L. *těres,* tapering as a tree, *těrětis,* of tapering as a tree), in *bot.,* nearly cylindrical ; having the transverse section nearly circular.

tergum, n., *těrg'·ŭm* (L. *tergum,* the back), among insects, the upper surface of the abdomen ; the dorsal arc in a somite.

Terminalia, n. plu., *těrm'·ĭn·āl'·ĭ·ă* (L. *termĭnus,* an end), a genus of plants, Ord. Combretaceæ, so named as having the leaves in bunches at the ends of the branches : **Terminalia bellerica,** *běl·lěr'·ĭk·ă* (Sp. *belěrico,* the fruit myrobalan), and **T. chebula,** *kěb'· ŭl·ă* (native name, E. I.), whose fruit, known as Myrobalans, is used as an astringent ; the fruit and galls used by dyers : **T. cattappa,** *kăt·ăp'·ă* (native name, E. I.), a species whose seeds are eaten as almonds ; the leaves and bark yield a black pigment, forming an Indian ink : **T. angustifolia,** *ăng·gŭst'·ĭ·fŏl'·ĭ·ă* (L. *angustus,* narrow ; *fŏlium,* a leaf), a species yielding a milky juice, which, dried, is fragrant, and is used as a kind of incense : **Terminalieæ,** n., *těrm'·ĭn·ăl·ĭ'ĕ·ē,* a tribe or Sub-order.

ternary, a., *těrn'·ăr·ĭ* (L. *ternus,* three each), threefold ; having parts arranged in threes: **ternate,** a., *tern'·ăt*, having compound leaves consisting of three leaflets; arranged by threes.

Ternstrœmiaceæ, n. plu., *tern'·strēm·ĭ·ā'·sĕ·ē* (after *Ternström*, a Swedish naturalist, 1745), the Tea family, an important Order of plants, yielding the various kinds of tea: **Ternstrœmia,** n., *tern·strēm'·ĭ·ă*, an interesting genus of plants.

tertian, a., *tĕr'·shĭ·ăn* (L. *tertĭus*, the third), occurring every third day, as a fever.

test, n., *tĕst* (L. *testa*, a shell), in *zool.*, the shell of Mollusca, thus sometimes called **testacea,** *tĕst·ā'·sĕ·ă*: **testa,** n., *tĕst'·ă*, **testæ,** n. plu., *tĕst'·ē*, in *bot.*, the outer covering of the seed; the shelly covering of certain animals: **testaceous,** a., *tĕst·ā'·shŭs*, having a hard, shelly covering.

testes, n. plu., *tĕst'·ēz* (L. *testis*, a witness), the organs in male animals which produce the semen or generative fluid; the testicles.

testicles, n. plu., *tĕst'·ĭk·ĭlz* (L. *testĭcŭlus*, a testicle, dim. of *testis*), the two male organs of generation: **testiculate,** a., *tĕst·ĭk'·ŭl·āt*, in *bot.*, having two oblong tubercules, as the roots in some Orchids.

Testudinaria, n., *tĕst·ŭd'·ĭn·ār'·ĭ·ă* (L. *testūdo*, a tortoise, *testūdĭnis*, of a tortoise), a genus of curious and interesting plants, Ord. Dioscoreaceæ, so named from the outside resemblance of the roots: **Testudinaria elephantipes,** *ĕl'·ĕf·ănt'·ĭp·ēz* (L. *elephas*, an elephant, *elephantis*, of an elephant; *pēs*, a foot, *pĕdis*, of a foot), the Tortoise plant, or elephant's foot, of the Cape, so named from its peculiar, thickened stem.

tetanus, n., *tĕt'·ăn·ŭs* (L. *tĕtănus*, Gr. *tĕtănos*, a stiffness or spasm of the neck—from Gr. *teinō*, I stretch), a diseased condition, characterised by painful and rigid contraction of the voluntary muscles, aggravated from time to time by very severe spasms; named **traumatic tetanus** when it comes on after wounds, and **idiopathic tetanus** when the symptoms exhibit themselves without any manifest cause: **tetanic,** a., *tĕt·ăn'·ĭk*, pert. to or affected by such symptoms as occur in tetanus; of or pert. to tetanus.

Tetrabranchiata, n. plu., *tĕt'·ră·brăngk·ĭ·āt'·ă* (Gr. *tetra*, four; *brangchia*, gills), an Order of the Cephalopoda, characterised by having four gills: **tetrabranchiate,** a., *-ĭ·āt*, having four gills.

tetradynamous, a., *tĕt'·ră·dĭn'·ăm·ŭs* (Gr. *tetra*, four; *dunămis*, power), in *bot.*, having four long stamens and two short, as in Cru ciferæ.

tetragonous, a., *tĕt·răg'·ŏn·ŭs*, also **tetragonal,** a., *tĕt·răg'·ŏn·ăl* (Gr. *tetra*, four; *gŏnĭa*, a corner), in *bot.*, having four angles, the faces being convex: **Tetragonia,** n., *tĕt'·ră·gōn'·ĭ·ă*, a genus of plants, Ord. Ficoideæ or Mesembryaceæ: **Tetragonia expansa,** *ĕks·păns'·ă* (L. *expansus*, spread apart, expanded), a species called New Zealand spinach.

tetragynous, a., *tĕt·rădj'·ĭn·ŭs* (Gr. *tetra*, four; *gūnē*, a female), in *bot.*, having four carpels or four styles.

tetramerous, a., *tĕt·răm'·ĕr·ŭs* (Gr. *tetra*, four; *mĕrŏs*, a part), in *bot.*, composed of four parts, or in fours, or in multiples of four.

tetrandrous, a., *tĕt·rănd'·rŭs* (Gr. *tetra*, four; *anēr*, a male, *andros*, of a male), in *bot.*, having four stamens.

Tetranthera, n., *tĕt'·rănth·ēr'·ă* (Gr. *tetra*, four; *anthērŏs*, flowery), a genus of plants, Ord. Lauraceæ: **Tetranthera laurifolia,** *lawr'·ĭ·fōl'·ĭ·ă* (L. *laurus*, the laurel; *fōlĭum*, a leaf), a species whose leaves and branches abound in a viscid juice, and the fruit yields a solid, rank-smelling fat, used for making candles.

tetrapetalous, a., *tĕt'·ră·pĕt'·ăl·ŭs* (Gr. *tetra*, four ; *petălŏn*, a leaf), in *bot.*, containing four distinct petals, flowers, or leaves.

tetrapterous, a., *tĕt·răp'·tĕr·ŭs* (Gr. *tetra*, four ; *ptĕrŏn*, a wing), having four wings.

tetraquetrous, a., *tĕt'·ră·kĕt'·rŭs* or *tĕt·răk'·trŭs* (Gr. *tetra*, four ; L. *quadra*, a square), in *bot.*, having four angles, the faces being concave ; see 'tetragonous.'

tetraspore, n., *tĕt'·ră·spōr* (Gr. *tetra*, four; *spora*, a seed), among the Algæ, reproductive bodies composed of four spores or germs: **tetrasporous**, a., *tĕt'·ră·spōr'·ŭs*, bearing tetraspores.

tetrathecal, a., *tĕt'·ră·thĕk'·ăl* (Gr. *tetra*, four ; *thēkē*, a case), in *bot.*, having four loculaments or thecæ.

tetter, n., *tĕt'·tĕr* (Icel. *titra*, to tremble ; Ger. *zitter*, a tetter), a skin-disease, often appearing on the face and side of the mouth ; herpes.

Thalamifloræ, n. plu., *thăl'·ăm·ĭ·flōr'·ē* (L. *thălămus*, a receptacle ; *flos*, a flower, *flōris*, of a flower), in *bot.*, a Sub-class of the class of plants Dicotyledones or Exogenæ : **thalamifloral**, a., *thăl'·ăm·ĭ·flōr'·ăl*, and **thalamiflorous**, a., *·flōr'·ŭs*, having the petals and stamens inserted on the thalamus or receptacle.

thalamium, n., *thăl·ăm'·ĭ·ŭm* (L. *thălămus*, a receptacle), in *bot.*, the layer of reproductive cells in the apothecia of Lichens : **thalamus**, n., *thăl'·ăm·ŭs*, the receptacle of a flower.

thalamus opticus, *thăl'·ăm·ŭs ŏp'·tĭk·ŭs*, **thalami optici**, plu., *thăl'·ăm·ī ŏp'·tĭs·ī* (L. *thălămus*, a bedroom, a receptacle ; *optĭcus*, optic), the posterior ganglia of the brain, which are of an oval shape, and rest on the correspond-

ing cerebral crura, which they in a manner embrace.

Thalassa-collida, n. plu., *thăl·ăs'·ă·kŏl'·lĭd·ă* (Gr. *thalassa*, the sea ; *kolla*, glue), a division or group of Protozoa.

thalline, n., *thăl'·ĭn* (Gr. *thăllŏs*, L. *thallus*, a young shoot or branch), in *bot.*, of the same substance as the thallus : **thallodal**, a., *thăl'·ŏd·ăl*, in same sense.

Thallogenæ, n. plu., *thăl·lŏdj'·ĕn·ē* (Gr. *thăllŏs*, a young shoot; *gĕnŏs*, birth, *gĕnnăō*, I beget), a Subclass of Cryptogamic plants: **thallogens**, n. plu., *thăl'·lō·jĕnz*, also **thallophytes**, n. plu., *·fīts* (Gr. *phutŏn*, a plant), plants producing a thallus ; plants bearing their fructification on a thallus ; also **thallophyta**, n. plu., *thăl·ŏf'·ĭt·ă*, in same sense.

thallus, n., *thăl'·lŭs*, also **thallome**, *thăl'·lŏm·ĕ* (Gr. *thăllŏs*, a young shoot, a frond), in *bot.*, a solid mass of cells, consisting of one or more layers, usually in the form of a flat stratum or expansion, or in the form of a lobe, leaf, or frond ; any structure having no morphological distinction of stem and leaves, and from which true roots are absent ; the vegetative system of Lichens.

Thea, n., *thē'·ă* (new L. *thēă*, the tea plant—said to be from *tcha*, the Chinese name for tea), a genus of valuable plants, Ord. Ternstrœmiaceæ : **Thea viridis**, *vir'·ĭd·ĭs* (L. *virĭdis*, green), the species artificially cultivated in Britain ; but there is said to be only one species, the different teas depending upon their mode of treatment, and their preparation for the market : **T. Bohea**, *bō·hē'·ă* (said to be from *Buoy* or *Booy*, a mountain in China), the common black tea, also called **T. Cantoniensis**, *kăn·tōn'·ĭ·ĕns'·ĭs* (of or from *Canton*): **T. Assamica**, *ăs·săm'·ĭk·ă* (of or from *Assam*), the Assam tea-plant : **theine**, n., *thē'·ĭn*, a bitter principle found in tea.

theca, n., *thēk'·ă*, **thecæ**, n. plu., *thēs'·ē* (Gr. *thēkē*, a sheath or case), the case containing the reproductive matter in some flowerless plants; spore cases of Mosses, and such like plants; in *anat.*, an organ or a part enclosing another, or which contains something : **thecaphore**, n., *thēk'·ă·fōr* (Gr. *phorĕō*, I bear), in *bot.*, the roundish stalk on which the ovary of some plants is elevated : **thecasporous**, a., *thēk·ăs'·pōr·ŭs* (Gr. *spora*, seed), applied to Fungi which have their spores placed in thecæ : **theciferous**, a., *thēs·ĭf'·ĕr·ŭs* (L. *fero*, I bear), bearing thecæ or asci.

Thecosomata, n. plu., *thēk'·ŏ·sŏm'·ăt·ă* (Gr. *thēkē*, a sheath ; *sōmă*, a body, *sōmătŏs*, of a body), a division of Pteropodous Molluscs, in which the body is protected by an external shell.

thenar, n., *thēn'·ăr* (Gr. *thĕnar*, the palm of the hand), in *anat.*, the fleshy mass which forms the ball of the thumb, consisting of four muscles : **thenal**, a., *thēn'·ăl*, pert. to or connected with the thenar.

Theobroma, n., *thē'·ō·brŏm'·ă* (Gr. *thĕŏs*, a god ; *brōma*, food), a genus of plants, Ord. Bittneriaceæ : **Theobroma cacao**, *kăk·ā'·ō* (said to be from Mexican *cacanatl*), a species producing the seeds or beans which are the chief ingredient in chocolate, and from which the best cocoas are wholly manufactured : **theobromine**, n., *thē'·ō·brŏm'·ĭn*, a crystalline principle, analogous to caffeine, obtained from the cacao beans.

Theophrasta, n., *thē'·ŏf·răst'·ă* (after *Theophrastus*, the father of natural history), a genus of plants, Ord. Myrsinaceæ : **Theophrasta Jussiæi**, *jŭs'·st·ē'·ī* (after *Antoine de Jussieu*, a botanist of Paris), a prickly-leaved shrub,

called Coco in St. Domingo, whose seeds are eatable, and made into a kind of bread.

therapeutic, a., *thĕr'·ăp·ūt'·ĭk*, also **therapeutical**, a., *-ĭk·ăl* (Gr. *therapeutikos*, having the power of healing — from *therapeuō*, I heal), pert. to the healing art ; curative : **therapeutics**, n. plu., *-ūt'·ĭks*, that department of medicine relating to the discovery and application of remedies for the cure of diseases.

thorax, n., *thōr'·ăks* (L. and Gr. *thōrax*, the breast, defensive armour for the breast), the chest; that part of the trunk situated between the neck and the abdomen, containing the heart, lungs, etc.: **thoracic**, a., *thōr·ăs'·ĭk*, of or relating to the thorax : **thoracic duct**, the common trunk of nearly all the lymphatic vessels of the body, which conveys the great mass of the lymph and chyle into the blood.

thrombus, n., *thrŏm'·bŭs* (Gr. *thrombos*, a clot of blood), in *surg.*, a plug formed in a vessel during life, or some time before death, generally in veins, but may occur in an artery, or even in the heart : **thrombosis**, n., *thrŏm·bōz'·ĭs* (Gr. *thrŏmbōsis*, a curdling or coagulation), the process of the coagulation of blood in the vessels during life.

thrum-eyed, a., *thrŭm·īd'* (Ger. *trumm*, a short, thick piece), in *bot.*, having short styles in flowers, as when the stigma does not appear at the upper part of the tube of the corolla, as seen in Primula.

thrush, n., *thrŭsh* (a corruption of Eng. *thrust*, a breaking out), an affection of the intestinal tract, usually met with in children, as a result of imperfect nutrition, and recognisable by the appearance of white specks and patches in the mouth and throat, which on examination are found to consist of a vegetable fungus, viz. the Oidium albicans.

Thuja, n., *thū'jă* (Gr. *thŭŏn*, incense, perfume), a genus of aromatic plants, Ord. Coniferæ, Sul -ord. Cupressineæ: **Thuja articulata**, *ărt·ĭk'ŭl·āt'ă* (L. *artĭcŭlātus*, furnished with joints, distinct), the Arar tree, which supplies a solid resin called Sandarach or Pounce, used to strew over MSS.: **T. occidentalis**, *ŏk'sĭd·ĕnt·āl'ĭs* (L. *occidentālis*, western), the common Arbor vitæ of gardens—so named from its supposed medicinal qualities: **T. orientalis**, *ōr'ĭ·ĕnt·āl'ĭs* (L. *orientālis*, eastern), a species also cultivated: **Thus**, *thŭs* (L. *thŭs*, or *tŭs*, incense), common frankincense, yielded by Pinus palustris and P. tæda, Ord. Coniferæ.

Thymelæaceæ, n. plu., *thĭm·ĕl'ē·ā'sĕ·ē* (L. *thymĕlœă*, the flax-leaved Daphne plant), the Daphne family, an Order of plants, the bark of many of which is acrid and irritant, and the fruit is often narcotic; see 'Daphne.'

thymus, n., *thĭm'ŭs* (Gr. *thŭmŏn*, a fleshy excrescence on the skin), a temporary organ of childhood, consisting of two lateral lobes, placed partly in the neck, and extending from the fourth costal cartilage upwards, as high as the border of the thyroid gland; it attains its full size at the end of the second year, after which it gradually dwindles, and almost disappears at puberty.

thyro-, *thĭr'.ō*, and **thyreo-**, *thĭr'.ĕ.ō* (Gr. *thŭrĕŏs*, a shield), a prefix in anatomical terms denoting connection with the thyroid cartilage: **thyroid**, a., *thĭr'.ōÿd* (Gr. *eidos*, resemblance), applied to one of the cartilages of the larynx, so named from its shield-like form; applied also to a glandular body lying in front of this cartilage, or to the arteries supplying the part; denoting the large bone at the bottom of the trunk, from its shield-like shape: **thyro-arytenoid**, *-ăr'.ĭt·ēn'.ōÿd*, a ligament, consisting of a thin band of elastic tissue, attached in front to the angle of the thyroid cartilage below the epiglottis, and named the superior or false vocal cords: **inferior thyro-arytenoid**, two strong fibrous bands, each consisting of a band of yellow elastic tissue, attached in front to the depression between the two alæ of the thyroid cartilage, and behind to the anterior angle of the base of the arytenoid, and named the inferior or true vocal cords.

thyrohyal, a., *thĭr'.ō·hĭ'.ăl* (Gr. *thurĕŏs*, a shield, the U-shaped bone; see 'hyo'), applied to two ossifications of the *hyoid* in the lower vertebrata; homologue of the larger horn of the hyoid bone in man.

thyrsus, n., *thĕrs'.ŭs*, also **thyrse**, n., *thĕrs* (L. *thyrsus*, Gr. *thursos*, a stalk, a stem), in *bot.*, a species of inflorescence; a very compact pannicle, as the flowers of the lilac, or as having the appearance of a bunch of grapes.

Thysanura, n. plu., *thĭs'ăn·ūr'.ă* (Gr. *thusanoi*, tassels, fringes; *oura*, a tail), an Order of Apterous Insects: **thysanurous**, a., *thĭs'.ăn·ūr'.ŭs*, having fringed tails.

tibia, n., *tĭb'.ĭ·ă* (L. *t nia*, a pipe or flute), the larger of the two bones of a leg, so called from its supposed resemblance to an ancient flute—the upper part resembling the expanded or trumpet-like end, and the lower the flute end; the shin-bone: **tibialis anticus**, *tĭb'.ĭ·āl'.ĭs ănt·ĭk'.ŭs* (L. *tibĭālis*, of or pert. to the shin-bone; *anticus*, in front), the fore part of the tibial muscle; one of two muscles of the tibia which bend the foot by drawing it upwards, etc.: the other is **tibialis posticus**, *tĭb'.ĭ·āl'.ĭs pŏst·ĭk'.ŭs* (L. *posticus*, behind the back part), the back part of the third muscle.

tic, n., *tĭk*, the common and short name for tic douloureux, *tĭk dŏō·lōōr·ŏ'* (F. *tic*, a spasm; *douloureux*, painful), that form of neuralgia affecting specially the fifth or sensory nerve of the face.

tigellus, n., *tĭdj·ĕl'lŭs*, tigella, n., *tĭdj·ĕl'lă*, and tigelle, n., *tĭdj·ĕl'* (new L. *tigellus*—from F. *tigelle*, from F. *tige*, a stem), the portion of the embryo between the radicle and cotyledons; the young embryonic axis : tigellary, a., *tĭdj·ĕl'lăr·ĭ*, having the sheathing portion of a leaf incorporated with the stem.

Tiliaceæ, n. plu., *tĭl'ĭ·ā'sĕ·ē* (L. *tĭlĭa*, the lime or linden tree), the Lime tree family, an Order of plants, many possessing mucilaginous properties, others furnish cordage : Tilia, n., *tĭl'ĭ·ă*, a genus of lofty, ornamental trees : Tilia Europæa, *ūr'·ōp·ē'ă* (of or from *Europe*), a species whose inner bark, called bast or bass, is tough and fibrous, and is made into Russian mats.

Tillandsia, n., *tĭl·lănd'sĭ·ă* (after *Tillands*, a physician and professor), a genus of interesting epiphytal plants, Ord. Bromeliaceæ: Tillandsia usneoides, *ŭs'·nē·ōȳd'·ēz* (*achneh*, the Arabic name for lichens ; Gr. *eidos*, resemblance), a species which has the appearance of the Beard moss, and is used for stuffing cushions, etc.; Tillandsias are hung from balconies in S. America as air-plants.

tinea, n., *tĭn'·ĕ·ă* (L. *tĭnĕa*, a gnawing worm), a general name for parasitic disease of the scalp ; pustular inflammation at the eyelashes : tinea sycosis, *sĭk·ōz'·ĭs* (Gr. *sūk'ȳn*, a fig, a fleshy tumour on the eyelids), a pustular inflammation affecting the roots of the hair of the eyelashes, beard, etc.

Tinospora, n., *tĭn·ŏs'·pŏr·ă* (L. *tĭnĕa*, a gnawing worm; Gr. *spora*, seed), a genus of plants, Ord. Menispermaceæ, consisting of climbing Indian shrubs, having extreme vitality: Tinospora cordifolia, *kŏrd'·ĭ·fōl'·ĭ·ă* (L. *cor*, the heart, *cordis*, of the heart ; *folium*, a leaf, *folĭa*, leaves), a species whose young shoots are used as emetics : T. cordifolia, and T. crispa, *krĭsp'·ă* (L. *crĭspŭs*, curled, wrinkled), species from which a bitter principle, called Guluncha, is obtained, considered a specific for the bites of poisonous insects and for the cure of ulcers ; administered also as a diuretic and tonic in fever, and also for snake-bites.

tissue, n., *tĭsh'·ū* (F. *tissu*, woven —from F. *tisser*, L. *texere*, to weave), in *anat.* or *bot.*, the minute elementary structures of which organs are composed, whether of animals or of plants.

tobacco, n., *tō·băk'·kō* (*tabaco*, Indian name for the pipe or tube in which they smoked, and transferred by the Spaniards to the plant itself; Sp. *tabaco*, F. *tabac*), an annual plant, Order Solanaceæ, having dingy-red, infundibuliform flowers, and large viscid leaves ; employed medicinally as a tincture, infusion, and wine, its oil is one of the most deadly of known poisons, acts medicinally as a sedative.

Toddalia, n., *tŏd·dāl'·ĭ·ă* (a native Malabar name), a genus of plants, Ord. Xanthoxylaceæ: Toddalia aculeata, *ăk·ūl'·ĕ·āt'·ă* (L. *aculĕātus*, thorny, prickly), a prickly, climbing plant of the Indian peninsula, etc., whose root furnishes a pungent aromatic, used in the cure of remittent fevers.

tomentose, a., *tōm'·ĕnt·ōz'* (L. *tōmentum*, a stuffing for cushions), in *bot.*, covered with hairs so close as scarcely to be discernible; having a whitish down like wool:

tomentum, n., *tōm·ĕnt'·ŭm*, in *bot.*, the closely matted hair or downy nap covering the leaves or stems of some plants; in *anat.*, the minutely-divided vessels on the surface of the brain, having a very flocculent appearance.

tomiparous, a., *tōm·ĭp'·ăr·ŭs* (Gr. *tomē*, a cutting; *tŏmŏs*, a slice; L. *părĭo*, I bring forth), in *bot.*, producing spores by division.

tone, n., *tōn* (Gr. *tŏnŏs*, a stretching, a tension), the state of the body in regard to the healthy performance of its animal functions: **tonic**, a., *tŏn'·ĭk*, imparting vigour to the bodily system: n., a medicine or agent which imparts vigour and strength to the body; a stomachic: **tonicity**, n., *tŏn·ĭs'·ĭt·ĭ*, a state of healthy tension of muscular fibres while at rest.

tonsils, n. plu., *tŏns'·ĭls* (L. *tonsillæ*, the tonsils of the neck—from *tonsilis*, shorn or clipt), two oblong glands situated on each side of the fauces at the base of the tongue: **tonsillitis**, n., *tŏns'·ĭl·ĭt'·ĭs*, inflammation of the tonsils, a form of sore throat: **tonsillitic**, a., *tŏns'·ĭl·lĭt'·ĭk*, related to or connected with the tonsils.

topical, a., *tŏp'·ĭk·ăl* (Gr. *topĭkos*, pert. to a place—from *tŏpŏs*, a place), in *med.*, pert. to an external local remedy, as a poultice, a blister, and the like.

torcular Herophili, *tŏrk'·ūl·ăr hĕr·ŏf'·ĭl·ī* (L. *torcŭlar*, a wine-press; *Herophili*, of *Herŏphĭlus*), the wine-press of Herophilus; the confluence or common point to which the venous sinuses converge, which are contained in the several processes or folds of the dura-mater of the brain.

tormentil, n., *tŏr·mĕnt'·ĭl* (L. *tormentum*, torture, anguish), the root of the **Potentilla tormentilla**, Ord. Rosaceæ, used as an astringent: **tormentilla erecta**, *tŏr·mĕnt·ĭl'·lă ēr·ĕkt'·ă* (L. *ērĕctus*, raised or set up), a species whose roots are used in the western isles of Scotland for tanning.

torrefaction, n., *tŏr'·rĕ·făk'·shŭn* (L. *torrĕŏ*, I dry or burn; *făcĭŏ*, I make), the operation of drying or scorching by fire, as in roasting or drying drugs.

Torula, n., *tŏr'·ūl·ă* (L. *torŭlus*, a tuft of hair), the yeast plant, a genus of Fungi: **torulose**, a., *tŏr'·ūl·ōz'*, exhibiting a succession of rounded swellings, as in the pods of some cruciferous plants.

torus, n., *tŏr'·ŭs* (L. *tŏrus*, a round swelling, a couch), in *bot.*, the axis on which all the parts of the floral whorls within the calyx are seated; a thalamus.

Totipalmatæ, n. plu., *tōt'·ĭ·păl·māt'·ē* (L. *tōtus*, whole; *palma*, the palm of the hand), in *zool.*, a group of Wading birds, having the hallux united to the other toes by a membrane in such a manner that the feet are completely webbed.

tourniquet, n., *tŏr'·nĭ·kĕt* (F. *tourniquet* — dim. from *tourner*, to turn), an instrument for the mechanical compression of a vessel for the prevention of hæmorrhage.

toxicology, n., *tŏks'·ĭk·ŏl'·ŏ·jĭ* (Gr. *toxĭkon*, poison; *logos*, discourse), the branch of medical science which relates to poisons, their effects, detection, and antidotes.

Toxicophlœa, n., *tŏks'·ĭk·ŏf·lē'·ă* (Gr. *toxĭkon*, poison; *phloios*, the bark of a tree), a genus of plants generally poisonous, Ord. Apocynaceæ: **Toxicophlœa Thunbergii**, *tŭn·bĕrj'·ĭ·ī* (after *Thunberg*, a botanist), a species whose bark is used to poison fish with at the Cape.

trabecula, n., *trăb·ĕk'·ūl·ă*, trabeculæ, n. plu., *-ūl·ē* (L. *trăbĕcŭla*, a little beam or rafter—from *trabs*, a beam), in *anat.*, numerous fibrous bands proceeding from the inner surface of the corpus cavernosum; two thick bars presented by the

2 D

blastema in the membranous condition of the primordial cranium ; a reticular framework of whitish elastic bands, forming a portion of the substance of the spleen ; in *bot.*, fibrous bands crossing from the ventral to the dorsal side of the microsporangia and macrosporangia of certain Lycopodiaceæ, and causing an imperfect segmentation : **trabecular,** a., *trăb·ĕk'·ŭl·ăr*, of or pert. to the bands or fibres forming a connecting or bounding medium in bodies or organs : **trabeculate,** a., *trăb·ĕk'·ŭl·āt*, in *bot.*, having horizontal cross bars, as on the inner surface of the teeth of the peristome.

trabs cerebri, *trăbz sĕr'·ĕb·rĭ* (L. *trabs*, a beam or rafter ; *cĕrĕbrum*, the brain, *cĕrĕbrĭ*, of the brain), the corpus callosum of the brain.

trachea, n., *trăk·ē'·ă* (Gr. *trachus*, masc., *tracheia*, fem., rough, rugged ; *tracheia*, the windpipe), the windpipe ; the common air passage of both lungs, consisting of an open tube commencing at the larynx above, and dividing below into two smaller tubes, the right and left bronchus—one for each lung : **tracheæ,** n. plu., *trăk·ē'·ē*, the breathing tubes of Insects and other articulate animals ; in *bot.*, spiral vessels in plants : **tracheitis,** *trăk'·ē·ĭt'·ĭs*, or **trachitis,** n., *trăk·ĭt'·ĭs*, inflammation of the trachea.

Trachearia, n. plu., *trăk'·ē·ār'·ĭ·ă* (Gr. *tracheia*, the windpipe), the Division of Arachnida which breathe by means of tracheæ : **tracheides,** n. plu., *trăk'·ē·ĭd'·ĕs* (Gr. *eidos*, resemblance), in *bot.*, vessels which serve as air-conducting tubes, after the protoplasm and cell-sap have disappeared.

trachenchyma, n., *trăk·ĕng'·kĭm·ă* (Gr. *tracheia*, the windpipe ; *engchuma*, what is poured in—from *cheuma*, juice, tissue), the

trachea or spiral vessels of plants ; tissue composed of spiral vessels.

tracheotomy, n., *trăk'·ē·ŏt'·ŏm·ĭ,* (Gr. *tracheia,* the windpipe ; *tomē,* a cutting), the operation of making an opening in the windpipe.

tragicus, a., *trădj'·ĭk·ŭs* (Gr. *trăgŏs,* a goat), a short, flattened band of muscular fibres, situated upon the outer surface of the tragus, the direction of its fibres being vertical : **antitragicus,** a., *ănt'·ĭ·trădj'·ĭk·ŭs,* this muscle arises from the outer part of the 'antitragus,' which see.

Tragopogon, n., *trăg'·ō·pōg'·ŏn* (Gr. *trăgŏs,* a goat ; *pōgŏn,* a beard), a genus of ornamental plants, Ord. Compositæ, Sub-ord. Cichoraceæ, so named from the long silky beard of the seeds : **Tragopogon porrifolius,** *pŏr'·rĭ·fōl'·ĭ·ŭs* (L. *porrum,* a leek ; *fŏlĭum,* a leaf), a species whose root produces Salsify, and is called the 'oyster plant' in America.

tragus, n., *trăg'·ŭs* (Gr. *trăgŏs,* a goat), a small pointed eminence in front of the concha of the ear, and projecting backwards over the meatus, so named because generally covered, on its under surface, with a tuft of hair resembling a goat's beard.

trama, n., *trăm'·ă* (L. *trāma,* the woof or filling of a web), in *bot.*, the central tissue of the lamellæ of gill-bearing fungi.

trance, n., *trăns* (old F. *transi,* fallen into a swoon ; F. *transe,* a swoon ; L. *transĕo,* I pass over), a cataleptic condition of the body of peculiar symptoms.

transudation, n., *trăns'·ūd·ā'·shŭn* (L. *trans,* through ; *sudo,* I sweat), the act or process of a fluid or vapour oozing through a porous substance or tissue.

transversalis abdominis, *trănz'·vĕrs·āl'·ĭs ăb·dŏm'·ĭn·ĭs* (L. *trănsversālis,* lying across, directed crosswise — from *trans,* across, *versus*, turned; *abdōmen*, the lower belly, *ab lŏmĭnĭs*, of the lower belly), the transverse muscle of the abdomen, a muscle so named from the direction of its fibres, is the most internal flat muscle of the abdomen, and supports and compresses the bowels : **transversalis cervicis**, *sĕrv·ĭs'ĭs*, or **colli**, *kŏl'lī* (L. *cervix*, the neck, *cervīcis*, of the neck ; *collum*, the neck, *collī*, of the neck), the transverse muscle of the neck ; a muscle on the inner side of the longissimus dorsi, arising by long tendons from the summits of the transverse processes of the dorsal vertebræ (3–6), and inserted into the five lower cervical vertebræ.

transversus auriculæ, *trăns·vĕrs'ŭs aŭr·ĭk'ŭl·ē* (L. *transversus*, directed across or athwart—from *trans*, across, *versus*, turned ; *aurĭcŭla*, the external ear, *aurĭcŭlœ*, of the external ear), a muscle placed on the cranial surface of the pinna of the ear, consisting of radiating fibres partly tendinous, partly muscular : **transversus pedis**, *pĕd'ĭs* (L. *pēs*, a foot, *pĕdis*, of a foot), a narrow, flat, muscular faciculus, stretched transversely across the heads of the metatarsal bones, between them and the flexor tendons : **t. perinæi**, *pĕr'ĭn·ē'ī* (Gr. *perinaion*, the space between the anus and the scrotum : new L. *pĕrĭnœum*, the perinæum, *perinœī*, of the perinæum), a narrow muscular slip, which passes more or less transversely across the back part of the perinæal space.

Trapa, n., *trăp'ă* (an adaptation of L. *trĭbŭlus*, a caltrop, a kind of thorn), a genus of aquatic plants, Ord. Myrtaceæ, the fruit of some of the species being furnished with four spines : **Trapa natans**, *năt'ănz* (L. *nătans*, swimming), the water-chestnut, which is eaten : **T. bicornis**, *bĭ·kŏrn'ĭs*

(L. *bĭcornis*, having two horns—from *bis*, twice, *cornu*, a horn), a species remarkable for its horned fruit, which is edible : **T. bispinosa**, *bĭs'·pĭn·ōs'ă* (L. *bis*, twice ; *spinōsus*, full of thorns), a species whose seeds are large and edible, largely cultivated ; in the East, Singhara nuts.

trapezium, n., *trăp·ēs'ĭ·ŭm* (Gr. *trapēzĭŏn*, a small table or counter), in *anat.*, one of the wristbones ; a bone of very irregular form, situated at the external and inferior part of the carpus, between the scaphoid and first metacarpal bone : **trapezius**, n., *trăp·ēz'ĭ·ŭs*, a broad, flat, triangular muscle, immediately beneath the skin, and covering the upper and back part of the neck and shoulders : **trapezoid**, n., *trăp'·ĕz·oўd* (Gr. *eidos*, resemblance), the smallest in the second of the wrist-bones, having a wedge-shaped form.

traumatic, a., *traŭom·ăt'ĭk* (L. *traumătĭcus*, Gr. *traumătĭkŏs*, fit for healing wounds — from Gr. *trauma*, a wound), applied to symptoms arising from wounds or local injuries : n., a medicine for the cure or alleviation of wounds.

trefoil-tendon, a., *trēf'·oўl* (L. *trĭfŏlĭum*, three-leaved grass—from *tres*, three ; *fŏlĭum*, a leaf : and *tendon*), in *anat.*, a strong aponeurosis, forming the central and highest part of the diaphragm, consisting of three lobes or alæ ; also named 'central tendon,' 'cordiform tendon,' or 'phrenic centre.'

Tremandraceæ, n. plu., *trĕm'·ăn·drā'·sĕ·ē* (Gr. *trēma*, a pore ; *anēr*, a male, *andrŏs*, of a male), an Order of Heath-like shrubs of Australia.

Trematoda, n. plu., *trĕm·ăt'·ŏd·ă* (Gr. *trēma*, an opening or pore, *trĕmătos*, of an opening ; *eidos*, resemblance), in *zool.*, an Order of Scolecida, intestinal worms comprising the fluke-worm, which are furnished with suctorial pores: **trematode**, n., *trĕm'-ăt-ōd*, one of the Trematoda or sucking worms.

tremelloid, a., *trĕm'-ĕl-ōyd* (L. *trĕmŭlŭs*, a shaking, a quaking : Gr. *trĕmō*, I tremble ; *eidos*, resemblance), in *bot.*, jelly-like in substance or appearance.

trepan, n., *trĕp-ăn'* (Gr. *trupănŏn*, a borer, an auger ; F. *trépan*), a circular saw employed for removing portions of the bone of the skull, when the skull is injured.

trephine, n., *trĕf-ēn'* or *trĕf'-ĭn* (also from Gr. *trupănŏn*, a borer, an auger), an improved circular saw, with a moveable centre pin, now used instead of the *trepan* for perforating the cranium, and removing circular pieces of bone from it.

triadelphous, a., *trĭ-ăd-ĕlf'-ŭs* (Gr. *treis*, three ; *adelphos*, a brother), in *bot.*, having stamens united in three bundles by their filaments.

triandrous, a., *trĭ-ănd'-rus* (Gr. *treis*, three; *anĕr*, a male, *andrŏs*, of a male), in *bot.*, having three stamens in a flower.

triangularis sterni, *trĭ-ăng-gŭl-ār'-ĭs stĕrn'-ĭ* (L. *trĭăngŭlāris*, triangular ; *sternum*, the breast-bone, *sternĭ*, of the breast-bone), in *anat.*, a thin plane of muscular and tendinous fibres, situated upon the inner wall of the front of the chest.

tribe, n., *trĭb* (L. *trĭbus*, a tribe—from *trĭs*, three), in *bot.*, a group of genera more nearly related in particular characters than others under the same Order ; a division between Order and Genus.

Tribulus, n., *trĭb'-ŭl-us* (L. *trĭbŭlus*, an instrument of four prongs, a caltrop), a genus of plants, Ord. Zygophyllaceæ, having each carpel of the species armed with three or four prickly points : **Tribulus terrestris**, *tĕr-rĕst'-rĭs* (L. *terrĕstris*, terrestrial—from

terra, land), a prickly plant of the East, found in Palestine; the supposed thistle of the New Test.

triceps, a., *trĭ'-sĕps* (L. *trĭceps*, having three heads—from *tris*, three ; *caput*, the head), having three heads: n., the three-headed muscle : **triceps extensor**, *trĭ'-sĕps ĕks-tĕns'-ŏr* (L. *extensor*, that which extends or stretches — from *extensus*, stretched out), the name of several muscles which extend a limb: **t. extensor cubiti**, *kŭb'-ĭt-ĭ* (L. *cubĭtum*, the elbow or bending of the arm), the stretching muscle of the arm having three heads ; a muscle arising by three heads from the scapula, and from the humerus, and inserted into the olecranon : **t. e. cruris**, *krōr'-ĭs* (L. *crus*, a leg, *cruris*, of a leg), the stretching-out muscle of the leg having three heads ; this muscle extends the leg, and consists of the three parts, 'vastus externus,' 'vastus internus,' and 'crureus,' which see : **t. longus adductor femoris**, *lŏng'-gus ăd-dŭkt'-ŏr fĕm'-ŏr-ĭs* (L. *triceps*, three-headed ; *longus*, long ; *adductor*, that which brings one part towards another ; *fĕmur*, the thigh, *fĕmŏris*, of the thigh), the long triceps adductor muscle of the thigh-bone.

Trichadenia, n., *trĭk'-ăd-ēn'-ĭ-ă* (Gr. *thrix*, hair, *trĭchŏs*, of hair ; *ădĕn*, an acorn), a genus of plants, Ord. Bixaceæ : **Trichadenia Zeylanica**, *zĭ-lăn'-ĭk-ă* (of or from *Ceylon*), a large tree of Ceylon, called Tettigass, which yields an oil used for burning in lamps, etc.

trichiasis, n., *trĭk-ĭ'-ăs-ĭs* (Gr. *thrix*, hair, *trĭchŏs*, of hair), a disease of the eye in which the eyelash turns in upon the eyeball and produces irritation.

Trichilia, n., *trĭk-ĭl'-ĭ-ă* (Gr. *trĭcha*, in three ways or parts), a genus of plants, Ord. Meliaceæ, having three-lobed stigmas, and their capsules three-celled, and three-valved: **Trichilia speciosa**, *spĕsh'-ĭ-ōz'-ă* (L. *spĕciōsus*, showy, handsome), a species from whose fruit a warm, pleasant-smelling oil is procured, valued in India for chronic rheumatism and paralytic affections, applied externally: **T. emetica**, *ĕm-ĕt'-ĭk-ă* (Gr. *ĕmĕtĭkŏs*, L. *ĕmĕtĭca*, an incitement to vomit), the Koka of the Arabs, a large tree, the fruit possesses emetic properties; mixed by the Arabians with the perfumes with which they wash their hair; also used for itch.

Trichina, n., *trĭk'-ĭn-ă*, **Trichinæ**, n. plu., *trĭk'-ĭn-ē* (Gr. *trichĭnos*, made of hair—from *thrix*, hair), animal parasites found in the muscles of the human body: **Trichina spiralis**, *spĭr-āl'-ĭs* (L. *spĭrālis*, spiral—from *spĭra*, a twist), a parasite of the human body, a bisexual and viviparous worm: **trichiniasis**, n., *trĭk'-ĭn-ī'-ăs-ĭs*, a formidable febrile disease, caused by the presence in large numbers of Trichinæ in the body, due to eating trichinous pork.

Trichocephalus dispar, *trĭk'-ō-sĕf'-ăl-ŭs dĭs'-păr* (Gr. *thrix*, hair; *kephalē*, the head; L. *dispar*, dissimilar), a thin filiform parasitic worm, found chiefly in the cæcum and large intestines: **Trichocysts**, n. plu., *trĭk'-ō-sĭsts* (Gr. *kustis*, a cyst), peculiar cells found in certain Infusoria.

trichogynium, n., *trĭk'-ō-jĭn'-ĭ-ŭm*, also **trichogyne**, n., *trĭk'-ō-jĭn* (Gr. *thrix*, hair, *trĭchos*, of hair; *gŭnē*, a female), in *bot.*, among the red sea-weeds, called Rhodospermeæ or Florideæ, a peculiar hair-like body surmounting a cell, which, after fertilization, is transformed into the cystocarp; a long, thin, hair-like, hyaline sac, forming a receptive organ for the spermatozoids.

trichome, n., *trĭk'-ōm* (Gr. *thrix*, hair; F. *trichome*), in *bot.*, any structure originating as an outgrowth of the epidermis.

trichophore, n., *trĭk'-ō-fōr* (Gr. *thrix*, hair; *phorĕō*, I bear), in *bot.*, the cellular body supporting the cystocarp, among some Florideæ; a group of cells from which the trichogynes spring.

Trichophyton tonsurans, *trĭk-ŏf'-ĭt-ŏn tŏns-ūr'-ăns* (Gr. *thrix*, hair; *phuton*, a plant; L. *tonsūrans*, clipping or pruning), a fungus, generally seen as spores, which affects the hair and skin, producing ringworm on the scalp, and the eruption, 'herpes circinatus,' on the body.

trichotomous, a., *trĭk-ŏt'-ŏm-ŭs* (Gr. *trĭchē*, in three parts; *tŏmē*, a cutting), divided into three parts; in *bot.*, divided successively into three branches: **trichotomy**, n., *trĭk-ŏt'-ŏm-ĭ*, division into three parts.

tricoccous, a., *trĭ-kŏk'-ŭs* (Gr. *treis*, three; *kokkos*, a kernel), in *bot.*, having three one-seeded cells; applied to a fruit having three elastically dehiscing cocci.

tricostate, a., *trĭ-kŏst'-āt* (L. *tris*, three; *costātus*, having ribs), in *bot.*, having three ribs; having ribs from the base.

tricuspid, a., *trĭ-kŭsp'-ĭd* (L. *tris*, three; *cuspis*, a point, *cuspĭdĭs*, of a point), having three summits or points: **tricuspidate**, a., *trĭ-kŭsp'-ĭd-āt*, having three long points.

tridactyle, a., *trĭ-dăk'-tĭl* (Gr. *treis*, three; *dăktŭlos*, a finger), having three fingers.

tridentate, a., *trĭ-dĕnt'-āt* (L. *trĭdens*, having three teeth — gen., *trĭdĕntis*—from *tris*, three; *dens*, a tooth), in *bot.*, having three tooth-like divisions.

trifacial, a., *trĭ-făsh'-ĭ-ăl* (L. *tris*, three; *făcĭes*, the face, the surface), a name for the fifth cranial nerve, forming the great sensitive nerve of the head and face.

trifarious, a., *trĭ-făr'-ĭ-ŭs* (L. *trĭ-*

fārĭus, of three sorts or ways, triple), in *bot.*, in three rows.

trifid, a., *trĭ'.fĭd* (L. *trĭfĭdus*, cleft in three parts—from *trĭs*, three ; *fĭdi*, I have cleft), in *bot.*, thrice cleft, midway to the base.

trifoliate, a., *trĭ.fŏl'.ĭ.āt*, also **trifoliolate**, a., *trĭ.fŏl'.ĭ.ŏl.āt* (L. *trĭs*, three ; *fŏlĭum*, a leaf), in *bot.*, having three leaves or leaflets growing from the same point.

Trifolium, n., *trĭ.fŏl'.ĭ.ŭm* (L. *trĭs*, three ; *fŏlĭum*, a leaf), an extensive genus of plants, Ord. Leguminosæ, Sub-ord. Papilionaceæ, known as clovers or trefoils, and having trifoliolate leaves : **Trifolium pratense**, *prăt.ĕns'.ē* (L. *prătĕnsis*, growing in meadows), the common red clover : **T. repens**, *rĕp'.ĕnz* (L. *rĕpĕns*, creeping), white Dutch clover—the shamrock of Ireland : **T. Alpinum**, *ălp.ĭn'.ŭm* (L. *Alpĭnus*, of or from the *Alps*), a species from whose leaves and roots the peculiar sweet principle called Glycyrrhizin is obtained : **T. incarnatum**, *ĭn'.kăr.nāt'.ŭm* (L. *incarnātŭs*, clothed in flesh—from *in*, in ; *carnātus*, fleshy — from *căro*, flesh), an annual species ; the carnation clover.

trigonal, a., *trĭg'.ŏn.ăl*, also **trigonous**, a., *trĭg'.ŏn.ŭs* (Gr. *treis*, three ; *gōnĭa*, an angle, a corner), in *bot.*, having the parts arranged in an alternating manner, as in the lily ; having three angles with three convex faces ; applied to stems.

trigone, n., *trĭg'.ŏn.ē*, or **trigonum**, n., *trĭ.gŏn'.ŭm* (Gr. *trĭgōnŏs*, a triangle—from *treis*, three ; *gōnĭa*, an angle), a triangular, smooth surface without rugæ, immediately behind the urethral orifice, the apex of which is directed forwards.

trigynous, a., *trĭdj'.ĭn.ŭs* (Gr. *treis*, three ; *gunē*, a woman), in *bot.*, having three carpels or three styles.

trijugate, a., *trĭ'.jŏŏg.āt* (L. *tris*,

three ; *jŭgum*, a yoke), in *bot.*, having three pairs of leaflets.

trilamellar, a., *trĭ.lăm'.ĕl.lăr* (L. *tris*, three ; *lămĕlla*, a small plate of metal), in *bot.*, applied to a compound stigma having three divisions flattened like bands.

Trilliaceæ, n. plu., *trĭl'.lĭ.ā'.sĕ.ē* (L. *trĭlĭx*, triple-twilled, *trĭlĭcis*, of triple - twilled), the Trillium family, an Order of plants, some acrid, others narcotic : **Trillium**, n., *trĭl'.lĭ.ŭm*, a genus of plants, so named from the calyx having three sepals, the corolla three petals, the pistil three styles, and the stem three leaves : **Trillium cernuum**, *sĕrn'.ū.ŭm* (L. *cernŭus*, stooping or bowing forward), a species whose rhizome is used as an emetic ; the juice of the berries with alum gives a blue colour.

trilobate, a., *trĭ.lŏb'.āt* (Gr. *treis*, three ; *lŏbos*, a lobe), having three lobes.

trilocular, a., *trĭ.lŏk'.ŭl.ăr* (L. *tris*, three ; *locŭlŭs*, a little place), in *bot.*, having three cells, or loculaments.

trimerous, a., *trĭm'.ĕr.ŭs* (Gr. *treis*, three ; *mĕrŏs*, a part), in *bot.*, composed of three parts, as a flower ; having its envelopes in three, or multiples of three.

trimorphic, a., *trĭ.mŏrf'.ĭk* (Gr. *treis*, three ; *morphē*, form, shape), in *bot.*, taking three forms of flowers in one species, each on a different plant, and having stamens and pistils.

trinervis, a., *trĭ.nĕrv'.ĭs* (L. *tris*, three ; *nervus*, a nerve), in *bot.*, having three ribs springing together from the base.

trioecious, a., *trĭ.ē'.shĭ.ŭs* (Gr. *treis*, three ; *oikos*, a house), in *bot.*, producing male, female, and hermaphrodite flowers, each on separate plants : **trioeciously-hermaphrodite**, another name for trimorphic.

tripartite, a., *trĭp'.ȧrt.ĭt* (L. *tris*, three ; *partītus*, divided), in *bot.*, parted into three divisions nearly to the base.

tripetalous, a., *trī·pĕt'.ăl·ŭs* (Gr. *treis*, three ; *petălŏn*, a leaf), in *bot.*, having three petals or flower leaves, as a corolla.

tripinnate, a., *trī·pĭn'nāt* (L. *tris*, three ; *pinna*, a feather), in *bot.*, divided three times in a pinnate manner, as a compound leaf ; having the pinnæ of a bipinnate leaf again pinnate : **tripinnatifid**, a., *trī'pĭn·nǎt'·ĭ·fĭd* (L. *findo*, I divide, *fĭdī*, I have divided), having a pinnatifid leaf with the segments divided twice in a pinnatifid manner.

triplicostate, a., *trĭp'.lĭ·kŏst'·āt* (L. *trĭplex*, threefold ; *costa*, a rib), in *bot.*, having three ribs proceeding from above the base of the leaf.

triploblastic, a., *trĭp'.lō·blǎst'.ĭk* (Gr. *triplŏos*, threefold ; *blastos*, a germ), in *zool.*, having ova in which the blastoderm separates into three parts.

triquetrous, a., *trĭk.ēt'.rŭs*, also **triquetral**, a., *trĭk.ēt'.rǎl* (L. *trĭquĕtrus*, three-sided), in *bot.*, having three angles with three concave faces ; in *anat.*, three-sided, or three-cornered, as a bone : **ossa triquetra**, *ŏs'.sǎ trĭk.ēt'.rǎ* (L. *ŏs*, a bone, *ŏssĭs*, of a bone), supernumerary ossicles found in a great number of skulls, interposed between the cranial bones, like islets in the sutures, and of irregular shape.

trisepalous, a., *trī·sĕp'.ǎl·ŭs* (L. *tris*, three ; Eng. *sepal*), in *bot.*, having three sepals, as a calyx.

triseptate, a., *trī·sĕpt'.āt* (L. *tris*, three ; *sĕptus*, hedged or fenced in), having three partitions or septa in an ovary or fruit.

trismus, n., *trĭz'.mŭs* (Gr. *trizō*, I gnash), a tetanic spasm affecting the muscles of the jaw ; lock-jaw :

trismus neonatorum, *nĕ·ŏn'.ǎt·ōr'.ŭm* (new L. *nĕonātōrum*, of the newly born—from Gr. *nĕŏs*, new ; L. *nātŭs*, born), a form of tetanus attacking infants within a few weeks after birth, characterised by congestion of the spinal arachnoid, with an effusion of blood or serum into its cavity — also called t. **nascentium**, *nǎs·sĕn'.shĭ·ŭm* (L. *nascens*, being born, *nascentĭum*, of those born) : t. **traumaticus**, *trǎwm·ǎt'.ĭk·ŭs* (Gr. *traumătikos*, fit for healing wounds —from *trauma*, a wound), tetanus, arising at all ages, arising from cold or a wound.

tristichous, a., *trĭst'.ĭk·ŭs* (Gr. *treis*, three ; *stichos*, a row), in *bot.*, in three rows.

triternate, a., *trī·tĕrn'.āt* (L. *tris*, three ; *terni*, three each), in *bot.*, divided three times in a ternate manner.

Triticum, n., *trĭt'.ĭk·ŭm* (L. *trĭtĭcum*, wheat—from *trĭtus*, a rubbing or wearing), the most important genus of the Order Gramineæ, producing the cereal grains : **Triticum vulgare**, *vŭlg·ār'.ĕ* (L. *vulgāris*, common), wheat : **T. æstivum**, *ĕst·īv'.ŭm* (L. *æstīvus*, pert. to summer), the varieties of spring wheat : **T. hybernum**, *hĭb·ĕrn'.ŭm* (L. *hybĕrnus*, pert. to winter, wintry), the varieties of winter wheat : **T. spelta**, *spĕlt'.ǎ* (AS. *spelt*, Ger. *spelt* or *spelz*, grain, wheat), spelt, an inferior kind of wheat, grown on the Continent, in the Bible called rye : **T. compositum**, *kŏm·pŏz'.ĭt·ŭm* (L. *compŏsĭtus*, placed or laid together), Egyptian or mummy wheat : **T. repens**, *rĕp'.ĕnz* (L. *rēpens*, creeping), couch-grass, or quitch-grass : **T. junceum**, *jŭn'.sĕ·ŭm* (L. *juncĕus*, made of rushes—from *juncus*, a rush), a species used in mucous discharges of the bladder.

tritozooid, n., *trĭt'.ō·zō'.ŏyd* (Gr. *trĭtŏs*, third ; *zoŏn*, an animal ;

eidos, resemblance), in *zool.*, a zooid of the third generation.

trituration, n., *trĭt·ūr·ā'·shŭn* (L. *trītūra*, a rubbing or wearing out), the act of reducing a substance to a fine powder by rubbing.

trivial names, *trĭv'·ĭ·ăl nāmz* (L. *trĭvĭālis*, that may be found everywhere — from *trĭvĭum*, a crossroad), the names added to the names of genera, which double or binomial names constitute the names of species, as *Triticum*, the generic name of certain cereals, while *Triticum vulgare* is the specific name of one of the genus, viz. common wheat : see Appendix on Specific Names.

trochal, a., *trōk'·ăl* (Gr. *trŏchŏs*, a wheel), wheel - shaped — applied to the ciliated discs of the Rotifera.

trochanter, n., *trōk·ănt'·ĕr* (Gr. *trochanter*, a runner—from *trochăō*, I roll or run round), one of the two processes or prominences at the upper part of each thighbone, named respectively the *major* and the *minor ;* they receive the large muscles which bend and extend the thigh, and turn it upon its axis, thus forming, as it were, a shoulder to each thigh-bone : **trochanteric**, a., *trōk'·ănt·ĕr'·ĭk*, of or pert. to the trochanters.

trochar, n., *trōk'·ăr* (F. *trocar*, a trocar; *trois-quarts*, three-fourths —from L. *tris*, three ; *quartus*, the fourth), a surgical instrument for taking off fluids from parts of the body, as in dropsy, so named from its triangular point.

trochlea, n., *trōk'·lĕ·ă* (L. *trochlea*, a case containing one or more pulleys ; Gr. *trŏchŏs*, a wheel), a pulley - like cartilage through which the tendon of the trochleary muscle passes : **trochlearis**, n., *trōk'·lĕ·ăr'·ĭs*, one of the projections of bones over which parts turn as ropes over pulleys : **troch-**

lear, a., *trōk'·lĕ·ăr*, shaped like a pulley : **trochleary**, a., *trōk'·lĕ·ăr'·ĭ*, of or pert. to the trochlea : **trochlear surface**, the smooth surface of the trochlea.

trochoid, a., *trōk'·ŏyd* (Gr. *trŏchŏs*, a wheel ; *eidos*, resemblance), conical, with a flat base, as the shells of Foraminifera.

Tropæolaceæ, n. plu., *trŏp·ē'·ōl·ā'·sĕ·ē* (Gr. *trŏpaiŏn*, a trophy), the Indian-Cress family, an Order of plants, having showy flowers, and more or less pungency, used as a cress : **Tropæolum**, n., *trŏp·ē'·ōl·ŭm*, a showy genus of plants, so named from their leaves resembling a buckler, and their flowers a helmet : **Tropæolum majus**, *mādj'·ŭs* (L. *mājor*, and *mājus*, greater), common Indian-Cress, or Garden Nasturtium, whose unripe fruit has been pickled, and used as capers : **T. tuberosum**, *tūb'·ĕr·ōz'·ŭm* (L. *tūbĕrōsus*, having fleshy knots—from *tūber*, a protuberance), a species whose roots are eaten in Peru.

trophi, n. plu., *trŏf'·ĭ* (Gr. *trŏphŏs*, rearing, nursing), the parts of the mouth in insects concerned in the acquisition and preparation of food : **trophic**, a., *trŏf'·ĭk*, connected with nourishment ; nourishing ; nutritious : **trophosome**, n., *trŏf'·ō·sŏm* (Gr. *sŏma*, body), the collective assemblage of the nutritive zoöids of any Hydrozoön.

trophosperm, n., *trŏf'·ŏ·spĕrm* (Gr. *trŏphŏs*, rearing, nursing ; *sperma*, seed), in *bot.*, a name applied to the placenta.

truncate, a., *trŭngk'·āt*, also **truncated**, a., *trŭngk'·āt·ĕd* (L. *truncātum*, to maim, to mutilate — from *truncus*, the bole or trunk of a tree), in *bot.*, terminating very abruptly, as if cut off at the end ; in *zool.*, abruptly cut off, as univalve shells, whose apex breaks off, the shells thus becoming decollated : **truncus**, n., *trŭngkʹŭs*, in *bot.*, the trunk or bole of a tree.

truss, n., *trŭs* (F. *trousser*, to pluck up; *trousse*, a bundle; L. *tortus*, a twisting, a wreath), a mechanical contrivance, usually for the support of parts concerned in abdominal rupture or hernia; also for the support or for the prevention of the protrusion of any viscus.

tryma, n., *trĭmʹă* (Gr. *truma*, a hole, an opening), in *bot.*, a fruit resembling a drupe, as the walnut, having a coriaceous or fleshy epicarp, and mesocarp, one-celled and one-seeded; a two-valved bony endocarp, having partitions on the inner concave surface, as the walnut.

tuber, n., *tūbʹĕr* (L. *tŭber*, a hump, a knob or excrescence), in *bot.*, a solid fleshy mass attached to many fibrous rooted plants; a thickened underground stem or branch, as the potato; in *anat.*, the rounded projection of a bone: **tuber annulare,** *ănʹnŭl·ārʹĕ* (L. *ānnŭlāris*, pert. to a ring—from *ānnŭlus*, a ring), in *anat.*, another name for 'pons Varolii,' see under 'pons': t. **calcis,** *kălʹsĭs* (L. *calx*, the heel, *calcis*, of the heel), the large posterior extremity of the os calcis, or largest bone of the foot, presenting inferiorly two tubercles which rest upon the ground when walking: t. **cinereum,** *sĭn·ērʹĕ·ŭm* (L. *cĭnĕrĕus*, ash-coloured—from *cinis*, ashes), a layer of grey matter at the base of the cerebrum: t. **cochleæ,** *kŏkʹlĕ·ē* (L. *cŏchlĕa*, a snail, *cŏchlĕæ*, of a snail), in the tympanum of the ear, the first turn of the cochlea.

tubercle, n., *tŭbʹĕr·kl* (L. *tuberculum*, a small hump or protuberance—from *tŭber*, a hump), a little tuber; in *med.*, a term of varied and wide application, generally a small tumour in any part; a new growth, composed of primitive cells and nuclei, and having a tendency to caseous or calcareous degeneration; a morbid, yellow, or caseous material, generally contained in cysts, of the size of a hemp seed, or of a pea, or loose in the structure of organs; in *bot.*, a swollen simple root, as in some Orchids; in *anat.*, a small protuberance, as the *tubercle* of the tibia: **tubercled,** a., *tŭbʹĕr·kld*, in *bot.*, covered with warts: **tubercular,** a., *tŭb·ĕrkʹŭl·ăr*, full of small knobs or tubercles; caused by tubercles; applied to morbid matter at one time compact and yellowish, at another calcareous, and sometimes becoming pultaceous, semi-fluid, and caseous.

tubercula quadrigemina, *tŭb·ĕrkʹŭl·ă kwŏdʹrĭ·jĕmʹĭn·ă* (L. *tuberculum*, a small hump; *quadrans*, a fourth; *gemini*, twins), in *anat.*, four rounded eminences in the cerebrum, separated by a crucial depression, and placed two on each side of the middle line, one before another—also called 'corpora quadrigemina': **tuberculum pharyngeum,** *fărʹĭn·jĕʹŭm* (new L. *phăryngēus*, pert. to the pharynx), the tubercle from which the mesial band attaching the pharynx to the skull principally springs.

tuberculosis, n., *tŭb·ĕrkʹŭl·ōzʹĭs* (L. *tuberculum*, a small hump), a form of fever accompanied by the formation of small bodies, called tubercles, in various tissues of the body: **tubercular meningitis,** a name given to the disease caused by the deposition of tubercles in the membranes of the cerebrum — also called 'acute hydrocephalus.'

tuberous, a., *tŭbʹĕr·ŭs* (L. *tŭber*, a small knob), in *bot.*, connected into a bunch by rootlets, as in the potato: **tuberosity,** n., *tŭbʹĕr·ŏsʹĭ·tĭ*, in *bot.*, a kind of projection or knob on a bone, generally forming an attachment to muscles.

Tubicola, n., *tŭb·ĭk'·ŏl·ă* (L. *tŭba*, a pipe ; *cŏlo,* I inhabit), the Order of Annelida which construct tubular cases, in which they protect themselves : **tubicolous,** a., *tŭb·ĭk'·ŏl·ŭs,* inhabiting a tube.

tubular, a., *tŭb'·ūl·ăr* (L. *tŭbŭlus,* a small pipe—from *tŭbus,* a pipe), hollow and cylindrical ; in *bot.,* applied to the regular florets of Compositæ.

tubuli contorti, *tŭb'·ūl·ī kōn·tŏrt'·ī* (L. *tŭbŭlus,* a small pipe ; *contortus,* twisted), twisted or convoluted tubules, which form the greater part of the cortical substance of the kidneys : **tubuli recti,** *rĕkt'·ī* (L. *rectus,* straight), straight tubules ; a name applied to the seminal ducts of the testis when they assume a comparatively straight course—also called 'vasa recta'; also denoting the straight portion of the tubules of the kidneys, which convey the urine : **t. seminiferi,** *sĕm'·ĭn·ĭf'·ĕr·ī* (L. *sĕmĭnĭf'ĕr,* bearing semen — from *semen,* seed ; *fĕro,* I bear), the small convoluted tubes in which the seminal fluid is secreted : **t. uriniferi,** *ūr'·ĭn·ĭf'·ĕr·ī* (L. *urĭnĭf'·ĕr,* bearing urine—from *ūrĭna,* urine ; *fĕro,* I bear), the tubules which bear or collect the urine of the kidneys.

tumour, n., *tŭm'·ĕr* (L. *tŭmor,* a swelling), a morbid growth on a part of the body in the form of a swelling or enlargement; a growth which may either be innocent or malignant—the former comprise 'sarcomata,' or fleshy growths; the latter, true cancers, or 'carcinomata.'

tunica adiposa, *tŭn'·ĭk·ă ăd'·ĭp·ōz'·ă* (L. *tŭnĭca,* a coating, a membrane; *adĭpōsus,* fatty—from *adeps,* fat), loose areolar tissue, usually containing much dense fat, which assists in maintaining the kidneys in their position : **tunica adven-**

titia, *ăd'·vĕn·tĭsh'·ĭ·ă* (L. *advĕntītīus,* foreign—from *ad,* to ; *vĕnĭo,* I come), the foreign or outside tunic ; the external coat of the arteries : **t. albuginea,** *ălb'·ū·jĭn'·ĕ·ă* (L. *albūgĭnĕus,* of a white appearance — from *albūgo,* the white of the eye), a strong capsule which encloses the testis proper: **t. choroidea,** *kŏr·ŏyd'·ĕ·ă* (Gr. *chŏrĭŏn,* skin or leather ; *eidos,* resemblance), the choroid coat of the eye, consisting of a dark - brown membrane lying between the sclerotic and the retina : **t. chorio-capillaris,** *kŏr'·ĭ·ō·kăp'·ĭl·lār'·ĭs* (L. *căpĭllāris,* pert. to the hair—from *căpĭllus,* the hair of the head), the inner part of the choroid coat of the eye, formed by the capillaries of the choroidal vessels: **t. vaginalis,** *vădj'·ĭn·āl'·ĭs* (L. *vagīnālis,* of or pert. to a *vāgīna* or sheath), a serous membrane whose visceral portion closely invests the great part of the body of the testis as well as the epididymis : **t. vasculosa testis,** *văsk'·ūl·ōz'·ă tĕst'·ĭs* (L. *vascŭlōsus,* full of small vessels — from *vascŭlum,* a small vessel ; *testis,* the testicle), the vascular tunic of the testicle ; the vascular network, together with its connecting areolar tissue, which surrounds the testicle.

tunicated, a., *tŭn'·ĭk·āt'·ĕd* (L. *tŭnĭca,* an under garment, a membrane), in *bot.,* covered by thin external scales, as the onion: **Tunicata,** n. plu., *tŭn'·ĭk·āt'·ă,* a class of Molluscoida, or headless Molluscs, which are enveloped in a tough, leathery case or test: **tunicle,** n., *tŭn'·ĭ·kl,* a natural covering ; an integument.

Turbellaria, n. plu., *tĕrb'·ĕl·lār'·ĭ·ă* (L. *turbellæ,* a bustle, a stir—from *turbo,* I disturb), in *zool.,* an Order of Scolecida, so named from the currents they cause in the water in which they exist.

turbinate, a., *tĕrb'·ĭn·āt,* also **turb-**

inated, a., *těrb'ĭn·ăt'ĕd* (L. *turb-ĭnātus*, pointed like a cone—from *turbo*, a whirl, a whipping-top), in *bot.*, in the form of a top; conical, with a round base; in *anat.*, applied to certain twisted bones of the nasal and olfactory chambers.

turio, n., *tūr'ĭ·ō* (L. *turĭō*, the tendril, a shoot), in *bot.*, a young shoot covered with scales sent up from an underground stem, as in asparagus; the early stage of a sucker when invested with leaf scales.

turmeric, n., *těrm'ěr·ĭk* (L. *terra-merĭta*, valuable earth), the branches of the rhizome or root-stock of the 'Carcuma longa,' Ord. Zingiberaceæ, reduced to a powder, of a lemon-yellow colour; see 'Carcuma.'

Turneraceæ, n. plu., *těrn'ěr·ā'sĕ·ē* (after *Rev. W. Turner*, an English botanist), the Turnera family, an Order of plants, natives of W. Indies and S. America: **Turnera, n.**, *těrn'ěr·ă*, a genus of elegant plants when in flower: **Turnera opifera**, *ŏp·ĭf'ěr·ă* (L. *opĭfěr*, bringing power—from *ops*, aid, power; *fěro*, I bear), an astringent, used in Brazil for dyspepsia: **T. ulmifolia**, *ŭl'mĭ·fōl'ĭ·ă* (L. *ulmus*, an elm tree; *fōlĭum*, a leaf), a species considered tonic and expectorant.

turpentine, n., *těrp'ĕnt·ĭn* (L. *terebinthus*, Ger. *terpentin*, the turpentine tree), a mixture of oil and resin obtained from various species of pine; in the form called 'oil of turpentine,' used as a stimulant, diuretic, cathartic, and anthelmintic.

Tussilago, n., *tŭs'sĭ·lāg'ō* (L. *tussis*, a cough, as used in relieving coughs), a genus of plants, Ord. Compositæ, Sub-ord. Corymbiferæ: **Tussilago farfara**, *fär'fär·ă* (L. *fărfărus*, the white poplar, as its leaves resemble those of the white poplar), the plant colt's-foot, has been used as a demulcent.

tutamina oculi, *tūt·ăm'ĭn·ă ŏk'ŭl·ī* (L. *tūtāmen*, a defence or protection; *ocŭlŭs*, the eye, *ŏc-ŭlī*, of the eye), the defences of the eye, a name applied to the eyelids.

Tylophora, n., *tĭl·ŏf'ŏr·ă* (Gr. *tŭlŏs*, a protuberance, a hardening; *phŏrĕō*, I bear), a genus of plants, Ord. Asclepiadaceæ, referring to its ventricose pollen masses: **Tylophora asthmatica**, *ăst·măt'ĭk·ă* (L. *asthmătĭcŭs*, afflicted with shortness of breath—from Gr. *asthma*, shortness of breath), an Indian plant, used instead of ipecacuanha.

tylosis, n., *tĭl·ōz'ĭs* (Gr. *tŭlŏs*, a protuberance, a callosity), in *med.*, a kind of ichthyosis or psoriasis of the tongue; in *bot.*, the development of irregular cells in the interior of pitted vessels, as in the Walnut, Oak, and Elm.

tympanum, n., *tĭm'păn·ŭm* (L. *tympănum*, Gr. *tumpănŏn*, a drum, a timbrel), the drum-like cavity which constitutes the middle ear, familiarly called the drum of the ear; in *bot.*, a membrane closing the thecæ in urn-mosses: **tympanic, a.**, *tĭm·păn'ĭk*, of or pert. to: **tympanites, n.**, *tĭm'păn·ĭt'ēz*, also **tympany, n.**, *tĭm'păn·ĭ*, a flatulent distension of the abdomen, in which the bowels swell up and resound like a drum when percussed: **tympanitis, n.**, *tĭm'păn·ĭt'ĭs*, inflammation of the lining membrane of the tympanum.

tynea sycosis, see 'tinea.'

type, n., *tīp* (L. *typus*, Gr. *tupos*, a figure, an image), the perfect representation or idea of anything; the peculiarity in the form of a disease; the primary model: **typical, a.**, *tĭp'ĭk·ăl*, an individual having pre-eminently the characteristics of the species; applied to a species or genus exhibiting in a marked degree the characteristics of the order.

Typha, n., *tȳf'·ă* (Gr. *tŭphŏs*, a marsh, a fen), a genus of plants, Ord. Araceæ, so named from their habitat : **Typha latifolia**, *lăt'·ĭ·fōl'·ĭ·ă* (L. *lātus*, broad ; *fōl'ĭum*, a leaf), the Great Reed Mace, the pollen of which is so abundant and easily collected and inflammable, that it is used as Lycopodium spores : **T. Shuttleworthii**, *shŭt'·il·wĕrth'·ĭ·ī* (after *Shuttleworth*, a botanist), the rhizomes of the species are used for food by certain natives of Australia : **T. latifolia**, and **T. angustifolia**, *ăng·gŭst'·ĭ·fōl'·ĭ·ă* (L. *angustus*, narrow ; *fōlĭum*, a leaf), species whose young shoots are eaten like asparagus by the Cossacks ; and the large, fleshy rhizomes are eaten by the Calmucks.

typhlitis, n., *tĭf·lĭt'·ĭs* (Gr. *tuphlŏs*, blind), inflammation of the cæcum.

typhoid, a., *tĭf'·ōȳd* (Gr. *tuphos*, smoke or stupor ; *eidos*, resemblance), applied to a form of continued fever, the causal germs of which are never found apart from the products of fæcal fermentation —characterised by an eruption of rose-coloured spots in successive crops, not always present : **typhus**, a., *tĭf'·ŭs*, a highly contagious, continued fever. occurring generally in an epidemic form in periods of famine and destitution —characterised by great languor and prostration, and a persistent eruption of a measly character, rarely absent.

Tyson's glands, *tī'·sŭnz* (after their discoverer, *Tyson*, the anatomist), numerous sebaceous glands collected round the cervix of the penis and corona — also called **glandulæ odoriferæ**, *glănd'·ŭl·ē ŏd'·ŏr·ĭf'·ĕr·ē*, the odoriferous glands, from the peculiar odour of their secretion.

ulcer, n., *ŭls'·ẽr* (L. *ulcus*, a sore, *ulcĕris*, of a sore), a dangerous running sore, arising from some constitutional disorder : **ulceration**, n., *ŭls'·ẽr·ā'·shŭn*, the process of forming into an ulcer, or becoming ulcerous : **ulcerous**, a., *ŭls'·ẽr·ŭs*, having the character of an ulcer.

Ulmaceæ, n. plu., *ŭl·mā'·sĕ·ē* (L. *ulmus*, an elm tree), the Elm family, an Order of trees or shrubs : **Ulmeæ**, n. plu., *ŭl'·mĕ·ē*, a Sub-order, constituting the true elms : **Ulmus**, n., *ŭl'·mŭs*, genus of fine forest trees : **Ulmus campestris**, *kăm·pĕst'·rĭs* (L. *cămpĕstris*, belonging to a field— from *campus*, a field), the English or small-leaved elm, producing a compact and durable timber ; its inner bark is bitter, mucilaginous, and astringent : **U. montana**, *mŏnt·ān'·ă* (L. *montānus*, of or belonging to a mountain—from *mons*, a mountain), the mountain wych, or Scotch elm : **U. fulva**, *fŭlv'·ă* (L. *fŭlvus*, deep-yellow, tawny), the red or slippery elm, used as a demulcent: **ulmaceous**, a., *ŭl·mā'·shŭs*, pert. to trees of the elm kind: **ulmic acid**, *ŭl'·mĭk*, a vegetable acid naturally exuding from the elm, oak, chestnut, etc. : **ulmin**, n., *ŭl'·mĭn*, the brown substance which exudes from the bark of the elm, and several other trees ; ulmic acid ; the brown matter found in decayed leaves and wood resembling ulmin.

ulna, n., *ŭl'·nă* (L. *ulna*, Gr. *ōlĕnē*, the elbow, the arm), in *anat.*, that bone of the forearm which, with the humerus, forms the elbow joint; the outermost of the two bones of the forearm, corresponding with the fibula of the hind limb: **ulnaris**, a., *ŭl·năr'·ĭs*, applied to two muscles of the forearm, a flexor muscle, and an extensor muscle : **ulnar**, a., *ŭl'·năr*, relating to the ulna, as *ulnar* artery.

Ulva, n., *ŭlv'ă* (L. *ulva*, sedge), a genus of Algæ, distinguished by their green colour : **Ulva latissima**, *lăt·ĭs'sĭm·ă* (L. *lătissĭmus*, broadest — from *lătus*, broad), a familiar species, frequently attached to oysters, and called oyster-green : **U. lactuca**, *lăk·tūk'ă* (L. *lăctūca*, lettuce), a species eaten under the name Green Laver.

umbel, n., *ŭm'bĕl* (L. *umbella*, a sun-shade, a parasol—from *umbra*, a shadow), in *bot.*, a particular arrangement of the flowers of certain plants, in which the peduncles, springing from a common centre, rise till they form a flat tuft, as in the familiar example of the inflorescence of the carrot, or hemlock. NOTE.— In the *corymb*, the flowers form a flat head, but they do not, as in the *umbel*, spring from a common centre. **umbellate**, a., *ŭm'bĕl·lāt*, in *bot.*, having the flowers arranged in a round, flat head, with the peduncles springing from a common centre ; in *zool.*, having a number of nearly equal ṛadii, all proceeding from a common centre.

Umbelliferæ, n. plu., *ŭm'bĕl·lĭf'·ĕr·ē* (L. *umbella*, a sun-shade, a parasol ; *fero*, I bear), the Umbelliferous family, an Order of plants, having various properties, some used as articles of diet, some yield gum, resinous, and other substances, and some are highly poisonous ; the Apiaceæ of Lindley : **umbelliferous**, a., *ŭm'bĕl·lĭf'·ĕr·ŭs*, producing or bearing umbels : **umbellule**, n., *ŭm'bĕl·ŭl*, a small umbel, seen in the compound umbellate flowers of many Umbelliferæ.

umbilicus, n., *ŭm'·bĭl·ĭk'·ŭs* (L. *umbĭlĭcŭs*, the navel), the central spot of the abdomen, marked by a depression ; the navel ; in *bot.*, the scar by which a seed is attached to the placenta, more usually called the hilum ; in *zool.*, the aperture at the base of the axis of certain univalve shells, when so seen they are said to be umbilicated or perforated : **umbilical**, a., *ŭm'·bĭl·ĭk'·ăl*, of or pert. to the navel : **umbilical cord**, in *anat.*, a cord-like substance which extends from the placenta to the navel of the fœtus ; the extremity of the malleus towards which the fibres of the membrana tympani converge ; in *bot.*, the prolongation by which the ovule is attached to the placenta : **umbilicate**, a., *ŭm·bĭl'·ĭk·āt*, in *bot.*, having a central depression ; fixed to a stalk by a point in the centre : also **umbilicated**, a., *-āt·ĕd*, in same sense.

umbo, n., *ŭm'·bō* (L. *umbo*, the boss of a shield), in *bot.*, a protuberant part or elevation on a surface, like the boss of an ancient shield ; in *zool.*, the beak of a bivalve shell : **umbonate**, a., *ŭm'·bōn·āt*, having a knob in the centre ; having a central elevation like the boss of an anc. shield.

umbraculiferous, a., *ŭm·brăk'·ŭl·ĭf'·ĕr·ŭs* (L. *ŭmbrācŭlum*, a sunshade, an umbrella—from *umbra*, a shade ; *fero*, I bear), in *bot.*, having the form of an expanded umbrella : **umbraculiform**, a., *ŭm'·brăk·ŭl'·ĭ·fŏrm* (L. *forma*, shape), in same sense as preceding: **umbraculum**, n., *ŭm·brăk'·ŭl·ŭm*, in *bot.*, the cap borne on the seta of Marchantia, Ord. Hepaticæ.

umbrella, n., *ŭm·brĕl'·lă* (It. *ombrella*, L. *umbella*, an umbrella—from L. *umbra*, a shade), in *zool.*, the contractile disc of one of the Lucernarida.

uncinate, a., *ŭn'·sĭn·āt* (L. *uncĭnus*, a hook, a barb), in *bot.*, provided with a hooked process ; in *zool.*, furnished with hooks or bent spines ; in *anat.*, a process of the ethmoid bone : **unciform**, a., *ŭn'·sĭ·fŏrm* (L. *forma*, shape), having a curved or hooked form ;

applied to one of the bones of the carpus, distinguished by its large process projecting forwards, and curved slightly outwards, on its anterior surface.

unequally-pinnate, in *bot.*, pinnate with a single terminal leaflet ; impari-pinnate.

ungual, a., *ŭng'gwăl* (L. *unguis,* a nail), pert. to a nail, claw, or hoof; having a nail, claw, or hoof attached : **ungual phalanges,** in *anat.*, the terminal phalanges of the digits, so named as provided with nails or claws: **unguiculate,** a., *ŭng·gwĭk'·ūl·āt,* furnished with claws ; in *bot.*, applied to petals having a claw : **unguis,** n., *ŭng'gwĭs,* in *anat.*, the bone like a nail ; the lachrymal bone, being a thin scale of bone placed at the anterior and inner part of the orbit ; in *bot.*, the claw, or narrowed part of a petal ; the stalk of a petal : **ungues,** n. plu., *ŭng'gwēz,* the pointed claws terminating the legs in insects.

Ungulata, n. plu., *ŭng'·gūl·āt'·ă* (L. *ungŭla,* a hoof or claw — from *unguis,* a nail), the Order of Mammals comprising the hoofed quadrupeds : **ungulate,** a., *ŭng'·gūl·āt,* having expanded nails constituting hoofs : **unguligrade,** n., *ŭng·gūl'·ĭ·grād* (L. *grădus,* a step), animals which walk upon hoofs.

unicostate, a., *ŭn'·ĭ·kŏst'·āt* (L. *ūnus,* one ; *costa,* a rib), in *bot.*, having a single rib or costa in the middle ; the midrib.

unijugate, a., *ŭn·ĭdj'·ŭg·āt* (L. *ūnus,* one ; *jŭgŭm,* a yoke), in *bot.*, applied to a pinnate leaf having one pair of leaflets.

unilateral, a., *ŭn'·ĭ·lăt'·ĕr·ăl* (L. *ūnus,* one ; *lătus,* a side, *lătĕris,* of a side), in *bot.*, arranged on one side, or turned to one side.

unilocular, a., *ŭn'·ĭ·lŏk'·ūl·ăr* (L. *ūnus,* one ; *lŏcŭlus,* a little place —from *lŏcus,* a place), having a single division or cavity ; one-celled.

union by first intention, see 'intention.'

uniparous, a., *ŭn·ĭp'·ăr·ŭs* (L. *ūnus,* one ; *părĭo,* I bear or bring forth), producing only one at a birth ; in *bot.*, having a cymose inflorescence in which the primary axis produces one bract, and so on, the cyme being elongated according to its development ; having a scorpioidal cyme.

unipetalous, a., *ŭn'·ĭ·pĕt'·ăl·ŭs* (L. *ūnus,* one ; Gr. *pĕtălon,* a leaf), in *bot.*, having a corolla consisting of one petal, which depends upon the abortion or non-development of others; 'unipetalous,' as a term, is quite distinct from 'mono-petalous.'

unipolar, a., *ŭn·ĭp'·ōl·ăr* (L. *ūnus,* one ; *pōlus,* a pole), in *anat.*, having a single pole, as ganglionic nerve-cells ; having but one radiating process.

uniseptate, a., *ŭn'·ĭ·sĕpt'·āt* (L. *ūnus,* one ; *septum,* a hedge), having but one septum.

uniseriate, a., *ŭn'·ĭ·sēr'·ĭ·āt* (L. *ūnus,* one; *sĕrĭēs,* a row, a series), in *bot.*, arranged in a single line or row.

unisexual, a., *ŭn'·ĭ·sĕks'·ū·ăl* (L. *ūnus,* one ; *sexus,* a sex), in *bot.*, having one sex only ; applied to plants having separate male and female flowers.

univalve, n., *ŭn'·ĭ·vălv* (L. *ūnus,* one ; *valvæ,* folding doors), a shell composed of a single piece or valve : **univalvular,** a., *ŭn'·ĭ·vălv'·ūl·ăr,* having one valve only.

unlining, n., *ŭn·lĭn'·ĭng* (L. *ŭn,* not ; *lĭnĕd,* a line), in *bot.*, the separation of parts originally united.

uovoli, n. plu., *ū·ŏv'·ōl·ī* (It. *uovolo,* a mushroom, a joint), knaurs on the olive tree from which roots and leaf-buds are produced.

Upas-tree, *ŭp'·ăs* (Malay *puhn-upas,* the poison tree—from *puhn,* tree ; *upas*, poison), a tree in Java whose shade and juices are poisonous ; see ' Antiaris.'

urachus, n., *ūr'·ăk·ŭs* (Gr. *ouŏn*, urine ; *ĕchō*, I have, I hold), the fibro-muscular cord which extends between the summit of the bladder and the umbilicus.

uræmia, n., *ūr·ēm'·ĭ·ă* (Gr. *ouŏn*, urine; *haima*, blood), a poisoning of the blood from retention of the products of retrograde metamorphosis, occurs in those diseases of the urinary organs which interfere with the secretions of the kidneys : **uræmic,** a., *ūr·ēm'·ĭk*, of or pert. to uræmia : **uræmic poisoning,** same as *uræmia*, that is, when the secretion products of the kidneys are no longer carried out of the system, but remain in, and contaminate the blood.

Urania, n., *ūr·ān'·ĭ·ă* (Gr. *ourăn·ĭŏs*, sublime, lofty — from the stateliness of the tree), a genus of splendid plants, Ord. Musaceæ : **Urania speciosa,** *spēsh'·ĭ·ōz'·ă* (L. *spĕciōsus*, full of beauty or display —from *spĕcies*, look, view, a sort), the Water-tree of the Dutch ; the Traveller's-tree of Madagascar, so named from the great quantity of water which flows from its stem or leaf-stalk when cut across; the juice of the fruit used for dyeing.

urates, n. plu., *ūr'·ātz* (Gr. *ouŏn*, urine), the most common of those deposits in the urine known as sand or gravel, usually of a pink or drab colour, and consisting of uric acid in combination with potash, soda, and ammonia: **urate,** n., a salt of uric acid.

Urceola, n., *ĕrs·ē'·ŏl·ă* (L. *urcĕŏlŭs*, a little water-pot), a genus of plants, Ord. Apocynaceæ : **Urceola elastica,** *ĕl·ăst'·ĭk·ă* (mid. L. *elastĭcus*, elastic), one of certain species which supply caoutchouc: **urceolate,** a., *ĕrs·ē'·ŏl·āt*, shaped like a pitcher.

urea, n., *ūr·ē'·ă* (Gr. *ouŏn*, urine),

a nitrogenous substance forming one of the chief constituents of the urine : **uric acid,** *ūr'·ĭk*, one of the constituents of urine.

Uredo, n., *ūr·ēd'·ō* (Gr. *ūro*, I scorch or burn), a genus of microscopic Fungi, usually known as mildew and blight, and which give to the part of the plant infested by them a burnt appearance : **Uredo fœtida,** *fēt'·ĭd·ă* (L. *fœtĭdus*, fetid, stinking), the Fungi called *pepper - brand*, having a peculiar fetid odour, which attacks grain: **U. segetum,** *sĕg'·ĕt·ŭm* (L. *Sĕgĕtĭa*, the goddess that protects the standing crops), the Fungus called *smut*, a sooty powder, having no odour, which attacks the flower of the grain : **U. rubigo,** *rŭb·ĭg'·ō* (L. *rubīgo*, rust, mildew), the Fungus called *rust* which attacks the leaves and chaff of the grain: **U. caries,** *kār'·ĭ·ēz* (L. *cărĭēs*, rottenness, decay), one of the Fungi which cause smut or blight.

ureter, n., *ūr·ēt'·ĕr* (Gr. *ourētĕr*, the passage through which the urine flows—from *ourĕō*, I make water), in *anat.*, a narrow tube or duct passing down from each kidney, which conveys the urine into the bladder : **ureteritis,** n., *ūr·ēt'·ĕr·īt'·ĭs*, inflammation of the ureter.

urethotomy, n., *ūr'·ĕth·ŏt'·ŏm·ĭ* (Gr. *ourēthra*, the passage through which the urine flows ; *tŏmē*, a cutting), the operation of opening the urethra.

urethra, n., *ūr·ēth'·ră* (Gr. *ourēthra*, the passage through which the urine flows—from *ourĕō*, I make water), the tube which allows the passage of the urine from the bladder, and conducts the semen of the male : **urethritis,** n., *ūr'·ĕth·rīt'·ĭs*, inflammation of the urethra : **urethral,** a., *ūr·ēth'·răl*, of or pert. to the urethra: **uretic,** n., *ūr·ĕt'·ĭk*, a medicine which increases the secretory action of the kidneys : **ureous diuresis,** *ŭr·ē'·ŭs dĭ'·ŭr·ēz'·ĭs,* among animals, a diuresis characterised by a high colour of the urine, with a peculiar slimy character, and strong odour.

uric, a., see under 'urea.'

urine, n., *ūr'·ĭn* (L. *urīna,* Gr. *ourŏn,* urine), the fluid secreted from the kidneys : **urina cibi,** *ŭr·īn'·ă sĭb'·ĭ* (L. *cĭbus,* food, *cĭbī,* of food), the urine of food; the urine passed shortly after partaking of food : **urina potus,** *pōt'·ŭs* (L. *pōtŭs,* drink, *pōtūs,* of drink), the urine of drink; the urine passed shortly after drinking freely of a fluid : **urina sanguinis,** *săng'·gwĭn·ĭs* (L. *sanguis,* blood, *sănguĭnis,* of blood), the urine of the blood ; the urine passed after a fast, as in the morning : **urinal,** n., *ŭr'·ĭn·ăl,* a vessel for receiving urine into ; a public or private place constructed for urinating in : **urinary,** a., *ŭr'·ĭn·ăr·ĭ,* of or pert. to urine : **urinate,** v., *ŭr'·ĭn·āt,* to pass urine : **uriniferous,** a., *ŭr'·ĭn·ĭf'·ĕr·ŭs* (L. *fero,* I bear), carrying or conveying urine.

Urodela, n. plu., *ŭr'·ō·dēl'·ă* (Gr. *ourŏn,* urine ; *dēlŏs,* visible, apparent), in *zool.,* the Order of the tailed Amphibians, as newts, etc.

urohyal, a., *ūr'·ō·hī'·ăl* (Gr. *oura,* the stern, the tail ; *hyoīdes,* the hyoid-bone), in most fishes, the constituent bone of the hæmal spine, extending backwards.

uroscopy, n., *ŭr·ŏs'·kŏp·ĭ* (Gr. *ourŏn,* urine ; *skŏpĕō,* I view), the determination of diseases from the inspection of the urine.

Urticaceæ, n. plu., *ĕrt'·ĭk·ā'·sĕ·ē* (L. *urtīca,* a stinging nettle—from *ŭro,* I `burn), the Nettle family, an Order of plants : **Urtica,** n., *ĕrt·īk'·ă,* a genus of plants, so named in reference to the stinging properties of most of the species : **Urtica dioica,** *dĭ·ŏyk'·ă* (Gr. *dis,* twice ; *oikĭa,* a house), the common stinging

nettle, a very ancient textile plant, young tops in spring eaten when cooked as a vegetable, and a colouring matter is obtained from its roots : **U. urens,** *ūr'·ĕnz* (L. *urens,* burning) ; and **U. pilulifera,** *pĭl'·ŭl·ĭf'·ĕr·ă* (L. *pĭlŭla,* a little ball; *fero,* I bear), are British species of stinging nettles, the last named having capitate female flowers, and the root is astringent and diuretic: **U. crenulata,** *krĕn'·ŭl·āt'·ă* (mid. L. *crēnŭla,* a little notch) ; and **U. stimulans,** *stĭm'·ŭl·ănz* (L. *stĭmŭlans,* pricking or goading on), Indian species, stinging powerfully : **U. urentissima,** *ūr'·ĕnt·ĭs'·sĭm·ă* (L. *urens,* burning, *urentis,* of burning), an Indian species, stinging so powerfully as to be called Devil's leaf, sometimes causing death : **U. cannabina,** *kăn·năb'·ĭn·ă* (L. *cannăbis,* the hemp ; *cannăbĭnus,* hempen) ; **U. tenacissima,** *tĕn'·ăs·ĭs'·sĭm·ă* (L. *tĕnax,* holding fast, tenacious, *tenācis,* of tenacious), are species which furnish fibres fit for cordage : **U. gigas,** *jīg'·ăs* (L. *gigas,* a giant), a species in Australia, was found to be 42 feet in circumference, forming a large tree.

urticaria, n., *ĕrt'·ĭk·ār'·ĭ·ă* (L. *urtīca,* a stinging nettle—from *ūro,* I burn), the nettle-rash, a troublesome cutaneous eruption, giving rise to a sensation similar to that felt after being stung by nettles : **urticating cells,** *ĕrt'·ĭk·āt'·ĭng,* the Cnidæ or thread-cells, by whose possession certain Cœlenterata obtain their power of stinging : **urtication,** n., *ĕrt'·ĭk·ā'·shŭn,* the act of whipping a limb with nettles.

ustulate, a., *ŭst'·ŭl·āt* (L. *ŭstŭlātum,* to burn a little, to scorch), in *bot.,* blackened as if burned : **ustulation,** n., *ŭst'·ŭl·ā'·shŭn,* the process of roasting or drying moist substances to prepare them for pulverising.

uterus, n., *ūt'·ĕr·ŭs* (L. *ŭtĕrus*, the womb, the matrix), the womb or organ of gestation, situated in the cavity of the pelvis, between the bladder and the rectum : **uterine**, a., *ūt'·ĕr·ĭn*, of or pert. to the uterus, or proceeding from it : **uteritis**, n., *ūt'·ĕr·it'·ĭs*, inflammation of the womb : **uterogestation**, *ūt'·ĕr·ŏ-*, the period of pregnancy.

utricle, n., *ūt'·rĭk·ŭl* (L. *ūtrĭcŭlus*, a small bag or bottle—from *ūtĕr*, a bag or bottle made from an animal's hide), in *bot.*, a thin-walled cell ; an air-bladder or cell ; a membranous one - seeded fruit ; in *anat.*, the larger of the two sacs of the vestibular portion of the ear : **utricular**, a., *ūt·rĭk'·ŭl·ăr*, containing vessels like small bags: **utriculus**, n., *ūt·rĭk'·ŭl·ŭs*, a kind of fruit with an inflated covering ; among Algæ, any loose cellular envelope containing spores ; a little bladder filled with air, attached to certain aquatic plants: **Primordial utricle**, within the cell-wall, and distinct from it, a delicate membrane or film immediately inclosing the cell contents.

Utricularia, n. plu., *ūt·rĭk'·ŭl·ăr'·ĭ·ă* (L. *ūtrĭcŭlus*, a small bag or bottle—from *ūtĕr*, a bag or bottle made from an animal's hide), a genus of plants, called Bladderworts, Order Lentibulariaceæ, so named from the utricles or bladders connected with the leaves, in which there exists a mucous fluid having cellular projections in the form of hairs : **Utricularia nelumbifolia**, *nĕl·ūm'·bĭ·fōl'·ĭ·ă* (said to be from *Nelumbo*, a Cingalese name ; L. *fŏlium*, a leaf), a singular plant which grows in the water collected at the bottom of the leaves of a large Tillandsia in Brazil, even sending out runners and shoots, and possessing a flowering stem two feet long.

Uvaria, n., *ūv·ār'·ĭ·ă* (L. *ūva*, a grape), a genus of climbing plants covered with star-shaped hairs, Ord. Anonaceæ : **Uvaria narium**, *nār'·ĭ·ŭm* (L. *nāris*, a nose, *nārĭum*, of noses; Gr. *nārŏs*, fluid), a species whose roots are fragrant and aromatic, used in India for fevers and liver complaints ; by distillation yield a fragrant greenish oil : **U. triloba**, *trĭl'·ŏb·ă* (Gr. *treis*, three; *lobos*, a lobe), a species containing a powerful acid, the leaves are used as an application for boils and abscesses, and the seeds are emetic : **U. febrifuga**, *fĕb·rĭf'·ūg·ă* (L. *febris*, a fever ; *fugo*, I drive away), a species to whose flowers the Indians ascribe febrifugal properties.

uvea, n., *ūv'·ĕ·ă* (L. *ūva*, a grape), in *anat.*, the posterior layer of the iris, which resembles the skin of a black grape.

uvula, n., *ūv'·ūl·ă* (L. *ūva*, a grape ; old F. *uvule*), in *anat.*, a muscular conical prominence projecting from the centre of the soft palate, and hanging down like a tongue—*Scotticè*, the pap of the hawse (Ger. *hals*, the throat); a small projection in the cerebellum : **uvula vesicæ**, *vĕs·is'·ē* (L. *vēsica*, the urinary bladder), a slight elevation of the mucous surface which projects from below into the urethral orifice of the urinary bladder.

vaccination, n., *văk'·sĭn·ā'·shŭn* (L. *vaccĭnus*, of or from cows—from *vacca*, a cow), the process by which the cow-pox or *vaccinia* is introduced into the human system, as a powerful protection against an attack of the smallpox : **vaccine**, a., *văk'·sĭn*, of or pert. to vaccinia or vaccination : **vaccinia**, n., *văk·sĭn'·ĭ·ă*, an eruptive vesicular disease, originally of the cow, now introduced into the human system as a protection against an attack of small-pox : **vaccinin**, n., *văk'·sĭn·ĭn*, the specific matter of cowpox.

Vacciniaceæ, n. plu., *văk·sĭn'·ĭ·ā'·sĕ·ē* (L. *vaccinĭum*, the bilberry), the Cranberry family, an Order of plants, some are astringent, and others yield sub-acid edible fruits : **Vaccinium**, n., *văk·sĭn'·ĭ·ŭm*, a genus of plants : **Vaccinium oxycoccus**, *ŏks'·ĭ·kŏk'·ŭs* (Gr. *oxus*, acid; *kokkos*, a berry); and **V. macrocarpum**, *măk'·rō·kârp'·ŭm* (G. *makros*, great ; *karpos*, fruit), are species which produce cranberries : **V. vitis-Idæa**, *vīt'·ĭs·ĭd·ē'·ă* (L. *vītĭs*, a vine ; *ĭdēă*, Idæan—from *Mount Ida* of Crete), the Idæan vine ; the Red Whortleberry or Cowberry, whose fruit or berries are often used instead of cranberries : **V. uliginosum**, *ŭl·ĭdj'·ĭn·ōz'·ŭm* (L. *ūlĭgĭnōsus*, full of moisture or wet—from *ūlĭgō*, moisture), the Black Whortleberry, found in Alpine countries : **V. myrtillus**, *mĕrt·ĭl'·lŭs* (L. *myrtus*, the myrtle), produces the Bilberry or Blaeberry.

vacuolæ, n. plu., *văk·ū'·ōl·ē*, also **vacuoli**, n. plu., *-ōl·ī*, and **vac'·uoles**, n. plu., *-ōlz* (L. dim. of *văcŭus*, void, empty), in *bot.*, and *animal histology*, clear spaces of indefinite size and arrangement in the protoplasm of a cell ; in *zool.*, little cavities found in the interior of many of the Protozoa, caused by the presence of little particles of food ; clear spaces often found in the tissues of the Cœlenterata.

vagina, n., *vădj·ĭn'·ă* (L. *vāgĭna*, a scabbard, a sheath), the canal or passage which leads from the external orifice of the female genitals to the uterus ; in *bot.*, a sheath formed by the petiole around the stem ; a sheath : **vaginal**, a., *vădj·ĭn'·ăl*, pert. to the vagina ; resembling a sheath: **vaginate**, a., *vădj'·ĭn·āt*, sheathed; invested as with a sheath : vag-

initis, n., *vădj'·ĭn·īt'·ĭs*, inflammation of the vagina.

vaginula, n., *vădj·ĭn'·ūl·ă*, also **vaginule**, n., *vădj'·ĭn·ūl* (L. *văgĭnŭlă*, a little sheath — from *vāgĭna*, a sheath), in *bot.*, a sheath surrounding the basal portion of the Archegonium, in Mosses.

vagus, n., *văg'·ŭs* (L. *văgus*, roaming, wandering), one of the three divisions of the eighth pair of cranial nerves having a more extensive distribution than any of the others.

Vahea, n., *vă·hē'·ă* (probably Sp. *vahear*, to emit steam or vapour), a genus of plants, Ord. Apocynaceæ, which yield caoutchouc : **Vahea gummifera**, *gŭm·mĭf'·ĕr·ă* (L. *gummi*, gum ; *fero*, I bear); and **V. Madagascariensis**, *măd'·ă·găsk'·ăr·ĭ·ĕns'·ĭs* (of or from *Madagascar*), are two large climbing shrubs or trees of Madagascar, yielding abundance of caoutchouc.

Valerianaceæ, n. plu., *văl·ēr'·ĭ·ăn·ā'·sĕ·ē* (after the anc. Roman *Valĕrius*, who first used it ; or L. *valĕo*, I am in health, from its virtues), the Valerian family, an Order of plants, which are generally strong scented or aromatic, some used as bitter tonics, anthelmintics, and antispasmodics : **Valeriana**, n., *văl·ēr'·ĭ·ăn'·ă*, a genus of plants, most of the species being ornamental in flower borders : **Valerianus officinalis**, *ŏf·fĭs'·ĭn·āl'·ĭs* (L. *offĭcĭnālis*, officinal, by authority —from *offĭcĭna*, a workshop), the common medicinal Valerian, having a bitter, acrid taste, and peculiar odour, disagreeable in the dry state, prescribed for hysteria: **Valerianus Celtica**, *sĕlt'·ĭk·ă* (L. *celtĭcus*, Celtic, pert. to Gaul), and others, possess similar properties : **Valerianella**, n., *văl·ēr'·ĭ·ăn·ĕl'·lă* (a dim. of *valerian*), a genus of plants : **Valerianella olitoria**, *ŏl'·ĭt·ōr'·ĭ·ă* (L. *ŏlĭtōrius*, belonging to a kitchen gardener—from *ŏlĭtor*, a kitchen gardener), a species whose young leaves are eaten as a salad, called by the French *Mâche*, and by the English *Lamb's lettuce*, and *corn salad*: **valerianic** or **valeric acid**, *văl·ĕr'·ĭ·ăn'·ĭk*, *văl·ĕr'ĭk*, an acid forming the leading ingredient of the volatile oil obtained from the Valerian root.

Vallisneria, n., *văl'·lĭs·nēr'·ĭ·ă* (after *Vallisnéri*, an Italian botanist), a genus of aquatic plants, Ord. Hydrocharidaceæ : **Vallisneria spiralis**, *spĭr·ăl'ĭs* (L. *spīrālis*, spiral—from *spīra*, a twist), a diœcious aquatic plant, the female flower developing along the spiral peduncle by which it reaches the surface of the water in order to receive the pollen ; Vallisneria, as well as Anacharis, show under the microscope the rotation of protoplasm in their cells: **Vallisnerieæ**, n. plu., *văl·lĭs'·nēr·ĭ'·ē·ē*, a tribe or Sub-order.

valve, n., *vălv* (L. *valvæ*, folding doors, *valvātus*, having folding doors), a cover or lid opening in one direction, and shutting in another ; in *bot.*, one of the pieces into which a pericarp or fruit separates, when separating naturally ; in *anat.*, folds of membrane guarding certain orifices and channels: **valvate**, a., *vălv'·āt*, in *bot.*, united or applied to each other by the margins only, as leaves in flower-buds or leaf-buds, the former being called *valvate æstivation*, the latter *valvate vernation ;* opening by valves, like the parts of certain seed-vessels : **valvular**, a., *vălv'·ŭl·ăr*, of or containing valves : **valvulitis**, n., *vălv'·ŭl·īt'·ĭs*, inflammation of valves.

valvulæ conniventes, *vălv'·ŭl·ē kŏn'·nĭv·ĕnt'·ēz* (L. dim. *valvŭlus* — from *valvæ*, folding doors ; *cŏnnīvens*, winking, *connīvĕntes*, plu.), in *anat.*, the permanent folds which exist in the lining membrane of the small intestine.

Vanilla, n., *văn·ĭl'·lă* (Sp. *vainilla*, a small pod or husk—from *vaina*, a scabbard or sheath), a genus of delightfully aromatic plants, Ord. Orchidaceæ : **Vanilla planifolia**, *plăn'·ĭ·fōl'·ĭ·ă* (L. *plānus*, flat ; *fŏlium*, a leaf) ; and **V. aromatica**, *ăr'·ōm·ăt'·ĭk·ă* (L. *ărōmătĭcus*, aromatic, fragrant—from *ărōma*, a spice), two species whose fleshy pod-like fruit, as well as that of other species, constitute the fragrant substance called *Vanilla*, employed to flavour confectionary, chocolate, etc.

vapours, n. plu., *văp'·ĕrz* (L. *vapor*, steam, exhalation ; F. *vapeur*), a disease characterised by nervous debility and depression of spirits ; hysteria.

varicella, n., *văr'·ĭs·ĕl'·lă* (a dim. from *vărĭŏla*, small-pox — from *vărĭus*, variegated, spotted), the chicken-pox or glass-pock.

varices, n. plu., *văr'·ĭs·ēz* (L. *vārix*, a dilated vein, *vārĭcis*, of a dilated vein, *vărĭces*, dilated veins), dilatations of veins ; in zool., the ridges or spinose lines marking a former position of the mouth in certain univalve shells : **varicose**, a., *văr'·ĭk·ōz*, denoting veins in a permanent state of dilatation, with an accumulation of dark-coloured blood : **varicocele**, n., *văr'·ĭk·ō·sēl'* (Gr. *kēlē*, a tumour), a swelling of the veins of the scrotum ; also of the spermatic cord : **varix**, n., *văr'·ĭks*, **varices**, n. plu., *văr'·ĭs·ēz*, a dilatation and convoluted state of the veins, accompanied with an accumulation of dark-coloured blood, due generally to an obstruction of the current of the blood towards the heart.

variety, n., *văr·ĭ'·ĕt·ĭ* (L. *vărĭĕtas*, diversity—from *vărĭus*, different, changing), a minor difference, as in form, colour, size, etc., existing in an individual of the same species, among animals or plants:

permanent **varieties or races,**
permanent minor differences,
among individuals of the same
species, arising from cultivation
and civilization, as well as from
natural causes.

variola, n., *văr·ĭ·ŏl·ă* (dim. from
L. *vărĭus*, varying, spotted), the
small-pox : **variolous,** a., *văr·ĭ·*
ŏl·ŭs, dotted with numerous small
impressions like those of the
small-pox ; relating to the small-
pox : **variolin,** n., *văr·ĭ·ŏl·ĭn*,
the specific matter of small-pox.

varix, see under 'varices.'

vasa aberrentia, *văz·ă ăb·ĕr·rĕn·*
shĭ·ă (L. *vas*, a vessel, *văsa*,
vessels ; *ăbĕrrentĭa*, participle,
plu., deviating from, wandering),
in *anat.*, long slender vessels
connecting the brachial or axillary
arteries with one of the arteries
of the forearm : **vasa afferentia,**
ăf·fĕr·ĕn·shĭ·ă (L. *ăffĕrĕns*, bringing or conveying to ; *ăffĕrĕntĭa*,
the plu. of the participle to
agree with *văsa*), lymphatics or
lacteals which enter a gland—also
called, **v. inferentia,** *ĭn·fĕr·ĕn·*
shĭ·ă (L. *ĭnfĕrens*, carrying or
bringing into ; *ĭnfĕrĕntĭă*, plu.) :
v. brevia, *brĕv·ĭ·ă* (L. *brĕvĭs*,
short ; *brĕvĭa*, plu.), from five to
seven small blood-vessels which
issue from the trunk of the splenic
artery, and reach the left extremity
of the stomach : **v. efferentia,** *ĕf·*
fĕr·ĕn·shĭ·ă (L. *ĕffĕrens*, bringing
or carrying out ; *ĕffĕrĕntĭă*, plu.),
small vessels which are straight
as they emerge from the testicle,
but become convoluted as they
proceed towards the epididymis,
forming a series of small conical
masses : **v. lactea** or **chylifera,**
lăkt·ĕ·ă or *kĭl·ĭf·ĕr·ă* (L. *lactĕus*,
pert. to milk—from *lac*, milk ; Gr.
chŭlos, juice ; L. *fero*, I bear), the
lacteals commencing in the coats
of the intestines, and extending
to the thoracic duct, in which they
terminate : **v. recta,** *rĕkt·ă* (L.
rectus, straight), small straight
blood-vessels lying between the
uriniferous tubes of the kidneys
and within the medullary substance ; straight seminal ducts of
the testicles, which pass through
their fibrous tissue, and end in
a close network of tubes : **v.
vasorum,** *văs·ōr·ŭm* (L. *răsa*,
vessels, *văsōrum*, of vessels),
small vessels, both venous and
arterial, on the coats of arteries,
veins, and lymphatics, which
serve for their nutrition : **v. vort-
icosa,** *vŏrt·ĭk·ōz·ă* (L. *vŏrtĭcōsus*,
full of vortices or eddies—from
vortex, a whirlpool), veins of the
choroid coat of the eye, so named
from their whorl-like arrangement.

vas aberrans, *văs ăb·ĕr·ănz* (L.
vas, a vessel ; *ăbĕrrans*, wandering), in *anat.*, a long narrow
tube, or diverticulum, leading off
from the lower part of the canal
of the epididymis, and ending by
a closed extremity : **vas deferens,**
dĕf·ĕr·ĕnz (L. *dēfĕrens*, bearing
or carrying away), the excretory
duct of the testis : **v. spirale,**
spīr·āl·ĕ (L. *spīrālis*, spiral —
from *spīra*, a fold, a coil), a small
single or branched blood-vessel
running along the under surface
of the membranous zone of the
internal ear.

vascular, a., *văsk·ŭl·ăr* (L. *vascŭlum*, a small vessel—from *vas*,
a vessel), consisting of or containing vessels, as arteries or
veins ; connected with the
circulatory system ; in *bot.*,
applied to tissue somewhat long ;
containing vessels like the tissue
of flowering plants, as distinguished from *cellular*.

vasculum, n., *văsk·ŭl·ŭm* (L.
vascŭlum, a small vessel—from
vas, a vessel), in *bot.*, a pitcher-
shaped leaf ; an Ascidium : **vasc-
uliform,** a., *văsk·ŭl·ĭ·fŏrm* (L.
forma, shape), having the form
of a pitcher or vasculum.

vasiform, a., *văs·ĭ·fŏrm* (L. *vas*,

a vessel ; *forma*, shape), in *bot.*, applied to a vegetable tissue called dotted vessels ; shaped like a blood-vessel.

vaso-motor, *văs'·ō-mōt'·ŏr* (L. *vas*, a vessel ; *mōtŏr*, a mover—from *mōto*, I keep moving), applied to nerves which govern the motions, and regulate the calibre of the blood-vessels ; the nerve fibres supplying the muscular coats of the blood-vessels.

vastus, n., *văst'·ŭs*, **vasti,** n. plu., *văst'·ī* (L. *vastus*, immense), a name applied to two portions of the 'triceps extensor cruris,' thus —**vastus externus,** *văst'·ŭs ĕks·tĕrn'·ŭs* (L. *externus*, outward) ; and **v. internus,** *ĭn·tĕrn'·ŭs* (L. *internus*, inward), the names designating a fleshy mass upon each side of it : **v. externus cruris,** *krŭr'·ĭs* (L. *crūs*, the leg or shin, *crūris*, of the leg), the full name of the vastus externus.

Vateria, n., *văt·ēr'·ĭ·ă* (after *Vater*, a German botanist), a genus of Indian trees, Ord. Dipterocarpaceæ : **Vateria Indica,** *ĭnd'·ĭk·ă* (L. *Indĭcus*, of or from India), a species which yields a gum resin, known as Indian Copal or Piney resin, used as a varnish, and in the manufacture of candles, and as incense ; in *med.*, used for rheumatic and other affections.

veil, n., *vāl* (old F. *veile*, a veil ; L. *vēlum*, a covering, a curtain), in *bot.*, the partial covering of the stem or margin of the cap among Fungi ; also said of the indusium of ferns.

vein, n., *vān* (F. *veine*, a vein ; L. *vēna*, a blood-vessel), in *anat.*, one of the vessels of the body which convey the blood back to the heart ; in *bot.*, one of the small branching ribs of a leaf.

vellus, n., *vĕl'·lŭs* (L. *vellus*, a fleece), in *bot.*, the stipe of certain Fungi.

velum, n., *vēl'·ŭm* (L. *vēlum*, a vail), in *zool.*, the membrane which surrounds and partially closes the mouth of the disc of the Medusæ, etc. ; in *bot.*, the cellular covering of the gills of an agaric in its early state : **velum interpositum,** *ĭn'·tĕr·pŏs'·ĭt·ŭm* (L. *intĕrpŏsĭtus*, a putting between, interposed), a vascular membrane reflected from the pia - mater, into the interior of the brain through the transverse fissure : **v. pendulum palati,** *pĕnd'·ŭl·ŭm păl·āt'·ī* (L. *pĕndŭlus*, hanging down, pendulous ; *pălātum*, the palate, *pălātī*, of the palate), in *anat.*, the soft palate, a moveable fold suspended from the posterior border of the hard palate, forming an incomplete septum between the mouth and pharynx.

velutinus, a., *vĕl'·ūt·īn'·ŭs* (mid. L. *velūtīnus*, velvety—from L. *vellus*, a fleece), in *bot.*, velvety ; applied to plants having a dense covering of short down like velvet : **velutinous,** a., *vĕl·ūt'·ĭn·ŭs*, having a velvety appearance.

vena azygos major, *vēn'·ă ăz'·ĭg·ŏs mādj'·ŏr* (L. *vēna*, a vein ; Gr. *azŭgos*, unyoked ; L. *mājor*, greater), in *anat.*, a vein on the right side of the body, commencing in the lumbar region : **vena cava,** *kāv'·ă* (L. *căvus*, hollow), a name applied to each of the two large veins which convey the blood back to right side of the heart—the *vena cava inferior* returning the blood from the lower limbs, and from the viscera of the abdomen and pelvis ; the *vena cava superior* returns the blood from the head, the neck, the upper limbs, and the thorax : **v. cordis magna,** *kŏrd'·ĭs măg'·nă* (L. *cor*, the heart, *cordis*, of the heart ; *magnus*, great), the great cardiac vein which coils round the left side of the base of the ventricle, and returns the blood from the substance of the heart to the right auricle :

v. hemiazygos, *hĕm'·ĭ·ăz'·ĭg·ŏs* (Gr. *hemi*, half; *azŭgos*, unyoked), the left or small azygos, a left intercostal vein, which crosses to join the main azygos near the seventh dorsal vertebra : **v. portæ,** *pŏrt'·ē* (L. *porta*, a gate, *portæ*, of a gate), the large vein which conveys the blood from the intestines into the liver, so named because it enters the porta or gate of the liver.

venæ basis vertebrarum, *vēn'·ē băs'·ĭs vĕrt'·ĕb·rār'·ŭm* (L. *vēna*, a vein, *venæ*, veins ; *băsĭs*, a base, or of a base ; *vĕrtebra*, a joint, *vĕrtebrārum*, of joints), the veins belonging to the bodies of the vertebræ ; comparatively large vessels or veins contained in the canals within the bodies of the vertebræ : **venæ comites,** *kŏm'·ĭt·ēz* (L. *cŏmĕs*, a companion, *cŏmĭtes*, companions), two or more deep veins accompanying an artery and its branches, following the distribution of such arteries : **v. cordis minimæ,** *kŏrd'·ĭs mĭn'·ĭm·ē* (L. *cor*, the heart, *cordis*, of the heart ; *mĭnĭmus*, least), the very small veins of the heart ; very small veins which open directly into the right auricle, and return the blood from the substance of the heart : **v. Galeni,** *găl·ēn'·ĭ* (L. *Gălēnus*, a celebrated anc. physician), two veins formed by the union of the minute veins of the choroid plexus of the brain : **v. parvæ,** *părv'·ē* (L. *parvus*, little, small), the small or anterior cardiac veins ; several small branches of veins which commence upon the anterior surface of the right ventricle, and open separately into the right auricle of the heart.

venation, n., *vĕn·ā'·shŭn* (L. *vēna*, a vein), in *bot.*, the arrangement of the veins or framework in leaves.

venery, n., *vĕn'·ĕr·ĭ* (L. *Venus*, the goddess of love, *vĕnĕris*, of Venus), sexual intercourse : **venereal,** a., *vĕn·ēr'·ĕ·ăl*, pert. to or connected with sexual intercourse.

venesection, n., *vĕn'·ĕ·sĕk'·shŭn* (L. *vēna*, a vein ; *sĕco*, I cut), the operation of bleeding from a vein, generally one at the head of the elbow.

venous, a., *vēn'·ŭs* (L. *vēnōsus*, full of veins — from *vēna*, a vein), pert. to or contained in a vein.

venter, n., *vĕnt'·ĕr* (L. *venter*, the belly), applied to the part of the internal surface of the ilium, which presents anteriorly a large smooth concave surface, lodging the iliacus muscle : **venter of the scapula,** *skăp'·ŭl·ă* (L. *scăpŭla*, the shoulder-blade), the anterior surface of the scapula, presenting a broad concavity, called the sub-scapular fossa : **ventral,** a., *vĕnt'·răl*, abdominal ; relating to the inferior surface of the body ; the opposite of dorsal ; in *bot.*, applied to the part of the carpel nearest the axis, or in front.

ventricle, n., *vĕnt'·rĭk·l* (L. dim. *ventriculus*, the belly or stomach — from *venter*, the belly), a small cavity in an animal body, as in the brain or heart ; applied to the cavities of the heart, which receive blood from the auricles : **ventricose,** a., *vĕnt'·rĭk·ōz*, distended ; swelling out in the middle, or unequally on one side : **ventricular,** a., *vĕnt·rĭk'·ŭl·ăr*, pert. to a ventricle or small cavity ; bellied.

Veratreæ, n. plu., *vĕr·āt'·rĕ·ē* (L. *vĕrātrum*, hellebore : *vērĕ*, truly ; *āter*, black), a Suborder of the Ord. Melanthaceæ : **Veratrum,** n., *vĕr·āt'·rŭm*, a genus of elegant plants when in flower, so named from the black colour of the root : **Veratrum album,** *ălb'·ŭm* (L. *albus*, white), a species whose rhizome or roots, the white hellebore of the Greeks, is an irritant, narcotic poison : **V. viride,** *vĭr'·ĭd·ē* (L.

vĭrĭdis, green), is an acrid, emetic, and powerful stimulant, followed by sedative effects : **ver-atrin**, n., *vĕr·āt'·rĭn*, or **veratria**, n., *vĕr·āt'·rĭ·ă*, an alkaloid, to whose presence is due the properties of Veratrum, used as an emetic and purgative, and in gout : **veratric acid**, *vĕr·āt'·rĭk*, an acid found in the seeds of Asagræa officinalis, formerly called Veratrum Sabadilla.

Verbascum, n., *vĕrb·ăsk'·ŭm* (L. *vĕrbăscum*, lungwort, the plant mullein), a genus of strong plants, producing an abundance of showy flowers, Ord. Scrophulariaceæ : **Verbascum Thapsus**, *thăp'·sŭs* (*Thapsus*, said to be an island where it grew), a species whose woolly leaves are emollient, and slightly narcotic, used in some pectoral affections ; also called Great Mullein.

Verbenaceæ, n. plu., *vĕrb'·ĕn·ā'·sĕ·ē* (L. *verbēnæ*, the boughs or branches of laurel, or other sacred boughs), the Vervain family, an Order of plants, many of which are fragrant and aromatic, some bitter and tonic, and some acrid : **Verbena**, n., *vĕrb·ēn'·ă*, a genus of extremely beautiful and ornamental plants in flower: **Verbena officinalis**, *ŏf·fĭs'·ĭn·āl'·ĭs* (L. *officīnālis*, official, by authority —from *officīna*, a workshop), the Vervain, a sacred plant among the Greeks, and received from them the name Holywort : **V. camædrifolia**, *kăm·ēd'·rĭ·fōl'·ĭ·ă* (Gr. *chamai*, on the ground ; Gr. *drus*, an oak ; L. *fōlium*, a leaf), a species from which the varieties of Verbenas of the gardens are chiefly obtained: **V. Teucrioides**, *tŭk'·rĭ·ōyd'·ēz* (after *Teucer*, its discoverer ; Gr. *eidos*, resemblance), a species whose flowers have a delightful jasmine-like odour.

Vermes, n., *vĕrm'·ēz* (L. *vermis*, a worm, *vermes*, worms), employed in nearly the same sense as Annaloida and Anarthropoda.

vermicular, a., *vĕrm·ĭk'·ŭl·ăr* (L. *vermicŭlŭs*, a little worm—from *vermis*, a worm), of or pert. to a worm ; that resembles the movements of a worm : **vermiculate**, a., *vĕrm·ĭk'·ŭl·āt*, also **vermiform**, a., *vĕrm'·ĭ·fŏrm* (L. *forma*, shape), resembling a worm ; shaped like a worm.

vernation, n., *vĕrn·ā'·shŭn* (L. *vernātio*, a renewal—from *ver*, spring), in *bot.*, the arrangement of the nascent leaves in the leaf-bud.

Veronica, n., *vĕr·ŏn'·ĭk·ă* (said to be a corruption of Arabic *viroo-nikoo*, beautiful remembrance ; It. and Sp. *veronica*), an extensive genus of plants, producing beautiful flowers, Ord. Scrophulariaceæ : **Veronica officinalis**, *ŏf·fĭs'·ĭn·āl'·ĭs* (L. *officīnālis*, official, by authority—from *officīna*, a workshop), a species whose leaves are bitter and astringent, sometimes used as tea.

verrucæ, n. plu., *vĕr·rōs'·ē* (L. *verrūca*, a wart, an excrescence, *verrucæ*, warts), in *bot.*, collections of thickened cells on the surface of plants, assuming a rounded form, and containing starch and other matters : **verrucæform**, a., *vĕr·rōs'·ē·fŏrm* (L. *forma*, shape), shaped like warts: **verrucose**, a., *vĕr'·rŏk·ōz'*, covered with wart-like excrescences.

versatile, a., *vĕrs'·ăt·ĭl* (L. *versāt·ĭlis*, that turns round, moveable —from *verso*, I turn much and often), in *bot.*, attached by one point to the filament, and so very easily turned round, as an anther.

vertebra, n., *vĕrt'·ĕb·ră*, vertebræ, n. plu., *vĕrt'·ĕb·rē* (L. *vertebra*, a joint—from *verto*, I turn), a bone of the spine or backbone, so called from its moving upon the adjoining one: **cervical vertebræ** are those of the neck, and are seven in number: **dorsal vertebræ**

are those of the back, and are twelve in number : **lumbar vertebræ** are those of the loins, and are five in number : **vertebral,** a., *vért'.ĕb·răl,* pert. to the joints of the spine or backbone.

vertebrate, a., *vért'.ĕb·rāt,* also **vertebrated,** a., *vért'.ĕb·rāt'.ĕd,* having a backbone or vertebral column : **Vertebrata,** n. plu., *vért'.ĕb·rāt'.ă,* the Division of the animal kingdom characterised by the possession of a backbone or vertebræ : **vertebra dentata,** *dĕnt·āt'.ă* (L. *dentātus,* toothed—from *dens,* a tooth), the second vertebra or axis, which forms a pivot on which the head with the first vertebra rotates : **vertebra prominens,** *prŏm'.ĭn·ĕnz* (L. *prōmĭnĕns,* standing out, prominent), the seventh cervical vertebra, so named because being so long it is readily felt beneath the skin.

vertex, n., *vért'.ĕks,* **vertices,** n. plu., *vért'.ĭs·ēz* (L. *vertex,* that which revolves about itself, the top or crown of the head—from *verto,* I turn), in *anat.,* the top or crown of the head : **vertical,** a., *vért'.ĭk·ăl,* perpendicular to the horizon ; standing upright.

verticil, n., *vért'.ĭs·ĭl* (L. *verticĭllus,* the whirl of a spindle, a little vertex—from *vertex,* a whirl, the top), in *bot.,* a whorl or form of inflorescence, having the flowers arranged in a circle around an axis: **verticillate,** a., *vért'.ĭs·ĭl·lāt,* having parts arranged in a whorl, or like the rays of a wheel ; **verticillaster,** n., *vért'.ĭs·ĭl·lăst'.ér* (L. *aster,* a diminutive termination), a false whorl or verticil, formed of two nearly sessile cymes, placed in the axils of opposite leaves.

vertigo, n., *vért·ĭg'.ō* (L. *vertĭgo,* a turning or whirling round—from *verto,* I turn about), giddiness, in which the patient feels that he is standing still, while the objects near him are running round.

verumontanum, *vĕr·ū·mŏnt'.ăn·ŭm* (L. *vĕru,* a spit, a dart ; *mons,* a mountain, *montis,* of a mountain), in *anat.,* a narrow longitudinal ridge of the urethra, formed by an elevation of the mucous membrane and its subjacent tissue.

vesica, n., *vĕs·ĭk'.ă* (L. *vēsīca,* the bladder), in *anat.,* the urinary bladder : **vesical,** a., *vĕs·ĭk'.ăl,* pert. to or in relation with the bladder: **vesicant,** n., *vĕs'.ĭk·ănt,* any external application which can raise a blister on the skin, as Spanish fly, acetic acid, etc.

vesicle, n., *vĕs'.ĭk·l,* also **vesicule,** n., *vĕs'.ĭk·ūl* (L. *vēsīcŭla,* a little bladder—from *vēsīca,* a bladder), a small bladder-like blister on an animal body ; a little sac or cyst; a small bladder-like cavity: **vesicula,** n., *vĕs·ĭk'.ūl·ă,* in *bot.,* composed of cells: **vesico-uterine,** *vĕs'.ĭk·ō·ūt'.ér·ĭn,* applied to folds of peritoneum extending from the uterus to the urinary bladder: **vesicula prostatica,** *prŏs·tăt'.ĭk·ă* (Gr. *prostātēs,* one who stands before, a leader), a depression at the forepart of the verumontanum in its middle line : **vesiculæ seminales,** *vĕs·ĭk'.ūl·ē sĕm•ĭn·āl'.ĕs* (L. *sēmen,* seed, *semĭnis,* of seed), the seminal vesicles in which the semen lodges.

vestibule, n., *vĕst'.ĭb·ūl* (L. *vestĭbŭlum,* a forecourt), a small oval cavity of the internal ear, forming an entry to the cochlea, etc.; a small cavity in the ventricle of the heart ; the angular interval between the nymphæ.

vestigium foraminis ovalis, *vĕst·ĭdj'.ĭ·ŭm fŏr·ăm'.ĭn·ĭs ŏv·āl'.ĭs* (L. *vestigĭum,* a trace, a vestige ; *fŏrāmen,* an aperture, *fŏrāminis,* of an aperture ; *ōvālis,* oval—from *ōvum,* an egg), the vestige of the foramen ovale of the fœtal heart, which indicates the original place of communication between the two auricles : **vestigial,** a., *vĕst·ĭdj'·ĭ·ăl*, pert. to a trace or vestige ; applied to a fold of the pericardium.

veterinary, a., *vĕt'·ĕr·ĭn·ăr'·ĭ* (L. *vĕtĕrīnārĭŭs*, belonging to beasts of burden—from *veterīnæ*, draught cattle, or beasts of burden), pert. to or connected with the art of treating the diseases of domestic animals.

vexillum, n., *vĕks·ĭl'·lŭm* (L. *vexillum*, a standard or banner), in *bot.*, the upper or posterior petal of a papilionaceous or pea flower : **vexillary,** a., *vĕks'·ĭl·lăr·ĭ*, denoting a form of æstivation in which the vexillum or upper petal is folded over the other.

vibices, n. plu., *vĭb·ĭs'·ēz* (L. *vībex*, the mark of a blow or stripe, *vībĭcis*, of the mark of a blow), patches of hæmorrhage, occurring in the skin in purpura; also known as 'ecchymosis'; called 'petechiæ' when very small.

vibracula, n., *vĭb·răk'·ŭl·ă* (L. *vībro*, I shake, I quiver), long, filamentous appendages found in many Polyzoa.

vibrio, n., *vĭb'·rĭ·ō*, **vibriones,** n. plu., *vĭb'·rĭ·ōn'·ēz* (L. *vībro*, I quiver or shake), minute thread - like animalcules found in many organic infusions.

vibrissæ, n. plu., *vĭb·rĭs'·sē* (L. *vibrissæ*, hairs in the nose of man —from *vĭbro*, I shake),hairs found growing at the entrance of the nostrils, and other outlets ; the whiskers in cats.

Viburnum, n., *vĭb·ĕrn'·ŭm* (L. *viburnum*, the wayfaring shrubs), a genus of elegant flowering shrubs, Ord. Caprifoliaceæ : **Viburnum lantana,** *lănt·ān'·ă* (the anc. name of viburnum), the pliant mealy tree, a species whose bark and berries are acrid: **V.opulus,** *ŏp'·ŭl·ŭs* (L. *cpŭlŭs*, a kind of maple tree), the Gueldres-rose,also called snowball, from its globular head of abortive leaves : **V. tinus,** *tĭn'·ŭs* (L. *tĭnus*, a plant, supposed V.

tinus), a species, the Laurustinus of gardeners.

Victoria,n.,*vĭk·tōr'·ĭ·ă* (after *Queen Victoria* of Britain and Ireland), a genus of noble aquatic plants, inhabiting the tranquil rivers of S. America, Ord. Nymphæaceæ, the seeds and root-stocks of many of the plants containing much starch, used as food : **Victoria regia,** *rēdj'·ĭ·ă* (L. *rēgĭus*, royal), one of the largest aquatics known, its very large flowers have a fine odour.

Vidian, a., *vĭd'·ĭ·ăn* (after *Vidius*, a professor at Paris), a name applied to (1) a small branch of the inferior maxillary artery ; (2) a canal which passes through the sphenoid bone horizontally ; (3) a nerve arising from the spheno-palatine ganglion, and - passing through the Vidian canal.

villi, n. plu., *vĭl'·ī* (L. *villus*, wool or hair, *villī*, hairs), in *anat.*, minute projections on the mucous lining of the intestinal canal, which are made up of blood-vessels, nerves, and absorbents ; in *bot.*, projections or papillæ on the surface of the epidermis of a plant, when these assume an elongated or conical form ; jagged leafy processes, covering the stem, amongst the leaves : **villose,** a., *vĭl·ōz'*, also **villous,** a., *vĭl'·ŭs*, in *bot.*, covered with long weak hairs or down ; in *anat.*, downy ; velvety : **villus,** n., *vĭl'·ŭs*, in *anat.*, one of the conical projections of the mucous membrane of the small intestines.

vincula accessoria tendinum, *vĭngk'·ŭl·ă ăk'·sĕs·sōr'·ĭ·ă tĕnd'·ĭn·ŭm* (L. *vĭncŭlum*, a chain, a fetter ; *accĕssōrĭus*, accessory, added—from *accĕssĭo*, an increase, an addition ; *tendo*, a tendon, *tĕndĭnum*, of tendons), the accessory fetters of the tendons ; also **vincula vasculosa,** *văsk'·ŭl·ōz'·ă* (L. *vascŭlōsus*, full of little vessels —from *văscŭlum*, a little vessel), the vascular fetters or fibres —are slender and loose bands, forming accessory fibres to the sheaths of the flexor tendons of the fingers.

Violaceæ, n. plu., *vī′·ŏl·ā′·sĕ·ē* (L. *vĭŏla*, the violet), the Violet family, an Order of plants, distinguished by the emetic properties of their roots : **Violeæ**, n. plu., *vī·ŏl′·ĕ·ē*, a tribe of plants : **Viola**, n., *vī′·ŏl·ă*, a genus of plants, esteemed for the beauty and scent of their flowers : **Viola odorata**, *ŏd′·ŏr·āt′·ă* (L. *odorātus*, having a scent or smell—from *ŏdŏr*, scent, smell), the sweet or March Violet, whose roots have been used as an emetic, and the petals as a laxative : **V. tricolor**, *trĭ·kŏl′·ŏr* (L. *trĭs*, three ; *cŏlor*, colour), Heart's-ease, which, with other species, have been used as demulcent expectorants ; all the cultivated varieties of the pansies have originated from V. tricolor : **V. canina**, *kăn·īn′·ă* (L. *canīnus*, of or pert. to a dog—from *cănĭs*, a dog), said to be good in cutaneous diseases.

Viperina, n. plu., *vĭp′·ĕr·īn′·ă* (L. *vīpĕrīnŭs*, pert. to an adder— from *vīpĕra*, an adder, a snake), in *zool.*, a group of the snakes : **viperine**, a., *vĭp′·ĕr·īn*, of or pert. to a snake.

virescence, n., *vĭr·ĕs′·ĕnz* (L. *virescens*, growing green), in *bot.*, the production of green in petals instead of the usual colouring matter : **virescent**, a., *vĭr·ĕs′·sĕnt*, approaching a green hue.

virgate, a., *vĕrg′·āt* (L. *virga*, a rod), in *bot.*, long and straight like a wand.

virus, n., *vīr′·ŭs* (L. *vīrus*, poison), a morbid poison, as of an ulcer ; the agent which transmits infectious diseases.

vis-a-fronte, *vĭs′·ă·frŏnt′·ē* (L. *vĭs*, strength, force ; *a*, from ; *frons*, the front, *frŏntĕ*, from the front), in *anat.*, one of the forces, called the 'Aspiratory force,' which

tend to produce a regular flow of blood ; in *bot.*, the evaporation of the leaves, which assists the ascent of the sap in plants : **vis-a-tergo**, *-tĕrg′·ō* (L. *tergum*, the back, *tĕrgo*, from the back), in *anat.*, the constant pressure from behind which causes the flow onwards of the blood towards the veins ; in *bot.*, the pushing force from below upwards by which the ascent of the sap in plants is assisted, caused by the absorption of moisture by the rootlets : **vis nervosa**, *nĕrv·ōz′·ă* (L. *nervōsus*, nervous — from *nervus*, a nerve), the property of nerves by which they convey stimuli to muscles either directly or circuitously.

viscera, n. plu., *vĭs′·ĕr·ă* (L. *viscus*, a bowel, *viscĕra*, the bowels), in *anat.*, the bowels, situated in the abdomen ; the contents of the thorax ; the contents of the cranium : **viscus**, n., *vĭsk′·ŭs*, any internal organ of the body : **visceral**, a., *vĭs′·sĕr·ăl*, pert. to the viscera.

viscous, a., *vĭsk′·ŭs* (L. *viscum*, the mistletoe, a sticky substance called birdlime made from it), glutinous ; clammy, like birdlime.

Viscum, n., *vĭsk′·ŭm* (L. *viscum*, the mistletoe ; *viscus*, birdlime, from the sticky nature of the berries), a genus of plants, Ord. Loranthaceæ : **Viscum album**, *ălb′·ŭm* (L. *albus*, white), the Mistletoe, a parasitic plant, chiefly found on apple trees, but was esteemed most by the Druids when found on oaks.

Vitaceæ, n. plu., *vĭt·ā′·sĕ·ē* (L. *vītĭs*, a vine), the Vine family, an Order of plants, also named 'Ampelideæ,' which see : **Vitis**, n., *vīt′·ĭs*, a very valuable and interesting genus of fruit-bearing plants : **Vitis vinifera**, *vĭn·ĭf′·ĕr·ă* (L. *vinum*, vine ; *fero*, I bear), the grape vine, whose unripe fruit contains the harsh acid juice called verjuice, the leaves are astringent, and are used in diarrhœa, its sap in France is the popular remedy for chronic ophthalmia; *raisins* are dried grapes : **V. vulpina**, *vŭlp·ĭn'·ă* (L. *vulpinus*, of or belonging to a fox—from *vulpes*, a fox), a species which yields the fox-grapes of Rhode Island.

vitelline, n., *vĭt·ĕl'·lĭn* (L. *vitellus*, the yolk of an egg), the albuminous substance of the yolk of eggs ; in *bot.*, the colour of the yolk of an egg : adj., applied to a membrane which encloses the yolk of the ovum : **vitellus**, n., *vĭt·ĕl'·lŭs*, in *anat.*, the contents of the ovum ; in *bot.*, the thickened sac within the nucleus which contains the amnios ; the embryo-sac, remaining distinct from the nucleus in the seeds, and forming a covering.

viticula, n., *vĭt·ĭk'·ŭl·ă* (dim. of L. *vitis*, a vine), in *bot.*, a trailing stem, as of a cucumber.

vitiligo, n., *vĭt'·ĭl·ĭg'·ō* (L. *vĭtĭlĭgo*, a cutaneous eruption, leprosy), a cutaneous disease, consisting of white patches on the skin, caused by loss of the usual colouring matter: **vitiligoidea**, *vĭt'·ĭl·ĭg·ōyd'·ĕ·ă* (Gr. *eidos*, resemblance), yellow patches sometimes met with round the eyelids, and elsewhere on the skin.

vitreous, a., *vĭt'·rĕ·ŭs* (L. *vĭtrĕus*, glassy, clear), applied to the glutinous, semi-fluid substance which fills up the central portion of the eye, and is quite transparent.

Vittæ, n. plu., *vĭt'·ē* (L. *vitta*, a band or fillet worn round the head among the ladies of anc. Rome), in *bot.*, narrow elongated receptacles of aromatic oil, occurring in the fruits of Umbellifers, appearing as brown dots between the pericarp and albumen in a transverse section of the fruit : **vittate**, a., *vĭt'·āt*, striped.

Vivianiaceæ, n. plu., *vĭv'·ĭ·ăn'·ĭ·ă'·sĕ·ē* (L. after *Viviana*, a botanist of Genoa), the Viviania family, an Order of plants of Chili : **Viviana**, n., *vĭv'·ĭ·ăn'·ă*, a genus of very pretty plants.

viviparous, a., *vĭv·ĭp'·ăr·ŭs* (L. *vivus*, alive ; *parto*, I produce), in *zool.*, bringing forth young alive ; in *bot.*, producing young plants in place of seeds ; attached in some unusual way to the parent, as young plants.

vivisection, n., *vĭv'·ĭ·sĕk'·shŭn* (L. *vivus*, alive.; *sectus*, cut), the dissection of an animal while living ; anatomical and surgical experiments on a living animal.

Vochysiaceæ, n. plu., *vŏk·ĭz'·ĭ·ā'·sĕ·ē* (from *Vochy*, the Guiana name of a species), the Vochysia family, an Order of plants, inhabiting the warmer parts of America : **Vochysia**, n., *vŏk·ĭz'·ĭ·ă*, a genus of plants, whose flowers are very sweet, and some yield a resinous juice.

volar, a., *vōl'·ăr* (L. *vŏla*, the palm of the hand), a branch of the radial artery, arises near the place where the radial leaves the front of the forearm, and passes onwards into the hands.

volubile, a., *vŏl·ūb'·ĭl·ĕ*, also *vŏl'·ūb·ĭl* (L. *volūbĭlĭs*, that turns itself round, twining—from *vŏlvo*, I turn round), in *bot.*, applied to stems, leaf-stalks, and the like, which have the property of twisting around some other body ; twining spirally.

volute, a., *vŏl·ūt'* (L. *volūtus*, turned round, twisted), in *bot.*, rolled up or twisted in any direction : **volution**, n., *vŏl·ū'·shŭn*, a spiral turn or wreath.

volva, n., *vŏlv'·ă* (L. *volva*, a wrapper—from *volvo*, I roll or turn about), in *bot.*, the involucrum-like base of the stipes of agarics, which was originally the bag enveloping the whole plant ; a general wrapper in Fungi.

volvulus, n., *vŏlv'·ŭl·ŭs* (new L. *vŏlvŭlus*, a little roll or wrapper—from *volvo*, I roll or turn about), in *surg.*, the passing of one portion of an intestine into another, commonly · the upper into the lower part.

vomer, n., *vōm'·ĕr* (L. *vōmer*, a ploughshare), in *anat.*, the slender thin bone separating the two nostrils, so named from its fancied resemblance to a ploughshare.

vomica, n., *vŏm'·ĭk·ă*, vomicæ, n. plu., *vŏm'·ĭs·ē* (L. *vŏmĭca*, a sore, a tumour), in *surg.*, the cavities formed in the destruction of the lungs; the collection of purulent matter in the lungs, forming cavities, constituting one of the most constant and important of morbid changes in chronic phthisis.

vulva, n., *vŭlv'·ă* (L. *vulva*, a womb), the external and visible parts of the female genitals : **vulvular**, a., *vŭlv'·ŭl·ăr*, pert. to, or in relation with the vulva.

warts, n., *wǎwrts* (Dut. *werte*, Ger. *warze*, a wart), in *med.*, dry excrescences of different forms, found on the skins of animals; verrucæ or papillary tumours; in *bot.*, firm glandular excrescences on the surfaces of plants.

wen, n., *wĕn* (AS. *wenn*, a swelling, a wart), an encysted tumour, affecting the head, face, or neck.

wheal, n., *hwēl* (AS. *walan*, a wheal; Goth. *valus*, Icel. *völr*, a rod, a stick), the raised streak on the skin left by a stripe, as with a cane ; red and white marks on the skin, seen in cases of nettle-rash.

whites, n. plu., *hwītz*, the popular name for 'leucorrhœa,' which see.

whitlow, n., *hwĭt'·lō* (Prov. Eng. *whickflaw* — from Prov. Eng. *whick*, quick, alive ; Eng. *flaw*), a flaw or sore about the quick of the nail ; an abscess beneath the

periosteum of the distal ph of any finger ; paronychia.

whooping-cough, *hŏp'·ing-* (a imitative of the sound), an tious disease, principally of hood, characterised by conv paroxysms of coughing, frequ fatal ; pertussis.

whorl, n., *hwŏrl* (Dan. *hver* turn ; Dut. *worwel*, a whi eddy), the spiral turn of a un shell ; any set of organs or a dages arranged in a circle a an axis ; leaves arranged regular circumference rou stem ; in *bot.*, a verticil.

wing, n., *wĭng*, in *bot.*, one two lateral petals of a pap aceous flower ; the broad edge of any organ : winge *wĭng'·ĕd*, furnished with l membranous expansions.

womb, n., *wŏm* (AS. *wamb*, *vomb*, belly, womb), the l organ in the female anin which the young is conceive nourished till birth.

woorali, see ' wourali.'

wornil, n., *wĕrn'·ĭl* (a dimi of *worm*), the larva or mag an insect found on the ba cattle.

wourali or **woorali**, n., *wi* (from a native name, *oura* arrow poison prepared b S. American Indians fror plant Strychnos toxifera, Guianensis, Ord. Loganiace

Wrightia, n., *rīt'·ĭ·ă* (afte *Wright*, a Scotch botani genus of plants, Ord. Apocy **Wrightia tinctoria**, *tĭngk-* (L. *tinctŏrĭus*, of or belong dyeing—from *tingo*, I d species from whose leaves ferior kind of indigo is pre **W. antidysenterica**, *ănt'·ĭ-* *tĕr'·ĭk·ă* (Gr. *anti*, against badly ; *entĕra*, the bowe species whose bark is the C bark of the Materia M valued as a tonic, a febrifug in dysentery.

:anthelasma, n., *zănth'.ĕl·ăz'·mă* (Gr. *xanthos*, yellow ; *ĕlasma*, a plate of metal hammered out), a disease of the skin characterised by yellow, slightly-raised patches, most common around the eyelids.

:anthic, a., *zănth'.ĭk* (Gr. *xanthos*, yellow), tending towards a yellow colour : **xanthine**, n., *zănth'.ĭn*, the yellow, insoluble, colouring matter in certain plants and flowers.

:anthophyll, n., *zănth'.ō·fĭl* (Gr. *xanthos*, yellow; *phullon*, a leaf), the yellow colouring matter of plants.

:anthorrhæa, n., *zănth'.ŏr·rē'.ă* (Gr. *xanthos*, yellow ; *rhĕō*, I flow), a genus of plants, Ord. Liliaceæ, to which belong the Black-boy, grass-gum trees of Australia : **Xanthorrhæa hastile**, *hăst·ĭl'.ĕ* (L. *hastĭlĕ*, the shaft of a spear), the grass-tree of New S. Wales, yields a yellow gum-like substance ; the leaves afford good fodder for cattle, and the natives eat the tender white centre of the top.

:anthoxylaceæ, n. plu., *zănth·ŏks'.ĭl·ā'·sĕ·ē* (Gr. *xanthos*, yellow; *zulŏn*, wood), the Xanthoxylon family, an Order of plants, some of which yield a volatile oil, aromatic and pungent, some are diaphoretics, others febrifugal and tonic : **Xanthoxylon**, n., *zănth·ŏks'.ĭl·ŏn*, a genus of plants, from their pungency sometimes called peppers : **Xanthoxylon fraxineum**, *frăks·ĭn'.ĕ·ŭm* (L. *fraxĭnĕus*, of ash - wood — from *frăxĭnus*, the ash - tree), the prickly ash, acts as a sialogogue : **X. caribæum**, *kăr'.ĭb·ē·ŭm* (of or from the *Caribbean Islands*), a W. Indian species having a bitter and febrifugal bark : **X. piperitum**, *pĭp'.ĕr·ĭt'.ŭm* (L. *pĭpĕrĭtus*, of or pert. to pepper — from *pĭper*, pepper), a Japanese species called Japan pepper : **xanthopicrine**, n., *zănth·ŏp'.ĭk·rĭn* (Gr. *xanthos*,

yellow; *pikros*, sweet), the bitter principle secreted by many species of the Order.

xeroderma, n., *zĕr'.ō·dĕrm'.ă* (Gr. *xĕros*, dry ; *dĕrma*, skin), a skin which is dry, hard, and rough ; also termed ichthyosis, or fish-skin disease.

xerophiles, n. plu., *zĕr'.ō·fĭlz* (Gr. *xĕros*, dry ; *phĭlĕō*, I love), in *bot.*, plants which require a large amount of heat and but little moisture : **xerophilous**, a., *zĕr·ŏf'.ĭl·ŭs*, of or pert. to such plants.

xiphisternum, n., *zĭf'.ĭ·stĕrn'.ŭm* (Gr. *xiphos*, a sword ; *stĕrnŏn*, the breast), in *zool.*, the inferior or posterior segment of the sternum, corresponding to the xiphoid cartilage of human anatomy.

xiphoid, a., *zĭf'.ōyd* (Gr. *xiphos*, a sword ; *eidos*, resemblance), in *anat.*, sword - shaped ; a term applied to the cartilage of the sternum.

xiphophyllous, a., *zĭf'.ō·fĭl'.ŭs* (Gr. *xĭphŏn*, a corn flag—from *xiphos*, a sword; *phullon*, a leaf), in *bot.*, having ensiform leaves.

Xiphosura, n., *zĭf'.ōz·ūr'.ă* (Gr. *xiphos*, a sword ; *oura*, a tail), an Order of Crustacea, charac-terised by their long, sword-like tails, as in the King-crabs.

xylem, n., *zĭl'.ĕm* (Gr. *xulŏn*, wood), bast-fibre or flax, procured from the inner bark of the stalk of Linum usitatissimum ; woody tissue.

xylocarp, n., *zĭl'.ō·kărp* (Gr. *xulŏn*, wood ; *karpos*, fruit), in *bot.*, a hard and woody fruit : **xylocarp-ous**, a., *zĭl'.ō·kărp'.ŭs*, having fruit becoming hard and woody.

xylophagous, a., *zĭl·ŏf'.ăg·ŭs* (Gr. *xulŏn*, wood ; *phago*, I eat), eat-ing or feeding on wood ; in *zool.*, applied to certain Mollusca.

Xylopia, n., *zĭl·ōp'.ĭ·ă* (Gr. *xulŏn*, wood ; *pĭkrŏs*, bitter), a genus of ornamental plants, Ord. Anon-aceæ, the wood of some species being extremely bitter : **Xylopia aromatica,** *ăr'.ōm.ăt'.ĭk·ă* (L. *ărōmătĭcus,* aromatic, fragrant— from *arōma,* a spice), a species commonly called Ethiopian pepper: **X.glabra,** *glăb'.ră* (L. *glăber,* smooth, without hair),. a species called Bitter-wood in W. Indies : **X. frutescens,** *frŏt·ĕs'.ĕns* (L. *frŭtĕx,* a shrub or bush), a native of Cayenne, the seeds used instead of spices: **X. grandiflora,** *grănd'. ĭ·flōr'.ă* (L. *grăndĭs,* great ; *flos,* a flower, *floris,* of a flower), a Brazilian species, esteemed for its carminative fruits, and febrifugal properties.

Xyridaceæ, n. plu., *zĭr'.ĭd·ā'.sĕ·ē* (Gr. *xurŏs,* sharp, razor-like), the Xyris family, an Order of plants, whose leaves terminate in sharp points : **Xyris,** n., *zĭr'.ĭs,* a genus of plants.

yaws, n. plu., *yăwz* (African *yaw,* a berry), a contagious disease, common in Africa, characterised by eruptions resembling strawberries.

yeast, n., *yēst* (Ger. *gascht,* froth of beer ; AS. *gist,* yeast), the froth in the working of beer ; the matter which separates from a liquid during the vinous fermentation ; yeast itself consists of a mass of minute cryptogamic plants : **yeast plant,** the popular name for the fungus or vinegar plant called Penicillium glaucum.

Zamia, n., *zăm'.ĭ·ă* (L. *zămĭa,* hurt, damage), a genus of very remarkable plants, nearly related both to ferns and palms, Ord. Cycadaceæ: **Zamia pumila,** *pŭm'. ĭl·ă* (L. *pŭmĭlus,* dwarfish, little), a species which supplies an amylaceous matter, has been sold as arrowroot : **Z. tenuis,** *tĕn'.ū·ĭs* (L. *tĕnŭis,* thin, fine); and **Z. furfuracea,** *fĕr'.fŭr·ā'.sĕ·ă* (L. *furfur,* bran or husks of wheat),

also produce a kind of arrowroot.

Zanthoxyllaceæ, n. plu.,*zănth·ŏks'. ĭl·lā'.sĕ·ē,* see ' Xanthoxyllaceæ.'

Zea, n., *zē'.ă* (Gr. *zeia,* Sansc. *zeva,* a species of corn), a genus of plants, Ord. Gramineæ, so named in reference to the nutritive qualities of the plants: **Zea mays,** *mā'.ĭz* (the Indian name), maize or Indian corn.

Zingiberaceæ, n. plu., *zĭnj'.ĭb·ĕr· ā'.sĕ·ē* (Gr. *zinggĭbĕris,* the ginger plant—from a native name), the Ginger family, an Order of plants, whose seeds and roots possess aromatic, stimulant properties— Order is also called Scitamineæ : **Zingiber,** n., *zĭnj'.ĭb·ĕr,* a genus of aromatic plants : **Zingiber officinale,** *ŏf.fĭs'.ĭn·āl'.ĕ* (L. *offĭcĭnālis,* officinal, by authority— from officĭna, a workshop), a species whose rhizomes constitute the ginger of commerce, imported from E. and W. Indies, roots used as preserves in their young state, used as a carminative and tonic in powder, syrup, or tincture.

Zizania, n., *zĭz·ān'.ĭ·ă* (Gr. *zizănĭŏn,* a weed growing among corn, darnel), a genus of plants, natives of America, Ord. Gramineæ : **Zizania aquatica,** *ăk·wăt'.ĭk·ă* (L. *ăquătĭcus,* growing or found in water — from *ăqua,* water), a species which supplies a kind of rice in Canada.

Zizyphus, n., *zĭz'.ĭf·ŭs* (L. *zĭzyphum,* Gr. *zĭzŭphŏn,* the jujube), a genus of pretty plants, Ord. Rhamnaceæ, the fruit of many being edible : **Zizyphus jujuba,** *jŏ'.jŏb·ă* (Gr. *zĭzŭphŏn,* Ar. *zĭfzŭf,* the jujube tree), a tree which supplies the fruit called jujube, and a kind of Scinde lac is found on it : **Z. lotus,** *lŏt'.ŭs* (Gr. *lotos,* L. *lotus,* the lotus), the Lotus or Lote-bush of the classics.

zona denticulata, *zōn'.ă dĕnt·ĭk'. ŭl·āt'.ă* (L. *zōna,* Gr. *zōnē,* a belt or girdle ; L. *dĕntĭcŭlātus,* furnished with small teeth—from *dens*, a tooth), in *anat.*, the toothed belt: zona glomerulosa, *glŏm·ĕr'·ŭl·ōz'·ă* (dim. from L. *glŏmĕrōsus*, like a ball, round—from *glŏmus*, a ball), the outer layer of the cortical part of the supra-renal bodies : z. pectinata, n., *pĕk'·tĭn·āt'·ă* (L. *pectĭnātus*, combed—from *pectin*, a comb), the comb-like belt ; the outer zone of the membrana basilaris: z. pellucida, *pĕl·lōs'·ĭd·ă* (L. *pellūcĭdus*, transparent), the external covering of the ovum ; a thick, colourless, transparent envelope which surrounds the substance of the yelk: z. reticularis, *rĕt·ĭk'·ŭl·ār'·ĭs* (L. *retĭculāris*, net-like—from *rētĕ*, a net), the inner layer of the cortical part of the supra-renal bodies.

zonate, a., *zōn'·āt* (L. *zōna*, a belt, a girdle), in *bot.*, marked with concentric undulations, bands, or zones.

zooid, a., *zō'·ŏyd* (Gr. *zoŏn*, an animal ; *eidos*, resemblance), an organism, more or less independent, produced by gemmation or fission : zoology, n., *zō·ŏl'·ō·jĭ* (Gr. *logos*, a discourse), that branch of natural history which treats of the structure, habits, classification, etc., of all animals.

zoophilous, a., *zō·ŏf'·ĭl·ŭs* (Gr. *zoŏn*, an animal ; *phĭlĕō*, I love), in *bot.*, applied to plants fertilised by the agency of insects.

zoophyte, n., *zō'·ŏf·īt* (Gr. *zoŏn*, an animal; *phŭtŏn*, a plant), in *zool.*, applied to many plant - like animals, such as sponges, corals, sea-anemones, etc.

zoosperms, n. plu., *zō'·ō·spĕrmz* (Gr. *zoŏn*, an animal ; *sperma*, seed), in *bot.*, the locomotive spores of some Algæ and Fungi ; zoospores; in *zool.*, animal semen.

zoospores, n. plu., *zō'·ō·spōrz* (Gr. *zoŏn*, an animal ; *spora*, seed), in *bot.*, moving spores provided with cilia ; zoosperms ; in *zool.*, the ciliated locomotive germs of some of the lowest forms of plants—the Protophyta.

zootheca, n., *zō'·ō·thēk'·ă* (Gr. *zoŏn*, an animal; *thēkē*, a case), in *bot.*, a cell containing a spermatozoid.

zootomy, n., *zō·ŏt'·ŏm·ĭ* (Gr. *zoŏn*, an animal ; *tomē*, a cutting), the dissection of the lower animals.

zygapophyses, n. plu., *zĭg'·ă·pŏf'·ĭs·ēs* (Gr. *zugon*, a yoke; *apŏphŭsĭs*, the process of a bone), the yoke-pieces ; the articulating processes of the vertebræ.

zygoma, n., *zĭg·ōm'·ă* (Gr. *zugōma*, a bolt or bar — from *zugon*, a yoke), in *anat.*, a bony arch at the upper part of the side of the face ; the arch formed by the zygomatic process of the temporal and cheek bones : zygomatic, a., *zĭg'·ōm·ăt'·ĭk*, pert. to the zygoma, or to the cheek-bone : zygomatic fossa, *fŏs'·să* (L. *fossa*, a ditch), an irregularly - shaped cavity, situated below and on the inner side of the zygoma : z. process, a thin, narrow projection of bone at the base of the squamous portion of the temporal bone : zygomaticus, a., *zĭg'·ōm·ăt'·ĭk·ŭs*, applied to two muscles — the *major*, arising from the cheek-bone, and inserted into the angle of the mouth ; the *minor*, arising higher on the cheek-bone, and inserted into the upper lip.

Zygophyllaceæ, n. plu., *zĭg'·ō·fĭl·lā'·sĕ·ē* (Gr. *zugon*, a yoke; *phullon*, a leaf, the leaves being in pairs), the Guiacum family, an Order of plants, some abound in stimulant resin, some are bitter and acrid, others are sudorifics : Zygophylleæ, n. plu., *zĭg'·ō·fĭl'·lĕ·ē*, a section or Sub-order, having albuminous seeds : Zygophyllum, n., *zĭg'·ō·fĭl'·lŭm*, a genus of plants : Zygophyllum fabago, *făb·āg'·ō* (L. *făbāgĭnus*, of or pert. to beans —from *făba*, a bean), the Bean-caper, so named from its flowers being used as capers, said to act as a febrifuge.

zymosis, n., *zĭm·ōz'·ĭs* (Gr. *zumōs-is*, fermentation, *zumōtŏs*, fermented—from *zumŏŏ*, I cause to ferment), in *med.*, a morbid action or condition, as of the blood, supposed to be allied to fermentation : zymotic, a., *zĭm·ŏt'·ĭk*, pert. to or caused by fermentation : zymotic diseases, that large class of contagious diseases supposed to be caused by the reception into the system of a virus which acts as a ferment ; the entire class of epidemic, endemic, and contagious diseases, which are in a great measure preventible.

THE TRIVIAL OR SECOND TERMS OF SPECIFIC NAMES.

In the nomenclature of plants, living and fossil, and in the nomenclature of sciences in general, *specific names* are *binomial*, that is, made up of two names, the first being the name of the *genus*, and the second indicating some peculiarities or properties characteristic of certain individuals, and distinguishing them from all others of the same *genus*. Thus, *Prunus* is a genus of plants, and *Prunus domestica*, a species, is the Plum tree and its varieties, which, when dried, constitute *Prunes*—the second name, *domestica*, meaning for *house-use ; Prunus spinosa* is the Sloe, *spinosa* meaning ' thorny,' ' prickly,' referring to the prickly character of the tree.

Such names, however derived, appear in Latin forms and terminations. In Latin nouns and adjectives, the terminations vary in order to express gender. The name of the genus is of course always a noun and the second or *trivial* name is always an adjective, or a noun used as an adjective. Whatever, therefore, the gender—masc., fem., or neut.—of the name of the genus may be, the termination of the second or *trivial* name must indicate the same gender. This explains why the same *trivial* name terminates sometimes with one letter, sometimes with another. Thus we have the specific names Helleborus *niger*, Morus *nigra*, Piper *nigrum*, in which the *trivial* name *niger* appears in the masculine, feminine, and neuter terminations respectively, in order to agree in gender with the *generic* names Helleborus, Morus, and Piper. Similarly we write Linaria *vulgaris*, Hordeum *vulgare ;* and Lathrus *sativus*, Latuca *sativa*, Lepidium *sativum*. Linnæus calls the second part of the specific name the *trivial* name. Sometimes the *trivial* name is not an adjective, but a noun, and used as such, in which case it is not made to agree in gender with the generic name, but governed by it in the genitive case ; thus, *Hydrangea Thunbergii*, the Hydrangea of Thunberg, *Musa sapientium*, the Musa trees of the Wise : or it is employed simply as an indeclinable adjective, and therefore united to a *generic* name of any gender.

For the benefit of those not familiar with the Latin language, the following Latin nouns and adjectives are declined, after studying which, the reader will be able to examine the list of specific names with greater advantage. The Latin words are marked with symbols as a guide to their pronunciation, and the meaning of each case follows. N.—Nominative, G.—Genitive, D.—Dative, A.—Accusative, V.—Vocative, Ab.—Ablative.

LATIN NOUNS.

FIRST DECLENSION.

SINGULAR NUMBER. FEM.

N. *pĭnnă*, or *pĕnnă*, a feather.
G. *pĭnnæ*, of a feather, or feather's.
D. *pĭnnæ*, to a feather.
A. *pĭnnăm*, a feather.
V. *pĭnnă*, O feather.
Ab. *pĭnnâ*, with or from a feather.

PLURAL NUMBER. FEM.

N. *pĭnnæ*, feathers.
G. *pĭnnăr'ŭm*, of feathers.
D. *pĭnnĭs*, to feathers.
A. *pĭnnăs*, feathers.
V. *pĭnnæ*, O feathers.
Ab. *pĭnnĭs*, with feathers.

SECOND DECLENSION IN ER.

SING. MASC.

N. *ăgĕr*, a field.
G. *ăgrī*, of a field.
D. *ăgrō*, to a field.
A. *ăgrŭm*, a field.
V. *ăgĕr*, a field.
Ab. *ăgrō*, with a field.

PLU. MASC.

N. *ăgrī*, fields.
G. *ăgrŏr'ŭm*, of fields.
D. *ăgrĭs*, to fields.
A. *ăgrŏs*, fields.
V. *ăgrī*, O fields.
Ab. *ăgrĭs*, with fields.

SECOND DECLENSION IN US.

SING. MASC.

N. *hŏrtŭs*, a garden.
G. *hŏrtī*, of a garden.
D. *hŏrtō*, to a garden.
A. *hŏrtŭm*, a garden.
V. *hŏrtĕ*, O garden.
Ab. *hŏrtō*, with or from a garden.

PLU. MASC.

N. *hŏrtī*, gardens.
G. *hŏrtŏr'ŭm*, of gardens.
D. *hŏrtĭs*, to gardens.
A. *hŏrtŏs*, gardens.
V. *hŏrtī*, O gardens.
Ab. *hŏrtĭs*, with gardens.

SECOND DECLENSION IN UM.

SING. NEUT.

N. *pōmŭm*, an apple.
G. *pōmī*, of an apple.
D. *pōmō*, to an apple.
A. *pōmŭm*, an apple.
V. *pōmŭm*, O apple
Ab. *pōmō*, with an apple.

PLU. NEUT.

N. *pōmă*, apples.
G. *pōmŏr'ŭm*, of apples.
D. *pōmĭs*, to apples.
A. *pōmă*, apples.
V. *pōmă*, O apples.
Ab. *pōmĭs*, with apples.

THIRD DECLENSION IN O.

SING. FEM.

N. *ĭmāg'ō*, an image.
G. *ĭmăg'ĭnĭs*, of an image.
D. *ĭmăg'ĭnī*, to an image.
A. *ĭmăg'ĭnĕm*, an image.
V. *ĭmāg'ō*, O image.
Ab. *ĭmăg'ĭnē*, with an image.

PLU. FEM.

N. *ĭmăg'ĭnēs*, images.
G. *ĭmăg'ĭnŭm*, of images.
D. *ĭm'ăgĭn'ĭbŭs*, to images.
A. *ĭmăg'ĭnēs*, images.
V. *ĭmăg'ĭnēs*, O images.
Ab. *ĭm'ăgĭn'ĭbŭs*, with images.

THIRD DECLENSION IN OR.

SING. NEUT.

N. *cŏr*, the heart.
G. *cŏrdĭs*, of the heart.
D. *cŏrdī*, to the heart.
A. *cŏr*, the heart.
V. *cŏr*, O the heart.
Ab. *cŏrdē*, with the heart.

2 F

PLU. NEUT.

N. *cŏrdă*, hearts.
G. *cŏrd'ĭŭm*, of hearts.
D. *cŏrd'ĭbŭs*, to hearts.
A. *cŏrdă*, hearts.
V. *cŏrdă*, O hearts.
Ab. *cŏrd'ĭbŭs*, with hearts.

THIRD DECLENSION IN **EX.**

SING. MASC.

N. *cŏrtĕx*, bark.
G. *cŏrt'ĭcĭs*, of bark.
D. *cŏrt'ĭcī*, to bark.
A. *cŏrt'ĭcĕm*, bark.
V. *cŏrtĕx*, O bark.
Ab. *cŏrt'ĭcē*, with bark.

PLU. MASC.

N. *cŏrt'ĭcēs*, barks.
G. *cŏrt'ĭcŭm*, of barks.
D. *cŏrtĭc'ĭbŭs*, to barks.
A. *cŏrt'ĭcēs*, barks.
V. *cŏrt'ĭcēs*, O barks.
Ab. *cŏrtĭc'ĭbŭs*, with barks.

THIRD DECLENSION IN **S.**

SING. MASC.

N. V. *flōs*, a flower.
G. *flōrĭs*, of a flower.
D. *flōrī*, to a flower.
A. *flōrĕm*, a flower.
Ab. *flōrē*, with a flower.

PLU. MASC.

N. A. V. *flōrēs*, flowers.
G. *flōrŭm*, of flowers.
D. Ab. *flōr'ĭbŭs*, to or with flowers.

THIRD DECLENSION IN **NS.**

SING. FEM.

N. V. *gĕns*, a clan.
G. *gĕntĭs*, of a clan.
D. *gĕntī*, to a clan.
A. *gĕntĕm*, a clan.
Ab. *gĕntē*, with a clan.

PLU. FEM.

N. A. V. *gĕntēs*, clans.
G. *gĕn'tĭŭm*, of clans.
D. Ab. *gĕnt'ĭbŭs*, to or with clans.

THIRD DECLENSION IN **E.**

SING. NEUT.

N. A. V. *mārē*, the sea.
G. *mārĭs*, of the sea.
D. Ab. *mārī*, to or with the sea.

PLU. NEUT.

N. A. V. *mār'ĭă*, seas.
G. *mār'ĭŭm*, of seas.
D. Ab. *mār'ĭbŭs*, to or with seas.

THIRD DECLENSION IN **EN.**

SING. NEUT.

N. A. V. *nŏmĕn*, a name.
G. *nŏm'ĭnĭs*, of a name.
D. *nŏm'ĭnī*, to a name.
Ab. *nŏm'ĭnē*, with a name.

PLU. NEUT.

N. A. V. *nŏm'ĭnă*, names.
G. *nŏm'ĭnŭm*, of names.
D. Ab. *nŏmĭn'ĭbŭs*, to or with names.

THIRD DECLENSION IN **O.**

SING. MASC.

N. V. *ŏrdō*, order, rank.
G. *ŏrd'ĭnĭs*, of order.
D. *ŏrd'ĭnī*, to order.
A. *ŏrd'ĭnĕm*, order.
Ab. *ŏrd'ĭnē*, with order.

PLU. MASC.

N. A. V. *ŏrd'ĭnēs*, orders or ranks.
G. *ŏrd'ĭnŭm*, of orders.
D. Ab. *ŏrdĭn'ĭbŭs*, to or with orders.

THIRD DECLENSION IN **S.**

SING. NEUT.

N. A. V. *ōs*, the mouth, the face.
G. *ōrĭs*, of the mouth.
D. *ōrī*, to the mouth.
Ab. *ōrē*, with the mouth.

PLU. NEUT.

N. A. V. *ōră*, mouths, faces.
G. *ōrŭm*, of mouths.
D. Ab. *ōr'ĭbŭs*, to or with mouths.

THIRD DECLENSION ALSO IN S.

SING. NEUT.

N. A. V. ŏs, a bone.
G. ŏssĭs, of a bone.
D. ŏssī, to a bone.
Ab. ŏssē, with a bone.

PLU. NEUT.

N. A. V. ŏssă, bones.
G. ŏs'sĭŭm, of bones.
D. Ab. ŏs'sĭbŭs, to or with bones.

THIRD DECLENSION IN US.

SING. NEUT.

N. A. V. vŭlnŭs, a wound.
G. vŭl'nĕrĭs, of a wound.
D. vŭl'nĕrī, to a wound.
Ab. vŭl'nĕrĕ, with a wound.

PLU. NEUT.

N. A. V. vŭl'nĕră, wounds.
G. vŭl'nĕrŭm, of wounds.
D. Ab. vŭlnĕr'ĭbŭs, to or with wounds.

FOURTH DECLENSION IN US.

SING. FEM.

N. mānŭs, a hand.
G. mānŭs, of a hand.
D. mān'-ūī, to a hand.
A. mānŭm, a hand.
V. mānŭs, O hand.
Ab. mānŭ, with a hand.

PLU. FEM.

N. mānŭs, hands.
G. mān'-ŭŭm, of hands.
D. mān'-ĭbŭs, to hands.
A. mānŭs, hands.
V. mānŭs, O hands.
Ab. mān'-ĭbŭs, with hands.

FOURTH DECLENSION ALSO IN US.

SING. MASC.

N. V. mŏrsŭs, a bite.
G. mŏrsŭs, of a bite.
D. mŏrs'-ūī, to a bite.
A. mŏrsŭm, a bite.
Ab. mŏrsū, with a bite.

PLU. MASC.

N. A. V. mŏrsŭs, bites.
G. mŏrs'-ŭŭm, of bites.
D. Ab. mŏrs'-ĭbŭs to or with bites.

LATIN ADJECTIVES.

Latin adjectives have their terminations of the *first* and *second* declensions, or of the *third* only. Adjectives have their terminations masc., fem., or neut., and are always made to agree with the gender of the nouns which they qualify. The following adjectives are first declined in their separate declensions, and then with nouns. Every vowel is sounded as marked, except in the case of diphthongs.

Albus, a., white.

SING.

	MASC.	FEM.	NEUT.	
N.	ălbŭs,	ălbă,	ălbŭm,	white.
G.	ălbī,	ălbœ, (ē)	ălbī,	of white.
D.	ălbō,	ălbœ, (ē)	ălbō,	to white.
A.	ălbŭm,	ălbăm,	ălbŭm,	white.
V.	ălbĕ,	ălbă,	ălbŭm,	O white.
Ab.	ălbō,	ălbā,	ălbō,	with white.

PLU.

MASC.	FEM.	NEUT.	
N. *ălbĭ,*	*ălbæ,*	*ălbă,*	white.
G. *ălbōr'-ŭm,*	*ălbār'-ŭm,*	*ălbōr'-ŭm,*	of white.
D. *ălbĭs,*	*ălbĭs,*	*ălbĭs,*	to white.
A. *ălbŏs,*	*ălbăs,*	*ălbă,*	white.
V. *ălbĭ,*	*ălbæ,*	*ălbă,*	O white.
Ab. *ălbĭs,*	*ălbĭs,*	*ălbĭs,*	with white.

Dulcis, a., sweet.
SING.

N. *dŭlcĭs,*	*dŭlcĭs,*	*dŭlcĕ,*	sweet.
G. *dŭlcĭs,*	*dŭlcĭs,*	*dŭlcĭs,*	of sweet.
D. *dŭlcī,*	*dŭlcī,*	*dŭlcī,*	to sweet.
A. *dŭlcĕm,*	*dŭlcĕm,*	*dŭlcĕ,*	sweet.
V. *dŭlcĭs,*	*dŭlcĭs,*	*dŭlcĕ,*	O sweet.
Ab. *dŭlcī,*	*dŭlcī,*	*dŭlcī,*	with sweet.

PLU.

N. *dŭlcēs,*	*dŭlcēs,*	*dŭl'-cĭă,*	sweet.
G. *dŭl'-cĭŭm,*	*dŭl'-cĭŭm,*	*dŭl'-cĭŭm,*	of sweet.
D. *dŭl'-cĭbŭs,*	*dŭl'-cĭbŭs,*	*dŭl'-cĭbŭs,*	to sweet.
A. *dŭlcēs,*	*dŭlcēs,*	*dŭl'-cĭă,*	sweet.
V. *dŭlcēs,*	*dŭlcēs,*	*dŭl'-cĭă,*	O sweet.
Ab. *dŭl'-cĭbŭs,*	*dŭl'-cĭbŭs,*	*dŭl'-cĭbŭs,*	with sweet.

Magnus, a., great.
SING.

N. *măgnŭs,*	*măgnă,*	*măgnŭm,*	great.
G. *măgnĭ,*	*măgnæ,*	*măgnĭ,*	of great.
D. *măgnō,*	*măgnæ,*	*măgnō,*	to great.
A. *măgnŭm,*	*măgnăm,*	*măgnŭm,*	great.
V. *măgnĕ,*	*măgnă,*	*măgnŭm,*	O great.
Ab. *măgnō,*	*măgnă,*	*măgnō,*	with great.

PLU.

N. V. *măgnĭ,*	*măgnæ,*	*măgnĭ,*	great.
G. *măgnōr'-ŭm,*	*măgnār'-ŭm,*	*măgnōr'-ŭm,*	of great.
D. Ab. *măgnĭs,*	*măgnĭs,*	*măgnĭs,*	to or with g.

Ingens, a., huge.
SING.

N. V. *ĭngĕns,*	*ĭngĕns,*	*ĭngĕns,*	huge.
G. *ĭngĕnt'-ĭs,*	*ĭngĕnt'-ĭs,*	*ĭngĕnt'-ĭs,*	of huge.
D. *ĭngĕnt'-ī,*	*ĭngĕnt'-ī,*	*ĭngĕnt'-ī*	to huge.
A. *ĭngĕnt'-ĕm,*	*ĭngĕnt'-ĕm,*	*ĭngĕns,*	huge.
Ab. *ĭngĕnt'-ē* or *ĭngĕnt'-ī,* etc.,			with huge.

PLU.

	MASC.	FEM.	NEUT.	
N. A. V.	*ĭngĕnt'·ēs,*	*ĭngĕnt'·ēs,*	*ĭngĕnt'·ĭă,*	huge.
G.	*ĭn·gĕn'·tĭŭm,*	*ĭn·gĕn'·tĭŭm,*	*ĭn·gĕn'·tĭŭm,*	of huge.
D. Ab.	*ĭngĕnt'·ĭbŭs,*	*ĭngĕnt'·ĭbŭs,*	*ĭngĕnt'·ĭbŭs,*	with huge.

The following is one adjective and nouns in full, corresponding in gender, as an example :—

SING.

N.	*ălbŭs āgĕr,*	*ălbă pĭnnă,*	*ălbŭm cŏr,*
	a white field ;	a white feather ;	a white heart.
G.	*ălbī āgrī,*	*ălbœ pĭnnœ,*	*ălbī cŏrdĭs,*
	of a white field ;	of a white feather ;	of a white heart.
D.	*ălbō āgrō,*	*ălbœ pĭnnœ,*	*ălbō cŏrdī,*
	to a white field ;	to a white feather ;	to a white heart.
A.	*ălbŭm āgrŭm,*	*ălbăm pĭnnăm,*	*ălbŭm cŏr,*
	a white field ;	a white feather ;	a white heart.
V.	*ălbē āgĕr,*	*ălbă pĭnnă,*	*ălbŭm cŏr,*
	O white field ;	O white feather ;	O white heart.
Ab.	*ălbō āgrō,*	*ălbă pĭnnă,*	*ălbō cŏrdē,*
	with a white field ;	with a white feather ;	with a white heart.

PLU.

N.	*ălbī āgrī,*	*ălbœ pĭnnœ,*	*ălbă cŏrdă,*
	white fields ;	white feathers ;	white hearts.
G.	*ălbōrŭm āgrōrŭm,*	*ălbārŭm pĭnnārŭm,*	*ălbōrŭm cŏrd'·ĭŭm,*
	of white fields ;	of white feathers ;	of white hearts.
D.	*ălbĭs āgrĭs,*	*ălbĭs pĭnnĭs,*	*ălbĭs cŏrdĭbŭs,*
	to white fields ;	to white feathers ;	to white hearts.
A.	*ălbōs āgrōs,*	*ălbăs pĭnnăs,*	*ălbă cŏrdă,*
	white fields ;	white feathers ;	white hearts.
V.	*ălbī āgrī,*	*ălbœ pĭnnœ,*	*ălbă cŏrdă,*
	O white fields ;	O white feathers ;	O white hearts.
Ab.	*ălbĭs āgrĭs,*	*ălbĭs pĭnnĭs,*	*ălbĭs cŏrdĭbŭs,*
	with white fields ;	with white feathers ;	with white hearts.

Latin adjectives, in grammars of that language, are usually inflected with only the masculine spelt in full, followed by the proper change of the terminations for the fem. and neut. In the subjoined list of the *trivial* or second part of specific names, which occur in the body of the present work, this plan will be adopted, each word being followed by its English meaning ; of course, only the nominatives in the three genders are given.

Trivial names often consist of nouns in any gender, which are made to do duty as indeclinable adjectives, that is, they remain unchanged in their terminations whatever the gender of the 'generic names' may be ; as *Triticum spelta,* spelt, or an inferior kind of wheat. The trivial noun-adjective is frequently found in the genitive case, as *Theophrasta Jussæi,* that is, the plant 'Theophrasta' of Jussæus, the botanist.

MASC.	FEM.	NEUT.	
ăb'·rŏt·ăn'·ŭs,	-ă,	-ŭm,	of or pert. to southernwood.
ăc'·ĕr,	-rĭs,	-rĕ,	sharp.
ăc'·ĕr·ĭ·fŏl'·ĭ·ŭs,	-ă,	-ŭm,	having a leaf like the maple tree.
ăc·ĕt'·ō·cĕl'·lŭs,	-ă,	-ŭm,	having slightly the taste of vinegar.
ăc'·ĕt·ōs'·ŭs,	-ă,	-ŭm,	having a taste like the sorrel.
ăc·ŭl'·ĕ·āt'·ŭs,	-ă,	-ŭm,	thorny ; prickly.
ăc·ūm'·ĭn·āt'·ŭs,	-ă,	-ŭm,	pointed ; sharpened.
ăc·ūt'·ĭ·fŏl'·ĭ·ŭs,	-ă,	-ŭm,	having sharp, pointed leaves.
æg'·ĭl·ŏps,	-ŏps,	-ŏps,	the oak, bearing acorns.
æs·tīv'·ŭs,	-ă,	-ŭm,	of or pert. to summer.
Af'·rĭc·ān'·ŭs,	-ă,	-ŭm,	of or belonging to Africa ; African.
ăl·āt'·ŭs,	-ă,	-ŭm,	furnished with wings.
ălb'·ĭc·ăns,	-ăns,	-ăns,	growing white.
ălb'·ĭd·ŭs,	-ă,	-ŭm,	whitish.
ălb'·ĭ·flōr'·ŭs,	-ă,	-ŭm,	having a white flower.
ălb'·ŭs,	-ă,	-ŭm,	white.
ăl'·cĕ·ŭs,	-ă,	-ŭm,	of or like mallows.
ăl'·lĭ·āc'·ĕ·ŭs,	-ă,	-ŭm,	like garlic.
ălp·ĕst'·rĭs,	-rĭs,	-rĕ,	of or from the Alps.
ălp·īn'·ŭs,	-ă,	-ŭm,	of or from the Alps.
ălt·ĭs'·sĭm·ŭs,	-ă,	-ŭm,	very high.
ăm'·ăr·ĕl'·lŭs,	-ă,	-ŭm,	somewhat bitter.
ăm·ār'·ŭs,	-ă,	-ŭm,	bitter.
Am·ĕr'·ĭc·ān'·ŭs,	-ă,	-ŭm,	of or from America.
ăm'·mō·dĕnd'·rŏn, n., a tree growing among sand.			
ăm'·mŏn·ĭ'·ăc·ŭs,	-ă,	-ŭm,	yielding a gum resin ; ammoniac.
ăm·yg'·dăl·īn'·ŭs,	-ă,	-ŭm,	of or from almonds.
ăn'·ă·cărd'·ĭ·ŭs,	-ă,	-ŭm,	shaped like a heart.
ăn·gŭs'·tĭ·fŏl'·ĭ·ŭs,	-ă,	-ŭm,	narrow-leaved.
ăn'·ĭs·āt'·ŭs,	-ă,	-ŭm,	of or like the anise plant.
ăn·ĭs'·ō·phyl'·lŭs,	-ă,	-ŭm,	having a leaf like the anise.
ăn·ĭs'·ŭs,	-ă,	-ŭm,	of the anise.
ăn'·nūl·īn'·ŭs,	-ă,	-ŭm,	of or like a ring.
ăn'·nū·ŭs,	-ă,	-ŭm,	annual ; that lasts a year.
ănt·ărct'·ĭc·ŭs,	-ă,	-ŭm,	found in the antarctic regions.
ănth'·ĕl·mĭnt'·ĭc·ŭs,	-ă,	-ŭm,	that which expels worms.
ănth·ĕl'·mĭ·ŭs,	-ă,	-ŭm,	good against worms.
ănt'·ĭ·dys·ĕn·tĕr'·ĭc·ŭs,-ă,	-ă,	-ŭm,	good against dysentery.
ăn'·tĭ·scŏr·būt'·ĭc·ŭs,	-ă,	-ŭm,	good against scurvy.
ăph'·ăc·ŭs,	-ă,	-ŭm,	of a kind of pulse.
Ăp'·ŏl·līn'·ĕ·ŭs,	-ă,	-ŭm,	of or pert. to Apollo.
ăq·uăt'·ĭc·ŭs, (wŏt)	-ă,	-ŭm,	growing in water.
ăq·uăt'·ĭl·ĭs, (wăt)	-ĭs,	-ĕ,	growing in or found near water.
ăq'·uĭ·fŏl'·ĭ·ŭs, (wĭ)	-ă,	-ŭm,	having needle-like leaves.
ăq'·uĭl·īn'·ŭs, (wĭl)	-ă,	-ŭm,	of or like an eagle.
ăq'·uĭl·ŭs, (wĭl)	-ă,	-ŭm,	of or pert. to an eagle; dark-coloured.
Ar·ăb'·ĭc·ŭs,	-ă,	-ŭm,	of or from Arabia.
ăr'·bŏr·ĕs'·cĕns,	-ĕns,	-ĕns,	growing into a tree.
ăr·bŏr'·ĕ·ŭs,	-ă,	-ŭm,	tree-like.
ăr'·bŏr·trĭst'·ĭs,	-ĭs,	-ĕ,	the sad tree.
ărc'·tĭc·ŭs,	-ă,	-ŭm,	of or from the north.
ăr'·ĕn·ăr'·ĭ·ŭs,	-ă,	-ŭm,	adapted to sandy soils.

MASC.	FEM.	NEUT.	
ăr·gĕnt'·ĕ·ŭs,	-ă,	-ŭm,	silvery.
ăr'·ĭs·tāt'·ŭs,	-ă,	-ŭm,	having a long ridged spine ; awned.
Ar'·mŏr·āc'·ĭ·ŭs,	-ă,	-ŭm,	of or from Brittany.
ăr'·ōm·ăt'·ĭc·ŭs,	-ă,	-ŭm,	aromatic ; fragrant.
ărt·ĭc'·ŭl·āt'·ŭs,	-ă,	-ŭm,	furnished with joints.
ăr·ŭnd'·ĭn·āc'·ĕ·ŭs,	-ă,	-ŭm,	having the nature of a reed.
ărv·ĕns'·ĭs,	-ĭs,	-ē,	field-inhabiting.
ăs'·ă·fœt'·ĭd·ŭs,	-ă,	-ŭm,	yielding a fetid gum resin.
ăs·cĕnd'·ĕns,	-ĕns,	-ĕns,	ascending ; growing upwards.
As'·ĭ·ăt'·ĭc·ŭs,	-ă,	-ŭm,	of or from Asia.
ăs'·pĕr·ŭs,	-ă,	-ŭm,	rough ; uneven.
As·săm'·ĭc·ŭs,	-ă,	-ŭm,	of or from Assam.
ăsth·măt'·ĭc·ŭs,	-ă,	-ŭm,	good for those afflicted with asthma.
ăŭc'·ŭp·ăr'·ĭ·ŭs,(ăŭk)·ă,	-ă,	-ŭm,	having power to catch birds.
ăŭr·āt'·ŭs, (ăŭr)	-ă,	-ŭm,	overlaid with gold.
ăŭr·ĭc'·ŭl·ŭs, (ăŭr)	-ă,	-ŭm,	having little ears.
Aŭs·trāl'·ĭs, (ăŭs)	-ĭs,	-ē,	of or from Australia.
ăŭ'·tŭm·nāl'·ĭs, (ăŭ)	-ĭs,	-ē,	autumnal.
Av'·ĕl·lān'·ŭs,	-ă,	-ŭm,	of or from Avella, a town of ancient Campania, where hard trees and nuts were numerous.
ăv·ĭc'·ŭl·ăr'·ĭs,	-ĭs,	-ē,	pert. to avicula, a small bird.
āv'·ĭ·ŭs,	-ă,	-ŭm,	of or belonging to birds.
Băb'·yl·ŏn'·ĭc·ŭs,	-ă,	-ŭm,	of or from Babylon.
băc·cāt'·ŭs,	-ă,	-ŭm,	furnished with berries or pearls.
băc'·cĭf·ĕr,	-cĭf'·ĕr·ă,	-ŭm,	bearing berries.
băl·săm'·ĭf·ĕr,	-ĭf'·ĕr·ă,	-ŭm,	bearing or producing balsam.
băm'·bŭs·ŏĭd'·ĭs, (ōўd)	-ĭs,	-ē,	resembling the bamboo.
Bârb'·ă·dĕns'·ĭs,	-ĭs,	-ē,	of or from Barbadoes.
bĕn'·ĕ·dĭct'·ŭs,	-ă,	-ŭm,	praised ; commended ; blessed.
Bĕr·mūd'·ĭ·ān'·ŭs,	-ă,	-ŭm,	of or from Bermuda.
bĕt'·ŭl·ĭn'·ŭs,	-ă,	-ŭm,	of or like the birch tree.
bĭ·cŏrn'·ĭs,	-ĭs,	-ē,	having two horns ; forked.
bĭ·ĕn'·nĭs,	-ĭs,	-ē,	lasting two years.
bĭ·fār'·ĭ·ŭs,	-ă,	-ŭm,	divided into two parts.
bĭj'·ŭg·ŭs,	-ă,	-ŭm,	yoked two together.
bĭs'·pĭn·ōs'·ŭs,	-ă,	-ŭm,	doubly full of thorns.
bĭs·tŏrt'·ŭs,	-ă,	-ŭm,	twice twisted.
bŏn'·ŭs,	-ă,	-ŭm,	good.
bŏr'·ĕ·āl'·ĭs,	-ĭs,	-ē,	of or from the north ; northern.
Brăs·ĭl'·ĭ·ĕns'·ĭs,	-ĭs,	-ē,	of or from Brazil.
brĕv'·ĭs,	-ĭs,	-ē,	short.
brŏn'·chĭ·āl'·ĭs,	-ĭs,	-ē,	of or belonging to the windpipe.
bŭlb'·ŏ·căst'·ăn·ŭs,	-ă,	-ŭm,	having a bulb like the chestnut.
bŭl·bōs'·ŭs,	-ă,	-ŭm,	having a bulbous root.
bŭt'·yr·āc'·ĕ·ŭs,	-ă,	-ŭm,	having the appearance or consistence of butter, as a product.
bŭt'·yr·ōs'·ŭs,	-ă,	-ŭm,	full of a buttery substance.

cœl'·ĭ·rōs'·ă, n., (sēl) the rose of the sky.

cœs'·pĭt·ōs'·ŭs,	-ă,	-ŭm,	of or pert. to a turf.

MASC.	FEM.	NEUT.	
căl'·yc·in'·ŭs, (*ĭs*)	*-ă,*	*-ŭm,*	having a flower-cup.
căm·pĕst'·rĭs,	*-rĭs,*	*-rĕ,*	of or belonging to a field.
căm'·phŏr·ŭs,	*-ă,*	*-ŭm,*	yielding camphor ; like camphor.
Căn'·ăd·ĕns'ĭs,	*-ĭs,*	*-ĕ,*	of or from Canada.
Căn·ār'·ĭ·ĕns'·ĭs,	*-ĭs,*	*-ĕ,*	from the Canary Islands.
căn'·dĕl·āb'·rŭs,	*-ă,*	*-ŭm,*	like a branched candlestick.
căn·īn'·ŭs,	*-ă,*	*-ŭm,*	of or pert. to a dog.
căn·năb'·ĭn·ŭs,	*-ă,*	*-ŭm,*	of or pert. to hemp.
Căn·tōn'·ĭ·ĕns'·ĭs,	*-ĭs,*	*-ĕ,*	of or from Canton.
căp·ræ'·ŭs, (*rē*)	*-ă,*	*-ŭm,*	pert. to a wild goat.
căp'·sŭl·ār'·ĭs,	*-ĭs,*	*-ĕ,*	having capsules.
cărd'·ăm·ōm'·ŭs,	*-ă,*	*-ŭm,*	pert. to *cardamon,* a kind of cress.
cărd·ŭnc'·ŭl·ŭs,	*-ă,*	*-ŭm,*	like a thistle ; pert. to the teasel.
Căr'·ĭb·bœ'·ŭs, (*bē*)	*-ă,*	*-ŭm,*	from Caribbean Islands.
Căr'·ĭc·ŭs,	*-ă,*	*-ŭm,*	of or from Caria.
cār'·ĭ·ēs, n., rottenness ; decay.			
cărn'·ĕ·ŭs,	*-ă,*	*-ŭm,*	having the appearance of flesh.
căr·nōs'·ŭs,	*-ă,*	*-ŭm,*	of or like flesh.
căr·ōt'·ă, n., a carrot.			
cărp'·ŭs, n., the wrist—from Gr. *kărp'·ŏs,* n., fruit.			
cărt'·ĭl·ăg·ĭn'·ĕ·ŭs,	*-ă,*	*-ŭm,*	of or like cartilage.
căr'·y·ō·phyl'·lŭs,	*-ă,*	*-ŭm,*	having leaves shaped like nuts.
căs'·ŭ·ăr·īn'·ŭs,	*-ă,*	*-ŭm,*	pert. to the cassowary tree.
căth·ărt'·ĭc·ŭs,	*-ă,*	*-ŭm,*	purifying or cleansing.
căŭl'·ĭ·flōr'·ŭs, (*kăŭl*)	*-ă,*	*-ŭm,*	having bright shining stems.
cē'·ĭb·ă, n., in Spain, the silk-cotton tree.			
Cĕlt'·ĭc·ŭs,	*-ă,*	*-ŭm,*	of or pert. to Gaul.
cĕnt·ăŭr'·ĕ·ŭs, (*ăŭr*)	*-ă,*	*-ŭm,*	pert. to the plant centaury.
cĕn'·tĭ·fōl'·ĭ·ŭs,	*-ă,*	*-ŭm,*	having a hundred leaves.
cĕph'·ă·ĕl'·ĭs,	*-ĭs,*	*-ĕ,*	having flowers disposed in heads.
cēp'·ŭs,	*-ă,*	*-ŭm,*	of or pert. to an onion.
cĕr·ē'·ĭf·ĕr,	*-ĭf'·ĕr·ă,*	*-ŭm,*	bearing wax.
cĕrn'·ŭ·ŭs,	*-ă,*	*-ŭm,*	sloping or bending forwards ; with the face downwards.
cĕr·ŭl'·ĕ·ŭs,	*-ă,*	*-ŭm,*	cerulean ; dark blue.
Chăl'·cĕd·ŏn'·ĭc·ŭs,	*-ă,*	*-ŭm,*	of or from Chalcedon.
chăm·œd'·rĭ·fōl'·ĭ·ŭs, (*ēd*)	*-ă,*	*-ŭm,*	low-growing and oak-leaved.
chăm·æm'·ŏr·ŭs, (*ēm*)	*-ă,*	*-ŭm,*	growing on the ground, and appearing like the mulberry.
chĭc'·ă, n., *tshĭk'·ă,* the Indian name for a beauty ; a pretty girl.			
chĭn'·ă, n., China ; from China.			
Chĭn·ĕns'·ĭs,	*-ĭs,*	*-ĕ,*	of or from China.
Chĭr·ōn'·ĭŭs,	*-ă,*	*-ŭm,*	of or belonging to Chiron, one of the fathers of medicine.
chlōr·ŏph'·ŏr·ĭs,	*-ĭs,*	*-ĕ,*	bearing a red colour.
chrys·ănth'·ŭs,	*-ă,*	*-ŭm,*	producing golden-coloured flowers.
cĭc'·ĕr·ŭs,	*-ă,*	*-ŭm,*	like the chick-pea.
cĭl'·ĭ·ār'·ĭs,	*-ĭs,*	*-ĕ,*	pert. to the eyelids ; ciliary.
cĭn'·ĕr·ār'·ĭ·ŭs,	*-ă,*	*-ŭm,*	having the colour of ashes.
cĭn·ĕr'·ĕ·ŭs,	*-ă,*	*-ŭm,*	ash-coloured.
cĭr'·cĭn·āl'·ĭs,	*-ĭs,*	*-ĕ,*	encircled ; whorled.
cĭr'·cŭm·scĭs'·sŭs,	*-ă,*	*-ŭm,*	torn or cut off around.

MASC.	FEM.	NEUT.	
cĭst·ŏĭd′·ĕ·ŭs, (ŏy̆d)	-ă,	-ŭm,	resembling a box or chest.
cĭt′·rĭ·fŏl′·ĭ·ŭs,	-ă,	-ŭm,	citron-leaved.
cĭt′·rĭ·ō·dōr′·ŭs,	-ă,	-ŭm,	having the nature of the citron.
cĭt·rŭl′·lŭs, n., the Sicilian water melon.			
clăv·āt′·ŭs,	-ă,	-ŭm,	furnished with nails.
cnē·ōr′·ŭs, (nē)	-ă,	-ŭm,	like a kind of nettle.
cŏc·cĭn′·ĕ·ŭs,	-ă,	-ŭm,	having a scarlet colour.
cŏc′·cŭl·ŭs, n., a little berry—from coccus, a berry.			
cŏch′·ĭn·ĕl′·lĭf·ĕr,	-ĭf′·ĕr·ă,	-ŭm,	bearing wood lice.
cæc·ŭt′·ĭ·ĕns, (sēk)	-ĕns,	-ĕns,	blinding.
cŏl′·ō·cynth′·ĭs, (sĭnth)	-ĭs,	-ē,	pert. to the wild gourd.
cŏl′·ŭb·rīn′·ŭs,	-ă,	-ŭm,	of or like a serpent.
cŏm·mūn′·ĭs,	-ĭs,	-ē,	common.
cōm·ōs′·ŭs,	-ă,	-ŭm,	like the hair of the head.
cŏm·păct′·ŭs,	-ă,	-ŭm,	pressed.
cŏm·pŏs′·ĭt·ŭs,	-ă,	-ŭm,	composite.
cŏn′·dĕns·āt′·ŭs,	-ă,	-ŭm,	made very dense.
cŏn·fĕrt′·ŭs,	-ă,	-ŭm,	thick ; dense.
cŏn·glŏm′·ĕr·āt′·ŭs,	-ă,	-ŭm,	rolled together.
cŏn′·ĭc·ŭs,	-ă,	-ĭm,	cone-like.
cŏn·nāt′·ŭs,	-ă,	-ŭm,	born with.
cŏn·sŏl′·ĭd·ŭs,	-ă,	-ŭm,	made very solid or firm.
cŏn′·tră·yĕrv′·ŭs,	-ă,	-ŭm,	pert. to Paraguayan tea.
cŏr′·ăl·lŏĭd′·ĭs, (lŏy̆d)	-ĭs,	-ē,	of or like red coral.
cŏrd·āt′·ŭs,	-ă,	-ŭm,	heart-shaped.
cŏrd′·ĭ·fŏl′·ĭ·ŭs,	-ă,	-ŭm,	leaves shaped like hearts.
cŏr′·ĭ·ă′·cĕ·ŭs,	-ă,	-ŭm,	resembling leather.
cŏr′·ĭ·ār′·ĭ·ŭs,	-ă,	-ŭm,	leathery.
cŏr′·ĭ·ŏph′·ŏr·ŭs,	-ă,	-ŭm,	bearing corianders.
cŏr·nĭc′·ŭl·āt′·ŭs,	-ă,	-ĭm,	having horns ; horned.
cŏr′·ŏn·ār′·ĭ·ŭs,	-ă,	-ŭm,	having a wreath or crown.
cŏrt′·ĭc·ōs′·ŭs,	-ă,	-ŭm,	full of bark.
cŏst′·ŭs, n., the Arabic name of plant Kasta, an aromatic plant.			
crăs′·sĭ·fŏl′·ĭ·ŭs,	-ă,	-ŭm,	having thick leaves.
crĕn·āt′·ŭs,	-ă,	-ŭm,	having notches.
crĕn′·ŭl·āt′·ŭs,	-ă,	-ŭm,	slightly notched.
crĕp′·ĭt·ăns,	-ăns,	-ăns,	creaking ; crackling.
Crēt′·ĭc·ŭs,	-ă,	-ŭm,	of or from Crete.
crĭsp′·ŭs,	-ă,	-ŭm,	curled or wrinkled.
crĭst·āt′·ŭs,	-ă,	-ŭm,	having crests.
crōc·āt′·ŭs,	-ă,	-ŭm,	saffron-yellow.
crŭm·ĕn′·ĭf·ĕr,	-ĭf′·ĕr·ă,	-ŭm,	carrying a bag.
crys·tăl′·lĭn·ŭs,	-ă,	-ŭm,	crystalline.
cŭn′·ē·ĭ·fŏl′·ĭ·ŭs,	-ă,	-ŭm,	having wedge-like leaves.
cūr′·ăs·săv′·ĭc·ŭs,	-ă,	-ŭm,	healing ; curing.
Cyd·ōn′·ĭ·ă, n., (sĭd) a town in the island of Crete.			
cym·ōs′·ŭs, (sĭm)	-ă,	-ŭm,	producing many shoots.
dăc·tyl′·ĭf·ĕr,	-ĭf′·ĕr·ă,	-ŭm,	bearing fingers.
dăc′·tyl·ŭs,	-ă,	-ŭm,	of or like a finger.
Dăm′·ăs·cēn′·ŭs,	-ă,	-ŭm,	of or from Damascus.
dĕc·ănd′·rŭs,	-ă,	-ŭm,	having ten stamens.

MASC.	FEM.	NEUT.	
dĕ·cŭmb'ĕns,	-ĕns,	-ĕns,	lying down.
dĕns'·căn·ĭn'·ŭs,	-ă,	-ŭm,	dog-toothed.
dĕns'·ĭ·flōr'·ŭs,	-ă,	-ŭm,	having dense flowers.
dĕ'·ŏd·ār'·ŭs,	-ă,	-ŭm,	producing sacred timber.
dĕ·scĕnd'·ĕns,	-ĕns,	-ĕns,	descending.
dĭch·ŏt'·ŏm·ŭs, (dĭk)	-ă,	-ŭm,	cut in halves.
Dĭc·tăm'·nŭs, n., the plant dittany—from Dĭctĕ, in Crete.			
dĭf·fūs'·ŭs,	-ă,	-ŭm,	spread out.
dĭg'·ĭt·āt'·ŭs,	-ă,	-ŭm,	having fingers or toes.
dĭl'·ăt·āt'·ŭs,	-ă,	-ŭm,	spread out ; extended.
dī·ŏĭc'·ŭs, (ōyk)	-ă,	-ŭm,	having a double house.
dĭst'·ĭch·ŭs,	-ă,	-ŭm,	consisting of two rows.
dōd'·ĕc·ănd'·rŭs,	-ă,	-ŭm,	having twelve stamens.
dōm·ĕst'·ĭc·ŭs,	-ă,	-ŭm,	of or belonging to the house.
drăc'·ō, n., a species of serpent.			
drăc·ŭnc'·ŭl·ŭs,	-ă,	-ŭm,	like a small serpent.
dŭl'·căm·ār'·ŭs,	-ă,	-ŭm,	bitter-sweet.
dŭl'·cĭs,	-ĭs,	-ē,	sweet.
dūm·ōs'·ŭs,	-ă,	-ŭm,	bushy ; abounding in bushes.
ĕb·ĕn'·ĭf·ĕr,	-ĭf'·ĕr·ă,	-ŭm,	bearing or producing ebony.
ĕb'·ĕn·ŭs,	-ă,	-ŭm,	of or like the ebon tree.
ĕch'·ĭn·āt'·ŭs,	-ă,	-ŭm,	prickly.
ĕd·ūl'·ĭs,	-ĭs,	-ē,	that may be eaten.
ĕf·fūs'·ŭs,	-ă,	-ŭm,	poured out ; shed.
E'·gyp·tĭ'·ăc·ŭs, (jĭp)	-ă,	-ŭm,	of or from Egypt.
ĕ·lăst'·ĭc·ŭs,	-ă,	-ŭm,	elastic.
ĕl'·ăt·ĕr'·ĭ·ŭs,	-ă,	-ŭm,	that drives out or expels.
ĕl·āt'·ĭ·ŏr,	-ŏr,	-ŭs,	more lofty ; more productive.
ĕl·āt'·ŭs,	-ă,	-ŭm,	lofty ; productive.
ĕl'·ĕph·ănt'·ĭp·ĭs,	-ĭs,	-ē,	like an elephant's foot.
ĕl'·ĕ·phănt'·ŭs,	-ă,	-ŭm,	like an elephant in size.
ĕl·lĭp'·tĭc·ŭs,	-ă,	-ŭm,	like an oval.
ĕl'·ŏng·āt'·ŭs,	-ă,	-ŭm,	made long.
ĕm'·bry·ŏp'·tĕr·ĭs,	-ĭs,	-ē,	having a germ appearing like a fern.
ĕm'·ĕr·ŭs,	-ă,	-ŭm,	not wild ; cultivated.
ĕm·ĕt'·ĭc·ŭs,	-ă,	-ŭm,	inciting to vomit.
ĕn·dĭv'·ĭ·ŭs,	-ă,	-ŭm,	used as a salad.
ĕnt'·ŏm·ō·rhĭz'·ŭs,	-ă,	-ŭm,	insect roots.
ē·pĭth'·ym·ŭs,	-ă,	-ŭm,	like the flower of thyme.
ĕ·quīn'·ŭs, n. masc., a horse ; ĕ·quīn'·ă, n. fem., a mare.			
ĕr'·ĭn·ā'·cĕ·ŭs,	-ă,	-ŭm,	like a hedgehog.
ĕr·ōs'·ŭs,	-ă,	-ŭm,	eaten away ; corroded.
ĕr'·yth·rīn'·ŭs,	-ă,	-ŭm,	having a red colour.
ĕsc·ŭl·ĕnt'·ŭs,	-ă,	-ŭm,	fit for food.
Eŭr'·ŏp·œ'·ŭs, (ē)	-ă,	-ŭm,	of or belonging to Europe.
ĕx·cĕls'·ĭ·ŏr,	-ŏr,	-ŭs,	more elevated ; loftier.
ĕx·cĕls'·ŭs,	-ă,	-ŭm,	elevated ; lofty.
ĕx·pāns'·ŭs,	-ă,	-ŭm,	spread out ; expanded.
ĕx'·plăn·āt'·ŭs,	-ă,	-ŭm,	spread out ; flattened.

făb·āg'·ō, n., the bean caper.

MASC.	FEM.	NEUT.	
făr'·făr·ŭs,	-ă,	-ŭm,	pert. to the white poplar.
făr·ĭn'·ă,	n., flour ; meal.		
făr·ĭn'·ĭf·ĕr,	-ĭf'·ĕr·ă,	-ŭm,	bearing or producing food.
făr'·ĭn·ōs'·ŭs,	-ă,	-ŭm,	mealy ; like meal.
făst·ĭg'·ĭ·āt'·ŭs, (ĭdj)	-ă,	-ŭm,	pointed at the top.
făst'·ū·ōs'·ŭs,	-ă,	-ŭm,	full of pride.
făv·ōs'·ŭs,	-ă,	-ŭm,	honeycombed.
fĕb·rĭf'·ūg·ŭs,	-ă,	-ŭm,	driving away fever.
fēl'·ĭx,	-ĭx,	-ĭx,	fruitful ; fertile.
fĕn'·ĕs·trāl'·ĭs,	-ĭs,	-ē,	belonging to windows or openings.
fĕn'·ĕs·trāt'·ŭs,	-ă,	-ŭm,	furnished with openings
fēr'·ŏx,	-ŏx,	-ŏx,	wild ; fierce.
fĕr'·rē·ŭs,	-ă,	-ŭm,	made of iron.
fĕr'·rŭg·ĭn'·ĕ·ŭs,	-ă,	-ŭm,	of an iron-rust colour.
fī·cār'·ĭ·ŭs,	-ă,	-ŭm,	of or like a fig.
fī·ĕn'·ĭf·ĕr,	-ĭf'·ĕr·ă,	-ŭm,	bearing hay.
fĭl'·ĭx,	n., a fern ; fĭl'·ĭx-măs, the male fern.		
fīl'·ŭm,	n., a string ; a cord.		
fĭm'·brĭ·āt'·ŭs,	-ă,	-ŭm,	having fringes.
fĭst'·ūl·ōs'·ŭs,	-ă,	-ŭm,	full of holes ; porous.
fĭst'·ūl·ŭs,	-ă,	-ŭm,	like a hollow reed.
flăb'·ĕl·lōĭd'·ĭs, (lōȳd)	-ĭs,	-ē,	resembling a fly-flap.
flăg·ĕl'·lĭ·fŏrm'·ĭs,	-ĭs,	-ē,	shaped like whips.
flăm'·mŭl·ŭs,	-ă,	-ŭm,	appearing a little flame.
flăv·ĕs'·cĕns,	-cĕns,	-cĕns,	becoming a golden-yellow.
flāv'·ŭs,	-ă,	-ŭm,	golden-yellow.
flĕx'·ū·ōs'·ŭs,	-ă,	-ŭm,	bent.
Flōr'·ĕnt·ĭn'·ŭs,	-ă,	-ŭm,	of or from Florence.
flōr'·ĭ·bŭnd'·ŭs,	-ă,	-ŭm,	abounding in flowers.
flŏr'·ĭd·ŭs,	-ă,	-ŭm,	flowery ; gay.
fœn·ĭc'·ŭl·ŭs, (fĕn)	-ă,	-ŭm,	like fennel.
fœt'·ĭd·ŭs, (fĕt)	-ă,	-ŭm,	fetid ; stinking.
fŏl·lĭc'·ŭl·ār'·ĭs,	-ĭs,	-ē,	having little bags inflated with air.
Fŏr'·mōs·ān'·ŭs,	-ă,	-ŭm,	of or from Formosa.
fŏr·mōs'·ŭs,	-ă,	-ŭm,	finely formed ; handsome.
fŏrt'·ĭs,	-ĭs,	-ē,	strong.
frăg'·ĭl·ĭs,	-ĭs,	-ē,	easily broken.
frăg'·răns,	-ăns,	-ăns,	sweet-smelling.
frăg'·rănt·ĭs'·sĭm·ŭs,	-ă,	-ŭm,	very sweet-smelling.
frăng'·ŭl·ŭs,	-ă,	-ŭm,	easily broken ; brittle.
frăx·ĭn·ĕl'·lŭs,	-ă,	-ŭm,	resembling the ash tree in the leaves
frăx·ĭn'·ĕ·ŭs,	-ă,	-ŭm,	of or pert. to ash wood.
frĭg'·ĭd·ŭs,	-ă,	-ŭm,	cold.
frŏnd·ōs'·ŭs,	-ă,	-ŭm,	abounding in leaves.
frŭt·ĕs'·cĕns,	-ĕns,	-ĕns,	like a shrub or bush.
frŭt'·ĭc·ăns,	-ăns,	-ăns,	putting forth shoots.
frŭt'·ĭc·ōs'·ŭs,	-ă,	-ŭm,	shrubby.
fūc'·ĭ·fŏrm'·ĭs, (fūs)	-ĭs,	-ē,	shaped like sea-weed.
fŭlg'·ĕns,	-ĕns,	-ĕns,	flashing ; shining.
fŭl·lōn'·ĭ·ŭs,	-ă,	-ŭm,	pert. to a fuller.
fŭl·lōn'·ŭm,	n. plu., of fullers.		
fŭl'·vŭs,	-ă,	-ŭm,	deep-yellow ; tawny.

MASC.	FEM.	NEUT.	
fŭn′·ĭf·ĕr,	*-ĭf′·ĕr·ă,*	*-ŭm,*	bearing cords or fibres.
fŭr·cāt′·ŭs,	*-ă,*	*-ŭm,*	forked.
fŭr′·fŭr·ā′·cĕ·ŭs,	*-ă,*	*-ŭm,*	having the appearance of bran.
găl′·băn·ĭf′·lŭ·ŭs,	*-ă,*	*-ŭm,*	flowing with resinous sap.
Găl′·lĭc·ŭs,	*-ă,*	*-ŭm,*	of or pert. to Gaul.
gĕm·mĭp′·ăr·ŭs,	*-ă,*	*-ŭm,*	producing buds.
Gĕr·măn′·ĭc·ŭs,	*-ă,*	*-ŭm,*	of or from Germany.
Gĭb·sōn′·ĭ, n., of Gibson.			
gĭg·ănt·ē′·ŭs,	*-ă,*	*-ŭm,*	gigantic.
gĭg′·ăs,	*-ăs,*	*-ăs,*	like a giant.
Gĭl′·ē·ăd·ĕns′·ĭs,	*-ĭs,*	*-ē,*	of or from Gilead.
glăb′·ĕr,	*-ră,*	*-rŭm,*	without hairs or bristles.
glā′·cĭ·āl′·ĭs,	*-ĭs,*	*-ē,*	icy ; frozen.
glănd·ŭl′·ĭf·ĕr,	*-ĭf′·ĕr·ă,*	*-ŭm,*	bearing glands.
glăŭc′·ŭs,	*-ă,*	*-ŭm,*	bluish-grey.
glŏb·ōs′·ŭs,	*-ă,*	*-ŭm,*	round ; globular.
glōb′·ŭs,	*-ă,*	*-ŭm,*	of or like a globe.
glŭt′·ĭn·ōs′·ŭs,	*-ă,*	*-ŭm,*	glutinous ; gluey.
gnăph′·ăl·ĭ·ŏĭd′·ēs, adj. plu., (năf) (ōўd)			like the plant cudweed.
gnĭd′·ĭ·ŭs, (nĭd)	*-ă,*	*-ŭm,*	of or from the laurel, from its anc. name.
Grœc′·ŭs, (grēk)	*-ă,*	*-ŭm,*	of or from Greece.
grăm·ĭn′·ĕ·ŭs,	*-ă,*	*-ŭm,*	of or pert. to grass.
grăn·āt′·ŭs,	*-ă,*	*-ŭm,*	having many grains or seeds.
grănd′·ĭ·flōr′·ŭs,	*-ă,*	*-ŭm,*	bearing large flowers.
grănd′·ĭs,	*-ĭs,*	*-ē,*	great.
grăn′·ŭl·āt′·ŭs,	*-ă,*	*-ŭm,*	having little grains.
grăt·ĭs′·sĭm·ŭs,	*-ă,*	*-ŭm,*	very pleasing.
grăv·ē′·ōl·ĕns,	*-ĕns,*	*-ĕns,*	strong-smelling.
grŏs·sŭl·ār′·ĭ·ŭs,	*-ă,*	*-ŭm,*	having the appearance of a gooseberry.
Guĭ·ăn·ĕns′·ĭs, (gwī)	*-ĭs,*	*-ē,*	of or from Guiana in America.
Guĭn′·ē·ĕns′·ĭs, (gwĭn)	*-ĭs,*	*-ē,*	of or from Guinea in Africa.
gŭm′·mĭ·fĕr,	*-mĭf′·ĕr·ă,*	*-ŭm,*	bearing or producing gum.
gŭt·tāt′·ŭs,	*-ă,*	*-ŭm,*	spotted ; speckled.
gŭt′·tŭs,	*-ă,*	*-ŭm,*	drop-like ; in drops.
gyr′·ăns,	*-ăns,*	*-ăns,*	turning in a circle.
gyr·ŏĭd′·ēś, adj. plu., (ōўd)			resembling a circular course.
hăst·āt′·ŭs,	*-ă,*	*-ŭm,*	spear-like.
hăst·ĭl′·ĭs,	*-ĭs,*	*-ē,*	like a spear shaft.
hĕd′·ĕr·æ, n. plu., (ē)			ivy plants.
Hĕl·ēn′·ĭ·ŭs,	*-ă,*	*-ŭm,*	after Helen of anc. Troy.
hĕl′·ĭx, n., a winding or spiral body.			
hĕm′·ĭs·phĕr′·ĭc·ŭs,	*-ă,*	*-ŭm,*	like a half globe.
Hĕn·rĭc′·ŭs,	*-ă,*	*-ŭm,*	of or pert. to Henry.
hĕp·tăg′·ōn·ŭs,	*-ă,*	*-ŭm,*	seven-angled.
Hĕr·ăc′·lē·ŏt′·ĭc·ŭs,	*-ă,*	*-ŭm,*	of or pert. to Heraclea, a city of Pontus.
hĕrb·ăc′·ē·ŭs,	*-ă,*	*-ŭm,*	grass-green ; herbaceous.
hĕrb·ār′·ŭm, n. plu., of herbs.			
hĕx·ănd′·rŭs,	*-ă,*	*-ŭm,*	having six stamens.
hĕx·ăsĭ′·ĭch·ŭs, (ĭk)	*-ă,*	*-ŭm,*	having six rows or ranks.

MASC.	FEM.	NEUT.	
hĭr·cīn'·ŭs,	-ă,	-ŭm,	of or like a goat.
hĭrt'·ŭs,	-ă,	-ŭm,	rough ; hairy.
Hĭs·păn'·ĭc·ŭs,	-ă,	-ŭm,	of or from Spain.
hĭsp'·ĭd·ŭs,	-ă,	-ŭm,	shaggy ; hairy.
hŏm'·ĭn·ĭs, n., of a man ; hŏm'·ō, n., a man.			
hŏrt·ĕns'·ĭs,	-ĭs,	-ē,	pert. to a garden.
Hŏŭs'·tŏn·ī, n., (hŏŭs) a proper name, of Houston.			
hŭm'·ĭl·ĭs,	-ĭs,	-ē,	lowly ; small.
hy·bérn'·ŭs,	-ă,	-ŭm,	of or pert. to winter.
hyb'·rĭd·ŭs,	-ă,	-ŭm,	of or relating to a hybrid.
hy'·drō·lăp'·ăth·ŭs,	-ă,	-ŭm,	of or pert. to the water-dock.
hyd'·rō·pĭp'·ér, n., a water plant having qualities like pepper.			
hy'·ĕm·āl'·ĭs,	-ĭs,	-ē,	of or belonging to winter.
hy'·pō·gœ'·ŭs, (jē)	-ă,	-ŭm,	under the earth.
hy·pŏs'·tŏm·ŏs,	-ŏs,	-ŏs,	situated under the mouth.

īl'·ĕx, n., the holm-oak.			
ĭn·cān'·ŭs,	-ă,	-ŭm,	hoary ; quite grey.
ĭn'·cår·nāt'·ŭs,	-ă,	-ŭm,	clothed in flesh.
ĭn·cīs'·ŭs,	-ă,	-ŭm,	notched ; indented.
Ĭn'·dĭc·ŭs,	-ă,	-ŭm,	of or from India.
ĭn·dĭg'·ŏf'·ér,	-ŏf'·ĕr·ă,	-ŭm,	producing indigo.
ĭn'·dĭg·ŏt'·ĭc·ŭs,	-ă,	-ŭm,	producing a blue-colouring matter.
ĭn·érm'·ĭs,	-ĭs,	-ē,	without weapons ; unarmed.
ĭn'·fĕc·tōr'·ĭ·ŭs,	-ă,	-ŭm,	that serves for dyeing.
ĭn·flāt'·ŭs,	-ă,	-ŭm,	puffed up ; inflated.
ĭn'·ō·phyl'·lŭs,	-ă,	-ŭm,	fibre-leaved.
ĭn·tĕg'·rĭ·fōl'·ĭ·ŭs,	-ă,	-ŭm,	having their leaves undivided.
ĭn'·tyb·ŭs,	-ă,	-ŭm,	having hollow tubes.
ĭp'·ĕ·căc·ŭ·ăn'·hă, n., having the qualities of the ipecacuanha.			
ĭr'·ĭd·ĭ·flōr'·ŭs,	-ă,	-ŭm,	iris-leaved.
Is·lănd'·ĭc·ŭs,	-ă,	-ŭm,	of or from Iceland.

Jăc·quĭn'·ĭ·ī, n., of Jacquinius, a botanist.			
jăl'·ăp·ŭs,	-ă,	-ŭm,	of or like the purgative plant jalap.
Jăp·ŏn'·ĭc·ŭs,	-ă,	-ŭm,	of or from Japan.
jŏn·quĭl'·lŭs, (kwĭl)	-ă,	-ŭm,	of or pert. to the jonquille, a daffodil species.
jū'·jŭb·ŭs,	-ă,	-ŭm,	of or like the jujube tree.
jŭn'·cĕ·ŭs,	-ă,	-ŭm,	like a rush or bulrush.

lăb·ŭrn'·ŭs,	-ă,	-ŭm,	of or pert. to the laburnum.
lăc·ĭn'·ĭ·āt'·ŭs,	-ă,	-ŭm,	jagged ; indented.
lăc'·rym·ăns,	-ăns,	-ăns,	weeping ; lamenting.
lăc'·tĕ·ŭs,	-ă,	-ŭm,	containing milk ; milky.
lăc'·tĭf·ér,	-tĭf'·ĕr·ă,	-ŭm,	producing milk.
lăc·tūc'·ă, n., lettuce.			
lănc'·ĕ·œ·fōl'·ĭ·ŭs, (ē)	-ă,	-ŭm,	having lance-shaped leaves.
lănc'·ĕ·ān'·ŭs,	-ă,	-ŭm,	like a lance or spear.
lănc'·ĕ·ŏl·āt'·ŭs,	-ă,	-ŭm,	lance-shaped.
lăngs·dŏrf'·ĭ·ī, n., of Langsdorf, a botanist.			
lān'·ĭg·ér,	-ĭg'·ĕr·ă,	-ŭm,	wool-bearing ; like wool.

MASC.	FEM.	NEUT.	
lăp'·pŭs,	-ă,	-ŭm,	having burs.
lāt'·ĭ·fōl'·ĭ·ŭs,	-ă,	-ŭm,	having broad leaves.
lăt·ĭs'·sĭm·ŭs,	-ă,	-ŭm,	very broad or wide.
lāŭr'·ĕ·fōl'·ĭ·ŭs, (lāwr)	-ă,	-ŭm,	laurel-leaved.
lāŭr·ē·ŏl·ŭs, (lāwr)	-ă,	-ŭm,	pert. to a small laurel.
lāŭr'·ō·cĕr'·ăs·ŭs, (lāwr)-	-ă,	-ŭm,	pert. to the cherry laurel.
lăx'·ĭ·ŭsc'·ŭl·ŭs, (ŭsk)-	-ă,	-ŭm,	somewhat wide or loose.
lĕnt·ĭsc'·ŭs,	-ă,	-ŭm,	of or pert. to the mastich tree.
lĕnt'·ŭs,	-ă,	-ŭm,	tough ; hard.

leūc'·ă·dĕnd'·rŏn, n., (lŏk) a white tree.

leū'·cō·rhĭz'·ŭs, (lŏ)	-ă,	-ŭm,	having a white root.
lĭv'·ĭg·āt'·ŭs,	-ă,	-ŭm,	softened ; macerated well.
lēv'·ĭs,	-ĭs,	-ē,	light ; not heavy.

Lĭb'·ăn·ĭ, n., of Mount Libanus, in Syria.

lĭch'·ĕn·ŏĭd'·ĭs, (ōyd)	-ĭs,	-ē,	resembling the lichen.

lĭn'·guă, n., (gwă) a tongue.

lĭnt'·ē·ār'·ĭ·ŭs,	-ă,	-ŭm,	of or pert. to linen.
lŏn'·gĭ·crūr'·ĭs,	-ĭs,	-ē,	having a long leg or limb.
lŏn'·gĭ·flōr'·ŭs,	-ă,	-ŭm,	having long flowers.
lŏn'·gŭs,	-ă,	-ŭm,	long.

lōt'·ŭs, n., the water-lily of the Nile.

lūc'·ĭd·ŭs,	-ă,	-ŭm,	clear ; bright.
lŭp'·ŭl·īn'·ŭs,	-ă,	-ŭm,	of or like the hop plant.
lŭp'·ŭl·ŭs,	-ă,	-ŭm,	like a little wolf ; of or like the hop plant.
Lŭs'·ĭt·ān'·ĭc·ŭs,	-ă,	-ŭm,	of or from Portugal.
lŭt·ē·ŏl·ŭs,	-ă,	-ŭm,	yellowish.
lŭt'·ē·ŭs,	-ă,	-ŭm,	of a yellow colour.
Lyc'·ĭ·ŭs,	-ă,	-ŭm,	of or from Lycia, Asia Minor.

măc·rănth'·ŭs,	-ă,	-ŭm,	having great flowers.
măc'·rō·cârp'·ŭs,	-ă,	-ŭm,	having very large fruit.
măc'·rō·pŭs,	-ă,	-ŭm,	having long feet.
măc'·ŭl·āt'·ŭs,	-ă,	-ŭm,	having spots or stains.
Măd'·ă·găsc'·ăr·ĭ·ĕns'·ĭs,	-ĭs,	-ē,	of or from Madagascar.
māj'·ŏr,	-ŏr,	-ŭs,	greater.
māj'·ŏr·ān'·ŭs,	-ă,	-ŭm,	flowing in Maius or May.

māj'·ŭs, adj. neut. (see major), greater.

māl'·ŭs,	-ă,	-ŭm,	bad ; sour.
măm·mōs'·ŭs,	-ă,	-ŭm,	having large breasts.
măn'·gĭf·ėr,	-gĭf'·ĕr·ă,	-ŭm,	yielding mango fruit.
măn'·nĭf·ėr,	-nĭf'·ĕr·ă,	-ŭm,	bearing manna.
măr'·gĭn·āt'·ŭs,	-ă,	-ŭm,	furnished with a border.
măr·ĭt'·ĭm·ŭs,	-ă,	-ŭm,	of or belonging to the sea.
măr·sŭp'·ĭ·ŭs,	-ă,	-ŭm,	having pouches.

măr'·tĭ·ĭ, n., of the month of March.

măsc'·ŭl·ŭs,	-ă,	-ŭm,	male.
Māŭr·ĭt'·ĭ·ān'·ŭs, (māwr)-	-ă,	-ŭm,	after Maurice of Nassau ; or or from Mauritius.
măx'·ĭm·ŭs,	-ă,	-ŭm,	the greatest ; the highest.
mĕd'·ĭc·ŭs,	-ă,	-ŭm,	used in medicine ; medicinal.
mēd'·ĭ·ŭs,	-ă,	-ŭm,	middle ; midst.

MASC.	FEM.	NEUT.	
mĕd·ŭl·lār′·ĭs,	-ĭs,	-ē,	having the nature of marrow or pith.
mĕg′·ăl·ō·cĕph′·ăl·ŭs,	-ă,	-ŭm,	having a very large head.
mĕl′·ăn·ō·cŏc′·cŭs,	-ă,	-ăm,	having black seeds or berries.
mĕl′·lĭf·ĕr,	-lĭf′·ĕr·ă,	-ŭm,	producing honey.
mēl′·ō, n., a melon.			
mĕl·ŏn′·gĕn·ŭs,	-ă,	-ŭm,	producing apples.
mĕth·ĭs′·tĭc·ŭs,	-ă,	-ŭm,	producing intoxication.
Mēz′·ĕr·ē′·ŭs,	-ă,	-ŭm,	from a town or district of Persia.
mĭc′·rŏ·kŭs,	-ă,	-ŭm,	being a little world.
mĭl′·ĭ·ā′·cĕ·ŭs,	-ă,	-ŭm,	of or pert. to millet.
mĭl′·ĭt·ār′·ĭs,	-ĭs,	-ē,	soldier-like.
mĭn′·ŏr,	-ŏr,	-ŭs,	less.
mīr·āb′·ĭl·ĭs,	-ĭs,	-ē,	wonderful ; marvellous.
mŏl′·lĭs,	-lĭs,	-lē,	pliant ; supple.
mŏn′·ō·spĕrm′·ŭs,	-ă,	-ŭm,	one-seeded.
Mŏns′·pĕl·ĭ′·ăc·ŭs,	-ă,	-ŭm,	from Montpelier in France.
mŏnt·ān′·ŭs,	-ă,	-ŭm,	of or belonging to a mountain.
mŏr·ĕl′·lŭs,	-ă,	-ŭm,	dark ; blackish.
mōr′·ĭ·ō, and mōr′·ĭ·ŏn, n., a dark-brown gem ; deadly nightshade ; male mandrake.			
mŏs·chāt′·ŭs,	-ă,	-ŭm,	smelling like musk.
mŭlt·ĭf′·ĭd·ŭs,	-ă,	-ŭm,	cleft or split into many parts.
mŭl′·tĭ·flōr′·ŭs,	-ă,	-ŭm,	having many flowers.
mŭl·tĭj′·ūg·ŭs,	-ă,	-ŭm,	having many yokes.
mūr′·ĭc·āt′·ŭs,	-ă,	-ŭm,	shaped like the murex shell ; pointed.
mŭs·cār′·ĭs,	-ĭs,	-ē,	of or pert. to flies, or to hair brooms.
mŭs·cĭp′·ŭl·ŭs,	-ă,	-ŭm,	pert. to a mouse-trap.
mūt·ăb′·ĭl·ĭs,	-ĭs,	-ē,	changeable.
myr′·rhă, n., myrrh.			
myr′·tĭ·fōl′·ĭ·ŭs,	-ă,	-ŭm,	having leaves like the myrtle.
myr·tĭl′·lŭs, .	-ă,	-ŭm,	pert. to the myrtle.
mys′·tăx, n., the upper lip ; the moustache.			
myx′·ă, n., mucus ; mucilage.			
năp·ĕl′·lŭs, n., a little turnip.			
nāp′·ŭs, n., a turnip.			
nār′·ĭs, n., a nose ; nār′·ĭ·ŭm, of noses.			
nāt′·ăns,	-ăns,	-ăns,	swimming.
nĕl·ŭm′·bĭ·fōl′·ĭ·ŭs,	-ă,	-ŭm,	having a leaf like the nelumbo.
nĕm′·ŏr·ōs′·ŭs,	-ă,	-ŭm,	pert. to a grove.
nĕp′·ĕnth·ŏĭd′·ĭs, (ŏȳd)-ĭs,	-ē,		producing a magic potion.
nēr′·ē·ĭ·fōl′·ĭ·ŭs,	-ă,	-ŭm,	a sea-god leaf.
nĭg′·ĕr,	-ră,	-rŭm,	black.
nĭt′·ĭd·ŭs,	-ă,	-ŭm,	shining.
nĭv·āl′·ĭs,	-ĭs,	-ē,	of or like snow.
nĭv′·ĕ·ŭs,	-ă,	-ŭm,	snowy.
nŏb′·ĭl·ĭs,	-ĭs,	-ē,	famous ; renowned.
nōd·ōs′·ŭs,	-ă,	-ŭm,	full of knots ; knotty.
nŏv′·ŭs,	-ă,	-ŭm,	new ; recent.
nŭb·ĭg′·ĕn·ŭs,	-ă,	-ŭm,	cloud-born.
nŭc′·ĭf·ĕr,	-ĭf′·ĕr·ă, -ŭm,		producing nuts.

MASC.	FEM.	NEUT.	
nŭt'·ăns,	-ăns,	-ăns,	nodding ; tottering.
nyc·tĭc'·ăl·ŭs,	-ă,	-ŭm,	night-calling.
ŏb·lĭq'·uŭs, (wŭs)	-ă,	-ŭm,	oblique ; slanting.
ŏb'·lŏng·āt'·ŭs,	-ă,	-ŭm,	extended in length.
ŏb'·ōv·āt'·ŭs,	-ă,	-ŭm,	inversely egg-shaped.
ŏb'·tūs·āt'·ŭs,	-ă,	-ŭm,	blunted.
ŏc'·cĭd·ĕnt·āl'·ĭs,	-ĭs,	-ē,	Western.
ŏd'·ŏr·āt·ĭs'·sĭm·ŭs,	-ă,	-ŭm,	very fragrant.
ŏd'·ŏr·āt'·ŭs,	-ă,	-ŭm,	having a scent or smell.
ŏd·ōr'·ŭs,	-ă,	-ŭm,	sweet-smelling.
ŏf'·fĭc·īn'·ă, n., a workshop.			
ŏf·fĭc'·ĭn·āl'·ĭs,	-ĭs,	-ē,	officinal ; by authority.
ŏl'·ĕ·ănd'·ĕr,	-ĕr·ă,	-ŭm,	a corruption of the *rhododen-* *dron.*
ŏl·ē'·ĭf·ĕr,	-ĭf'·ĕr·ă,	-ŭm,	bearing oil.
ŏl'·ĕr·āc'·ĕ·ŭs,	-ă,	-ŭm,	herb-like.
ŏl'·ĭt·ōr'·ĭ·ŭs,	-ă,	-ŭm,	belonging to vegetables.
ŏl·lār'·ĭ·ŭs,	-ă,	-ŭm,	of or like a pot.
ŏn·ĭt'·ēs, plural of *onitis,* which see.			
ŏn·ĭt'·ĭs,	-ĭs,	-ē,	belonging to sweet marjoram.
ŏp'·ĭf·ĕr,	-ĭf'·ĕr·ă,	-ŭm,	bearing aid or power.
ŏp·pŏs'·ĭt·ĭ·fōl'·ĭ·ŭs,	-ă,	-ŭm,	having opposite leaves.
ŏp'·ŭl·ŭs, n., a kind of maple tree.			
ŏr'·ĕl·lān'·ă, n. (Sp.), arnatto.			
ŏr'·ĭ·ĕnt·āl'·ĭs,	-ĭs,	-ē,	of or from the East ; Eastern.
ōv·āl'·ĭ·fōl'·ĭ·ŭs,	-ă,	-ŭm,	having oval leaves.
ōv·āl'·ĭs,	-ĭs,	-ē,	oval.
ōv·āt'·ŭs,	-ă,	-ŭm,	egg-shaped.
ōv'·ĭf·ĕr,	-ĭf'·ĕr·ă,	-ŭm,	bearing eggs.
ōv'·ĭg·ĕr,	-ĭg'·ĕr·ă,	-ŭm,	bearing eggs.
ŏx'·y·cŏc'·cŭs,	-ă,	-ŭm,	having acid berries.
păb'·ŭl·ār'·ĭ·ŭs,	-ă,	-ŭm,	producing food.
păl'·lĭd·ŭs,	-ă,	-ŭm,	of a pale or pallid colour.
păl·māt'·ŭs,	-ă,	-ŭm,	palm-leaf shaped.
păl·mĭt'·ŭs,	-ă,	-ŭm,	having young branches.
păl·ŭs'·trĭs,	-ĭs,	-ē,	marshy ; swampy.
păn·ĭc'·ŭl·āt'·ŭs,	-ă,	-ŭm,	having a tuft or panicle.
păp·ĭl'·ĭ·ō, n., a butterfly.			
păp·ĭl'·ĭ·ōn·āc'·ĕ·ŭs,	-ă,	-ŭm,	of or pert. to a butterfly.
păp'·yr·āc'·ĕ·ŭs,	-ă,	-ŭm,	of or like the paper-reed.
păp·yr'·ĭf·ĕr,	-ĭf'·ĕr·ă,	-ŭm,	bearing a paper-reed.
păp·yr'·ŭs, n., the paper-reed.			
păr'·ăd·ĭs·ĭ'·ăc·ŭs,	-ă,	-ŭm,	of or belonging to Paradise.
păr'·ă·dŏx'·ŭs,	-ă,	-ŭm,	marvellous ; strange.
Păr'·ă·guĕns'·ĭs, (guĕns)	-ĭs, -ē,		of or from Paraguay.
păr'·ĭ·ĕt·īn'·ŭs,	-ă,	-ŭm,	of or belonging to old walls.
părvĭ·fōl'·ĭ·ŭs,	-ă,	-ŭm,	having small leaves.
pĕd'·ĭ·cĕl·lāt'·ŭs,	-ă,	-ŭm,	furnished with foot stalks.
pĕd·ŭnc'·ŭl·āt'·ŭs,	-ă,	-ŭm,	having little feet.
pĕl·lūc'·ĭd·ŭs,	-ă,	-ŭm,	transparent.

MASC.	FEM.	NEUT.	
pĕlt·āt'·ŭs,	-ă,	-ŭm,	armed with small shields, half-moon shaped.
pĕnt·ănd'·rŭs,	-ă,	-ŭm,	having five stamens.
pĕp·lŏĭd'·ĭs, (lŏȳd)	-ĭs,	-ē,	resembling a covering.
pĕp'·ō, n., a large melon; a pumpkin.			
pĕr·ĕn'·nĭs,	-ĭs,	-ē,	lasting the whole year; never failing.
pĕr·fŏl'·ĭ·āt'·ŭs,	-ă,	-ŭm,	completely leaved.
pĕrf'·ŏr·āt'·ŭs,	-ă,	-ŭm,	bored or pierced through.
Pĕrs'·ĭc·ār'·ī, n. plu. (mod. L.), Persia.			
Pĕrs'·ĭc·ŭs,	-ă,	-ŭm,	of or from Persia.
pĕr·tūs'·ŭs,	-ă,	-ŭm,	perforated.
Pĕr·ŭv·ĭ·ān'·ŭs,	-ă,	-ŭm,	of or from Peru.
phĕl·lănd'·rŭs,	-ă,	-ŭm,	having leaves like the ivy.
phĕl'·lŏs, n., the cork tree.			
phŏs·phŏr'·ĕ·ŭs,	-ă,	-ŭm,	bearing or bringing light.
phym'·ăt·ōd'·ĭs,	-ĭs,	-ē,	affected with hard swellings.
pĭc·tōr'·ĭ·ŭs,	-ă,	-ŭm,	like a painting.
pĭl·ōs'·ŭs,	-ă,	-ŭm,	hairy.
pĭl·ŭl'·ĭf·ĕr,	-ĭf'·ĕr·ă,	-ŭm,	bearing little balls.
pĭm·ĕnt'·ŭs,	-ă,	-ŭm,	of or like Indian pepper.
pĭn·ăst'·ĕr, n., a wild pine.			
pĭn'·ĕ·ŭs,	-ă,	-ŭm,	of or like the pine.
pĭn'·gŭĭs (gwĭs),	-gŭĭs,	-guē,	fat.
pĭn'·năt·ĭf'·ĭd·ŭs,	-ă,	-ŭm,	divided in segments in a feathery manner.
pĭn·năt'·ŭs,	-ă,	-ŭm,	winged; feathered.
pĭp'·ĕr·ĭt'·ŭs,	-ă,	-ŭm,	of or pert. to pepper.
plān'·ĭ·fŏl'·ĭ·ŭs,	-ă,	-ŭm,	having flat leaves.
plăt'·ăn·ŏĭd'·ĕs, plu., (ŏȳd) having the appearance of the plane tree.			
plăt'·y·cărp'·ŭs,	-ă,	-ŭm,	having broad fruit.
plĭc·āt'·ŭs,	-ă,	-ŭm,	folded; having folds.
pō·ĕt'·ĭc·ŭs,	-ă,	-ŭm,	poetical.
pōl·ār'·ĭs,	-ĭs,	-ē,	of or pert. to the pole.
pŏl'·y·phyl'·lŭs,	-ă,	-ŭm,	having many leaves.
pŏm'·ĭf·ĕr,	-ĭf'·ĕr·ă,	-ŭm,	bearing apples.
Pŏnt'·ĭc·ŭs,	-ă,	-ŭm,	of or from the Black Sea or Pontus.
pōp·ŭl'·nĕ·ŭs,	-ă,	-ŭm,	of or belonging to the poplar.
pŏr'·rĭ·fŏl'·ĭ·ŭs,	-ă,	-ŭm,	having leaves like leeks.
pŏr'·rŭm, n., a leek.			
prăt·ĕns'·ĭs,	-ĭs,	-ē,	growing in meadows.
prō·cēr'·ŭs,	-ă,	-ŭm,	high; tall.
prō·cŭmb'·ĕns,	-ĕns,	-ĕns,	leaning or bending forwards.
prō·fūs'·ŭs,	-ă,	-ŭm,	spread out; extended.
prŏl'·ĭf·ĕr,	-lĭf'·ĕr·ă,	-ŭm,	bearing offspring.
prūr'·ĭ·ĕns,	-ĕns,	-ĕns,	itching; producing an itching.
prūr·ĭt'·ŭs,	-ă,	-ŭm,	itched.
pseūd·ăc'·ŏr·ŭs, (sūd) -ă,		-ŭm,	of the false sweet flag.
pseūd'·ō·chĭn'·ă, n., (sūd) false Peruvian bark.			
psĭt'·tă·cīn'·ŭs, (sĭt) -ă,		-ŭm,	of or pert. to a parrot.
ptĕr'·ĭg·ō·spĕrm'·ŭs,(tĕr) -ă,-ŭm,			having winged seed.
pūb'·ĕns,	-ĕns,	-ĕns,	exuberant; juicy.
pūd·ĭc'·ŭs,	-ă,	-ŭm,	bashful; modest.

2 G

MASC.	FEM.	NEUT.	
pŭl·ĕg'·ĭ·ŭs,	-ă,	-ŭm,	pert. to flea-bane or penny royal.
pŭm·ĭl'·ĭ·ō, n., a dwarf ; a pigmy.			
pŭm'·ĭl·ŭs,	-ă,	-ŭm,	dwarfish ; little.
pŭnc·tāt'·ŭs,	-ă,	-ŭm,	having punctures.
pūn·ĭc'·ĭ·fōl'·ĭ·ŭs,	-ă,	-ŭm,	having African leaves.
pŭrg'·ăns,	-ăns,	-ăns,	cleaning or clearing out.
pŭrg'·ŭs,	-ă,	-ŭm,	cleaning out ; purging.
pŭr·pŭr'·ĕ·ŭs,	-ă,	-ŭm,	purple-coloured.
pyr·ēth'·rŭs,	-ă,	·ŭm,	of or like fire.
pyr'·ĭf·ĕr,	-ĭf'·ĕr·ă,	-ŭm,	bearing pears.
pyr'·ĭ·fŏrm'·ĭs,	-ĭs,	-ē,	shaped like a pear.
quăd·răng'·ŭl·ār'·ĭs,	-ĭs,	-ē,	having four corners ; four-square.
quăd·rāt'·ŭs,	-ă,	-ŭm,	in the form of a square.
quăd'·rĭ·fōl'·ĭ·ŭs,	-ă,	-ŭm,	four-leaved.
quăd'·rĭ·vălv'·ĭs,	-ĭs,	-ē,	having square folding doors.
quæs·ĭt'·ŭs, (kwĕs)	-ă,	-ŭm,	sought out ; select.
quĭn'·quĕ·fōl'·ĭ·ŭs,(kwĭn)	-ă,	-ŭm,	having five leaves.
răc'·ĕm·ōs'·ŭs,	-ă,	-ŭm,	full of clusters ; clustering.
rād'·ĭ·āt'·ŭs,	-ă,	-ŭm,	having rays ; rayed.
răd·ĭc'·ăns,	-ăns,	-ăns,	striking or taking root.
Răf·flēs'·ĭ·ān'·ŭs,	-ă,	-ŭm,	after Sir Stamford Raffles.
răm·ōs'·ŭs,	-ă,	-ŭm,	branchy ; ramose.
Rănd'·ĭ·ŭs,	-ă,	-ŭm,	after Rand, botanist.
răng·ĭf'·ĕr·ĭn'·ŭs,	-ă,	-ŭm,	of or pert. to the reindeer.
răp·ŭnc'·ŭl·ŭs,	-ă,	-ŭm,	like a little turnip.
rāp'·ŭs,	-ă,	-ŭm,	of or like a turnip.
rĕct'·ŭs,	-ă,	-ŭm,	straight.
rĕ·gīn'·ă, n., a queen ; rĕ·gīn'·æ, of a queen.			
rēg'·ĭ·ŭs,	-ă,	-ŭm,	royal.
rĕ·lĭg'·ĭ·ōs'·ŭs,	-ă,	-ŭm,	sacred ; religious.
rĕ·pănd'·ŭs,	-ă,	-ŭm,	bent backward ; turned up.
rēp'·ĕns,	-ĕns,	-ĕns,	creeping.
rĕs·ĭn'·ĭf·ĕr,	-ĭf'·ĕr·ă,	-ŭm,	producing resin.
rĕt·ĭc'·ŭl·āt'·ŭs,	-ă,	-ŭm,	net-like ; reticulated.
rĕv'·ōl·ūt'·ŭs,	-ă,	-ŭm,	rolled back.
rhăm·nōïd'·ĭs, (nŏÿd)	-ĭs,	-ē,	resembling the white thorn.
Rhă·pŏnt'·ĭc·ŭs,	-ă,	-ŭm,	of or from the Rha or the river Volga.
rhœ'·ăs, (rē)	-ăs,	-ăs,	flowing as juice.
rĭv·āl'·ĭs,	-ĭs,	-ē,	belonging to a small stream or brook.
rōb·ŭst'·ŭs,	-ă,	-ŭm,	of or like oak wood ; strong.
Rōm·ān'·ŭs,	-ă,	-ŭm,	of or from Rome.
rōs'·ă-sĭn·ĕns'·ĭs,	-ĭs,	-ē,	like a Chinese rose.
rōs'·ĕ·ŭs,	-ă,	-ŭm,	like a rose.
rōs'·măr·ĭn'·ŭs,	-ă,	-ŭm,	like marine dew ; rosemary.
rōt·ŭnd'·ĭ·fōl'·ĭ·ŭs,	-ă,	-ŭm,	having round leaves.
Rŏx·bŭrgh'·ĭ·ī, n. plu., (bĕrg) after Roxburgh, a county of Scotland.			
rŏÿ·āl'·ĭs,	-ĭs,	-ē,	royal.
rub'·ĕr,	-ră,	-rŭm,	red.
rŭb·īg'·ō, n., rust ; mildew.			
rŭb'·rĭ·câul'·ĭs,(kâwl)	-ĭs,	-ē,	having a red stem.

MASC.	FEM.	NEUT.	
rŭd·ĕnt'·ŭs,	-ă,	-ŭm,	pert. to ropes ; like a rope.
rŭs'·cĭ·fōl'·ĭ·ŭs,	-ă,	-ŭm,	having leaves of the colour of the ruscus.
rŭst'·ĭc·ŭs,	-ă,	-ŭm,	rustic ; country.

săb·ĭn'·ă, n., employed by the anc. Sabine priests.			
săc·chăr'·ĭf·ér,	-ă,	-ŭm,	bearing or producing sugar.
săc'·chăr·ĭn'·ŭs,	-ă,	-ĭm,	producing sweet juice.
săc'·cĭd·ōr'·ŭs,	-ă,	-ŭm,	furnishing sacs or bags.
săc'·cĭf·ér,	-cĭf'·ĕr·ă,	-ŭm,	bearing bags.
săg·ĭt'·tă, n., an arrow.			
săl·ĭs'·ĭ·fōl'·ĭ·ŭs,	-ă,	-ŭm,	leaved like the willow
sălv·āt'·rĭx, n., a saviour.			
sănc'·tŭs,	-ă,	-ŭm,	sacred ; holy.
săn·guĭn'·ĕ·ŭs,(gwĭn)-ă,		-ŭm,	of or like blood.
sănt·āl·ĭn'·ŭs,	-ă,	-ŭm,	of or pert. to santal wood.
săp'·ĭd·ŭs,	-ă,	-ŭm,	tasting ; savouring.
săp'·ĭ·ĕnt'·ŭs,	-ă,	-ŭm,	having a good taste.
săp'·ōn·āc'·ĕ·ŭs,	-ă,	-ŭm,	of or pert. to soap.
săp'·ōn·ār'·ĭ·ŭs,	-ă,	-ŭm,	of or like soap.
sârc'·ō·kŏl'·lŭs,	-ă,	-ŭm,	producing flesh-glue.
sāt·īv'·ŭs,	-ă,	-ŭm,	that may be sown or planted.
scāb'·ér,	-ră,	-rŭm,	rough ; scabby.
scăm·mŏn'·ĭc·ŭs,	-ă,	-ŭm,	like scammony.
scănd'·ĕns,	-ĕns,	-ĕns,	climbing.
scĕl'·ĕr·āt'·ŭs,	-ă,	-ŭm,	polluted.
schōl·ār'·ĭs,	-ĭs,	-ē,	pert. to a school.
scĭp'·ĭ·ōn'·ŭs, (sĭp)	-ă,	-ŭm,	like a staff, or a grape-stalk.
scŏl'·ym·ŭs, (ĭm)	-ă,	-ŭm,	like an edible kind of thistle.
scōp·ār'·ĭ·ŭs,	-ă,	-ŭm,	of or pert. to a broom.
Scŏt'·ĭc·ŭs,	-ă,	-ŭm,	of or from Scotland.
scūt·āt'·ŭs,	-ă,	-ŭm,	armed with shields.
sĕb'·ĭf·ér,	-ĭf'·ĕr·ă,	-ŭm,	bearing fat or tallow.
sĕg'·ĕt·ŭs,	-ă,	-ŭm,	of or belonging to the goddess of standing crops.
sĕl'·ăg·ĭn·ŏĭd'·ĭs, (ŏyd)	-ĭs,	-ē,	resembling the upright club moss.
sĕm'·pér·vīr'·ĕns,	-ĕns,	-ĕns,	always flourishing or verdant.
Sĕn'·ĕg·ŭs,	-ă,	-ŭm,	of or from Senegal.
sĕn·īl'·ĭs,	-ĭs,	-ē,	aged.
sĕns'·ĭt·īv'·ŭs,	-ă,	-ŭm,	having the power of feeling.
sēp'·ĭ·ŭs,	-ă,	-ŭm,	pert. to shell of cuttle-fish.
sēpt·ăng'·ŭl·ār'·ĭs,	-ĭs,	-ē,	seven-angled.
sĕr·ĭc'·ĕ·ŭs,	-ă,	-ŭm,	silky.
sĕrp'·ĕnt·ār'·ĭ·ŭs,	-ă,	-ŭm,	of or like a serpent.
sĕr'·răt·ĭ·fōl'·ĭ·ŭs,	-ă,	-ŭm,	having saw-shaped leaves.
sĕs'·sĭl·ĭ·flōr'·ŭs,	-ă,	-ŭm,	having dwarf flowers.
sĕt·ōs'·ŭs,	-ă,	-ŭm,	having coarse hair or bristles.
sĭl'·ĭq·uŭs (wŭs),	-ă,	-ŭm,	having pods ; like the carob.
sĭm'·ŭl·ār'·ĭs,	-ĭs,	-ē,	making or looking like.
Sĭn·ĕns'·ĭs,	-ĭs,	-ē,	of or from Sina or China.
sŏb·ŏl'·ĭf·ér,	-ĭf'·ĕr·ă,	-ŭm,	bearing sprouts or shoots
Sŏc'·ōt·rĭn'·ŭs,	-ă,	-ŭm,	of or from Socotra.

MASC.	FEM.	NEUT.	
sŏm·nĭf·ĕr,	-nĭf′·ĕr·ă,	-ŭm,	sleep-bringing.
sŏn·ŏr′·ŭs,	-ă,	-ŭm,	sounding.
sŏrb′·ĭl·ĭs,	-ĭs,	-ē,	that may be sucked up or supped.
sŏrb′·ŭs,	-ă,	-ŭm,	of the sorb or service tree.
spăr′·tŭm, n., a plant from Spain.			
spĕc′·ĭ·ōs′·ŭs,	-ă,	-ŭm,	full of beauty or display.
spĕlt′·ă, n., grain or wheat.			
spĭc·ăt′·ŭs,	-ă,	-ŭm,	furnished with spikes.
spĭn·ōs·ĭs′·sĭm·ŭs,	-ă,	-ŭm,	very thorny or prickly.
spĭn·ōs′·ŭs,	-ă,	-ŭm,	thorny ; prickly.
spĭr·āl′·ĭs,	-ĭs,	-ē,	spiral.
splĕnd′·ĕns,	-ĕns,	-ĕns,	bright ; shining.
squăm·ăr′·ĕ·ŭs,	-ă,	-ŭm,	full of scales ; scaly.
squăm·ăt′·ŭs, (skwăm) -ă,		-ŭm,	having scales.
squăm·ōs′·ŭs,	-ă,	-ŭm,	scaly.
stăph′·ys·āg′·rĭ·ŭs,	-ă,	-ŭm,	of or connected with country raisins.
stĭm′·ŭl·ăns,	-ăns,	-ăns,	pricking or goading on.
străm·ĭn′·ĕ·ŭs,	-ă,	-ŭm,	⎫ having straw-like or fibrous roots.
străm·ōn′·ĭ·ŭs,	-ă,	-ŭm,	⎭
styr′·ă·cĭf′·lŭ·ŭs,	-ă,	-ŭm,	abounding with the liquid resinous gum storax.
suăv′·ĭs,	-ĭs,	-ē,	sweet ; pleasant.
sūb′·ĕr, n., the cork tree.			
sūb′·tŏm·ĕnt·ōs′·ŭs,	-ă,	-ŭm,	having an inferior pubescence.
sŭc′·cĕd·ān′·ĕ·ŭs,	-ă,	-ŭm,	that supplies the place of something.
sŭc′·cĭ·rŭb′·ĕr,	-ră,	-rŭm,	having red juice.
sŭc·cīs′·ŭs,	-ă,	-ŭm,	cut off or down.
Sŭĕs′·ĭc·ŭs, (swĕs)	-ă,	-ŭm,	of or from Sweden.
sū·ĭll′·ŭs,	-ă,	-ŭm,	of or belonging to a swine.
sū·pĕrb′·ŭs,	-ă,	-ŭm,	proud.
Swī′·ĕt·ēn′·ĭ·ŭs,	-ă,	-ŭm,	after Swieten, a Dutch botanist.
syc′·ō·mōr′·ŭs,	-ă,	-ŭm,	of the mulberry tree.
syl·văt′·ĭc·ŭs,	-ă,	-ŭm,	living in the woods.
syl·vĕst′·rĭs,	-ĭs,	-ē,	woody.
Syr·ĭ′·ăc·ŭs,	-ă,	-ŭm,	of or from Syria.
tœd′·ă, n., (tēd) the pitch-pine tree.			
Tăr·tăr′·ĭ·ŭs,	-ă,	-ŭm,	of or belonging to the infernal regions ; or from Tartary.
tĕn′·ăc·ĭs′·sĭm·ŭs,	-ă,	-ŭm,	holding very fast.
tēn′·ăx,	-ăx,	-ăx,	holding fast.
tĕn′·ŭ·ĭ·fōl′·ĭ·ŭs,	-ă,	-ŭm,	having thin leaves.
tĕn′·ŭ·ĭs,	-ĭs,	-ē,	thin ; fine.
tĕn′·ŭs, n., Latin name of a tree, unknown.			
tĕr·ĕb·ĭnth′·ŭs, n., the turpentine tree.			
tĕrm′·ĭn·āl′·ĭs,	-ĭs,	-ē,	terminal or bounding, as planted for hedges in India.
tĕr·rĕst′·rĭs,	-ĭs,	-ē,	terrestrial.
tĕt′·răl·ĭx,	-ĭx,	-ĭx,	of or belonging to a heath plant.
tĕt·răn′·drŭs,	-ă,	-ŭm,	having four stamens.
tĕxt′·ĭl·ĭs,	-ĭs,	-ē,	woven ; wrought.
Thē·bā′·ĭc·ŭs,	-ă,	-ŭm,	of or from Thebes in Egypt.

MASC.	FEM.	NEUT.	
thŭr'·ĭf·ẽr,	-ĭf'·ẽr·ă,	-ŭm,	bearing frankincense.
thym'·ĭ·fōl'·ĭ·ŭs,	-ă,	-ŭm,	thyme-leaved.
tĭl'·ĭ·āc'·ĕ·ŭs,	-ă,	-ŭm,	pert. to the linden tree.
tĭnc·tōr'·ĭ·ŭs,	-ă,	-ŭm,	having the quality of dyeing.
tĭng'·ĕns,	-ĕns,	-ĕns,	dyeing.
Tō·bāc'·ŭs,	-ă,	-ŭm,	of or from Tobago, W. Indies.
Tōl·ū'·ĭf·ẽr,	-ĭf'·ẽr·ă,	-ŭm,	bearing Tolu balsam.
tŏm'·ĕnt·ōs'·ŭs,	-ă,	-ŭm,	woolly ; downy.
tŏns·ūr'·ăns,	-ăns,	-ăns,	clipping or pruning.
tŏr'·mĕnt·ĭl'·lŭs,	-ă,	-ŭm,	relieving pain or torment.
tŏrt'·ĭl·ĭs,	-ĭs,	-ē,	twined ; twisted.
tŏx'·ĭc·ār'·ĭ·ŭs,	-ă,	-ŭm,	producing poison for arrows.
tŏx'·ĭc·ō·dĕnd'·rŏn, n., a tree that produces poison.			
tŏx'·ĭf·ẽr,	-ĭf'·ẽr·ă,	-ŭm,	producing poison.
trâŭm·ăt'·ĭc·ŭs, (trāŭm)	-ă,	-ŭm,	fit for healing wounds.
trēm'·ĕns,	-ĕns,	-ĕns,	shaking ; quivering.
trĕm'·ŭl·ŭs,	-ă,	-ŭm,	trembling.
trī·ănd'·rŭs,	-ă,	-ŭm,	having three stamens.
trī·cōl'·ŏr,	-ŏr,	-ŏr,	having three colours.
trī'·dĕnt·āt'·ŭs,	-ă,	-ŭm,	having three teeth or tines.
trī·fōl'·ĭ·āt'·ŭs,	-ă,	-ŭm,	three-leaved.
trĭg'·ŏn·cĕph'·ăl·ŭs,	-ă,	-ŭm,	having triangular heads.
trĭl'·ŏb·ŭs,	-ă,	-ŭm,	having three lobes.
Trī·pōl'·ĭ·ŭs,	-ă,	-ŭm,	of or from Tripoli in Africa.
trĭs'·tĭs,	-ĭs,	-ē,	sad ; mournful.
trŭnc·āt'·ŭs,	-ă,	-ŭm,	lopped off ; truncated.
tūb'·æ·fŏrm'·ĭs, (ē)	-ĭs,	-ē,	shaped like a trumpet.
tūb'·ĕr·ōs'·ŭs,	-ă,	-ŭm,	having fleshy knots; having humps.
tūl·ĭp'·ĭf·ẽr,	-ĭf'·ẽr·ă,	-ŭm,	producing tulips.
tŭrb'·ĭn·āt'·ŭs,	-ă,	-ŭm,	cone-shaped.
typh·ĭn'·ŭs,	-ă,	-ŭm,	of or like spelt or German wheat.
ūl·ĭg'·ĭn·ōs'·ŭs,	-ă,	-ŭm,	full of moisture.
ŭl'·mĭ·fōl'·ĭ·ŭs,	-ă,	-ŭm,	having leaves like the elm.
ŭm'·bĕl·lāt'·ŭs,	-ă,	-ŭm,	forming little shadows ; bearing umbels.
ŭm'·brăc·ŭl'·ĭf·ẽr,	-ĭf'·ẽr·ă,	-ŭm,	supplying a shade.
ŭn'·cĭn·āt'·ŭs,	-ă,	-ŭm,	furnished with hooks.
ŭn'·dŭl·āt'·ŭs,	-ă,	-ŭm,	undulated ; like waves.
ŭn'·ĕd·ō, n., the arbute or strawberry tree.			
ŭrb·ān'·ŭs,	-ă,	-ŭm,	belonging to the city or town.
ūr'·ĕns,	-ĕns,	-ĕns,	parched ; dried up.
ūr'·ĕnt·ĭs'·sĭm·ŭs,	-ă,	-ŭm,	very acrid or burning.
ūs'·ĭt·ăt·ĭs'·sĭm·ŭs,	-ă,	-ŭm,	very often used ; very common.
ūs'·ĭt·āt'·ŭs,	-ă,	-ŭm,	used often ; common.
ūt'·ĭl·ĭs,	-ĭs,	-ē,	useful ; profitable.
ūt·rĭc'·ŭl·āt'·ŭs,	-ă,	-ŭm,	having a small skin or leathern bottle.
ūv'·ă·ŭrs'·ĭ, n., the grape of the bear.			
ūv'·ĭf·ẽr,	-ĭf'·ẽr·ă,	-ŭm,	bearing grapes.
vāg'·ăns,	-ăns,	-ăns,	wandering about.

MASC.	FEM.	NEUT.	
vār′·ĭ·ĕg·āt′·ŭs,	-ă,	-ŭm,	having various colours.
vār′·ĭ·ōl·ār′·ĭs,	-ĭs,	-ē,	diversified.
vār′·ĭ·ŭs,	-ă,	-ŭm,	changing ; varying.
vĕn′·ĕn·āt′·ŭs,	-ă,	-ŭm,	furnished with poison.
vĕn′·ĕn·ōs′·ŭs,	-ă,	-ŭm,	very poisonous.
vĕn′·ŭl·ōs′·ŭs,	-ă,	-ŭm,	full of small veins.
vēr′·ĭs,	-ĭs,	-ē,	pert. to spring ; n., of spring.
vėrn·ĭc′·ĭf·ėr,	-ĭf′·ėr·ă, -ŭm,		bearing or bringing spring.
vėrn′·ĭc·ĭf′·lŭ·ŭs,	-ă,	-ŭm,	yielding varnish.
vėrn′·ŭs,	-ă,	-ŭm,	of or belonging to spring.
vēr′·ŭs,	-ă,	-ŭm,	real ; genuine.
vĕsc′·ŭs,	-ă,	-ŭm,	small ; feeble ; fine.
vĕs·ĭc′·ŭl·ōs′·ŭs,	-ă,	-ŭm,	having little vesicles.
vĕsp′·ėr·tĭl·ĭ·ōn′·ĭs,	n., of the animal called a bat.		
vīn′·ĭ·fėr,	-ĭf′·ėr·ă, -ŭm,		producing wine.
vīn·ōs′·ŭs,	-ă,	-ŭm,	having the taste of wine.
vĭ′·ōl·āc′·ĕ·ŭs,	-ă,	-ŭm,	violet-coloured.
vĭrg·āt′·ŭs,	-ă,	-ŭm,	made of twigs or osiers.
vĭrg′·ĭn·ĭ·ān′·ŭs,	-ă,	-ŭm,	pert. to a virgin ; of Virginia.
vĭrg·ĭn′·ĭc·ŭs,	-ă,	-ŭm,	virgin.
vĭr′·ĭd·ĭs,	-ĭs,	-ē,	green.
vĭr·ōs′·ŭs,	-ă,	-ŭm,	slimy ; poisonous.
vĭt′·ĕl·lĭn′·ŭs,	-ă,	-ŭm,	of a yellow colour.
vīt′·ĭs,	n., a vine.		
vŏl′·ĭt·āns,	-āns,	-āns,	flying to and fro.
vŏm′·ĭt·ōr′·ĭ·ŭs,	-ă,	-ŭm,	that provokes vomiting.
vŭl·gār′·ĭs,	-ĭs,	-ē,	common ; vulgar.
vŭl·gāt′·ŭs,	-ă,	-ŭm,	made common.
vŭlp·ĭn′·ŭs,	-ă,	-ŭm,	of or pert. to a fox.
zĕb·rīn′·ŭs,	-ă,	-ŭm,	striped like a zebra.
zĕy·lăn′·ĭc·ŭs, (zī)	-ă,	-ŭm,	of or from Ceylon.

PREFIXES.

NOTE.—A prefix is a significant particle placed before a word, or a root, in order to modify its meaning. As the constituent part of a word, a prefix can be readily separated and defined. *Note.*—In the examples the *prefixes* are printed in italics.

In medical compound terms, a *prefix* is very frequently formed from an independent word, and made to end in *o*, followed by a hyphen, which *prefix* then indicates ‘connection or association with,’ or ‘relation to,’ the second term of the compound; thus, *cerebro-spinal* is an adjective which indicates ‘connection or association with’ the *brain* and *spine.*

The *word-prefixes* in *o* will be generally found in their proper places in the body of the work. Only a few omitted ones are given in the following list of prefixes.

a (AS.), at ; in ; on : *a*head, *at* the head ; *a*sleep, *in* sleep ; *a*ground, *on* ground ; *a*ware = *ge*ware (AS. *ge*).

a, with its forms ab, abs (L.), from ; away from : *a*void, to part from ;

*a*vert, to turn away from : *ab*solve, to loose from : *ab*stract, to draw from.

a, also **an** (Gr.), without ; not : *aby*ss, a place without a bottom ; *a*theist, a man without God : *an*archy, a society without a government ; *an*omalous, not similar : tonic, having tone ; *a*tonic, without tone.

ad, assuming for the sake of euphony the various forms of **a, ac, af, ag, al, an, ap, ar, as, at,** according to the commencing letter of the primitive or root (L.), to ; towards : *ad*here, to stick to ; *ad*duce, to lead to : **ad** becomes **a** before **s,** as in *a*scend, to climb to : **ac** before **c,** as in *ac*cede, to yield to ; *ac*crue, to grow to : **af** before **f,** as in *af*fix, to fix to ; *af*fiance, to give faith to : **ag** before **g,** as in *ag*gregate, to collect into one mass ; *ag*gravate, to make heavy to : **al** before **l,** as in *al*lot, to apportion to ; *al*locate, to give a place to : **an** before **n,** as in *an*nex, to tie to ; *an*nounce, to tell to : **ap** before **p,** as in *ap*pend, to hang to ; *ap*plaud, to clap the hands to : **ar** before **r,** as in *ar*rive, to come to the shore ; *ar*range, to put into a row : **as** before **s,** as in *as*sign, to allot to ; *as*sist, to stand to : **at** before **t,** as in *at*tract, to draw to ; *at*test, to bear witness to.

adeno-, *ăd·ēn'·ō* (Gr. *adēn,* an acorn, a gland), denoting connection with glands ; as *adeno*-cele, a glandular tumour.

al, *ăl* (Ar.), an Arabic prefix signifying 'the' ; or used to denote 'eminence' or an 'essence' ; as *al*chemy, that is, *al kimia,* the secret art.

am, amb, also **ambi,** and **amphi** (L. *ambo,* both ; Gr. *amphi,* about, on both sides), both ; round ; about : *am*putate, to cut off round about, as a leg : *am*bition, a going round : *am*bidextrous, using both hands as right : *amphi*bious, able to live in both elements ; *amphi*theatre, a theatre on all sides ; *amphi*gens, plants which increase by growth on all sides.

an, see **a** (Gr.).

ana (Gr.), up ; up through ; back ; again : *ana*tomy, a cutting up through ; *ana*logy, a reasoning back ; *ana*lysis, a loosening up through ; *ana*chronism, a dating up or back : denoting also, throughout ; an increase or repetition ; see **āā** in medical abbrev.

ante, in one case **anti** (L.), before, in time or place : *ante*chamber, a chamber before the principal one ; *ante*cedent, going before : *anti*cipate, to take before, to foresee.

anti, also **ant** (Gr.), against ; opposite : *anti*dote, something given as good against ; *anti*pathy, a feeling against : *ant*arctic, opposite the arctic or north.

apo (Gr. *apo ;* Sans. *apa,* off, away), away ; from : *apo*stasy, a standing away from ; *apo*stle, one sent from.

arterio-, *ăr·tēr'·ĭ·ō* (Gr. *artēria,* an artery), of or connected with an artery ; as *arterio*-phlebotomy, blood-letting as by leeches, the scarificator, or lancet.

auriculo-, *aur·ĭk'·ŭl·ō* (L. *auricula,* the flap of the ear), denoting connection with the ear, or with the auricles of the heart ; see under 'auricle.'

be (AS. *be,* sometimes *ge*), to make ; to take from : **be** prefixed to a noun forms a verb, as in *be*calm, to make calm ; *be*dim, to make dim ; *be*friend, to act as a friend to ; *be*head, to take the head from : **be** prefixed to a verb signifies 'about' ; over ; for ; as *be*gird, to gird about ; *be*daub, to daub over ; *be*speak, to speak for : **be** as the first element in an adverb, a preposition, or a conjunction, signifies 'by or in': *betimes*, in time; *behind*, in the rear of; *before*, in front of; *because*, by cause of.

bi, also **bis** (L. *bis*, twice; another form of **dis**), twice; two; double; in two: *bisect*, to cut into two equal parts; *bicipital*, having a double head: *biscuit*, bread twice baked: **bis** becomes, for sake of euphony, **bin**, as in *bin*oxalate.

brachio-, *brăk'ĭ·ō* (L. *brachium*, an arm), denoting a connection with the arm; as *brachio*-cephalic, connected with the arm and head.

broncho-, *brŏngk'ō* (Gr. *bronchos*, the windpipe), denoting relation to, or connection with, the brachea or windpipe; as *broncho*-pneumonia, inflammation of the bronchia, and the substance of the lungs.

bucco-, *bŭk'kō* (L. *bucca*, the cheek), denoting connection with the cheek or its muscles.

calcareo-, *kăl·kār'ĕ·ō* (L. *calcarius*, pert. to lime—from *calx*, lime), having calcareous matter or lime in the composition of the compound; as *calcareo*-silicious, consisting of calcareous and silicious earth.

carpo-, *kărp'ō* (*carpus*, a Latinised form of Gr. *karpos*, the wrist), denoting connection with the wrist; as *carpo*-meta-carpal, pert. to the hand and wrist, including the fingers.

cata, also **cat**, and **cath** (Gr.), down; downwards; under; against; completeness: *cata*combs, hollow places underground; *cata*logue, consisting of words put down as in a list: *cate*chise, to speak down to others; *cat*optrics, the science of light reflected downwards; *cath*-olic, the whole, in completeness.

chloro-, *klōr'ō*, also **chlor-**, *klōr* (Gr. *chlōros*, grass-green), denoting that chlorine is one of the components of the substance; of a grass-green, or deep-yellow.

chondro-, *kŏn'drō* (Gr. *chondros*, cartilage), denoting connection with the cartilage; as *chondro*-xiphoid, connected with the xiphoid cartilage.

circum, also **circu** (L.), around; round about: *circum*ference, that which goes round; *circum*scribe, to write around, to limit: *circuit*, a moving or passing round.

cis (L.), on this side: *cis*alpine, on this side the Alps.

cleido-, *klĕid'ō* (Gr. *kleis*, a key, a clavicle, *kleidos*, of a key), denoting connection with the clavicle; as *cleido*-costal, connected with the clavicle and ribs.

con, assuming the various forms **co**, **cog**, **col**, **com**, **cor**, according to the commencing letter of the word or root (L. *cum*, with), together; with; together with: *con*cede, to yield together; *con*tract, to draw together: **con** becomes **co** before a vowel or **h**, as *co*alesce, to grow together; *co*erce, to force together; *co*herent, sticking together: **cog** before **n**, as *cog*nate, born together; *cog*nition, knowledge together: **col** before **l**, as *col*lect, to gather together; *col*late, to bring together: **com** before **m**, **b**, or **p**, as *com*merce, a trading together; *com*bustion, a burning together; *com*pose, to put together: **cor** before **r**, as *cor*rect, to make straight with; *cor*rode, to gnaw together.

contra, also its forms **counter** and **contro** (L. *contra*, F. *contre*, against), against; in opposition to: *contra*dict, to speak against; *counter*act, to act against; *contro*vert, to contend against in words or writing.

crico-, *krĭk'ō* (Gr. *krĭkŏs*, a ring), denoting attachment to or connection with the cricoid cartilage; as *crico*-thyroid, denoting a membrane, forming one of the three ligaments which connect the cricoid and thyroid cartilages.

cysto-, *sĭst'·ō* (Gr. *kustis,* a bladder), denoting connection with the bladder; as *cysto-*lithiasis, urinary calculus disease.

dacryo-, *dăk'·rĭ·ō* (Gr. *dakrū,* a tear, *dakrŭŏs,* of a tear), denoting connection with the lachrymal apparatus; as *dacryo-*adenalgia, pain or disease of the lachrymal gland.

de (L.), down; from; separation: *de*cide, to cut down; *de*grade, to put a step down; *de*mand, to order from; *de*pose, to put down.

deca, *děk'·ă* (Gr. *deka*), ten; as *deca*gon, a figure having ten equal angles and sides.

dermo-, *dĕrm'·ō,* **dermat-,** *dĕrm'·ăt,* and **dermato-** (Gr. *derma,* skin), denoting connection with the skin; as *dermato-*pathia, a suggested term for disease of the skin.

deut-, *dūt,* and **deuto-,** *dūt'·ō* (Gr. *deutĕros,* second), denoting 'two' or 'double,' as the combinations of two equivalents of oxygen with a metal: *deut*oxide, a substance in the second degree of oxydation, that is, a substance containing two equivalents of oxygen to one of another body.

dia, *dī'·ă* (Gr. *dia,* through—from *duo,* two), two; through; asunder: *dia*logue, a conversation between two; *dia*phanous, letting light through; *dia*meter, the measure through the centre.

dis, *dĭs,* with its forms **di** and **dif** (L. and Gr. *dis,* twice, in two parts), not; the opposite of; asunder or apart; two: *dis*agree, the opposite of agree; *dis*pel, to drive asunder; *dis*pose, to place asunder; *dis*relish, not to relish; *dis*syllable, a word of two syllables; *dis*annul, to render null—*dis* being only intensive: **dis** becomes **di** before **s, v,** etc., as *di*sperse, to spread asunder; ˉ*di*vert, to turn aside or apart: **dif** before **f,** as *dif*fuse, to pour apart; *dif*fer, to bear apart.

dorso-, *dŏrs'·ō* (L. *dorsum,* the back), denoting connection with the back; as *dorso-*cervical, designating a region situated at the back part of the neck.

duo-, *dŭ'·ō* (L. *duo,* two), denoting the second or duplicate; as *duo-*sternal, denoting the second bone or gladiolus of the sternum.

dys, *dĭs* (Gr. *dus,* with difficulty, bad), an inseparable prefix, denoting badly; with difficulty; hard; opposed to Gr. *eu,* well: *dys*crasia, an ill habit of body.

e is a form of L. **ex,** and **ec** a form of Gr. **ex,** which see.

electro-, *ĕ·lĕk'·trō* (Gr. *elektron,* amber), denoting connection with the phenomena of electricity or galvanism; as *electro-*biology, the doctrine which treats of the influence of electricity on life.

en (AS.; F.), to make; to surround: *en*able, to make able; *en*noble, to make noble: **en** becomes **em** before **b** or **p,** as *em*bezzle, to make as one's own what belongs to another; *em*ploy, to make use of; *em*brace, to surround as with the arms.

en (F. *en;* L. *in;* Gr. *en;* AS. *em,* in), in; on; into: *en*cage, to put into a cage; *en*close, to close in; *en*kindle, to set on fire: **en** becomes **em** before **b** or **p,** as *em*balm, to put into balsam; *em*bosom, to hold or enclose in the bosom; *em*pale, to drive a stake into: **en** or **em** from the Greek, and used as a prefix in words derived from the Greek, as *en*demic, on the people; *en*ergy, work or power in: *em*phasis, a speaking with the force of the voice on: some words are written indifferently with *en* or *in,* as *en*close or *in*close.

endeca-, *ĕn'.dĕk.ă* (Gr. *endeca*, eleven), eleven ; in L. *undecim.*

endo-, *ĕn'.dō* (Gr. *endon*, within), within or inwards ; as *endo*-**skeleton**, an inner or internal skeleton.

ennea-, *ĕn'.nĕ.ă* (Gr. *ĕnnĕă*, nine), having nine ; in L. *novem ;* as *ennea*ndrous, having nine stamens.

ens, *ĕnz* (L. *ens*, being), any being or substance ; in *chem.*, an essence containing the whole qualities or virtues of a compound substance.

entero-, *ĕn'.tĕr.ō* (Gr. *ĕntĕrŏn*, an intestine), denoting connection with the intestines ; as *entero*colitis, inflammation of the small intestine and colon.

epi, with its forms **ep** and **eph** (Gr.), on ; upon ; during : **ep** is used before a vowel, **eph** with an aspirate, and **epi** before a consonant : *epi*dermis, a skin upon a skin; *epi*taph, a writing upon a tombstone: *ep*och, a point of time fixed on : *ep*hemeral, existence only upon a day.

ex, with its forms **e**, **ef** (L.), from ; out ; out of : *ex*haust, to draw out ; *ex*pire, to breathe out : *e*merge, to rise out of : *ef*fect, to work out ; *ef*fulgence, a shining out.

ex or **ek**, also **ec** (Gr.), out ; out of ; from : *ex*odus, a going out : *ec*stasy, a standing out of the body : *ec*centric, out of the centre.

excito-, *ĕks·sīt'.ō* (L. *excito*, I stir up), denoting power to rouse or stimulate to action ; as *excito*-**motory**, the function of the nervous system by which the impressions conveyed to the brain result in muscular action without sensation or volition.

exo-, *ĕks'.ō* (Gr.), without : *exo*tic, that which is introduced from without.

extra (L.), on the outside; beyond; in excess; additional: *extra*vagant, wandering beyond limits ; *extra*vasate, to let or force out beyond the proper vessel ; *extra*-judicial, on the outside of ordinary court procedure.

ferro-, *fĕr'.rō* (L. *ferrum*, iron), denoting connection with iron ; as *ferro*-cyanic, denoting an acid compounded of cyanogen, iron, and hydrogen.

fibro-, *fib'.rō* (L. *fibra*, a fibre, a band), denoting a fibrous state of the substance ; as *fibro*-**cartilage**, a texture consisting of white, fibrous tissue and cartilage.

for, sometimes **fore** (Ger. *ver*, Goth. *fair*, away), not ; against ; forth ; away : *for*bid, to bid a thing away ; *for*get, to away-get ; *for*swear, to swear against : *for*ego, to go without.

fore (Ger. *vor*, before ; AS. *for*, for), before ; in front of : *fore*ordain, to ordain beforehand ; *fore*tell, to tell before ; *fore*ground, ground in front. NOTE.—The prep. *for* and the prefixes *for* and *fore* are radically connected.

gain (AS.), against : *gain*say, to speak against.

gastero-, *găs'.tĕr.ō*, **gastro-**, *găs'.trō*, and **gastr-**, *găs'.tr* (Gr. *gastĕr*, the stomach), prefixes denoting relation to, or connection with, the stomach ; as *gastro*-**cephalitis**, inflammation of the stomach and head.

hæm-, *hēm*, **hæma-**, *hēm'.ă*, **hæmat-**, *hēm'.ăt*, and **hæmato-**, *hēm'.ăt.ō* (Gr. *haima*, blood, *haimătŏs*, of blood), different forms, signifying blood ; having a reference to, or connected with, blood : *hæm*alopia, an effusion of blood into the ball of the eye ; a blood-shot eye : *hæmato*metra, retention of blood in the womb.

hemi-, *hĕm'.ĭ* (Gr. *hēmĭ*, half), half ; in L. *semi ;* as *hemi*crania, pain on one side of the head only.

hepato-, *hĕp'ăt·ō* (Gr. *hēpar*, the liver), denoting connection with the liver ; as *hepato*-cystic, denoting connection between the liver and gall-bladder.

hepta-, *hĕp'tă* (Gr. *hepta*, seven), seven ; in L. *septem ;* as *hepta*gynous, having seven styles.

hetero-, *hĕt'ĕr·ō* (Gr. *hĕtĕros*, opposite, different), denoting difference ; dissimilarity ; as *hetero*dromous, having spirals running in opposite directions.

hexa-, *hĕks'ă* (Gr. *hexa*, six), six ; in L. *sex ;* as *hex*androus, having six stamens.

holo-, *hōl'ō* (Gr. *hŏlŏs*, entire), entire ; complete ; as *holo*petalous, having entire petals.

homo-, *hōm'ō*, **homœo-,** *hōm·ē'ō* (Gr. *hŏmŏs*, alike ; *hŏmoiŏs*, similar, like); *homo* signifies equality or sameness ; *homœo*, similarity : *homo*carpous, having all the fruits of a flower-head alike *: homœo*meric, having similarity of parts.

hyper, *hĭp'ĕr* (Gr. *huper*), above ; over; beyond : *hyper*borean, beyond the north ; *hyper*critical, judging over-exactly.

hypo, *hĭp'ō* (Gr. *hupo*), under ; beneath ; indicating a less quantity : *hypo*crite, one who keeps his real character under ; *hypo*tenuse, the line extended under the right angle : **hyp,** *hĭp*, slightness, or incompleteness, as *hyp*algia, slight pain.

iatro-, *ī·ăt'rō* (Gr. *iătros*, a physician), denoting connection with the healing art ; as *iatro*physics, physics as applied to medicine.

icos-, *ĭk'ŏs*, and **icosi-,** *ĭk·ōz'ī* (Gr. *eikosi*, twenty), twenty ; in L. *viginti.*

ideo-, *ĭd'ĕ·ō* (Gr. *idea*, idea, abstract notion), denoting connection with ideas or mind ; as *ideo*logy, the science of ideas or mind.

idio-, *ĭd'ĭ·ō* (Gr. *idĭŏs*, peculiar), denoting something peculiar to the person, thing, or part spoken of ; as *idio*spasm, spasm or cramp occurring in one part only.

in, also its forms **il, im, ir** (L. *in*, in, within), in ; into ; on—in verbs and nouns ; as *in*clude, to shut in ; *in*cision, a cutting into : in becomes il before l, as *il*luminate, to throw light on : im before b, p, or m, as *im*bibe, to drink in ; *im*port, to carry in ; *im*mure, to put within walls : ir before r, as *ir*rigate, to let water flow on : in sometimes becomes en—see en 2.

in, also its forms **ig, il, im, ir** (L. *in*, not), signifies 'not' before adjectives : *in*correct, not correct ; *in*capable, not able to take : in becomes ig before n, as *ig*noble, not noble ; *ig*nominious, not of a good name : il before l, as *il*licit, not permitted ; *il*liberal, not free or generous : im before m or p, as *im*mature, not ripe ; *im*prudent, not prudent : ir before r, as *ir*regular, not according to rule : *ir*religious, not religious.

infra, *ĭn'frä* (L. *infra*, beneath), denoting under or beneath, as *infra*-orbital, situated underneath the orbit, as an artery.

inter-, *ĭn'tĕr* (L.), between ; among or amongst ; in the midst : *inter*cede, to go between ; *inter*fere, to strike amongst ; *inter*pose, to place amongst : **intel,** as in *intel*ligence, understanding among.

intro-, *ĭn'trō* (L.), within ; into ; in : *intro*duce, to lead within ; *intro*mit, to send in.

iod-, *ī·ŏd'*, and **iodo-,** *ī·ŏd'ō* (Gr. *iōdĕs*, resembling a violet in colour ; new L. *iōdĭŭm*, iodine), denoting iodine as an element of a compound; as *iodo*form, denoting a saffron-coloured substance containing iodine.

irido-, *ĭr'ĭd·ō* (L. *iris*, the rainbow, *ĭrĭdis*, of the rainbow), denoting connection with the iris of the eye; as *irido-*dialysis, an operation for an artificial pupil of the eye.

iso-, *ĭs'ō* (Gr. *isos*, equal, similar), denoting equality, likeness, or similarity; as *iso*cheimal, having the same, or a similar winter temperature.

juxta, *jŭks'tă* (L.), close to; near to; nigh: *juxta*position, a position close to.

leuco-, *lōk'ō* (Gr. *leukos*, white), denoting 'whiteness'; as *leuco-*derma, a cutaneous disease characterised by white patches on healthy skins.

litho-, *lĭth'ō* (Gr. *lithos*, a stone), having reference to a stone, or a calculus; as *litho*lysis, the treatment for the solution of stone in the bladder.

macro-, *măk'rō* (Gr. *makros*, long), denoting largeness or length; as *macro*carpous, having large fruit.

magneto-, *măg·nēt'ō* (Gr. L., *magnes*, the loadstone, L. *magnētis*, of the loadstone), connected with magnetism; as *magneto-*electricity, the electric phenomena produced by magnetism.

medico-, *mĕd'ĭk·ō* (L. *mĕdĭco*, I cure or heal), denoting connection with medicine; as *medico-*legal, pert. to law as affected by medical facts.

mega-, *mĕg'ă*, and **megalo-**, *mĕg'ăl·ō* (Gr. *mĕgas*, great), large; of great size; as *mega*therium, a fossil creature of enormous size.

mercurio-, *mĕr·kūr'ĭ·ō* (L. *mercŭrĭus*, mercury), denoting a connection with mercury; as *mercurio-*syphilitic, resulting partly from the effects of mercury, and partly from syphilis.

mes-, *mĕs*, and **meso-**, *mĕz'ō* (Gr. *mĕsŏs*, middle), denoting the middle; as *meso*phlœum, the middle layer of the bark.

meta, *mĕt'ă*, also its form **met** (Gr.), beyond; after; over; a change or transference: *meta*phor, that which carries a word beyond its usual meaning; *meta*morphosis, a change of form: *met*onymy, that which changes one word or name for another related to it; *met*hod, after a settled way.

micro, *mĭk'rō* (Gr. *mikrŏs*, small), denoting of small size; as *micro*meter, an instr. for measuring minute objects under the microscope.

mis (Goth. *mis*, implying error, separation; AS. *mis*, defect), divergence; error; defect; wrong: *mis*apply, to apply wrongly; *mis*lay, to lay in a wrong place; *mis*behaviour, ill-behaviour; *mis*conduct, defect in conduct.

mon-, *mŏn*, and **mono-**, *mŏn'ō* (Gr. *mŏnŏs*, one, single), one; in L. *unus;* as *mon*androus, having one stamen.

muco-, *mūk'ō* (L. *mūcŭs*, nasal secretion), denoting connection with mucus; as *muco-*enteritis, inflammation of the mucous coat of the intestines.

mult-, *mŭlt*, and **multi-**, *mŭlt'ĭ* (L. *multus*, many, much), many in number; much; as *mult*angular, having many corners or angles.

myelo-, *mĭ'ĕl·ō* (Gr. *muĕlos*, marrow), denoting connection with the brain or spinal marrow; as *myelo-*meningitis, inflammation of the spinal cord.

myo-, *mĭ'ō* (Gr. *mŭs*, a muscle), denoting connection with a muscle; as *myo*carditis, inflammation of the muscular substance of the heart.

neo-, *nē'·ō* (Gr. *něŏs*, new), recent ; new; as *neo*plasm, a new formation or growth.

nitro-, *nīt'·rō*, and **nitr-**, *nīt'·r* (Gr. *nītrŏn*, L. *nītrum*, a mineral alkali), denoting the presence of nitre, or nitric acid ; as *nitr*ification, the process of converting into nitre.

non-, *nŏn* (L.), not ; reversing the sense ; as *non*-**ability**, want of ability.

ob, with its forms **oc, of, o, op** (L.), in the way of ; against ; out : *ob*ject, something cast in the way of ; *ob*solete, grown out of use : **ob** becomes **oc** before **c**, as in *oc*casion, a falling in the way of : **of** before **f**, as in *of*fend, to strike against : **o** before **m**, as in *o*mit, to leave out : **op** before **p**, as in *op*pose, to place against : in *bot.*, reversed, or contrariwise, as *ob*compressed, flattened in front and behind, not laterally ; *ob*ovate, inversely ovate.

occipito-, *ŏk·sĭp'·ĭt·ō* (L. *occiput*, the back part of the head), denoting connection with the occipital bone, or 'os occipitis' ; as *occipito*-**front-alis**, a thin, flat muscle which arises from the transverse ridge of the occipital bone.

oleo-, *ōl'·ĕ·ō* (L. *ŏlĕum*, oil), combined with oil, or containing it ; as *oleo*-**albuminous**, consisting of oil and albumen.

omo-, *ōm'·ō* (Gr. *ōmŏs*, a shoulder), denoting attachment to, or connection with, the scapula ; as *omo*-**hyoid**, denoting a muscle between the scapula and the hyoid bone.

ortho-, *ŏrth'·ō* (Gr. *ŏrthŏs*, straight), straight ; upright ; as *ortho*pnœa, inability to breathe except in the upright position.

osteo-, *ŏst'·ĕ·ō* (Gr. *ŏstĕŏn*, a bone), denoting connection with, or reference to, a bone ; as *osteo*-**dentine**, a substance intermediate in structure between dentine and bone.

oxy-, *ŏks'·ĭ* (Gr. *oxus*, sour, acid), denoting the presence of oxygen or an acid ; acute ; sharp : also assumes the forms **oxi** and **oxu** : as *oxy*mel, a mixture of vinegar and honey.

pachy-, *păk'·ĭ* (Gr. *păchus*, thick), thick ; dense ; as *pachy*-dermatous, having a thick skin.

pan-, *păn*, **pant-**, *pănt*, and **panto-**, *pănt'·ō* (Gr. *pan*, all), all ; everything : *pan*demonium, the place of all the demons : *panto*mime, a theatrical dumb show of all sorts of actions and characters.

para, *păr'·ă*, also **par** (Gr. *para*, by, along), side by side as if for comparison ; like ; unlike ; contrary to : *para*dox, that which is contrary to received opinion : *parody*, a poetical composition, *like* in substance, but *unlike* in sense, to another.

penta-, *pĕnt'·ă*, and **pente-**, *pĕnt'·ē* (Gr. *pĕntĕ*, five), five ; in L. *quinque;* as *penta*phyllous, having five leaves.

per, with its form **pel** (L.), through ; thoroughly ; by ; for : *per*ennial, lasting through the year ; *per*fect, done thoroughly : **per** becomes **pel** before **l**, as in *pel*lucid, thoroughly clear.

peri, *pĕr'·ĭ* (Gr.), round ; about : *peri*meter, the measure round about ; *peri*od, a way round.

pharyngo-, *făr·ĭng'·gō* (Gr. *pharungx*, the gullet or windpipe), denoting connection with the pharynx or windpipe ; as *pharyngo*-**glossal**, pert. to the pharynx and tongue.

phyllo-, *fĭl'·lō* (Gr. *phyllon*, a leaf), a leaf ; in L. *folium;* as *phyllo*taxis, the arrangement of leaves on the stem.

platy-, *plăt'·ĭ* (Gr. *platus*, broad), broad ; in L. *latus;* as *platy*phyllous, having broad leaves.

pleuro-, *plŏr'.ō* (Gr. *pleura*, the side), denoting connection with the pleura, a side, or a rib ; as *pleuro*dynia, rheumatic or spasmodic pain in the side.

pluri-, *plŏr'.ĭ* (L. *plūrēs*, many), many ; several ; as *pluri*partite, separated into many distinct divisions.

pneumato-, *nūm'.ăt·ō*, **pneumo-**, *nūm'.ō*, and **pneumon-**, *nūm'.ŏn* (Gr. *pneuma*, air), connected with air or breath ; as *pneumato*-cyst, an air sac or float of certain Hydrozoa.

podo-, *pŏd'.ō* (Gr. *pous*, a foot, *pŏdŏs*, of a foot), a foot or stalk ; in L. *pes*, a foot, *pĕdĭs*, of a foot ; as *podo*carp, a stalk supporting the fruit.

poly-, *pŏl'.ĭ* (Gr. *polus*, many), many ; in L. *multus ;* as *poly*spermal, containing many seeds.

post (L.), behind ; after ; afterwards : *post*fix, that which is put after ; *post*script, that which is written afterwards.

præ or **pre** (L. *præ*), before ; priority of time, place, or rank : *pre*cede, to go before ; *pre*dict, to say or tell before.

preter-, *prĕt'.ĕr* (L. *præter*), beyond ; more than : *preter*natural, beyond the course of nature ; *preter*-imperfect, more than imperfect.

pro, with its forms **por** and **pur** (L. *pro*, for ; Gr. *pro*, before), for ; forward ; forth : *pro*ceed, to go forward ; *pro*voke, to call forth : *por*tend, to indicate events forward ; *pur*sue, to follow forward.

proto-, *prōt'.ō* (Gr. *prōtŏs*, first), first ; lowest ; in *chem.*, a first degree of combination, as of oxygen with metals ; as *proto*plast, the thing first formed.

pseud-, *sūd*, and **pseudo-**, *sūd'.ō* (Gr. *pseudēs*, false), false or spurious ; as *pseudo*-membrane, a false membrane.

pyr-, *pĕr*, and **pyro-**, *pīr'.ō* (Gr. *pur*, fire, *puros*, of fire), denoting relation to, or connection with, fire or heat ; as *pyro*genous, produced or formed by fire.

radio-, *rād'.ĭ.ō* (L. *rădĭus*, a spoke, a ray), denoting connection with the smaller bone of the forearm ; as *radio*-carpal, applied to the joint at the wrist which unites the 'radius' with the 'carpus.'

re (L.), back or again ; anew or a second time : *re*affirm, to firm again ; *re*commence, to begin anew.

retro (L.), back ; backward : *retro*spect, a looking back.

rhino-, *rĭn'.ō* (Gr. *rhĭn*, the nose, *rhĭnŏs*, of the nose), denoting connection with the nose ; as *rhino*dynia, pain of the nose.

sarco-, *sărk'.ō* (Gr. *sarx*, flesh), denoting 'flesh or fleshy' ; as *sarco*carp, the fleshy part of certain fruits.

sclero-, *sklēr'.ō* (Gr. *sklērŏs*, hard), denoting hardness ; as *sclero*derma, a disease in which the skin hardens and indurates.

se (L.), aside ; a separating from : *se*cede, to go aside, to separate from ; *se*duce, to lead aside.

semi-, (L.), half ; in part : *semi*circle, half a circle ; in Gr. *hemi*, half.

septem-, *sĕpt'.ĕm* (L. *septem*, seven), seven ; in Gr. *hepta*, seven ; as *septem*nervine, having seven nerves.

sesqui-, *sĕs'.kwĭ* (L. *sesqui*, more by a half), in *chem.*, denoting that 1½ equivalents of one constituent is united to one equivalent of another, or in the proportion of 'three to two' ; a whole and a half; as *sesqui*-carbonate, a salt composed of 1½ equivalents of carbonic acid and 1 equivalent of any base.

sex-, *sĕks* (L. *sex*, six), six ; in Gr. *hexa*, six ; as *sex*digitism, the condition of having six fingers on a hand, and six toes on a foot.

sex-, sĕks (L. *sexus*, sex, *sexūs*, of sex), sex; as *sexi*ferous, provided with sexual organs.

sine (L.), without : *sine*cure, an office which has an income but not employment.

stomato-, stŏm'ăt·ō, and **stomo-, stŏm'ō** (Gr. *stŏma*, the mouth), denoting connection with the mouth ; as *stomato*-**gastric**, connected with the mouth and stomach.

stylo-, stil'ō (L. *stylus*, Gr. *stūlŏs*, a column, a style or pen), denoting connection with the styloid process of the temporal bone ; as *stylo*-**glossus**, the shortest of three muscles which spring from the styloid process of the temporal bone, situated partly under the tongue.

sub, with its forms **suc, suf, sug, sum, sup, sus** (L.), under ; below ; beneath : *sub*scribe, to write under ; *sub*side, to settle under : **sub** becomes **suc** before **c**, as in *suc*ceed, to follow under or in order : **suf** before **f**, as in *suf*fer, to bear up under : **sug** before **g**, as in *sug*gest, to carry or lay under : **sum** before **m**, as in *sum*mon, to warn beneath or secretly : **sup** before **p**, as in *sup*plant, to trip up beneath : **sus** before **c, p, t**, etc., as in *sus*ceptible, capable of being laid hold of beneath ; *sus*pend, to hang beneath.

subter (L.), beneath ; under : *subter*fuge, a flying under or beneath.

super, with its form **sur** (L.), above ; over ; in excess : *super*human, above human ; *super*sede, to sit or be above : **super** assumes the French form **sur**, as in *sur*charge, to charge in excess.

syn, with its forms **sy, syl, sym** (Gr.), with ; together ; united : *syn*tax, a putting together in order : **syn** becomes **sy** before **s**, as in *sy*stem, that which is formed of parts placed together : **syl** before **l**, as in *syl*lable, several letters taken together to form a single sound : **sym** before **b, p**, or **m**, as in *sym*pathy, feeling with another ; *sym*bol, that which is thrown together with something else ; *sym*metry, state of having the parts of the same measure with.

ter-, tĕr (L. *tĕr*, thrice), in *chem.*, denoting three atoms of acid combined with one of base.

tetra-, tĕt'ră (Gr. *tetra*, four), four ; in L. *quatuor ;* as *tetra*gynous, having four carpels, or four styles.

trachelo-, trăk·ēl'·ō (Gr. *trachēlŏs*, the neck), denoting connection with the throat or neck ; as *trachelo*-**mastoid**, a muscle which passes from the neck to the mastoid process of the skull.

tracheo-, trăk'·ē·ō (Gr. *tracheia*, the windpipe), denoting connection with the trachea or windpipe ; as *tracheo*-**bronchitis**, inflammation of the trachea and bronchi.

trans, with its form **tra** (L.), across ; over ; beyond ; through : *trans*act, to carry or drive through ; *trans*gress, to go over or beyond : **trans** is contracted into **tra**, as in *tra*verse, to turn or lie across.

tri-, trī (L. *tris*, Gr. *treis*, thrice), three ; in threes ; as *tri*adelphous, in *bot.*, having stamens united into three bundles by their filaments ; *tri*angle, a figure of three sides and angles ; *tri*sect, to cut into three equal parts.

ultra (L.), beyond ; on the other side ; extreme : *ultra*montane, on the other side of the mountain.

un (AS. *un*, a privative or negative particle), not ; the opposite of — used in these senses before adjectives, or nouns derived from adjectives : *un*fruitful, not fruitful ; *un*fruitfulness, the state of not being

fruitful ; *un*able, not able : **un** before a verb signifies ' to deprive of';
to undo ; *un*dress, to take off clothes ; *un*crown, to deprive of a
crown : **un** is equivalent to the Latin prefix **in** when it signifies
' not ' : **in** and **un** are often used indifferently before adjectives—see
in.

undecim-, *ŭn'.dĕs·ĭm* (L. *undĕcĭm*, eleven), eleven ; in Gr. *endeka*.

under (Goth. *undar*, Ger. *unter*, under), that which is less than right
or ordinary ; lower in rank or degree ; beneath : *under*coat, a coat
beneath ; *under*-clerk, an inferior clerk.

uni-, *ūn'.ĭ* (L. *ūnus*, one), one ; in Gr. *mono* ; as *uni*parous, having only
one at a birth.

utero-, *ūt'.ĕr·ō* (L. *ŭtĕrŭs*, the womb), denoting connection with the
womb ; as *utero*-abdominal, pert. to the uterus and the abdomen.

vegeto-, *vĕdj'.ĕt·ō* (L. *vegĕtus*, lively, vigorous—from *vegĕŏ*, I quicken),
denoting connection with vegetable life; as *vegeto*-**animal**, having the
nature of both vegetable and animal life.

viginti-, *vĭdj·ĭn'.tĭ* (L. *viginti*, twenty), twenty ; in Gr. *icosi*.

xylo-, *zĭl'.ō* (L. *xulŏn*, wood), denoting some connection with wood.

POSTFIXES.

A postfix is a particle, generally significant, placed after a word, or a root, to
modify its meaning. *Note.*—There are many postfixes or terminations which are
not now significant. These are letters or syllables in present use which apparently
serve only to lengthen the words, though once significant. The postfixes are placed
in groups according to their signification. In the examples the root-parts are printed
in black type, and the postfixes in italics.

Those postfixes only are given in the following list which affect scientific terms.

-aceæ, *ā'.sĕ·ē* (L. *ācĕus*), in *bot.*, a postfix which terminates the names
of Orders ; as **Drocer**aceæ, the Sundew family of plants ; **Oxalid**aceæ,
the Wood-sorrel family of plants.

-aceous, *ā'.shŭs*, and **-ous**, *ŭs*,—*aceous* denotes resemblance to a substance ; as **membran**aceous, resembling a membrane, having the
consistence or structure of membrane ; **carbon**aceous, partaking of
the qualities or appearance of carbon : *ous* denotes the substance
itself ; as **membran**ous, belonging to, or consisting of, membranes.

-adæ, *ăd'.ē*, the same as **idæ**, which see.

-agoga, *ăg·ōg'.ă*, and **-agogue**, *ăg·ōg'* (Gr. *agōgos*, a leader—from *agō*,
I lead or drive), denoting substances which expel others ; as **emmen**-
agogue, a medicine which has the power of promoting the menstrual
discharge.

-agra, *ăg'.ră* (Gr. *agra*, a seizure), denoting a seizure of pain ; as **pod**agra,
gout of the foot.

-algia, *ălj'.ĭ·ă* (Gr. *algos*, pain), denoting the presence of pain ; as **nephr**-
algia, pain, or neuralgia, in the kidney : same as **-odynia**.

-ana, *ăn'.ă* (L. *ānus*), denoting a collection of memorable sayings
or loose thoughts ; as **Johnson**iana, a collection of the sayings, etc., of
Johnson.

-ate, *āt* (L. *ātus*), in *chem.*, a postfix which, substituted in the name of
an acid ending in **ic**, expresses a combination of that acid with a

salifiable base; as **nitra*te*** of silver, that is, a combination of nitric acid with the salifiable base silver.

-cele, *sĕl* (Gr. *kēlē*, a tumour), denoting a tumour caused by the protrusion of some soft part ; denoting the swelling of a part ; denoting an enlargement by a contained fluid ; as **entero*cele***, abdominal hernia, containing intestine only.

-cle, *kl*, and **-cule**, *kūl*, etc. (L. *culus*), denoting 'little'; 'diminution'; as **animal*cule***, a very little creature : **pil*ule***, a little pill.

-colla, *kŏl'lă* (Gr. *kolla*, glue), denoting glue, or a resemblance to it; as **sarco*colla***, flesh glue.

-eæ, *ĕ·ē*, in *bot.*, a postfix terminating names of Sub-orders; as **Phytolaccœæ**, a Sub-order of the Order Phytolaccaceæ.

-form, *fŏrm* (L. *forma*, form, shape), denoting 'resemblance'; in Gr. *id* or *ide;* as **fili*form***, having the form or shape of a thread.

-fuge, *fŭdj* (L. *fŭgo*, 1 drive away or expel), denoting one substance which expels another, or a disease ; as **febri*fuge***, a medicine which expels or cures a fever.

-gen, *jĕn*, **-geny**, *jĕn'ĭ*, **-genesis**, *jĕn'ĕs·ĭs*, and **-genous**, *jĕn'ŭs*, etc. (Gr. *gĕnĕsis*, generation ; *gĕnŏs*, birth ; *gennŏō*, I produce), denoting 'production'; 'generation'; as **organo*genesis***, or **organo*geny***, the production or generation of organs : **caprig*enous***, produced or generated by a goat.

-graphy, *grăj'ĭ* (Gr. *grăphō*, I write), denoting the description of a thing, either in writing or by means of diagrams; as **atmo*graphy***, a description or history of vapours.

-ia, *ĭ·ă* (L. *ĭŭs*), a postfix which forms the termination of medical terms denoting 'a diseased state or condition' ; as **leucim*ia***, a condition of the blood in which there is a deficiency of colouring matter ; **dipsomani*a***, a condition in which there is an irresistible longing for alcoholic liquors : also in *bot.*, terminating many names of genera : **-ious**, *ĭ·ŭs*, the terminations of the adjectives formed from them.

-ic, *ĭk* (L. *icus*, Gr. *ikos*), in *chem.*, denoting the acid containing most oxygen, when more than one is formed; as **nitr*ic***, **sulphur*ic*** : in *phys.* and *path.*, expressing the condition of being excited ; see ode.

-ida, *ĭd'ă*, see **idæ**.

-idæ, *ĭd'ē*, **-adæ**, *ăd'ē*, and **-ides**, *ĭdz* (Gr. *ides*), a postfix signifying 'descent'; denoting a family or group exhibiting some points of likeness ; as **can*idæ***, the Dog family, including dogs, foxes, and wolves.

-ide, *ĭd*, and **-ides**, *ĭdz* (Gr. *eidos*, resemblance), a postfix of such terms as *oxygen*, *chlorine*, *fluorine*, and *iodine*, used to indicate combinations with each other, or with simple combustibles or metals, in proportions not forming acids ; as **ox*ide*** of chlorine, **chlor*ide*** of sulphur, **iod*ide*** of iron, etc.

-idea, *ĭd'ĕ·ă*, **-idean**, *ĭd'ĕ·ăn*, and **-ideus**, *ĭd'ĕ·ŭs* (L. *ideus*—from Gr. *eidos*, resemblance), that which bears resemblance, or related to such ; as **aryteno*idean***, *ăr·ĭt'ĕn·ŏyd'ĕ·ăn*, pert. to that which is arytenoid or funnel-shaped.

-ides, *ĭdz*, as if **-eides** and **-oides**, *ŏydz* (Gr. *eidos*, resemblance), a postfix preceded by *o*, denoting 'resemblance or likeness to an object'; as **alkal*oides***, substances having a likeness or resemblance to alkaloids.

-ides, see **idæ**.

2 H

-ine, *ĭn*, or -in, *ĭn* (L. *inus*), a common termination in chemical terms, but varying much in signification ; as hæmat*in*, the colouring matter resulting from the decomposition of *hæmoglobin* by heat : hæmat*ine*, the colouring matter of logwood : stear*in*, the solid fatty principle of animal fat : mul*in*, a modification of starch : e is now pretty generally omitted in the terminations of such words.

-ine has been usually applied to the alkaloids produced from vegetable substances, and the compounds possessing the closest analogies to them ; as quin*ine*, atrop*ine*, anil*ine*, etc., but we now say quin*ia*.

-ite, *ĭt* (L. *itus*), a postfix which, in the name of an acid, substituted for *ous* expresses combination of that acid with a salifiable base ; as sulph*ite* of potash, that is, a combination of sulphur*ous* acid with the base potash.

-ite, *ĭt* (Gr. *lĭthos*, a stone), in *geol.*, an abbreviation of *lite*, meaning 'stone'; 'resembling stone'; as quartz*ite*, granular quartz ; ammon*ite*, a certain fossil shell.

-itis, *ĭt'ĭs* (Gr. *ĭēmi*, I discharge, I set against), in *med.*, a postfix in Gr. names of organs, denoting inflammation of the organ indicated, as card*itis*, inflammation of the heart ; laryng*itis*, inflammation of the larynx.

-lite, *līt* (Gr. *lithos*, a stone), in *geol.*, stone ; as mel*lite*, honey-stone.

-logy, *lŏdj'ĭ* (Gr. *logos*, a word, a description), denoting a description of, or a treatise on, a subject ; as laryngo*logy*, a treatise on the larynx.

-lysis, *lĭs'ĭs* (Gr. *lŭsĭs*, a loosening, a release), denoting 'a solution'; 'a resolution'; etc. ; as para*lysis*, a loosening of nervous energy.

-meter, *mĕt'ĕr* (Gr. *mĕtrŏn*, a measure), denoting 'measure,' or 'measurer'; as baro*meter*, a measurer of weight.

-o, a common terminating vowel of the first part of binomial compounds, denoting intimate 'connection or association,' either friendly or hostile, or otherwise, with the second part ; thus, Angl*o*-Indian, that is, India as associated with, or influenced by England ; metall*o*-chemistry, the branch of chemistry which treats specially of metals.

-ode, *ōd* (Gr. *ōdēs*, excess or fulness), in *med.*, denoting 'an unexcited condition'; as tetan*ode*, tetanus without excitability, as distinguished from tetanic, denoting the excited state of tetanus.

-odes, *ōdz* (Gr. *ōdēs*, excess or fulness), in scientific terms, 'plenty or fulness'; as hæmat*odes*, full of blood.

-œcium, *ē'shĭ-ŭm*, and -œcious, *ē'shŭs* (Gr. *oikos*, a house or family), in *bot.*, denoting the arrangement of stamens and pistils in flowers ; as andr*œcium*, the staminal organs : mon*œcious*, possessing two kinds of unisexual flowers on the same individual.

-ops, *ŏps*, -opsia, *ŏps'ĭ-ă*, and -opia, *ŏp'ĭ-ă* (Gr. *ŏps*, the eye, *ŏpsĭs*, sight), denoting connection with the eye and vision ; as my*opia*, shortness of sight.

-ous, *ŭs*, and -ose, *ōz* (L. *osus*), in *chem.*, denoting that compound which has a smaller quantity of oxygen than the one which ends in ic ; thus, nitr*ous* acid, the acid which contains a smaller quantity of oxygen than nitr*ic* acid.

-pathy, *păth'ĭ*, and -pathia, *păth'ĭ-ă* (Gr. *păthŏs*, suffering, disease), denoting 'feeling or suffering with'; 'affection'; as deutero*pathy*, a secondary or sympathetic disease.

-phore, *fōr*, -phorum, *fōr'ŭm*, and -phorus, *fōr'ŭs* (Gr. *phŏrĕō*, I bear,

I carry), denoting 'bearing'; 'producing'; in L. *fer* and *ferus*; as galacto*phorous*, milk-bearing.

-rhœa, *rē'ā* (Gr. *rhoa*, a stream—from *rhĕō*, I flow), denoting 'a discharge'; as diarrhœa, a discharge from the bowels: the r is doubled after a vowel.

-scope, *skōp*, and -scopy, *skōp'ī* (Gr. *skŏpŏs*, an inspector—from *skopĕō*, I view), denoting examination; indication, etc.; as stetho*scope*, an instr. for assisting in the examination of the sound of the chest.

-tome, *tōm*, -tomia, *tōm'ī-ā*, -tomy, *tōm'ī* (Gr. *tŏmē*, a cutting), denoting 'a cutting'; 'incision'; as ana*tomy*, the art of cutting up a dead animal for scientific purposes.

-ula, *ŭl'ā*, -ule, *ŭl*, and -ulus, *ŭl'ŭs* (L. *ulus*), diminution; littleness; as glob*ule*, a little globe.

-uret, *ūr'ĕt* (L. *ūro*, I burn), denoting the combination of simple inflammable bodies with one another, or with a metal; as sulph*uret*, the combination of sulphur with an alkali or a metal; phosph*uretted*, combined with phosphorus, etc.

ABBREVIATIONS

USED BY MEDICAL MEN IN WRITING OUT PRESCRIPTIONS.

The abbreviations begin with small letters, as the form in which they are usually found. Of course, when commencing a sentence, an abbreviation will commence with a large or capital letter. Only those in common use have been given.

a or **āā** (Gr. *ănă*), of each; denotes that an equal quantity of each ingredient named is to be taken.

abs. febr. (L. *absĕntĕ*, absent; *fĕbrĕ*, fever), fever being absent; in the absence of fever.

add. (L. *ăddĕ*, add), add; also (L. *addantur*, they may be added), add; let there be added.

ad lib. or **ad libit.** (L. *ad*, to or at; *lĭbĭtum*, one's pleasure), at pleasure.

admov. (L. *ădmŏvĕ*, apply), apply; also (L. *admŏvĕātŭr*, it may be applied, or *admŏvĕăntur*, they may be applied), let there be applied.

altern. horis (L. *altĕrnĭs*, in alternate; *hōrĭs*, in hours), in alternate hours; every other hour.

alt. noct. (L. *altĕrnĭs*, in alternate; *nŏctĭbus*, in nights), in alternate nights; every other night.

alvo adst. (L. *ălvō*, with the belly; *ădstrĭcta*, with bound or pressed close), with the belly bound; when the bowels are costive.

aq. bull. (L. *ăquă*, water; *bŭllĭĕns*, boiling), boiling water.

aq. dest. (L. *ăquă*, water; *dēstĭllătă*, distilled), distilled water.

aq. ferv. (L. *ăquă*, water; *fĕrvĕns*, boiling), boiling water.

aq. font. (L. *ăquă*, water; *fŏntānă*, of or from a spring), spring water; also (L. *ăquă*, water; *fŏntĭs*, of the fountain), the water of a spring.

bib. (L. *bĭbĕ*), drink thou.

bis ind. (L. *bĭs*, twice; *ĭndĭĕs*, from day to day), twice a day.

b. c. (L. *bălnĕŭm*, a bath; *călĭdum*, warm), a warm-water bath for the patient.

b. m. (L. *bălnĕŭm*, a bath; mid. L. *mărĭæ*, of the sea—from *mărĭa*, the sea ; L. *măris*, of the sea—from *măre̅*, the sea), a bath of sea-water ; a water bath.

b. v. (L. *bălne̅um*, a bath ; *văpŏrōsum*, full of steam or vapour), a vapour bath for the patient.

bull. (L. *bŭllĭăt*, it may boil, or *bŭllĭănt*, they may boil), let it boil, or let them boil.

cap. (L. *căpĭăt*, he may take), let him take ; let the patient take.

c. c. or č. (L. *cŭcŭrbĭtŭlă*, the bitter gourd, a cupping-glass ; *crŭĕntă*, bloody—from *crŭŏr*, blood), the cupping-glass with the scarificator.

cochleat. (L. *cŏchlĕātĭm*, spirally, by spoonfuls—from *cŏchlĕă*, a snail-shell), by spoonfuls.

coch. ampl. (L. *cŏchlĕārĕ*, a spoon ; *ămplŭm*, large), a large spoonful.

coch. infant. (L. *cŏchlĕārĕ*, a spoon ; *ĭnfăntĭs*, of an infant), a child's spoonful.

coch. magn. (L. *cŏchlĕārĕ*, a spoon ; *măgnŭm*, large), a large spoonful.

coch. med. (L. *cŏchlĕārĕ*, a spoon ; *mĕdĭŭm*, middle), a middling spoonful : coch. mod. (L. *mŏdĭcŭm*, moderate), a moderate spoonful, that is, a dessert spoonful.

coch. parv. (L. *cŏchlĕārĕ*, a spoon ; *părvŭm*, small), a small spoonful.

coq. (L. *cŏque̅*, *kŏk-we̅*), boil thou.

col. (L. *cōlă*), do thou strain or filter it : col. (L. *cōlātŭs*), strained or filtered.

colat. (L. *cōlātŭr*, it may be strained or filtered), let it be strained or filtered : colet. (L. *cōlētŭr*, it may be strained or filtered), let it be strained or filtered : colent. (L. *cōlĕntŭr*, they may be strained or filtered), let them be strained or filtered.

color. (L. *cŏlōrētŭr*, it may be coloured), let it be coloured.

comp. (L. *cŏmpŏsĭtŭs*), compounded.

cong. (mid. L. *cŏngĭŭs*), a gallon.

c. n. (L. *crăs*, to-morrow ; *nŏcte̅*, in the night), to-morrow night : c. v. (L. *crăs*, to-morrow ; *vĕspĕrĕ*, in the evening), to-morrow evening : cras mane (L. *mănĕ*, the morning), to-morrow morning.

cuj. (L. *cŭjŭs*, of which—from *qui*, who), of which.

cyath. thee (L. *cyăthō*, in a cup ; new L. *thĕæ*, of tea—from *thĕă*, the tea-plant), in a cup of tea.

cyath. vinar. (L. *cyăthŭs*, a cup, a liquid measure ; *vīnărĭŭs*, of or belonging to wine—from *vīnŭm*, wine), a wine-glassful.

det. (L. *dētŭr*, it may be given—from *do*, I give), let it be given.

dieb. alt. (L. *dĭēbŭs*, on or in days—from *dĭēs*, a day ; *altĕrnĭs*, on alternate), on alternate days ; every other day : dieb. tert. (L. *tĕrtĭĭs*, on third—from *tĕrtĭŭs*, third), on every third day.

dim. (L. *dĭmĭdĭŭm*), a half ; one half.

div. in p. æq. (L. *dīvĭdātŭr*, it may be divided ; *in*, into ; *părtēs*, parts—from *pars*, a part ; *æquālēs*, equal), let it be divided into equal parts.

donec alv. bis dej. (L. *dōnĕc*, until ; *ălvŭs*, the belly ; *bĭs*, twice ; *dējĭcĭātŭr*, it may be thrown or cast down, it may be purged—from *dē*, down, and *jăcĭō*, I throw), until the belly is twice evacuated ; until two stools have been obtained.

donec alv. sol. fuer. (L. *dōnĕc*, until ; *ălvŭs*, the belly ; *sŏlūtă*, loosened, unbound ; *fŭĕrĭt*, it may have been), until the belly has been loosened ; until a stool has been obtained.

dos. or **d.** (Gr. *dŏsĭs*, that which is given—from *dĭdŏmĭ*, I give), a dose.

ejusd. (L. *ĕjŭsdĕm*, of the same—from *ĭdĕm*, the same), of the same.

f. or **fac.** (L. *fĭăt*, it may be made, or *fĭănt*, they may be made—from *făcĭŏ*, I make), let it be made ; let them be made.

f. pil. xii. (L. *făc*, make thou ; *pĭlŭlăs*, little balls; *dŭŏdĕcĭm*, twelve), make twelve pills.

feb. dur. (L. *fĕbrē*, in or with the fever—from *fĕbrĭs*, a fever; *dŭrăntĕ*, with continuing), the fever continuing ; while the fever continues.

ft. haust. or **f. h.** (L. *fĭăt*, it may be made ; *haustŭs*, a drink, a draught), let a draught be made.

ft. mist. (L. *fĭăt*, it may be made; *mĭstŭră*, a mixture), let a mixture be made.

f. s. a. (L. *fĭăt*, it may be made; *sĕcŭndŭm*, according to ; *ărtĕm*, art), let it be made according to art.

garg. (L. *gărgărĭsma*), a gargle.

grana or **gr.** (L. *grānŭm*, a grain, *grānă*, grains), a grain ; grains.

gtt. (L. *gŭttă*, a drop, *gŭttæ*, drops), a drop ; drops : **guttat.** (L. *gŭttātĭm*), by drops.

h. s. or **hor. som.** (L. *hŏrā*, at the hour ; *somnī*, of sleep), at the hour of sleep ; on retiring to rest.

ind. or **indies** (L. *ĭndīĕs*, from day to day—from *in*, into ; *dīĕs*, a day), from day to day ; daily.

inject. (L. *ĭnjĕctĭŏ*), an injection.

lat. dol. (L. *lătĕrī*, on or to the side—from *lătŭs*, aside ; *dŏlĕntī*, to suffering or feeling pain—from *dŏlĕŏ*, I suffer pain), to the side which suffers pain.

lot. (L. *lŏtĭŏ*, a washing, a bathing), a lotion.

mane pr. (L. *mānē*, at or in the morning ; *prīmŏ*, in the first), very early in the morning.

manip., see under **mis.**

mic. pan. (L. *mĭcă*, a crumb, a morsel ; *pānĭs*, of bread), a crumb or morsel of bread.

min. (L. *mĭnĭmum*), a minim, the 60th of a drachm measure.

mis. or **misce** (L. *mĭscĕ*, mix thou), mix : **misce mensura** (L. *mĕnsŭrā*, by measure), mix by measure : **misce manipulus** (L. *mănĭpŭlŭs*, a handful), mix a handful.

mist. (L. *mĭstūra*), a mixture.

mitt. or **mitte** (L. *mĭttĕ*, send thou), send ; (L. *mĭttātŭr*, it may be sent), let it be sent ; (L. *mĭttăntŭr*, they may be sent), let them be sent.

mod. præsc. (L. *mŏdŏ*, in the manner or way ; *præscrĭptŏ*, in præscribed or directed), in the manner præscribed.

oct. or **o.** (L. *ŏctārĭŭs*, a pint—from *octo*, eight), a pint ; a pint of wine.

omn. bid. (L. *ŏmnĭ*, on every ; *bĭdŭŏ*, in a period of two days), every two days : **omn. bih.** (L. *bĭhŏrĭŏ*, in a period of two hours—from *bis*, twice ; *hora*, an hour), every two hours ; **omn. hor.** (L. *hŏrā*, in the hour), every hour.

omn. man. (L. *ŏmnĭ*, on every ; *mānē*, on the morning), on every morning : **omn. noct.** (L. *nŏctĕ*, on the night), on every night.

omn. quadr. hor. (L. *ŏmnī*, on every ; *quădrăntĕ*, in a fourth part ; *hōræ*, of an hour), every quarter of an hour.

P. B. (L. *phărmăcōpœiă*, the Pharmacopœia ; *Brĭtănnĭæ*, of Britain), the British Pharmacopœia. NOTE.—Before the recent Medical Amendment Act, there used to exist separate Pharmacopœias for London, Edinburgh, and Dublin, respectively. These are now merged into the one named above. **P. U. S.** is the Pharmacopœia of the United States.

pocal. (L. *pōcŭlŭm*), a cup ; a tea-cup : **pocill.** (L. *pōcĭllŭm*), a little cup.

post sing. sed. liq. (L. *post*, after ; *sĭngŭlăs*, each ; *sēdēs*, seats, stools ; *lĭquĭdăs*, liquids), after each loose stool.

ppt. or **prep.** (L. *præpărătă*), prepared ; made ready.

p. r. n. (L. *prō*, on account of ; *rē*, for or with a thing ; *nātă*, born), according as circumstances require ; occasionally.

pulv. (L. *pŭlvĭs*, dust, powder), a powder ; (L. *pŭlvĕrīzātŭs*), powdered.

q. s. (L. *quăntŭm*, as much as ; *suffĭcĭăt*, it may be sufficient), as much as may be sufficient.

quaq. (L. *quăqŭĕ, kwā:kwĕ*), from every one : **quisq.** (L. *quĭsquĕ, kwĭs: kwĕ*), every one.

quor. (L. *quōrŭm*, of which (things)—from *quŏd*, which), of which medicines or ingredients.

℞. (L. *rĕcĭpĕ*), take thou. NOTE.—The ℞ with the down stroke is said to be simply an accommodation or corruption of the anc. heathen symbol ♃ , an invocation to Jupiter for his blessing on the formula as a curative agent: ♃ is used also as the astronomical symbol for the planet Jupiter.

red. in pulv. (L. *rĕdŭctŭs*, brought or reduced ; *in*, to ; *pŭlvĕrĕm*, powder—from *pŭlvĭs*, dust, powder), reduced to powder.

repet. (L. *rĕpĕtātŭr*, it may be repeated), let it be repeated or continued ; (L. *rĕpĕtăntŭr*, they may be repeated), let them be repeated or continued.

s. a. (L. *sĕcŭndŭm*, according to ; *ărtĕm*, art—from *ars*, art), according to art.

semidr. (L. *sĕmĭdrăchmă*, a half drachm—from *sĕmĭ*, half, and *drachma*), half a drachm.

semih. (L. *sĕmĭhōră*, a half hour—from *sĕmĭ*, half ; *hōră*, an hour), half an hour.

sescunc. (L. *sĕscŭncĭă*, one and a half unciæ—from *sĕsquĭ*, one half more ; *ŭncĭă*, a twelfth part, an ounce), an ounce and a half.

sesquih. (L. *sĕsquĭhōră*—from *sĕsquĭ*, one half more ; *hōră*, an hour), an hour and a half.

sign. n. pr. (L. *sĭgnĕtŭr*, it may be marked or stamped ; or *signa*, mark thou ; *nōmĭnĕ*, with the name—from *nōmĕn*, the name ; *prŏprĭŏ*, with proper), let it be stamped or marked with its proper name, that is, no special directions are required. NOTE.—The directions for the patient are always written in English, but for the druggist they are written in Latin.

signat. (mid. L. *sĭgnătūră*), a label.

sing. (L. *sĭngŭlōrŭm*, of one to each—from *sĭngŭlī*, one to each, single), of each.

sol. (L. *sŏlūtĭŏ*), a solution.

ss. (L. *sēmĭ*), half.

st. or stet. (L. *stĕt*, it may stand), let it stand ; (L. *stĕnt*, they may stand), let them stand.

suc. (L. *succus*), juice.

sum. (L. *sūmĕ*), take thou ; (L. *sūmăt*, he may take), let him take ; (L. *sūmātŭr*, it may be taken), let it be taken ; (L. *sūmăntŭr*, they may be taken), let them be taken ; (L. *sūmĕndŭs*), to be taken.

s. v. (L. *spĭrĭtŭs*, spirit ; *vīnōsŭs*, having the flavour or quality of wine), ardent spirit ; diluted spirited of wine : **s. v. r.** (L. *spĭrĭtŭs*, spirit ; *vīnī*, of wine ; *rĕctĭfĭcātŭs*, rectified—from *rĕctŭs*, straight, right, and *făcĭō*, I make), rectified spirit of wine : **s. v. t.** (L. *spĭrĭtŭs vīnōsŭs ;* *tĕnŭĭs*, thin), diluted spirit ; half and half spirit of wine and water.

tinct. (L. *tĭnctūră*), a tincture.

troc. (mid. L. *trōchĭscŭs*, n. sing. ; *trōchĭscī*, n. plu.), troches or lozenges.

vom. urg. (L. *vŏmĭtĭōnĕ*, with vomiting ; *ŭrgĕntĕ*, with pressing or urgent), the vomiting being troublesome ; when the vomiting begins.

---

gr. is the symbol denoting a grain.

Ә is the symbol denoting a scruple = 20 grains troy.

ᴣ is the symbol denoting a drachm = 3 scruples, or = 60 grs. troy.

ᴣ is the symbol for an ounce troy = 8 drachms = 24 scruples = 480 grs. troy; also = 16th part of a wine pint; or 20th part of an imperial pint.

i means one of the quantity indicated by the symbol : **ij** = two of the quantity named : **iij** = three of the quantity named : **iv** = four of the quantity named : **v** = five of the quantity named : **vj** = six of the quantity named : **vij** = seven of the quantity named : **ss.** (for *semi*, half)=half of the quantity named, and so on.

Thus we have **gr. vj** = six grains ; **gr. iij** = 3 grains : **Әj** = one scruple ; **Әiij** = three scruples ; **Әjss** = one and a half scruples ; **Әij** = two scruples : **ᴣiv** = four drachms ; **ᴣijss** = two drachms and a half ; **ᴣj** = one ounce : **ᴣviij** = eight ounces ; **ᴣss** = half an ounce : **i** and **j** are symbols used in same sense as above, **j** always terminating.

O. (L. *octārĭŭs*, a pint), is a less common symbol, and denotes a pint, as **Oj** = one pint ; **Oij** = two pints ; **Oss** = half a pint.

gtt. (L. *gutta*, a drop), as **gtt. xx.** = twenty drops.

I. APOTHECARIES' WEIGHT—*For Medical Prescriptions.*

20 grains (grs. xx) = 1 scruple (Әj).
3 scruples (Әiij) = 1 dram (ᴣj) = 60 grs.
8 drams (ᴣviij) = 1 ounce (ᴣj) = 480 grs.
12 ounces (ᴣxij) = 1 lb.

NOTE.—For ordinary purposes the avoirdupois ounce and pound are used, but for prescriptions the above are still in use.

NEW APOTHECARIES' WEIGHT.

437½ grains = 1 oz.
16 oz. = 1 lb.

II. LIQUID MEASURE—*For Water, Spirits, Wines, etc.*
>   4 gills   = 1 pint (pt.).
>   2 pints = 1 quaɪɟ (qt.).
>   4 quarts = 1 gallon (gal.).

III. APOTHECARIES' FLUID MEASURE.
>   60 minims (mm.LX)   = 1 fluid dram (f.ʒj).
>   8 fluid drams (ff.ʒviij) = 1 fluid ounce (f.ʒj).
>   20 fluid ounces (ff.ʒxx) = 1 pint (Oj).
>   8 pints (Oviij)   = 1 gallon.

1 ounce of distilled water weighs 1 ounce avoirdupois.
The pint and gallon are the same as the imperial pint and gallon.

NOTE.—m=1 minim ; mm=2 or more minims : f=fluid ; ff=2 or more of the fluid thing named.

––––––––

NOMENCLATURE OF CLASSIFICATION IN THE VEGETABLE AND ANIMAL KINGDOMS.

I. **Species,** an assemblage of individuals, resembling each other in their essential characters, and having a community of descent, comprising *varieties* and *races.*

II. **Genus,** a group of species, possessing a community of essential details of structure.

III. **Family,** a group of genera, agreeing in their general characters.

IV. **Order,** a group of families related to one another by structural characters common to all.   𝓍

NOTE.—Family and Order are often used synonymously, especially in botany.

V. **Class,** a very large division, comprising animals or plants which are formed upon the same fundamental plan of structure, but which differ in the method in which the plan is executed.

Sub-kingdom, a primary division of the animal or vegetable kingdom.

Example of the Animal Kingdom.

Sub-kingdom, *Vertebrata,* or vertebrate animals.
Class, *Mammalia,* or animals which suckle their young.
Order, *Carnivora,* or beasts of prey.
Family, *Canidæ,* or genera of the dog kind.
Genus, *Canis,* the dog kind; as the dog, wolf, and jackal.
Species, *Canis familiaris,* the dog and its varieties.

The usual Sub-divisions in botany are

I. **Species.**
>   (1) Varieties.

II. **Genus.**
>   (1) Sub-genus or Section.

III. **Order** or **Family.**
>   (1) Sub-order.
>   (2) Tribe.
>   (3) Sub-tribe.

IV. **Class.**
>   (1) Sub-class.